Lecture Notes in Computational Vision and Biomechanics

Volume 30

The research related to the analysis of living structures (Biomechanics) has been a source of recent research in several distinct areas of science, for example, Mathematics, Mechanical Engineering, Physics, Informatics, Medicine and Sport. However, for its successful achievement, numerous research topics should be considered, such as image processing and analysis, geometric and numerical modelling, biomechanics, experimental analysis, mechanobiology and enhanced visualization, and their application to real cases must be developed and more investigation is needed. Additionally, enhanced hardware solutions and less invasive devices are demanded.

On the other hand, Image Analysis (Computational Vision) is used for the extraction of high level information from static images or dynamic image sequences. Examples of applications involving image analysis can be the study of motion of structures from image sequences, shape reconstruction from images, and medical diagnosis. As a multidisciplinary area, Computational Vision considers techniques and methods from other disciplines, such as Artificial Intelligence, Signal Processing, Mathematics, Physics and Informatics. Despite the many research projects in this area, more robust and efficient methods of Computational Imaging are still demanded in many application domains in Medicine, and their validation in real scenarios is matter of urgency.

These two important and predominant branches of Science are increasingly considered to be strongly connected and related. Hence, the main goal of the LNCV&B book series consists of the provision of a comprehensive forum for discussion on the current state-of-the-art in these fields by emphasizing their connection. The book series covers (but is not limited to):

- Applications of Computational Vision and Biomechanics
- Biometrics and Biomedical Pattern Analysis
- Cellular Imaging and Cellular Mechanics
- Clinical Biomechanics
- Computational Bioimaging and Visualization
- Computational Biology in Biomedical Imaging
- Development of Biomechanical Devices
- Device and Technique Development for Biomedical Imaging
- Digital Geometry Algorithms for Computational Vision and Visualization
- Experimental Biomechanics
- Gait & Posture Mechanics
- Multiscale Analysis in Biomechanics
- Neuromuscular Biomechanics
- Numerical Methods for Living Tissues
- Numerical Simulation
- Software Development on Computational Vision and Biomechanics

- Grid and High Performance Computing for Computational Vision and Biomechanics
- Image-based Geometric Modeling and Mesh Generation
- Image Processing and Analysis
- Image Processing and Visualization in Biofluids
- Image Understanding
- Material Models
- Mechanobiology
- Medical Image Analysis
- Molecular Mechanics
- Multi-Modal Image Systems
- Multiscale Biosensors in Biomedical Imaging
- Multiscale Devices and Biomems for Biomedical Imaging
- Musculoskeletal Biomechanics
- Sport Biomechanics
- Virtual Reality in Biomechanics
- Vision Systems

More information about this series at http://www.springer.com/series/8910

Durai Pandian · Xavier Fernando
Zubair Baig · Fuqian Shi
Editors

Proceedings of the International Conference on ISMAC in Computational Vision and Bio-Engineering 2018 (ISMAC-CVB)

Volume 2

 Springer

Editors
Durai Pandian
SCAD Institute of Technology
Palladam, India

Xavier Fernando
Department of Electrical
and Computer Engineering
Ryerson University
Toronto, ON, Canada

Zubair Baig
School of Computer and Security
Science
Edith Cowan University
Joondalup, WA, Australia

Fuqian Shi
Wenzhou Medical University
Wenzhou, China

ISSN 2212-9391   ISSN 2212-9413 (electronic)
Lecture Notes in Computational Vision and Biomechanics
ISBN 978-3-030-00664-8   ISBN 978-3-030-00665-5 (eBook)
https://doi.org/10.1007/978-3-030-00665-5

Library of Congress Control Number: 2018954619

Contents

Relation Extraction Using Convolutional Neural Networks

V. Hariharan, M. Anand Kumar and K. P. Soman

Abstract Identifying the relationship between the entities plays a key role in understanding any natural language. The relation extraction is a task, which finds the relationship between entities in a sentence. The relation extraction and named entity recognition are the subtasks of information extraction. In this paper, we have experimented and analyzed the closed-domain relation extraction using three variants of temporal convolutional neural network on SemEval-2018 and SemEval-2010 relation extraction corpus. In this approach, the word-level features are formed from the distributed representation of text and the position information of entity are used as the feature for the model.

1 Introduction

Ever since the Internet outbreak, there has been a tremendous rise in the amount of information that has been found on the various online platforms like Reddit, Stack Exchange, etc. To study and understand this information on a large scale in various platforms, it is necessary to build an information extraction system for qualitative data analysis. The information extraction is a task of extracting the factual information from the unstructured text. This task can be done jointly or can be divided into two subtasks such as Named Entity Recognition (NER) and relation extraction. In the NER task, named entities in the text are identified and each entity is assigned to a single category [1]. After recognizing the entities, finding the semantic relation between them is called relation extraction. This work focuses toward solv-

V. Hariharan (✉) · M. Anand Kumar (✉) · K. P. Soman (✉)
Center for Computational Engineering and Networking (CEN), Amrita School
of Engineering, Amrita Vishwa Vidyapeetham, Coimbatore, India
e-mail: cb.en.p2cen16007@cb.students.amrita.edu

M. Anand Kumar
e-mail: m_anandkumar@cb.amrita.edu

K. P. Soman
e-mail: kp_soman@amrita.edu

© Springer Nature Switzerland AG 2019
D. Pandian et al. (eds.), *Proceedings of the International Conference on ISMAC in Computational Vision and Bio-Engineering 2018 (ISMAC-CVB)*, Lecture Notes in Computational Vision and Biomechanics 30,
https://doi.org/10.1007/978-3-030-00665-5_90

ing the closed-domain relation extraction using deep learning methods. In closed-domain relation extraction, the number of entities and the relation between them are fixed. Previous machine learning based information extraction system employs handcrafted features and uses existing Natural Language Processing (NLP) pipelines. These crafted features may fail to capture all the relevant information. The errors present in the pre-existing pipelines will also propagate through the system which hinders the performance of these systems. The deep learning based approach for relation extraction is resilient and more independent of manual feature engineering, thereby relying on pre-existing NLP pipelines decreases.

2 Background

Relation extraction is a key task for building knowledge systems, extending existing ontology like BabelNet, and acting as a knowledge base for question answering systems. Relation extraction is done via various methods like hand-built patterns, bootstrapping with a small highly annotated data, distant supervision method which efficiently uses the existing knowledge bases like freebase, wikidata to extract relations on a bigger corpus [2], and supervised methods. In supervised methods, getting a large quantity of annotated data for the training is key for the high performance of the model. In this work, we are focused toward the supervised relation extraction methods.

2.1 Convolutional Neural Network

The convolutional neural network (CNN) is a kind of feed-forward neural network which is obtained by a slight variation from the multilayer perceptron. In CNN, in order to retain the spatial information of the image, convolution is performed over the image with the different sized filters instead of connecting each unit in the hidden layer to all the units in the previous layer [3]. After convolution operation, a nonlinear activation function like tanh squashes each element to a value between -1 and 1. After applying nonlinearity, a max pooling is done over the output with a fixed size filter which downsamples the information along the spatial dimension. The max pooling operation is done with an intent to capture the most prominent feature in the image, in spite of the slight variation of the spatial features caused by rotation, scaling, and translation. In general, a convolutional neural network will have several stacks of convolution followed by a nonlinear activation before making it to fully connected layer for classification. Several convolutional layers are generally used since the filters in the initial layer learn to detect edge features in the images and the later layers use these low-level features to learn higher level features of images.

2.2   Temporal Convolutional Neural Network

The phenomenal success of the convolutional neural network in the image domain has inspired many researchers to experiment CNN in various domains. In text domain, in order to best understand the meaning, it is necessary to consider the text as a temporal sequence (i.e., each word occurring in the space of time) rather than considering them as an independent feature. The convolution is seen as the operation of blending of features, and it is inferred that convoluted representation has information of nearby words. In NLP, operations are performed over representations given to the text. These representations are given either in the word or character level. In word level, each word is given a vector representation, which is either randomly initialized or assigned through the distributed representations of words [4]. Vector dimension or length of the representations is empirically fixed; this corresponds to the amount of information contained in it. In temporal convolution, the convolution is performed over the temporal sequence [5]. Here, the filter dimension must be equal to the vector dimension of the word embedding. So during convolution, filters slide over the words. In convolution operation, over a window of h words, a filter w is applied to produce new features.

$$o_i = f(w.x_{i:i+h-1} + b) \tag{1}$$

Here, o_i is new feature obtained for the words $x_{i:i+h-1}$, b is the bias term, and f is a nonlinear function like Relu. Similarly, features are obtained for all possible word windows.

$$o = \{o_1, o_2, \ldots, o_{i:i+h-1}\} \tag{2}$$

Then max-over time pooling is done over every feature map of all the available filter, which gives $\hat{o} = \max\{o\}$. $\hat{o}$ is the feature of the corresponding filter. The idea of max-over time pooling is to capture the most significant feature in each feature map. The max-over time pooling naturally deals with different length sequences and decreases the temporal dimension. The pooling operation is generally connected to a dense layer and output layer. The number of classes is equal to the number of neurons in the output layer.

3   Relation Extraction Corpus

In this work, we have used two openly available corpora for relation extraction. The first corpus is taken from SemEval-2010 Task-8 is multi-way classification of semantic relations between pairs of nominals [6], and the second corpus taken from the SemEval-2018 Task-7 is semantic relation extraction and classification in scientific papers [7]. The number of training instances, number of relation classes, and

Table 1 My caption

Sl. No	Corpus	No of instances train–test split	No. of relation classes
1	SemEval-2010 task-8 10717 sentences	75–25	10
2	SemEval-2018 task-7 350 documents	80–20	6

Table 2 SemEval-2010 corpus statistics

Sl. No	Relation	Freq (%)
1	Entity–origin	9.1
2	Instrument–agency	6.2
3	Product–producer	8.8
4	Component–whole	11.7
5	Entity–destination	10.6
6	Cause–effect	12.4
7	Message–topic	8.4
8	Content–container	6.8
9	Member–collection	8.6
10	Other	17.4

Table 3 SemEval-2018 corpus statistics

Sl. No	Relation type	Example	Freq (%)
1	Usage	x is used for y	39
2	Part–whole	x is found in y	19
3	Result	x gives result as y	5.8
4	Topic	x deal with topic y	1.4
5	Compare	x is compared to y	7.7
6	Model	x is model of y	26.5

train–test split of the two corpora are given in Table 1. The class distribution frequency of the two corpora is given in Tables 2 and 3. For the SemEval-2010 corpus shown in Table 2, the frequency of class distribution except "other" class is pretty evenly distributed. For the SemEval-2018 corpus shown in Table 3, the "usage", "model", and "part–whole" have more occurrences when compared with the rest of the other classes.

A preprocessed sample instance from the two corpora is given below: SemEval-2010 task-8 corpus

The (e1) cancer (e1) was caused due to radiation (e2) exposure (e2).

SemEval-2018 task-7 corpus

The key features of the system include (i) robust efficient (e1) parsing (e1) of (e2) korean (e2) (e3) verb final language (e3) with (e4) overt case markers (e4), relatively (e5) free word order (e5), and frequent omissions of (e6) arguments (e6); (ii) high-quality (e7) translations (e7) via (e8) word sense disambiguation (e8) and accurate (e9) word order generation (e9) of the (e10) target language (e10); and (iii) (e11) rapid system development (e11) and porting to new (e12) domains (e12) via (e13) knowledge-based automated acquisition of grammars (e13).

4 Methodology

In this approach, the closed-domain relation extraction is a problem that is modeled as a multi-class classification problem. Some of the constraints of relation extraction are loosened since the entities in the text are already known. In this approach, temporal convolution neural network is employed for relation extraction. The temporal convolution neural network typically models the sequence information present in the text but in order for convolution network to be aware of the entities in sentence or context, the position information about the entities are fed into the network [8]. This is done by feeding an additional array per entity into the network along with the word embedding information as shown in Fig. 1. These arrays have the distance information of each word with respect to the index of the entities.

For a cause–effect relation example as shown in Fig. 1, $e1$ is the effect and $e2$ is the cause. The position array of the first entity $e1 = [-1, 0, 1, 2, 3, 4, 5, 6]$ and the second entity $e2 = [-7, -6, -5, -4, -3, -2, -1, 0]$ is formed; this is done by calculating the distance of each word from the entity. The length of position array will be equal to the length of sentence. While training, the sequence is sufficiently zero padded wherever necessary. To avoid the negative values in the position array, it is converted into an index which starts from zero. The position indices along with the word representation are fed into the temporal convolutional neural network. The word

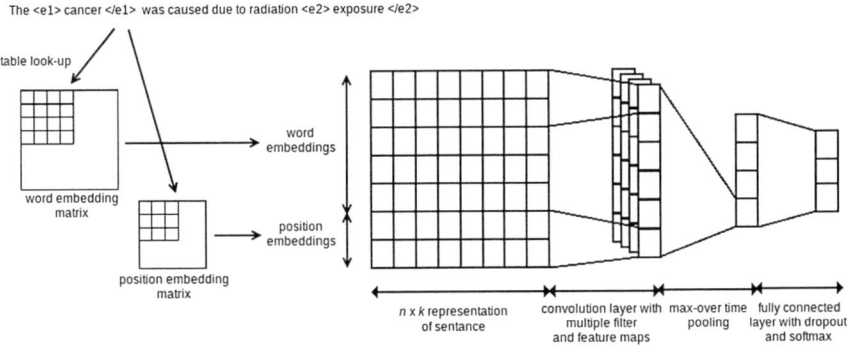

Fig. 1 CNN-based relation extraction architecture

representation is initialized with the word vector from distributed representation [9] word embedding models like word2vec [10] and GloVe [11].

The three variants of temporal CNN used for relation extraction are listed below.

Temporal CNN: This variant uses the word representation initialized through the distributed representation of words by using popular distributed representation framework like word2vec and position embedding information as features.

Temporal CNN with word channel: In this variant in addition to the word and position embedding features, a separate temporal CNN is trained on the word; this forms the local representation for the words with respect to the sentences. This channel is called as word channel and is concatenated with the temporal CNN with word and position embedding features after max-over time pooling. Note that temporal CNN's word embeddings are initialized from the pretrained word embeddings while word channel embeddings are locally learned.

Multichannel temporal CNN: This variant uses two channels of temporal CNN where each channel is initialized with the word vector from different distributed representation frameworks like word2vec and GloVe. Then these two channels are concatenated before max-over time pooling. This idea this model is inspired by the work [12].

5   Result and Discussion

In this work, temporal convolutional neural network based relation extraction model is applied on SemEval-2018 task-7 and SemEval-2010 task-8 corpus. On these two corpora, we have applied three variants of temporal CNN. The first model is a temporal CNN architecture with pretrained word embeddings and position embedding as features; this is considered as the baseline. In the second model, in addition to a temporal CNN with word and position embedding, an additional word channel is introduced; it is seen that having an additional word channel improves the accuracy and F1 score of the classification task. Then, in the final model, temporal CNN with multichannel word embeddings is used. It is observed that using different distributed word representation frameworks significantly helps the model to learn better representation for sentences, thereby increasing the overall performance of the classification task. Tables 4 and 5 show the accuracy and F1 score of our various approaches on the two corpora. So, it is seen that model which uses additional word representation channel improves the overall accuracy of the classification task. The state-of-the-art (SOTA) performance reported for the SemEval-2010 task-8 corpus is using bidirectional gated recurrent unit with attention [13].

Table 4 Results for SemEval-2010 corpus

Sl. No	Architecture	Filter length	Accuracy	F1 (macro)
1	Temporal CNN	3	0.77	0.740
2	Temporal CNN	4	0.76	0.741
3	Temporal CNN with word channel	3	0.78	0.761
4	Temporal CNN with word channel	4	0.78	0.764
5	Multichannel temporal CNN	3	0.79	0.769
6	Multichannel temporal CNN	4	0.79	0.771
7	BiGRU with ATT (SOTA)	–	–	0.84

Table 5 Results for SemEval-2018 corpus

Sl. No	Architecture	Filter size	Accuracy	F1 (macro)
1	Temporal CNN	3	0.5512	0.291
2	Temporal CNN	4	0.5441	0.280
3	Temporal CNN with word channel	3	0.5657	0.327
4	Temporal CNN with word channel	4	0.5749	0.316
5	Multichannel temporal CNN	3	0.5896	0.325
6	Multichannel temporal CNN	4	0.6196	0.313

6   Conclusion and Future Work

In this paper, the task of relation extraction is modeled as a multi-class classification problem. For classifying relations, temporal convolution neural network is employed and the word representation is done using distributed representation framework like word2vec and GloVe. In this work, three variants based on temporal convolutional neural network are applied on SemEval 2018 and 2010 relation extraction corpus and it is observed that model which has multiple channels of word representation helps and improves the overall accuracy of the relation extraction task. For the future work, we are planning to implement a multi-level attention model which uses attention to learn a better representation for the words and also employs attention mechanism for classification. In our next work, we are planning to make an end-to-end network which does named entity recognition and relation extraction together.

References

1. Devi GR, Veena PV, Anand Kumar M, Soman KP (2018) Entity extraction of Hindi–English and Tamil–English code-mixed social media text. In: Forum for information retrieval evaluation. Springer, pp 206–218
2. Mintz M, Bills S, Snow R, Jurafsky D (2009) Distant supervision for relation extraction without labeled data. In: Proceedings of the joint conference of the 47th annual meeting of the ACL and the 4th international joint conference on natural language processing of the AFNLP, vol 2, pp 1003–1011. Association for Computational Linguistics
3. LeCun Yann, Bengio Yoshua et al (1995) Convolutional networks for images, speech, and time series. Handb Brain Theor Neural Netw 3361(10):1995
4. Barathi Ganesh HB, Anand Kumar M, Soman KP (2018) From vector space models to vector space models of semantics. In: Forum for information retrieval evaluation. Springer, pp 50–60
5. Kim Y (2014) Convolutional neural networks for sentence classification. arXiv preprint arXiv: 1408.5882
6. Hendrickx I, Kim SN, Kozareva Z, Nakov P, Séaghdha DÓ, Padó S, Pennacchiotti M, Romano L, Szpakowicz S (2009) Semeval-2010 task 8: multi-way classification of semantic relations between pairs of nominals. In: Proceedings of the workshop on semantic evaluations: recent achievements and future directions, pp 94–99. Association for Computational Linguistics
7. Ga´bor K, Buscaldi D, Schumann A-K, QasemiZadeh B, Zargayouna H, Charnois T (2018) Semeval-2018 Task 7: semantic relation extraction and classification in scientific papers. In: Proceedings of international workshop on semantic evaluation (SemEval-2018), New Orleans, LA, USA
8. Zeng D, Liu K, Lai S, Zhou G, Zhao J (2014) Relation classification via convolutional deep neural network. In: Proceedings of COLING 2014, the 25th international conference on computational linguistics: technical papers, pp 2335–2344
9. Barathi Ganesh HB, Anand Kumar M, Soman KP (2016) Distributional semantic representation for information retrieval. In: Working notes of FIRE, pp 1–7
10. Mikolov T, Sutskever I, Chen K, Corrado GS, Dean J (2013) Distributed representations of words and phrases and their compositionality. In: Advances in neural information processing systems, pp 3111–3119
11. Pennington J, Socher R, Manning C (2014) Glove: global vectors for word representation. In: Proceedings of the 2014 conference on empirical methods in natural language processing (EMNLP), pp 1532–1543
12. Shi H, Ushio T, Endo M, Yamagami K, Horii N (2016) A multichannel convolutional neural network for cross-language dialog state tracking. In: 2016 IEEE Spoken language technology workshop (SLT). IEEE, pp 559–564
13. Kim J, Lee J-H (2017) Multiple range-restricted bidirectional gated recurrent units with attention for relation classification. arXiv preprint arXiv:1707.01265

Multi-Object Detection Using Modified GMM-Based Background Subtraction Technique

Rohini Chavan, S. R. Gengaje and Shilpa Gaikwad

Abstract Detection of objects is the most important and challenging task in video surveillance system in order to track the object and to determine meaningful and suspicious activities in outdoor environment. In this paper, we have implemented novel approach as modified Gaussian mixture model (GMM) based object detection technique. The object detection performance is improved compared to original GMM by adaptively tuning its parameters to deal with the dynamic changes that occurred in the scene in outdoor environment. Proposed adaptive tuning approach significantly reduces the overload experimentations and minimizes the errors that occurred in empirical tuning traditional GMM technique. The performance of the proposed system is evaluated using open source database consisting of seven video sequences of critical background condition.

1 Introduction

Human requirement of automated detection system in personal, commercial, industrial, and military areas leads to development of video analytics which will make lives easier and enable us to compete with future technologies [1]. On the other hand, it pushes us to analyze the challenges of automated video surveillance scenarios. Humans have an amazing capacity for decision-making but are notoriously poor at maintaining concentration levels. A variety of studies has shown that after 20 min of watching, up to 90% of the information being shown on monitors will be missed. As closed-circuit television (CCTV) culture continues to grow, humans

R. Chavan (✉) · S. Gaikwad
Bharati Vidyapeeth University College of Engineering, Pune, India
e-mail: chavanrohini10@gmail.com

S. Gaikwad
e-mail: spgaikwad@bvucoep.edu.in

S. R. Gengaje
Walchand Institute of Technology, Solapur, India
e-mail: srgengaje@rediffmail.com

© Springer Nature Switzerland AG 2019
D. Pandian et al. (eds.), *Proceedings of the International Conference on ISMAC in Computational Vision and Bio-Engineering 2018 (ISMAC-CVB)*, Lecture Notes in Computational Vision and Biomechanics 30,
https://doi.org/10.1007/978-3-030-00665-5_91

would require to observe feed from hundreds of camera 24×7 [2]. It shows requirement of automatic system that analyzes and stores video from 100s of cameras and other sensors, detecting events of interest continuously and browsing of data through sophisticated user interface. It is simply known as video analytics [3, 4].

Recent research in computer vision is giving more stress on developing system for monitoring and detecting humans. It is helpful for people in personal, industrial, commercial, and military areas to develop innovations in video analysis, to compete with future technologies and to accept the challenges in automatic video surveillance system. Video surveillance tries to detect, classify, and track objects over a sequence of images and help to understand and describe the object behavior by human operator. This system monitors sensitive areas such as airport, bank, parking lots, and country borders. The processing framework of an automated video surveillance system includes stages like object detection, object classification, and object tracking. Almost every video surveillance system starts with motion detection. Motion detection aims at segmenting regions of interest corresponding to moving objects from remaining image. Subsequent processes such as object classification and tracking performances are greatly dependent on it. If there is significant fluctuations in color, shape, and texture of moving object, it causes difficulty in handling these objects. A frame of video sequence consists of two groups of pixels. The first group represents foreground objects and second group belongs to background pixels. Different techniques such as frame differencing, adaptive median filtering, and background subtraction are used for extraction of objects from stationary background [5]. The most popular and commonly used approach for detection of foreground objects is background subtraction. The important steps in background subtraction algorithm are background modeling and foreground detection [6]. Background modeling gives reference frame which represents statistical description of entire background scene. The background is modeled to extract interested object from video frames. It is designed with first few frames of video sequence. But, in case of quasi-stationary background such as wavering of trees, flags, and water, it is more challenging to extract exact moving object. In this situation, single model background frame is not enough to accurately detect the moving object but adaptive background modeling technique is used for exact detection of objects from dynamic background [7].

2 Object Detection Using Adaptive Gaussian Mixture Model

2.1 Basic Gaussian Mixture Model

A Gaussian mixture model (GMM) is parametric probability density function presented as a weighted sum of K Gaussian component densities [8, 9]. It is represented by the following equation:

$$P(x_t) = \sum_{i=1}^{k} \left(\omega_{i,t} \, n(X_t; \mu_{i,t}, \sum i, t) \right) \tag{1}$$

$$\sum_{i=1}^{k} \left(\omega_{i,t} \right) = 1 \tag{2}$$

where x is a D dimensional data vector and ω is weight of ith Gaussian component. Here, k is the number of Gaussian distributions, t represents time, μ is mean value of the ith Gaussian mixture at time t, and $\sum i, t$ is the covariance matrix. The entire GMM is scaled by mean vectors, covariance matrices, and mixture weights of all component densities. The mean of such mixture is represented by following equation:

$$\mu_t = \sum_{i=1}^{k} \omega_{i,t} \mu_{i,t} \tag{3}$$

There are several variants on the GMM and covariance matrices constrained to be diagonal. The selection of number of components and full or diagonal covariance matrix is often determined by the availability of data for estimating GMM parameters.

Background subtraction object detection technique is popular as it is less complex, simple and easy to implement. It takes the difference between current frame (I_t) and reference frame. The reference frame is denoted by (B_{t-1}). Hence, difference image (D_t) is given by

$$D_t = |B_{t-1} - I_t| \tag{4}$$

Foreground mask (F_t) is given by applying threshold to difference image

$$F_t = 1, \text{ when } D_t > \text{Th}$$
$$F_t = 0, \text{ when } D_t < \text{Th}$$

2.2 GMM Model Initialization and Maintenance

For stationary process pixels, EM algorithm is applicable. K-means algorithm is an alternative to EM [10]. Using K-means approximation, every new pixel value X_t is checked against existing K Gaussian distribution until match is found. A match is given by

$$\text{Sqrt}\left((X_{t+1} - \mu_{i,t}) T . \sum_{i,t}^{-1} (X_{t+1} - \mu_{i,t}) \right) < k\sigma_{i,t} \tag{5}$$

where k is constant threshold value which is selected as 2.5. If K distribution is not matched with current pixel value then least probable distribution is replaced with current distribution value as its mean, weight, and variance. Prior weights of K distributions at time t are adjusted as follows:

$$\omega_{k,t} = (1 - \alpha)\omega_{k,t} - 1 + \alpha(M_{k,t}) \tag{6}$$

where α is learning rate and $M_{k,t}$ is 1 for model which is matched and 0 for other models. After operating this approximation, weights are again normalized. The μ and σ parameters remain same for unmatched distributions. The parameters of distribution which matches new observations are updated as follows:

$$\mu_t = (1 - \rho)\mu_t - 1 + \rho X_t \tag{7}$$

$$\sigma t = (1 - \rho)\sigma 2t - 1 + \rho(X_t - \mu t)T(X_t - \mu t) \tag{8}$$

$$\sigma_t = (1 - \rho)\sigma_{t-1} + \rho(X_t - \mu t)^{\mathrm{T}}(X_t - \mu t) \tag{9}$$

where

$$\rho = \alpha.n(X_t|\mu_k, \sigma) \tag{10}$$

One advantage of this technique is that when new thing is added in the model then it will not completely destroy the previous background model but it can update the model.

3 Object Detection Using Adaptive GMM

The system is implemented using two steps such as GMM-based object detection and noise removal using morphological operations. Implementation is done using MATLAB 2014v with the help of computer vision system toolbox. The small detected regions whose area is less than moving object and which are not part of foreground object can be removed using noise removal algorithm. Finally, output binary image is compared with ground truth image for performance evaluation to determine accuracy.

Background modeling is adaptive to accommodate all the changes occurring in the background scene. It is very sensitive to dynamic changes that have occurred in the scene which causes consequent need of adaptation of background as per the variations in background. The research has progressed toward improving robustness and accuracy in background subtraction method for complex background condition like sudden and slow illumination change. A common attribute of BS algorithm is learning rate, threshold, and constant parameter K which can be empirically adjusted to get desired accuracy. However, tuning process for these parameters has been less attentive due to lack of awareness. Stauffer and Grimson [7] suggested that selection of learning rate and threshold value is important among all other parameters. Tuning process for these parameters requires time intense repeated experimentation to achieve optimum results. It is very challenging to set the parameters because it requires understanding of background situation and common setting for different scenarios may not produce accurate result. All these aspects put limitations on effective use of background subtraction algorithm and demand improvement and extension of original GMM.

Recent years, researchers are focused on developing innovative technology to improve performance of IVS in terms of accuracy, speed, and complexity. To design novel approach for GMM parameter tuning based on extraction of statistical features and map with GMM training parameters [11]. Learning rate parameter is very important which determines the rate of change of background. Large amount of experimentation is required to set the value of learning rate for exact detection of foreground object. It is required to develop the system which tunes the parameter automatically for satisfactory performance of GMM.

GMM modeling is able to handle multimodal background scene. Performance of GMM-based background subtraction is decided by pixel-wise comparison of ground truth and actual foreground mask. Performance of the system is evaluated with the help of primary metrics such as true positive (TP), true negative (TN), false positive (FP), and false negative (FN) and secondary metrics like sensitivity, accuracy, miss rate, recall, and precision. Precision reflects false detection rate and recall gives accuracy of detection. Precision and recall are the two important measures in order to estimate detection algorithm systematically and quantitatively [12, 13].

$$\text{Precision}(\%) = \frac{TP}{TP + FP} * 100 \tag{11}$$

Our proposed system brings innovation in original GMM-based object detection system through tuning and adaptation of important parameters such as number of component, learning rate, and threshold.

3.1 Video Database

Wallflower database is open source database [9]. It includes seven sets of video sequence with different critical situations in background. Video frames of size 160×120 pixel, sampled at 4 Hz. Data set provider also gives one ground truth image and text file having description of all video sequences. Ground truth is binary image representing foreground mask of specific frame in video sequence. Table shows all test sequences along with their ground truth (Table 1).

Table 1 Wallflower dataset of seven different video sequences along with ground truth

Sr. No.	Name	Test sequence	Ground truth
1	Moved object (MO)		
2	Time of day (TOD)		
3	Light switch (LS)		
4	Waving tree (WT)		
5	Camouflage (C)		
6	Foreground aperture (FA)		
7	Bootstrap (B)		

3.2 Experimental Setup

The main focus of research is based on appropriate selection of GMM training parameters like K, α, and T. The selection of K Gaussian component value is function of complexity of background scene. If the background is simple and unimodal, the value of K must be selected as 1 or 2. For complex multimodal background, value of K is more than 2 and less than 5 so as to improve the accuracy in detection process. Various pairs of α and T are evaluated on Wallflower dataset. After lots of experimentation best pair of α and T is identified based on performance analysis on various Wallflower videos [9]. Parameter initialization, training, and testing are the three important steps for object detection process.

3.3 Parameter Initialization

Object detection system includes various GMM parameters like number of training frames, initial variance, and training parameters (K, α, and T). They are initialized as follows:

Number of training frames: 200 (given by data set provider),
Number of component: 4,
Initial variance: 0.006, and
Threshold: Adjusting value empirically (0.5, 0.6, 0.7, 0.8, 0.9).

4 Experimental Results

GMM-based object detection system is evaluated using various settings of α and T for each sequence. After this experimentation, for all videos, appropriate setting of α and T is decided based on lowest value of total error. Performance metrics are calculated for each sequence by comparing detected mask with ground truth.

Results are as follows (Fig. 1):

GMM-based background subtraction technique gives best overall detection performance at $\alpha = 0.001$ and $T = 0.9$. These parameter settings improve the accuracy of foreground mask which is almost matching with ground truth. Learning rate and threshold have enough power to tune object detection performance. Best overall performance setting has less probability to give best result at individual level. Best individual performance may be obtained by different settings of parameters for some of the sequences. Performance analysis for best α and T can be done at pixel level. Empirically selection of higher threshold value gives merging of foreground objects with background. It leads to increase in false negative and decrease in true positive. For faster changing background, empirical selection of lower value of α is too low to adapt such background changes. Empirical setting of threshold value, $T = 0.9$, is

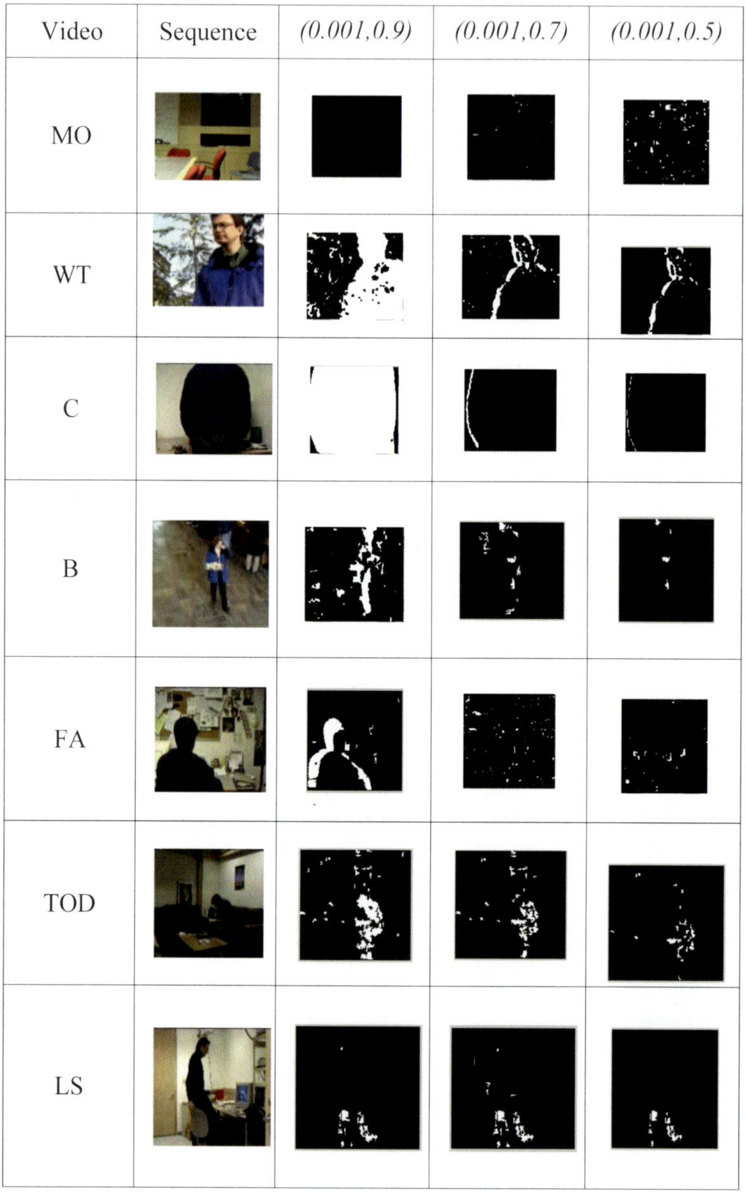

Video	Sequence	(0.001,0.9)	(0.001,0.7)	(0.001,0.5)
MO				
WT				
C				
B				
FA				
TOD				
LS				

Fig. 1 Foreground mask obtained using GMM for different values of α and T

being so high that all foreground pixels are merged into the background. It gives an increase in false negative and decrease in true positive pixels. Same way, empirical selection of learning rate, $\alpha = 0.001$, is being too low for rapid changing background. Thus, misclassification is higher and accuracy is lower for those video sequences in

Table 2 Performance evaluation of proposed system on Wallflower dataset

Video	True positive (TP)	True negative (TN)	False positive (FP)	False negative (FN)	Precision (%)	Accuracy (%)
MO	0	19,200	0	0	100	100
WT	12,558	766	306	4570	90	80.34
C	3399	1039	770	8231	81.53	83.68
B	2546	15,927	295	432	89.61	85.49
TOD	125	17,762	0	1313	90.16	91.43
FA	3754	13,718	745	983	83.44	84.69
LS	718	12,670	3359	2453	69.72	77.87

which sudden change of illumination is occurring. This experimentation also suggests that various settings of (α and T) may result in more improved performance for different video sequences than fixed selection of (α and T) (Table 2).

5 Conclusion

Proposed research emphasizes on proper tuning of important GMM parameter leading to improvement in the performance accuracy of GMM-based object detection system. GMM parameters mainly include number of mixture component (K), learning rate (α), and threshold (T). We have implemented two approaches for tuning these parameters such as traditional empirical tuning and automated adaptive tuning based on background dynamics. Traditional empirical tuning method is implemented using different settings of α and T, while K is kept constant to high value for complex scene. After large number of experimentation, appropriate pair of α and T is selected based on low performance error.

Proposed adaptive tuning method involves adaptation of α and keeping T and K constant to appropriate value. Unique EIR concept is used to extract background dynamics for current frame. Learning rate is tuned depending on EIR. This modified approach improves the result of GMM compared to the original GMM. This result strongly emphasizes the strength of learning rate adaptation. Performances of GMM with these tuning methods are evaluated based on foreground mask obtained using GMM and ground truth image in database. The analysis of performance can be done using primary metrics such as TP, TN, FP, and FN as well as secondary metrics like precision and accuracy. Our proposed system performance is compared with traditional empirical method and other existing techniques. Our research is implemented on MATLAB 2014 platform. Different functions from MATLAB computer vision toolbox are used for implementation of algorithm.

References

1. Hsieh J-W, Yu S-H, Chen Y-S (2006) An automatic traffic Surveillance system for vehicle tracking and classification. IEEE Trans Intell Transp Syst 7
2. Chauhan AK, Krishan P (2013) Moving object tracking using Gaussian mixture model and optical flow. Int J Adv Res Comput Sci Softw Eng 3
3. Picardi M (2004) Background subtraction technique: a review. In: IEEE international conference on systems, man and cybermetics
4. Viola P, Jones M (2005) Detecting Pedestrians using patterns of motion and appearance. Int J Comput Vision 63(2):153–161
5. Bo W, Nevatia R (2007) Detection and tracking of multiple partially occluded Humans by Baysian combination of edgelet based part detector. Int J Comput Vision 75(2):247–266
6. Ran Y, Weiss I (2007) Pedestrian Detecting via periodic motion analysis. Int J Comput Vision 71(2):143–160
7. Stauffer C, Grimson WEL (1999) Adaptive background mixture models for real time tracking. In: International conference on computer vision and pattern recognition 2
8. Cucchiaria R, Grana C, Piccardi M, Prati A (2003) Detecting moving objects, ghosts and shadows in video streams. IEEE Trans PAMI 25(10):1337–1342
9. Toyama K, Krumm J, Brumitt B (1999) Wallflower: Principles and practice of background maintenance. In: International conference of computer vision, pp 255–261
10. Zhang, LZ, Hou Z, Wang H, Tan M (2005) An adaptive mixture Gaussian background model with online background reconstruction and motion segmentation. ICIT, pp 23–27
11. White B, Shah M (2007) Automatically tuning background subtraction parameters using Particle swarm optimization. In: IEEE international conference on multimedia and Expo, China, pp 1826–1829
12. Harville M, Gordon G, Woodfill J (2001) Foreground segmentation using adaptive mixture models in color and depth. In: Proceeding of the IEEE workshop on detection and recognition of events in Video, Canada
13. Elgammal A, Harwood D, Davis L (2000) Non parametric model for background subtraction. In: European conference on computer vision, pp 751–767

Motion Detection Algorithm for Surveillance Videos

M. Srenithi and P. N. Kumar

Abstract Locality Sensitive Hashing (LSH) is an approach which is extensively used for comparing document similarity. In our work, this technique is incorporated in a video environment for finding dissimilarity between the frames in the video so as to detect motion. This has been implemented for a single point camera archiving, wherein the images are converted into pixel file using a rasterization procedure. Pixels are then tokenized and hashed using minhashing procedure which employs a randomized algorithm to quickly estimate the Jaccard similarity. LSH finds the dissimilarity among the frames in the video by breaking the minhashes into a series of band comprising of rows. The proposed procedure is implemented on multiple datasets, and from the experimental analysis, we infer that it is capable of isolating the motions in a video file.

1 Introduction

Smart video surveillance plays a vital role in ensuring safety and security in public and private spaces. Particularly in vast spaces and big organizations, possession of an inherent capability to detect motion without human intervention assumes importance (where large number of cameras may be deployed). The pertinent challenging factors in the detection of motion from surveillance videos are fatigue for a human to review the video; manifold increase in computation cost for further processing and the requirement of large storage space. Thus, as a part of any intelligent video surveillance system, inclusion of video motion detection capability that can automatically detect the events and render further support in the ensuing video processing techniques, is a force multiplier.

M. Srenithi · P. N. Kumar (✉)
Department of Computer Science and Engineering, Amrita School of Engineering,
Amrita Vishwa Vidyapeetham, Coimbatore, India
e-mail: pn_kumar@cb.amrita.edu

M. Srenithi
e-mail: cb.en.p2cse16018@cb.students.amrita.edu

© Springer Nature Switzerland AG 2019
D. Pandian et al. (eds.), *Proceedings of the International Conference on ISMAC in Computational Vision and Bio-Engineering 2018 (ISMAC-CVB)*, Lecture Notes in Computational Vision and Biomechanics 30,
https://doi.org/10.1007/978-3-030-00665-5_92

2   Related Work

Traditional motion detection approaches are background subtraction; frame differencing; temporal differencing and optical flow. Background subtraction method
detects motion by pixel-by-pixel subtraction of the current image from a reference
background image [1]. Frame differencing identifies the existence of moving object
by considering the variation between two successive frames [2]. In these methods,
the performance is adversely affected when the background is dynamic. In temporal
differencing [3], the current frame is compared with the previous frame, and the
compared frame is then threshold to fragment out the foreground objects [4]. This
approach fails when the pixel values are uniformly distributed. Optical flow [5] is the
motion of objects in a visual sight caused by relative motion between the camera and
the frame. This method, however, is computationally expensive. Gregory presented
a content-based retrieval method for long-surveillance videos in which archival of
video is based on low-level spatio-temporal extraction [5]. They are hashed into an
inverted index using Locality Sensitive Hashing (LSH) for query flexibility. Partial matches are extracted to user-created queries and [5] assembled subsequently
into full matches using dynamic programming that assembles the indexed low-level
features into a video segment that matches the query route by exploiting causal-
ity [5]. Ling et al. [6] presented a contour based object tracking approach. Using
multi-feature fusion strategy, the rough location of an object is found. Extracting the
contours with the help of region-based object contour extraction, accurate and robust
object contour tracking have been demonstrated. Zhao et al. [7] presented a method
in which the average values of a continuous multi-frame gray image are calculated,
and through statistical averaging of the continuous image sequence the background
image is obtained.

3   Background

Hashing has gained much popularity for quick estimate of similarity computation.
Among the diverse hashing methods, Locality Sensitive Hashing is extensively used
in high dimensional databases. In LSH [8], a group of arbitrary hash functions called
Minhash are associated with the banding technique to eliminate pairs with high
similarity from the pairs with less similarity. Minhash transforms large files into a
matrix of signatures which preserves the similarity of record pairs and LSH finds
identical pairs which are most probably similar in the same bucket, thus abstaining
assimilation of each and every combination of files.

3.1 Minhashing

Minhashing is a technique which uses the random hash function for fast Jaccard similarity computation. Amid Minhash and Jaccard Similarity there exists a significant relationship, which is the probability that the Jaccard similarity for two sets equals the value given by the Minhash function for those sets. To apprehend how, it is essential to visualize the columns of those sets A and B, therefore rows are separated into three categories: category X rows contain 1 in both columns, category Y rows contain 1 and 0 alternatively, and category Z rows contain 0 in both. Meanwhile, most of the rows are of category Z; Jaccard similarity (A, B) is determined by the ratio of category X and Y [9]. Thus Jaccard similarity $(A, B) = x/x+y$ where x denotes rows of category X and y denotes rows of category Y. As the size of $A \cap B$ is x and the size of $A \cup B$ is $x + y$ [9].

Assume for probability $h(A) = h(B)$ that the rows are randomly permuted, and when we traverse from the top $x/(x + y)$ is probability to see category X before a category Y. When the first row from the top is a category X, then $h(A) = h(B)$ [9]. Otherwise, if the first row is a category Y, then the Minhash value is acquired by set with a 1. Thus, $h(A) \neq h(B)$ if category Y row is meet first [9]. Thus concluding, the probability $h(A) = h(B)$ is $x/(x + y)$, which is the Jaccard similarity of A and B.

3.2 Minhash Signatures

To represent sets as a characteristic matrix M, a random n number of permutations of the rows of the matrix are chosen. Minhash functions are determined as $h_1, h_2, \ldots, h_n$. The minhash signature for S is constructed from the column representing set S, that is the $[h_1(S), h_2(S), \ldots, h_n(S)]$ vector [9]. This hash-values list is represented as a column [9]. Thus, a signature matrix is formed from matrix M, in which the minhash signature replaces the ith column of M for the set of the ith column.

3.3 Locality Sensitive Hashing

Although the concept of minhash function can be employed to compress bulk files into small signatures and conserve the required similarity, efficient similarity computation is still impossible [9]. Since the total number of pairs of files will be too large. Hence, minhash is associated with the banding technique Locality Sensitive Hashing to facilitate efficient similarity computation. In Locality Sensitive Hashing, files are hashed numerous times, and such that similar files are probably hashed to the same bucket than dissimilar files are. Any pair of files that is hashed to the same bucket for any hash function becomes a candidate pair. Candidate pairs are considered for similarity. This intends that dissimilar pairs by no means hash to the same bucket.

Having the minhash signatures, an efficient way to further handle the hashing is to divide the signatures into series of bands b comprising of rows r [9]. There is a hash function for each band that takes a column's segment inside that band and hashes them to some extensive number of buckets. The equivalent hash function can be utilized for all the bands, but for each band we use a separate bucket array, so columns with the same vector in different bands will not hash to the same bucket [9].

4   Proposed Method

In this work, we consider the problem of detecting motion in sequence of frames within a video and propose a procedure based on minhashing and LSH. We use rasterizing procedure [10] in which frames of the video are chosen, raster layer objects created, and finally these raster layer objects are aggregated to store the pixels in a text file. Each frame in the video is modelled to pixel files of type text. In order to detect motion, each pair of frame needs to be compared. This is not only time consuming, but also is prohibitively expensive as well. To surmount this problem we employ minhashing and LSH algorithms [11].

Minhash function tokenizes the pixels into a set of hash integers, from which the minimum value is selected, which is similar to random selection of a token. In LSH, [11] minhashes are shattered into a sequence of bands consisting of series of rows. Each band is hashed to a bucket. If two continuous frames have different minhashes in a band, they are hashed to the same bucket and considered as candidate pairs from which motion is detected. The outline workflow of the proposed method is shown in Fig. 1.

The procedure to implement the LSH algorithm is depicted in the architecture diagram given in Fig. 2.

5   Implementation and Results

5.1   Rasterization of Frames in the Video

Video is considered as frames per second for rasterization and a raster layer object is created from the image object. Raster layer object [10] is a dot matrix data structure, representing a rectangular grid of pixels of a frame as a rectangular grid. The procedure for calculating the pixel values of the frames in the video is as follows:

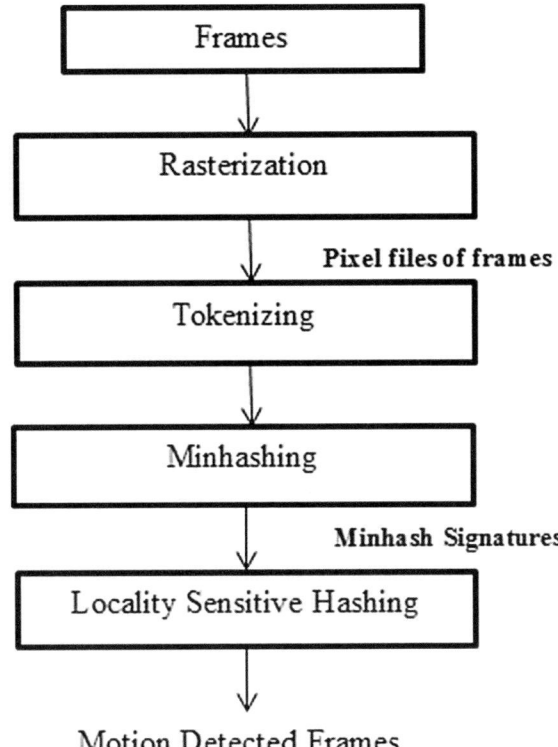

Fig. 1 Proposed workflow of motion detection

Motion Detected Frames

Input: Set of frames extracted from the video
Output: Set of text files containing pixel values of each frame.
Steps:

1. Convert each frame as Raster Layer Object.
2. Assign a variable to the extent co-ordinates of frame.
3. Calculate the pixel value of the frames.
4. Pixel values are aggregated and stored in a text file.

5.2 Minhashing

The pixels of the frame are tokenized using a tokenizer, converted into a set of hash integers using a minhash function, and we select the minimum value from the set of integers. This is similar to selecting a random token. Minhash function [11] does this repeatedly by effectively selecting n random token using various hash functions. For random hash functions, minhash generator takes seed for re-creating equivalent

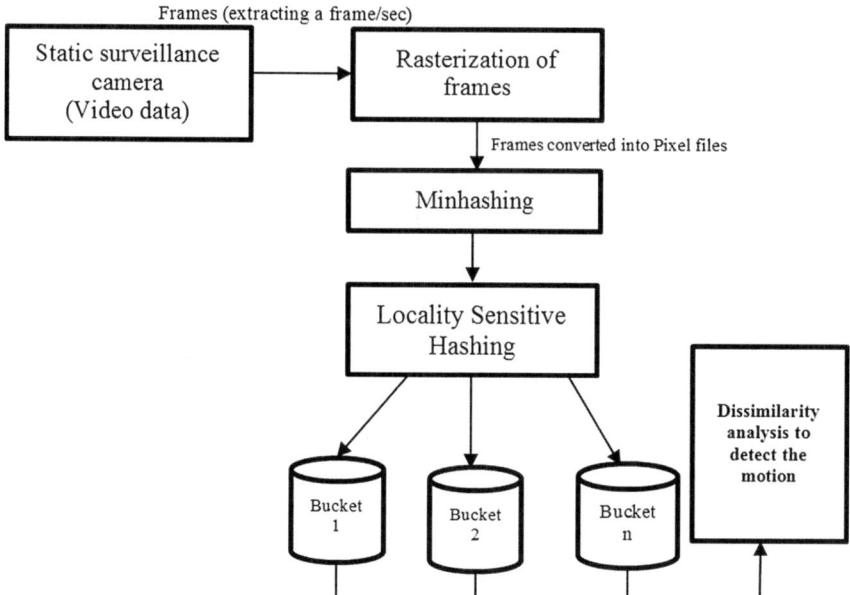

Fig. 2 The LSH implementation architecture

minhash function. Thus minhash function converts the set of tokenized pixels into n
randomly selected and hashed tokens.

5.3 Locality Sensitive Hashing

The LSH [11] function solves the problem of complexity. The minhashes are broken
down into series of bands comprising of rows. Each band is hashed to a bucket.
Frames having exact minhashes in a band are hashed into the different bucket, and
frames were motions have occurred that is with different minhashes are hashed into
same buckets so as to detect the motions. The procedure for LSH is as follows:

Input: Set of text files containing pixel values of each frame.
Output: Frames containing abnormal events
Steps:

1. Initialize the minhash generator#n and the seed value.
2. Load the pixel files.
3. Tokenize the pixel files.
4. Apply hash functions to convert the tokens into set of hash integers.
5. Select the minimum value token.
6. Break the minhashes into bands and rows.

7. Each band is hashed into buckets.
8. Similar bands are hashed into separate buckets.
9. Dissimilar bands are found in same buckets.

5.4 Datasets

We have considered two datasets for our experiment, viz. (1) A real-time static surveillance dataset and (2) Video Surveillance Online Repository (VISOR). The real-time static surveillance dataset includes approximately one hour long videos captured from a university campus; each consisting of 500 to 600 frames per second with a resolution of 352×288. The video has been captured both during the day (Fig. 3) and at night (Fig. 4). Movement of pedestrians and vehicles has been captured to illustrate motion detection. The VISOR [12] video (Fig. 5) contains one minute long, 60 frames per second data with a resolution of 360×288. It is a static camera traffic surveillance video taken from Video Surveillance Online Repository.

Fig. 3 An illustrative frame of a real-time dataset (Daylight)

Fig. 4 An illustrative frame of a real-time dataset (At night)

Fig. 5 Examples of detected motions in VISOR Dataset

5.5 *Results*

The results obtained in terms of Precision, Recall and F-Measure employing the min-hash and LSH techniques are tabulated below as a confusion matrix for both data sets used by us. In F-measure twice as much weight is given to precision since accurately detecting the motions is essential, thus an $F_{0.5}$ score is used. In motion detection using this technique, we have achieved an accuracy of over 85% for detection of motion in videos (Table 1).

A comparison of the results achieved in our experiment of employing LSH, with other techniques reported viz., MODE [13], DP-GMM [14], and Culibrk [15] is given in Table 2. From the table, we can infer that the accuracy of the method proposed which is experimented on real-time dataset is significant and is higher than few of the existing methods.

As shown in the Table 3 storing the detected frames reduces the amount of storage up to 90% for real-time and VISOR videos. Also saving the detected frames lessen the fatigue of security personnel to sit and review the surveillance videos.

The Receiver Operating Characteristic (ROC) curve obtained for the motion detection experiment for the real-time static surveillance video and VISOR static surveillance video is as shown in Fig. 6.

Table 1 Results of motion detection employing LSH

Dataset	Precision	Recall	$F_{0.5}$ Measure
Real-time	0.865	0.93	0.88
VISOR	0.84	0.82	0.835

Table 2 Comparison of accuracy obtained/reported

Method	Accuracy reported
Motion detection using LSH	0.87
DP-GMM [14]	0.7567
Culibrk [15]	0.5256
MODE [13]	0.7628
Belhani [16]	0.85

Table 3 Compression ratio of the videos

Video	Actual size (Mb)	Compressed size (Mb)
Video 1 (Real-time)	24.2	2.57
Video 2 (Real-time)	33.3	2.1
Video 3 (Real-time)	27	2.3
VISOR video	15.3	1.5

Fig. 6 Receiver operating characteristic of LSH Method

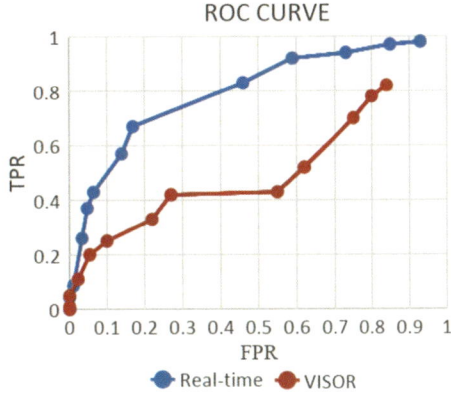

6   Conclusion

In this work, we have developed a motion detection system for static surveillance videos based on minhashing, followed by the application of Locality Sensitive Hashing (LSH) technique. The extracted frames are rasterized and minhashed. The minhashed frames are then broken into a sequence of bands and hashed into buckets employing the LSH technique. Experimental results on the real-time static surveillance and in the VISOR dataset demonstrates the improved accuracy and effectiveness of the proposed method and also out-performs the few of the existing methods. Such a system can be deployed in the monitoring station of surveillance videos in large establishments. This method is not only accurate and efficient but also would reduce the fatigue to the security personnel nominated to screen and monitor large sized recordings of videos in installations and open spaces.

References

1. Joshan J, Suresh P (2012) Systematic survey on object tracking methods in video. Int J Adv Res Comput Eng Technol (IJARCET)
2. Papageorgiou, Oren, Poggio (1998) A general framework for object detection. In: IEEE international conference on computer vision (ICCV)
3. Astha, Manoj, Kailash (2016) Survey on video object detection and tracking. Int J Current Trends Eng Technol
4. Singh T, Sanju, Vijay (2014) A new algorithm designing for detection of moving objects in video. Int J Comput Appl
5. Gregory C, Elgharib M, Saligrama V, Jodoin P-M (2015) Retrieval in long surveillance videos using user-described motion and object attributes. IEEE Trans Circuits Syst Video Technol
6. Lu X, Song L, Yu S, Ling N (2012) Object contour tracking using multi-feature fusion based particle filter. In: IEEE conference on industrial electronics and applications (ICIEA)
7. Yang K, Cai Z, Zhao L (2013) Algorithm research on moving object detection of surveillance video sequence. Opt Photonics J

8. Sowmya K, Kumar PN (2018) Traffic density analysis employing locality sensitive hashing on GPS data and image processing techniques. In: Computational vision and bio inspired computing

9. Rajaraman A, Ullman J (2011) Mining of massive datasets, Cambridge University Press, New York. chap. 3

10. https://www.rdocumentation.org/packages/raster/versions/2.5–8/topics/raster

11. Mullen L Minhash and locality-sensitive hashing. https://cran.r-project.org/web/packages/textreuse/vignettes/textreuse-minhash.html

12. Vezzani R, Cucchiara R (2010) Video surveillance online repository (ViSOR): an integrated framework in Multimedia Tools and Applications, Kluwer Academic Press

13. Singh B, Singh D, Singh G, Sharma N, Sibbal V (2014) Motion detection for video surveillance. In: International conference on signal propagation and computer technology (iCSPCT)

14. Haines TSF, Xiang T (2012) Background subtraction with dirichlet processes. In: 12th European conference on computer vision

15. Culibrk D, Marques O, Socek D, Kalv H, Furht B (2007) Neural network approach to background modeling for video object segmentation. IEEE Trans Neural Networks

16. Belhani H, Guezouli L (2016) Automatic detection of moving objects in a video surveillance. In: Global Summit on Computer and Information Technology (GSCIT)

6D Pose Estimation of 3D Objects in Scenes with Mutual Similarities and Occlusions

Tiandi Chen and Lifeng Sun

Abstract Estimation of six degrees of freedom (6DoF) attitude of rigid bodies is an essential issue in such fields as robotics and virtual reality. This paper proposed a method that could accurately estimate 6DoF attitude of known rigid bodies with RGB and RGB-D input image data. Additionally, the method could also work well in the scenes in which objects exhibit mutual similarities and occlusion. As one of the contributions made by this paper, a modularized assembly line method was proposed, which integrated deep learning and multi-view geometry method. At first, a neural network for instance segmentation was used to identify the general locations of known objects in the images and give the bounding boxes and masks. Then 6DoF attitude was estimated roughly according to the local features of RGB-D images and templates. Finally, purely geometric method was used to refine the estimation. Another contribution of this paper was the correction of misclassification with the help of some information reserved in the process of training the network. The proposed method achieves a superior performance on a challenging public dataset.

1 Introduction

Estimation of 6DoF attitude of 3D objects has long been an issue arousing study interests in the field of computer vision. This problem is also highly relevant to robotics and virtual reality. With the development of the fields and the improvement of calculation ability, this problem has been gradually developing from simple 2D image detection to complete 6D altitude estimation of objects, including 3DoF location and 3DoF rotation direction. Methods in early period mainly calculated 3D altitude by calculating the feature points of objects according to their surface textures and carrying out matching [1]. The past decade has seen the rapid development of depth

T. Chen · L. Sun (✉)
Department of Computer Science and Technology, Tsinghua University, Beijing, China
e-mail: sunlf@mail.tsinghua.edu.cn

T. Chen
e-mail: chentd11@gmail.com

© Springer Nature Switzerland AG 2019
D. Pandian et al. (eds.), *Proceedings of the International Conference on ISMAC in Computational Vision and Bio-Engineering 2018 (ISMAC-CVB)*, Lecture Notes in Computational Vision and Biomechanics 30,
https://doi.org/10.1007/978-3-030-00665-5_93

camera technique, and post-estimation of texture-less objects becomes more easily. The present studies mainly focus on altitude estimation in difficult scenes with heavy occlusion and other clutter. The solving of this problem will improve the robustness of object recognition algorithm.

For texture-less objects, template-based methods, such as Point-Pair Features and LINEMOD were once popular. The core thought is to obtain the CAD model of a known object and generate the point cloud templates of the object presenting different altitudes. The templates are one-to-one corresponding to poses. Then, local features of point clouds are extracted from the input scenes and compared with the existing templates. Then, ICP algorithm is used for refining the pose, with the final results being output.

The latest study results are already able to address the pose estimation problem of most scenes. Reference [2] improved the point-pair matching method used before and delivered encouraging pose estimation results of rigid bodies in the case of RGB-D input. Reference [3] introduced a new algorithm architecture, which got great recognition accuracy in LINEMOD datasets. The latest dataset T-LESS [4] contained more similar low texture objects and more serious blocking, making the problem more challenging.

There are some similarities between the work in this paper and the work before. Reference [3, 5, 6] all used image data to initialize object coordinate system, while this paper adopted depth data. In addition, the masks of all objects in this paper were known after the first phase, which offered a great advantage.

We summarize our contributions as follows:

- The test result of public datasets showed that our method achieved a good performance, which is much better than the previous methods.
- We proposed a modularized assembly line system, in which modules were not coupled and could work independently, making it possible to improve individual parts.
- We implement a tool to generate new scenes and render it as point cloud data, which generate training and test data randomly and automatically.
- Taking the advantage of this tool, we developed a method to quantitatively measure the similarity between two categories of 3D object, which helps correct the classification error.

The remainder of the paper is organized as follows. Relevant work in this field was introduced in the second part of this paper, detailed introduction to our method was given in the third part, our experiment results were provided in the fourth part, and a summary was made in the fifth part.

2 Related Work

Traditional 2D object detection methods are mainly to match sparse features [1] or templates [7]. Between the two methods, template-based method can better process cases involving low texture objects. Generally speaking, data collected by cameras have small projected areas and texture cannot provide enough feature points. Thus, template-based method usually performs better than feature-based methods. A major drawback of the method completely based on templates is the sensitivity to the occlusion which would reduce the accuracy of recognition and pose estimation.

With the launch of Kinect [8] and other consumer depth transducers, template-based methods can be used more easily. LINEMOD template-based method [9] is a good template-based method. This method combines the features of the normal of point cloud and the normal of RGB image boundary, and generates a descriptor based on the two features, with comparison being made with all template descriptors in the template library. Later, some work optimized the speed of template-based method [10]. Other voting methods have also delivered fine effects, especially point-pair features [11, 2] was also based on point-pair features and made some improvements for the blocking problem that template methods were difficult to address.

The development of deep learning method also leads to breakthroughs in the field of computer vision recognition. Learning-based method is widely applied in the 6D altitude estimation of rigid bodies as well. Reference [12] used neural network to conduct direct regression on pose of objects. But a method completely relying on learning cannot produce highly accurate altitude estimation results. Reference [13] proposed to optimize results by classifying views of discrete objects but did not deliver any test result of the method in challenging public datasets. It seems difficult for pure learning method to achieve a high accuracy of pose estimation. Learning method was only used to figure out the rough locations and masks of the objects rather than directly obtaining 6D pose in an end-to-end network.

Another way to solve the problem was to combine voting and learning together. Reference [14] adopted 6D voting on each image block of RGB-D images using Hough Forest. Reference [15] extracted features with auto-encoder, calculated one feature for each altitude, carried out feature matching on the input RGB-D images, and found out the optimal matches by using the voting method. Random forest, as an improvement of the voting method, can work out the correspondence between images and object coordinates as well. Reference [16] used random forest to match image blocks to the 3D coordinate system and then made corrections by RANSAC.

Reference [17] BB8 was a method proposed recently, which took RGB images as the only input and conducted altitude estimation only based on sparse fixed points of several objects. The pose was estimated by finding the exact points having highly obvious features on object surface. But our experiment results showed that the accuracy could be improved when point cloud information relied on more if there was point cloud data.

All in all, it's difficult to estimate 6D pose of objects in a scene with occlusion. And there were few previous studies that could solve the problem of pose estimation of objects with mutual similarity. However, our method achieved greater accuracy in these cases when RGB-D data input was used.

3   Proposed Method

The framework of our method is shown as follows. First, we used a network (Multi-task Network Cascade, [18]) to conduct instance segmentation for the input images, then found several most similar categories for each object and calculated the similarities of all templates, and weighted the similarities according to the classification error rate obtained in the training of the last step. We choose several most similar templates, which correspond to pose hypothesis. At last, we verify and optimize the pose hypothesis obtained from the template matching of the last step, with the refined results being output (Fig. 1).

The core of the method proposed in this paper was to decompose the problem into three cascade and low-coupling tasks. First, the general location of the object was given according to a detection conducted on 2D image. Then the general pose was provided through comparison between local features of the point cloud and the template point cloud. After that, 6D pose was got through optimization using a pure geometric method. In the first step, the existing instance segmentation method was used to train a deep neural network, as well as classify objects and give the boundary boxes, with the masks of objects being provided at the same time. In the meanwhile, during the training in the first step, some training data and testing data that contained the blocking of objects were generated randomly and automatically, with the error rate of classification being worked out. To some degree, the error rate of classification indicated the similarity level of objects. Based on this, an algorithm could be designed to search among templates of similar object categories in the subsequent template matching process, thus reducing the error rate. In the second step, LINEMOD was used to search among the point cloud templates for the most similar templates of the objects already having masks and falling into some categories. Then, a rough pose estimation value was obtained. In the third step, verifications and refine were made.

In summary, the method in this paper combined the learning method and the traditional geometric method together, which was an advantage. Our goal was to obtain accurate pose estimation, and the mathematic model of the problem is known if the category of the object was determined, so we believe that end-to-end learning

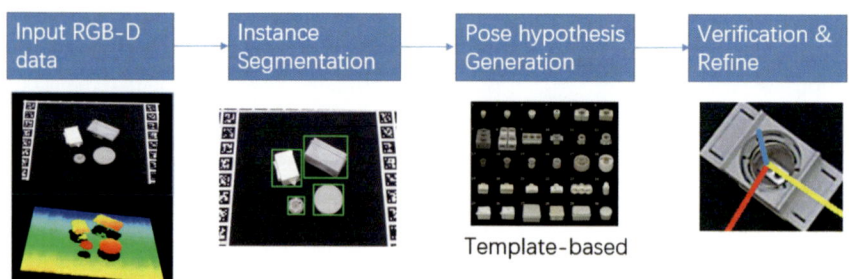

Fig. 1 Architecture of our system

would compromise accuracy. But it was hard for traditional methods to perform a recognition task very well, while the learning method had advantages in this issue. The initial recognition in step one gave the categories and masks of objects, which both reduced the matching times in the template matching stage and mitigated the impacts of noise because of the existence of determined masks. In the subsequent altitude optimization, if the error was found to be large, the smallest error would be found among several alternative results provided in step two and be output, which reduced the calculation work. In this way, both the accuracy and the calculation efficiency were guaranteed.

3.1 Instance Segmentation and Similarity Calculation Between Categories

The main purpose of this step was to find the classification results and masks of all known objects in a scene. The scene was segmented with Multi-task Network Cascade (MNC). If there were K known objects, with an input image being given, the output of the network would contain the masks of all K instances and their categories.

To handle the problem of misclassification, we automatically and randomly generated many scenes with some object was blocked and render them to point cloud data. Then we input the patch of object RGB image to the classifier, and the classification results of which were summarized. The frequency that object i was classified into j was recorded as C_{ij}. $i \neq j$ meant incorrect classification. If the C_{ij} is large, we may need to verify the category of this instance, even calculating the ICP error in the last step to determine which category it belongs to.

3.2 Generating Pose Hypothesis

Here we chose Linemod algorithm to find out the most similar template of the actual pose. Generating templates was a key step in the algorithm. Since 3D CAD models of some objects were not easy to be obtained in actual situations, we developed a method of generating templates of objects having different poses based on the collected RGB-D data. As for the view points, we selected all vertexes of the regular icosahedron on the sphere, as well as the midpoints of all edges of the regular icosahedron. At each view point, we could create a template by rotating through a certain angle for each time. In the experiment, we created a template by moving through every $5°$. As a result, $42 \times 72 = 3024$ templates were obtained in total.

After the templates were created, an altitude assumption could be obtained through comparison among similarities between objects in the scenario and templates. Linemod algorithm gave scores to the similarities between each template and the input RGB-D patch of each instance. Generally speaking, a template having the highest score could be taken as the most similar one, with its corresponding pose being taken as the initial value in the following optimization step. However, since there might be incorrect classification, our strategy was to detect all templates in all categories of $C_{ij} > 0.1$, take several templates with the highest similarity scores given by Linemod, and make further judgments at the time of refining the pose in the next step.

3.3 Pose Refine

For an object that was detected in the scene, its mask could help to locate the corresponding part in the point cloud. Those pose hypothesis obtained in the last step were recorded. Iterative Closest Point (ICP) algorithm was performed on the corresponding point cloud Po of each pose assumption, and the result with the smallest matching error was taken as the final output result.

4 Experiments

To test the performance of our system, we conducted a series of experiments. In this section, we first introduce the public dataset we conduct our test on and then show our results. Finally, we show the runtime performance of the system.

4.1 Test Dataset

T-LESS [4] is a new public dataset for testing system performance for 6D pose estimation of objects. This dataset contains 30 types of test objects, most of which are industrial parts without obvious texture. Most of these parts have a single color, so color-based methods cannot work well. The challenge of this data set is that objects are symmetric and mutually similar in shape and size. The test image has 20 different complex scenes, varying from a simple scene composed of several isolated objects to a large number of different objects and other unrelated objects with occlusion.

4.2   Performance

4.2.1   Accuracy of Recognition

We test all 20 scenes of the dataset, and the output of the recognition phase is as follows (Fig. 2):

Because there are many indistinguishable obstructions in some scenes, we just evaluate the instances which have a visible surface area of more than 10% of the total surface area. To figure out the accuracy of recognition, we calculate intersection over union (IoU) of our result and ground truth bounding box, and if the IoU is greater than a threshold (we set the threshold to 0.5), we apply this result as a correct recognition. The accuracy of recognition is as follows (Table 1).

We also figure out the accuracy of recognition of objects belonging to different categories, and we compare our results with BB8 [17]. It should be noted that BB8 only uses RGB data and our method needs depth data (Table 2).

4.2.2   Precision of Pose Estimation

We just sum up 3D location error of instances which is recognized correctly. The average location errors of each scene are as follows (Table 3).

This result achieves a high precision and is not related to the complexity of scenes. So It seems that the bottleneck of the performance is object detection and recognition.

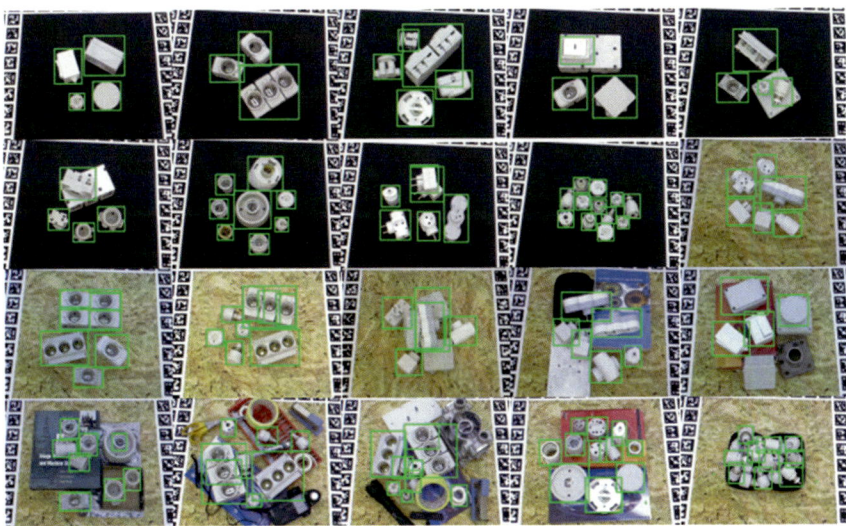

Fig. 2 Recognition result of the instance segmentation network

Table 1 Accuracy of recognition in each scenes

Scene no.	Total accuracy (%)	Scene no.	Total accuracy (%)
1	96.6	11	67.0
2	98.6	12	62.9
3	93.5	13	72.5
4	94.0	14	85.2
5	89.3	15	93.5
6	91.3	16	81.3
7	71.8	17	58.1
8	91.9	18	65.0
9	77.2	19	59.2
10	86.4	20	49.5

Table 2 Compared with BB8

Sense ID: [Obj. IDs]	Accuracy (BB8, RGB only)	Accuracy (Ours, RGB-D)
1: [2, 30]	50.8, 55.4	90.0, 92.0
2: [5, 6]	56.5, 55.6	84.6, 89.3
4: [5, 26, 28]	68.7, 53.3, 40.6	87.3, 76.4, 72.9
5: [1, 10, 27]	39.6, 69.9, 50.1	63.9, 89.3, 88.7
7: [1, 3, 13–18]	42.0, 61.7, 64.5, 40.7, 39.7, 45.7, 50.2, 83.7	77.4, 81.8, 79.5, 71.4, 52.3, 73.0, 83.1, 94.2

Table 3 location error in each scenes. Note that only correct recognition results are applied

Scene no.	Avg. location error (m)	Scene no.	Avg. location error (m)
1	0.0077	11	0.0095
2	0.0059	12	0.0099
3	0.0116	13	0.0173
4	0.0112	14	0.0210
5	0.0085	15	0.0114
6	0.0128	16	0.0106
7	0.0288	17	0.0094
8	0.0167	18	0.0139
9	0.0093	19	0.0127
10	*0.0120*	20	0.0242

Table 4 Time spent of each module

Module	Avg. time for one input scene (ms)
Instance segmentation	11
Pose hypothesis generation	130
Pose refine	6

4.2.3   Runtime Benchmark

At present, the instance segmentation module is implemented by GPU, and other two modules are implemented by CPU with multithread. A computer with an Intel i5-4210H CPU, an NVIDIA GeForce GTX1060 GPU, and an 8 GB memory is used to test. The time spent on every module of the system is as follows (Table 4).

It should be noted that the time required for the second phase is linear with the number of template matches required. So reducing the number of generated templates will speed up the system. If the objects exhibit less mutual similarity, the time spent will also reduce.

5   Conclusion

We proposed a modular system for 6D pose estimation of known objects. Combining deep learning methods and traditional geometry methods, the system achieves a high accuracy of recognition and high precision of pose estimation. And with the help of quantitatively measure the similarity between two categories of 3D object, the system is able to correct the misclassification, which enhances the robustness of the system. Additionally, each module of this system is independent, which makes it easy to modify and improve. From the experiment results, it seems that the bottleneck of performance is still object detection and recognition. The future work will construct a more reasonable neural network for object detection, and optimize the implement for real-time tasks.

References

1. Lowe DG (2001) Local feature view clustering for 3D object recognition. In: CVPR
2. Hinterstoisser S, Lepetit V, Rajkumar N, Konolige K (2016) Going further with point pair features. In: ECCV
3. Michel F, Kirillov A, Brachmann E, Krull A, Gumhold S, Savchynskyy B, Rother C (2017) Global hypothesis generation for 6D object pose estimation. In: CVPR
4. Hodan T et al (2017) T-LESS: An RGB-D dataset for 6D pose estimation of texture-less objects. In: 2017 IEEE Winter Conference on Applications of Computer Vision (WACV), IEEE
5. Brachmann E, Krull A, Michel F, Gumhold S, Shotton J, Rother C (2014) Learning 6D object pose estimation using 3D object coordinates. In: ECCV

6. Brachmann E, Michel F, Krull A, Yang MY, Gumhold S, Rother C (2016) Uncertainty-driven 6D pose estimation of objects and scenes from a single RGB image. In: CVPR
7. Huttenlocher D, Klanderman G, Rucklidge W (1993) Comparing images using the Hausdorff distance. IEEE Trans PAMI
8. WA MCR Kinect for Xbox 360
9. Hinterstoisser S, Holzer S, Cagniart C, Ilic S, Konolige K, Navab N, Lepetit V (2011) Multi-modal templates for real-time detection of texture-less objects in heavily cluttered scenes. In: ICCV
10. Rios-Cabrera R, Tuytelaars T (2013) Discriminatively trained templates for 3D object detection: a real time scalable approach. In: ICCV
11. Drost B, Ulrich M, Navab N, Ilic S (2010) Model globally, match locally: Efficient and robust 3D object recognition. In: CVPR
12. Kendall A, Grimes M, Cipolla R (2015) PoseNet: a convolutional network for real-time 6-DoF camera relocalization. In: ICCV
13. Kehl W, Manhardt F, Tombari F, Ilic S, Navab N (2017) SSD6D: making RGB-based 3D detection and 6D pose estimation great again. In: ICCV
14. Tejani A, Tang D, Kouskouridas R, Kim T-K (2014) Latent Class Hough forests for 3D object detection and pose estimation. In: ECCV
15. Kehl W, Milletari F, Tombari F, Ilic S, Navab N (2016) Deep learning of local RGB-D patches for 3D object detection and 6D pose estimation. In: ECCV
16. Brachmann E, Krull A, Michel F, Gumhold S, Shotton J, Rother C (2014) Learning 6D object pose estimation using 3D object coordinates. In: ECCV
17. Rad M, Lepetit V (2017) BB8: A scalable, accurate, robust to partial occlusion method for predicting the 3D poses of challenging objects without using depth. In: ICCV. 1
18. Dai J, He K, Sun J (2016) Instance-aware semantic segmentation via multi-task network cascades. In: CVPR

A Simple and Enhanced Low-Light Image Enhancement Process Using Effective Illumination Mapping Approach

Vallabhuni Vijay, V. Siva Nagaraju, M. Sai Greeshma, B. Revanth Reddy, U. Suresh Kumar and C. Surekha

Abstract When an image is captured in low-light, it gets the low visibility. To overcome the low visibility of the image, some operations are to be performed. But in this paper, image enhancement is introduced using illumination mapping. First, R, G, B maximum values in each pixel of the considered image are to be calculated and then convert it into a grey scale image by applying the formulae. Some filters are used to remove the noise, the choice of filter depends on the type of noise, and then the image is preprocessed. The logarithmic transformation helps to increase the brightness and contrast of the image with a certain amount. Earlier there were some methods to enhance the low-light image, but illumination map existence is chosen. In this illumination, the image will be enhanced with the good quality and efficiency. The illumination technique will be the more efficient and more quality. The illumination corrects the R, G, B values to get the desired image, then Gamma Correction is applied. The Gamma Correction is a non-linear power transform, it helps to increase or decrease the brightness of the desired image when a low value of gamma is taken, the brightness will be increased and when a high value of gamma is taken, and the brightness will be decreased. The proposed system is implemented using MATLAB software. When different types of images are applied, different contrast and brightness levels that depend on the type of image are observed.

1 Introduction

An image captured in low-light conditions is to be enhanced for a better visibility. Image enhancement plays a vital role in improving the digital image quality. Histogram equalization (HE) in [1] is widely used to adjust the image intensities and to enhance the contrast of the image in a variety of applications due to its simple

V. Vijay (✉) · V. Siva Nagaraju · M. Sai Greeshma · B. Revanth Reddy · U. Suresh Kumar
C. Surekha
Department of Electronics and Communication Engineering, Institute of Aeronautical
Engineering, Dundigal, Hyderabad 500043, India
e-mail: v.vijay@iare.ac.in

© Springer Nature Switzerland AG 2019   975
D. Pandian et al. (eds.), *Proceedings of the International Conference on ISMAC in Computational Vision and Bio-Engineering 2018 (ISMAC-CVB)*, Lecture Notes in Computational Vision and Biomechanics 30,
https://doi.org/10.1007/978-3-030-00665-5_94

function and its effectiveness. Contextual and variational contrast enhancement in [2, 3] proposes an algorithm that enhances the contrast of input image using interpixel contextual information. It maps the elements of one histogram to elements of another histogram diagonally. Layered difference representation is used to enhance the image contrast by mapping the input grey levels to the output grey levels. The grey-level difference between the adjacent pixels is amplified. Naturalness preserved enhancement algorithm preserves the naturalness of the image by using three methods, namely, lightness-order-error measure, bright pass filter and bi-log transformation. These are used for preserving naturalness and also enhances the image. Applications for the above-mentioned methods are that these are applied for thresholding, normalization of magnetic resonance imaging (MRI) images. For comparing two or more images on a particular basis, histograms of the images are to be normalized.

The major drawbacks of existing methods are an increase in the contrast of noise by decreasing the quality of the image. The results are computationally intensive. When we increase the brightness of the image, the brighter particles in the image will get brighter. One major drawback observed is with rapid change in technology, techniques that are used in real-time applications always keeps on changing less than a decade, i.e., digital image processing's real-time products can only survive in the market for a maximum of 3–5 years only and in this span other technologies arises. The initial cost of the technique is high depending on the type of system.

Proposed work mainly describes input image, that is, the low-light captured image is chosen and is to be enhanced to obtain the desired modified image as said in [4]. The file path of the image to be enhanced is considered and is preprocessed so that noise can be reduced easily. Noise present in the image causes blurriness. These noise or distortions are filtered using preprocessing step. The uneven illumination caused by sensors, default, non-uniform illumination of the scene or orientation of objects can be rectified using illumination correlation. Prospective correlation, retrospective correlation, and other using low pass filtering as from [5] is also done using illumination correction [6]. In illumination correction brightness and contrast can be applied to the image whereever required non-linear distribution is done. The greyscale levels between black and white colours are enhanced using gamma correction [7]. Gamma correction also enables us to compare grey levels between the adjacent pixels and stop adjust whichever colour is required. Pixel intensity enhancement has to be carefully adjusting brightness or contrast levels to larger extend may cause irregularities in the image. These pixels enhancement can be applied to both grey and colour scale image. Illumination map estimation is done using RGB channels individually to the input image and then is converted to grey scale image to check the improvements in the image. Enhancement is done until the desired output is obtained. Advantages and applications are also explained in this paper.

This paper is well organized as it clearly states step to step process of this project. Image enhancement is the major part as it is a basic need when any image is considered. The drawbacks in previous methods and the possibility to overcome them are discussed in the above paragraphs. Reduction of noise as explained in [8, 9], better version of HE and other steps has been taken. In the above-mentioned theory, it clearly states that what the steps are done from the beginning input image to obtaining the desired output. Section 2 describes how the input image is taken, reduction of noise, deblurring the blurred image, etc., are explained individually. Adjusting the brightness and contrast of the image is done accordingly. Applying different filters individually for the considered input image. Gamma correction enhances the pixel intensity. Gaussian and median filters [10] are used for noise reduction. Recovering the desired image from noise affected or blurred features can also be achieved [11]–[17]. In continuation with Sects. 2, 3 deals with the simulation results and comparison table of the input and output images.

2 Design Methodology and Observation

The design methodology briefly describes how image enhancement, preprocessing, transformation, illumination correction and gamma correction are performed on the image which is taken under low-light conditions. The enhanced image is obtained by using the below-explained methods.

2.1 Image Enhancement

Image enhancement is the process by which the digital images can be adjusted for better quality and for further use. Here the noise can be reduced, the images are sharpened and brightened.

The various methods involved in image enhancement are morphological operations, HE for removal of noise wiener filter is used, linear contrast adjustment, median filtering, unsharp mask filtering and decorrelation stretch [1].

Advantages of image enhancement are manipulating the pixel values in the image and it is easy for visual interpretation [2].

The flow chart is given in Fig. 1 describes each and every step of the techniques which are used for enhancement of the image. The algorithm shown below gives the information about how an image is taken and processed in a step by step manner. The first is to reduce the noise and enhancing the image by using some of the transforms and in the final step the image contrast levels are increased by using gamma correction.

Algorithm: Image Enhancement
Input: The input image is given by $F(x, y)$.
Initialization:
Update the value of $R(x, y)$ using equation (1).
Update the value of $g(x, y)$ using equation (2).
Update the value of $T_w(f)$ and $T_b(f)$ using $g(x, y)$.
Output: The solution is s.

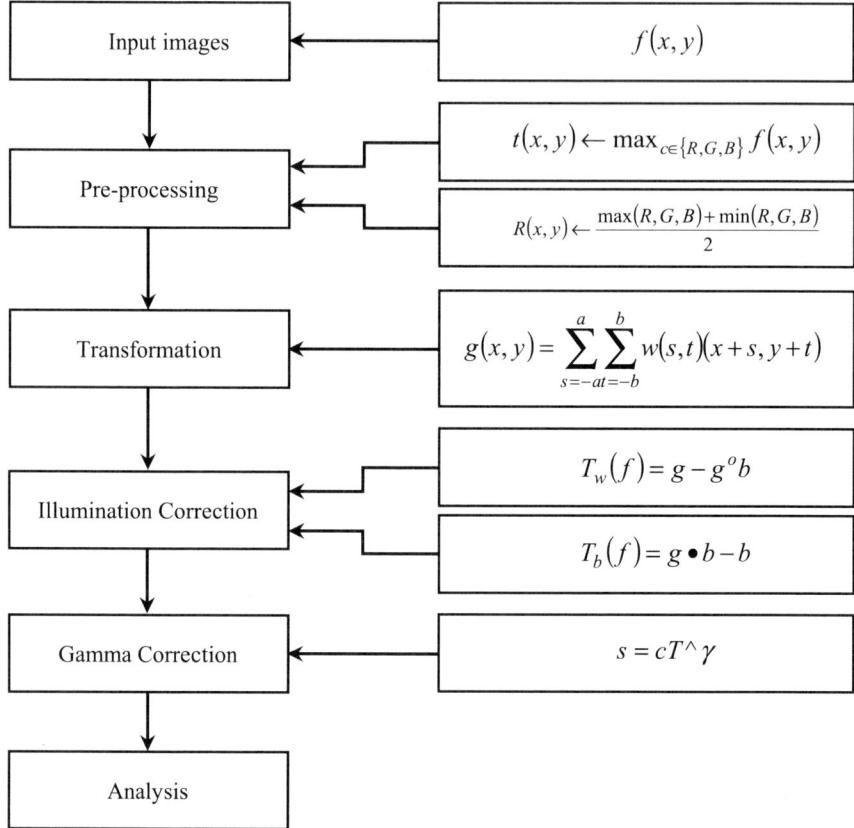

Fig. 1 Flow chart

2.2 Preprocessing

Preprocessing is a technique used to suppress the noise and distortion which are present in the image and it enhances the information in the image. The median filter is used in reducing the noise. It is a non-linear digital filtering technique. Reduction of noise is a major solution in preprocessing step for improvement of further use. Figure 2b shows how noise is reduced. By using a median filter, it preserves edges when noise is reduced. The main advantage of preprocessing is to reduce the noise [4].

$$t(x, y) \leftarrow \max_{c \in \{R,G,B\}} f(x, y) \qquad (1)$$

$$R(x, y) \leftarrow \frac{\max(R, G, B) + \min(R, G, B)}{2} \qquad (2)$$

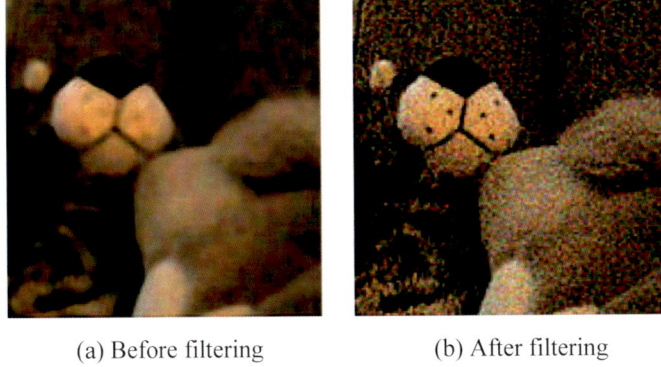

(a) Before filtering　　　　　　　　(b) After filtering

Fig. 2 Removal of noise using a median filter

Fig. 3 Image after transformation

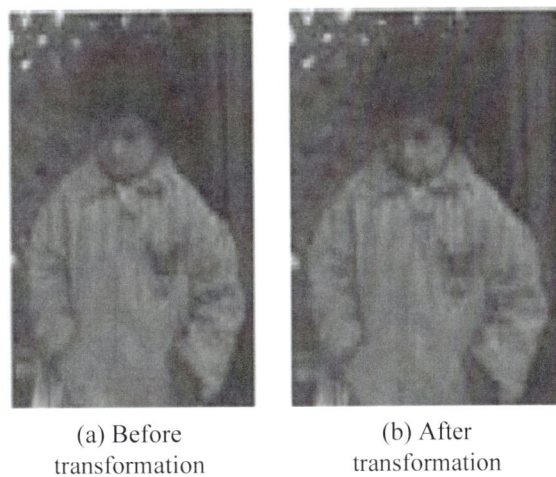

(a) Before transformation　　　　　(b) After transformation

The two Eqs. (1) and (2) represent the input image and the image converted into a grey image. Here f (x,y) represents the input image and R (x,y) is the grey image.

2.3　Transformation

Transformation is a technique used for operations in the image. First, the image is transformed into spatial domain then it removes the noise and again convert it into the frequency domain. The image before and after transformation is shown in Fig. 3a and b. The mathematical representation of the spatial domain is given in Eq. (3).

$$g(x, y) = \sum_{s=-a}^{a} \sum_{t=-b}^{b} w(s, t)(x + s, y + t) \tag{3}$$

where $w(s, t)$ is the image which is a preprocessed image, $R(x, y)$ is the grey image and $g(x, y)$ is the transformed image.

Advantages of transformation are noise reduction. In the frequency domain, fast convolution is carried out [5].

2.4 Illumination Correction

It is used to adjust the brightness and contrast level of the image. Illumination correction is done by two transforms, namely, top-hat and bot-hat transform. Top-hat transform is used in digital image processing and mathematical morphology. It is an operation used for the extraction of small elements that are given in Fig. 4.

It is also used for extracting the details from the given image. There are two types of top-hat transforms, namely, (1) white top-hat transform, (2) black top-hat transform. The white top-hat transform is known as the difference between the input image and the structuring element by opening it. White top-hat transform is represented by $T_w(f)$. The black top-hat transform is known as the difference between the closing of structuring element and input image and it is represented by $T_b(f)$. In various applications, top-hat transform is used [7].

$$T_w(f) = g - g^o b \tag{4}$$
$$T_b(f) = g \bullet b - b \tag{5}$$

The above two Eqs. (4) and (5) are top-hat and bot-hat transforms, where g is transformed image and b is the structuring element.

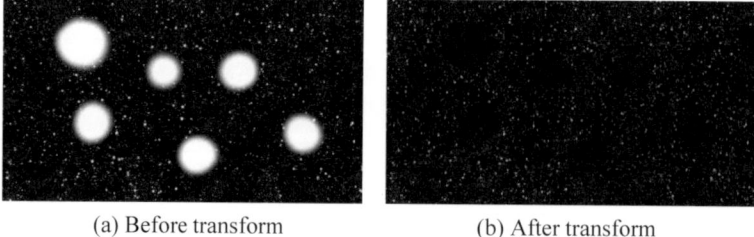

(a) Before transform   (b) After transform

Fig. 4 Illumination correction

(a) Before applying gamma   (b) After applying gamma

Fig. 5 Gamma correction

2.5 *Gamma Correction*

Gamma correction is also known as power law transformation. The nth power and nth root transformations are two types of power law transformations. These transformations are given by the expression presented in Eq. (6).

$$s = cT \wedge \gamma \tag{6}$$

This γ symbol is called gamma, by reflecting this symbol name, it is also known as gamma transformation. By varying the γ values the image can be enhanced that is shown in Fig. 5. Here c is a colour component value ranging from 0 raised to some power γ and T is the image. Different devices or monitors have different display settings, by this, the intensity also changes. The advantage of gamma is that the image can be displayed on CRT and LCD [8].

3 Simulation Results of the Proposed Design

Figure. 6a-l represents each and every method in the proposed model about how the image is taken and enhanced. First, the input image is converted into RGB channels and then converted into a grey image. The grey image is then preprocessed using a median filter and the image is transformed into frequency domain [5]. The image in [7] frequency domain is enhanced by changing the contrast levels using top-hat and bot-hat transform and next to the image is corrected by using gamma correction; in this, the image is enhanced by changing the gamma values. Thus, the output image is obtained shown in Fig. 6(l).

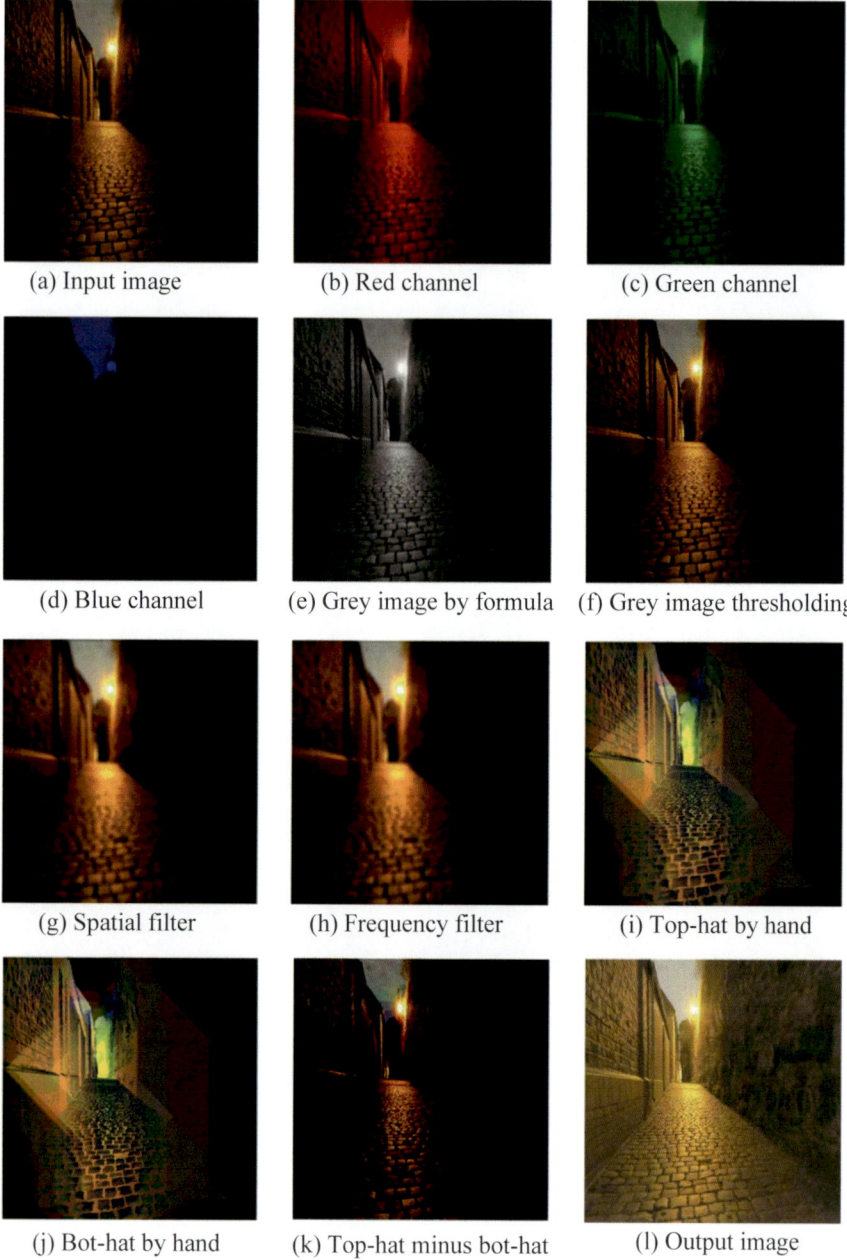

(a) Input image (b) Red channel (c) Green channel

(d) Blue channel (e) Grey image by formula (f) Grey image thresholding

(g) Spatial filter (h) Frequency filter (i) Top-hat by hand

(j) Bot-hat by hand (k) Top-hat minus bot-hat (l) Output image

Fig. 6 Simulation Results

Table 1 Comparison table

Type of image	MSE	SNR
Input image	1107.9	0.6948
Red channel	1223.9	0.8791
Green channel	1242.9	0.8463
Blue channel	1278.3	0.7873
Grey image by formulae	1148.5	1.0286
Grey image threshold	1228.7	0.9052
Spatial filter	1217.4	0.9286
Frequency filter	1215.7	0.9281
Top-hat transform	1211.1	0.9055
Output image	1133.05	1.1090
Input image	1107.9	0.6948

The comparison of different types of images is done by calculating the mean square error (MSE) and signal to noise ratio (SNR). The MSE and SNR of the input image are 1107.9, 0.6948. The MSE and SNR of the output image are 411.90, 1.55, respectively. It is as shown in Table 1.

4 Conclusion

In this paper, a low-light image enhancement is proposed. By decomposing a low-light image into the illumination component, it offers a solution to expand illumination and enhances image details separately. Specifically, the illumination component is processed using median image filter in gradient domain, followed by a non-linear logarithmic transform. Then the illumination component is enhanced by the gamma transform. This leads to enhance the low-light images and effectively reduce the distortions in colour image and also reduces noise and image blurring. Then, the final result is analyzed by the output of the enhanced illumination component. Experimental results show that the enhanced images produced by the proposed method are visually clear and more effective by the performance of the proposed method outperforms the existing methods in terms of both HE and Adaptive histogram equalization (AHE) assessments. Moreover, the proposed algorithm is efficient because the computation complexity is not related to filter size. The proposed method has great potential to implement in a real-time low-light video processing.

References

1. Ghitta O, Ilea DE, Whelan PF (2013) Texture enhanced histogram equalization using TV-L1 image decomposition. IEEE Trans Image Process 22(8):3133–3144
2. Celik T, Tjahjadi T (2011) Contextual and Variational Contrast Enhancement. IEEE Trans Image Process 20(12):3431–3441
3. Luo Y, Guan Y-P (2015) Structuralcompensation enhancement method fornonuniform illumination images. Appl Opt 54(10):2929–2938
4. Guo X, Li Y, Ling H (2017) LIME: Low-Light Image Enhancement via Illumination Map Estimation. IEEE Trans Image Process 26(2):982–993
5. Low pass filters, https://www.picosecond.com/objects/AN
6. Yang J, Zhong W, Miao Z (2016) On the Image enhancement histogram processing. In: 3rd international conference on informative and cybernetics for computational social systems (ICCSS). Jinzhou, China, pp 252–255
7. Kubinger W, Vincze M, Ayromiou M (1998) The role of gamma correction in colour image processing. In: 9th European signal processing conference (EUSIPCO 1998). Vienna, Austria, pp 1–4
8. Noise reduction filters for image processing, https://www.sciencedirect.com/science/article/pii/S1875389212005494
9. Filters for noise reduction, http://www.radiomuseum.org/forumdata/users/4767/file/Tektronix_VerticalAmplifierCircuits_Part1.pdf
10. Huang T, Yang G, Tang G (2014) A fast two-dimensional median filtering algorithm. IEEE Trans Acoust Speech Signal Process 27(1):13–18
11. Gehler P, Rother C, Kiefel M, Zhang L, Scholkopf B (2011) Recovering intrinsic images with a global sparsity prior on reflectance. In: Neural information processing systems. California, United States, pp 765–773
12. Enhancement methods in image processing, https://in.mathworks.com/discovery/image-enhancement.html
13. Analysis of image enhancement, http://acharya.ac.in/aigs/firstissuepapers/paper7.pdf
14. Image pre-processing, https://www.slideshare.net/ASHI14march/image-pre-processing
15. Image transformations, https://www.tutorialspoint.com/dip/image_transformations.htm
16. Top-hat transform, https://en.wikipedia.org/wiki/Top-hat_transform
17. Image processing algorithms part 6: gamma correction, http://www.dfstudios.co.uk/articles/programming/image-programming-algorithms/image-processing-algorithms-part-6-gamma-correction/

Comparison of Particle Swarm Optimization and Weighted Artificial Bee Colony Techniques in Classification of Dementia Using MRI Images

N. Bharanidharan and Harikumar Rajaguru

Abstract Numerous soft computing techniques are used nowadays to analyze medical images, and diagnosis of disease is computerized. This paper compares the performance of Weighted Artificial Bee Colony and Particle Swarm Optimization in the diagnosis of dementia using MRI images. For analysis, cross-sectional MRI of 235 subjects collected from OASIS is used. By adjusting the weights for both optimization techniques in a proper manner, optimized results can be reached. These techniques classify the cross-sectional image into three categories and give almost equal Goodness Detection Ratio of 78% along with different regression ratios.

1 Introduction

Magnetic Resonance Imaging (MRI) is a popular and robust visualization technique which helps to characterize the internal structures of the human body in a secured way [1]. In the diagnosis of various brain disorders, MRI plays a vital role due to the potential of recording data in a distinguished manner for dissimilar soft tissues. The usefulness of MRI images is appreciably exaggerated when automated, precise classification is possible.

Dementia is an overall term which denotes the group of symptoms which causes gradual declination of memory or thinking/reasoning skills. Other symptoms of dementia include notable difference in personality and emotional regulation. This disease is so common to aged peoples and the people affected with dementia are presumed to grow from 35 million today to 65 million by the year 2030 [2].

In dementia diagnosis, radiologists have to integrate brain MRI scans along with their medical knowledge to know the severity of dementia and to decide the treat-

N. Bharanidharan (✉) · H. Rajaguru
Department of ECE, Bannari Amman Institute of Technology, Sathyamangalam, India
e-mail: bharanidharan@bitsathy.ac.in

H. Rajaguru
e-mail: harikumarr@bitsathy.ac.in

© Springer Nature Switzerland AG 2019
D. Pandian et al. (eds.), *Proceedings of the International Conference on ISMAC in Computational Vision and Bio-Engineering 2018 (ISMAC-CVB)*, Lecture Notes in Computational Vision and Biomechanics 30,
https://doi.org/10.1007/978-3-030-00665-5_95

ment method [3]. But the greater difficulty in MRI-based dementia detection will be large number of scanned images for each patient which causes diagnosing dementia manually a tedious process. Hence, computer-aided dementia diagnosis methods are essential to increase the diagnostic correctness. Numerous algorithms have evolved in the past three decades to solve this problem, but still there is a vacuum for an accurate and computerized technique to help dementia detection.

The paper is organized as follows: Second section presents Particle Swarm Optimization (PSO) algorithm and third section explains the original and weighted Artificial Bee Colony (ABC) technique and fourth section deals with classification of dementia and results are analyzed in the last section.

2   Particle Swarm Optimization Technique

Evolutionary Algorithm (EA) imitates principles of uninhibited evolution to perform optimization process in numerous methods. Replicating the natural character of bird flocks, this population-based swarm intelligence optimization technique was modelled. Group of random particles (solutions) are assigned for PSO initially and then the optimum solution is found iteratively by renovating its position and velocity [4–6]. Every particle will have both social learning and cognitive learning, i.e. they change its present position depending on its distance from the global best and its personal best. Depending on the problem of optimization, assessment of closeness between the particle and global optimum, i.e. fitness function is chosen. Each particle i carries the following data during flight: x_i, ith particle's current position (x-vector), v_i, ith particle's current velocity (v-vector). p_i (p-vector).

The personal best (pbest) position of a particular particle is the finest position which the particle i, visited so far, i.e. a position of particular particle which gave maximum fitness value so far. Consider f as the fitness function, and then the personal best position of ith particle at a time step t is updated as

$$p_i(t+1) = \begin{cases} p_i(t) & if \ f(x_i(t+1)) \geq f(p_i(t)) \\ x_i(t+1) & if \ f(x_i(t+1)) < f(p_i(t)) \end{cases} \tag{1}$$

Information sharing among swarm members is the important characteristics of PSO. This sharing of information helps to calculate the best position that enables social learning. In PSO, most of the times, either ring or star topology will be used to share information among neighbours.

Let gbest be the postion of global best, then

$$gbest \in \{p_0(t), p_1(t), \ldots, p_m(t)\} = \min\{f(p_0(t)), f(p_1(t)), \ldots, f(p_m(t))\} \tag{2}$$

The velocity and position of ith particle are renovated using the following equations:

$$v_i(t+1) = w * v_i(t) + c_1 * r_1 * (p_i(t) - x_i(t) + c_2 * r_2 * (gbest - x_i(t)) \quad (3)$$

$$x_i(t+1) = x_i(t) + v_i(t+1) \quad (4)$$

Here w is inertia weight while c_1 and c_2 are cognitive learning constant and social learning constant, respectively. The parameters r_1 and r_2 are arbitrary numbers with value ranging from 0 to 1 to avoid local minima problem. To reduce the effect of the previous velocity, inertia term is used. A very little inertia term generally pulls exploitation, while a very huge inertia term may pull exploration. The performance of PSO is extremely depending on the parameters $c1$, $c2$ and w. Iteratively, PSO changes velocity and position and repeats until velocity changes are nearing zero or a particular number of maximum iterations have been reached.

3 Artificial Bee Colony

This algorithm categorizes bees in three groups: employed bees, scouts and onlookers. Employed bees will search for food and pass their food information to onlookers. Onlookers select good food source using the information shared by employed bees. If the eminence of the particular food source is not enhanced, then it is abandoned. Every food source (x_i) corresponds to a feasible solution of the optimization problem and fitness value $(f(x_i))$ is used to know the nectar amount of a food source [7, 8]. The count of employed bees is chosen equal to the number of food sources. Based on the optimization problem, the food sources are initialized.

The nectar information is evaluated by onlooker bees and the food source with higher quality will be chosen by onlookers. The food quality can be calculated using the following equation:

$$p_i = \frac{f(x_i)}{\sum_{i=1}^{n} f(x_i)} \quad (5)$$

3.1 Methodology for ABC

When a food source x_i is chosen using any one selection method, update on the solution in her memory will be done based on the equation:

$$x_i' = x_i + \alpha(x_i - x_k) \quad (6)$$

where x_i' is a new feasible solution produced from its previous solution x_i and the randomly selected neighbouring solution x_k; α is a random number between $[-1, 1]$.

3.2 Weighted ABC

In the original ABC algorithm, random number (α) is used to eradicate local minima problem by introducing some randomness. But there should be controlling weight parameter for this randomness in a classification problem. In proposed weighted ABC, position is updated using equation given as

$$x_i(t+1) = x_i(t) + a_1 * r_1 * (x_i(t) - x_k(t)) + b_1 * r_2 * (x_i(t-1) - x_k(t-1)) \quad (7)$$

where t denotes the iteration step, r_1 and r_2 are random numbers in the range $[0, 1]$, k denotes the neighbour chosen randomly, a_1 and b_1 are the control weights.

4 Classification of Dementia

OASIS furnishes brain MRI images related to dementia comprising cross-sectional brain images of 416 patients with age ranging between 18 and 96. Clinical Dementia Rating (CDR) of 235 patients is accessible, out of 416 patients. Each subject under evaluation will be categorized as Non-Dementia (ND), Very Mild Dementia (VMD) or Mild Dementia (MD). Targets are fixed for each class during the training and testing phase as shown in Table 1.

During both training and testing, MRI brain image of every patient is split into 16 sub-images and statistical features like mean, variance, skewness, kurtosis features are obtained (Fig. 1).

4.1 PSO Based Classification

For each image 64 normalized features are initialized as 64 particles (solutions). Velocity of all solutions is arbitrarily assigned as 0.5 and maximum number of iterations are used as stopping criterion. PSO nears the target precisely with more number of iterations and so optimal value selection of maximum number of iterations is required. The optimal values for w, c_1 and c_2 are selected while training and kept as constant during valuation.

Table 1 Data set and its target		ND	VMD	MD
	Training	63	35	15
	Testing	62	35	15
	Target	0.1	0.6	0.9

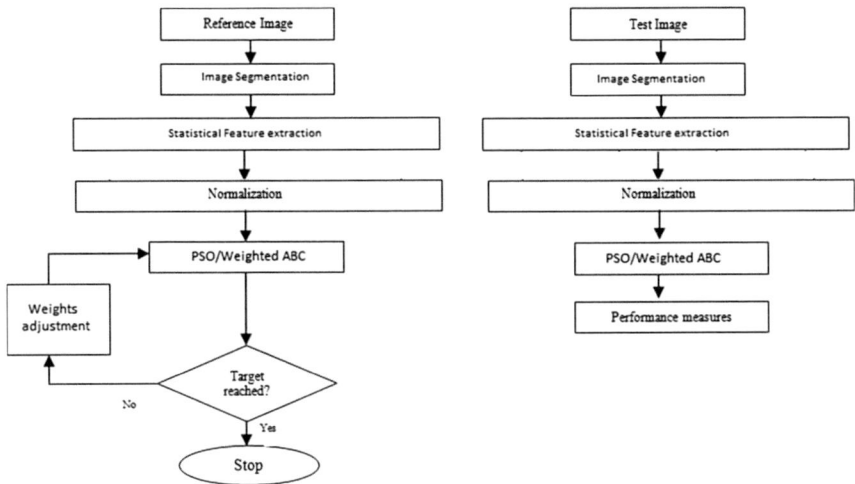

Fig. 1 Flow chart of training and testing the dementia classifier based on PSO/Weighted ABC

4.2 Weighted ABC Based Classification

For each image, 64 normalized features are initialized as 64 food sources. Similar to PSO, maximum number of iterations are used as a stopping criterion. Roulette wheel based selection approach is used for selecting the food source with high nectar quantity. The optimal values for a_1 and b_1 are chosen while training and kept as constant during valuation (Fig. 2).

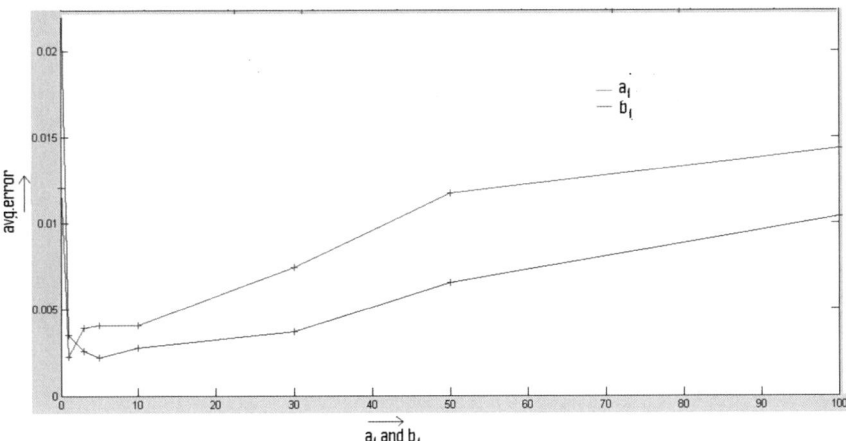

Fig. 2 Average error response of weighted ABC for variation in a_1 and b_1 weights

5 Results and Discussion

To examine the performance of the classifier, Ratios of Regression (R_1 and R_2) and Goodness Detection Ratio (GDR) are determined from the confusion matrix. A classifier is named as good if the GDR crosses 75%.

$$GDR = \frac{PC - MC}{PC + FA} \times 100\% \tag{8}$$

$$R_1 = \frac{MC}{FA} \tag{9}$$

$$R_2 = 1 - \frac{FA}{MC} \tag{10}$$

In the above equations, PC means Perfect Classification, MC means Missed Classification and FA means False Alarm.

From Fig. 3, it is observed that convergence speed of weighted ABC is high compared to the PSO. Hence the computational time of weightedABC will be less due to reduced number of iterations. On the other hand, average error will not be equal to zero for weighted ABC even after 1000 iterations while PSO approaches zero average error with 100 iterations (Table 2).

From the above results, GDR will be equal to 77.57% and Ratios of regression, $R_1 = 0.714$, $R_2 = -0.4$ for PSO-based classifier. For Weighted ABC based classifier, GDR $= 77.22\%$, $R_1 = 2.28$ and $R_2 = 0.562$.

Thus, the overall classification accuracy is same for PSO and Weighted ABC based classifier if optimal weights are chosen. But there is a huge difference in regression ratio between the two types. It is due to high false alarm rate and low missed classification rate of PSO-based classifier; high missed classification rate and low false alarm rate of Weighted ABC based classifier.

Table 2 Confusion matrix of PSO and weighted ABC based classifier

Actual/ predicted class	PSO based classifier			Weighted ABC based classifier		
	ND	VMD	MD	ND	VMD	MD
ND	61	6	0	55	12	0
VMD	4	23	8	3	28	4
MD	0	6	9	0	4	11

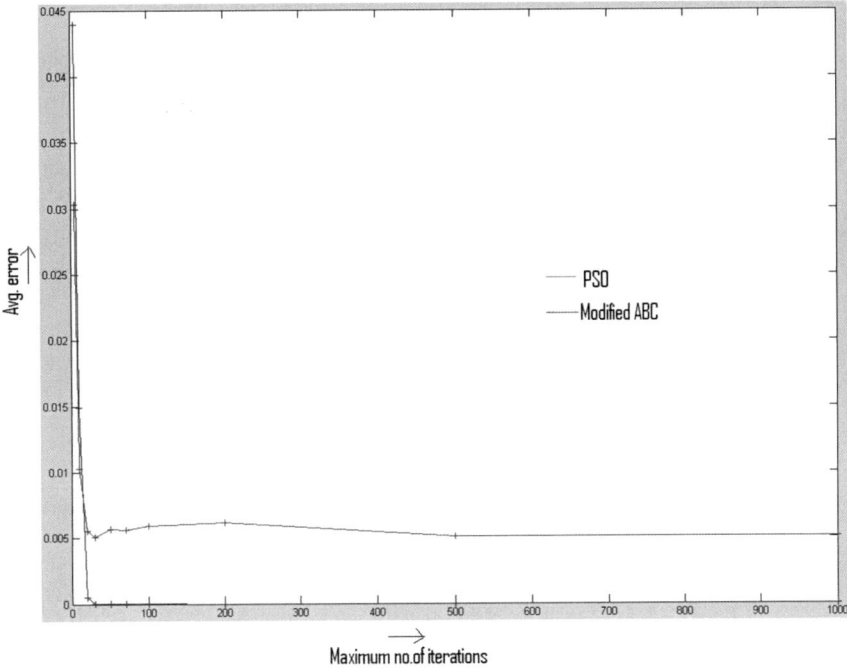

Fig. 3 Comparison of PSO and weighted ABC using average error response for variation in maximum number of iterations

6 Conclusion

The performance of PSO and weighted ABC based classifier in dementia classification is analyzed. GDR can be increased by tuning the values of weights of PSO and weighted ABC optimally. Improper selection of weights may result in low GDR. Performance of weighted ABC and PSO based classification methods can be enhanced taking into consideration other parameters like age, whole brain volume, total intracranial volume, etc. Hybrid classifiers such as ABC-PSO and ACO-PSO will be the direction of future research work.

References

1. Rodriguez AO (2004) Principles of magnetic resonance imaging. Rev Mex Fis 50:272–286
2. Korolev IO (2014) Alzheimer's disease: a clinical and basic science review. Med Stud Res J 04 (2014)
3. Kapse RS, Salankar SS, Babar M (2015) Literature survey on detection of brain tumor from MRI images. IOSR J Electron Commun Eng (IOSR-JECE) 10:80–86

4. Kennedy J, Eberhart R (1995) Particle swarm optimization. IEEE Int Conf Neural Netw Australia 4:1942–1948
5. Mohsen F, Hadhoud M, Mostafa K, Amin K (2012) A new image segmentation method based on particle swarm optimization. Int Arab J Inf Technol 9
6. Omran M, Engelbrecht A, Salman AA (2005) Particle swarm optimization method for image clustering. Int J Pattern Recogn Artif Intell. https://doi.org/10.1142/S0218001405004083
7. Chun-Feng W, Kui L, Pei-Ping S (2014) Hybrid artificial bee colony algorithm and particle swarm search for global optimization. Math Prob Eng Article ID 832949
8. Cao L, Xue D (2015) Research on modified artificial bee colony clustering algorithm. In: International conference on network and information systems for computers. https://doi.org/10.1109/ICNISC.2015.62

Automatic Segmentation of Malaria Affected Erythrocyte in Thin Blood Films

Komal B. Rode and Sangita D. Bharkad

Abstract In today's world, a highly précised diagnostic method needs to be improved for management of feverish sickness and ensure that medicines are prescribed when necessary. The proposed algorithm is applied to giemsa-stained thin blood films. Using triangle's thresholding technique, the erythrocytes are segmented, HSV color space based feature extraction is applied on the segmented erythrocytes. Features extracted are given to SVM classifier to identify whether the query sample is affected with a parasite or a normal sample. The performance of this algorithm is evaluated on the database collected from CDC website and 20 samples taken manually from a local hospital.

1 Introduction

Malaria is a feverish illness and a main public health issue worldwide. It is evoked by a parasitic protozoan of the genre plasmodium transferred through the bite of infected female Anopheles mosquito. It has an impingement on children less than 5 years of age and pregnant women contributing significantly to maternal deaths, low-birth weights in infants [1]. There are five different malaria species *p. falciparum*, *p. vivax*, *p. malariae*, *p. ovale*, and *p. knowlesi* which are harmful to human. *p. falciparum* is one of the deadly species which if not detected before time may lead to serious health issues. Blood smear image of a normal person consists of as RBCs, WBCs, platelets, and the liquid plasma as main constituents. When a person is suffering from malaria the parasites dwell in RBCs. Figure 1 shows the microscopic malaria blood sample image which shows the ring-like structure of malaria parasites and the normal red blood cells.

K. B. Rode (✉) · S. D. Bharkad
Department of Electronics and Telecommunication, Government College
of Engineering, Aurangabad, India
e-mail: komal18oct@gmail.com

S. D. Bharkad
e-mail: sangita.bharkad@gmail.com

© Springer Nature Switzerland AG 2019   993
D. Pandian et al. (eds.), *Proceedings of the International Conference on ISMAC
in Computational Vision and Bio-Engineering 2018 (ISMAC-CVB)*, Lecture Notes
in Computational Vision and Biomechanics 30,
https://doi.org/10.1007/978-3-030-00665-5_96

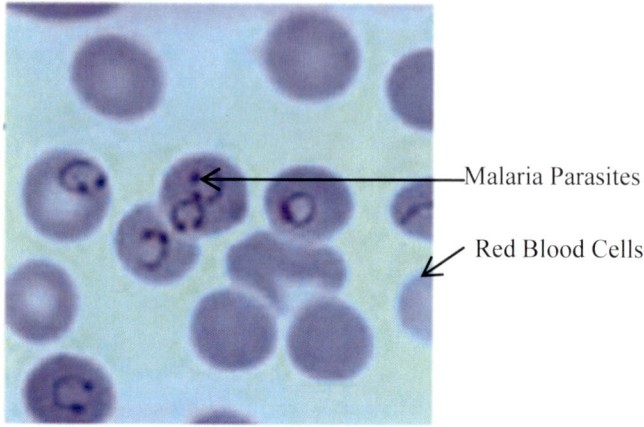

Fig. 1 Microscopic blood sample

Table 1 Pathology methods for malaria test [3]

Sr. no	Name of test	Time required (min)
1	PBS	30
2	QBS	10–15
3	RDT	10–15
4	PCR	45

Traditionally, to validate the presence of parasites different stains such as Giemsa, Leishman, AB are used depending on the environmental conditions. The peripheral smear preparation consists of two approaches thick smear and thin smear. The pathologist uses thick smear for parasitemia estimation and thin smear for species estimation. Staining is a process which highlights the parasites, making it easier for the expertise to identify the presence of parasites in the RBCs. The parasites appear dark (the color depends upon the stain) in color due to the presence of chromatin dots while erythrocytes are light in color (as the nucleus is absent), the background contains plasma lightly colored [2, 3]. The pure standard for malaria parasite estimation is smear preparation which is based on microscopic examination. Different clinical diagnosis methods such as PBS (peripheral blood smear), serological tests, QBS (quantitative buffy coat), and RDT (Rapid diagnostic test) are used for examining the presence of malaria parasites in blood samples. The efficiency of laboratory techniques mainly depends on factors like sensitivity, a number of parasites per microlitre of blood, time required, and cost. Table 1 describes the clinical malaria diagnosis methods and time required to perform the respective tests.

However, clinical diagnostic tests are based on visual inspection by the medical experts. The visual analysis can be hard. Hence, the outcome of diagnostic results on a visual assumption due to human interactions may sometimes be prone to error.

1.1 Motivation

With the advance in technology, many computer-aided techniques have been successfully developed for the reason of facilitating and accelerating the clinical diagnosis test. Today, reducing the time factor is a major concern for malaria diagnosis and making it manual interpretation independent. Proper medication can be given only when the clinical test is a success. New kits available suffer from flaws like temperature tolerance, shows false negative count, and their sensitivity vary considerably [3]. Reducing the flaws in kits available and a new automatic diagnosis tool is a vital factor in finding diseases. By reducing the manual interruption, medical expertise can focus their attention on the patients with a positive test. Therefore, we aim to propose an automatic method which identifies the affected RBC's with maximum accuracy. The key points of our algorithm are the triangle's thresholding method [4] which separates the foreground and background to identify the region of interest.

1.2 Related Work

Many researchers have proposed different image processing related work for parasite identification. Yashasvi et al. [5] used chan-vese segmentation method for boundary extraction of the cells. Hough transform to count the RBCs with k-means clustering for parasite detection. Gloria et al. [6] proposed a method which consists of the preprocessing step for luminance correction. A segmentation technique by inclusion tree representation is used to separate the erythrocytes from the background. Recognition of erythrocytes is done using color pixel classification. Feature extraction using the color histogram, Tamura texture, and saturation histogram is performed. For classification of parasites, radial basis and polynomial kernel function are considered. The amount of parasitemia present is obtained using a trained bank of the classifier. This technique has 99.7% specificity and 94% sensitivity. Boray et al. [7] have given a brief review of the computer vision techniques with partial solutions related to diagnostic techniques for malaria parasite detection. Minh-Tam et al. [8] used Zack's thresholding for extraction of the nucleated portion. All the solid constituents are identified by comparing the malaria image with an empty field image. The size of erythrocytes is obtained. Based on isolated erythrocytes area, the leukocytes and gametocytes are detected and isolated accordingly. Kareem et al. [9] mentioned a method that differentiates gametocytes using modified annular ring ratio. Kareem et al. [10] described a method in which depending on the geometry and color features of the erythrocytes the parasites are identified with good accuracy results. Literature mentioned above mainly focus on identifying the parasites present in the erythrocytes; working in unsupervised conditions are not addressed. Understanding the manual diagnosis method need of algorithm is to reduce the medical expertise intervention and time factor. Hence, with focus on above factors, an automatic algorithm is proposed.

Rest of the paper describes the method implemented which is organized as follows. Section 2 describes the proposed algorithm. Section 3 presents the experimental results of the proposed approach. Finally, the implementation is concluded in Sect. 4.

2 Proposed Approach

Figure 2 presents the flow of the proposed work. The approach is a simple algorithm which segments the cells present in each image and classifies whether the cell is infected with malaria parasites or normal RBC.

2.1 *Image Acquisition*

For image acquisition, first step is slide preparation by an expert for manual examination. Giemsa staining technique is used, dilution is made in alkaline buffer with pH 7.2. After staining slide is rinsed under tap water and dried. Slides are prepared to detect the count of parasitemia under a microscope by lab expertise. Images were acquired by placing the high-definition digital camera with an extension to a microscope. For this work, a 12-megapixel resolution camera was used for acquiring the sample images.

Fig. 2 Proposed approach

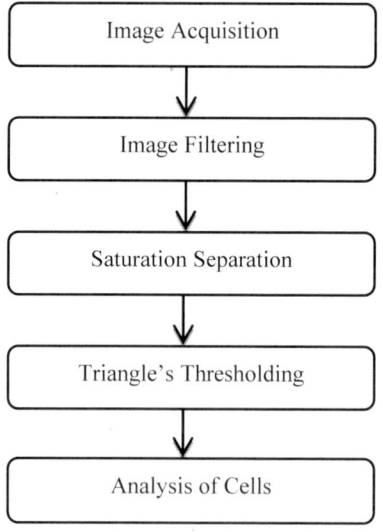

2.2 Average Filtering

Filtering is a preprocessing method which aims in the suppression of image noise, detect edges, etc. Averaging filters smoothens the image by reducing intensity variation in neighbor pixels. It reduces noise, suppresses irrelevant detail in the image, and blurs the image which makes the transition smooth from one color to another. It works replacing each pixel to pixel by its average value with a 3 by 3 filter mask. The mask gives equal weights to all the pixels. Average filtering is chosen since it blurs the image noise components introduced, i.e., artifacts which are smaller in size. These artifacts should be blended because they increase chances of false prediction as parasites. Equation (1) gives the mathematical average filtering operation.

$$g_{ij} = \sum_{w=-m}^{m} \sum_{l=-m}^{m} k_{wl} f_{i+w,j+l} \tag{1}$$

where g_{ij} the average filtered image. k is an average filter mask of size 3 by 3. $w = -m \leq 0 \leq m, l = -m \leq 0 < m$, f_{ij} is original image. The parasites appear dark colored objects (purplish) in the microscopic image; however, the visual extent of staining varies due to lighting and imaging conditions. Based on color features, the parasites are very well distinguished in hue-saturation-value color space as compared to the gray level image. Hence the original image is converted to HSV space. Figure 3a–d shows the image after average filtering, hue component image, saturation component image, and value component image, respectively.

2.3 Saturation Separation

The RBCs are lightly stained as mature RBCs do not have nuclei. The parasites consist of chromatin dots which appear as dark objects in the sample image. Saturation channel gives us the degree of color diluted by white light. Considering the images based on intensity, the brighter proportion of the parasites is visible on the light background. Extracting the saturation channel in an image describes color similarities to human eye precipitance. Therefore, to detect a dark object within an RBC, selecting saturation component of the original image gives better understanding of parasites. Histogram of saturation channel helps in finding the threshold value for further processing. Equation (2) denotes the mathematical expression for H component in terms of R, G, B values and Eq. (3) gives a mathematical expression for S component in terms of R, G, and B component, respectively. Figure 4 shows the histogram plot of saturation channel.

$$H = tan \frac{(3(G - B))}{(R - G) + (R - B)} \tag{2}$$

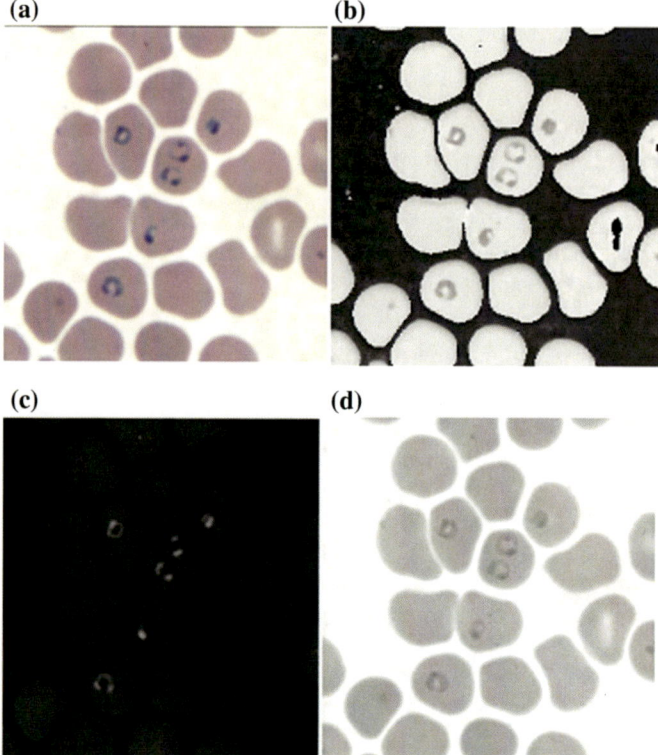

Fig. 3 **a** Filtered image. **b** Hue image. **c** Saturation image. **d** Value image

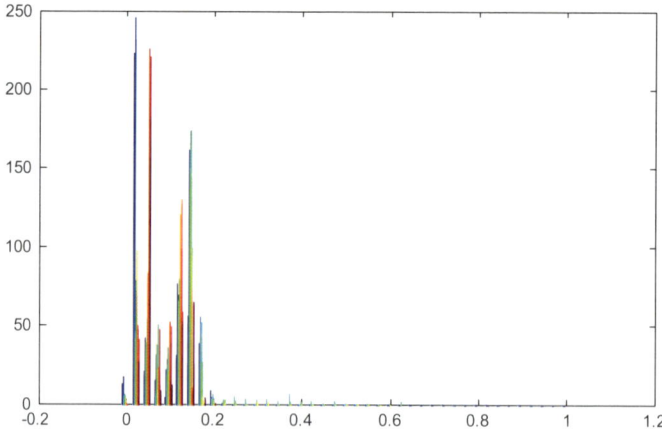

Fig. 4 Histogram plot of saturation channel

Fig. 5 Triangle's
thresholding

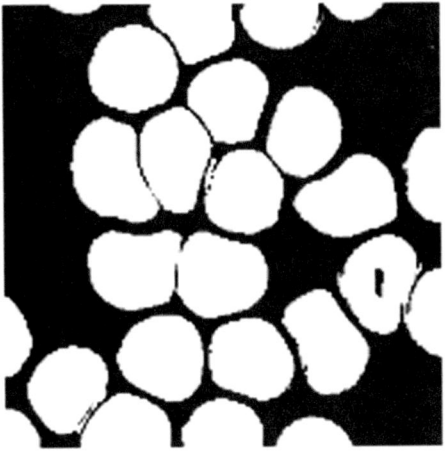

$$S = 1 - \frac{\min(R, G, B)}{V} \tag{3}$$

where R, G, B is Red, Green, and Blue component images, respectively.

2.4   Triangle's Thresholding

Triangle's thresholding method is performed on the saturation channel. By plotting a histogram of saturation channel, the optimal threshold value is obtained. The triangle's thresholding [11] method constructs a line between the highest and the lowest value of the histogram. Further, the distance is measured between line constructed and the values of the histogram between the minimum and maximum values for all the values denoted by h, where h is the value through which the distance K reaches its highest brightness. The point h at which maximum distance obtained is considered as a threshold. Figure 5 shows the image after triangle's thresholding.

2.5   Analysis of Cells

Applying triangle's thresholding algorithm gives a binary image followed by an opening operation which removes the small unwanted component. The statistical features average, standard deviation of HSV color space and the gray scale are extracted.Support Vector Machine (SVM), a supervised learning algorithm is considered for classification problem of erythrocytes. Binary classification is done to classify whether the random erythrocyte is affected or not. SVM with optimal separating hyperplane plots each point in the n-dimensional space and classifies based on finding the hyperplane which separates the two classes. The mathematical expression for average and standard deviation are mentioned in Eqs. (4) and (5), respectively. The Eq. (6) denotes the expression for linear classification of SVM classifier.

Average:

$$\mu_{H,S,V} = \frac{1}{IJ} \sum_{k=1}^{I} \sum_{l=1}^{J} P_{kl}^c \tag{4}$$

Standard deviation:

$$\sigma_{H,S,V} = \left[\frac{1}{IJ} \sum_{k=1}^{I} \sum_{l=1}^{J} \left(P_{kl}^c - \mu_c \right)^2 \right]^{\frac{1}{2}} \tag{5}$$

For linear classifier

$$f(x) = w^T x + b \tag{6}$$

where w, b, and x is normal weight, b bias, and support vector, respectively.

3   Experimental Results

The experimentation is conducted on 40 images with different stained images. The samples were obtained from the center for disease control and prevention [12] and Medicare hospital, Ahmednagar. The sample size 256×256 is considered as input image for implementation of proposed algorithm. The experimentation performed is implemented in MATLABv2013a environment. Despite the poor slide quality, the system identifies the erythrocytes properly.

$$A = \frac{T_p + T_n}{T_p + F_p + T_n + F_n} \tag{7}$$

$$S = \frac{T_p}{T_p + F_n} \tag{8}$$

$$SP = \frac{T_n}{T_n + F_p} \tag{9}$$

where A, S, SP is overall accuracy, sensitivity, and specificity, respectively. T_p is true positive the classifier identifies an affected cell as an affected cell, T_n is true negative identifies normal as a normal cell; F_p is false positive the affected sample is not identified, F_n is false negative false normal identified. Each and every cell in the original image is checked whether the parasite is present or not. Table 2. shows

Table 2 Comparsion of manual and automatic technique

Patient list	Manual detection of affected cells	Detection by proposed approach
Patient 1	3	3
Patient 2	11	10
Patient 3	5	5
Patient 4	18	18
Patient 5	2	2
Patient 6	2	2
Patient 7	1	1
Patient 8	1	1
Patient 9	2	2
Patient 10	1	1
Patient 11	1	1
Patient 12	1	1
Patient 13	1	1
Patient 14	2	2
Patient 15	4	3
Patient 16	2	2
Patient 17	1	1
Patient 18	1	1
Patient 19	2	2
Patient 20	3	2

Table 3 Comparison of proposed method with existing methods

Method	Specificity	Sensitivity	Accuracy
Diaz et al. [6]	99.7	94	–
Purwar et al. [5]	50–88	100	–
Proposed approach	80	100	85

comparison of results obtained by manual technique and proposed algorithm for identification of infected samples.

Table 3 shows a comparison of the existing methods with the proposed method in terms of specificity, sensitivity, and accuracy.

4   Conclusion

This paper describes the segmentation of RBCs using triangle's method which achieved a good accuracy in identifying an infected RBC and a normal RBC. The proposed work provides a fully automated technique which would help the lab expertise in counting number of parasitic cells, identification of affected cells and will improve management of this febrile illness to ensure the antimalarial medication. Early detection of malaria will be life-saving. Further, we can extend the work by identifying the species and the stages of the affected samples which would definitely help in improving the diagnosis method.

Acknowledgements We are heartily grateful to the Medicare hospital, Ahmednagar for providing us the blood slides images which lead to the implementation of our project and some images were taken from the CDC website.

References

1. World Malaria Report: World Health Organization (2017)
2. Wassmer E (2017) Grau: Severe Malaria: What's new on the pathogenesis front? Elsevier Int J Parasitol 47:145–152
3. Azikiwe, Ifezulike, Siminialayi, Amazu, Enye, Nwakwunite (2012) A comparative laboratory diagnosis of malaria: microscopy versus rapid diagnostic test kits. Elsevier Asian Pac J Trop Biomed 4:307–310(2012)
4. Nugroho HA, son of Ali Akbar, Elsa Herdiana Murhandar E (2015) Feature extraction and classification for detection of malaria parasites in a thin blood smear. In: Proceeding of IEEE conference on information technology, computer, and electrical engineering. pp 197–201
5. Yashasvi, Shah, Clarke, Almugairi A, Muehlenbachs A (2011) Automated and unsupervised detection of malarial parasites in microscopic images. Malar J 1–10
6. Gloria, Gonzalez, Romero (2009) A semi-automatic method for quantification and classification of erythrocytes infected with malaria parasites in microscopic images. Elsevier J Biomed Inf 42:296–307
7. Boray, Dempster, Kale (2009) Computer vision for microscopy diagnosis of malaria. Malar J 9:1–12
8. Minh-Tam Le, Bretschneider, Kuss, Preiser (2008) A novel semi-automatic image processing approach to determine Plasmodium falciparum Parasitemia in Giemsa-Stained thin blood smears. In: Research article BMC cell biology, pp 1–12
9. Kareem S, Morling RCS, Kale I (2011) A novel method to count the RBCs in thin blood films. In: Proceedings of IEEE conference on circuits and systems, pp 1021–1024
10. Kareem S, Morling RCS, Kale I (2012) Automated malaria parasites detection in thin blood films—a hybrid illumination and color constancy insensitive, morphological approach. In: Proceedings of IEEE conference on circuits and systems, pp 240–243
11. Sadeghian, Seman Z, Ramli AR, Kahar BHA, M-Iqbal Saripan (2009) A framework for white blood cell segmentation in microscopic blood images using digital image processing.11:196–206
12. https://www.cdc.gov (centre for disease control and prevention)

Detection and Recognition of Vehicle Using Principal Component Analysis

Kolandapalayam Shanmugam Selvanayaki, Rm. Somasundaram and J. Shyamala Devi

Abstract The idea of detection of moving objects and the concept of classification of moving objects is considered to be the important part of research in video processing and in real-time applications for surveillance and tracking of vehicles. In scientific terms, image processing is said to be any form of signal processing, where the input is an image and the output of image processing may taken be either an image or a set of characteristics or parameters related to the image. The proposed work of the paper is to detect and classify vehicles in a given video. It consists of two modules, first one uses GLOH algorithm for feature extraction and feature reduction. The second module classifies the vehicle from an input video frame using PCA. The final result is obtained by integrating the above-said modules.

1 Introduction

Many of the existing image-processing techniques treat the image as a two-dimensional signal and applies the standard signal-processing techniques to the image. The detection of an object is done by either one method, where the object gets detected in every frame or the object is detected when it appears first in the video [1]. The most predominantly used approach in object detection is using the information exists in a single frame. A common approach for object detection is to

K. S. Selvanayaki (✉)
Department of Computer Science, Concordia University, Chicago,
River Forest 60305, Illinois, USA
e-mail: selvanayaki.ks@cuchicago.edu

Rm. Somasundaram
Department of Computer Science & Engineering, SNS College of Engineering, Coimbatore,
Tamil Nadu, India
e-mail: ilakiasomu@gmail.com

J. Shyamala Devi
Department of Computer Science, SRM University, Chennai, Tamil Nadu, India
e-mail: shyamaladevi@gmail.com

© Springer Nature Switzerland AG 2019
D. Pandian et al. (eds.), *Proceedings of the International Conference on ISMAC in Computational Vision and Bio-Engineering 2018 (ISMAC-CVB)*, Lecture Notes in Computational Vision and Biomechanics 30,
https://doi.org/10.1007/978-3-030-00665-5_97

use information in a single frame. Several features like the shape, color, logo, etc., could be used to detect and classify the vehicle in the video.

In computer vision tasks, GLOH (Gradient Location and Orientation Histogram) is a robust image descriptor which is mostly used. Principal Components Analysis (PCA) reduces the higher dimensionality identified by the descriptor. Therefore, GLOH is a combination of PCA and the SIFT algorithms. GLOH algorithm is performed in two steps initially the keypoints and features are extracted using SIFT algorithm and further dimensionality reduction is achieved using the PCA algorithm. Scale Invariant Feature Transform (SIFT) features are features which are extracted from images that helps to match the different views of the same object. The extracted features are identified to be invariant in considering to scale and orientation and are mostly distinctive of the image. SIFT is not just scale invariant but also rotation, illumination, viewpoint invariant.

2 Principal Component Analysis (PCA)

Principal components analysis (PCA) is one of the classical methods. It provides a sequence of best linear approximations to a given high-dimensional observation; with all the existing techniques, PCA has to be taken as one of the most popular techniques for reduction in dimensionality.

2.1 Principal Component Analysis for Classification

Classification is the process of identifying the set of categories, a new observation belongs by considering the training set of data containing observations for known category membership. The algorithm gives the classification in a concrete implementation is called as a classifier. Sometimes, "classifier" refers to mathematical functions given by classification algorithms, which is mapping input data for this category. Some of the commonly used classification techniques are State Vector Machines (SVM), Neural networks, and Principal Component Analysis (PCA). Some of these techniques are discussed below.

2.2 Existing Approaches for Classification

SVM: Support Vector Machine (SVM) is a discriminative classifier formally defined by a separating hyperplane. It is a supervised learning model which analyze data and recognize patterns that can be used for regression analysis.

Neural Networks: A Neural network consists of a closely connected group of artificial neurons, and process the information in a connectionist approach for computation.

Principal Component Analysis: Principal component analysis is an algorithm which achieves dimensionality reduction. In our proposed system, PCA is used for classification on cars in a frame of video based on its brand. PCA classification aims to maximize between-class separation and aims at accurate classification.

3   Overall System Design

The main focus of this paper is to detect and classify vehicles (cars) in a given video. Gradient Location Orientation Histogram (GLOH) algorithm and [SIFT-PCA] are used for feature extraction and feature reduction. PCA is used for further classification of the vehicle from the input video frame. In this system, there are basically three modules. The SIFT module which consists of key point extraction using SIFT and feature extraction. The next module consists of PCA reduction. The final module is the classification of the vehicle using the PCA algorithm. All the three modules are then integrated to provide the final result and the overall system design is given in Fig. 1.

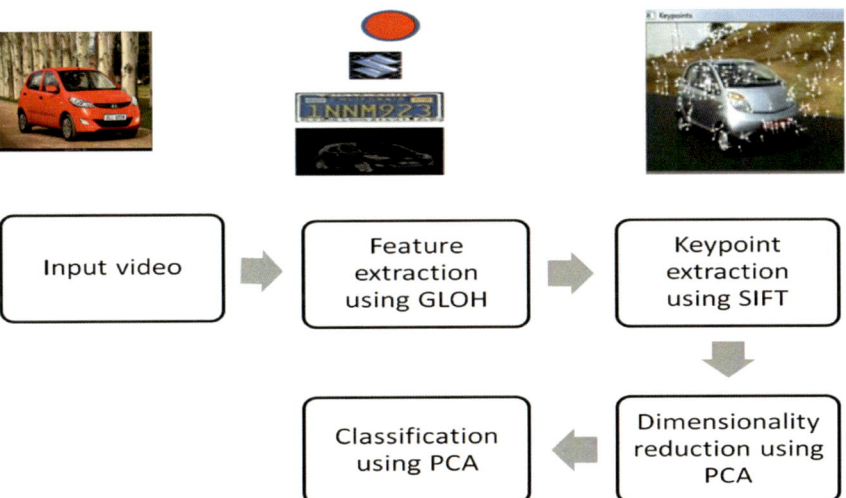

Fig. 1 Overall system design

4 Number Plate Detection and Character Recognition

The aim is to extract the vehicle registration numbers from car images and recognize the same without the use of human intervention for ease of authentication system. The first step is to exactly locate and extract the license plate from car video and the next is to extract the text for further authentication purpose.

Number plate detection is done in two steps.

- Extracting number plate from a video using BLOBs library
- Extracting the characters using Aostsoft Image to Text OCR Converter.

BLOBs library is used for doing computer vision by finding BLOBs on an image, for areas whose brightness is above or below a particular value. It also provides functions and manipulates filtering and extracting results from the extracted BLOBs. In order to extract the numbers in the license plate, Aostsoft Image to Text OCR Converter is used.

4.1 Overall Architecture

The overall architecture for Number Plate Detection is given in Fig. 2.

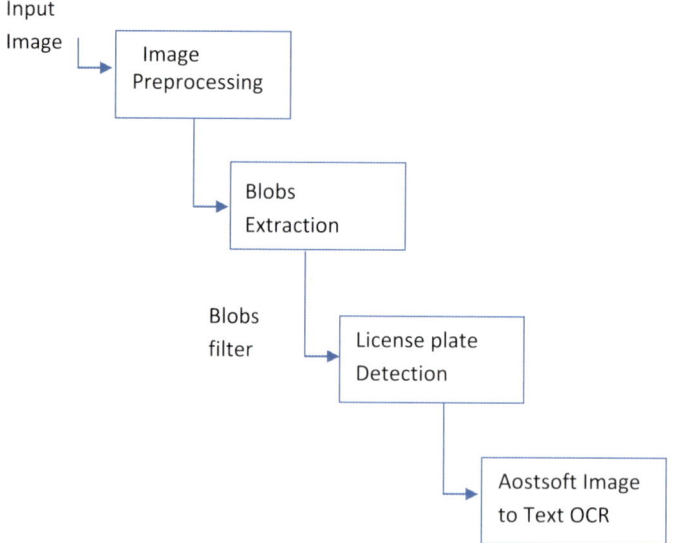

Fig. 2 Overall architecture for Number Plate Detection

Input	Smoothing	TOP_HAT	Threshold

Fig. 3 Sequence of preprocessing techniques

4.2 Extracting Number Plate

Number plate extraction involves two steps, namely, preprocessing techniques to reduce noise, BLOBs extraction, and text extraction.

4.3 Preprocessing Techniques

The most important part of image processing is just not detection but the preprocessing techniques that is done before the operation which helps in removing noise and thus reduce errors. Smoothing, morphological operation, and threshold are the three techniques chosen. Input image is first smoothened to remove noise.

Next technique used is TOP HAT, a morphological operation. TOP HAT exaggerates those portions which are lighter than the surroundings. In our case, number plate is lighter region. Next is to apply threshold which makes lighter regions more prominent for BLOBs extraction. OTSU threshold technique is used, which is based on moving the average value of pixels within the image (Fig. 3).

4.4 BLOBs Extraction

BLOB detection is focussed at the detection of points and/or regions in the image which differs in properties like brightness or color compared to their surroundings. Using BLOBs library such regions are extracted. This includes a large collection of BLOBs area of various sizes. The library gives the two basic functionalities namely (1) Extracts eight-connected components also referred as BLOBs in binary or grayscale image and (2) Filters the BLOBs which are obtained to get the interest objects in the image, which used Filter method from CBLOBResult (Fig. 4).

As a result, insignificant BLOB that is either too large or too small is also detected. These BLOBs are filtered based on area and ratio constraints. Ratio is calculated as width/height. It is observed that ratio calculated for the number plate of a car falls in

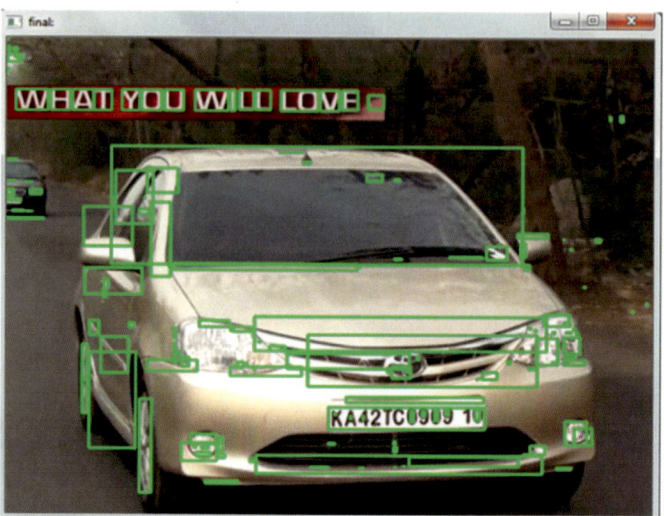

Fig. 4 Result of BLOB detection

Fig. 5 Result of BLOB detection after filtering

the range of 1.8–8 and area less than 8000. Those BLOBs which do not satisfy these conditions are eliminated. Thus, the number plate part is detected (Fig. 5).

4.5 Text Extraction

Now only the number plate is detected. Further, the text is to be extracted and checked whether it is a government car or not. Aostsoft Image to Text OCR Converter is used to extract the text in the number plate. The localized region is given as an input, from which individual characters are extracted which is then saved into a text file. The easy and flexible Image OCR converter is Aostsoft Image to Text OCR Converter. It is much helpful in extracting text from various types of images such as JPG, JPEG, TIF, TIFF, BMP, GIF, PNG, PSD, EMF, WMF, DCX, PCX, JP2, J2K, etc. The authentication is based on the occurrence of the letter "G" or "g". In case if the letter G/g is present then it is classified as Government car else not a government car.

4.6 Color Detection

The one of the important feature is color, which is chosen to uniquely label a group of cars. This system combines two methods to increase the efficiency of color detection over a range of conditions.

Haar cascade classifier to localize the car in the frame or image: A Haar Cascade Classifier which is trained to localize the car region alone in the frame is used to find the car position in the frame. This car position is given as the input for the color detection algorithm which estimates the color of the car.

Find color using the center point of localized region: This module has considered middle pixel and six neighboring pixels and estimates the color of the car using the RGB model. For a video, the color values of the points are averaged to get the exact color [2].

Condition where algorithm fails	Condition where algorithm works properly
(1) When the pixel color represented by the midpoint of the Region of Interest (ROI) is not the proper color of the car	(1) When the pixel color represented by the midpoint of the Region of Interest (ROI) is the proper color of the car
(2) When there is too much of light illuminating the front view, which may give a white dominant shade and not the original color	(2) The system's reliability is not marred by the background like in other methods

4.7 Shape Extraction

This system combines three different methods to extract the shape of the object, namely, BLOBs detection, Finding Contours, and Contour Comparison. Once the

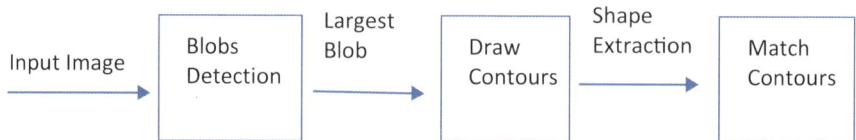

Fig. 6 Flow of shape extraction technique

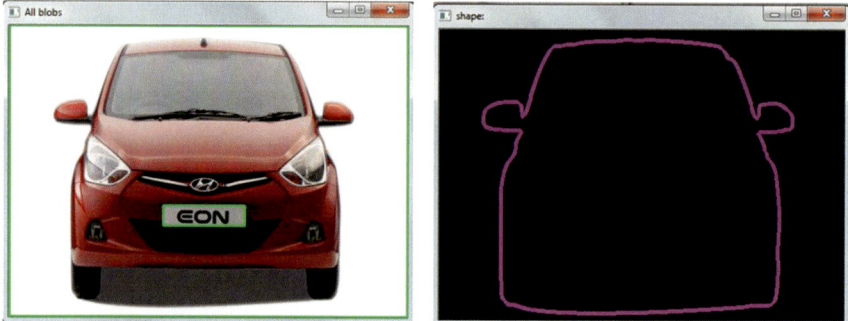

Fig. 7 Result of BLOB detection

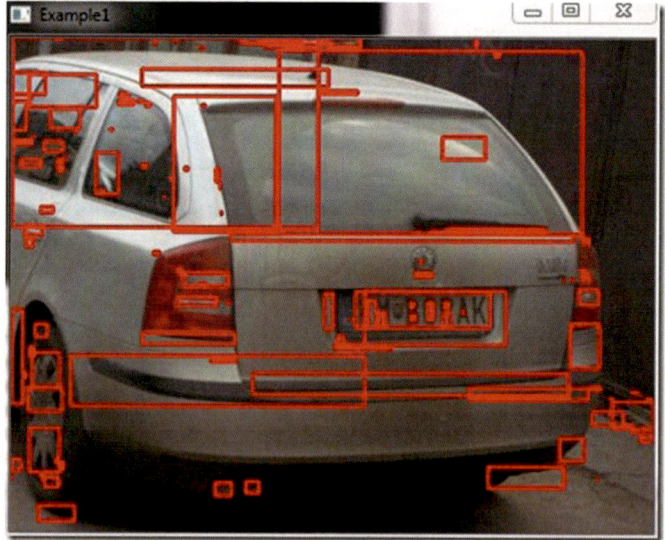

Fig. 8 BLOB detection

shape is extracted, they are compared with templates in order to find the match percentage value (Figs. 6, 7, 8, 9, 10 and 11).

The above algorithms were used to extract user-defined features from the given car in the video. At the end of this phase, a feature vector was formed which described

```
f:\project\Final working code\Final working code without svm\siftttt+ all features+video+release - ...

Feature Matching Count--

Hyundai Count: 2
Toyota Count: 0
Suzuki Count: 376
Feature 1: 3
................................
...MODULE 2: COLOUR DETECTION...
Colour Count --

White Count: 708
Black Count: 24Gray Count: 40
Wrong Count: 236
Feature 2: 1
................................
...MODULE 3: NUMBER PLATE DETECTION...
Number Plate --
Feature 3: 1
................................
...MODULE 4: SHAPE DETECTION...
Confidance Value -- 62.575756
Feature 4: 1
................................

Press any key to continue . . .
```

Fig. 9 Output of feature extraction

Fig. 10 Output of number plate detection

the features of the given car in the frame. The feature vector has three feature values among which the first value is for the logo the next feature value is for the color and

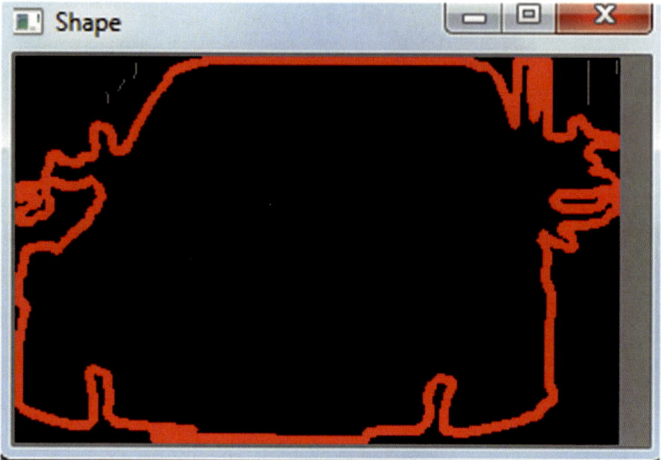

Fig. 11 Output of shape extraction

the last feature value is for the shape. It was earlier decided that SVM will be used
for classification. But since it was suggested that classification could be done using
PCA this module was completed at this stage and classification was done using PCA
on the reduced frame from the previous phase.

4.8 Steps in PCA Classification

Loading images for training

The identified steps in PCA Classification is as follows:

- Finding the PCA subspace
- Thesising the training images
- Saving the learned face model
- The recognition phase
- Finding the nearest neighbor (Figs. 12 and 13).

5 Conclusion

The classification is done during this phase and classification is done based on the
nearest neighbor by finding out the minimum distance. This stage has a training and
testing phase where the classification model is learned during the training phase and
actual classification is done in the testing phase. This model can be improved by

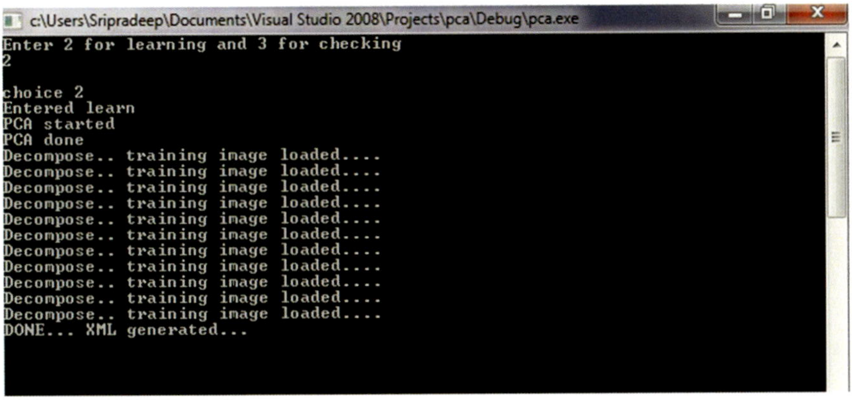

Fig. 12 Training phase output

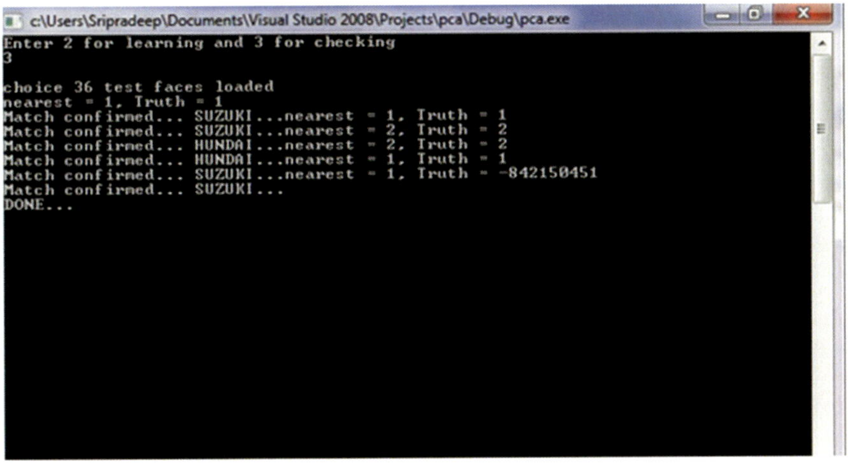

Fig. 13 Test recognition phase

taking other different distance measuring approached to find out the distance. This classification is faster than the already existing method because it is a pixel-based method whereas most of the other methods are feature-based.

There are a few more conditions under which the algorithm misclassifies or fails, which are given as follows

Condition where the algorithm works	Condition where the algorithm fails
(1) The factors like width and height of each input image (parameter 2 of cvCalcEigenObjects(), input image) need to be identical. Also, these images should be single channel	(1) The factors like width and height of each input image (parameter 2 of cvCalcEigenObjects(), input image) need not to be identical. Also these images are not single channel
(2) The factors like width and height of the average image (parameter 9 of cvCalcEigenObjects(), average image) needs to be the same as the width and height of the input images	(2) The factors like width and height of the average image (parameter 9 of cvCalcEigenObjects(), average image) need not the same as the width and height of the input images
(3) The width and height of each eigenvector image (parameter 3 of cvCalcEigenObjects(), output) must also be the same as the width and height of the input images	(3) The width and height of each eigenvector image (parameter 3 of cvCalcEigenObjects(), output) need not be the same as the width and height of the input images
(4) All the images probably need to be single channel	(4) All the images probably need not be single channel

Table 1 Results of various brands of car after classification

Brand	Positive images used	Negative images used	Results obtained for image
Suzuki	10	Zero	It is been to get accepted for the trained set of images
Suzuki	50	10	Worked well for Suzuki cars
Hyundai	5	10	Car detected as some other brand
Hyundai	10	2	Car detected as Hyundai

For a given query image, let's assume:

There are M relevant images of "correct matches" in the collection. Upon execution of the query, it is possible to retrieve N images.

- In that, only R are relevant.
  So, Recall $= R/M =$ Number of retrieved images that are relevant/Total number of relevant images.
- *Precision* $= R/N =$ Number of retrieved images that are relevant/Total number of retrieved images.

The recall is the answer to the question: How close am I to getting *all* good matches? The precision is the answer to the question: How close am I to getting *only* good matches?

In PCA classification precision and recall are inversely proportional.

The classifier works well for Suzuki and Hyundai cars and classifies those cars correctly when there was more number of positive images. The classifier can be extended further for classification of other brands of cars as well. The following are the results obtained for various brands of cars after classification (Table 1).

It can be seen from the results that principal component analysis gives better results with good classification measures.

References

1. Anagnostopoulos CNE, Anagnostopoulos IE, Psoroulas ID, Loumos V, Kayafas E (2008) License plate recognition from still Images and video sequences: a survey. Intell Trans Syst IEEE Trans 9:377–391. https://doi.org/10.1109/tits.2008.922938
2. Butzke M, Silva AG, Hounsell MS, Pillon MA (2008) Automatic recognition of vehicle attribute-color classification and logo segmentation. Hifen, Urugaiana, pp 32–62
3. Cao X, Wu C, Yan P, Li X (2011) Linear svm classification using boosting hog features for vehicle detection in low-altitude airborne videos. In: Proceedings of the 18th IEEE international conference on image processing
4. Zhang C, Chen X, Chen W (2006) A PCA-based vehicle classification framework. In: 22nd international conference on data engineering workshops
5. Lowe DG (2004) Distinctive image features from scale-invariant keypoints. Int J Comput Vision 60(2):91–110
6. Bagarinao E, Kurita T, Higashikubo M, Inayoshi H (2009) Adapting SVM image classifiers to changes in imaging conditions using incremental SVM: an application to car detection. In: Proceeding of 9th Asian conference on computer vision (ACCV), pp 363–372
7. Liu H (2009) Skew detection for complex document images using robust borderlines in both text and non-text regions. Sci Dir, Patt Recogn Lett
8. Kim KK, Kim KI, Kim JB, Kim HJ (2000) Learning-based approach for license plate recognition. In: Proceeding of IEEE workshop on neural networks for signal processing, vol 2, pp 614–623
9. Mikolajczyk K, Schmid C (2005) A performance evaluation of local descriptors. IEEE Trans Patt Anal Mach Intell 27(10):1615–1630
10. Juan L, Gwun O (2010) A comparison of SIFT, PCA-SIFT and SURF. Int J Image Process (IJIP) 3(4)
11. Sivaraman S, Manubhai Trivedi M (2010) A general active-learning framework for on-road vehicle recognition and tracking. IEEE Trans Intell Transport Syst 11(2):267–276
12. Vapnik V, Golowich S, Smola A (1997) Support vector method for function approximation, regression estimation, and signal processing. Advances in Neural Information Processing Systems, Cambridge, MA, MIT Press, pp 281–287
13. Shao Y, Lunetta RS (2010) Comparison of support vector machine, neural network, and CART algorithms for the land cover classification using limited training data points. Environmental Protection Agency, USA
14. Yasser Arafat S, Saleem M, Afaq Hussain S (2009) Comparative analysis of invariant schemes for logo classification. In: Proceedings of international conference of emerging technologies (ICET), pp 256–261
15. Chen Z, Pears N, Freeman M, Austin J (2009) Road vehicle classification using support vector machines. Intell Comput Intell Syst

Convolution Neural Networks: A Case Study on Brain Tumor Segmentation in Medical Care

Jayanthi Prisilla and V. Murali Krishna Iyyanki

Abstract Image segmentation is dividing of medical imaging into parts and extracting the regions of interest. The study involves the images of brain tumors where the tumor part is segmented from the image and analyzed accurately and efficiently. Convolution Neural Network (CNN) has made a tremendous progress in the field of the Medical and Information Technology. With CNN model, one may not be able to reorganize higher risk patients to get immediate aid they require but also communicate through the network to the clinicians, surgeons, eventually improving the standard of patient care in the medical system.

1 Introduction

The convolutional neural networks are making progress in image recognition, image processing, and image segmentation which are essential for medical imaging. A medical database consists of very huge images consisting of X-rays, MRI, and CT-Scans of various ailments. The manual segmentation is time-consuming process and is tedious, and is prone to human error. Automatic segmentation is cheap, reliable, and is scalable; works with large networks with more accurate and better at classification. This segmentation provides a faster and a well-planned treatment for the large-scale research.

J. Prisilla (✉)
Hyderabad, India
e-mail: prisillaj28@gmail.com

V. M. K. Iyyanki
Defence Research and Development Organization, Hyderabad, India
e-mail: iyyanki@gmail.com

© Springer Nature Switzerland AG 2019
D. Pandian et al. (eds.), *Proceedings of the International Conference on ISMAC in Computational Vision and Bio-Engineering 2018 (ISMAC-CVB)*, Lecture Notes in Computational Vision and Biomechanics 30,
https://doi.org/10.1007/978-3-030-00665-5_98

2   Previous Study

Case-Based Reasoning Clinical Decision Support System (CBR CDSS) developed by Yin et al. [1] for diagnosing the Probable Migraine (PM) and Probable Tension-Type Headache (PTTH) with much better accuracy. An AI technique, case-based reasoning is used to solve very near similarity solutions and is accepted to be effective methods of managing implicit knowledge. The study states it comprises two entities: symptoms and diagnosis. The symptoms are of 74 types for which diagnosis is the solution for each case. Genetic algorithm is a search heuristics that is used for optimization and searching problems.

PLA General Hospital China, an International Headache Center from where the testing data sets were collected consisted of 222 former cases. The proposed CBR CDSS provides the approval grounded on the similarity measurement between fresh cases and former cases. The KNN method was considered to be more effective over other algorithms for its simplicity and higher accuracy. The equation is given as similarity $(A, B) = \sum f (A_i, B_i) \times w_i$, where A denotes problem case, in every case B indicates the current case, and i is the distinct features from 1 to n, the quantity of attributes is given by n and the similarity function is f of features i in the case A and B, w is the weight of each distinct feature. The efficacy of the proposed system in the work was evaluated using the recall rate, precision rate, f-score, and accuracy. The recall rate and precision is a quantity of completeness and exactness, respectively.

The high degree of exactness in identifying several primary headache; when symptoms overlapped it was impotent to analyze accurately PM and PTTH. Hence the weighted CBR CDSS method was developed to differentiate the two types of probable primary headaches to improve the diagnosis and treatment of headaches [1].

Computer-aided diagnoses developed by Allen J. Moses et al. (2006) enhances the capability of a person's mind, integrates more information (unseen by physicians) and searches instantly far more information of chronic head pain. The major argument over headache whether it is a symptom of a disease or an illness is still unknown. This led to the development computer software, computer-aided diagnoses, which help the physician to diagnose the headache accurately and give the treatment accordingly.

The results of the data analysis emphasize the need for thorough exploration of computer-aided diagnoses. A computerized database appears to be reasonable and useful for data gathering. A large user database is needed for giving individual diagnoses statistically. Statistical analysis of databases is as large as the Internet is capable of generating relevant information on more headache sufferers. However, currently such evidence-based standards do not exist in statistics, but data mining and information technology offer evidence-based and potential for exploration [2].

Diagnostic prediction model for the vestibular disease developed Grill et al. [3] and was implemented into a mobile device application that may assist PHP with their clinical decisions. The most common complaints are vertigo and dizziness. The signs and symptoms of vertigo and their complex interactions are based on well-performing prediction model. The data is collected from over 3000 patients per year at German Center for Vertigo and Balance Disorders. Patients are put to a series of questions to determine the type of vertigo and dizziness. Accuracy is defined by the ratio of all correct classifications, i.e., true positives (t_p) and true negatives (t_n) over all collected samples. **Overall accuracy** is equated to $(\mathbf{t_p + t_n})/(\mathbf{t_p + f_p + t_n + f_n})$.

Four steps are involved in machine learning workflow:

1. Cleansing and preparing data for training, checks redundant features and replaces missing data.
2. Selection of model, used to predict the accurate vertigo diagnosis from a number of independent variables based on any one algorithm—multi-class prediction task, namely, multi-class decision forests, an artificial neural network with one hidden layer and a multi-class logistic regression algorithm.
3. Model implementation and training, the modules are manually placed, connected, and configured. Each model provides different configuration parameters.
4. Evaluation, selection, fine tuning of models and publication of each model provides different parameters to optimize the accuracy. The multi-class decision forest is most suitable model for that task based on varied with the parameters of this model, number of decision trees, depth of the trees and the number of random splits of the trees.

The study involves 688 patients for sociodemographic and diagnostic information. With the overall accuracy of the multi-class decision forest was 0.62, of the multi-class neural network 0.33, and of the multi-class logistic regression 0.18. The cloud machine learning platform combines a high usability with breadth of the evaluation results and their visualization [3].

Mirarchi et al. [4] suggested a framework that implements pathology distribution and classification, it involves gathering and investigating data inflowing from vestibular system to attain supplementary information and support physicians in diagnosing. The study of vestibular functions is to recognize and distinguish vestibular disorders that are liable in human pathologies such as head tilt, asymmetrical ataxia, or nystagmus. The test named Vestibular Evoked Myogenic Potentials (VEMPs) is applied with external electrodes to evaluate the vestibular system that can be performed in cervical and ocular zone to measure the myogenic potentials called, respectively, cervical VEMP (cVEMPs) and ocular VEMP (oVEMPs) potentials. The methods such as (1) neural networks, (2) decision trees, (3) genetic algorithms, and (4) nearest neighbor method are adopted for healthcare. The analysis of health data is done by Bayesian-based methods. The statistical dependencies for gene data are carried out by Bayesian networks. The Entity Relational Diagram was used for clinical information such as the database of patient, examination, treatment, diagnosis and device data and information. PostGreSQL, with spatial extension, is used for database instance and further the system was developed by using a web application and graphical user interface. The probability of an experiment was analyzed based on Bayesian methods. The libraries include R software for statistical computing and graphics and using Weka module, the results are compared for auto-learning.

The proposed study enhances to get related and valuable information of vestibular pathology to the sex. The development of vestibular pathology was carried out with the analysis performed, improving the quality and the quantity of data [4].

Maximal frequent item-set Algorithm (MAFIA) is an algorithm used by Nishara Banu and Gomathy [5] for mining highest frequent itemsets from a database. The transactional database is stored efficiently as a series of vertical bitmaps using MAFIA. Decision trees are powerful for classification and prediction using C4.5 algorithm. The method of operation of k-means algorithm helps achieve its name. In data mining, clustering is used to identify the interesting patterns in a specified dataset. In medical imaging, the k-means algorithm is used for unordered method of defining clusters.

The study involves predicting heart diseases and the heart disease database is preprocessed effectively by deleting corresponding records and inserting missing values. The well-ordered preprocessed dataset is collected by K-means algorithm further processed with the K value of 2. The predicted values of heart attack and prescription ID with their levels are calculated. The data is trained to foresee the heart attack level and shows information gain for the efficient heart attack level using the C4.5. High accuracy, high precision, and recall metrics are achieved [5].

Health care is unstructured and huge. Hence, C4.5 decision tree has high values over k-mean-based MAFIA and k-mean MAFIA with ID3. The study shows efficient fragmenting and extracting significant forms from the heart attack database warehouses for the prediction of heart attack.

Halicek [6] has developed an automated classification of normal and cancerous, head and neck tissue using deep convolutional neutral networks (CNNs). The deep learning has the potential to be implemented into a tissue classifier, fully trainable on a database of hyperspectral images from tissue specimens that can produce near real-time tissue labeling for intraoperative cancer detection. A CNN was implemented using tensor flow to classify the spectral patches as either normal or cancer tissue. The neural network architecture consisted of six convolutional layers and three fully connected layers. The patch size used was 10×10, and the kernel size used for convolutions was 3×3. The output of each convolutional layer is $10 \times 10 \times N$, where N is the number of filters in the convolutional layer. The final layer, i.e., softmax, generates a probability of the pixel belonging to either class. The CNN classification performance was evaluated using leave-one-patient-out external-validation to calculate the sensitivity, specificity, and accuracy.

The experimental results show that the CNN has the potential for use in the automatic labeling of cancer and normal tissue using hyperspectral images 81% sensitivity, 78% specificity, and 80% accuracy, which could be useful for intraoperative cancer detection. The proposed technique is fast and does not require any further post-processing to enhance the results [6].

Prisilla et al. [7] discusses about the role of the deep learning has an application of the artificial intelligence helping to predict the early detection of the disease. The outstanding results in the image and speech recognition of the deep learning have proved over other techniques at predicting the potential drug molecules and analyzing particle accelerator data. The manual data analysis coupled with few methods is used for the unpredictable growth of medical databases for efficient computer-assisted analysis. The powerful tools for quality control like artificial neural networks (ANNs) are used in modeling complex relationships in multifaceted processes [7]. The results showed that the deep learning technique can reduce the error in diagnosing the disease, predict the result faster and provide indispensable treatment for the patient's life-saving.

Prisilla [8] identifies that the quality transmission is required to maintain the standardization, reliability, and correctness in the medical network with the packet loss. This medical network is referred to as butterfly network. The concept of butterfly network is to bring all the physicians and engineers to a unique platform where any clinic can access the services of the equipment by sharing among them. Thus, cloud computing helps in improving diagnoses. The butterfly network topology is based on mesh topology, where nodes take a square form. This butterfly network topology includes $(K+1)2^K$ nodes organized in $K+1$ ranks, each $n=2^K$ nodes. Any specific length is not mentioned for the edges, as the nodes keep increases and then the physical distance between them increases. In 3D network, communication time increases as the network grows and more over the edge length increases with number of processors. The source sends the packets to destination by avoiding the interference or packet loss with the help of network coding. Encoding the contents and forwarding the packets to help to achieve maximum bandwidth. The key purpose of networking coding is it delivers efficiently, ease of management, and resilience to network dynamics. Every node is anticipated to be a medical care center linked with each other. The information is passed in the form of packets from one medical care to another using cloud-based services.

In this study, the performance of the network is increased with the reduction of packet loss and the packet loss is identified very easily. The three causes for packet loss are (1) network link failure (2) when the buffer overflow occurs (3) the packet does not arrive at its destination due to congestion.

The butterfly network has high security with few resources and aims in reducing the cost and network diameter [8].

3   Convolution Neural Network

Captivating an image as input and labeling it as output is referred as image recognition. When a machine reads an image, it reads as an array of pixel values, each value between ranges of 0–255 shown in Fig. 1.

Convolution is a process of combining two functions by multiplying them. To convolve means to roll together. Basically, a convolutional network implements a kind of search. Convolutional networks are used for image detection and take many searches over a single image in horizontal lines, diagonal ones, vertically several times as a visual element is required. A kernel is defined as a matrix of numbers which is used in image convolutions. Various sized kernels comprising different patterns of numbers produce different results and the size of a kernel is arbitrary but 3×3 is used maximum.

In Fig. 2, the INPUT image [$32 \times 32 \times 3$] will hold the raw pixel values, an image of width 32, height 32, and with three color channels red, green, and blue. The result

Fig. 1 How machine reads
an image in pixels

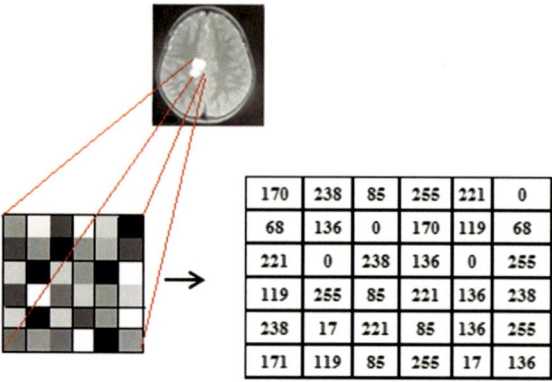

170	238	85	255	221	0
68	136	0	170	119	68
221	0	238	136	0	255
119	255	85	221	136	238
238	17	221	85	136	255
171	119	85	255	17	136

Fig. 2 $32 \times 32 \times 3$ image

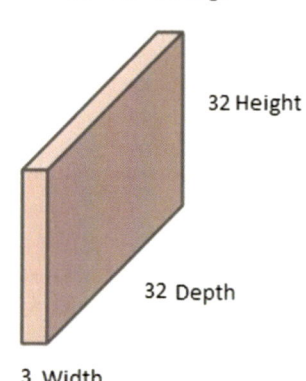

32 * 32 * 3 Image

32 Height

32 Depth

3 Width

of taking a dot product between the filter and small $5 \times 5 \times 3$ chunk of the image $(3 \times 3 \times 3 = 75$ − dimension dot product + bias) as in Fig. 3. When the filter slides around the input image or convolving, it is multiplying the values in the filter with the original pixel values of the image, i.e., element-wise multiplications.

A Conv Net has many layers. The three principal layers of Convolution Neural Network are as follows:

1. CONV Layers
2. ReLU Layer (Rectified Linear Unit) in Fig. 5
3. Subsampling Layers (Fig. 4).

The output of neurons are connected to the inputs and are calculated by CONV layer, each computing a dot product between their weights and a small region connected to the input volume.

ReLU layer applies an element-wise activation function, such as the max(0, x). Thresholding at zero leaves the size unchanged [$32 \times 32 \times 12$]. POOL layer performs a down-sampling operation along the spatial dimensions (width, height), resulting in volume such as [$16 \times 16 \times 12$] and used widely in CNN.

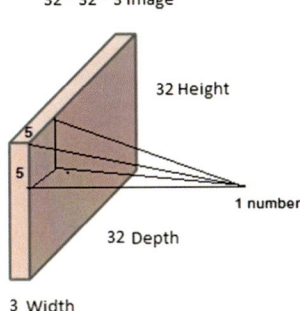

Fig. 3 Applying 5 × 5 filters

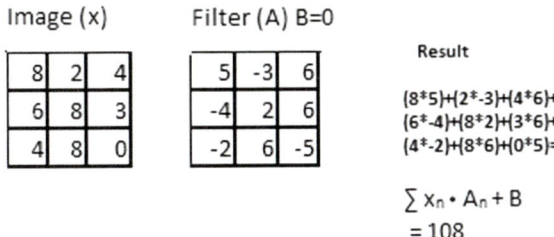

Fig. 4 Filtering

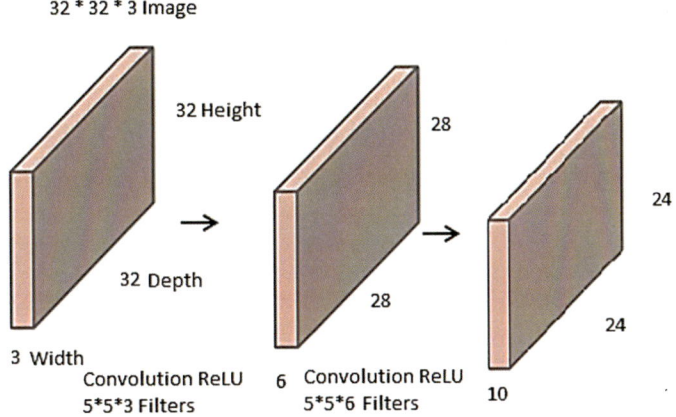

Fig. 5 Conv–ReLU layers in sequence

Max pooling is a technique of subsampling, the subsampling is to get an input representation by reducing its parameters of the network, which helps in reducing overfitting. Max pooling takes the highest pixel value from a region depending on its size. The pooling size is 2 × 2 pixels as in Fig. 6 [9].

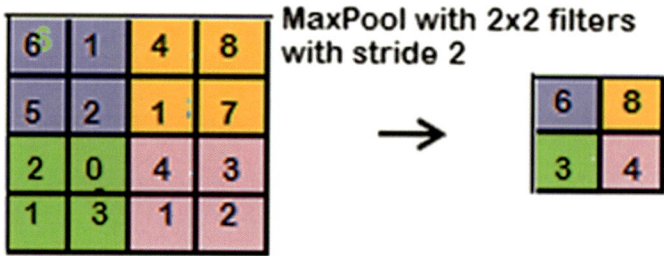

Fig. 6 Maxpooling

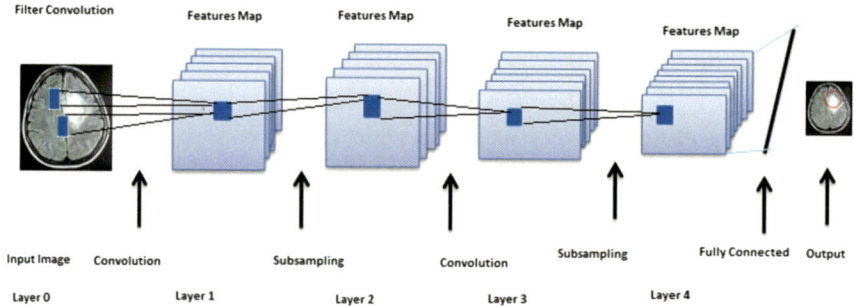

Fig. 7 Fully convolutional neural network

FC, fully connected layer of neurons at the end of CNN will compute the resulting size $[1 \times 1 \times 10]$ and it is multi-layer perceptron that uses a softmax activation function in the output layer. Every neuron in the previous layer in the fully connected layers is connected to every neuron on the next layer. The image is segmented to the proper labels after retrieving all of the features from each image as shown in Fig. 7.

4 Results and Discussion

4.1 Image Processing

The elements of image analysis are divided into three areas first, low-level processing, second, intermediate level processing, and last, high-level processing. This research concentrates on the intermediate-level processing that deals with extracting and describing the features in an image resulting from a low-level processing. It involves segmentation and description.

Gray-level slicing, the simple intensity transformation deals with the application which includes enhancing features like flaws in X-rays images, MRI, and CT-scans. In two approaches, one approach deals with the display of high values for all gray levels

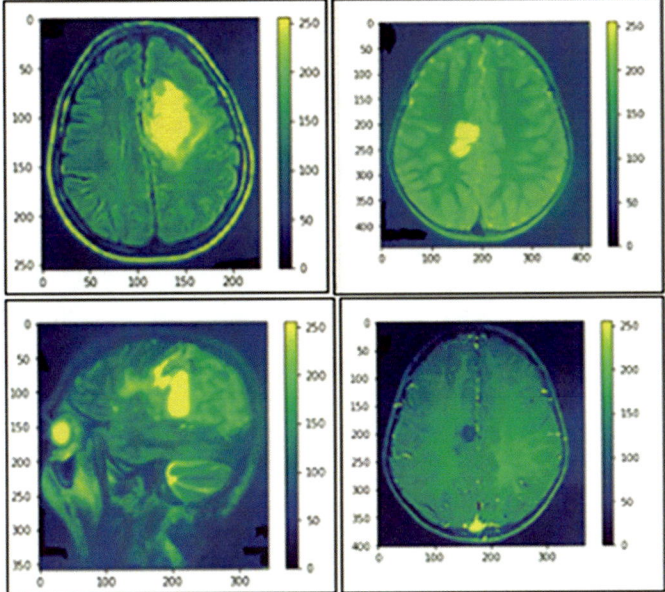

Fig. 8 Tumor part is highlighted

in the range of interest and low value for all the other gray levels. This transformation results in a binary image. Based on the transformation, the second approach brightens the desired range of the gray levels of the image but preserves the background and gray-level tones in the image.

Image subtraction, the difference between two images say $p(x, y)$ and $q(x, y)$ is represented as $R(x, y) = p(x, y) - q(x, y)$, the difference between all pairs of corresponding pixels from p and q. This important feature is used in segmentation and enhancement as displayed in Figs. 8 and 9 [10].

4.2 Image Segmentation

The process of partitioning a digital image to multiple segments, the aim of segmenting an image is to give meaningful information. The segmentation assigns a label to every pixel so that every pixel of same characteristics holds the same characteristics.

The proposed CNN experiment was conducted on the X-ray image of 100 patients having brain tumors which were collected from Gandhi Hospital, Secunderabad, India. The proposed CNN is implemented with the Spyder, a powerful interactive development environment for the Python language. The experiments are performed on a PC with an Intel Core i5-4300 M CPU @ 2.60 GHz processor and 8 GB RAM. On average, the CNN takes about 1 s processing time for each image.

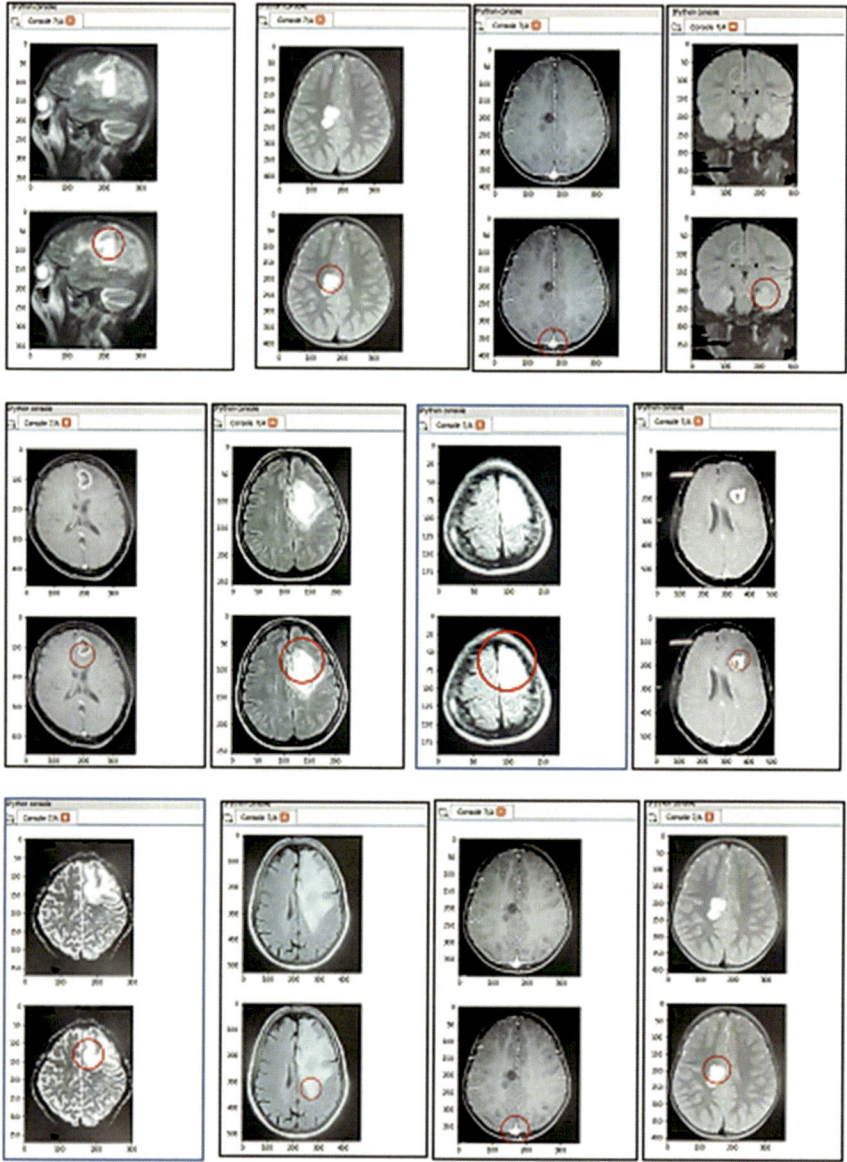

Fig. 9 Outputs of CNN for detecting the tumor part in the brain tumor

5 Conclusions

Fully convolutional networks have a rich essence of models, of which classification convnets are most significant. Extending these classification nets to segmentation, and improving the architecture with multi-resolution layer combinations improves simplifying and speeds up learning. The integration of image segmentation is such that the network can learn certain specific feature than any macroscopic features, and moreover inclusion of more abnormal examples which are very variant and unusual. Hence, the CNN gives more accurate segmentation result which helps in understanding the tumor part in the images of the brain thus helping the surgeon to diagnose it faster and for further analysis. The research is still in progress to recognize the technology growth in the field of medicine.

Conflicts of Interest Prisilla Jayanthi is the principal investigator of the CNN study and involved in script making. Dr. I. V. Murali Krishna is a keen guide in directing the research work and giving the novel ideas. The X-ray reports were collected from the Gandhi Hospital, Secunderabad, India and thanks for their support.

References

1. Yin Z et al (2015) A clinical decision support system for the diagnosis of probable migraine and probable tension-type headache based on case-based reasoning. J Headache Pain 16:29
2. Moses AJ et al (2006) Computer-aided diagnoses of chronic head pain: explanation, study data, implications, and challenges. J Craniomandibular Pract
3. Grill E et al (2016) Developing and implementing diagnostic prediction models for vestibular diseases in primary care. In: Exploring complexity in health: an interdisciplinary systems approach
4. Mirarchi D et al (2016) Applying mining techniques to analyze vestibular data. Elsevier Procedia Comput Sci 98(2016):467–472
5. Nishara Banu MA, Gomathy B (2013) Disease predicting system using data mining techniques. Int J Tech Res Appl. e-ISSN: 2320-8163
6. Halicek M et al (2017) Deep convolutional neural networks for classifying head and neck cancer using hyperspectral imaging. J Biomed Opt 22(6):060503
7. Prisilla J et al (2017) Deep learning in oncology—a case study on brain tumor. Int J Cancer Res Ther. ISSN: 2476-237
8. Prisilla J (2016) The butterfly network in medical care using clouds services. ICTBIG IEEE. https://doi.org/10.1109/ictbig.2016.7892655. Electronic ISBN: 978-1-5090-5515-9
9. Chen K, Seuret M (2017) Convolutional neural networks for page segmentation of historical document images
10. Gonzalez RC, Woods RE, Digital image processing. Pearson Education Asia

Vision-Based Algorithm for Fire Detection in Smart Buildings

Patel Abhilasha Paresh and Latha Parameswaran

Abstract Recent technological advancement has opened the space for a gradual increase in the number of smart buildings. Public safety and security has becomes a matter of concern with such a development, especially in areas of fire accidents. The conventional fire detection system usually worked on sensors and takes time for fire detection. This work presents an early fire detection system that unlike conventional fire detection system is cost-effective with high fire detection rate. The proposed algorithm uses features like color, increase in area and intensity flicker for early detection of fire. Segmentation of fire colored regions is done with the help of $L^*a^*b^*$, YCbCr, and RGB color space. Analysis of fire, i.e., fire area, its spread, temporal information, direction of the fire, and its average growth rate are measured using optical flow and blob analysis. Accuracy and F measure are used to evaluate the accuracy of the proposed system. Experimental results show that the average accuracy of the system is above 80% which is more promising in a video.

1 Introduction

The social progress, economic development, and technological advancements have increased the construction of buildings across a very large space. With such advancements, public safety and security have become a matter of concern especially in the case of fire accidents. Fire could cause huge economic and ecological damage and most importantly endangering people's life. Thus, such loss due to fire can be avoided or reduced with the help of an efficient and early fire detection system. Early detection of fire would limit the loss to minimum. Conventional fire detectors work on the basis of certain fire features like smoke, heat, or radiation. These fire detectors are

P. A. Paresh · L. Parameswaran (✉)
Department of Computer Science and Engineering, Amrita School of Engineering,
Amrita Vishwa Vidyapeetham, Coimbatore, India
e-mail: p_latha@cb.amrita.edu

P. A. Paresh
e-mail: cb.en.p2cvi16005@cb.students.amrita.edu

© Springer Nature Switzerland AG 2019
D. Pandian et al. (eds.), *Proceedings of the International Conference on ISMAC in Computational Vision and Bio-Engineering 2018 (ISMAC-CVB)*, Lecture Notes in Computational Vision and Biomechanics 30,
https://doi.org/10.1007/978-3-030-00665-5_99

usually based on infrared sensors, optical sensors, and ion sensors that respond to fire features. The reliability and the positional distribution of these sensors play an important role in early and efficient fire detection. Thus, early detection of fire in such a scenario seems practically impossible. The conventional system requires dense distribution of sensors, large detection time, costly, and thus fails to meet the needs for fire detection in large areas. Vision-based fire detection system not only overcomes these concerns but it also unlike conventional fire detection system facilitates earlier detection of fire and provides information regarding the location, direction, spread, and growth rate of fire.

This paper is structured as follows: Sect. 2 investigates the survey on the works in visual fire detection domain. The proposed work is described in Sect. 3. It discusses the architecture diagram and algorithms used in this paper. Architecture diagram discusses the workflow of the proposed work while algorithm section explains various techniques used in this paper. In Sect. 4, results and analysis of the proposed work based on different data set are discussed before drawing conclusion in Sect. 5.

2 Related Works

With the increasing concern of safety and security of life and property, there is a considerable increase in the development of visual fire detection systems that serve the need for fire detection in smart buildings. These systems detect fire with the help of color, shape, motion, geometry, and texture characteristics of fire. Color detection and dynamic feature analysis form the basis for most of the vision-based fire detection systems. A survey of key techniques is discussed here.

Celik and Demirel [1] proposed a color model that checks for the fire pixels using YCbCr color model which distinguishes luminance from chrominance than any other color model like RGB and $L^*a^*b^*$. Zhang et al. [2] considering the flicker characteristic of a fire, proposed a forest fire detection system based on contours of fire flame. This work combines two methods, i.e. wavelet which is basically used to detect fire, pixels is combined with FFT that describes the contour of the fire.

$L^*a^*b^*$ color model is used in [3] for fast and efficient fire detection system. Two improved features like color and motion are used in [4] to propose an improved probabilistic approach. In [5] color information, area size, coarseness of the surface, roughness of the boundary, and so on, are taken into consideration by Borges et al. to evaluate the behavioral change by observing the changes in between the frame with the help of low-level feature. Bayes classifier is used to detect fire. Considering the flicker property of a fire, Toreyin et al. [6] uses wavelet transform for detection of flickering pixels that indicate the presence of flames along with the Hidden Markov Models. Bohush and Brouka [7] use contrast and motion features to detect smoke while color and texture features are used for detection of flame in video.

3 Proposed Work

This work proposes an early fire detection system using color, flicker, area, and motion characteristic of fire for various online videos. Color segmentation is used to extract all the regions resembling the fire color and intensity. The color models used are RGB, YCbCr, and $L^*a^*b^*$. Morphological operations are used as post-processing to fill in for the fire pixels not detected due to the use of different color models. Flicker image is obtained by segmenting the high flicker feature regions. Optical flow is used to measure the speed and the direction of fire. Blob analysis, on the other hand, describes the height, width and the area of the detected fire.

3.1 Architecture Diagram

The architecture of the proposed algorithm is shown in Fig. 1. Two consecutive frames from input video are given to color segmentation module that extracts all colors ranging from yellow to red [1, 3, 8–10]. The empty spaces in extracted region are then filled using dilation operation. The rate of change in the area between the two consecutive frames is then measured. The frame is considered for fire detection if change in the area is greater than the threshold value. On the other hand, flicker image is measured by considering all the frames in 1s [11]. Both the images are then masked to get a fire image. The analysis of this fire is done with the help of Luckas Kanade optical flow and blob analysis [5, 12, 13].

3.2 Algorithm

Segmentation Using Flicker Intensity flickering feature is one of the important characteristics in fire detection. Since, positions of pixel around fire regions may or may not be a flame pixel in short time interval, thus having strong intensity differences. A pixel is considered to be fire if conditions in Eqs. (1) and (2) are satisfied [11].

$$D_{x,y} = \sum_{t=1}^{100} I_{x,y}(t) * \text{diff}_{x,y}(t) \tag{1}$$

$$D_{x,y} \geq \text{threshold} \tag{2}$$

High flicker increases the value of $D_{x,y}$ as in Eq. (1). The distinguishing factor that separates fire from non-fire is the second condition as in Eq. (2). A pixel is said to be fire if $D_{x,y}$ is greater than threshold otherwise it is just a disturbance pixel.

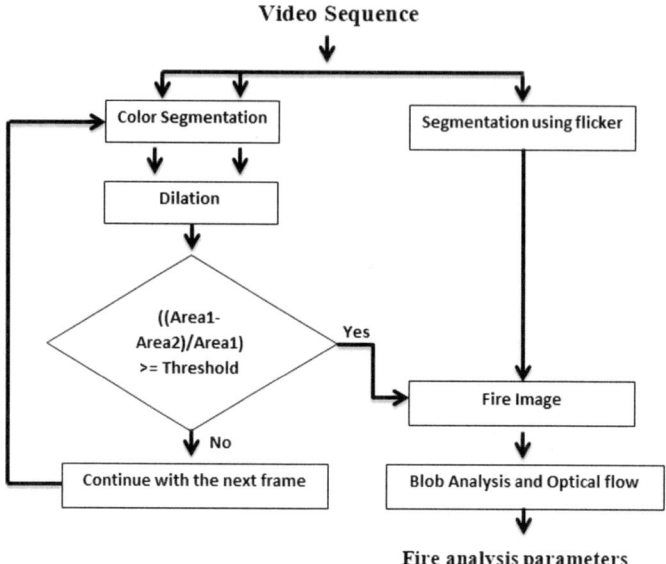

Fig. 1 Architecture of the proposed algorithm

Segmentation Using Color Fire can be characterized by color. Fire being high in illumination and color in the range of yellow to red, we can use this to characterize the fire pixels and detect its existence. Rule-based color segmentation for fire color regions is done using three color models, i.e. $L^*a^*b^*$, YCbCr, and RGB [14]. $L^*a^*b^*$ and YCbCr are chosen because of its ability to extract the illumination and chrominance separately for the segmentation of high-intensity regions. YCbCr is used to detect very high-intensity yellow color regions. $L^*a^*b^*$ color space, on the other hand, is used to detect orange and darker orange regions. RGB is used because most of the visible video cameras capture videos is in RGB format which makes our segmentation computational less complex. RGB is used to extract red regions of fire.

Lab Color Model Color of the fire usually ranges from yellow to red. $L^*a^*b^*$ color model is basically used to detect orange and darker orange fire. For an image to check for the existence of fire in it using a $L^*a^*b^*$ color model, its statistical measures like average can be calculated using the Eqs. (3), (4) and (5) [3, 9, 10].

$$L_m = \frac{1}{N} \sum_x \sum_y L^*(x, y) \tag{3}$$

$$a_m = \frac{1}{N} \sum_x \sum_y a^*(x, y) \tag{4}$$

$$b_m = \frac{1}{N} \sum_x \sum_y b^*(x, y) \tag{5}$$

where

L_m^*, a_m^* and b_m^* are average values of the L^* (luminance),
a^*, b^* (chrominance), respectively,
N is the total pixels in a frame and (x, y) is spatial pixel location in an image.

$L^*a^*b^*$ color model assumes that the regions of fire in a frame is the most luminous region and also is near to the color red. Thus, considering these conditions, fire rules defined in Eqs. (6), (7), (8) and (9) are formulated [3]

$$R1(x, y) = \begin{cases} 1, & L^*(x, y) \geq L_m^* \\ 0, & \text{Otherwise} \end{cases} \tag{6}$$

$$R2(x, y) = \begin{cases} 1, & a^*(x, y) \geq a_m^* \\ 0, & \text{Otherwise} \end{cases} \tag{7}$$

$$R3(x, y) = \begin{cases} 1, & b^*(x, y) \geq b_m^* \\ 0, & \text{Otherwise} \end{cases} \tag{8}$$

$$R4(x, y) = \begin{cases} 1, & b^*(x, y) \geq a_m^* \\ 0, & \text{Otherwise} \end{cases} \tag{9}$$

A pixel is said to be a fire pixel if it satisfies all the conditions by Eqs. (6), (7), (8) and (9).

RGB Color Model RGB color model extracts red color information of the image. Considering the RGB value of an image, a pixel is defined as a fire if its R channel intensity value is more than G channel value, G channel intensity value is more than B channel value. Since red color is more prominent than green and blue in fire, we can also say that for a pixel to be fire it needs to have more red intensity than its average red intensity. So mathematically, for a pixel to be a fire, it must satisfy Eqs. (10), (11) and (12) [8].

$$R_m = \frac{1}{N} \sum_x \sum_y R(x, y) \tag{10}$$

where

R_m is the mean of the R channel.

$$R1(x, y) = \begin{cases} 1, & R(x, y) > G(x, y) > B(x, y) \\ 0, & \text{Otherwise} \end{cases} \tag{11}$$

$$R2(x, y) = \begin{cases} 1, & R(x, y) \geq R_m \\ 0, & \text{Otherwise} \end{cases} \tag{12}$$

YCbCr Color Model YCbCr is an efficient color model that responds to high luminous yellow and brighter yellow regions. In YCbCr color model, for fire pixel, fol-
lowing relation between Y, Cb, and Cr channels holds true $Y(x, y) \geq Cr(x, y) \geq Cb(x, y)$ [1]. A pixel is said to be a fire if it satisfies Eqs. (13), (14), (15), and (16).

$$R1(x, y) = \begin{cases} 1, & (R(x, y) \geq G(x, y)), (G(x, y) > B(x, y)) \\ 0, & \text{Otherwise} \end{cases} \tag{13}$$

$$R2(x, y) = \begin{cases} 1, & (R(x, y) > (190)), (G(x, y) > 100), (B(x, y) < 140) \\ 0, & \text{Otherwise} \end{cases} \tag{14}$$

$$R3(x, y) = \begin{cases} 1, & Y(x, y) \geq Cb(x, y) \\ 0, & \text{Otherwise} \end{cases} \tag{15}$$

$$R4(x, y) = \begin{cases} 1, & Cr(x, y) \geq Cb(x, y) \\ 0, & \text{Otherwise} \end{cases} \tag{16}$$

Blob Analysis Blob is a patch of connected pixel having similar properties to form
a single structure. In case of binary image, background is denoted as zero whereas
every non zero value represents the object. The main aim of blob analysis is not to
detect fire rather analyze the detected fire, i.e. area covered by fire, its height and
width [13]. A bounding box is drawn over the detected fire regions. Time at which
fire is detected can also be obtained by considering the frame rate and frame for
which fire was detected.

Optical Flow Optical flow is used in this work to track the features and measure its
displacement between the two consecutive frames [5]. It measures the average rate
and the direction of fire. The optical flow algorithm (Lucas–Kanade) [12] is based
on three assumptions as follows:

Brightness Constancy It assumes that the brightness of each pixel remains alike from
frame to frame. Considering the grayscale images, the change in the intensity of the
pixels is always lesser than the threshold such that pixels are tracked properly frame
by frame.

$$f(x, t) = I(x(t), t) = I(x(t + 1), t + 1); \frac{\partial f(x)}{\partial t} = 0 \tag{17}$$

Temporal Persistence The motion of the moving pixel should be slow such that it is
tracked properly. It should not move fast.

$$\frac{\partial I}{\partial x}|_t \left(\frac{\partial x}{\partial t}\right) + \frac{\partial I}{\partial t}|_{x(t)} = 0; I_x v + I_t = 0; v = -\frac{I_t}{I_x} \tag{18}$$

Spatial Coherence It assumes that the neighbouring pixel has similar motion as that
of the moving pixel. It is assumed that all the pixels lie on the same plane. Hence, it
can be said that all pixel has similar motion.

$$\begin{bmatrix} I_x(p_1) & I_y(p_1) \\ I_x(p_{25}) & I_y(p_{25}) \end{bmatrix} \begin{bmatrix} u \\ v \end{bmatrix} = - \begin{bmatrix} I_t(p_1) \\ I_t(p_{25}) \end{bmatrix}; Ad = -B \tag{19}$$

Optical flow analysis. The output optical flow vector is obtained for each pixel from the above steps. The optical flow vectors can be defined as $p = [px, py]_i$, and $q = [qx, qy]_i$ where $i = 1, 2, \ldots, n$, p, and q are the position vectors in both frames. The velocity of the fire blob can be considered as the average velocity vector of each fire pixel and this is given by Eq. (20) as below:

$$V_{avg} = \frac{1}{n} \sum_{i=1}^{n} \sqrt{(p_{yi} - q_{yi})^2 + (p_{xi} - q_{xi})^2} \tag{20}$$

4 Experimental Results

The proposed system is evaluated with the help of various datasets. The dataset include fire and non-fire videos taken at different time (day, evening and night) and of different fire size (small, medium and large). The videos are of different format

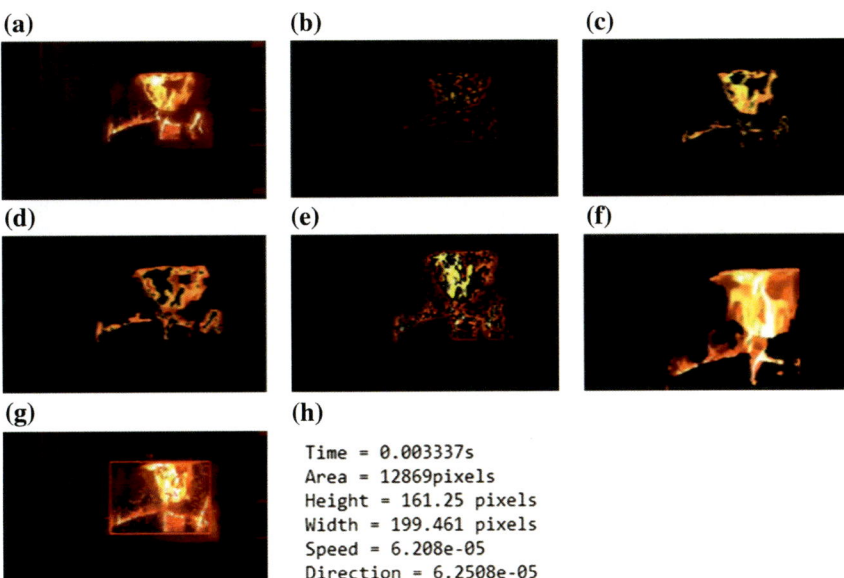

(a) **(b)** **(c)**

(d) **(e)** **(f)**

(g) **(h)**

```
Time = 0.003337s
Area = 12869pixels
Height = 161.25 pixels
Width = 199.461 pixels
Speed = 6.208e-05
Direction = 6.2508e-05
```

Fig. 2 **a** Input frame, **b** segmentation using RGB color model, **c** segmentation using YCbCr color model, **d** segmentation using $L^*a^*b^*$ color model, **e** dilation of color segmented image, **f** flicker image, **g** optical flow for fire detected regions, **h** fire analysis using optical flow and blob analysis

like.mp4 and .avi with varied length ranging from 2 to 14 s. Different resolution videos ranging from 240 * 320 to 720 * 1280 are considered. The proposed algorithm is implemented on Intel Core i7-4510U CPU @2.00 GHz with MATLAB R2016a. The intermediate results of the experiment are shown in Fig. 2.

The parameters used for evaluation are Accuracy and F measure. These can be calculated as discussed in Eqs. (21) and (22) referred from [13].

$$\text{Accuracy} = \frac{TP + TN}{TP + TN + FN + FP} \tag{21}$$

$$F \text{ measure} = \frac{2 * \text{Recall} * \text{Precision}}{(\text{Recall} + \text{Precision})} \tag{22}$$

TP   Fire image predicted as fire image
TN   Non-fire image predicted as non-fire image
FP   Non-fire image predicted as fire image
FN   Fire image is predicted as non-fire image.

Performance of the system is shown in Table 1. The average accuracy for the proposed fire detection system is 80.98%. The average F measure value is 89.95%.

5   Conclusion

This work proposes an early vision fire detection system that uses fire characteristics like color, area, intensity flicker, and motion for detection of fire. The fire is analyzed using optical flow and blob analysis. Blob analysis is to obtain details like location of fire, height, width while Lucas–Kanade optical flow measures the average velocity and direction of the fire. Experimental results show that the proposed work gives better results for fire videos. The average accuracy of the system is around 80.98%. Non-fire videos have less accuracy and enhancements needs to be done to improve upon the accuracy of non-fire videos. Videos with fire of very high intensity having color almost similar to white is not detected properly so enhancements for the same can be made in future to improve the false negatives of the system.

Table 1 Analysis of proposed visual fire detection system

Video	Duration (s)	No. of frames	No. of fire frames	No. of non-fire frames	TP	TN	FP	FN	Accuracy	F measure
video1	5.77	144	144	0	119	0	0	25	0.8264	0.9049
video2	3.00	61	61	0	45	0	0	16	0.826	0.8491
video3	9.00	265	265	0	110	0	0	55	0.6667	0.800
video4	3.00	59	59	0	46	0	0	13	0.7797	0.8762
video5	6.00	178	0	178	0	148	30	0	0.8315	–
video6	3.85	72	72	0	50	0	0	22	0.6944	0.8197
video7	4.00	76	76	0	51	0	0	25	0.6711	0.8031
video8	4.68	93	93	0	80	0	0	13	0.8602	0.9249
video9	3.00	52	52	0	49	0	0	3	0.9423	0.9703
video10	2.60	43	0	43	0	23	20	0	0.5349	–
video11	2.50	39	39	0	36	0	0	3	0.9231	0.9600
video12	3.64	67	67	0	62	0	0	5	0.9254	0.9612
video13	4.4	86	86	0	84	0	0	2	0.9767	0.9882
video14	3.88	73	73	0	72	0	0	1	0.9863	0.9931
video15	2.80	46	46	0	43	0	0	3	0.9348	0.9663
video16	3.45	61	61	0	58	0	0	3	0.9508	0.9758
video17	3.16	55	55	0	54	0	0	1	0.9818	0.9908
video18	5.68	118	0	118	0	68	50	0	0.5763	–
video19	8.00	212	212	0	134	0	0	78	0.6321	0.7746
video20	5.13	125	125	0	122	0	0	23	0.8414	0.9139
video21	6.2	131	131	0	65	0	0	66	0.4962	0.6633
video22	7.56	165	165	0	112	0	0	53	0.6788	0.8087
video23	8.28	183	183	0	161	0	0	22	0.8798	0.9360
video24	3.64	67	67	0	64	0	0	3	0.9552	0.9771
video25	2.24	32	32	0	28	0	0	4	0.8750	0.9333
Average									0.8098	0.8995

References

1. Celik T, Demirel H (2009) Fire detection in video sequences using a generic color model. Fire Saf J 44(2):147–158
2. Zhang Z, Zhao J, Zhang D, Qu C, Ke Y, Cai B (2008) Contour based forest fire detection using FFT and wavelet. In: 2008 International conference on computer science and software engineering, vol 1, pp 760–763
3. Celik T (2010) Fast and efficient method for fire detection using image processing. ETRI J 32(6):881–890
4. Zhang Z, Shen T, Zou J (2014) An improved probabilistic approach for fire detection in videos. Fire Technol 50(3):745–752
5. Kumar TS, Gautam KS, Haritha H (2016) Debris detection and tracking system in water bodies using motion estimation technique. In: Innovations in bio-inspired computing and applications, pp 275–284

6. Toreyin BU, Dedeoglu Y, Cetin AE (2005) Flame detection in video using hidden markov models. In: IEEE international conference on image processing, vol 2, pp II–1230
7. Bohush R, Brouka N (2013) Smoke and flame detection in video sequences based on static and dynamic features. In: Signal processing: algorithms, architectures, arrangements, and applications (SPA), pp 20–25
8. Chen T-H, Wu P-H, Chiou Y-C (2004) An early fire-detection method based on image processing. In: 2004 International conference on image processing, vol 3, pp 1707–1710
9. Yuan C, Liu Z, Zhang Y (2015) UAV-based forest fire detection and tracking using image processing techniques. In: International conference on unmanned aircraft systems, pp 639–643
10. Kecheril SS, Dr. Venkataraman D, Suganthi J, Sujathan K (2012) Segmentation of lung glandular cells using multiple color spaces. Int J Comput Sci Eng Appl 2(3):147–158
11. Stadler A, Windisch T, Diepold K (2014) Comparison of intensity flickering features for video based flame detection algorithms. Fire Saf J 66:1–7
12. Rinsurongkawong S, Ekpanyapong M, Dailey MN (2012) Fire detection for early fire alarm based on optical flow video processing. In: 9th International conference on electrical engineering/electronics, computer, telecommunications and information technology, pp 1–4
13. Kim Y-H, Kim A, Jeong H-Y (2014) RGB color model based the fire detection algorithm in video sequences on wireless sensor network. Int J Distrib Sens Netw 10(4)
14. Toulouse T, Rossi L, Akhloufi M, Celik T, Maldague X (2015) Benchmarking of wildland fire colour segmentation algorithms. IET Image Process 9(12):1064–1072

Comparative Performance Analysis of Local Feature Descriptors for Biomedical Image Retrieval

Suchita Sharma and Ashutosh Aggarwal

Abstract Biomedical imaging field is growing enormously from last decade. The medical images have been used and stored continuously for diagnosis as well as research purposes. There exist several methods that tend to provide real-time retrieval of medical images from such storage repositories. Therefore, in this paper, we strive to present an exhaustive performance comparison of existing and recently published state-of-the-art local feature descriptors for retrieval of CT and MR images. All the compared methods have been tested on two standard test databases, namely, NEMA CT and NEMA MRI. Additional experiments have been conducted to analyze the noise robustness ability of all the compared approaches. Lastly, the methods are also compared in terms of their computational complexity and total CPU time taken to retrieve images corresponding to the given query image.

1 Introduction

The usage of biomedical images during diagnosis of a patient in hospitals is increasing rapidly. The biomedical images can be in the form of X-ray, computed tomography (CT), magnetic resonance imaging (MRI), etc. Due to the rapid increment in biomedical images, it is becoming a more challenging task to diagnose a patient in clinics and hospitals. Hence, there is a need to have a more accurate and effective searching, indexing and retrieving techniques for a proper diagnosis. Moreover, it cannot be achieved without a proper arrangement of the biomedical images data. To cope up with such type of situations a new technique named content-based image retrieval (CBIR) system comes into the existence. CBIR systems have benefited the

S. Sharma (✉) · A. Aggarwal
Department of Computer Science and Engineering, Thapar Institute of Engineering and
Technology, Patiala, India
e-mail: ssharma_me16@thapar.edu

A. Aggarwal
e-mail: ashutosh.aggarwal@thapar.edu

© Springer Nature Switzerland AG 2019
D. Pandian et al. (eds.), *Proceedings of the International Conference on ISMAC
in Computational Vision and Bio-Engineering 2018 (ISMAC-CVB)*, Lecture Notes
in Computational Vision and Biomechanics 30,
https://doi.org/10.1007/978-3-030-00665-5_100

collection and management of continually increasing biomedical images as well as the medicine, R&D department, and education.

CBIR systems include a process of retrieval the most relevant images from an already existing image dataset on the basis of their primary (e.g., color, texture, and shape) or semantic features. To develop robust and efficient medical image retrieval (MIR) system, extraction of appropriate visual features like shape, color, and texture plays a significant role. These features (especially textures) provide the salient information from medical images in an efficient manner. The visual feature descriptors of an image can be categorized into two types as the local feature descriptors and the global feature descriptors.

In local feature descriptors, images present in the database are divided into the blocks and binary representation of each block is obtained by exploring the neighborhood in different directions. Then these blocks are scale to calculate local pattern and by concatenation of these local patterns, a feature vector is formed to retrieve the medical images. In global feature descriptors, the entire image is considered as a single unit for the purpose of feature extraction. The description of how the CBIR systems work with different feature extraction techniques is detailed in [1–5]. As biomedical images can be in the different formats, in this paper our aim is to retrieve the CT and MRI images. The work done in the initial days for the retrieval of MRI and CT images is mentioned in [6]. The PACS (Picture archiving and communication system), IRMA (Image retrieval for medical applications), SPIRS (Spine pathology and image retrieval system, etc., are the most popular content-based medical image retrieval systems (MIR).

Initially, local binary pattern (LBP) [7] has been proposed as a local feature descriptor for the purpose of the texture synthesis. LBP achieves very appreciable results in texture synthesis because of its simple working strategy. LBP has given a new insight into the research areas like medical, motion analyzing, face detection and image browsing and retrieval, etc. Because of its uniform direction working strategy, the results expecting under the difficult lighting conditions, noise conditions, pose conditions, aging of the images, etc., are not that much satisfactory. To deal with such type of conditions LBP is further extended to its various variants for serving the different application areas. To cope up the lightning conditions Tan and Triggs introduced a new local feature extraction technique named Local ternary patterns (LTP) [8] as a three-valued code formation method. With the help of LTP, it becomes easier to analyze the different body organs under different pose and lighting conditions. In the same trend, local quantized patterns (LQP) [9] and local tetra patterns (LTerPs) [10] are introduced for the retrieval of the biomedical images. The LTerPs have the extraordinary power to explain the spatial structure of the body organs in a more accurate and detailed way in a four coded-values format. The proposed extractor has the problem of high feature vector length. To overcome the feature vector length problem, Murala and Wu further introduced a new method of feature extraction named Local co-occurrence ternary patterns (LTCoP) [11] especially to retrieve out the MRI images. Later on, the local mesh edge patterns (LMeP) [12] for the retrieval of biomedical images has been introduced by the same authors. The working strategy in local mesh edge pattern (LMeP) is the little bit different than the already existing local methods, the LMeP simply encodes the relationship among the neighbors. After that, the LMeP has been further extended to a new feature descriptor method called Local Mesh Peak Valley edge patterns (LMePVPs) [13].

All the above-mentioned methods can work only for the two-dimensional view of the input medical grayscale images. To work with the 3D view of the medical images a new method called spherical symmetric three-dimensional local ternary patterns (SS-3D-LTP) [14] has been introduced. The SS-3D-LTP have the advantage to encode the relationship between the central pixel to its neighboring pixels in a selected five directions with an aim to convert the input 2D image to the 3D plane with the help of the Gaussian Filter Bank. From the 3D view of the body organs, it has become easier for the physicians to analyze the region of disorder more quickly and in an effective way. The LQPs are further extended to local quantized extrema patterns (LQEP) [15] for the better image retrieval. Later on, Dubey et al. introduce a new feature descriptor Local wavelet pattern (LWP) [16] especially for the CT images retrieval and the method has more discriminate power as compared to the already existing ones. The problem with LWP is to have the lesser dimensionality.

To cope with the dimensionality problem, the same author further proposed the new methods named local diagonal extrema pattern (LDEP) [17], local bit-plane decoded patterns (LBDP) [18] and local bit-plane dissimilarity patterns (LBDISP) [19] especially to achieve the better performance during the retrieval of CT images. After that, Deep et al. introduces the new methods called local mesh ternary patterns (LMeTP) [20], directional local ternary quantized extrema (DLTerQEPs) [21] and local quantized extrema quinary patterns (LQEQP) [22] to extend the standard LBP and LTP functionality in a 2D image to the selected directions especially for the retrieval of MRI and CT images. DLTerQEPs uses the ternary patterns from the horizontal, vertical, diagonal, anti-diagonal (HVDA$_7$) structure of the directional local extrema pattern values of an image to encode more spatial structure information for the better retrieval. LMeTP simply uses the ternary patterns from the mesh patterns of an image to encode the better spatial structural information of the medical images. In the same trend, Verma and Raman have been introduced a new method named local tri-diagonal patterns (LTDP) [23] to achieve the better retrieval performance rate over the LBP and all its variants by encoding all the possible information associated with a pixel in an image. LTDP has the capabilities to extract more accurate and systematic features from an input image in the comparison to the existing methods. Although the LTDP has been proved as a better feature descriptor, sometimes results may lead to the loss of discrimination information. To have a control over the loss of discrimination information the new method named local directional gradient pattern (LDGP) [24] has been introduced. In the same trend, a new feature descriptor method Local neighborhood Intensity Pattern (LNIP) [25] has been proposed. LNIP works on the idea to calculate the intensity difference between a particular pixel's adjacent neighbors. Recently, a new local descriptor method Local neighborhood difference pattern (LNDP) [26] by Verma and Raman has been introduced to find out the mutual relationship of all the neighboring pixels in a binary pattern with respect to each pixel in the image. The LNDP has the complementary results over LBP and all its variants.

The objective of the review presented in this paper is to identify those methods which are effective as well as efficient in accurate retrieval of different biomedical images.

The rest of the paper is organized as, Sect. 2 presents a brief overview of LBP. Section 3 presents the experimental framework used for testing the retrieval performances of all the compared approaches. Section 4 presents the computational complexity associated with all the compared approaches. Last, the paper is concluded in Sect. 5.

2   Existing Local Feature Extraction Methods

2.1   Local Binary Patterns

The local binary pattern (LBP) was proposed by Ojala et al. for the purpose of texture classification. LBP had success in terms of speed and there was no need to tune any parameter. For a given input grey scale image, the LBP pattern is calculated by constructing a local window around each pixel of the image to encode the intensity difference relationship between a particular central pixel (I_c) to all its neighboring pixels intensity values (I_i). The number of neighboring pixels is depends on the value of radius R. If $R = 1$ means that the eight neighbors will be lying on the circumference of the circle around the central pixel. The value obtained from the intensity difference are replaced with the binary number either 0 or 1. Thus, a binary code of a particular length is obtained for the selected central pixel, to get a LBP map the weights are assigned to the obtained binary values as described in Eq. (1)

$$LBP_{N,R} = \sum_{i=1}^{N} 2^{i-1} \times D_1(I_i, I_c) \qquad (1)$$

To construct a feature vector for the entire image a histogram concatenation is done for all possible obtained LBP maps as described in Eq. (3)

$$Hist\_LBP = \sum_{i=1}^{X} \sum_{J=1}^{Y} D_2(LBP(I_i, I_c), L); L \in \left[0, \left(2^N - 1\right)\right] \qquad (2)$$

$$D_2(a_1, a_2) = \begin{cases} 1, & a_1 = a_2 \\ 0, & \text{otherwise} \end{cases} \qquad (3)$$

2.2 Local Ternary Patterns

Tan and Triggs extended the LBP from the two-valued code to three-valued code. The mathematical equation to represent the LTP is as

$$\vec{F}(x, g_c, t) = \begin{cases} +1, & x \geq g_c + t \\ 0, & |x - g_c| < t \\ -1, & x \leq g_c - t \end{cases} \tag{4}$$

where t is a threshold value, values $g_c - t$ and $g_c + t$ are considered to be as the low and high range of values. The values lie in the lower range of the low are replaced as -1, values in the upper range to the high value replaced by $+1$ and the value lies between the high and low is considered to be as the 0. In LTP the threshold value is a user defined value. Hence, the obtained LTP values are more resistant to noise and no more effect of the invariance to the gray-level transformation. To construct a feature vector for the entire image a histogram concatenation is done for all possible obtained LTP maps as described earlier in LBP.

The readers are requested to refer to cited references in the introduction section to read about the other local feature descriptors compared in this paper.

3 Experiments and Results

In this section, we present the computational framework employed to test the performance of the existing local feature descriptors for biomedical image retrieval. We have considered the following mentioned methods in our comparative analysis. All the methods have been implemented in MATLAB 2013.

S. No.	Abbreviation	Method name	Year
1	LBP	Local Binary Pattern	2002
2	LTP	Local Ternary Pattern	2009
3	LQryP	Local Quinary Patterns	2010
4	LTrP	Local Tetra Patterns	2012
5	LTCoP	Local Ternary Co-occurrence Patterns	2013
6	LMeP	Local Mesh Patterns	2014
7	LMePVEP	Local Mesh Peak Valley Edge Patterns	2014
8	LDEP	Local Diagonal Extrema Pattern	2015
9	LWP	Local Wavelet Pattern	2015

(continued)

(continued)

S. No.	Abbreviation	Method name	Year
10	LQEP	Local Quantized Extrema Patterns	2015
11	SS-3D-LTP	Spherical Symmetric 3D Local Ternary Patterns	2015
12	LBDISP	Local Bit-Plane Dissimilarity Pattern	2016
13	LBDP	Local Bit-Plane Decoded Pattern	2016
14	DLTerQEP	Directional Local Ternary Quantized Extrema Pattern	2016
15	LMeTP	Local Mesh Ternary Patterns	2016
16	LNIP	Local Neighborhood Intensity Pattern	2017
17	LTDP	Local Tri-directional Patterns	2017
18	LNDP	Local Neighborhood Difference Pattern	2017
19	LDGP	Local Directional Gradient Pattern	2017
20	LQEQryP	Local Quantized Extrema Quinary Pattern	2017

In our experiments, we use Extended Canberra similarity measure to compare the feature vectors of the query image with that of database images. The mathematical expression for the Extended Canberra distance is as

$$D_{\mathrm{ECD}}(t, Q) = \sum_{\tau=1}^{\dim} \frac{\left| F^{Q}(\tau) - F^{t}(\tau) \right|}{\left(F^{Q}(\tau) + \mu^{Q} \right) + \left(F^{t}(\tau) + \mu^{t} \right)},$$

$$\mu^{Q} = \frac{1}{\dim} \sum_{\tau=1}^{\dim} F^{Q}(\tau)$$

$$\mu^{t} = \frac{1}{\dim} \sum_{\tau=1}^{\dim} F^{t}(\tau) \tag{5}$$

dim represents the dimension of the feature vectors, $F^{Q}(\tau)$ and $F^{t}(\tau)$ is the τth element of the feature vectors of query image Q and database image t, respectively. In our experiments, every image in the database is used as a query image. The average precision rate (APR) average retrieval rate (ARR), F_score and mean average precision (mAP) are used as the performance measures for each retrieval method.

3.1 Dataset Used

In this paper two datasets namely NEMA CT [27] and NEMA MRI [28] consisting of CT and MR images of various body organs are used for the experimental work. The NEMA CT consisting of 600 CT images categorized it into 10 categories having

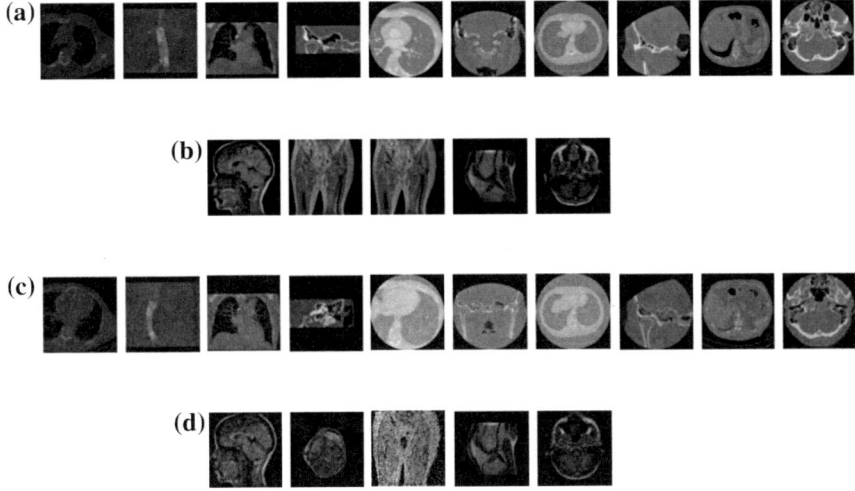

Fig. 1 Sample images from **a** NEMA CT database, **b** NEMA MRI database, **c** noisy NEMA CT database and **d** noisy NEMA MRI database

54, 70, 66, 50, 15, 60, 52, 104, 60, and 69 images. The second dataset is NEMA MRI containing 372 MR images and categorized it into 5 categories having 72, 100, 76, 59, and 65 images. Figure 1a, b shows sample image from each category of NEMA CT and NEMA MRI datasets, respectively. Besides experimenting with the above-mentioned databases, additional experiments have been performed to test the effectiveness and superiority of all the compared methods under noisy conditions. In this process, a new copy of each of the above-mentioned databases has been created consisting of noisy images. The noisy images are obtained by corrupting the original noise-free images with additive white Gaussian noise with zero mean and $\sigma \in [5, 50]$. Figure 1c, d shows sample noisy images.

3.2 Experimental Results

This section provides several experiments to demonstrate the effectiveness of existing local feature extraction methods and for all these approaches, we have taken a window size of 3×3 with $(P, R) = (8, 1)$.

3.2.1 Results on Noise-Free NEMA CT Database

Table 1 compares the retrieval performance of the all existing local feature extraction methods in terms of ARR, ARP, F_score, and mAP. The experiments have been performed to obtain the best 100 matches from the NEMA CT database. From Table 1,

Table 1 Comparison of performances of all existing local feature extraction techniques on the bases of ARR, APR, F_score and mAP for best $\eta = 100$ on noise-free NEMA CT database and all the mentioned values are in percentages (%)

	ARR	APR	F_score	mAP
LMePTerP	96.36	64.98	77.62	98.55
DLTerQEP	96.45	64.92	77.61	99.00
LQEP	96.43	64.70	77.44	98.96
LQEQerP	95.83	64.65	77.21	98.56
LMeP	96.19	64.64	77.32	98.12
LNDP	96.42	64.62	77.38	98.07
SS-3D-LTP	95.92	64.62	77.22	98.43
LTP	95.99	64.52	77.17	98.83
LTDP	96.18	64.50	77.21	98.50
LTrP	95.92	64.26	76.96	97.92
LQP	95.27	64.00	76.56	98.48
LDGP	95.52	63.98	76.63	97.27
LNIP	95.45	63.93	76.57	97.68
LDEP	95.02	63.72	76.28	97.23
LTCoP	95.47	63.55	76.31	98.36
LMePVEP	93.51	62.91	75.22	97.43
LBP	94.19	62.25	74.96	97.85
LWP	80.37	52.76	63.70	89.89
LBDP	76.91	50.06	60.64	84.45
LBDISP	71.14	45.20	55.28	82.41

it can be observed that the results obtained from DLTerQEP method are better than those obtained from the other existing methods. The compared method has the ARR, APR, F_score and mAP retrieval rates of 96.45% (ARR), 64.92% (APR), 77.61% (F_score), and 99.00% (mAP), with maximum gain of 0.79% (ARR), 3.11% (APR), 2.45% (F_score), and 0.12% (mAP) over other methods. The APR graph for all the compared methods has been shown in Fig. 2. The graph demonstrates the variation in the retrieval rates of all the compared methods with the number of top matches.

3.2.2 Results on Noise-Free NEMA MRI Database

Figure 3 shows the graphical comparison of the APR rate of all existing local feature extraction methods. The retrieval rates of all the compared methods for top 100 matches are shown in Table 2. The methods LBP, LTP, LTrP, LTCoP, LNIP, LTDP, LMeP, LMePVEP, LNDP, LDGP, LQEP, and DLTerQEP gave similar and maximum retrieval performance of 100% (ARR), 77.06% (APR), 87.04% (F_score), and 100% (mAP). Among all the compared methods, the retrieval performance of LWP has been

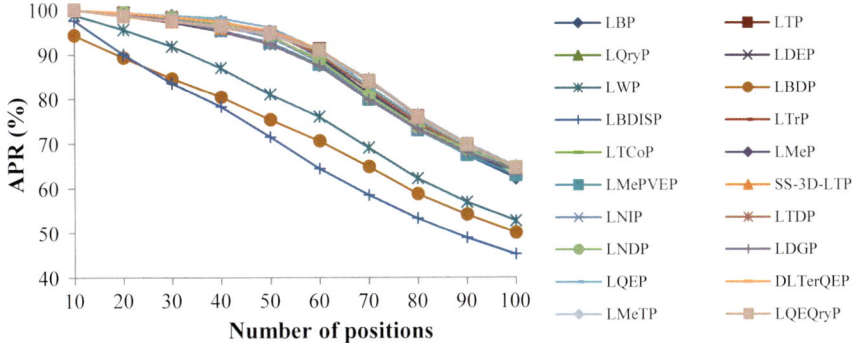

Fig. 2 Comparison between all existing local techniques on the bases of APR on noise-free NEMA CT database

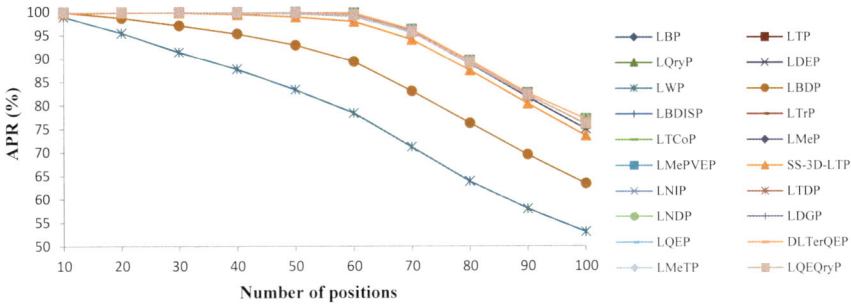

Fig. 3 Comparison between all the existing techniques on the bases of APR on noise-free NEMA MRI database

observed to be the lowest with the loss in retrieval rates of 28.49% (ARR), 27.26% (APR), 28.16% (F_score), and 13.32% (mAP).

3.2.3 Results on Noisy Databases

Tables 3 and 4 summarizes the retrieval performance of all the compared methods for top 100 matches and Figs. 4 and 5 shows the APR retrieval rates of all existing local feature extraction methods tested over noisy NEMA CT and noisy NEMA MRI databases, respectively. The method LBDP obtained the highest retrieval rates of 47.11% (ARR), 32.57% (APR), 38.51% (F_score), and 35.83% (mAP) on noisy NEMA CT database and 27.01% (ARR), 15.94% (APR), 20.13% (F_score), and 40.69% (mAP) and 60.54% (ARR), 46.20% (APR), 52.40% (F_score), and 50.17% (mAP) on noisy NEMA MRI database. From the results, it is observed that the method LBDP is highly robust to noise among all the compared methods and is the most suitable local feature extraction method for the retrieval of noisy CT and MR images.

Table 2 Comparison of performances of all existing local feature extraction techniques on the bases of ARR, APR, F_score and mAP for best $\eta = 100$ on noise-free NEMA MRI database and all the mentioned values are in percentages (%)

	ARR	APR	F_score	mAP
LBP	100.00	77.06	87.04	100.00
LTP	100.00	77.06	87.04	100.00
LTrP	100.00	77.06	87.04	100.00
LTCoP	100.00	77.06	87.04	100.00
LMeP	100.00	77.06	87.04	100.00
LMePVEP	100.00	77.06	87.04	100.00
LNIP	100.00	77.06	87.04	100.00
LTDP	100.00	77.06	87.04	100.00
LNDP	100.00	77.06	87.04	100.00
LDGP	100.00	77.06	87.04	100.00
LQEP	100.00	77.06	87.04	100.00
DLTerQEP	100.00	77.06	87.04	100.00
LBDISP	99.99	77.05	87.03	99.99
LQEQerP	98.83	76.06	85.96	99.86
LMePTerP	98.80	76.06	85.95	99.77
LQP	98.79	75.95	85.88	99.80
LDEP	97.90	74.96	84.91	99.63
SS-3D-LTP	96.22	73.51	83.34	98.87
LBDP	83.80	63.36	72.16	93.93
LWP	71.51	53.04	60.91	86.68

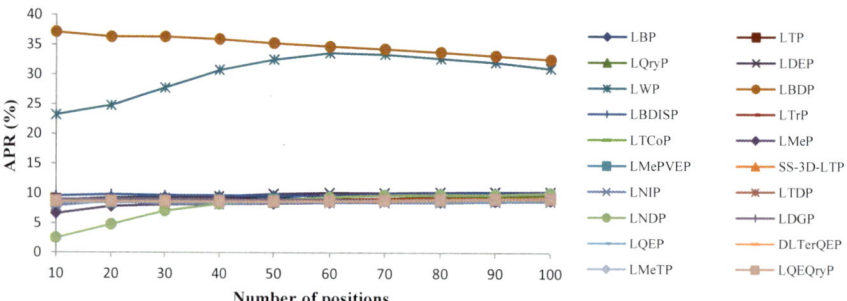

Fig. 4 Comparison between all the existing techniques on the bases of APR on noisy NEMA CT database

4   Computational Efficiency

Biomedical image retrieval requires methods which are effective as well as computationally efficient. Such methods should provide real-time response, i.e., the time

Table 3 Comparison of performances of all existing local feature extraction techniques on the bases of ARR, APR, F_score and mAP for best $\eta = 100$ on noisy NEMA CT database and all the mentioned values are in percentages (%)

	ARR	APR	F_score	mAP
LBDP	47.11	32.57	38.51	35.83
LWP	46.51	31.03	37.23	31.00
LDEP	18.39	10.27	13.18	9.48
LBDISP	19.05	10.11	13.21	9.71
LNDP	16.42	10.09	12.50	7.88
LQP	18.45	9.73	12.74	8.83
SS-3D-LTP	16.75	9.24	11.91	8.77
LTP	16.80	9.18	11.87	8.87
LMePVEP	16.81	9.18	11.87	8.69
LMePTerP	16.67	9.18	11.84	8.61
LQEQerP	16.67	9.18	11.84	8.68
LTrP	17.03	9.16	11.91	9.03
LTCoP	16.77	9.14	11.83	8.72
DLTerQEP	16.68	9.10	11.77	8.66
LDGP	17.28	9.05	11.88	8.99
LTDP	16.59	8.97	11.64	8.70
LQEP	16.55	8.93	11.60	8.52
LMeP	16.66	8.91	11.61	7.67
LBP	16.59	8.79	11.49	9.22
LNIP	16.24	8.67	11.31	8.01

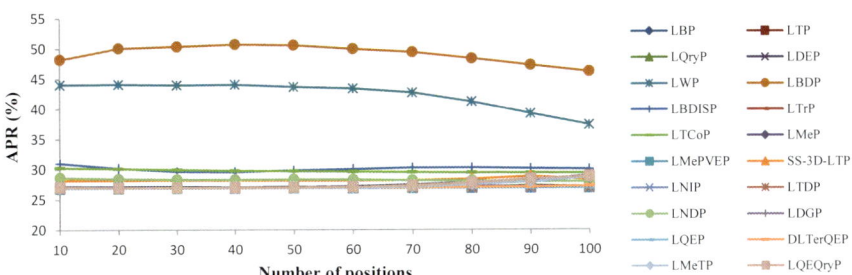

Fig. 5 Comparison between all the existing techniques on the bases of APR on noisy NEMA MRI Database

associated with the retrieval of medical images should be very less. The computational time for any retrieval system depends on two stages: (1) Feature Extraction Stage and (2) Retrieval (Classification) Stage. Table 5 shows the comparison of the feature vector length as well as the total CPU time of all the compared approaches for a given query image. The method LQEQerP has the highest number of features, i.e., 16,384 features. From table, it can be observed that the total CPU time for the

Table 4 Comparison of performances of all existing local feature extraction techniques on the bases of ARR, APR, F_score and mAP for best $\eta = 100$ on noisy NEMA MRI database and all the mentioned values are in percentages (%)

	ARR	APR	F_score	mAP
LBDP	60.54	46.20	52.40	50.17
LWP	46.34	37.33	41.35	42.04
LBDISP	34.91	29.98	32.26	30.05
LTCoP	30.36	29.33	29.84	29.92
LDEP	29.84	28.96	29.39	27.08
LQEQerP	30.04	28.83	29.42	27.04
LMePTerP	29.45	28.69	29.06	26.90
SS-3D-LTP	29.56	28.28	28.91	28.06
LNDP	29.90	27.92	28.88	28.14
LMeP	27.37	27.20	27.29	26.92
DLTerQEP	27.33	27.20	27.26	26.92
LBP	27.28	27.14	27.21	27.14
LQP	28.36	27.05	27.69	26.63
LDGP	26.96	26.94	26.95	26.95
LMePVEP	26.90	26.89	26.90	26.88
LTP	26.89	26.89	26.89	26.88
LNIP	26.89	26.88	26.88	26.88
LTDP	26.89	26.88	26.89	26.88
LTrP	26.88	26.88	26.88	26.88
LQEP	26.88	26.88	26.88	26.88

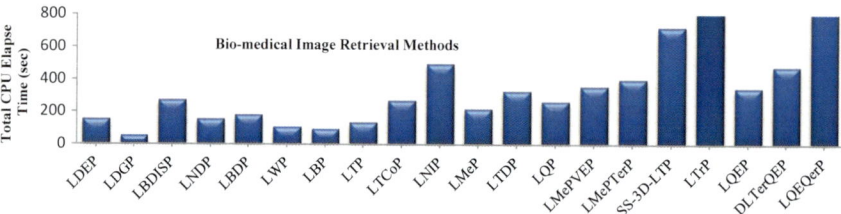

Fig. 6 Total CPU Elapse time (s) comparison of all the existing biomedical retrieval approaches over NEMA CT database

LDGP method is lowest among all other compared (existing) methods. Figures 6 and 7 show the total CPU elapse time (s) of all the compared approaches for the NEMA CT and NEMA MRI databases.

Table 5 CPU elapse time (s) taken for extracting the features and retrieved out the images from NEMA CT and NEMA MRI databases images using already existing local feature extraction approaches

Feature descriptors	No. of features	Feature extraction time (A)		Retrieval time (B)		Total CPU time (A + B)	
		NEMA CT	NEMA MRI	NEMA CT	NEMA MRI	NEMA CT	NEMA MRI
LDEP	24	156.13	24.20	1.86	1.42	157.99	25.62
LDGP	64	51.38	7.96	4.33	3.07	55.71	11.03
LBDISP	256	265.07	41.09	11.51	7.25	276.58	48.34
LNDP	256	146.43	22.70	11.51	7.25	157.94	29.95
LBDP	256	171.88	26.64	11.51	7.25	183.39	33.89
LWP	256	98.25	15.23	11.51	7.25	109.76	22.48
LBP	256	85.03	13.18	11.51	7.25	96.54	20.43
LTP	512	117.54	18.22	20.54	12.26	138.08	30.48
LTCoP	512	252.74	39.17	20.54	12.26	273.28	51.43
LNIP	512	477.18	73.96	20.54	12.26	497.72	86.22
LMeP	768	190.27	29.49	30.21	16.41	220.48	45.9
LTDP	768	302.54	46.89	30.21	16.41	332.75	63.3
LQP	1024	226.25	35.07	41.68	21.98	267.93	57.05
LMePVEP	1024	318.63	49.39	41.68	21.98	360.31	71.37
LMePTerP	1536	343.45	53.23	60.46	31.38	403.91	84.61
SS-3D-LTP	2560	629.75	97.61	93.82	43.03	723.57	140.64
LTrP	3328	7848.78	1216.56	121.70	51.80	7970.48	1268.36
LQEP	4096	202.86	31.44	147.91	63.01	350.77	94.45
DLTerQEP	8192	210.02	32.55	271.52	118.62	481.54	151.17
LQEQerP	16,384	385.55	59.76	510.18	252.26	895.73	312.02

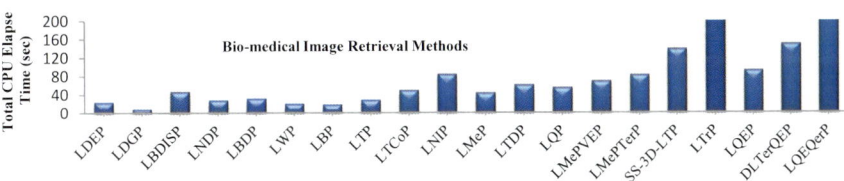

Fig. 7 Total CPU elapse time (s) comparison of all the existing biomedical retrieval approaches over NEMA MRI database

5 Conclusion

In this paper, a comparison between all the existing local biomedical image retrieval approach for CT and MR images has been presented. Experiments have been performed over both the noisy as well as noise-free images of two standard databases namely NEMA CT and NEMA MRI to check the performance comparison of existing and recently published state-of-the-art local feature descriptors. After the thorough analysis of the obtained results, it was observed that the LDGP approach is most suitable for retrieval of noise-free images. LDGP is found to be the most balanced approach in terms of its retrieval performance and the computational time requirements. However, in the case of noisy datasets, LBDP and LWP approach achieved maximum retrieval rates. The higher retrieval rates demonstrate their superior noise robustness capabilities in comparison to other compared approaches.

References

1. Rui Y, Huang TS (1999) Image retrieval: current techniques, promising directions and open issues. J Vis Commun Image Represent 10:39–62
2. Smeulders AWM, Worring M, Gupta SA, Jain R (2000) Content-based image retrieval at the end of the early years. IEEE Trans Pattern Anal Mach Intell 22:1349–1380
3. Kokare M, Chatterji BN, Biswas PK (2002) A survey on current content-based image retrieval methods. IETE J Res 48:261–271
4. Liu Y, Zhang D, Lu G, Ma WY (2007) A survey of content-based image retrieval with high-level semantics. Pattern Recogn 40:262–282
5. Muller H, Michoux N, Bandon D, Geisbuhler A (2004) A review of content-based image retrieval systems in medical applications—clinical benefits and future directions. Int J Med Inform 73:1–23
6. Manjunath KN, Renuka A, Niranjan UC (2007) Linear models of the cumulative distribution function for content-based medical image retrieval. J Med Syst 31:433–443
7. Ojala T, Pietikäinen M, Harwood D (1996) A comparative study of texture measures with classification based on featured distributions. Pattern Recogn 29(1):51–59
8. Tan X, Triggs B (2010) Enhanced local texture feature sets for face recognition under difficult lighting conditions. IEEE Trans Image Process 19(6):1635–1650
9. Hussain ul S, Triggs B (2012) Visual recognition using local quantized patterns. ECCV 2012. Part II, LNCS 7573, Italy, pp 716–729
10. Murala S, Maheshwari RP, Balasubramanian R (2012) Local tetra patterns: a new feature descriptor for content-based image retrieval. IEEE Trans Image Process 21(5):2874–2886
11. Murala S, Wu QMJ (2013) Local ternary co-occurrence patterns: a new feature descriptor for MRI and CT image retrieval. Neurocomputing 119:399–412
12. Murala S, Wu QMJ (2014) Local mesh patterns versus local binary patterns: biomedical image indexing and retrieval. IEEE J Biomed Health Inform 18(3):929–938
13. Murala S, Wu QMJ (2014) MRI and CT image indexing and retrieval using local mesh peak valley edge patterns. Sig Process Image Commun 29(3):400–409
14. Murala S, Wu QMJ (2015) Spherical symmetric 3D local ternary patterns for natural, texture and biomedical image indexing and retrieval. Neurocomputing 149:1502–1514
15. Rao LK, Rao DV (2015) Local quantized extrema patterns for content-based natural and texture image retrieval. Human-Centric Comput Inf Sci 5(1):26
16. Dubey SR, Singh SK, Singh RK (2015) Local wavelet pattern: a new feature descriptor for image retrieval in medical CT databases. IEEE Trans Image Process 24(12):5892–5903
17. Dubey SR, Singh SK, Singh RK (2015) Local diagonal extrema pattern: a new and efficient feature descriptor for CT image retrieval. IEEE Signal Process Lett 22(9):1215–1219
18. Dubey SR, Singh SK, Singh RK (2016) Local bit-plane decoded pattern: a novel feature descriptor for biomedical image retrieval. IEEE J Biomed Health Inform 20(4):1139–1147

19. Dubey SR, Singh SK, Singh RK (2016) Novel local bit-plane dissimilarity pattern for computed tomography image retrieval. Electron Lett 52(15):1290–1292
20. Deep G, Kaur L, Gupta S (2016) Directional local ternary quantized extrema pattern: a new descriptor for biomedical image indexing and retrieval. Eng Sci Technol Int J 19(4):1895–1909
21. Deep G, Kaur L, Gupta S (2016) Local mesh ternary patterns: a new descriptor for MRI and CT biomedical image indexing and retrieval. Comput Methods Biomech Biomed Eng: Imaging Vis 6(2):155–169
22. Verma M, Raman B (2016) Local tri-directional patterns: a new texture feature descriptor for image retrieval. Digit Signal Proc 51:62–72
23. Deep G, Kaur L, Gupta S (2017) Local quantized extrema quinary pattern: a new descriptor for biomedical image indexing and retrieval. Comput Methods Biomech Biomed Eng: Imaging Vis 1–17
24. Chakraborty S, Singh SK, Chakraborty P (2017) Local directional gradient pattern: a local descriptor for face recognition. Multimedia Tools Appl 76(1):1201–1216
25. Banerjee P, Bhunia AK, Bhattacharyya A, Roy PP, Murala S (2017) Local neighborhood intensity pattern. A new texture feature descriptor for image retrieval. arXiv preprint arXiv:1709.02463
26. Verma M, Raman B (2017) Local neighborhood difference pattern. A new feature descriptor for natural and texture image retrieval. Multimedia Tools Appl 1–24
27. NEMA CT Image Database [Online] ftp://medical.nema.org/medical/Dicom/Multiframe/CT
28. NEMA MR Image Database [Online] ftp://medical.nema.org/medical/Dicom/Multiframe/MR

Detection of Liver Tumor Using Gradient Vector Flow Algorithm

Jisha Baby, T. Rajalakshmi and U. Snekhalatha

Abstract Liver tumor also known as the hepatic tumor is a type of growth found in or on the liver. Identifying the tumor location can be a tedious, error-prone and need an experts study to identify it. This paper presents a segmentation technique to segment the liver tumor using Gradient Vector Flow (GVF) snakes algorithm. To initiate snakes algorithm the images need to be insensitive to noise, Wiener Filter is proposed to remove the noise. The GVF snake starts its process by initially extending it to create an initial boundary. The GVF forces are calculated and help in driving the algorithm to stretch and bend the initial contour towards the region of interest due to the difference in intensity. The images were classified into tumor and non-tumor categories by Artificial Neural Network Classifier depending on the features extracted which showed notable dissimilarity between normal and abnormal images.

1 Introduction

There are many types of liver malfunction found in day to day life one among them being liver tumors. Liver tumor is also known as the hepatic tumors are growths found in or around the area of liver. There are various types of liver tumors which can be formed as liver is made up of different cell types. These growths found in the liver can be benign or malignant which are cancerous growths. It can be found using various liver function test which helps in finding out the initial liver problem. Liver tumors can be found only using any of the imaging techniques. The radiologists carefully study the liver image from the scans to find the accurate location of the tumor, if present. Since these are done manually it takes time to find the location of tumor, they are also more prone to errors and requires an experts study to find

J. Baby (✉) · T. Rajalakshmi · U. Snekhalatha
Department of Biomedical Engineering, SRM Institute of Science and Technology,
Kattankulathur, Chennai 603203, Tamil Nadu, India
e-mail: jisha_babykutty@srmuniv.edu.in

T. Rajalakshmi
e-mail: abirajalakshmix@gmail.com

© Springer Nature Switzerland AG 2019   1055
D. Pandian et al. (eds.), *Proceedings of the International Conference on ISMAC
in Computational Vision and Bio-Engineering 2018 (ISMAC-CVB)*, Lecture Notes
in Computational Vision and Biomechanics 30,
https://doi.org/10.1007/978-3-030-00665-5_101

the exact location. The study done in this paper is about segmenting the region of interest from the CT imaging techniques using the Gradient Vector Flow (GVF) Snakes Algorithm.

A snakes algorithm is a type of Active contour which are used to initiate a curve in the image to detect boundaries of an object in facial recognition [1, 2], medical image processing [3–5], image segmentation [6] and tracking of objects [7, 8]. An initial contour is set on the object to create an initial boundary which goes through various iteration allowing the contour to converge into the true boundaries of the object. The algorithm can be used to find an organs outline from any imaging techniques or to identify a part form the unit being manufactured. From the instigation of Snakes algorithm by Kass et al. [9] in the year of 1988 there are two problems which could not be solved. One of the problems was that the initial contour could not move to the objects which were situated far away and the other was that it was not able to get into any concavities present in boundary. The introduction of GVF snake has proven to be much better than the previous methods and also has a much faster rate in segmenting the desired part. Gradient Vector Flow snakes algorithm was used by various researchers to segment as well as to find the location of the desired part or region of interest.

2   Related Work

Bahreini et al. in 2010 used GVF algorithm to segment the lesions present in the Breast from the MRI Images [10]. The study was conducted on 52 female who had different types of lesions on the breast. In these 52 females who had lesions, 33 were malignant and 19 were benign lesions. The study included selecting the required region, segmenting the region of interest using GVF algorithm and being verified by the manual segmentation done by the radiologist's. Gradient Vector Flow Snake segmentation technique was able to segment 97% of the malignant and 89.5% of benign lesion borders with a threshold overlap of 0.6 thus helping to make accurate segmentation on the borders of breast lesions.

Díaz-Parra et al. [11] used Gradient vector flow algorithm in 2014 on computed tomography images for a fully automated spinal canal segmentation for radiation therapy. This paper proposed a fully automated Gradient Vector Flow-based snake algorithm which is used for segmenting the canal of a spine. This method was conducted on three patients with the help of Dice coefficient and the results were inferred as 79.5, 81.8 and 83.7%.

Gradient Vector Flow Algorithm was used in medical image segmentation in 2012 by Yuhua et al. [12]. This paper proposed Sigmoid Gradient Vector Flow (SGVF) algorithm which was used to improve the performance in the contour. The proposed method can avoid leakage from the weak edges as the external energy of the algorithm is callous to the noises. The algorithm was used on both synthetic and real images of an ultrasound and MRI imaging techniques which removed the noises and also extracted the edges which were weak. The results of this method presented that the algorithm lead to much more precise segmentation.

Qiongfei et al. [13] used Gradient Vector Flow Model for infrared image segmentation in 2015. This paper used the extended neighborhood along with the Gradient Vector Flow algorithm which analyses a pixel of an image and then checks the neighboring pixel. If any of the neighboring pixel is bigger than the previous then it replaces the original pixel. This calculation technique uses lesser number of iterations and the results acquired is also better. The new calculation method only needs less iteration numbers and also obtained better effects.

Mahmoud MKA et al. [14] used Gradient Vector Flow snakes algorithm to segment skin cancer images in 2011. In this paper images of skin cancer is segmented using the GVF snakes algorithm. For the hair on the skin to be removed and the noises to be calloused an adaptive filter consisting of Wiener and Median filter is used. The boundary of the skin cancer can be tracked even with the presence of any other objects near the region of interest by extending the algorithm to one direction. This research was conducted on eight skin cancer patients and the algorithm was considered to be effective.

3 Proposed Work

The methodology of liver tumor segmentation goes forward in various stages using Matlab software. As per the diagram shown in Fig. 1, Stage one consists of the original input image of the liver which has been acquired from the databases [15–17].

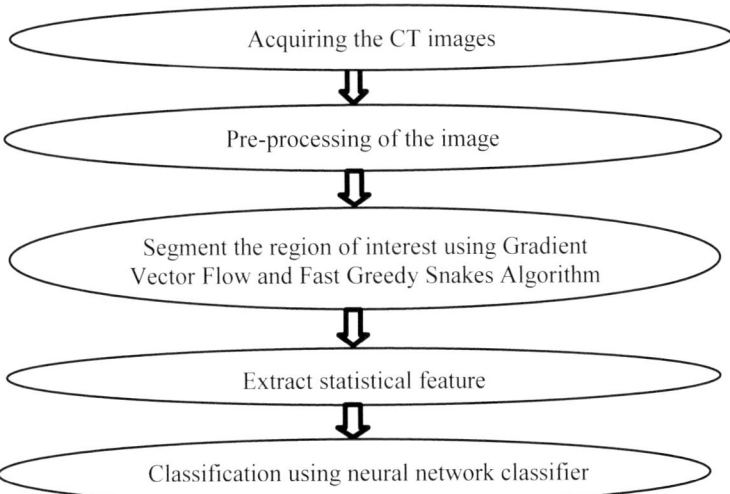

Fig. 1 Methodology for liver tumor segmentation

Stage two consists of preprocessing the images which includes converting the color image into greyscale image and also to resize the image of 0–255 size, as 0–125 and 135–255 can identify the difference between its pixel but that of in the between has minute changes which can be only detected if its resized to that scale. In stage three we propose use of Wiener filter, this filter is used to remove the unwanted noises present in the image. Stage three is used to initialize the Gradient vector flow GVF snakes algorithm to initialize the contour on the true image and segment it automatically. In stage four it is proposed to give the features extracted of tumor and non-tumor images as input to the Artificial Neural Network Classifier to train it to classify itself into tumor images or non-tumor images.

Preprocessing of the image removes distortions and enhances the image quality for better image and processing. The preprocessed image is then filtered and can be segmented using the proposed segmentation algorithm. The segmented image is given to the backpropagation Neural Network classifier to classify the images between tumor and non-tumor liver. The classifier used here has three different layers which are the input layer, hidden layer and the output layer which are added with weights for training purpose. These weights are modified during the training process to make an appropriate decision when it comes through similar input. The main reason for the use of this classifier was due to its ability to generate complex decision boundaries in any situation.

3.1 Preprocessing and Image Filtering

Preprocessing [15, 16] is used to improve the data which helps in subduing the difficult noises and also increases the quality of the images so that it can undergo further processing. Pre-processing of images in this section consists of resizing the image and then converting them to grayscale image using various Matlab commands.

The Wiener filter [16, 17] is used to generate an estimated signal by giving the input as the known signal and comparing it to the unknown signal by computing the unknown signal to give an output similar to the known signal. This filter can be used to remove any type of distortions present in an image and acquire only the interested signal. Some of the reasons why wiener filter was used over the other filter was because it did not require noncausal or causal filter and FIR(Finite Impulse Response) for filtering the images.

3.2 Image Segmentation

The process of slicing the images into various parts is called Image segmentation. Segmentation technique is used to gather peculiar information from a digital library or to locate objects of interest from a vast amount of objects. The method used here is Gradient Vector Flow snakes algorithm. Gradient vector flow (GVF) [18–21] snake algorithm, starts off by calculating the GVF forces. These forces help the initial contour to mold and stretch to the boundaries of the region of interest. The GVF forces are found using generalized diffusion equations to both components of the gradient of an image edge map. Energy function $L(c)$ of a GVF algorithm defined on the contour is given in Eq. (1):

$$L(c) = \text{Internal energy}(L_{int}) + \text{External energy}(L_{ext}) \tag{1}$$

$$L(c) = L_{int} + L_{ext} \tag{2}$$

where L_{int} and L_{ext} denote the internal and external energies, respectively. The initial contour converges to the contour smoothly with the help of the internal energy. The contour development is categorized using the external energy. The contour fits into the edges due to the image energy calculated. The external energy is said to be zero when the contour is on the boundary of an object as the position will not change. If there is any change in the position of the contour is due to the difference in the control points.

In this paper, the algorithm here uses GVF algorithm to segment the image by initializing an initial contour to the tumor liver CT image. The snake's algorithm consists of energy function as shown in Eq. (3)

$$L = \int_0^1 L_{int}(v(s)) + L_{image}(v(s)) + L_{con}(v(s))ds \tag{3}$$

In here we use the GVF field as the constraint energy on the original snake's energy equation as given above. To get the GVF field we need to extract the edge map function from the image function. The GVF field is calculated using Eq. (4)

$$L = \iint \mu\left(ux^2 + uy^2 + vx^2 + vy^2\right) + |\nabla f|^2 |g - \nabla f^2| dx dy \tag{4}$$

The smoothening term also known as the data term denoted as "μ" which is set as $1 * e^{-7}$ adjusts the trade off between the 1st and 2nd term as the value of the data term (μ) decreases simultaneously the noise too decreases. As the GVF field is found its applied to the snakes algorithm [10–15, 22].

3.3 Feature Extraction and Classification

This phase is important as it extracts various features of the images helping them to provide information to the classifier. Feature extraction is used to find data which can be built to be informative, help in further learning, better interpretations, and get derived values necessary. The features which are selected need to be accurate and in similar form for the classification to be correct. The selected features are required to contain information's acquired from the input data in order to undergo the desired work. Following features were found from the segmented image for the purpose of classification.

Mean—It finds the basic texture feature representing the average pixel value of the image useful for delivering the background.

$$\text{Mean} = \frac{1}{n} * \sum x \tag{5}$$

Standard Deviation—It indicates how a histogram is spread along the image. It indicates what other pixel values belong to the part of the background

$$\text{Standard Deviation} = \sqrt{\frac{1}{N} * \sum (x - \mu)^2} \tag{6}$$

Correlation—It allows the image to compare for two different signals allowing to detect any difference in this waveforms.

$$\text{Correlation} = \frac{N \sum xy - \left(\sum x\right)\left(\sum y\right)}{\sqrt{\left[N \sum x^2 - \sum x^2\right]\left[N \sum y^2 - \sum y^2\right]}} \tag{7}$$

Contrast—It is the difference between the maximum and minimum pixel intensity of an image. Viewing of the image is made easier with the help of this feature.

Kurtosis—It is interpreted in combination with the noise and resolution measurement.

$$\text{Kurtosis} = n \frac{\sum_{i=1}^{n} (x_i - x_{\text{avg}})^4}{\left(\sum_{i=1}^{n} (x_i - x_{\text{avg}})^2\right)^2} \tag{8}$$

Skewness—It helps in making a judgment about the surface of the image.

$$\text{Skewness} = \sqrt{n} \frac{\sum_{i=1}^{n} (x_i - x_{\text{avg}})^3}{\left(\sum_{i=1}^{n} (x_i - x_{\text{avg}})^2\right)^{3/2}} \tag{9}$$

Homogeneity—It is used to check if all the pixel in the image has same degree of value.

Energy—It is calculated using mean squared value of the image signal. It is a signal used to help in finding the frequency being distributed.

$$\text{Energy} = 1/2 \, mv^2 + mgh \tag{10}$$

Back Propagation Artificial Neural Network Classifier is used to classify the proposed system using the features extracted. The classifier consists of three layers input, hidden and output layer which are connected to each other using weights. These weights are modified during the training process to make an appropriate decision

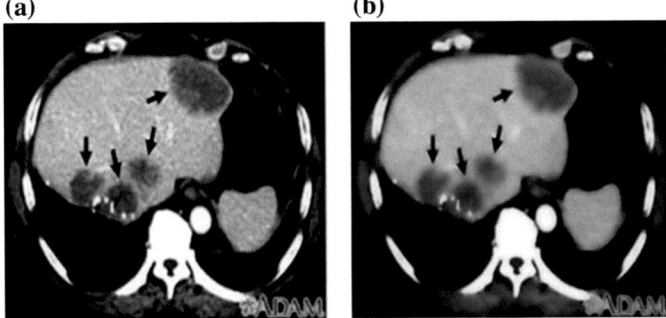

Fig. 2 **a** Input image **b** filtered image

when it comes through similar input. The main reason for the use of this classifier was due to its ability to generate complex decision boundaries in any situation. The features extracted were given as input and was trained by the classifier using the various layers present in the classifier to give the true positive, true negative, false positive, and false negative values. The images were classified as normal or abnormal CT images using this classifier. The formula for finding the confusion matrix is given by Eqs. (11)–(13):

$$\text{Accuracy} = \frac{\sum \text{True Positive} + \sum \text{False Positive}}{\sum \text{Total Population}} \qquad (11)$$

$$\text{False Discovery Rate} = \frac{\sum \text{False Positive}}{\sum \text{Predicted Condition Positive}} \qquad (12)$$

$$\text{Negative Predictive Value} = \frac{\sum \text{True Negative}}{\sum \text{Predicted Condition Negative}} \qquad (13)$$

4   Results Analysis

The experiment was carried out on liver CT images taken from different database, on 25 abnormal image and 25 normal image. The original liver CT image as shown in Fig. 2a is provided as input. These images were preprocessed to give a resized image and then converted to grayscale image. It reduces the complexity and increases the accuracy of the applied algorithm. The Wiener filter is used to reduce the noise and reconstruct an image to perform morphological operations. Figure 2b shows the Wiener-filtered images are segmented using Gradient Vector Flow Snake Algorithm where an initial contour is initiated as shown in Fig. 4a. With respect to the initial contour, the internal and external energy is calculated and helps the snake to move towards its desired region of interest which are shown in Fig. 4b, c, d. Figure 5 shows the final segmented image (Fig. 3).

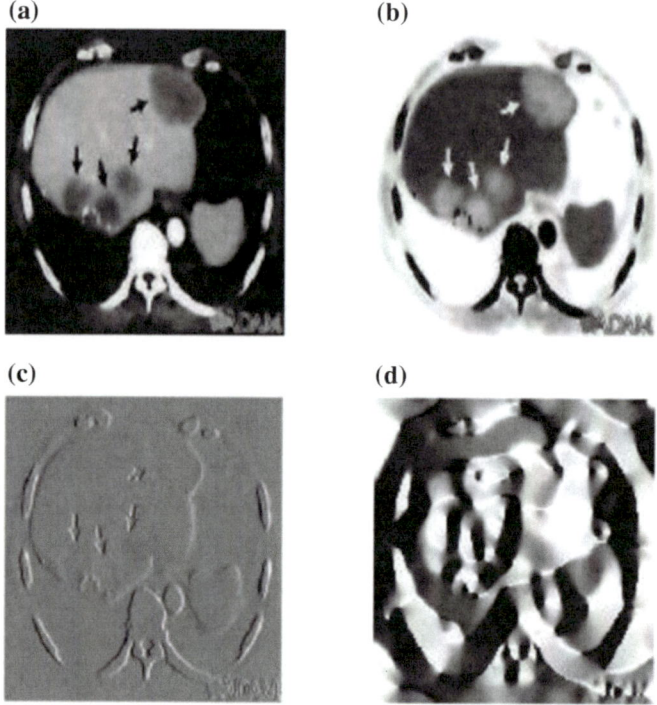

Fig. 3 Segmentation process **a** test image **b** edge map **c** edge map gradient **d** normalized GVF field

Fig. 4 Segmented image

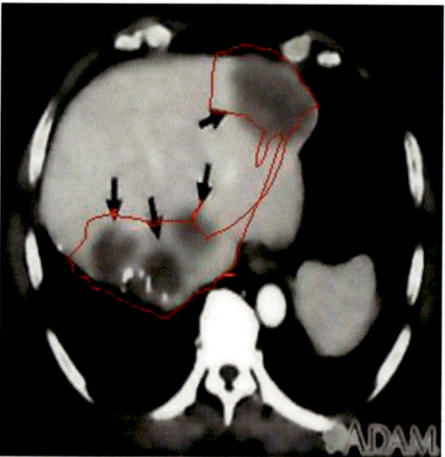

Various statistical features calculated are tabulated in Table 1 from which it can be inferred that mean of normal CT images has higher value than the abnormal CT images, kurtosis of the abnormal images should be closer or lesser than 3 and that of

Fig. 5 Regression curve

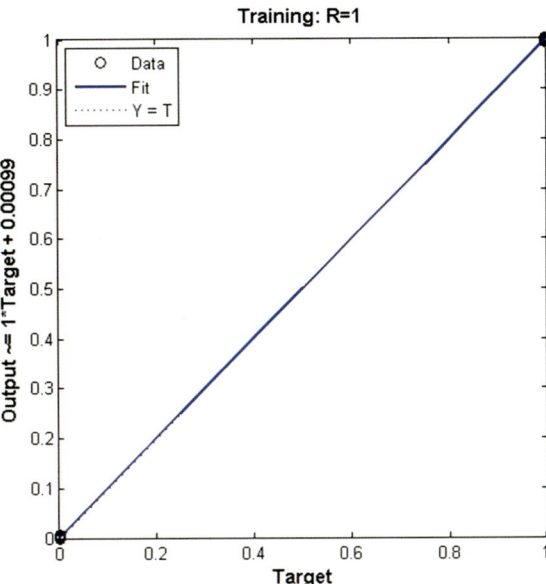

Table 1 Feature values of normal and abnormal CT image

S. No.	Feature extraction	Normal CT images	Abnormal CT images
1.	MEAN	9.508627e+01	7.670058e+01
2.	Kurtosis	4.282876e+00	3.045479e+00
3.	Skewness	3.861835e−01	7.492599e−01
4.	Correlation	9.734316e−01	9.437237e−01
5.	Homogeneity	9.331582e−01	9.921990e−01
6.	Contrast	2.003370e−01	4.368566e−01
7.	Energy	2.781193e−01	8.327415e−01
8.	Standard deviation	6.397086e+01	8.260880e+01

normal images will be greater than 3. Skewness is lesser in normal CT images than that of abnormal CT image. Correlation is slightly higher in the normal CT images. Homogeneity is higher in normal CT images. The value of contrast will be higher in abnormal CT images than the normal CT images. Energy difference is very high in abnormal CT image and the Standard deviation found is lesser in normal CT images.

The feature extracted is then given as input to the neural network classifier which classifies the image into tumor or non-tumor liver. The confusion matrix was calculated to find the accuracy of the system by calculating the true positive, negative predictive values, true negative, false discovery rate, false positive, predicted condition positive, predicted condition negative and false negatives values of the images and the accuracy was found to be 88% as shown in Table 2.

Table 2 Confusion matrix table

Total population	Condition positive	Condition negative	Accuracy = 88%
Positive predicted condition	True positive = 23	False positive = 4	False discovery rate = 14.8%
Negative predicted condition	False negative = 2	True negative = 21	Negative predictive values = 91.3%

The regression curve is used to study the linearity of the signals generated by studying the relationship between one or several signals. Regression curve is used for prediction and forecasting by learning in the Neural Networks is plotted and shown in Fig. 5. This helps in classifying the liver into normal or abnormal images.

5 Conclusion

This paper presented liver tumor image segmentation and its classification. A total of 50 images containing 25 normal and abnormal images each were used. The images were given as input and then the images were filtered using Wiener filter after going through the preprocessing. The filtered image went through the segmentation algorithm proposed in the paper where the region of interest was segmented and various features were extracted which was given as input to the neural network classifier to train itself to classify itself into tumor or non-tumor images whenever an image is given as input.

For the future work a better algorithm Fast Greedy Snakes Algorithm is used to segment the region of interest more accurately and faster than the current segmentation method.

References

1. Lam KM, Yan H (1994) Fast algorithm for locating head boundaries. J Electron Imaging 3(4):352–359
2. Waite JB, Welsh WJ (1990) Head boundary location using snakes. Br Telecom Technol J 8(3):127–135
3. McInernery T, Terzopolous D (1996) Deformable models in medical image analysis: a survey. Med Image Anal 1(2):91–108
4. Jain AK, Smith SP, Backer E (1980) Segmentation of muscle cell pictures: a preliminary study. IEEE Trans Pattern Anal Mach Intell PAMI-2(3):232–242
5. Fok YL, Chan JCK, Chin RT (1996) Automated analysis of nerve cell images using active contour models. IEEE Trans Med Imaging 15(3):353–368
6. Ginneken B, Frangi AF, Staal JJ, Haar Romeny BM, Viergever MA (2002) Active shape model segmentation with optimal features. IEEE Trans Med Imaging 21(8):924–933
7. Lie WN (1995) Automatic target segmentation by locally adaptive image thresholding. IEEE Trans Image Process 4(7):1036–1041

8. Couvignou PA, Papanikolopoulos NP, Khosla PK (1993) On the use of snakes for 3-D robotic visual tracking. In: Proceedings of IEEE CVPR, 1993, pp 750–751
9. Kass M, Witkin A, Terzopoulos D (1987) Snakes: active contour models. Int J Comput Vis 1(4):321–331
10. Bahreini L, Fatemizadeh E, Gity M (2010) Gradient vector flow snake segmentation of breast lesions in dynamic contrast-enhanced MR images. In: 17th Iranian conference of biomedical engineering (ICBME2010), 3–4 Nov 2010, pp 1–4
11. Díaz-Parra A, Arana E, Moratal D (2014) Fully automatic spinal canal segmentation for radiation therapy using a gradient vector flow-based method on computed tomography images: a preliminary study. In: 2014 36th annual international conference of the IEEE 26–30 Aug 2014, Engineering in Medicine and Biology Society (EMBC), pp 5518–5521
12. Yuhua Y, Lixiong L, Lejian L, Ming W, Jianping G, Yinghui L (2012) Sigmoid gradient vector flow for medical image segmentation. In: 2012 IEEE 11th international conference on signal processing (ICSP) 21–25 Oct 2012, pp 881–884
13. Qiongfei W, Yong Z, Zhiqiang Z (2015) Infrared image segmentation based on gradient vector flow model. In: 2015 sixth international conference on intelligent systems design and engineering applications (ISDEA), pp 460–462
14. Mahmoud MKA, Al-Jumaily A (2011) Segmentation of skin cancer images based on gradient vector flow (GVF) snake. In: 2011 IEEE international conference on mechatronics and automation, Aug 7–10, Beijing, China, pp 216–220
15. Cleary K (2007) Original datasets. Midas-original datasets. doi:hdl.handle.net/1926/587
16. Al-Kadi OS, Chung DYF, Carlisle RC, Coussios CC, Noble JA (2014) Quantification of ultra-sonic texture intra-heterogeneity via volumetric stochastic modeling for tissue characterization. Med Image Anal. https://doi.org/10.1016/j.media.2014.12.004
17. Grove O, Berglund AE, Schabath MB, Aerts HJ, Dekker A, Wang H, Gillies RJ (2015) Data from: quantitative computed tomographic descriptors associate tumor shape complexity and intratumor heterogeneity with prognosis in lung adenocarcinoma. Cancer Imaging Arch. https://doi.org/10.7937/k9/tcia.2015.a6v7jiwx
18. Gonzalez RC, Woods RE, Eddins SL (2009) Digital image processing, using MATLAB, 2nd edn. Gatesmark Publishing, USA
19. Pratt WK (2007) Digital image processing, 4th edn, p 16
20. Pratt WK (1972) Generalized wiener filtering computation techniques. IEEE Trans Comput c-21:636–641
21. Kass M, Witkin A, Terzopolous D (1987) Snakes: active contour models. Int J Comput 1(4):321–331
22. Kazerooni AF, Ahmadian A, Serej ND, Rad HS, Saberi H, Yousefi H, Farnia P (2011) Segmentation of brain tumors in MRI images using multi-scale gradient vector flow. In: 33rd annual international conference of the IEEE EMBS Boston, Massachusetts, USA, August 30–September 3, 2011, pp 7973–7976
23. Xu C, Prince JL (1998) Snakes, shapes, and gradient vector flow. IEEE Trans Image Process 7(3):359

Hyperspectral Image Classification Using Semi-supervised Random Forest

Sunit Kumar Adhikary and Sourish Gunesh Dhekane

Abstract In this paper, a hyperspectral image classification technique is proposed using semi-supervised random forest (SSRF). Robust node splitting in the random forest requires enormous training data, which is scarce in remote sensing applications. In order to overcome this drawback, we propose utilizing unlabeled data in conjunction with labeled data to assist the splitting process. Moreover, in order to tackle the curse of dimensionality associated with a hyperspectral image, we explore nonnegative matrix factorization (NMF) to remove redundant information. Experimental results confirm the efficacy of the proposed method.

1 Introduction

Recent developments in optics and imaging highlight the fact that materials when exposed to various wavelength intervals of electromagnetic spectrum show signature properties [1]. Trees show higher reflectance for wavelengths near infrared region compared with certain visible wavelengths which are its signature property. This behavior is in contrast to other materials like bricks, asphalt, and soil, which have a similar magnitude of reflectance for both ranges of wavelength. A hyperspectral image is basically a collection of images collected in several wavelengths, which often belong to a narrow bandwidth. Thus, it contains a tremendous amount of information and thus finds utility in several remote sensing applications like land cover mapping, change detection, mineral identification, crop analysis, and lot more. Classification of hyperspectral images is the key concept in these applications. However, their low spatial resolution leads to lack of photo-interpretability. This, in turn, hampers the possibility of using supervised learning methods which require labeled data. The acute scarcity of labeled patterns where the number of training samples is rela-

S. K. Adhikary (✉) · S. G. Dhekane
Indian Institute of Information Technology Guwahati, Guwahati, India
e-mail: sunitadhikary1409@gmail.com

S. G. Dhekane
e-mail: sourishdhekane@gmail.com

© Springer Nature Switzerland AG 2019   1067
D. Pandian et al. (eds.), *Proceedings of the International Conference on ISMAC in Computational Vision and Bio-Engineering 2018 (ISMAC-CVB)*, Lecture Notes in Computational Vision and Biomechanics 30,
https://doi.org/10.1007/978-3-030-00665-5_102

tively small compared to the number of features [2] and the curse of dimensionality (i.e., the Hughes phenomenon [3]) pose a serious challenge in the classification of hyperspectral images.

In order to deal efficiently with the problem of lack of labeled patterns, semi-supervised learning (SSL), and active learning (AL) is mostly explored in the remote sensing literature. Semi-supervised techniques try to exploit unlabeled data and extract valuable information from both labeled and unlabeled data during training. On the other hand, active learning techniques build a small set of extremely valuable labeled patterns. A number of SSL techniques have been applied to multispectral and hyperspectral image classification, including Transductive SVMs (TSVM) [4], Laplacian SVMs (LapSVM) [5], and Graph-based methods [6]. Ratle et al. [7] proposed semi-supervised neural networks (SSNN) to label scenes from hyperspectral data which were trained with gradient descent and were shown to outperform traditional margin-based SSL techniques. It has been pointed out in [8] that the effectiveness of margin maximization based methods for SSL depends heavily on specific data distribution which is usually difficult to be satisfied in many applications. Hence, it is desirable to find a method that can utilize the unlabeled data without losing its flexibility.

We explore random forest (RF) for this work. Node splitting in RF is the key issue. Our approach is to use the semi-supervised node splitting criterion stated in [11] to build decision trees. However, the density estimation based technique proposed in [11] seems not to be a good criterion for splitting nodes when there is an acute scarcity of labeled patterns. We have modified it with the incorporation of unsupervised information gain for node splitting. Moreover, nonnegative matrix factorization (NMF) has been used for band reduction and an additional constraint has been imposed in order to be in accordance with the linear mixture model.

2 Methodology

Hyperspectral images are rich in information attributed to its large number of spectral bands. But due to lack of photo-interpretability of hyperspectral images, it is difficult to collect labeled data which poses a serious challenge to the classification of hyperspectral images. Moreover, the spectral bands are in a narrow bandwidth, which leads to a high correlation between several spectral bands. Thus, it becomes absolutely necessary to transform the image into one of the reduced spectral bands. This reduction of spectral bands also considered as feature extraction can be carried out by selecting certain bands or by using a transform that produces features as combinations of bands. This reduction in hyperspectral image analysis can be linked to the hyperspectral unmixing problem. Usually, a pixel in a hyperspectral image contains the reflectance values of various materials because of its low spatial resolution. The observed values at any point can be considered to be a linear mixture of reflectance of materials corresponding to the pixel.

According to the linear mixture model (LMM), each l-dimensional observed pixel vector X present in the image cube can be expressed as a linear combination of "m" materials present in the pixel weighted by its fractional abundance given as

$$X = \sum_{i=1}^{m} u_i v_i + w = Vu + w \tag{1}$$

where V is the $1 \times m$ matrix of spectra $(v_1, \ldots, v_m)$ of the individual composing materials (also called endmembers), u is an m-dimensional vector describing the fractional abundances of the endmembers in the mixture (abundance vector) and w is the additive noise vector. The elements of the abundance vector are assumed to be positive and with unit sum:

$$u_i \geq 0 \; i = 1, \ldots, m \tag{2}$$

$$\sum_{i=1}^{m} u_i = 1 \tag{3}$$

The unmixing can, however, not directly be linked to feature extraction. In techniques like PCA, ICA the endmembers correspond to rows in the linear transform and are orthogonal. This may be a condition too strict for endmember extraction where the base materials may only slightly differ from each other. Moreover satisfying the additivity conditions as stated in Eqs. (2) and (3) cannot be ensured. Hence, we explore NMF in this work which is in coherence with the LMM.

2.1 NMF

NMF is a matrix factorization method which decomposes the given nonnegative matrix $X \in R^{L \times N}$ into a basis matrix $(U \in R^{L \times M})$ and a weight matrix $(V \in R^{M \times N})$ having nonnegative elements such that

$$X \approx UV \tag{4}$$

Appropriate values for U and V are found by formulating an optimization problem by minimizing the Euclidean distance between X and UV, keeping both U and V nonnegative

$$\text{minimize} \quad f(U, V) = \frac{1}{2} \| X - UV \|_F^2 \tag{5}$$

$$\text{subject to} \quad U \geq 0, \; V \geq 0 \tag{6}$$

It should be noted that the operator $\|.\|._F$ represents the Frobenius norm. The solution of the above optimization problem yields the multiplicative iterative update rules as stated in [9]

$$U_{i,j} \leftarrow U_{ij} \frac{\left(XV^T\right)_{ij}}{\left(UVV^T\right)_{ij}} \tag{7}$$

$$V_{i,j} \leftarrow V_{ij} \frac{\left(U^T X\right)_{ij}}{\left(U^T XV^T V\right)_{ij}} \tag{8}$$

While implementing NMF in this case, the parameter M acts as the hyperparameter. It represents the number of endmembers and is empirically determined. This formulation of NMF does not satisfy Eq. (3). In order to enforce this restriction, we normalize the columns of V after the above optimization problem converges.

2.2 Random Forest

The random forest (RF) is a collection of decision trees $F = \{t_1, t_2, \ldots, t_N\}$ grown on a subset of the training set drawn randomly with replacement also called as bagging [10]. The trees are usually grown until there are only pure samples in a node or until there exists a specified number of samples in it. Data in a node of a tree is split and sent to 2 sister nodes based on a partition criterion like information gain, Gini index, and Bayesian classification error. The splitting of the data can, however, be carried out in various manners [13]. In our work, we chose hyperplanes aligned to the axis. However, in order to invoke more randomness in the decision trees a subset of the feature space is considered in each node [13] to find the best node split. The information gain used to measure the quality of node split is defined as the difference of entropies before and after splitting given as

$$I_j = I_j^s = H(R) - \frac{|R_1|}{|R|} H(R_1) - \frac{|R_r|}{|R|} H(R_r) \tag{9}$$

where R is an internal node, R_1 and R_r are its left and right child, respectively. $H(R)$ is the Shannon entropy of R

$$H(R) = -\sum_{k=1}^{K} p_k \log p_k \tag{10}$$

where p_k is the probability distribution of the kth class in node R of a tree, similarly $H(R_r)$ and $H(R_1)$ should be calculated. Traditionally, we can use the law of total probability to calculate p_k.

$$p_k = \int_R p(k|x)\mathrm{d}p(x) \tag{11}$$

In the fully supervised training phase which contains labeled samples $p(k|x_{R_i})$, which represents the probability of the ith sample in node R having class k is either 1 or 0. So p_k can be readily obtained. While testing a test sample x, RF assigns probability estimate for each class is as follows:

$$p(k|x) = \frac{1}{N} \sum_{i=1}^{N} p_i(k|x) \tag{12}$$

where $p_i(k|x)$ is the probability estimation of class k given by the ith tree. The overall decision function of RF is defined as

$$\mathcal{F} = \operatorname{argmax}_{k \in y} p(k|x) \tag{13}$$

2.3 Semi-supervised Node Splitting

The problem with the fully supervised splitting is that, although the distribution $p(k|x_{R_i})$ is given by the labeled sample, the sparsely labeled data cannot give a good approximation of the marginal distribution which may lead to a worse choice of the separating hyperplane. Unfortunately, the insufficiency of labeled training data usually leads to a sparse distribution and a bad approximation. Motivated by the work done in [11] we explore unlabeled data to do a semi-supervised node split. A new problem arises, the posteriori distribution $\hat{p}(k|x_{R_i})$ of unlabeled data x of node R in the ith tree is unknown. It can be estimated as in [11] as the probability density ratio.

$$\hat{p}(k|x_{R_i}) = \frac{p(x_{R_i}|k)}{\sum_{j=1}^{K} p(x_{R_i}|j)} \tag{14}$$

For $p(x_{R_i}|k)$, we apply a kernel-based density estimation with Gaussian kernel [12]

$$K_h(u) = h^{-d}(2\pi)^{-d/2}\exp\left\{-\frac{1}{2h^2}u^{\mathrm{T}}u\right\} \tag{15}$$

where h is the bandwidth to be determined and d is the data dimension. We then have the following estimation for an unlabeled sample:

$$p(x_{R_i}|k) = \frac{1}{n_k} \sum_{y_j=k} K_h(x_{R_i} - x_j) \tag{16}$$

where n_k is the number of samples that are labeled k. Once $\hat{p}(k|x_{R_i})$ is determined for the unlabeled samples it can be treated like labeled patterns. Equation (9) can be used to split the nodes, p_k being calculated using Eq. (11). But when there is severely less labeled data in a node, the posterior probability calculated based on the density estimate by the above method cannot be relied upon. In that case, we obtain supervised information gain with the help of a few labeled patterns I_j^s as in Eq. (9) to which we add an unsupervised information gain for calculating the total information gain

$$I_j = I_j^s + \lambda\, I_j^u \tag{17}$$

Assuming multi-variate Gaussian distributions at the nodes. The unsupervised gain term I_j^u can be calculated with both labeled and unlabeled samples as in [13] is given as

$$I_j^u = \log(\|\Lambda(R)\|) - \frac{|R_1|}{|R|}\log(\|\Lambda(R_1)\|) - \frac{|R_r|}{|R|}\log(\|\Lambda(R_r)\|) \tag{18}$$

where $\|.\|$ represents determinant and Λ represents the $p \times p$ covariance matrix of the data in the node. The trees are grown on chunks of data bagged from the entire set of labeled and unlabeled samples to an extent until there are pure true labeled samples following the new splitting criterion stated above. The collection of decision trees grown following this scheme is what we refer to as semi-supervised random forest (SSRF).

3 Experiments

3.1 Dataset

A real-world hyperspectral image dataset called "University of Pavia" (UP) is used to validate the efficacy of the proposed method in our study. Reflective Optics System Imaging Spectrometer (ROSIS-3) sensor has been used to collect the dataset. It has 103 spectral bands collected within the wavelength range of 430–860 nm. The image consists of 610×340 pixels at a spatial resolution of 1.3 m per pixel. This dataset considers 9 land cover classes of interest. Table 1 shows a number of samples in each land cover class.

Table 1 Number of samples per class

Class labels	Number of samples
Asphalt	6631
Meadows	18,649
Gravel	2099
Trees	3064
Painted metal sheets	1345
Bare soil	5029
Bitumen	1330
Self-blocking bricks	3682
Shadows	947

3.2 Experimental Setup and Results

In our experiment, the given hyperspectral image is decomposed using the NMF scheme described above. We perform 300 iterations with the loss function converging to a global/local optima. The number of endmembers has been empirically chosen to be 10. For the classification process, labeled set X_L contains 5, 10, 15 samples per class, respectively, randomly selected from the labeled pixels on the hyperspectral image. The test set includes all the samples leaving aside the samples used for training the classifier. The decision trees of semi-supervised random forest (SSRF) are grown on the entire dataset (both training and test data). Each node in a tree is split with the semi-supervised node split criterion described above. When the number of labeled samples in a node is more than 4, we estimate the posterior probability of unlabeled data with the density estimation step as is stated in Eq. (16) and then use the supervised information gain stated in Eq. (9) to do the node split as if the entire data was a supervised data. But when the number of labeled samples is severely less in a particular node the density estimated with Gaussian kernel cannot be relied upon and hence we add an unsupervised information gain term to the supervised information gain as in Eq. (17). Several values of λ were experimented and was finally set to 0.3. The supervised information gain is calculated with the small amount of labeled patterns in the node but the entire labeled and unlabeled patterns are used to obtain the unsupervised information gain. The generated SSRF consist of a total of 300 trees.

The classification accuracy of the proposed technique is compared with TSSNN [7] and LapSSNN [7]. The results are shown in Table 2. The proposed technique can be seen to beat the neural network based semi-supervised schemes with 10 and 15 samples per class but gives a comparable result in 5 sample per class. Thus, it can be concluded that the proposed scheme is better able to leverage information present in unlabeled data.

Table 2 Classification accuracy for Pavia University dataset

Labeled examples per class	5	10	15
TSSNN [7]	**73.2 ± 2.79**	75.3 ± 2.09	76.3 ± 5.24
LapSSNN [7]	72.7 ± 2.82	74.1 ± 1.35	74.9 ± 4.07
Proposed technique	72.9 ± 3.72	**77.6 ± 1.66**	**79.2 ± 1.1**

4 Conclusion

In this paper, we proposed a semi-supervised random forest for classification. The improved semi-supervised node split criterion has successfully leveraged information from unlabeled data. The enhanced value of classification accuracy and its low variance bear testimony of the effectiveness of the proposed technique. Apart from the improvement in classification accuracy, the proposed technique is computationally efficient. Unlike traditional SSL techniques which are either iterative in nature or are optimized with gradient descent, the proposed semi-supervised technique has less time complexity when expressed as a function of the number of samples.

Acknowledgements The authors thank Prof. Gamba for providing the University of Pavia dataset.

References

1. Mather PM, Koch M (2010) Computer processing of remotely-sensed images, 4th edn. Wiley, New York
2. Baraldi A, Bruzzone L, Blonda P (2005) Quality assessment of classification and cluster maps without ground truth knowledge. IEEE Trans Geosci Remote Sens 43:857–873
3. Hughes GF (1968) On the mean accuracy of statistical pattern recognition. IEEE Trans Inf Theory 14:55–63
4. Bruzzone L, Chi M, Marconcini M (2006) A novel transductive SVM for semisupervised classification of remote-sensing images. IEEE Trans Geosci Remote Sens 44:3363–3373
5. Gomez-Chova L, Camps-Valls G, Munnoz-Mari J, Calpe J (2008) Semisupervised image classification with Laplacian support vector machines. IEEE Geosci Remote Sens Lett 5:336–340
6. Camps-Valls G, Bandos TV, Zhou D (2007) Semi-supervised graph-based hyperspectral image classification. IEEE Geosci Remote Sens Lett 45:3044–3054
7. Ratle F, Camps-Valls G, Weston J (2010) Semisupervised neural networks for efficient hyperspectral image classification. IEEE Trans Geosci Remote Sens 48(5):2271–2282
8. Zhang T, Oles F (2000) A probability analysis on the value of unlabeled data for classification problems. In: 17th International conference on machine learning. ACM Press, California, pp 1191–1198
9. Lee DD, Seung S (2000) Algorithms for non-negative matrix factorization. In: Neural information processing systems. MIT Press, Denver, pp 556–562
10. Breiman L (1996) Bagging predictors. Mach Learn 26:123–140
11. Liu X, Song M, Tao D, Liu Z, Zhang L, Chen C, Bu J (2013) Semi-supervised node splitting for random forest construction. In: Conference on computer vision and pattern recognition. IEEE Press, Oregon, pp 492–499

12. Silverman B (1986) Density estimation for statistics and data analysis. Chapman and Hall/CRC, London
13. Criminisi A, Shotton J, Konukoglu E (2011) Decision forests for classification, regression, density estimation, mainfold learning and semi-supervised learning. Technical report, Microsoft Research

Clustering of Various Parameters to Catalog Human Bone Disorders Through Soft Computing Simulation

S. Ramkumar and R. Malathi

Abstract Every minute nearly 20 fractures occur due to bone disorders in the world. People around the world could not able to differentiate the difference between bone disorders. This chapter is a novel approach toward differentiation of different bone disorders like osteoporosis and osteopenia with influences several parameters. Accordingly, five different parameters such as Calcium, Phosphate, Vitamin D3, Parathyroid hormone (PTH) level, and calcitonin level are considered for the study to categorize the bone disorders. The present approach is an attempt to combine the clinical data measured from each patient and their respective bone mineral density value for the better classification. This is a unique study to provide combined information of both clinical and image processing studies. For this purpose, the above-mentioned parameters and bone density values were observed from ten different patients. All these data were used as an input for soft computing using MATLAB for further processing the data. Initially, unsupervised mapping classifier is adopted to classify bone disorder, for which the clinical parameters are compared with bone density value using k-means clustering algorithm. The prime idea behind using of k-means technique is that the feasibility to classify the inputs based on the distance between the input seeds. With reference to the perpendicular distance between the seed inputs, the bone disorders have been cataloged. The repeated iterations lead to best clustering results.

S. Ramkumar (✉) · R. Malathi
Department of EIE, Annamalai University, Chidambaram, Tamil Nadu, India
e-mail: sram0829@gmail.com

R. Malathi
e-mail: vsmalu@gmail.com

S. Ramkumar
Department of EIE, Veltech University, Chennai, Tamil Nadu, India

© Springer Nature Switzerland AG 2019

D. Pandian et al. (eds.), *Proceedings of the International Conference on ISMAC in Computational Vision and Bio-Engineering 2018 (ISMAC-CVB)*, Lecture Notes in Computational Vision and Biomechanics 30,
https://doi.org/10.1007/978-3-030-00665-5_103

1 Introduction

Bone is an essential material of the human skeleton to enable locomotion and ensuring the protection of living organisms. The main function of skeleton bone is to provide mechanical support and also in maintaining calcium homeostasis and hematopoiesis of an organism. When compared to men, women are three times prone to osteoporosis that leads to a femur bone mass deficiency after menopause. The role of hormonal change after menopause in women is inevitable in inducing osteoporotic conditions [1, 2]. Femur bone is the longest and strongest bone in our body, extended from hip to the knee. Particularly, femur that is thigh bone will be affected more due to frequent physical activities such as walking, running, standing, etc.

The human body is made up of different kinds of tissues such as muscle tissue, nervous tissue, epithelial tissue, and connective tissue. The connective tissue is a type of tissue which is responsible for the formation of bones. The subdivisions of the connective tissue are proper connective tissue and specialized connective tissue. Particularly, the bone tissues belong to specialized connective tissue category and appeared in vascularization state with a hard consistency. Due to continuous mineral deposition (mainly calcium) and collagen fibers in extracellular matrix raised the hardness of bone. In general, the human body consists of 206 bones which are seen in different shapes and sizes. Around 80% of bone mass in the skeleton is composed of cortical bone. At the beginning of nineteenth century, the term osteoporosis was proposed to describe the porosity of an elderly human based on histological findings.

Osteoporosis is a quite common bone disease in elderly people of both genders mainly due to comprised bone strength. This has lead to an increased risk of fracture. In women, postmenopausal osteoporosis is highly responsible for such decrease in bone density and bone strength. BMD (bone mineral density) can be measured by following methods such as SPA (Single Photon Absorptiometry) QCT (Quantitative Computed Tomography), QUS (Qualitative Ultrasound), (DPA) Dual Photon Absorptiometry, DXR (Digital X-ray Radiogrammetry), SEXA (Single-energy X-ray Absorptiometry), and DEXA (Dual-energy X-ray Absorptiometry). These tests are carried out for various diseased conditions like Hypophosphatasia, Klippel Feil syndrome, Craniosynostosis, Fibrous dysplasia, and Osteoarthritis involved with bone mineral density reduction [3–6].

In the present study, DEXA method output has been selected to conduct a comparative assessment of associated clinical parameters [7] output for bone disease patients'. Based on the comparative assessment, the classification of the disease occurs into their sub levels. This study will give additional information on patient classification and fracture-prone area to prevent or beneficial treatment of patients. The most successful practice for measuring bone mass density (BMD) is dual-energy X-ray absorptiometry (DEXA) scanning [8, 9]. The output of the DEXA is represented as T-score system as derived from comparing the individual BMD measurement of patients with normal. Osteoporosis is predominantly found in women rather than men, minimum prone to bone fracture because of their high bone density. Based on the literature survey, osteoporosis is less or no symptomatic disease. It is quite difficult to identify this disease at an early stage. After clinical manifestation and occurrence of noticeable loss around 30–40% in bone mass of patients are considered as a diagnostic parameter. Due to limitations in early diagnostic procedures, there is no prevention of this disease at this stage. All the supportive treatment of osteoporosis patients can support in reducing the severity of bone density loss instead of curing the disease. Therefore, this disease is termed as "silent epidemic" disease. Especially, age and gender are non-modifiable risk factors responsible for both osteopenia and osteoporosis condition. Unless changes in bone density, this disease is unpredictable [10, 11].

Osteoporosis has been considered as a public health problem, its diagnosis and fractures rehabilitation are high and it shed an impact on treatment cost. It is such a burden for developing countries to meet their countries growth and health status. To avoid severe bone loss and fractures, early diagnosis method is the only solution for these types of bone diseases. This study focused to develop combine diagnostic methods using a computational algorithm and its application for identification and classification of bone disease patients into either osteopenia or osteoporosis condition using input values of clinical diagnosis and iDEXA scanning method. In the case of medical application, sensitivity in diagnostic procedures and medical problem detection and its accuracy all must be taken into consideration.

There are many types of fluids available in the human. In all those body fluids, calcium is the key electrolytes in body fluids. In general, 99% of calcium intake is from the food and it is utilized for bone mineralization. The remaining 1% is inevitable for neuromuscular excitability, cardiac activity, blood coagulation, and membrane permeability [12]. The calcium in the body (e, i) present in plasma either surrounded by albumin or it may be presented as in free ionized form. Serum calcium level normal ranges from 8.8 to 10.4 mg/dl in a healthy adult. The total free calcium present in three forms such as calcium ion account, protein-bound complex, and ionic complex. At the same time, the percentage of calcium contents is also varied as 51, 40, and 9%. Likewise normal physiological phosphate ranges from 0.8 to 3.74 mg/dl for healthy adult; [13] The vitamin D3 level ranges from 30 to 100 ng/mL; Parathyroid hormone (PTH) level ranges from 1.9 to 106 pg/mL; [14] and the normal range in calcitonin level is different for male and female that is for female (0.1–10.9) and for male (0.2–27.7) [15].

2 Research Methodology

The research methodology completely depends on the classification of bone disorders. The inputs are classified based on the training and further, it helps to cluster them into groups. In this classification, clustering is done using k-means clustering algorithm.

Table 1 Serum calcium, Ph, Vit-D3, PTH, Calcitonin, and DEXA level inputs [16]

S. No.	Sex	Age	Measured serum level (Ca) (mg/dL)	Measured serum level (Ph) (mg/dL)	Measured serum level (Vit-D3) (ng/mL)	Measured serum level (PTH) (pg/mL)	Measured serum level (Calci-tonin)	DEXA value
1	F	55	8.31	5.9	20.3	13	2.99	−1.893
2	F	43	7.90	4.07	21.4	28	3.11	−1.661
3	M	69	8.25	8.31	29.9	37	10.14	−1.931
4	M	58	8.19	6.60	28.5	33	9.33	−1.895
5	F	55	5.94	10.47	20.5	11	0.07	−2.574
6	F	52	6.33	9.55	15.8	19	2.29	−1.417
7	M	78	7.31	8.11	25.1	22	0.20	−1.768
8	F	51	8.22	9.39	27.0	14	0.88	−1.915
9	F	66	7.93	8.52	22.8	17	2.03	−1.355
10	F	72	5.65	7.99	19.9	12	0.05	−2.901

2.1 k-Means Clustering Algorithm

k-means clustering is a machine learning algorithm or data mining, which is used to cluster the data into clusters of related data without any information of those associations. It is one of the clustering methods and it is normally used in biometrics, medical imaging, etc. It is an evolutionary algorithm that the name set from its method of operation. The algorithm clusters data into k groups, where k is an input parameter. Then allocate every observation to clusters based on the observation's closeness to the mean of the cluster. The clusters mean is then re-estimated and the procedure initiates again. The k-means algorithm works as follows:

1. The algorithm randomly picks k points as the primary cluster centroids.
2. Every point in the dataset is allocated to the closed cluster based on the Euclidean distance between every point and cluster center.
3. Each cluster centroid is re-estimated as the average of the points in each cluster.
4. Steps 2 and 3 are repeated waiting for the clusters to diverge. Convergence is well defined depending upon the implementation, but it usually means that either no observations alteration clusters when steps 2 and 3 are repeated or that the variations do not make a material modification in the definition of the clusters [16] (Table 1).

3 Data Collection

The k-means clustering is done in order to find the nearest clusters for Calcium, Phosphate, Vitamin D3, PTH, and Calcitonin. In this, the number of datasets that is being adapted is 10, i.e., Calcium has 10 values which include the age, gender, serum

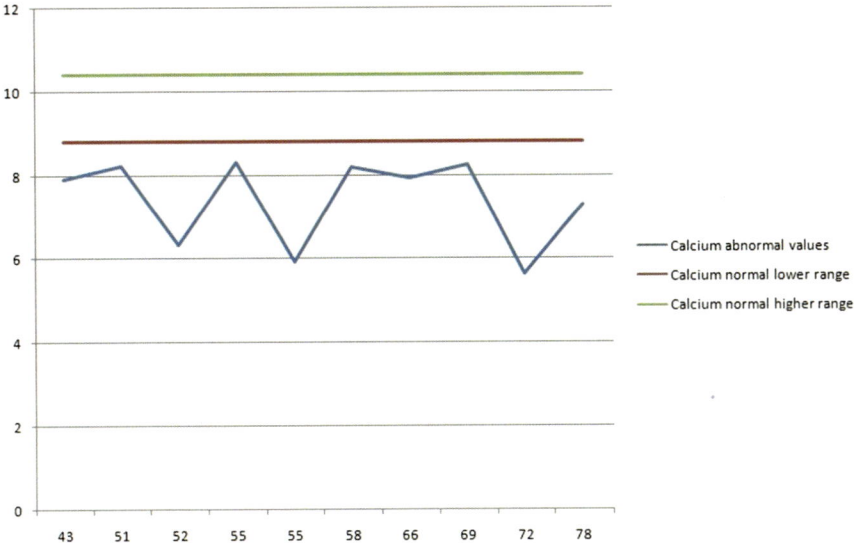

Fig. 1 Graph between abnormality and normal for Calcium

level, and T-score value similarly the values has been detected for Phosphate, Vitamin D3, PTH, and Calcitonin. The following figures were plotted for representation of the abnormalities of Calcium, Phosphate, Vitamin D3, PTH, and Calcitonin. The following plot representation the threshold values of parameters Figs. 1, 2, 3, 4, and 5 [17, 18].

4   Result with Discussion

The function k-means partitions data into k equally exclusive groups and proceeds the index of the cluster to which it has allotted every reflection [16]. Unlike hierarchical clustering, k-means clustering functions on actual observations and produces a single level of clusters. The partitions mean that k-means grouping is often extra fit than hierarchical clustering for huge amounts of data. k-means works for every observation in given data as an object having a location in space. It determines a partition in which substances within each cluster are as close to each other as possible and as far from substances in other clusters as possible. Every cluster in the partition is defined by its member objects and by its centroid or center [19]. The centroid for every cluster is the point to which the sum of distances from all substances in that cluster is reduced. This has been obtained in the Figs. 6, 7, 8, 9, and 10.

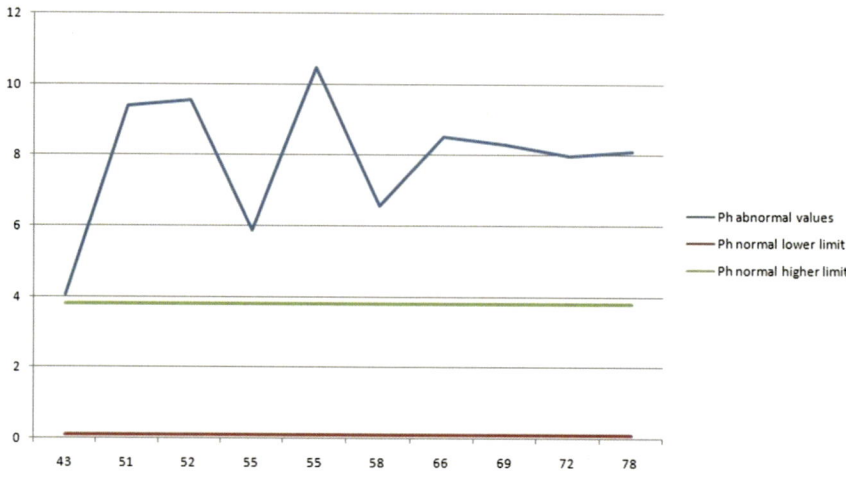

Fig. 2 Graph between abnormality and normal for Ph

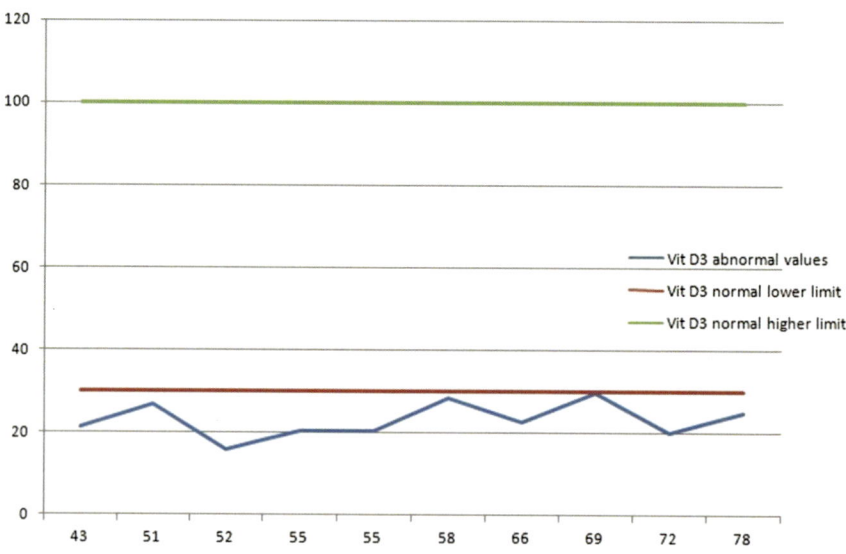

Fig. 3 Graph between abnormality and normal for Vitamin D3

4.1 Calcium

See Fig. 6.

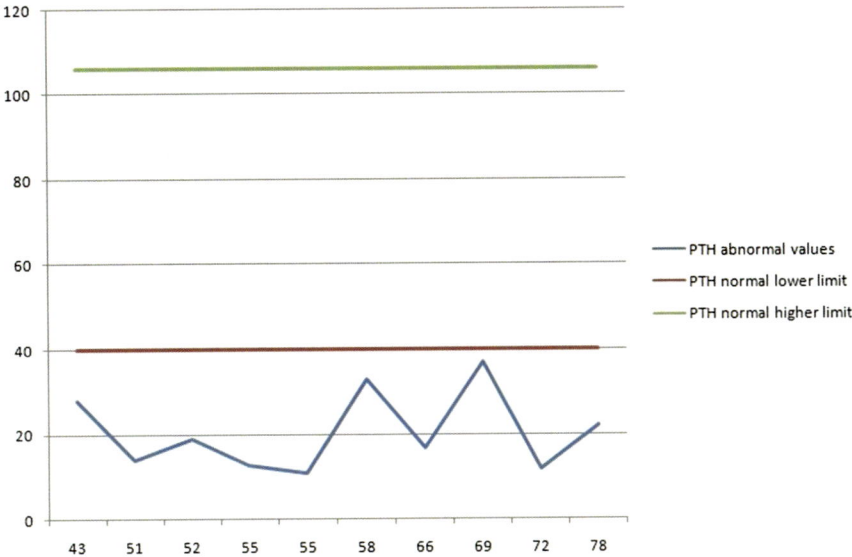

Fig. 4 Graph between abnormality and normal for PTH

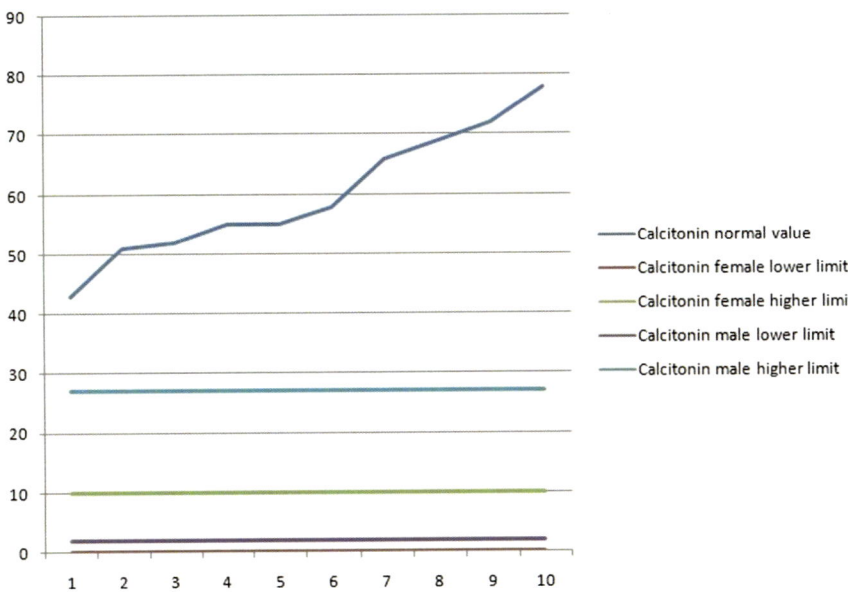

Fig. 5 Graph between abnormality and normal for Calcitonin

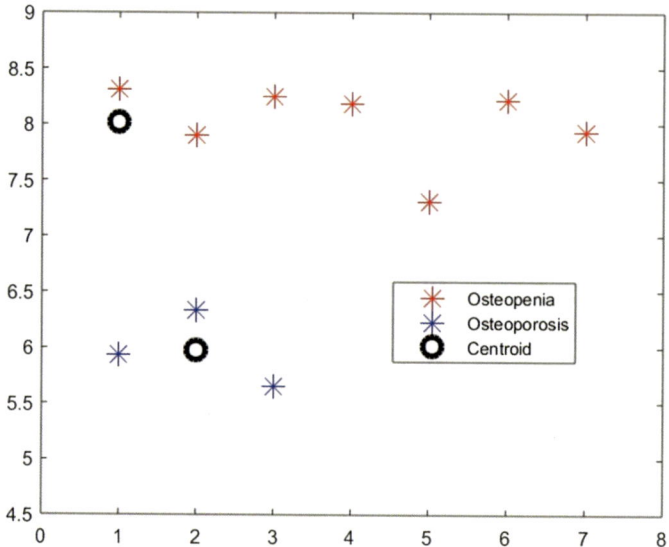

Fig. 6 Output of k-means clustering of Ca

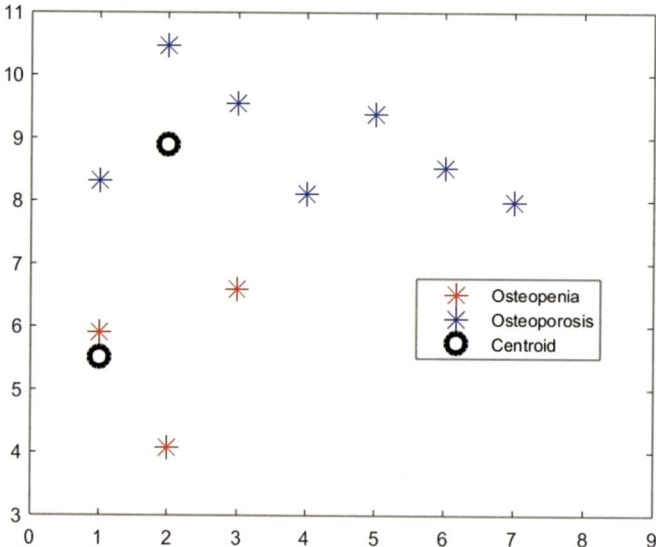

Fig. 7 Output of k-means clustering of Ph

4.2 Phosphate

See Fig. 7.

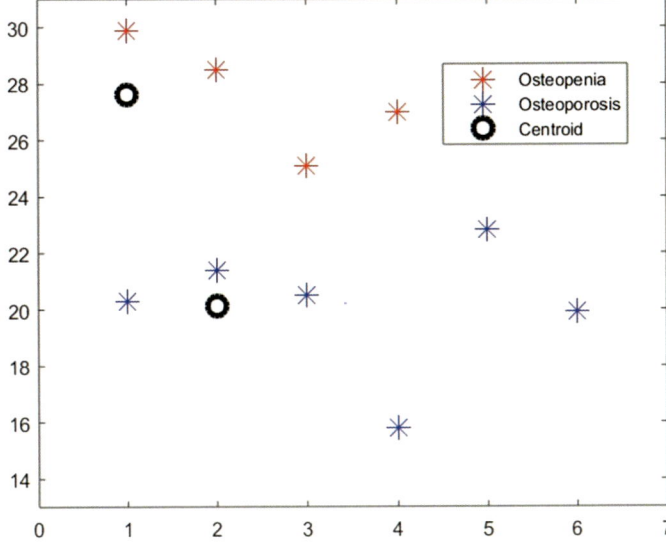

Fig. 8 Output of k-means clustering of Vitamin D3

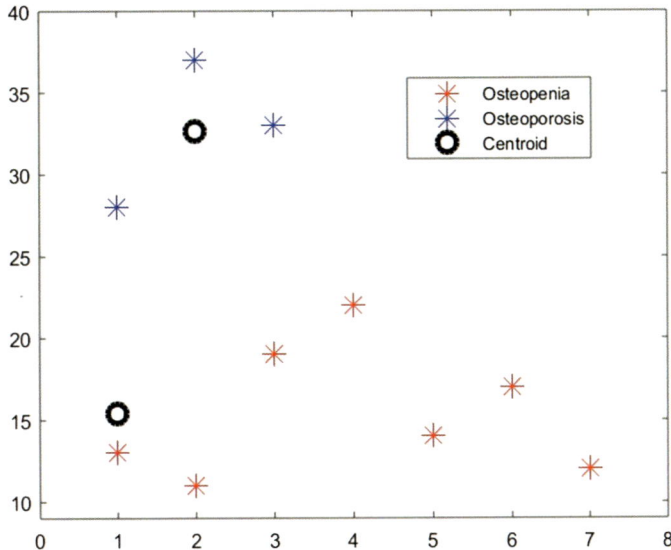

Fig. 9 Output of k-means clustering of PTH

4.3 Vitamin D3

See Fig. 8.

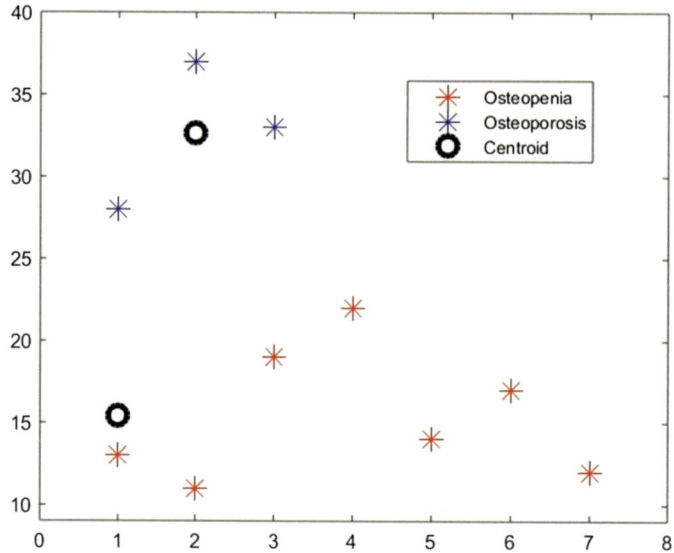

Fig. 10 Output of k-means clustering of PTH

4.4 PTH

See Fig. 9.

4.5 Calcitonin

The number of inputs in a two-dimensional plane will be clustered into different groups based on 'k' value. Select any two inputs from entire inputs. The selected inputs were named as seed inputs. A geometrical line is drawn between the seed inputs, which is bisected by a perpendicular line. The bisected line divides the entire inputs into two half, i.e., both sides of the perpendicular line form two different groups. To obtain optimal results, iterations were carried out. Hereby, the 'k' value assigned as 2, so the two groups of clusters are formed based on the bone diseases. Hence, the cluster's inputs were noted with different colors such as blue and red.

5 Conclusion

The present approach has attempted to combine clinical data measured from each patients and their respective DEXA value for classification of patients into different catalogues of bone disorder conditions. This approach has included 10 patients and their five influencing parameters such Calcium, Phosphate, Vitamin D_3, Parathyroid hormone (PTH) level and calcitonin level are taken into account along with respective DEXA value. The k-means clustering algorithm is a type of neural network. Particularly, this algorithm uses multilevel networks. The number of inputs in a two-dimensional plane will be clustered into different groups based on 'k' value. Select any two inputs from entire inputs. The selected inputs were named as seed inputs. A geometrical line is drawn between the seed inputs, which is bisected by a perpendicular line. The bisected line divides the entire inputs into two half, i.e., both sides of the perpendicular line form two different groups. To obtain optimal results, iterations were carried out. Hereby the 'k' value assigned as 2, so the two groups of clusters formed based on the bone diseases. Hence the cluster's inputs were noted with different colors such as blue and red. This results in the classification of bone disorders based on severities. This concludes that each one of the parameters creates an impact on the results. In case of any abnormalities at one of the parameters' values then it can be considered as the beginning stages of bone disorder diseases. Further testing this soft simulation with an increased number of population, this method could be improvised. Therefore, the application of this method can be extended. This approach creates awareness among the people toward bone disorder diseases.

References

1. Shatrugna V, Kulkarni B, Kumar PA, Rani KU, Balakrishna N (2005) Bone status of Indian women from a low-income group and its relationship to the nutritional status. Osteoporos Int 16:1827–1835. https://doi.org/10.1007/s00198-005-1933-1
2. Siris ES, Miller PD, Barrett-Connor E, Faulkner KG, Wehren LE, Abbott TA et al (2001) Identification and fracture outcomes of undiagnosed low bone mineral density in postmenopausal women. JAMA 286:2815. https://doi.org/10.1001/jama.286.22.2815
3. Park M, Kang B, Jin SJ, Luo S (2009) Computer aided diagnosis system of medical images using incremental learning method. Expert Syst Appl 36:7242–7251. https://doi.org/10.1016/j.eswa.2008.09.058
4. Ding F, Leow WK, Sen Howe T (n.d.) Automatic segmentation of femur bones in anterior-posterior pelvis X-ray images. In: Computer vision analysis of images patterns. Springer, Berlin, Heidelberg, pp 205–212. https://doi.org/10.1007/978-3-540-74272-2_26
5. Lim SE, Xing Y, Chen Y, Leow WK, Sen Howe T, Png MA (2004) Detection of femur and radius fractures in X-ray images. In: Proceedings of the 2nd international conference on advanced medical signal information processing, vol 1. pp 249–256
6. Armato III SG, Sensakovic WF (n.d.) Automated lung segmentation for thoracic CT: impact on computer-aided diagnosis 1. https://doi.org/10.1016/j.xacra.2004.06.005
7. Pulkkinen P, Jämsä T, Lochmüller E-M, Kuhn V, Nieminen MT, Eckstein F (2008) Experimental hip fracture load can be predicted from plain radiography by combined analysis of trabecular bone structure and bone geometry. Osteoporos Int 19:547–558. https://doi.org/10.1007/s00198-007-0479-9
8. Sapthagirivasan V, Anburajan M (2013) Diagnosis of osteoporosis by extraction of trabecular features from hip radiographs using support vector machine: an investigation panorama with DXA. Comput Biol Med 43:1910–1919. https://doi.org/10.1016/j.compbiomed.2013.09.002

9. Kavitha MS, Asano A, Taguchi A, Kurita T, Sanada M (2012) Diagnosis of osteoporosis from dental panoramic radiographs using the support vector machine method in a computer-aided system. BMC Med Imaging 12:1. https://doi.org/10.1186/1471-2342-12-1

10. Detection of osteoporosis and osteopenia using bone densitometer – simulation study. Materials Today: Proceedings (Elsevier) Volume 5, 1024–6

11. Marwaha RK, Tandon N, Garg MK, Kanwar R, Narang A, Sastry A et al (2011) Bone health in healthy Indian population aged 50 years and above. Osteoporos Int 22:2829–2836. https://doi.org/10.1007/s00198-010-1507-8

12. Wang L, Nancollas GH, Henneman ZJ, Klein E, Weiner S (2006) Nanosized particles in bone and dissolution insensitivity of bone mineral. Biointerphases 1:106–111. https://doi.org/10.1116/1.2354575

13. World Health Organization (2004) WHO scientific group on the assessment of osteoporosis at primary health care level. In: Summary meeting report, pp 5–7

14. Nalavade K, Meshram BB (2014) Evaluation of k-means clustering for effective intrusion detection and prevention in massive network traffic data. Int J Comput Appl 96(7):9–14

15. Kavitha MS, Asano A, Taguchi A, Kurita T, Sanada M (2012) Diagnosis of osteoporosis from dental panoramic radiographs using the support vector machine method in a computer-aided system. BMC Med Imaging 12(1):1

16. K*-means: an effective and efficient k-means clustering algorithm. IEEE Xplore. Retrieved from https://doi.org/10.1109/bdcloud-SocialComSustainCom.2016.46. Accessed on 31 Oct 2016

17. McCormick CC (2002) Passive diffusion does not play a major role in the absorption of dietary calcium in normal adults. J Nutr 132:3428–30. Retrieved from http://www.ncbi.nlm.nih.gov/pubmed/12421863

18. Kanis JA (2004) WHO scientific group on the assessment of osteoporosis at primary health care level. World Health Organisation, 5–7 May 2004. https://doi.org/10.1016/s0140-6736(02)08761-5

19. Vesanto J, Alhoniemi E (2000) Clustering of the self-organizing map. IEEE Trans Neural Netw 11:586–600. https://doi.org/10.1109/72.846731

IoT-Based Embedded Smart Lock Control Using Face Recognition System

J. Krishna Chaithanya, G. A. E. Satish Kumar and T. Ramasri

Abstract Smart home security and remote monitoring have become vital and indispensable in recent times, and with the advent of new concepts like Internet of Things and development of advanced authentication and security technologies, the need for smarter security systems has only been growing. The design and development of an intelligent web-based door lock control system using face recognition technology, for authentication, remote monitoring of visitors and remote control of smart door lock have been reported in this paper. This system uses Haar-like features for face detection and Local Binary Pattern Histogram (LBPH) for face recognition. The system also includes a web-based remote monitoring, an authentication module, and a bare-bones embedded IoT server, which transmits the live pictures of the visitors via email along with an SMS notification, and the owner can then remotely control the lock by responding to the email with predefined security codes to unlock the door. This system finds wide applications in smart homes where the physical presence of the owner at all times is not possible, and where a remote authentication and control is desired. The system has been implemented and tested using the Raspberry Pi 2 board, Python along with OpenCV are used to program the various face recognition and control modules.

1 Introduction

With the advent of various smart technologies, the need for better and more intelligent security and monitoring has been growing. In recent times, the need for security and surveillance has become vital in many areas such as homes, offices, banks, etc. In the recent past, various authentication techniques have been designed and implemented, passwords, patterns, RFID to name a few, these technologies have

J. Krishna Chaithanya (✉) · G. A. E. Satish Kumar
Department of ECE, Vardhaman College of Engineering, Hyderabad, India
e-mail: j.krishnachaitanya@vardhman.org

T. Ramasri
Department of ECE, SVUCE, SVU, Tirupati, India

D. Pandian et al. (eds.), *Proceedings of the International Conference on ISMAC in Computational Vision and Bio-Engineering 2018 (ISMAC-CVB)*, Lecture Notes in Computational Vision and Biomechanics 30,
https://doi.org/10.1007/978-3-030-00665-5_104

their advantages and disadvantages, the passwords and patterns once traced may compromise the security, the fingerprint-based systems are regularly prone errors arising due to external factors and mismatches. Hence, arises a need for a more efficient and effective way of authentication [1]. The authentication based on Face Recognition has played a pivotal role over the years providing unmatched levels of efficiency and accuracy [2, 3] but was limited to a few high-security establishments and large corporations as the design, and implementation costs are high. Today, thanks to ongoing research and development, the algorithms have become more accessible and find a broad range of applications [4]. Also, the present day home security systems have not been updated since years, the physical presence of the owner/key is mandatory to gain access to the house, the proposed system hence includes an intelligent web-based embedded server aimed at providing remote authentication and control of the door lock using email and basic IoT concepts.

1.1 Face Recognition

In the proposed system, we introduce a low-cost extendable framework for embedded smart home security system, which consists of a face recognition module. This system uses Haar-like features for face detection and Local Binary Pattern Histogram (LBPH) for face recognition [5, 6]. The system a cascade classifier in face detection and face recognition is carried out in three stages, namely feature extraction, matching, and classification. The distinctive and at the most important options are extracted, and the face image is compared with pictures in information throughout the last stage (classification). This native binary pattern for person's face recognition takes into account each form and texture information for analysis. The image given is segregated into little elements from which the Local Binary Patterns are adopted and clubbed into one vector feature. This feature vector helps in measuring similarities between pictures by forming an associate in nursing economical illustration of a face. The algorithms are implemented using the OpenCV library of Python programming language running on the Raspberry Pi 2.

1.2 Remote Monitoring Control

In addition to the face recognition-based authentication, the proposed system also includes an intelligent web-based embedded server for remote authentication and control of the door lock. This technology plays a crucial role in enabling remote access to a person/persons if required and if deemed necessary by the owner, who is not required to be present in-person. When an unauthorized person is detected by the system, it emails the owner a live picture of the individual along with as SMS notification and waits for his command. If the owner recognizes the person and would like to provide them with access, he can do so by sending a predefined security code to the server via email, the server, in turn, checks the authenticity of the code and unlocks the door accordingly [7, 8]. The server also maintains a web page which can only be accessed by the owner to bypass the face recognition system in case of error. This subsystem is implemented using Python programming language along with the Linux server and PHP scripts running on the Raspberry Pi 2.

2 System Design and Architecture

2.1 System Architecture

The proposed system is a combination of various modules namely, imaging module, core module and the door lock module. Figure 1 represent the block diagram. The imaging module is responsible for capturing the images of visitor/visitors and forwarding it to the core module for further processing; this is realized using a USB web camera. The door lock module is a combination of driver circuitry and a DC motor to drive a lock, responsible for locking and unlocking the door as required [4] (Fig. 2).

The heart of this system is the core module which is realized using the Raspberry Pi 2, its responsibilities include, acquiring images from the camera, processing the acquired image as required, maintaining the facial image database, comparing the acquired image with the database, sending commands to the door lock module, etc. The core module also acts as an embedded web server responsible for sending and receiving emails, sending out SMS notifications and backend access [7].

3 System Description

3.1 Imaging Module

The imaging module in the proposed system is realized using a USB web Camera, the main reason behind choosing USB Camera over the Pi camera is the cost effectiveness. The camera features a high-quality CMOS sensor, with an image resolution of 25 MP (Interpolated), an adjustable lens for focus adjustment, a frame rate of 30 fps and f2.0 lens. The USB camera also is equipped with night vision for low light photography. The camera interfaces with the Raspberry Pi via the USB 2.0 port and is responsible for capturing images when requested, the pictures are captured by using the command fswebcam (Fig. 3).

Fig. 1 Flow chart

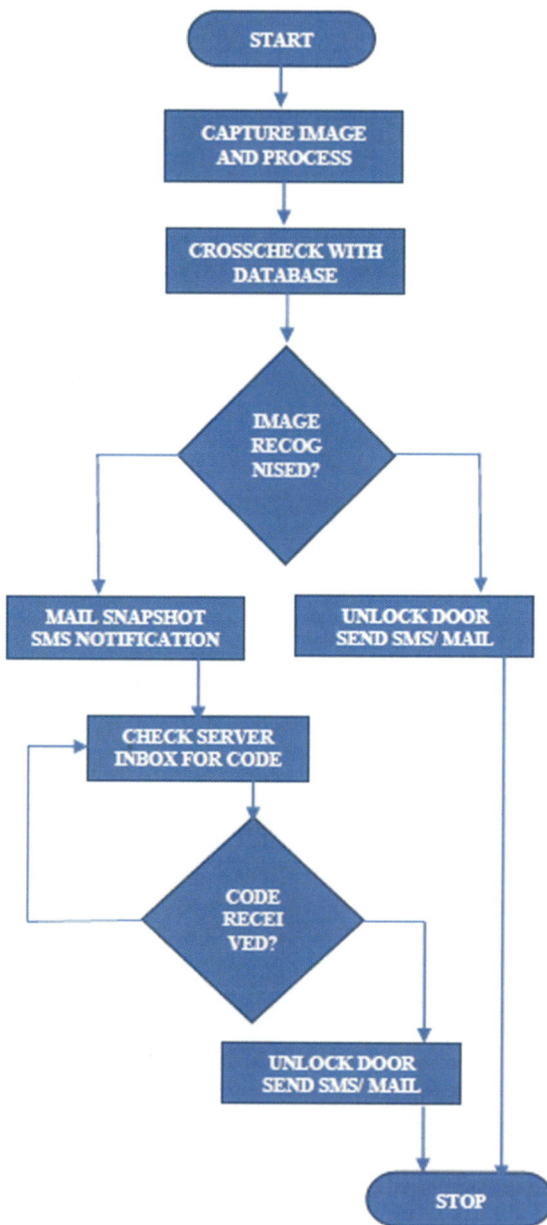

3.2 Raspberry Pi Core Module

The core module of the system is realized using a Raspberry Pi 2 board; it's a $35 bare-bones computer designed and developed by the Raspberry Pi Foundation, the Pi

Fig. 2 Block diagram

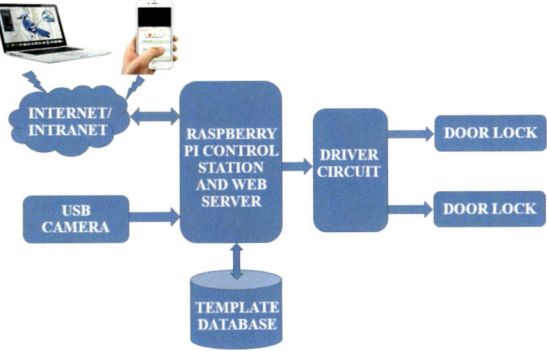

Fig. 3 USB camera

2 features a BCM 2836 System-on-Chip which includes a Quad-Core 32-Bit ARM Cortex A7 CPU clocked at 900 MHz paired with 1 GB of RAM. It also has VideoCore IV GPU for graphical processing applications, it also includes four USB ports for peripherals and 40 Pin General Purpose Input Output (GPIO) pins for interfacing the Pi with external electronic circuits, these GPIO pins are used to interface the Pi to the door lock module. The Raspberry Pi is designed to run various Linux based operating systems and has Raspbian as its official operating system and Python as its official programming language. In this system the core module plays a highly pivotal role and is responsible for various functions, the core module is responsible for acquiring the images from the camera, processing and storing. It's also responsible for maintaining the facial database which consists of pictures of all the authorized persons for reference. It is in charge of employing the face detection and recognition algorithms and has to decide whether is a person is authorized or not. It's responsible for controlling the door lock module by sending lock/unlock commands using Python code via GPIO to the motor driver (Fig. 4).

Fig. 4 Raspberry Pi 2 module

3.3   Embedded Server and IoT

Another crucial function of the core module is to act as an embedded web server, the primary responsibilities of this server include, transmitting the visitor/visitors images via email to the owner, SMS notifications, look for emails from the owner and find the security code from the emails for authorization. This system employs an embedded server approach for communicating with the user and with the internet/intranet. Python code is used to program certain aspects of this system such as sending and receiving emails and text messages. Standard Python libraries corresponding to the web such as urlib2, cookielib for online SMS service; imaplib, poplib, email, smtp, etc. for sending and receiving emails are imported and used accordingly. This system uses web-based SMS client way2sms to send SMS alerts to the owner; it's also configured with a dedicated gmail id to send and receive emails. The system is also configured using Apache to act as a server, which is useful to remotely monitor the conditions. The owner can log into the server using a dedicated static IP assigned to the Raspberry Pi, another important function of this server is to provide a secure back door to lock/unlock the door by bypassing the face recognition feature in case of a failure or emergency. This is a secret feature and is only accessible by the owner.

3.4   Face Detection and Recognition

Many kinds of face detection algorithms are used in many appliances, surveillance systems, gaming, human-computer interaction, etc. Paul Viola and Michael Jones devised a formula for object detection using Haar-feature based cascade classifiers. It's a machine learning based algorithm in which some positive and negative images are employed to train the cascade classifier. Once the classifier is trained, features are extracted which is in turn used for object/face detection [5]. The OpenCV library of Python provides support for using the Haar cascade classifiers for face detection, and it's equipped with both trainer and classifier. Thus the required XML classifiers

Fig. 5 Door lock module

are loaded using Python. For face recognition to be carried out, various face recognition algorithms such as Eigenfaces, Fisherfaces, LBPH Algorithm, etc. are available [6, 10]. This system uses Lower Binary Pattern Histograms method to perform face recognition. The Eigenfaces and Fisherfaces methods employ a holistic approach to recognitions, the data in these techniques are treated as a vector in high dimensional space which is not always ideal, hence in the LBPH algorithm the idea is to look at the lower dimensional subspace for useful information. In this method, the local structure in an image is summarized by comparing the pixels with their corresponding neighbors [7, 8]. Consider a pixel. It's surrounded by eight other pixels as neighbors. Now comparing the intensity of the center pixel with that of the neighbors if the intensity of the center pixel is higher than or equal to that of the neighbor under consideration then it's denoted by 1, else with 0. Thus, for each pixel a binary representation is formed, which leads to a total of 2^8 possible combinations, these combinations are termed as Local Binary Patterns. The OpenCV library of Python features a rich variety of face recognition algorithms through its FaceRecognizer class. The LBP algorithm is enabled by using the command createLBPHFaceRecognizer(). This system uses the LBP algorithm paired with the Yale Facedatabase or Yalefaces [6, 9] (Fig. 5).

3.5 Doorlock Module

The door lock module of this system is simulated using a DC motor to demonstrate the locking and unlocking function. This module is a combination of a relay driver circuit and a DC motor; this system uses an HFD27 Series 5V 1A 125 Ω (DPDT) Through Hole SubMiniature DIP Relay to control the DC motor. The driver circuit is also provided with leads for a 9 V battery to drive the motor when triggered. The driver is triggered by the core module through the GPIO pins.

Fig. 6 Intial setup

4   Design and Implementation

This section emphasizes on the actual hardware implementation of the proposed system, the various modules, components, peripherals and the interconnections between
them are discussed here. The first stage of the implementation is to prepare the Raspberry Pi 2 module for its first boot; this is done by downloading the latest version of
the Raspbian operating system from the official Raspberry Pi website. A microSD
card is the formatted using SD Formatter; it's then flashed with the Raspbian OS using
Win32 Disk Imager [10]. The first boot is then completed on the Raspberry Pi connecting the required peripherals, such as power supply, keyboard, mouse, Ethernet
cable, etc. (Fig. 6).

The Raspberry Pi for optimal operation requires a quality power supply; the Pi can
be driven by using any Micro USB based mobile phone chargers with a good current
rating, and this system is powered by a 5V 2A power bank for uninterrupted operation.
Since the Raspberry Pi doesn't natively support wireless internet a USB WiFi dongle
is used for connectivity; the Pi also has an Ethernet port which can be used to
gain wired internet access [9]. Using Python programming language preinstalled on
Raspbian the source code of the system is provided and tested appropriately. The
USB Camera is interfaced, the GPIO pins are programmed using commands in Linux
and Python in this stage. The camera is interfaced to the Pi via the USB port and the
door lock module is interfaced via the GPIO pins on the Pi (Figs. 7 and 8).

5   Experimental Results

This section emphasizes on the final results of the proposed system, the system has
been configured to recognize one of the author's face, and thus some face images

Fig. 7 GPIO and peripherals

Fig. 8 Final setup

are taken in varying light conditions and are added to the database which is already populated with faces from Yale database. The system correctly recognizes the face and unlocks the door which is simulated by the DC Motor along with an SMS alert. In the case of an unauthorized person, the algorithm reports non-availability of the face in the database to the core module, which in turn forwards the live snapshot to the owner's email address for manual authentication along with an SMS alert. The owner now has to manually provide access to the person if he/she recognizes the person, this is done by replying to the Pi's email with a secure code as its subject, this code can be changed by the owner. Once the Pi receives this code, it validates it and unlocks the door.

6   Conclusion

This paper presents the design and implementation of an intelligent home security system using a robust, low-cost, low power single chip approach with the Internet as its backbone. This paper also explores the immense potential of computer vision in general and face recognition, in particular, the possibilities of IoT in home security and automation. The versatility and prowess of Linux operating system, the Python programming language, and the OpenCV library have also been explored, in depth.

References

1. Ahonen T, Pietikäinen M, Hadid M, Mäenpää T (2004) Face recognition based on the appearance of local region. Machine Vision Group, InfoTech., University of Oulu, IEEE, Finland
2. Ahonen T, Hadid A, Pietikäinen M (2006) Face description with local binary patterns: application to face recognition. Draft
3. Faizi A (2008) Robust face detection using template matching algorithm. University of Toronto, Canada
4. Feng P (2004) Face recognition based on elastic template. Beijing University of Technology, China. Yang MH, Kriegman DJ, Ahuja N (2002) Detecting faces in images: a survey. IEEE Trans PAMI
5. Hadid A, Heikkilä M, Ahonen T, Pietikäinen M (2004) A novel approach to access control based on face recognition. Machine Vision Group, InfoTech Oulu and Department of Electrical and Information Engineering, University of Oulu, Finland
6. Rodriguez Y (2006) Face detection and verification using local binary patterns. Ph.D. thesis, Acole Polytechnique Federale de Lausanne
7. Nosaka R, Ohkawa Y, Fukui K (2012) Feature extraction based on co-occurrence of adjacent local binary patterns. In: Proceedings of the 5th Pacific Rim conference on advances in image and video technology, vol Part II, PSIVT 2011, pp 82–91
8. Zhang C, Zhang Z (2009) A survey of recent advances in face detection. In: Face recognition: face in video, age in variance, and facial marks. 2010 Unsang Park, Michigan State University
9. Zhang H, Zhao D (2004) Spatial histogram features for face detection in color images. In: IEEE 5th Pacific Rim conference on multimedia. Tokyo, Japan, pp 377–384
10. Brubaker S, Wu J, Sun J, Mullin M, Rehg J (2005) On the design of cascades of boosted ensembles for face detection. Technical report GIT-GVU-05-28, Georgia Institute of Technology

MediCloud: Cloud-Based Solution to Patient's Medical Records

G. B. Praveen, Anita Agrawal, Jainam Shah and Amalin Prince

Abstract Cloud computing has lately emerged as a new standard for hosting and delivering services over the internet. As per the Code of Federal Regulations, the hospitals are required to retain the radiological records for a duration of 5 years. Healthcare organizations are considering cloud computing as an attractive option for managing radiological imaging data. In this paper, we have proposed a public cloud-based "Infrastructure as a Service (IaaS)" model as a solution for maintaining the medical records of patients. New patient information is registered at the hospital with the help of a unique identification number. For each user, the bucket is created in the Amazon AWS cloud to store or retrieve the data. Data access is performed with the help of username and password provided by the web link which is embedded in the unique QR code. Two QR codes are used, the first code gives the access to the login page, whereas the latter one is used for accessing the corresponding user bucket.

1 Introduction

A prototype with a universal, on-demand access to a shared pool of resources which can be procured and supplied with nominal management effort or service provider interaction is defined as "Cloud Computing" [1, 2].

G. B. Praveen (✉) · A. Agrawal · J. Shah · A. Prince
Department of Electrical and Electronics Engineering, BITS Pilani – K.K Birla Goa Campus,
Goa, India
e-mail: p20120404@goa.bits-pilani.ac.in

A. Agrawal
e-mail: aagrawal@goa.bits-pilani.ac.in

J. Shah
e-mail: f2014023@goa.bits-pilani.ac.in

A. Prince
e-mail: amalinprince@goa.bits-pilani.ac.in

© Springer Nature Switzerland AG 2019
D. Pandian et al. (eds.), *Proceedings of the International Conference on ISMAC in Computational Vision and Bio-Engineering 2018 (ISMAC-CVB)*, Lecture Notes in Computational Vision and Biomechanics 30,
https://doi.org/10.1007/978-3-030-00665-5_105

Hospitals in the developing countries face funds shortage to invest in operation expansion. The healthcare industry is gradually adopting latest technology for storing digital images in the cloud database. Currently, the technology adoption is limited only to billing automation or back-office systems [3]. Digitization of medical records of patients is a vital and essential stage for quick and reliable storage and retrieval of patient's medical records [4]. However, it is not easy to maintain huge data centers, since heavy investment is required in human resources and technology for the same. The importance of cloud-based system is that data can be made accessed at any location and at any time. Cloud-based service provides pay as per use model so that massive investment such as buying and deploying expensive technologies can be avoided.

Rolim et al. [5] proposed an automated framework with the help of sensors to acquire vital data from the human body and to store the information in the cloud. A platform based on cloud computing framework is proposed for the management of mobile and wearable healthcare sensors [6]. Chang et al. [7] developed a cloud computing adoption framework (CCAF) and cloud storage design and deployment framework based on storage area network (SAN).

Client platform security is the major drawback of the current e-health solutions and standards. To overcome the security issue Lohr et al. [8] presented a secured framework for establishing privacy domains in e-health infrastructures. He et al. [9] proposed a private cloud platform architecture which consists of six layers and utilizes the message queue as a cloud engine. To overcome the drawbacks of the abovementioned frameworks, we propose a public cloud-based Infrastructure as a Service (IaaS) model as a solution for maintaining the medical records of patients. Characteristics and model deployment strategies of cloud computing are discussed in Sect. 2 and real-time challenges faced by the hospitals is described in Sect. 3. Section 4 describes the system model and proposed methodology is briefed in Sect. 5, concluding remarks are drawn in Sect. 6.

2 Characteristics and Deployment Models of Cloud Computing

The NIST has described the characteristics of the cloud as follows [1, 10, 11]:

- On-demand self-service: A user can demand computing capabilities, such as network storage, server time, etc., whenever required.
- Broad network access: Users can make use of the facilities over the network and access it through varied client platforms such as mobile phones, laptops, and tablets.
- Resource pooling: The different virtual and physical resources of the service provider are allocated and reallocated dynamically as per the customer demands. The computing resources are combined to serve numerous customers.

Fig. 1 Cloud deployment
models

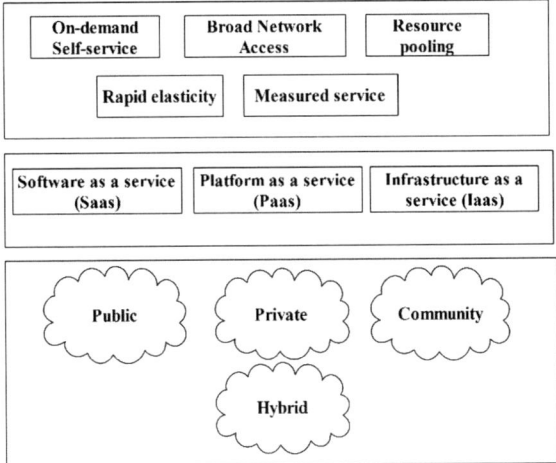

- Rapid elasticity: The computing capabilities often seem to be infinite and can be procured in required quantity as per the requirement. Capabilities can be demanded and released elastically. This process is generally automatic and it scales inward and outward proportionally with the demand.
- Measured service: Cloud systems provide a metering ability to control and optimize the usage of the resource.

The NIST has listed four categories of cloud deployment models as shown in Fig. 1 [1, 10–12]:

- Public cloud: The general public can use the services and resources via the internet. These services are typically low cost. Public cloud can be utilized by the general public, small and medium enterprise clients and even by the large organizations. Amazon AWS, Microsoft Azure, Google Cloud Platform are some of the providers of public cloud service.
- Private cloud: The cloud infrastructure is set up exclusively for an organization, although it may be managed by a third party or the organization itself. Management by third-party permits better controls over the infrastructure, security and management issues related to the cloud resources. The people outside the organization do not have access to it as it is restricted. It can be on the premises of the organization. This has less elasticity when compared with other deployment models.
- Community cloud: A group of organizations which share mutual interests and use the services. These services support a specific community which has similar concerns. Government institutions can share or members or partners of a common group can use community cloud.
- Hybrid cloud: It is a blend of two or more clouds (private, community, or public). This provides the most flexibility because it takes the benefits of all the other categories public, private, or community.

NIST definition of Cloud:

- Three service models of cloud computing was introduced by NIST [10, 13, 14]:
- Infrastructure as a Service (IaaS): The customer can demand processing capabilities, storage facilities, network abilities and also other fundamental computing resources which include operating systems, applications, and database systems. It is divided into compute clouds and resource clouds. Users can demand and access computational resources such as RAM, memory, CPU, processing power, virtual machines in compute cloud whereas, in resource cloud, users can access enhanced virtualization capabilities. They provide scalable resources as services to users.
- Platform as a Service (PaaS): The consumers can develop and host applications and software on a platform which provides computational resources. The behavior of the server can be controlled by making use of dedicated application programming interfaces (APIs) according to user requests.
- Software as a Service (SaaS): The provider's applications which run on the cloud are provided as a software service to the customers. The consumers can access the hosted applications or software over the internet.

3   Challenges

There are various challenges to the hospitals which are directly related to their Information Technology (IT) department.

3.1   Data Storage

The volume of digital data that is generated by healthcare systems has grown exponentially in the past decade. It is not an easy job to maintain this data. The medical regulations of country specify for how long the data must be stored and archived. For example, healthcare providers need to retain medical records archived for 6 years after discharge according to Health Insurance Portability and Accountability Act (HIPAA) [15]. Data backup and recovery has to be done as a measure to protect the data.

3.2   Data Sharing

Patients need to carry hard copies of scans and reports every time they visit a doctor. The data might be digitized but even though sharing and accessing it from any location is difficult.

3.3   Security

Patient's image scans and reports need to be maintained confidentially [16]. Secured data storage is an important task, hence these data should be encrypted and saved in the cloud. Encryption Algorithm implementation is a difficult task for the hospital's IT department.

3.4   Availability and Data Corruption

Availability of a system is a measure of how reliable the system is. The data must be available as and when required. The data stored may get corrupted over time due to bit-flip errors.

3.5   High Cost of Maintaining Infrastructure

It is not easy to maintain huge data centers. It requires a huge investment in human resources and technology [3]. Also, the current systems need upgradation which requires capital.

4   System Model

This section deals with the definition of our proposed model which consists of the below-mentioned entities.

- Object: An object consists of any files of types (e.g., PDF, JPEG, DOC, etc.) [17].
- Bucket: Buckets are the containers for objects [17].
- Identity and Access Management services (IAM): AWS identity and web management is a web service by which the AWS resource access can be controlled. The identity management portion of the IAM service is required for user authentication (sign in) and authorization. The access management portion of the IAM service defines the permission for a user or other entity in an account. Permissions are granted through policies that are created and then attached to users [18].
- Registered user: The patient who comes to a hospital and gets registered is called a registered user.
- Root user: The admin of the system is called as the root user. The root user has complete access to all the services and resources.

5 Methodology

The proposed framework uses public cloud such as Amazon AWS for maintaining the medical records of patients as depicted in Fig. 2.

When the patient is first registered at the hospital (registered user), the patient will be given a unique identification number which will be used to identify each patient uniquely. A bucket will be created for that registered user by the root user. Access Control Policy will be specified such that only the bucket corresponding to that registered user can be accessed by that registered user. No registered user can access any other registered user's bucket. For this purpose, identity access and management (IAM) services are used. Username and password are generated corresponding to each registered user. QR code which contains the web link for accessing the data will be given. QR code is generated by a Ruby Script which uses Barby gem. On scanning the QR code, the user will be redirected to a login page. Another QR code, which contains the username, password, and the steps to be followed will be given. On scanning this QR code, the text will appear. The user has to enter the username and password on the login page and then follow the steps that appeared in the QR code as per the requirements, i.e., to store new data or to retrieve previously stored data. QR code is printed and given so that the user does not have to remember the

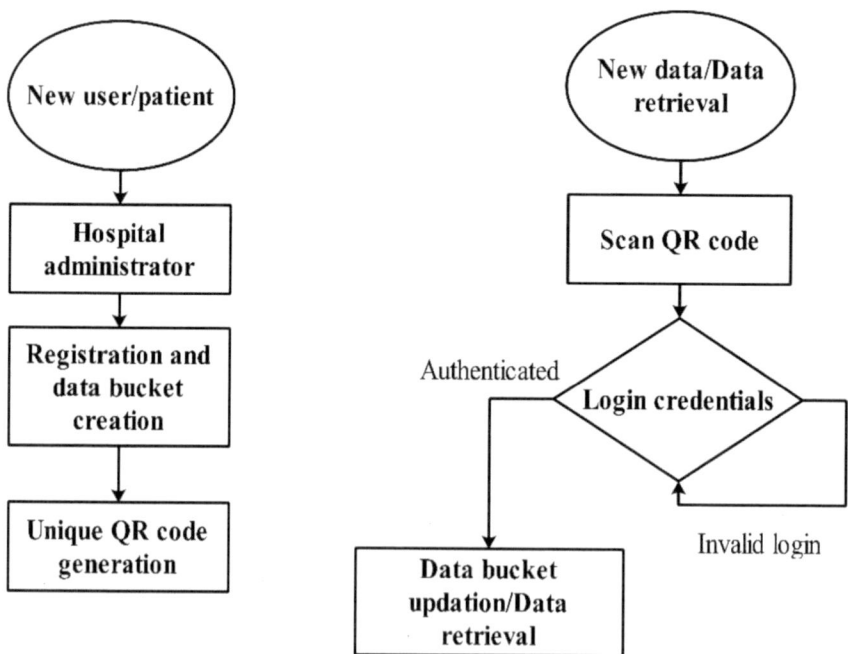

Fig. 2 Proposed methodology

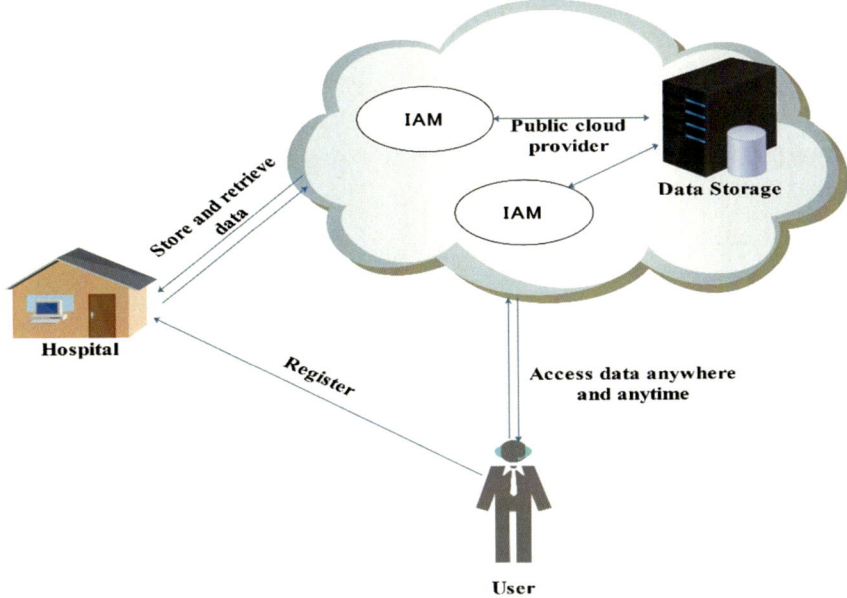

Fig. 3 Overview of medical cloud

links, username, password and the steps to be followed. A complete overview of the proposed system is shown in Fig. 3.

The abovementioned methodology solves the challenges listed in the previous section.

5.1 Durability

The objects are stored on multiple devices across multiple facilities by Amazon Simple Storage Service (S3) in Amazon S3 region redundantly. The service is designed in such a way that concurrent device failures are quickly detected and any lost redundancy is repaired. When a data storage request is received, the data gets stored in multiple data centers [19].

5.2 Accessibility

The data to be retrieved is accessible from any location anytime. It can be accessed on a computer, laptops, and even on smart mobile phones.

5.3 Data Encryption

Users are authenticated before access to the data. The data uploaded and downloaded to the cloud service provider is secured via SSL endpoints using the HTTPS protocol. The security of data stored in the cloud when at rest is achieved by using the security features provided by the cloud provider. Amazon S3 provides server-side encryption for the data at rest. The data is encrypted by one of the strongest block ciphers available, i.e., the AES-256 (Advanced Encryption Standard) [19].

5.4 Availability and Detection of Data Corruption

Amazon S3 checks for data corruption by applying algorithms such as a combination of MD5 checksums and cyclic redundancy checks (CRCs). These checksums are performed on data at rest and any corruption found is repaired using redundant data [19].

5.5 Scalability

Cloud provides the ability of scalability [20]. There is no need to guess the amount of data that will be generated as is the case with data centers. The storage size can be increased and decreased as and when required. The pay as per use model of cloud is very beneficial as you need to pay only for the services used.

6 Conclusion

Robust data management of patient information is facilitated by the cloud storage system. The image scans and reports of the patients get stored in the cloud database, which facilitates data sharing across any location and at any time. In this paper, we have proposed a public cloud-based infrastructure as a service (IaaS) model as a solution for maintaining the medical records of patients. Two QR codes are generated for the accessibility of the login page and bucket access respectively. Cloud storage provides a scalable environment for businesses to increase or decrease storage on-demand. Cloud storage provides durability since data is stored in multiple data centers, and failure of one data center will not affect the stored data in the other center. Cloud storage has transformed the capital-intensive system and saves costs due to its pay as per usage model. Secured data storage and reduction of operating risks are the main advantages of using cloud storage system. The healthcare organizations such as hospitals need cost-effective and innovative methods to deal with the challenges mentioned in the previous sections. We intend to apply the solution proposed above to real-world application in future works in order to assess the advantages of the proposed solution.

Appendix

A. QR Code

QR code is the acronym for the Quick Response code. QR Code is a kind of 2D barcode. Large capacity, Small printout size, high-speed scanning, etc., are some of the features of QR code [21]. It can encode up to 4296 alphanumeric characters in a single pattern [22]. In a QR code, there are three large square patterns (there is a small black square surrounded by white bars inside it). There is an additional square in newer versions. The other area of QR code is used to encode the embedded information and it consists of a number of small blocks. URLs and alphanumeric characters can be easily encoded in a QR code. Easy access to the website is possible by scanning the QR code. QR code scanning just requires a mobile phone with a camera.

B. AES-256 (Advanced Encryption Standard)

AES is a modern block cipher which supports key lengths of 128, 192, and 256 bits. As the name suggests, AES-256 uses a key length of 256 bits. AES is a symmetric key algorithm which uses the same key for encryption and decryption. The best-known attack till date against the AES is the brute force attack [23]. Practically, no one without the key can read the data encrypted by AES.

C. Cyclic Redundancy Checks (CRCs)

Cyclic Redundancy check is an error detecting code. Accidental changes to data caused by bit-flip errors can be detected by this algorithm. When the blocks of data enter the system, a short check value is attached which is calculated on the basis of the remainder of a polynomial division of the contents of data. The calculation is again done when retrieving data. If both the values do not match, then data is found to be corrupted. Thus, corrective measures can be taken. CRCs are so called because the algorithm is based on a cyclic code and the data verification value (check value) is a redundancy (it expands the message without adding information) [24].

References

1. Mell P, Grance T (2011) The NIST definition of cloud computing
2. Mell P, Grance T (2010) The NIST definition of cloud computing. Commun ACM 53(6):50
3. Chand R, Tripathi M, Mishra SK (2016) Cloud computing for medical applications and health-care delivery: technology, application, security and swot analysis

4. Aziz HA, Guled A (2016) Cloud computing and healthcare services
5. Rolim, CO et al (2010) A cloud computing solution for patient's data collection in healthcare institutions. In: Second IEEE international conference on eHealth, telemedicine, and social medicine, pp 95–99
6. Doukas C, Maglogiannis I (2012) Bringing IoT and cloud computing towards pervasive healthcare. In: Sixth IEEE international conference on innovative mobile and internet services in ubiquitous computing (IMIS), pp 922–926
7. Chang V, Walters RJ, Wills G (2012) Cloud storage and bioinformatics in a private cloud deployment: lessons for data-intensive research. In: International conference on cloud computing and services science. Springer, pp 245–264
8. Lohr H, Sadeghi A-R, Winandy M (2010) Securing the e-health cloud. In: Proceedings of the 1st ACM international health informatics symposium, pp 220–229
9. He C, Fan X, Li Y (2013) Toward ubiquitous healthcare services with a novel efficient cloud platform. IEEE Trans Biomed Eng 60(1):230–234
10. National Institute of Standards and Technology (2009). Retrieved from www.nist.gov
11. Wang L, Alexander CA (2013) Medical applications, and healthcare based on cloud computing. Int J Cloud Comput Serv Sci 2(4):217
12. Shadi JH, Tim B, Wei G, Dachuan H, Song W (2010) Cloud types and services. In: Handbook of cloud computing. Springer, pp 335–355
13. Vilaplana J, Solsona F, Abella F, Filgueira R, Rius J (2013) The cloud paradigm applied to e-Health. BMC Med Inform Decis Mak 35(13):1–10
14. Chang V, Roure DD, Wills G, Walters RJ, Barry T (2011) Organizational sustainability modeling for return on investment (ROI): case studies presented by a National Health Service (NHS) Trust UK. CIT J Comput Inf Technol 19(3):177–192
15. John N, Shenoy S (2014) Health cloud-healthcare as a service (HaaS). In: IEEE International conference on advances in computing, communications, and informatics (ICACCI), pp 1963–1966
16. Rolim CO, Koch FL, Westphal CB, Werner J, Fracalossi A, Salvador GS (2010) A cloud computing solution for patient's data collection in healthcare institutions. In: Second IEEE international conference on eHealth, telemedicine, and social medicine, pp 95–99
17. Amazon Simple Storage Service (S3) Introduction. Retrieved from http://docs.aws.amazon.com/AmazonS3/latest/dev/Introduction.html
18. Amazon Identity and Access Management (IAM) Services Introduction. Retrieved from http://docs.aws.amazon.com/IAM/latest/UserGuide/introduction.html
19. Amazon Simple Storage Service. Retrieved from https://aws.amazon.com/s3/faqs/
20. Marwan M, Kartit A, Ouahmane H (2016) Cloud-based medical image issues. Int J Appl Eng Res 11(5):3713–3719
21. Sun A, Sun Y, Liu C (2007) The QR-code reorganization in illegible snapshots taken by mobile phones. In: IEEE international conference on computational science and its applications (ICCSA), pp 532–538
22. Kan TW, Teng CH, Chou WS (2009) Applying QR code in augmented reality applications. In: ACM proceedings of the 8th international conference on virtual reality continuum and its applications in industry, pp 253–257
23. Paar C, Pelzl J (2009) Understanding cryptography: a textbook for students and practitioners. Springer, Berlin
24. Sarwate DV (1988) Computation of cyclic redundancy checks via table look-up. Commun ACM 31(8):1008–1013

A Trio Approach Satisfying CIA Triad for Medical Image Security

Sivasaranyan Guhan, Sridevi Arumugham, Siva Janakiraman,
Amirtharajan Rengarajan and Sundararaman Rajagopalan

Abstract Medical image security attains a great appeal due to the challenges in transmitting them through an open network channel. Even a small change in medical information leads to a wrong diagnosis. Hence, methods to prevent the attacks and tampering on the medical images in the open channel have a great demand. This work proposes such an attractor-assisted medical image watermarking and double chaotic encryption which preserves the confidentiality, integrity and authenticity of the medical image. Patient diagnosis details are compressed using lossless compression approach and embedded in the pixels of patient's DICOM image. Attractor-based selective watermarking is implemented using integer wavelet transform. Encryption will be carried out by employing chaotic maps using confusion and diffusion operations. The effectiveness of the work will be evaluated using standard analyses namely MSE, PSNR, SSIM, entropy, correlation, histogram and key space.

1 Introduction

Recent trends in technological growth are exponential. This exponential growth led to multimedia communication in medicine industry for telemedicine. Multimedia communication in telemedicine is established by the use of electronic health record technology. Electronic health records (EHRs) like health history data of a patient, demographic data, physical examination data, laboratory test solutions, treatment measures and prescriptions are extremely confidential in nature. Information security is achieved by ensuring that the (i) transmitted medical images cannot be retrieved by unauthorized parties (Confidentiality), (ii) received images are not tampered during transmission (Integrity), and (iii) images are from original sources and reach the original receivers (Authentication).

S. Guhan · S. Arumugham · S. Janakiraman · A. Rengarajan · S. Rajagopalan (✉)
School of Electrical and Electronics Engineering, SASTRA Deemed to be University,
Thanjavur 613401, India
e-mail: raman@ece.sastra.edu

© Springer Nature Switzerland AG 2019
D. Pandian et al. (eds.), *Proceedings of the International Conference on ISMAC
in Computational Vision and Bio-Engineering 2018 (ISMAC-CVB)*, Lecture Notes
in Computational Vision and Biomechanics 30,
https://doi.org/10.1007/978-3-030-00665-5_106

In this paper, we have proposed a hybrid model which includes three major security techniques like encoding, watermarking and encryption that ensures the aforementioned prerequisites. Many compression techniques have been used world-wide to reduce size and traffic. Using lossless compression, the received image is as same as the transmitted image that is, no information is lost [1]. Many previous researches have shown that arithmetic encoding is the most optimum technique and achieves higher compression techniques than other lossless techniques. In this paper, arithmetic encoding is used to encode the image. Arithmetic encoding is variable length encoding technique that achieves lossless compression [2]. Arithmetic encoding encodes the target image into binary bit stream.

Many researchers have proposed many digital watermarking techniques. Kavitha and Shan [3] proposed a watermarking scheme for DICOM image using IWT. Alpha blending watermarking requires the receiver to have the original host image. This can be disadvantageous in most cases. Hence, a lot of research is happening in the bit plane watermarking [4]. Watermarking can be classified into two major areas. Many researches on watermarking suggest that frequency domain watermarking provides higher robustness and imperceptibility [5]. Due to the rapid development of the Internet, the need for digital product protection is on the rise. This concern led to the need for more secure digital watermarking technologies [6]. Recent researches in frequency domain watermarking show that IWT is computationally faster and provides better results than DWT. Despite many researches in this technology, very few researchers have proposed selective watermarking. Singh et al. [7] proposed multiple watermarking algorithms by using selective DWT coefficients for embedding. Malonia and Agarwal [8] elucidated an algorithm where watermarking is done on the DWT coefficients selected by arithmetic progression. The results from selective embedding show that the watermark has higher robustness and imperceptibility.

Hybridization of watermarking with encryption has proved that it can significantly increase in the medical image security [9]. Though there are many algorithms available for encryption, chaotic maps have drawn a lot of attention. Because of the nonlinearity and unpredictability, the chaotic maps provide researchers have focussed on using it for the encryption purpose. Safi and Maghari [10] examined the competence of a combination of two chaotic maps. The results are better and less complicated than many other proposals with similar results. Shyamala [11] proposed chaotic confusion—diffusion algorithm. It makes use of the fact that mathematical equations generated by chaotic maps are reversible and efficient enough to cause diffusion on the image pixels. Huang and Yang [12] introduced two encryption algorithm to randomize transforms, chaotic scrambling, chaotic permutation and chaotic diffusion. Wang et al. [13] proposed an image encryption using a combination of 1D chaotic maps to confuse the image. The experimental results showed that the scheme is resistive to different attacks.

2 Preliminaries

2.1 Lorenz Attractor (LA)

A system may tend to develop towards a set of numerical values for a wide variety of initial conditions. These set of numerical values are called attractors. Lorenz attractor is one of the attractors that provides a chaotic solution. Lorenz attractor is a set of chaotic solutions of the Lorenz system. Lorenz system is modelled by three ordinary differential equations. For system that behaves chaotic when the constants $\sigma = 10$; $\rho = 28$; $\beta = 8/3$ [14].

$$\frac{dx}{dt} = \sigma(y - x) \tag{1}$$

$$\frac{dy}{dt} = x(\rho - z) - y \tag{2}$$

$$\frac{dz}{dt} = (xy) - (\beta z) \tag{3}$$

2.2 Logistic Map

Chaotic maps are functions that are highly sensitive to initial conditions and exhibit chaotic behaviour. Logistic map is one such 1D map which exhibits chaotic behaviour from a simple nonlinear dynamic equation. Logistic map is given by

$$x_{n+1} = r x_n (1 - x) \tag{4}$$

x_n should be a number between 0 and 1. Logistic map becomes chaotic when the constant $r \sim 3.5699$. For different constant values and seed value x_0, the map shows different characteristics.

2.3 Tent Map

Tent map is another such 1D chaotic map that exhibits chaotic behaviour from a simple nonlinear dynamic equation. Tent map is given by

$$x_{n+1} = \mu x_n \quad \text{for } x_n < 1/2 \tag{5}$$

$$x_{n+1} = \mu x_n (1 - x_n) \quad \text{for } x_n \geq 1/2 \tag{6}$$

x_n should be a number between 0 and 1. When constant $\mu = 2$, the system maps the interval [0, 1] onto itself. The periodic points are dense in [0, 1], so the map becomes chaotic. For different constant values and seed value x_0, the map shows different characteristics.

2.4 Integer Wavelet Transform

Integer wavelet transform is used for its finite integer coefficients which makes the transform lossless. Due to the fact that all the coefficients of IWT are integers, there is no rounding off errors so the watermark can be extracted without any loss. Lifting scheme is implemented to perform IWT. Lifting scheme consists of three stages. They are split, predict and update.

Split: The main signal is decomposed into even sequence and odd sequence.
Predict: The numbers from one sequence are predicted on the basis of other sequence.
Update: Even samples are updated from the input even samples and the updated odd samples.

3 Proposed Work

The proposed method introduces randomness while embedding and encrypts the watermarked image taking advantage of various chaotic maps. In this work, a medical EHR 1 of size 128×128 is taken as secret image and a host 8 bit DICOM of size 512×512 is taken for validations. The use of Lorenz attractors, logistic maps and tent map for watermarking and encryption is explained in the detailed block diagram Fig. 1.

3.1 Watermarking

i. Secret image is encoded using arithmetic encoding technique which gives binary bit stream of length N.
ii. Permutation operation is performed when binary sequence is left cyclic shifted by a seed value.
iii. The bit stream is divided into three arrays such that
$(1 - i), ((i + 1) - j)$ and $((j + 1) - N)$
where $1 \leq i, j \leq N$ and $i \neq j$.
iv. Host image of size $M \times N$ is decomposed using integer wavelet transform into its sub-bands LL, LH, HL and HH each of size $M/2 \times N/2$.
v. The HH sub-band is subdivided into three segments say HH_1, HH_2 and HH_3.

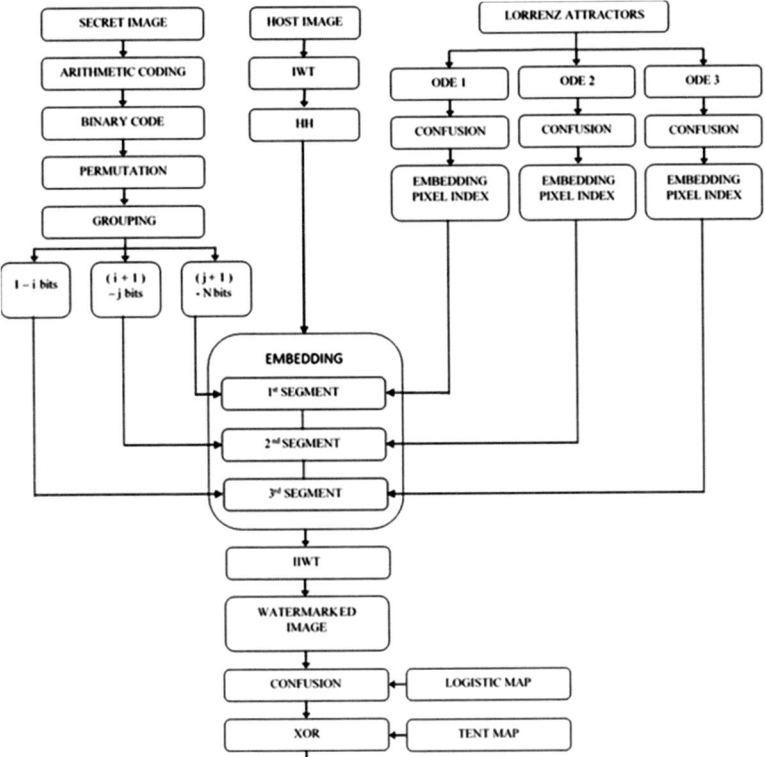

Fig. 1 Detailed block diagram of the proposed method

vi. Lorenz attractor (LA) ordinary differential Eqs. (1)–(3) are used to generate pseudo random number sequence

$$S_{LA1} = \left(S_1, S_2, \ldots, S_{M/6 \times N/6} \right)$$
$$S_{LA2} = \left(S_{M/6+1}, S_{M/6+2}, \ldots, S_{M/3 \times N/3} \right)$$
$$S_{LA3} = \left(S_{M/3+1}, S_{M/3+2}, \ldots, S_{M/2 \times N/2} \right)$$

vii. Perform sorting in ascending order on the pseudo random sequence S_{LA1}, S_{LA2} and S_{LA3} as follows:

$$[I_{s1}, Y_{s1}] = \text{sort}(S_{LA1})$$
$$[I_{s2}, Y_{s2}] = \text{sort}(S_{LA2})$$
$$[I_{s3}, Y_{s3}] = \text{sort}(S_{LA3})$$

Where $[\bullet, \bullet] = $ sort $(\bullet)$ is the sequence indexing function; I_s is the new index after sorting the data S and Y_s is the sorted sequence after sorting S.

viii. Each secret array is embedded into the first bit plane of the coefficients in the I_s position of each segments.

ix. Perform inverse integer wavelet transform on the LL, LH, HL and modified HH sub-bands to get the final watermarked image of size $M \times N$.

3.2 Encryption

i. Logistic chaotic map Eq. (4) is used to generate pseudo random number sequence

$$S_{\text{logistic}} = (S_1, S_2, \ldots, S_{M \times N})$$

which forms the key 1.

ii. Perform sorting in ascending order on the pseudo random sequence S_{logistic} such that

$$\left[I_{\text{logistic}}, Y_{\text{logistic}} \right] = \text{sort} \left(S_{\text{logistic}} \right)$$

where I_{logistic} is the new index after sorting the data S_{logistic} and Y_{logistic} is the sorted sequence after sorting S_{logistic}.

iii. Perform confusion on the pixels of watermarked image of size $(M \times N)$ such that

$$\text{Logistic}(i) \Rightarrow \text{Watermarked} \left(I_{\text{logistic}}, (i) \right),$$

where $1 \leq i \leq (M \times N)$

This results in first level chaotic encryption.

iv. Scale up Watermarked$_{\text{logistic}}$ by using, $X_{\text{scaled}} = \text{mod}(\text{Watermarked}_{\text{logistic}} \times 10^{14}, 256)$.

v. Tent chaotic map Eq. (5) is used to generate pseudo random number sequence

$$S_{\text{tent}} = (S_1, S_2, \ldots S_{M \times N})$$

which forms the key 2.

vi. Perform XOR to diffuse the confused image with key 2. This results in second level chaotic encryption and final encrypted image

$$\text{Image}_{\text{enc}} = X_{\text{scaled}} \oplus S_{\text{tent}}.$$

vii. Decryption and extraction are the reverse processes of encryption and watermarking, respectively.

4 Results and Discussion

The proposed work includes three modules namely compression, watermarking and encryption processes. The compression modules are validated by calculating compression ratio and average length. The watermarking module is validated by calculating MSE, PSNR and SSIM. The encryption module is validated by calculating correlation coefficients, entropy, histogram analysis, NPCR analysis and key space analysis. The simulation work of the proposed work is done using MATLAB R2016b in a system with 4 GB RAM, 500 GB Hard drive, Intel Core i3 processor and Windows 8.1 operating system. Figure 2 shows the final output image of each module.

The same secret image as shown in Fig. 2a is watermarked within different DICOM images and encrypted. The DICOM images and its corresponding encrypted outputs are shown in Fig. 3.

4.1 Compression Ratio

Compression Ratio is defined as ratio of number of bits in the original image and number of bits in the compressed image. Compression ratio is used to validate the efficiency of the used compression technique. Compression ratio is given by

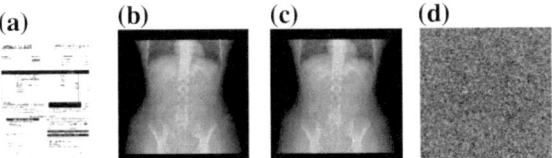

Fig. 2 **a** EHR 1, **b** original DICOM 1, **c** watermarked DICOM image, **d** encrypted image ENC 1

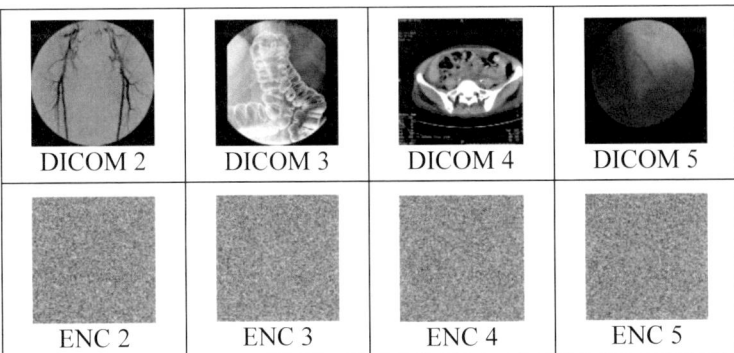

Fig. 3 Host DICOM images and its corresponding cipher images

Table 1 Compression ratio and average length

Image	Compression ratio	Average length
EHR 1	1.6753	4.7754
EHR 2	1.5039	5.3196
EHR 3	1.3930	5.7429
EHR 4	1.3905	5.7532
EHR 5	1.8143	4.4094

$$\text{Compression ratio} = \frac{\text{Number of bits in the original image}}{\text{Number of bits in the compressed image}}$$

4.2 Average Length

Average length is defined as the average of number of bits required to encode a symbol in a sequence. Arithmetic coding uses less number of bits for symbols with high probability. The compression ratio and average length of various secret images are validated and shown in Table 1.

4.3 Mean Square Error

Error metrics such are used to compare two image processing techniques. Mean square error is one such metric which is used to validate the proposed technique. Mean square error is the cumulative squared error between the original image and watermarked image. The ideal value of MSE between the two images is 0. Mean squared error is given by

$$\text{MSE} = \frac{1}{MN} \sum_{y=1}^{M} \sum_{x=1}^{N} \left[I(x, y) - I'(x, y) \right]^2$$

x Row
a Total number of rows
y Column
b Total number of columns.

Table 2 MSE, PSNR and SSIM values

Image	MSE	PSNR	SSIM
DICOM 1	0.0284	63.5999	0.8703
DICOM 2	0.0281	63.6386	0.9420
DICOM 3	0.0281	63.6504	0.9079
DICOM 4	0.0283	63.6122	0.9386
DICOM 5	0.0282	63.6251	0.8931

4.4 Peak Signal-to-Noise Ratio (PSNR)

Peak signal-to-noise ratio is the ratio of peak power of signal to power of noise in an image. PSNR is calculated between the watermarked image and the original image. The ideal value of PSNR is infinity. The mathematical formula for PSNR is

$$PSNR = 20 * \log_{10}\left(\frac{MAX}{\sqrt{(MSE)}}\right)$$

MAX   Maximum possible pixel value of the image.

4.5 Structural Similarity Index

Structural similarity index is a full reference index used to measure the similarity between two images. The ideal value of SSIM is 1. SSIM is measured between two images of same size using

$$SSIM = \frac{(2\mu_x\mu_y + c_1)(2\sigma_{xy} + c_2)}{(\mu_x^2 + \mu_y^2 + c_1)(\sigma_x^2 + \sigma_y^2 + c_2)}$$

μ_x average of x
σ_x variance of x
σ_{xy} covariance of x and y
μ_y average of y
σ_y variance of y
c_1, c_2 constants to stabilize the division.

From Table 2, the proposed method gives MSE nearly equal to 0, PSNR value is above the acceptable range of 40 dB and the SSIM value is nearly 1 showing that the proposed method is much better than the previous works.

Table 3 Correlation coefficients and entropy of DICOM images

Images	Entropy		Correlation coefficients			
	Original	Cipher	Direction	Original	Cipher	Ref. [15]
DICOM 1	6.4444	7.9907	H	0.9868	−0.0019	0.0014
			V	0.9892	0.0040	−0.0009
			D	0.9800	0.0004	0.0065
DICOM 2	4.9982	7.9778	H	0.9777	−0.0006	0.0081
			V	0.9835	0.0024	−0.0039
			D	0.9707	0.0006	0.0030
DICOM 3	5.7158	7.9536	H	0.9872	0.0032	−0.0032
			V	0.9859	−0.0017	0.0016
			D	0.9768	0.0020	−0.0029
DICOM 4	3.8657	7.9714	H	0.9131	−0.0007	0.0075
			V	0.9172	−0.0005	0.0142
			D	0.8713	0.0001	0.0061
DICOM 5	4.7728	7.9650	H	0.9972	0.0017	NA
			V	0.9958	0.0024	
			D	0.9962	−0.0027	

4.6 Entropy

Entropy measures the randomness among the pixels of an image. The ideal value of an 8-bit image is 8. The proposed method gives entropy value close to 8 as shown in Table 3.

$$H = -\sum_{i=1}^{n} P_i \times \log_2 P_i$$

H Entropy of the Image
N Grey level of an input image (0–255)
P_i Probability of the occurrence of symbol i.

4.7 Correlation Coefficient

Correlation coefficient is statistical metric used to validate the encrypted image. It is a numerical measure of relationship between two adjacent pixels of an image. For an original image, the coefficient is 1 which indicates that there is strong agreement between the pixels. And for an encrypted image, the value should be very less. Figure 4 shows the correlation graph for a test image. The statistical validation of the proposed method is proved from Table 3. The metrics are compared with an existing algorithm to prove its efficiency.

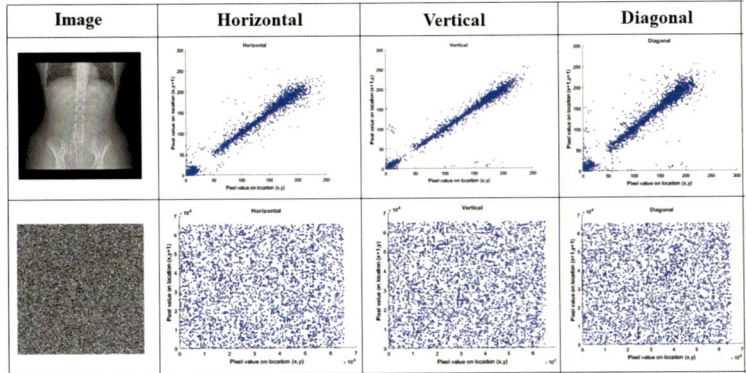

Fig. 4 Correlation graph of plain image and encrypted image

Fig. 5 Histogram analysis of plain and cipher image of DICOM 1

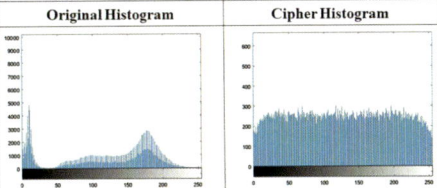

$$x = \frac{n\left(\sum ab\right) - \left(\sum a\right)\left(\sum b\right)}{\sqrt{\left[n\sum a^2 - \left(\sum a\right)^2\right]\left[n\sum b^2 - \left(\sum b\right)^2\right]}}$$

x	Correlation Coefficient
n	Total number of Pixels
a and b	Adjacent pixels
H	Horizontal
V	Vertical
D	Diagonal.

4.8 Histogram Analysis

Histogram is the accurate representation of distribution of pixels of an image. Histogram is used to measure the intensity levels and its frequency of an image. From Fig. 5, it is clear that the original image has different levels of intensity and the encrypted image obtained from the proposed method has a flat response.

Table 4 NPCR analysis

Image	NPCR	Ref. [15]
DICOM 1	99.9866	99.6240
DICOM 2	99.8497	99.2360
DICOM 3	99.7509	99.1382
DICOM 4	99.9943	99.2360
DICOM 5	99.7757	99.6076

4.9 NPCR Analysis

Number of pixel change rate (NPCR) is a measure used to test the influence of plain image over the encrypted image. A perfect encryption algorithm is highly resistive to differential attacks and possesses better diffusion property. These values are calculated for different test images and shown in Table 4.

4.10 Key Space Analysis

Key space analysis is to check the feasibility of brute force attack. When the cipher image is decrypted using all possible keys, it is called brute force attack. The key space should be large enough to withstand the brute force attack. For an encryption algorithm with 128-bit key, it requires 1.3736×10^{31} years to decrypt the algorithm. The secret keys used in this proposed method includes Lorenz attractors, logistic map, tent map and a seed value for cyclic shift. The total number of initial conditions required for Lorenz attractors (6), logistic map (2) and tent map (2) is 10. The seed value N used for cyclic shift operation also makes up for key space. Hence, the proposed method achieves a key space of $N \times 10^{170} > 2^{128}$ which is large enough to resist the brute force attack. Comparing to previous works [10] which has a key space of 2^{68} and [13] which has a key space of 2^{128} the proposed work is highly secure.

5 Conclusion

This paper is aimed to produce a comprehensive image hiding and image encryption taking advantage of all available algorithms. Based on the results obtained, it is clear that combining major image processing techniques like compression, watermarking and encryption to form a hybrid model gives more security than many previous models. This method provides a more secure and reliable means of communication in the telemedicine industry by ensuring the CIA triad.

References

1. Shahbahrami A, Bahrampour R, Rostami MS, Ayoubi M (2011) Evaluation of Huffman and arithmetic algorithms for multimedia compression standards. Int J Comput Sci Eng Appl 1(4):34–47
2. Mukesh PS, Pandya MS, Pathak S (2013) Enhancing AES algorithm with arithmetic coding. In: Proceedings of the 2013 international conference green computing, communication and conservation of energy, ICGCE 2013, pp 83–86
3. Kavitha KJ, Shan BP (2017) Implementation of DWM for medical images using IWT and QR code as a watermark. In: 2017 Conference on emerging devices and smart systems, ICEDSS 2017, Mar 2017, pp 252–255
4. Tyagi S, Singh HV, Agarwal R, Gangwar SK (2016) Digital watermarking techniques for security applications. In: 2016 International conference on emerging trends in electrical electronics and sustainable energy systems, pp 379–382
5. Zhang Y (2009) Digital Watermarking technology: a review. In: 2009 ETP international conference on future computer and communication, pp 250–252
6. Gupta P, Parmar G (2017) Image watermarking using IWT-SVD and its comparative analysis with DWT-SVD. In: 2017 International conference on computer, communication and electronics COMPTELIX 2017, pp 527–531
7. Singh AK, Kumar B, Dave M, Mohan A (2015) Multiple watermarking on medical images using selective discrete wavelet transform coefficients. J Med Imaging Health Inform 5(3):607–614
8. Malonia M, Agarwal SK (2016) Digital image watermarking using discrete wavelet transform and arithmetic progression technique. In: 2016 IEEE students' conference on electrical, electronics and computer science, pp 1–6
9. Aparna P, Kishore PVV (2018) An efficient medical image watermarking technique in e-healthcare application using hybridization of compression and cryptography algorithm. J Intell Syst 27(1):115–133
10. Safi HW, Maghari AY (2017) Image encryption using double chaotic logistic map. In: International conference on promising electronic technologies, pp 66–70
11. Shyamala N (2017) Reversible chaotic encryption techniques for images. pp 2–6
12. Huang H, Yang S (2016) Colour image encryption based on logistic mapping and double random-phase encoding. IET Image Process 11(4):211–216
13. Wang H, Xiao D, Chen X, Huang H (2018) Cryptanalysis and enhancements of image encryption using combination of the 1D chaotic map. Sig Process 144:444–452
14. https://en.wikipedia.org/w/index.php?title=Lorenz_system&oldid=833739483
15. Chandrasekaran J, Thiruvengadam SJ (2017) A hybrid chaotic and number theoretic approach for securing DICOM images. Secur Commun Netw 2017

A Novel Hybrid Method for Time Series Forecasting Using Soft Computing Approach

Arpita Sanghani, Nirav Bhatt and N. C. Chauhan

Abstract Improving the forecasting accuracy of time series is important and has always been a challenging research domain. From many decades, Auto-Regressive Integrated Moving Average (ARIMA) has been popularly used for statistic forecasting however it will solely forecast linear half accurately because it cannot capture the nonlinear patterns. Therefore here, we have projected a hybrid model of ARIMA and SVM. As Support Vector Machine (SVM) has demonstrated great outcomes in solving nonlinear regression estimation problems and to utilize the linear strength of ARIMA. Comparison with other models using different datasets has been done and the results are very promising.

1 Introduction

Time series prediction is an endlessly growing analysis domain. The accuracy of prediction is the main goal to realize. Prediction future is the best exploitation by time series prediction. With the statistic prediction, past data assortment of the constant variable is used to build a model to forecast long-term accessibility of those data. Then established model is utilized thus on extrapolate the statistic into the longer term. This modeling approach is very useful once little information is obtainable regarding past variables and no different things are thought [1–5].

A. Sanghani (✉)
Department of Computer Engineering, BVM Engineering College, Vallabh Vidyanagar 388120, Gujarat, India
e-mail: arpita.sanghani@gmail.com

N. Bhatt
Department of Information Technology, CSPIT, CHARUSAT, Anand 388421, Gujarat, India
e-mail: niravbhatt.it@charusat.ac.in

N. C. Chauhan
Department of Information Technology, A.D. Patel Institute of Technology, New V.V. Nagar, Anand 388121, Gujarat, India
e-mail: narendracchauhan@gmail.com

© Springer Nature Switzerland AG 2019
D. Pandian et al. (eds.), *Proceedings of the International Conference on ISMAC in Computational Vision and Bio-Engineering 2018 (ISMAC-CVB)*, Lecture Notes in Computational Vision and Biomechanics 30,
https://doi.org/10.1007/978-3-030-00665-5_107

First, only the ARIMA model was used and was popular for time series forecasting but it was observed that individual models were not capable to handle nonlinearity. So, combine modeling came into the picture. Many combine models were introduced. And the results of some of them were also better than the individual. ARIMA works well for linear forecasting but it has shown a drawback of not giving good results when applied to nonlinear data. And after that ARIMA + ANN hybrid model came into the picture with various versions and it started giving goods results for the different type of time series data. But neural network also has some drawback and as we know that SVM is a good alternative to ANN so there is still scope for forecasting results improvement [6–8].

Our purpose in this paper is to do build a novel hybrid model for time series forecasting using ARIMA and a unique soft computing and Support Vector Machines (SVM) based on machine learning. The idea for exploiting SVM came from literature survey [9]. This technique is used to accurately forecast statistic knowledge and the processes of the system are usually nonlinear, nonstationary and not highlighted earlier. SVMs additionally verified to beat alternative nonlinear ways along with neural network primarily focused on nonlinear prediction ways like multilayer perceptrons. The detailed paper is organized as follows. In Sect. 1, introduction to recent time series trends, Sect. 2 describes the traditional model ARIMA of time series. Section 3 describes regarding soft computing approach model for time series Support Vector Machine (SVM). Section 4 gives detail about the proposed model. Section 5 shows the dataset and results then concludes this work by the conclusion section of the proposed work and future goals.

2 Auto-Regressive Integrated Moving Average (ARIMA) Model

For statistic prediction initially of all ancient applied mathematics models were developed like moving average, exponential smoothing, and autoregressive integrated moving average. Among all, ARIMA became well liked for linear forecasting [1, 10–13].

In this strategy, the given measuring information is introductory checked for stationarity and if don't seem to be then the differencing operation is performed. On the off likelihood that the data are still nonstationary, differencing is a new performed until the information are eventually created stationary [1, 10]. On the other hand differencing is carried out d times, the combination request of the ARIMA philosophy said to be d. The resultant output is sculptural as associate autoregressive moving average (ARIMA) measuring as takes after [1, 10–12, 14, 15]. The information value at any denoted time t, say y_t, is taken under consideration as a function of the earlier p information values, say $y_{t-1}, y_{t-2}, \ldots, y_{t-p}$ and therefore the errors at times $t, t-1, \ldots, t-q$ say $n_t, n_{t-1}, \ldots, n_{t-q}$. The generated ARMA equation is shown in Eq. 1 [1, 2, 10, 12, 13, 16, 17]. In Eq. 1, a_1 to a_p be the autoregressive (AR) coefficients and b to b_q comprises the MA coefficients. So the statistic model is denoted as ARIMA (p, d, q) because it is a mixture of AR and MA. The ARMA model accept that the blunder arrangement n_t is common and that is Gaussian circulated, the difference of mistake is furthermore a model parameter [1, 10] The ARIMA displaying technique consists of three stages: (a) recognizing the model request, i.e., recognizing p and q; (b) assessing the model coefficients; (c) forecast the data [1, 10, 12–14, 16].

The model coefficients are assessed utilizing the Box–Jenkins strategy. At long last, the model coefficients are measurable and the future estimations of the measurement are anticipated utilizing the accessible past data values and hence the model coefficients. ARIMA models anticipate straight measurement data with wonderful accuracy [1, 10–14, 17–20].

$$y_t = a_1 y_{t1} + a_2 y_{t2} + \cdots + a_p y_{tp} + n_t + b_1 n_{t1} + \cdots + b_q n_{tq} \tag{1}$$

In the autoregressive integrated moving average model, the long-term value of a variable is thought to be a linear operator of numerous past observations and random errors [1, 2, 10, 12, 17, 18, 20–22]. ARIMA consolidates an assumption that the real information is regularly straight this suspicion is a reward and disadvantage each for ARIMA because it offers smart forecast result for linear statistic and it cannot deal with nonlinear data [1, 2, 10–12, 17, 18, 20, 21, 23–25].

3 Soft Computing Approach Models

Soft computing is becoming extremely popular and it is getting used in all the fields for obtaining sensible results. And as from literature survey, it proves that ancient models do not seem to be capable to handle nonlinear statistic knowledge properly. Thus, soft computing and intelligence system are a smart alternative for statistic forecasting [26–29].

3.1 Support Vector Machine (SVM) Model

Soft computing approach is Support Vector Machine here. After the introduction of an alternative called loss function, we have use SVM for time series forecasting also [6, 8, 20, 30–34]. The advantage of SVM to resolve nonlinear data problem proves it capable to use for statistic forecasting [32, 33]. The basic plan of SVM as shown in Fig. 1 plots the data x to high-dimensional feature space F by nonlinear plotting space so do linear regression [6, 8, 20, 30–38]. Suppose we have given set of training that consists of n data points $G = \{x_i, d_i\}_{i=1}^n$ [34] along with input data $x_i \in R^p$, p consists of a total number of patterns of data and data generated and $d_i \in R^p$ is the result [34].

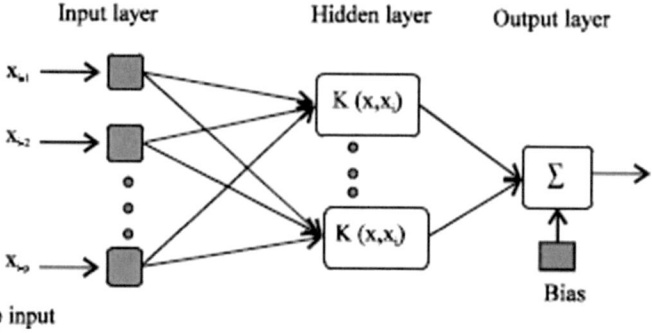

Fig. 1 The architecture of SVM (*Source* [33, p. 952])

The SVM regression as shown in the papers [6, 8, 30, 32, 34, 36] can be calculated approximately by the below function Eq. 2

$$f(x) = w\phi(x) + b, \quad \phi : R^n \rightarrow F, w \in F \, [32, 36] \tag{2}$$

where b denotes scalar threshold [36]; ϕ denotes the high-dimensional feature house that is nonlinearly mapped from the input space x. Thus, the regression inside the high-dimensional feature house corresponds to nonlinear regression in low dimension input house, which ignores the real number computation among w and ϕ in the high-dimensional feature house. The coefficients w and b calculable by minimizing the regularized perform as shown in Eq. 3 [8, 20, 31, 32, 36].

$$R(C) = \frac{1}{2}||w||^2 + \frac{C}{n}\sum_{i=1}^{n} L_\varepsilon(d_i, y_i) \tag{3}$$

where

$$L_\varepsilon(d_i, y_i) = \begin{cases} |d_i - y_i| - \varepsilon, & \text{if } |d_i - y_i| \geq \varepsilon \\ 0, & \text{otherwise} \end{cases} \tag{4}$$

To find out the assessment of w and b Eq. 3 is converted to the primal function generated by Eq. 5 by introducing the new positive slack variable ξ and ξ^* as given [5, 6, 34, 36]:

$$\text{Mininmize } R(w, \xi^*) = \frac{1}{2}||w||^2 + \frac{C}{n}\sum_{i=1}^{n}(\xi_i + \xi_i^*) \tag{5}$$

$$\text{subject to} \begin{cases} d_i - w\phi(x_i) - b_i \le \varepsilon + \xi_i \\ w\phi(x_i) + b_i - y_i \le \varepsilon + \xi_i^* & [39] \\ \xi_i \xi_i^* \ge 0 \end{cases}$$

The term $\frac{1}{2}||w||^2$ denotes the weights vector norm, d_i denotes the required output and C defined as regularized constant [33, 34, 36]. Finally, by introducing Lagrange multipliers as given in [30] and expanding the optimality constraints, the choice operate generated by the equivalent weight of Eq. 2 denotes the subsequent express form [40]

$$f\left(x, \alpha_i, \alpha_i^*\right) = \sum_{i=1}^{n} \left(\alpha_i - \alpha_i^*\right) K\left(x, x_i\right) + b \, [39] \tag{6}$$

In Eq. 6, α_i and α_i^* are called Lagrange multipliers [39]. They fulfill the $\alpha_i \times \alpha_i^* = 0$ and $\alpha_i \ge 0$ and $\alpha_i^* \ge 0$ where $i = 1, 2, \ldots, n$ [30, 35, 40] and are generated by maximizing the dual function of Eq. 7, and the maximal dual function generated in Eq. 7 that denotes the below form [36]:

$$R\left(\alpha_i, \alpha_i^*\right) = \sum_{i=1}^{n} d_i \left(\alpha_i - \alpha_i^*\right) - \varepsilon \sum_{i=1}^{n} \left(\alpha_i + \alpha_i^*\right)$$
$$- \sum_{i=1}^{n} \cdot \sum_{j=1}^{n} \left(\alpha_i - \alpha_i^*\right)\left(\alpha_j - \alpha_j^*\right) K\left(x_i, x_j\right) \tag{7}$$

With constraints, [36]

$$\sum_{i=1}^{n} \left(\alpha_i - \alpha_i^*\right) = 0, \quad \begin{matrix} 0 \le \alpha_i \le C & i = 1, 2, \ldots, n \\ 0 \le \alpha_i^* \le C & i = 1, 2, \ldots, n \end{matrix} \, [36]$$

$K\left(x_i, x_j\right)$ denotes kernel function [39]. The output of the kernel function is equal to join multiplication of two vectors x_i and x_j inside the feature house $\phi(x_i)$ and $\phi(x_j)$, i.e., $K\left(x_i, x_j\right) = \phi(x_i)$ times $\phi(x_j)$ [39–41]. Generally, Gaussian function denotes kernel function and the equation is shown as below [32, 36]

$$K\left(x_i, x_j\right) = -\exp\left(\frac{||(x_i - x_j)||^2}{2\sigma^2}\right) \, [36, 39] \tag{8}$$

4 Proposed Work

In this paper, a fresh novel hybrid model has been planned. The system includes ARIMA and SVM. Auto-Regressive Integrated Moving Average (ARIMA) is con-

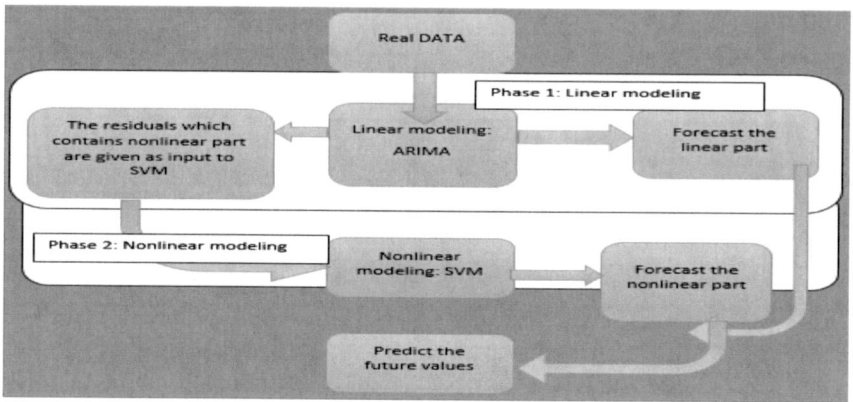

Fig. 2 Proposed model

sidered as one of the popular linear models in statistic prediction throughout the beyond three decades [21, 22]. Current analysis work in prediction along with Support Vector Machine (SVM) suggests that SVM are usually a promising completely different to the traditional linear ways. ARIMA models and SVM together typically compared along with mixed conclusions on the basis of the prevalence for predicting performance [21, 22]. In this paper, a hybrid approach that mixes each ARIMA and SVM model is planned to find out the benefit of the distinctive power of ARIMA and SVM models in linear and nonlinear modeling. Many ARIMA and ANN hybrid models have been planned early which they have given higher results but the accuracy of the model remains a research gap and research tells that SVM could also be a better completely different to ANN so we tend to use SVM in our system. The planned model diagram is shown below in Fig. 2. The proposed model is divided into two phase:

1. Linear Modeling
2. Nonlinear Modeling

Linear modeling

Here, real data is made up of both linear and nonlinear part is taken as input as shown in Eq. 9:

$$Y_{t=L_t+N_t} \tag{9}$$

where Y_t is real data, L_t denotes linear part, and N_t denotes the nonlinear part of data [42]. The assumption is made that real data is a combination of linear and nonlinear data as shown in Eq. 10. Then **Step 1:** First ARIMA is used to design the linear part of data and it produces two outputs: linear forecast values and the residuals.

$$L_t = \left[\sum_{i=1}^{p} \varphi_i Z_{t-i} - \sum_{j=1}^{q} \theta_j \varepsilon_{t-j} \right] + \varepsilon_t + e_t = L_t^{\wedge} + e_t \qquad (10)$$

where $L_t^{\wedge}$ is the linear part is forecast value and e_t is the residual part [36] which can be calculated using Eqs. 11 and 13:

$$e_t = Y_t - L_t^{\wedge} \qquad (11)$$

Step 2: This residual contains the nonlinear data is given as input to SVM for nonlinear modeling.

Nonlinear modeling

Step 3: Now SVM is used for nonlinear modeling as is best for nonlinear modeling. The residuals from ARIMA are given as input to SVM and it produces the output as forecast values for the nonlinear part by the Eq. 12 as shown below.

$$N_t^{\wedge} = f(e_{t-1}, \dots, e_{t-n}) + \varepsilon_t \qquad (12)$$

Step 4: Now to predict the future the forecast values from ARIMA and SVM are combined and given as input as shown in Eq. 13 and it predicts future values.

$$Y_t^{\wedge} = L_t^{\wedge} + N_t^{\wedge} \qquad (13)$$

Step 5: After that, the accuracy of the model is checked by performances measures like MSE.

5   Dataset and Result Analysis

Here, the derived model has been checked for different datasets and the outputs are compared with other models outputs too. The proposed model work is done using R programming.

5.1   Sunspot Data

The sunspot data [3, 10, 11, 21] have been used that consists of the yearly number of sunspots detected on the face of sun ranging from 1700 to 1987, consisting of a total of 288 observations. The plot of particular time series is shown in Fig. 3 says that there is a cyclical pattern that denotes a mean cycle of about 11 years [21, 22].

Fig. 3 Sunspot data R plot

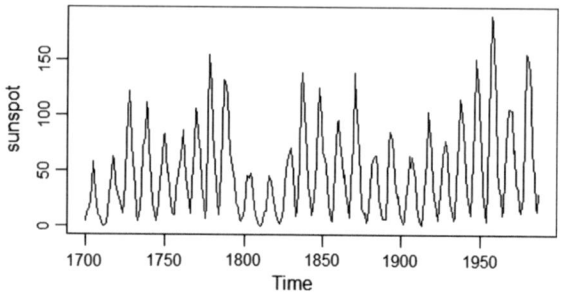

Fig. 4 Lynx data R plot

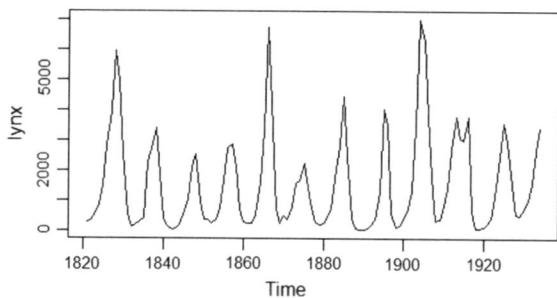

5.2 Canadian Lynx Dataset

The lynx series [10, 11, 13, 18, 22] consists of the total number of lynx collected per year in the Mackenzie River district of Northern Canada. The lynx data are plotted in Fig. 4 that denotes time period approximately up to 10 years. The datasets consist of total 114 observations, referring to the period of 1821–1934 [21, 22].

5.3 US Monthly Electricity Dataset

The US monthly electricity series [11, 17] contains monthly electricity net generation measured in billions of kilowatt hours (kWh) from January 1973 to October 2010. The dataset has total 454 observations. The plot of series is shown in Fig. 5.

And from the graphs Figs. 6, 7 and 8. And tables of different model comparison for 3 datasets, i.e., Tables 1, 2 and 3 we can see that the proposed model has shown promising accuracy over past models.

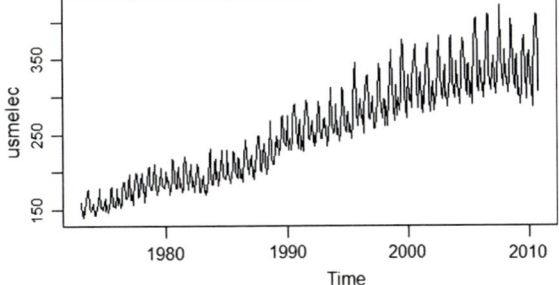

Fig. 5 US monthly electricity data R plot

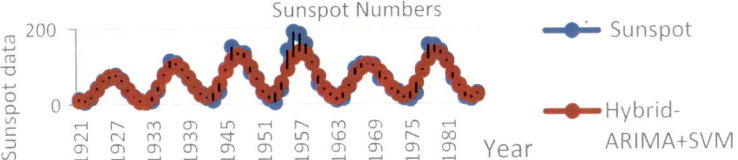

Fig. 6 Sunspot data proposed model R graph

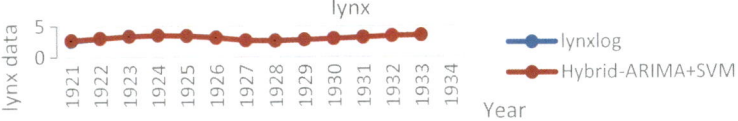

Fig. 7 Lynx data proposed model R graph

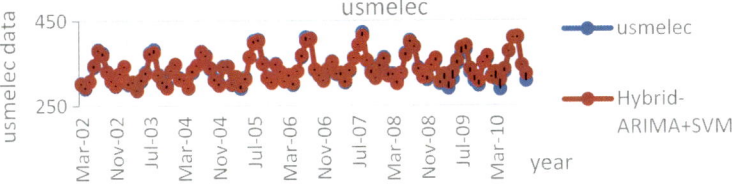

Fig. 8 US electricity data proposed model R graph

6 Conclusion and Future Work

And the results are outstanding for novel hybrid proposed model compared to previously proposed models. The proposed model works well with all type of time series datasets and also gives more accurate results as we can see in tables and graphs above. Thus, SVM can be used as a good alternate of various other models for a hybrid model to get best time series results. But still, SVM parameter selection is an open research gap and needs some good work for better selection process as if

Table 1 Sunspot data result in comparison

Sunspot [221:67] = 288

Model	MSE	MAE
ARIMA	253.87	17.66
ANN	315.79	12.26
ARIMA ANN [1]	290.19	12.78
ARIMA ANN [2]	240.20	12.10
SVM	285.91	12.00
ARIMA SVM	206.52	11.66

Table 2 Lynx data result in comparison

Lynx [100:14] = 114

Model	MSE	MAE
ARIMA	0.025	0.123
ANN	0.034	0.148
ARIMA ANN [1]	0.021	0.115
ARIMA ANN [2]	0.014	0.096
SVM	0.076	0.070
ARIMA SVM	0.012	0.019

Table 3 US electricity data result comparison

Usmelec [350:104] = 454

Model	MSE	MAE
ARIMA	44.80	4.939
ANN	49.74	5.427
ARIMA ANN [1]	78.95	6.696
ARIMA ANN [2]	79.98	7.654
SVM	119.70	6.451
ARIMA SVM	**36.63**	**4.376**

parameters are not selected properly SVM can lead to poor results. Further in the future, we can try other soft computing methods like ANFIS to achieve still more accurate results.

References

1. Babu CN, Reddy BE (2014) A moving-average filter based hybrid ARIMA–ANN model for forecasting time series data. Appl Soft Comput 23:27–38
2. Li C, Chiang T-W (2013) Complex neurofuzzy ARIMA forecasting—a new approach using complex fuzzy sets. Fuzzy Syst IEEE Trans 21:567–584
3. Zounemat-Kermani M, Teshnehlab M (2008) Using adaptive neuro-fuzzy inference system for hydrological time series prediction. Appl Soft Comput J 8:928–936

4. Khashei M, Bijari M (2010) An artificial neural network (p, d, q) model for timeseries fore-casting. Expert Syst Appl 37:479–489
5. Chen K-Y, Wang C-H (2007) A hybrid SARIMA and support vector machines in forecasting the production values of the machinery industry in Taiwan. Expert Syst Appl 32:254–264
6. Sapankevych N, Sankar R (2009) Time series prediction using support vector machines: a survey. IEEE Comput Intell Mag 4:24–38
7. Sanghani A, Bhatt N, Chauhan NC (2016) A review of soft computing techniques for time series forecasting. Indian J Sci Technol 9
8. Agrawal RKRA (2013) An introductory study on time series modeling and forecasting. CoRR 1302.6613
9. CiteULike Everyone's library. The underlying motivation for exploiting SVM is that the ability of this technique to accurately forecast statistic knowledge once the underlying system pro-cesses are usually nonlinear, non-stationary and not outlined a-priori. Retrieved from http://www.citeulike.org/
10. Babu CN, Reddy BE (2014) A moving-average filter based hybrid ARIMA–ANN model for forecasting time series data. Appl Soft Comput 23:27–38
11. Wang L, Zou H, Su J, Li L, Chaudhry S (2013) An ARIMA-ANN hybrid model for time series forecasting. Syst Res Behav Sci 30:244–259
12. Khashei M, Bijari M (2011) A novel hybridization of artificial neural networks and ARIMA models for time series forecasting. Appl Soft Comput 11:2664–2675
13. Zhang GP (2003) Time series forecasting using a hybrid ARIMA and neural network model. Neurocomputing 50:159–175
14. Li CS, Chiang TW (2013) Complex neurofuzzy ARIMA forecasting—a new approach using complex fuzzy sets. IEEE Trans Fuzzy Syst 21:567–584
15. Sakurai Y, Faloutsos C (2015) Mining and forecasting of big time-series data. 919–922
16. Sciences RB, Sato RC (2013) Disease management with ARIMA model in time series Geren-ciamento de doenças utilizando séries temporais com o modelo. ARIMA 11:128–131
17. Khashei M, Bijari M (2010) An artificial neural network (p, d, q) model for timeseries fore-casting. Expert Syst Appl 37:479–489
18. Khashei M, Bijari M (2011) A novel hybridization of artificial neural networks and ARIMA models for time series forecasting. Appl Soft Comput J 11:2664–2675
19. De Gooijer JG, Hyndman RJ (2006) 25 years of time series forecasting. Int J Forecast 22:443–473
20. Goel N, Singh S, Aseri TC, Goel N, Singh S, Aseri TC (2013) A review of soft computing techniques for gene prediction. ISRN Genomics 2013:1–8
21. Zhang GP (2003) Time series forecasting using a hybrid ARIMA and neural network model. Neurocomputing 50:159–175
22. Managed DNS | Dyn. Retrieved from https://dyn.com/dns/
23. Khashei M, Bijari M (2012) A new class of hybrid models for time series forecasting. Expert Syst Appl 39:4344–4357
24. Peng H-W, Wu S-F, Wei C-C, Lee S-J (2015) Time series forecasting with a neuro-fuzzy modeling scheme. Appl Soft Comput 32:481–493
25. Wen X, Academy C (2015) A comparative study of artificial neural network, adaptive neuro fuzzy
26. Mandal SN, Choudhury JP, Chaudhuri SRB, De D (2008) Soft computing approach in predic-tion of a time series data. J Theor Appl Inf Technol 8:1131–1141
27. Bautu E, Barbulescu A (2013) Forecasting meteorological time series using soft computing methods: an empirical study. Appl Math 7:1297–1306
28. Meryem O, Ismail J, Mohammed E-M (2014) A comparative study of predictive algorithms for time series forecasting. In: 2014 third ieee international colloquium in information science and technology (CIST), pp 68–73
29. Rojas I, Palomares H (2004) Soft-computing techniques for time series forecasting. In: Pro-ceedings of the European symposium on artificial neural networks, pp 93–102

30. Xiang L, Zhu Y, Tang G (2009) A hybrid support vector regression for time series forecasting. In: WRI world congress on software engineering, WCSE'09, pp 161–165
31. Gunn SR (1998) Support vector machines for classification and regression by
32. Xiang L, Tang G, Zhang C (2008) Simulation of time series prediction based on hybrid support vector regression. In: Fourth international conference on natural computation, 2008. ICNC'08, pp 167–171
33. Samsudin R (2010) A comparison of time series forecasting using support vector machine and artificial neural network model. J Appl Sci 10:950–958
34. Welcome to Science Alert. Retrieved from http://scialert.net/
35. Chen K-Y (2007) Forecasting systems reliability based on support vector regression with genetic algorithms. Reliab Eng Syst Saf 92:423–432
36. Xiang LXL, Zhu YZY, Tang GTG A hybrid support vector regression for time series forecasting. In: 2009 WRI world congress on software engineering, vol 4, pp 161–165
37. Gunn SR et al (1998) Support vector machines for classification and regression. ISIS technical report, vol 14
38. Vijayalaksmi DP, Babu KSJ (2015) Water supply system demand forecasting using adaptive neuro-fuzzy inference system. Aquat Procedia 4:950–956
39. Bao Y, Liu Z (2006) A fast grid search method in support vector regression forecasting time series. In: Corchado E, Yin H, Botti V, Fyfe C (eds) Intelligent data engineering and automated learning—IDEAL 2006: 7th international conference, Burgos, Spain, 20–23 Sept 2006. Proceedings. Springer, Berlin, Heidelberg, pp 504–511
40. Li Y, Fang T (2003) Rough set methods for constructing support vector machines. In: Wang G, Liu Q, Yao Y, Skowron A (eds) Rough sets, fuzzy sets, data mining, and granular computing: 9th international conference, RSFDGrC 2003, Chongqing, China, 26–29 May 2003 Proceedings. Springer, Berlin, Heidelberg, pp 334–338
41. Zhang X, Zhang T, Young AA, Li X (2014) Applications and comparisons of four time series models in epidemiological surveillance Data. PLoS ONE 9:e88075
42. Chen KY, Wang CH (2007) A hybrid SARIMA and support vector machines in forecasting the production values of the machinery industry in Taiwan. Expert Syst Appl 32:254–264

Medical Image Classification Using MRI: An Investigation

R. Merjulah and J. Chandra

Abstract The main objective of the paper is to review the performance of various machine learning classification technique currently used for magnetic resonance imaging. The prerequisite for the best classification technique is the main drive for the paper. In magnetic resonance imaging, detection of various diseases might be simple but the physicians need quantification for further treatment. So, the machine learning along with digital image processing aids for the diagnosis of the diseases and synergizes between the computer and the radiologist. The review of machine learning classification based on the support vector machine, discrete wavelet transform, artificial neural network, and principal component analysis reveals that discrete wavelet transform combined with other highly used method like PCA, ANN, etc., will bring high accuracy rate of 100%. The hybrid technique provides the second opinion to the radiologist on taking the decision.

1 Introduction

Over the earlier several decayed, ailments have dropped before the scythe of the intelligence of the human in the system of the biomedical developments with the understanding of the various diseases. A qualified radiologist visually examines the medical images and further recognizes the signs of the various diseases [1]. Image analysis supports the semiautomatic or automatic technique for the illustration of the acquired medical images. Since the generation of the clinical data is more, it is difficult to physically describe and classify data in suitable time. Despite the specialist's skills and experience, the qualitative manual analysis is based on the vision of the human system. The incapability of the human eye cannot able to differentiate

R. Merjulah (✉) · J. Chandra
Christ University, Bangalore, India
e-mail: r.merjulah@res.christuniversity.in

J. Chandra
e-mail: Chandra.j@christuniversity.in

© Springer Nature Switzerland AG 2019   1135
D. Pandian et al. (eds.), *Proceedings of the International Conference on ISMAC in Computational Vision and Bio-Engineering 2018 (ISMAC-CVB)*, Lecture Notes in Computational Vision and Biomechanics 30,
https://doi.org/10.1007/978-3-030-00665-5_108

the several tens of levels in the MRI. The inability leads in using the computer-aided machine as the second opinion for the in-depth and high-resolution of MRI images.

Machine learning (MI) is the procedure of image processing, natural language processing and computer vision are the key artificial intelligence (AI) technologies forming the pillars for the radiologists. Current AI an has accurate rate that surpasses the radiologists in some area of narrow study [2]. The strength of the AI is to make the clinical judgment through the accurate detection and classification of the MRI images. MI is for the automatic detection as well as the classification of various diagnosis using MRI. Combining the radiologist and the AI as the hybrid model will lead to promising accuracy in the diagnosis. Incorporation of the MI will become the driving force for the health care in the future and also increases the value of the Radiology in the medical field [3].

Classification algorithm is classified into supervised and unsupervised methods. To train the dataset in the supervised method, the available labels for every data points are used for the classifier. In unsupervised methods, the datasets are trained using the restricted prior information of the data. The unsupervised method which analyzes the arrangement of the data to infer labels. In the medical imaging, the label often initiates from the spatially determined histological examinations. In a classical classifier, every object is trained and tested by the vector of the features and a rule is applied to classify the tested vector. In the image classification problem, these feature vectors are obtained from the images using every pixel that corresponds to one feature. Although there are many classification techniques for the medical images, most of them are not suitable for various medical images. Finding out the best method for the medical image dataset is the challenging task.

The classification process work flows like image acquisition, preprocessing, feature extraction, classification, and evaluation. Image acquisition is nothing but the selection of images that ranges as either CT or MRI or X-rays, etc. Preprocessing is the process of executing the raw images for the best performance of the dataset as images. Preprocessing phase includes cropping, filtering, noise removal, etc. Feature extraction involves with the feature estimation and the selection of the features. The literature provides some features like gabor, texture, wavelet histogram, etc. Every aspect explains the content of the images. Feature extraction is to extract the prominent features that correspond to the different classes of objects. Therefore improved feature extraction brings high accuracy in the classification technique, where the classification is divided into three division namely neural network, texture classification, and data mining technique.

In general, the complete diagnosis process includes different stages such as preprocessing, segmentation, feature selection, feature extraction, classification, etc. The purpose of the medical image analysis is to automate the process of measuring and for meaningful interpretation. The main objective of the work is to identify the best classification technique for MRI on various diagnosis.

2 Related Works

Recently, a lot of researches based on the classification technique include either the direct implementation of the best image classification technique or some other research coming out with specific problem-solving technique using hybrid model. Since, the image processing for the medical images like computer tomography (CT), electrocardiogram (ECG), MRI, etc., is totally more challenging that the other image processing.

Kumar et al. [4] have proposed a hybrid model on the brain MRI tumor images which contains discrete wavelet transform (DWT) for feature extraction, genetic algorithm for decreasing the number of features and used support vector machine (SVM) for the brain tumor classification. The accuracy varies from 80 to 90% for 25 dataset of MRI brain images.

Zhang et al. [5] have proposed a hybrid model for the classification of MRI brain images as normal or abnormal. The feature extraction includes a DWT from images, the reduction of the features includes the principle component analysis (PCA), and the reduced features are passed through the forward neural network (FNN). An improved artificial bee colony (ABC) algorithm named scaled chaotic artificial bee colony (SCABC) is used for the optimization of the parameters. The experiment is tested with 66 brain images and achieved 100% accuracy on MSE for the classification.

Chaplot et al. [6] have proposed wavelets to neural network self-organized maps (SOM) and SVM for the classification of normal and abnormal brain MRI. The classification achieves 94% on SOM and 98% on SVM for 52 MRI brain image dataset.

Saritha et al. [7] have proposed an integrating wavelet entropy based spider web plots combined with the probabilistic neural network for the classification of the brain MRI. The classification yields 100% accuracy on the brain dataset.

Gupta et al. [8] have proposed DWT for the feature extraction and PCA as feature extraction and used various classification technique like SVM, K-nearest neighbor (KNN), classification and regression tree, and random forest (RA). The SVM without the PCA yields 88% accuracy on the classification of cerebral Tumor.

Duchesne et al. [9] have proposed SVM for the automatic classification of neurodegenerative diseases, specifically Alzheimer's dementia (AD). The classification is based on the separation of 75 probable AD and 75 age-matched normal controls. The result of classification achieves 92% based on the least squares optimization.

Sayed et al. [10] have proposed KNN and linear discriminant analysis (LDA) for the classification of two groups namely benign and malignant tumors. The examined classification of the breast tumors with the two methods reveals the significant accuracy compared with the pathological analysis and resubstitution error.

Gatidis et al. [11] have proposed a classification of the multidimensional medical images for the local prostate cancer. To train the SVM, a single dataset was applied to the spatial constrain fuzzy C-Means (SFCM) algorithm. The proposed model is verified with the false positive, false negative and accuracy where determined and compared with the manual tumor delineation. The hybrid model of the combination of SFCM and SVM yields a better result than the unsupervised SFCM alone.

Garro et al. [12] have proposed ANN with ABC for the classification of the DNA microarray. ABC is used to train the ANN with the reduced genes. ABC is used mainly for the dimensionality reduction for selecting the best set of genes.

Dandil et al. [13] have proposed a spatial fuzzy C-means technique for the segmentation and SVM for the classification of the brain tumor into malign and benign which yields the accuracy of 91.49%.

Dahshan et al. [14] have proposed feedback pulse-coupled neural network for the image segmentation and for the identification of the ROI. The further work on the feature extraction is done using the DWT. Further, for the reduction of the dimensionality features, the PCA technique is used and the reduced features are passed through the back-propagation neural network for the classification of abnormal or normal brain based on the selection of the feature parameters.

Kalbkhani et al. [15] have proposed a multi-cluster feature selection technique for the feature selection on the brain MRI. The features are obtained from the DWT. Further, the features are applied to K-nearest neighbor (KNN) for the classification of normal MRI or one of seven various illness.

Amien et al. [16] have proposed an automated brain tumor diagnosis for MRI. First step involves the reduction of the noise and increasing the contrast of the images. Second step involves the extraction of the texture features and for the dimensionality reduction, the PCA is used. Finally, the reduced features are passed through the back-propagation neural network (BPNN) for the classification of brain images as edema or normal or cancer and the result achieves the accuracy of 96.8%.

Through the above literature review, it is observed that the hybrid machine learning models bring out high accuracy for the successful classification.

3 Methodology

In the section, the algorithms for the analysis of the medical MRI images are explained based on the classification techniques like SVM, DWT, ANN, and PCA. The study is to investigate the best classification technique for the MRI and determines that DWT with other highly used algorithms like ANN and PCA yields high accuracy rate.

3.1 Support Vector Machine

SVM is the supervised machine learning algorithm which is used for the regression as well as for the classification. In the SVM classification, the plotting happens with every data item as point in the n dimensions where the number of features is n. Then the classification is performed using the hyperplane that differentiates the classes. Manually selecting the hyper plane is the challenging task. Choose the hyperplane that segregates the classes better. In SVM for the automatic hyperplane selection, a technique called Kernel trick is used. Kernel trick transforms the low dimension input space to the high dimension space. It is commonly used in the nonlinear separation problem.

Mathematical calculation falls into the following steps

Step 1: Dataset D is inserted into the SVM for the classification

The data is composed of n vectors x_i. The x_i value is associated with the y_i that indicates the elements which belong to the +1 class or -1 class. y_i belong to either -1 or +1.

$$\text{Dataset } D = \left((x_i, y_i) | x_i \in R_p, y_i \in \{-1, 1\}\right)_{i=1}^{n} \tag{1}$$

Step 2: Selecting the hyperplane with no points

The hyperplanes are defined as H such that

$$H_1 \text{ defines as } w.x_i + b = +1 \tag{2}$$
$$H_2 \text{ defines as } w.x_i + b = -1 \tag{3}$$
$$H_0 \text{ defines as } w.x_i + b = 0 \tag{4}$$

where H_1 and H_2 are the classification planes and H_0 is the in-between median. The weight vector is w, input vector is x, and bias is b.

$d+$ mentioned as the short distance to the close positive point.
$d-$ mentioned as the short distance to the close negative point.

3.2   Discrete Wavelet Transform

Wavelets are to transform the signal hooked on to the set of basic functions. The DWT is to transform a discrete time signal to discrete wavelet. DWT is the highly efficient and decomposition of signals. DWT is the multi-resolution exploration and it decomposes the images in scaling function and wavelet coefficient. Wavelet converts the images into series of wavelets which stores more efficiently than the pixel blocks. Digital filtering technique for the digital signal for DWT timescale representation. The signal is analyzed which is passed through the filters with cutoff frequency for various scales that reduce the time taken during the computation process [17]. In two dimensional DWT, the rows are first processed with one level of decomposition. The decomposition separates the first half that stores the details of the coefficients and the other half stores the average coefficients. The procedure is continued for the column results in sub-bands.

3.2.1 Haar Transform

Haar transform is a modest category of wavelet transform. In discrete, the Haar wavelets are linked to the scientific operation named as haar transform. This transform provides the prototype for the other wavelet transform. Similarly, other transforms decompose the discrete signal into two sub-signals of the length. The running average is one signal and the other is the fluctuation or running difference.

3.2.2 Daubechies Transform

The Daubechies transform defines the similar way as haar transform by processing the running average and the differences between the scalable product with the scaling signals and the wavelets [18]. The wavelet balances the occurrence responses but it is nonlinear phase response. Daubechies transform uses overlapping windows, so, high-frequency coefficient spectrum that reflects the high-frequency changes.

Basically, the medical images need the accuracy without the loss of the information. DWT is based on the discrete time signal that provides the multi-resolution without loss of information on the MRI data.

3.3 Artificial Neural Network

ANN is based on the supervised family. ANN acquires the knowledge through the connection of the network. The general steps of the neural network are to define the fixed number of nodes in the input, output, and in the hidden layers. The following steps are used for training the ANN:

Step 1: Push the input for the processing.
Step 2: Every input has to be weighted, which is nothing but multiplied by random value of either +1 or −1.
Step 3: Add all the weighted input values.
Step 4: Pass the sum through activation function for producing the output of the network.

ANN has the learning ability for various classification problem with back-propagation [19], and improved back-propagation algorithm [20].

3.4 Principal Component Analysis

PCA is the usual statistical model used for the reduction of the dimensionality of the data. There are many algorithms for the reduction of the features in the images and choosing important ones. The PCA is the best among that for the reduction of the features and improves the accuracy result in the classification task.

If there are n is the observation and p is the variable, then the number of PCA is

$$\min(n - 1, p) \tag{5}$$

PCA preserving as many variables that has been translated into the new variable that are the linear functions on the original dataset which increases the variance that are not overlapped with each other. The new variable reduces the eigenvector problem. The linear combination of matrix in

$$\sum_{j=1}^{p} a_j x_j = xa \tag{6}$$

where p is the data value, n is the dimension vectors of $x_1, x_2 \ldots, x_p$, $n \times p$ is the data matrix x, whose jth column is the x_j vector of the jth variable, constant vector is $a_1, a_2, a_3, \ldots, a_p$.

The steps in the PCA includes,

Step 1: Using the dataset "S" calculate the mean value
Step 2: Subtract the mean using "S". A new matrix "A" value is obtained.
Step 3: Process the covariance "c" from "A" using $C = AA^T$ where the computation process through

$$1, \ldots, l, \ A_k \in R,^N \sum_{k=1}^{l} A_k = 0,$$

$$\text{Covariance matrix is } c = \frac{1}{l} \sum_{i=1}^{l} A_i A_i^T \tag{7}$$

Step 4: The values of the eigen will be obtained through the covariance matrix where $[V_1, V_2, \ldots, V_n]$.
Step 5: Using the covariance matrix C eigenvectors are calculated.
Step 6: Linear combination of eigenvectors of vector S or $S - \bar{S}$

$$S = b_1 u_1 + b_2 u_2 + b_3 u_3 \cdots + b_n u_n \tag{8}$$

Since the covariance form is $[V_1, V_2, \ldots, V_n]$.
Step 7: The largest eigenvalue gives the lower dimensional dataset

$$S = \sum_{i=0}^{1} b_i u_i, \ 1 < N \tag{9}$$

The lower component dimensional space is the principal components. The principal components will be independent in the normal case distribution.

3.5 Self-organizing Map

SOM is developed by Kohonen [21], the SOM texture classification algorithm [22], and the steps involved in SOM are as follows

Step 1: Initialize the three nodes which includes connection weights, output nodes, and input nodes. Where the input vector is N, M is the two-dimensional map as output vector, the weight is w_{ij} from the input nodes to the M output nodes.

Step 2: Place every document in order. Set the value 1 for the document which has the corresponding term. Set value 0 if the document has no corresponding term. The document is placed several times.

Step 3: The distance is computed using the Euclidean distance d_j between the input vector and the output vector.

$$d_j = \sum_{i=0}^{N-1} (x_i(t) - w_{ij}(t))^2 \tag{10}$$

where the value of $x_i(t)$ can be either 0 or 1 that depends on the ith document term at the time t. The vector that represents j map node in the vector document space is w_{ij}. The interpretation of the weight from the input node is i to the output node j in the neutral net.

Step 4: Choose the winning node j and the weights to node j has to be updated and its neighbors to reduce the distance.

$$w_{ij}(t+1) = w_{ij}(t) + \eta(t)(x_i(t+1) - w_{ij}(t)) \tag{11}$$

where $\eta(t)$ is an error adjusting coefficient $(0 < \eta(t) < 1)$ which decreases the overall time.

Step 5: Label the map regions. After the training of the network, assign the term to every node by selecting the winning term. The similar winning term will be combined to form the group. The resulting map represents the documents that joined with them. The processing time increases because the Steps 2 to 4 are repeated several times.

4 Result Analysis

In the section, the dataset description and the performance evaluation of the DWT, SVM, ANN, KNN, and the hybrid approaches are shown to find the best classification

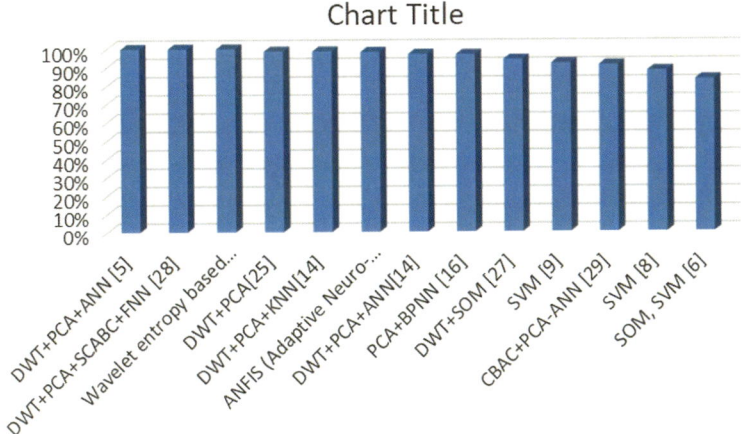

Fig. 1 Performance graph on various classification technique using MRI

model for the classification of the medical MRI data images. The result is compared base on the accuracy.

The whole brain atlas dataset is the information resources of the central nervous system imaging that has the clinical information of the magnetic resonance imaging, X-ray computed tomography, and the nuclear medicine images [23]. The atlas project is part of the radiology department and neurology from Brigham and women's hospital.

Allan Institute is another open source repository that currently collaborated with three entries namely Allen Institute of Cell Science, brain science, and finally the Paul G. Allen Frontiers group. The approaches for the bioscience ranges from the smallest molecular scale to level of the entire system [24].

Accuracy is the probability for the diagnosis testing which can be calculated as follows:

$$Accuracy = \frac{TP + TN}{TP + TN + FP + FN} \tag{12}$$

where TP is True Positive for the correct classification of positive cases, TN is True Negative for the correct classification of the negative cases, FP is the False Positives for the incorrect classification of negative cases, and FN is the False Negative for the incorrect classification of the positive cases.

Table 1 shows the performance evaluation of the classification technique and the hybrid approaches with the accuracy. To evaluate the effectiveness based on the accuracy, the machine learning technique has been compared with the existing models in the table to find out the high accuracy classification. Figure 1 shows the performance comparison of the existing system. Finally, after the investigation, it reveals that the DWT along with the highly used machine learning technique like ANN and PCA of hybrid model brings out high accuracy.

Table 1 Performance evaluation of the various classification techniques using MRI

S. No.	Dataset	Approach	Classification accuracy (%)
1	Brain	DWT + PCA + ANN [5]	100
2	Brain	SOM, SVM [6]	83.21
3	Brain	Wavelet entropy based spider web plot + PNN [7]	100
4	Cerebral tumor	SVM [8]	88
5	Alzheimer's dementia	SVM [9]	92
6	Brain	DWT + PCA + ANN [14]	97
7	Brain	DWT + PCA + KNN [14]	98.60
8	Brain	PCA + BPNN [16]	96.8
9	Brain	DWT + PCA [25]	98.75
10	Brain	ANFIS (Adaptive neuro-fuzzy inference system) [26]	98.25
11	Brain	DWT + SOM [27]	94
12	Brain	DWT + PCA + SCABC + FNN [28]	100
13	Brain	CBAC + PCA − ANN [29]	91

5   Conclusion

A study on the recent classification techniques based on the magnetic resonance imaging collectively gets the knowledge and its accuracy is the stepping stone for the researchers for the classification of medical images. The investigation based on the classification of medical images reveals that the discrete wavelet transform along with the highly used classification technique like artificial neural network, principal component analysis, etc., brings out high accuracy of 100%. The survey is designed towards the expansion of image processing and machine intelligence technique. The digital image processing methodologies along with the intelligence technique support the radiologist in effective diagnosis, and as hybrid methods provide the second opinion and the assistance to the radiologist.

References

1. Mohan G, Subashini MM (2018) MRI based medical image analysis: survey on brain tumor grade classification. Biomed Sig Process Control 39:139–161
2. Hinton G (2016) Machine learning and the market for intelligence. In: Machine learning and the market for intelligence conference
3. Dean BL, Drayer BP, Bird CR, Flom RA, Hodak JA, Coons SW, Carey RG (1990) Gliomas: classification with MR imaging. Radiology 174:411–415
4. Kumar S, Dabas C, Godara S (2017) Classification of brain MRI tumor images: a hybrid approach. Procedia Comput Sci 122:510–517. Springer
5. Zhang Y, Wu L, Wang S (2011) Magnetic resonance brain image classification by an improve artificial bee colony algorithm. Prog Electromagn Resolut 116:65–79
6. Chaplot S, Patnaik LM, Jagannathan NR (2006) Classification of magnetic resonance brain images using wavelets as input to support vector machine and neural network. Biomed Sig Process Control 1(1):86–92
7. Saritha M, Paul Joseph K, Mathew AT (2013) Classification of MRI brain images using combined wavelet entropy based spider web plots and probabilistic neural network. Pattern Recogn Lett 34(16):2151–2156
8. Gupta T, Gandhi TK, Gupta RK, Panigrahi BK (2017) Classification of patients with tumor using MR FLAIR images. Pattern Recogn Lett
9. Duchesne S, Caroli A, Geroldi C, Barillot C, Frisoni GB, Collins DL (2008) MRI-based automated computer classification of probable AD versus normal controls. IEEE Trans Med Imaging 27:509–520
10. Sayed AM, Zaghloul E, Nassef TM (2016) Automatic classification of breast tumors using features extracted from magnetic resonance images. Procedia Comput Sci 95:392–398
11. Gatidis S, Scharpf M, Martirosian P, Bezrukov I, Küstner T, Hennenlotter J, Kruck S, Kaufmann S, Schraml C, Fougere C, Schwenzer NF, Schmidt H (2015) Combined unsupervised—supervised classification of multiparametric PET/MRI data: application to prostate cancer. Biomedicine 28(7):26
12. Garro BA, Rodriguez K, Vazquez RA (2016) Classification of DNA microarrays using artificial neural network and ABC algorithm. Appl Soft Comput 38:548–560 Elsevier
13. E. Dandil., M. Cakiroglu., Z. Eksi.,: Computer-aided diagnosis of malign and benign brain tumors on MR images, in ICT Innovations, pp. 157–166, 2014
14. El-Dahshan ESA, Hosney T, Salem ABM (2010) Hybrid intelligent techniques for MRI brain images classification. Digital Sig Process 20:433–44
15. Kalbkhani H, Salimi A, Shayesteh MG (2015) Classification of brain MRI using multi-cluster feature selection and KNN classifier. In: Electrical engineering conference. IEEE, pp 2164–7054
16. Amien MB, Abd-elrehman A, Ibrahim W (2013) An intelligent model for automatic brain-tumor diagnosis based-on MRI images. Int J Comput Appl 72(23):21–24
17. Bute YS, Jasutkar RW (2012) Implementation of discrete wavelet transform processor for image compression. Int J Comput Sci Netw 1:1–5
18. Mahmoud MI, Dessouky MIM, Deyab S, Elfouly FH (2007) Comparison between Haar and Daubechies wavelet transformions on FPGA technology. In: Proceedings of world academy of science, in engineering and technology, vol 20
19. Heermann PD, Khazenie N (1992) Classification of multispectral remote sensing data using a back-propagation neural network. IEEE Trans Geosci Remote Sens 30(1):81–88
20. Nawi NM, Ransing RS, Salleh MNM, Ghazali R, Abdul Hamid N (2010) An improved back propagation neural network algorithm on classification problems. Database Theory Appl Bio-Sci Bio-Technol 118:177–188
21. Kohonen T (1995) Self-organization maps. Springer, Berlin, Heidelberg
22. Raghu PP, Poongodi R, Yegnanarayana B (1995) A combined neural network approach for texture classification. Neural Networks 8(6):975–987

23. Summers D (2003) Harvard whole brain atlas. J Neurol Neurosurg Psychiatry 74(3):288
24. Miller JA et al (2014) BrainSpan Atlas of the developing human brain, in Transcriptional land-scape of the prenatal human brain. Nature 508:199–206. https://doi.org/10.1038/nature13185
25. Zhang Y, Wang S, Wu L (2010) A novel method for magnetic resonance brain image classifi-cation based on adaptive chaotic PSO. Prog Electromagn Res 109:325–343
26. Roy S, Sadhu S, Bandyopadhyay SK, Bhattacharyya D, Kim T-H (2016) Brain tumor clas-sification using adaptive neuro-fuzzy inference system from MRI. Int J Bio-Sci Bio-Technol 8(3):203–218
27. Chaplot S, Patnaik LM, Jagannathan NR (2006) Classification of magnetic resonance brain images using wavelets as input to support vector machine and neural network. Biomed Sig Process Control 1:86–92
28. Zhang Y, Wu L, Wang S (2012) Magnetic resonance brain image classification by an improved artificial bee colony algorithm. Prog Electromagn Res 130:369–388
29. Sachdeva J, Kumar V, Gupta I, Khandelwal N, Ahuja CK (2013) Segmentation, Feature Extrac-tion and Multiclass brain tumor classification. J Digital Imaging 26(6):1141–1150

Tumor Detection and Analysis Using Improved Fuzzy C-Means Algorithm

R. Swathika, T. Sree Sharmila, M. Janani Bharathi and S. Jacindha

Abstract Formation of abnormal cells in brain serves the major cause of tumor. With estimated deaths of 229,000 as of 2015, it has become an issue to be dealt. The less awareness of brain tumor owes to lots of unaccounted deaths. Thus, we aim in developing an app which could serve the purpose of detecting the tumor and giving additional information related to the detected tumor. This app takes in an MRI image and does preprocessing followed by clustering, segmentation, and binarization. The preprocessing involves the conversion of the image into grayscale and noise filtering. We aim at using improved Fuzzy c-means algorithm for clustering and segmentation. Binarization mainly aims at calculating the tumor size useful for further analysis. The improved fuzzy c-means algorithm overcomes the various constraints of k-means algorithm such as time complexity, processing of noisy images, and memory space.

1 Introduction

Brain tumor segmentation and analysis is a field of interest for many research scholars. But an efficient method for analyzing the tumor is yet to be developed as the existing methods face one or the other drawbacks. As for now, the best segmentation technique is k-means [1] but the efficiency of k-means is limited by various constraints such as processing noisy images, time complexity, and memory space. Fuzzy c-means [FCM] was brought into picture to serve this purpose but neither of them could eradicate these problems in an efficient way. Thus, we propose improved fuzzy c-means algorithm for clustering and segmentation of the image. The IFCM takes the frequency of the pixel values instead of the actual values which had proved to greatly reduce the time complexity and memory space. Awareness about the brain tumor is very less among the people. In a developing nation like our country still a major population is uneducated and they have very little knowledge about tumor. If our system could analyze the tumor more efficiently, we definitely believe that we

R. Swathika (✉) · T. Sree Sharmila · M. Janani Bharathi · S. Jacindha
Department of Information Technology, SSN College of Engineering, Chennai, India
e-mail: swathikar@ssn.edu.in

© Springer Nature Switzerland AG 2019
D. Pandian et al. (eds.), *Proceedings of the International Conference on ISMAC in Computational Vision and Bio-Engineering 2018 (ISMAC-CVB)*, Lecture Notes in Computational Vision and Biomechanics 30,
https://doi.org/10.1007/978-3-030-00665-5_109

could save more lives. Today in this era of technological advancements, we aim in providing efficient and interesting technology which could take MRI image as an input and detect the tumor area and depending upon the size of the tumor detected, it could suggest the stage of the tumor. Localized detection of the tumor could give an output as benign and if it had spread over a number of regions, it could classify it as malignant or secondary. This technology could be used by doctors in their analyzation.

2 Review of Existing Work

Yisu Lu et al. proposed Dirichlet's process mixture model for Automatic Multimodal Brain Tumor Segmentation [2]. The time complexity of the work is $O(n^2)$. Hence, the processing of the high-resolution images could become a tedious and time-consuming job. 1024 by 1024 data matrix could take computational time of 10 M s. It also takes up lots of the memory space.

In automatic image segmentation using convolutional neural network [3], an optimal number of clusters for each unique image is predetermined and hence it enhances the clustered image and serves the best in terms of the accuracy. Our proposed algorithm could be used in combination with this work for the predetermination of number of clusters to be passed an argument to the function performing the IFCM.

Model-based brain and tumor segmentation [4] uses expectation maximization clustering algorithm. This is variant of k-means and hence suffers from the drawbacks of time complexity and memory space. The time complexity is in the order of $O(n * k)$. Thus, here also the time increases in proportion with the input size.

Telrandhe et al. [5] proposed detection of brain tumor from MRI images by using segmentation and SVM. The segmentation and preprocessing of image are done using k-means algorithm. The classification and detection of brain tumor are done using SVM technique. Ghassabeh et al. [6] used IFCM with genetic algorithm optimization for segmentation of brain tissue. Forghani et al. [7] proposed a segmentation of brain tissue using IFCM with particle swarm optimization.

Most of the other applications currently available use either k-means, FCM else hierarchical clustering models for the segmentation followed by the binarization and thus extracting the tumor from the MRI image.

2.1 Traditional FCM

Traditional FCM (fuzzy c-means clustering) is a segmentation technique which segments the input image into number of predefined clusters. It is developed by Dunn and changed by Bezdek [8]. Segmentation is done in accordance with the similarity of the pixel values. The advantage of FCM is that it is a soft clustering technique and hence is able to outperform hard clustering technique k-means. The performance evaluation of k-means and clustering algorithms with that of FCM is discussed in detail in Table 1. FCM allows a pixel to belong to two or more clusters and hence provides more accuracy in comparison with other techniques. The following are the steps of traditional fuzzy c-means algorithm.

Algorithm: fcm(x[l..n])
//Clustering by traditional FCM method
//Input: The data matrix x of dimension l by n
//Output: The membership matrix U
Step1: Choose random centroid at least 2 and put values to them randomly.
Step2: Compute membership matrix:

$$U_{ij} = \frac{1}{\sum_{k=1}^{c} \left[\frac{|x_i - c_j|}{x_i - c_k} \right]^{\frac{2}{m-1}}}, \quad \text{where } m > 1, c \text{ cluster's No.}$$

where U is the membership matrix,
C is the centroid matrix.
i, j, k are loop variables.
Step3: calculate the clusters centers:

$$C = \frac{\sum_{i=1}^{n} U^m_{ij} * x_i}{\sum_{i=1}^{n} U^m_{ij}}$$

Step4:
if $C^{k-1} - C^k < \varepsilon$
then Stop else go to Step2.

This traditional algorithm is an iterative algorithm that suffers from time and memory consumption because it computes membership value for each item in the data.

3 Proposed Work

Our app gets the image from the gallery and converts it into gray scale. After filtering, the image is clustered and segmented using IFCM. The IFCM output then undergoes binarization to isolate the tumor with which the size of the tumor is calculated. The steps used for tumor detection are clustering and thresholding. The process diagram is shown in Fig. 1.

Table 1 Comparison of performance evaluation of the proposed clustering algorithm

IMAGES	FCM	K-MEANS	IFCM
COMPUTATIONAL TIME	36.415 sec	3.041 sec	2.1764 sec
COMPUTATIONAL TIME	71.562 sec	6.692 sec	6.4719 sec
COMPUTATIONAL TIME	147.706 sec	10.586 sec	10.414 sec
COMPUTATIONAL TIME	290.432 sec	16.55 sec	13.687 sec

(continued)

Table 1 (continued)

IMAGES	FCM	K-MEANS	IFCM
COMPUTATIONAL TIME	10.689 sec	2.798 sec	1.728 sec
COMPUTIONAL TIME	34.837 sec	18.03 sec	3.153 sec
COMPUTATIONAL TIME	20.458 sec	3.065 sec	2.585 sec
COMPUTATIONAL TIME	7.824 sec	2.764 sec	2.014 sec

Fig. 1 Process flowchart

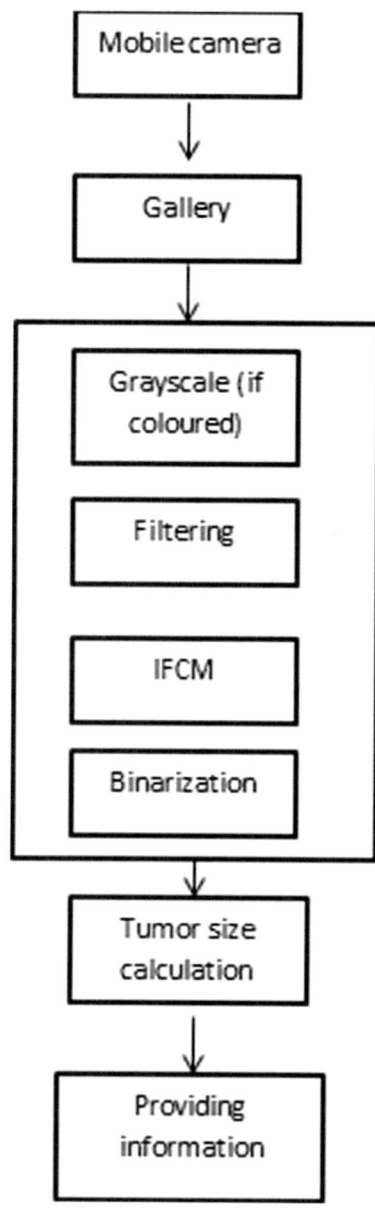

3.1 Preprocessing

Converting image into gray valued matrix

The input image in the form of an MRI scanned image is first converted into gray valued matrix. Conversion of image into grey valued matrix can be done by an inbuilt function in MATLAB.

Noise filtering—median filtering

It filters the image and gives a smooth appearance and also prepares it for further processing. Median filtering has proved to be very effective in the field of medical image processing (Figs. 2 and 3).

Fig. 2 Original image

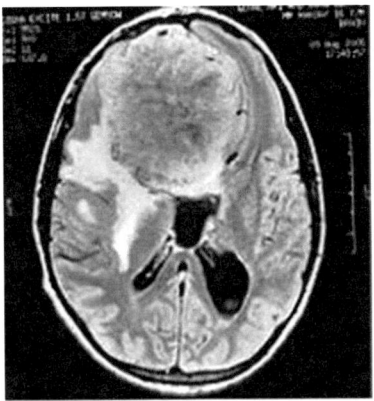

Fig. 3 After median filtering

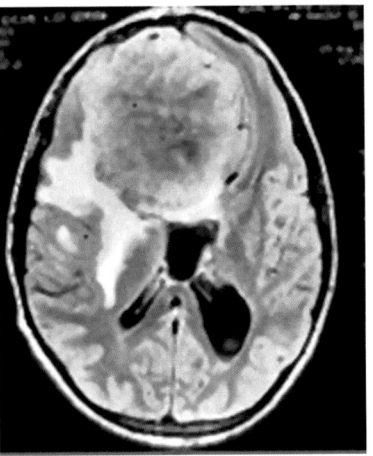

3.2 Improved Fuzzy C-Means

In the following section, we propose the improved fuzzy c-mean algorithm:

Algorithm: ifcm(data[1..n])
//IFCM method
//Input: data matrix data of dimension 1 by n
//Output: The membership matrix U
Step1: Let H represent the frequency of each item in Data.
Step2: create vector I=min(Data): max (Data)
Step3: Choose random centroid at least 2.
Step4: Compute membership matrix:

$$U_{ij} = \frac{1}{\sum_{k=1}^{c} \left[\frac{|I_i - c_j|}{I_i - c_k} \right]^{\frac{2}{m-1}}}$$

Where U is the membership matrix,
I is a vector which contains the unique elements in the data matrix I the ascending order from minimum to maximum,
m is a minimum exponential value and is ≥ 2
c is the centroid matrix
Step5: calculate the cluster center:

$$C = \frac{\sum_{i=1}^{n} U^m * H * 1}{\sum_{i=1}^{n} U^m * H}$$

Step 6: if $C^{(k-1)} - C^k < \varepsilon$ then stop else go to step

In traditional FCM, the membership matric raised to the power m is multiplied with a data points matrix whereas in IFCM [improved fuzzy c-means] frequency matrix H is multiplied. H is N by 1 matrix and $N \ll$ total no. of. data points and $0 < N < 256$. I is also a N by 1 matrix. Hence on the application of IFCM, the number of iterations for the calculation of new center and new membership matrix decreases rapidly thus reducing the overall time complexity to a great extent. The maximum number of elements to be processed here is 256 whereas in traditional FCM, it could be $256 * 256 = 65,536$. The minimum value of m is 2. The proposed algorithm does not depend on whole data of image; it actually depends on data that represent the frequency of each data item in original image's data. A number of frequencies at most are 256.

```
I=unique(data);
H1=[I,histc(data(:),I)];
H=H1(:,2);
tmp1=H.*I;

mf = U.^expo;          % MF matrix after exponential modification
center = mf*tmp1./(sum(mf,2)*ones(1,size(tmp1,1))*H);  %new center
dist = distfcm_f(center, data);        % fill the distance matrix
obj_fcn = sum(sum((dist.^2).*mf));     % objective function
tmp = dist.^(-2/(expo-1));             % calculate new U, suppose expo != 1
U_new = tmp./(ones(cluster_n, 1)*sum(tmp));
```

Distfcm_f is calculated as such

```
I=unique(data);
H1=[I,histc(data(:),I)];
H=H1(:,2);
tmp1=H.*I;

mf = U.^expo;          % MF matrix after exponential modification
center = mf*tmp1./(sum(mf,2)*ones(1,size(tmp1,1))*H);  %new center
dist = distfcm_f(center, data);        % fill the distance matrix
obj_fcn = sum(sum((dist.^2).*mf));     % objective function
tmp = dist.^(-2/(expo-1));             % calculate new U, suppose expo != 1
U_new = tmp./(ones(cluster_n, 1)*sum(tmp));
```

3.3 Binarization

Binarization is the process by which a binary mask is applied over the image obtained on segmentation. Segmentation is the process of extraction of tumorous area from the input MRI image and is done using the proposed method (IFCM). On application of binary mask, each pixel takes the value of either 1 or 0. Value of the pixel is determined based on the input threshold. Hence, this process is also called thresholding. Let f be the input image and has pixel values in the gray scale range of k. Let threshold value be T, which also lies in the gray scale range k. Each pixel of f is compared with T (as shown in the Eq. 1). In case of the pixel value being less than T, the pixel is assigned a value of 0 and vice versa. This process acts as primary step in tumor size calculation.

$$g(n) = \begin{cases} 0 & \text{if } (n) \geq T \\ 1 & \text{if } (n) < T \end{cases} \tag{1}$$

where $f(n)$ is the transform coefficient,
T is the threshold value and
$g(n)$ is the transform coefficient after the binary decision for the same value of n.

3.4 Calculation of Tumor Size

In the approximate reasoning step, the tumor area is calculated using the binarization method. That is from the image having only two values either black or white (0 or 1). Here 256 × 256 JPEG image is a maximum image size. The binary image can be represented as a summation of total number of white and black pixels.

The tumor size is the summation of total number of white pixels. The users of the app will be provided various information depending on the tumor size. The relevant information for an appropriate tumor size will already be available in the app's database.

4 Implementation and Result

On the application of the clustering algorithm IFCM, the tumor cells are clearly segmented in the IFCM image, which is shown in Fig. 5. The various clustering algorithms are applied on the images in Fig. 4 which were obtained from dataset of brain tumor images provided by the department of computer science, University of Cyprus [9] and evident results are taken. A dataset of tumor-MRI images of the brain was obtained and the algorithm was tested on sufficient number of images (sample shown in Fig. 5). Figure 6a representing the IFCM iteration count takes negligibly small number of iterations whereas Fig. 6b corresponding to FCM takes around 42 iterations with a computation time of 290 s. In the denoised image, the edges are smoothened and now the image is ready for the application of IFCM. One of the sample's FCM and IFCM computation time is compared below. Figure 7 shows the clustered and segmented sample image using FCM and IFCM.

Image is segmented and thresholded by the proposed IFCM. We calculated the frequency matrix using histc function. Instead of multiplying with each data point, the distance matrix calculated was multiplied with the frequency matrix. Hence the number of iterations reduced significantly and the time complexity was also greatly reduced. The objective function was defined and in each iteration, it was reduced till the condition to exit out of the loop was satisfied. If the condition fails in the worst case, the iterations may run 100 times but still, the time complexity remains small.

As interpreted from Fig. 8, the computation time comes around 6.4719 s if the summation is taken over the column total time.

From Fig. 9, it is clear that the time of computation comes around 71.562 s for traditional FCM segmentation and thresholding. A speedup of 5.3 times is achieved by using IFCM.

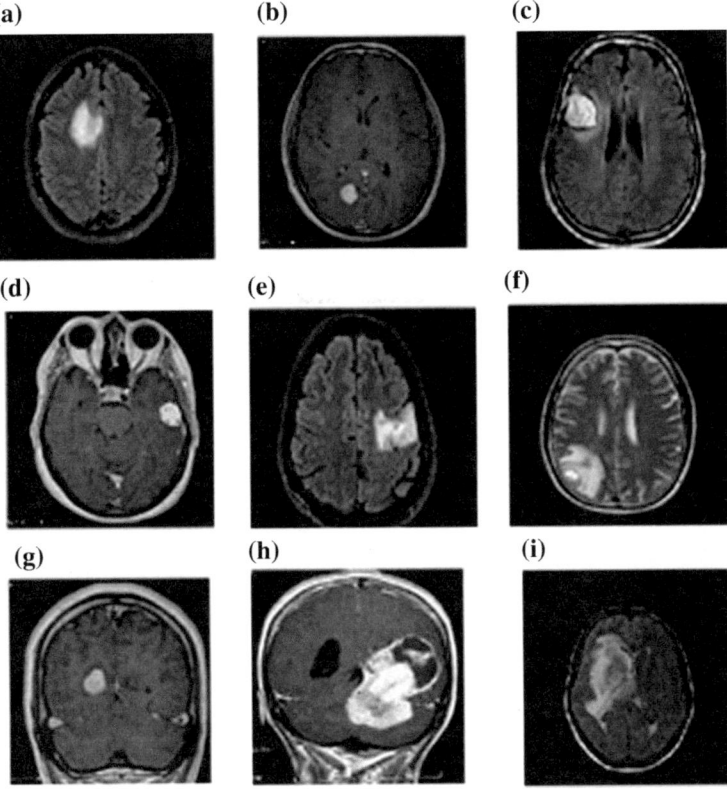

Fig. 4 Dataset of original images

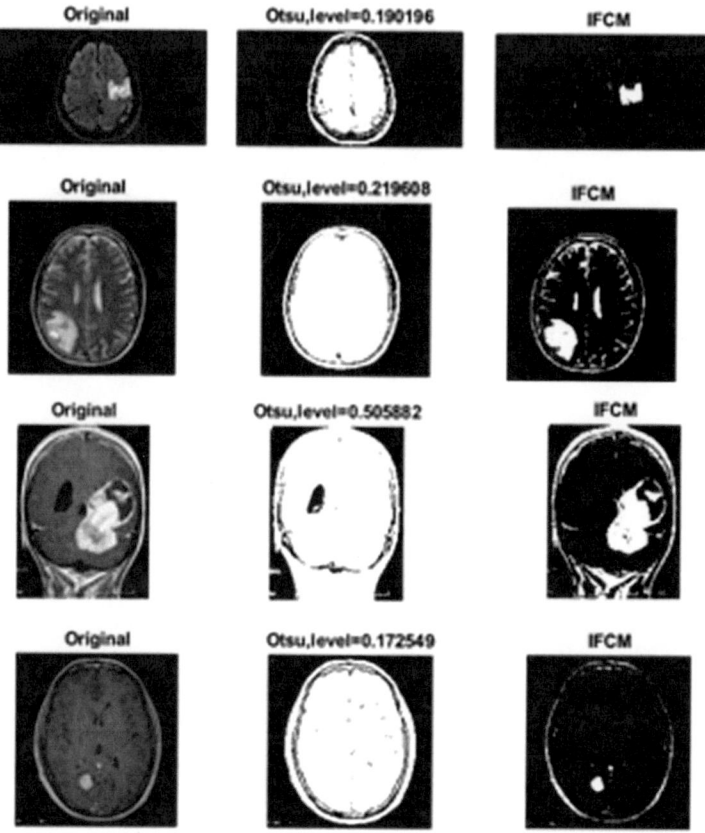

Fig. 5 IFCM samples

The computation time analysis for the best segmentation algorithms is discussed below. Their drawbacks are also mentioned with pictorial representations.

k-means algorithm is a hard clustering technique and hence a data point can belong to only one cluster, thus failing to give accurate results as shown in Fig. 10b.

Despite providing accurate results, FCM is a time-consuming process as the membership matrix of the size of the data matrix is calculated for each iteration. From the comparison table provided above, it is very clear that the computation time is nearly 10–20 times more than that of IFCM and k-means clustering techniques. Table 1 provides a comparative analysis of the dataset of 10 images in terms of output efficiency and computational time.

(a)
```
Iteration count = 1, obj. fcn = 34.403991
Iteration count = 2, obj. fcn = 27.267812
Iteration count = 3, obj. fcn = 27.267787
Iteration count = 4, obj. fcn = 27.267787
Iteration count = 1, obj. fcn = 34.614916
Iteration count = 2, obj. fcn = 27.267793
Iteration count = 3, obj. fcn = 27.267787
```
(b)
```
Iteration count = 37, obj. fcn = 4535.488914
Iteration count = 38, obj. fcn = 4535.488509
Iteration count = 39, obj. fcn = 4535.488368
Iteration count = 40, obj. fcn = 4535.488318
Iteration count = 41, obj. fcn = 4535.488300
Iteration count = 42, obj. fcn = 4535.488294
```

Fig. 6 **a** IFCM iteration, **b** FCM iteration

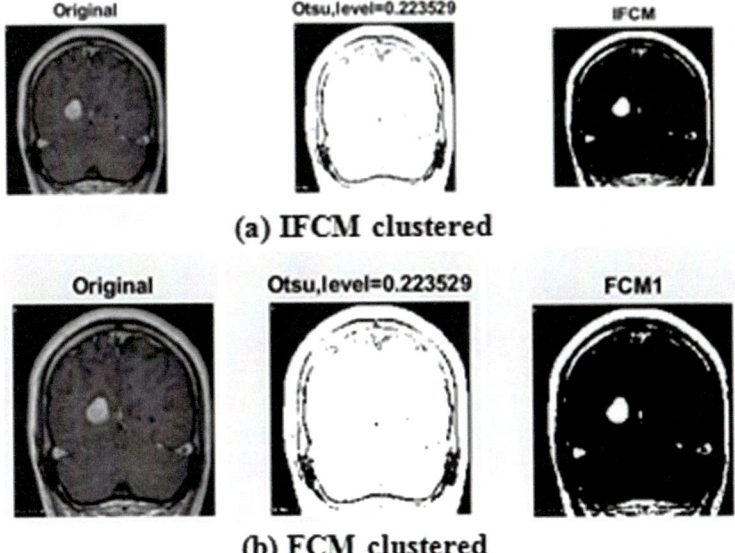

(a) IFCM clustered

(b) FCM clustered

Fig. 7 Clustered images

Function Name	Calls	**Total Time**	Self Time*	Total Time Plot (dark band = self time)
m_f	1	1.438 s	0.008 s	
fcmthresh_f	2	1.080 s	0.005 s	
fcm_f	2	1.066 s	0.009 s	
stepfcm_f	7	0.945 s	0.290 s	
unique	16	0.763 s	0.004 s	
unique>uniqueR2012a	16	0.759 s	0.759 s	
distfcm_f	7	0.330 s	0.003 s	

Fig. 8 Time of computation for IFCM for sample Fig. 7a

Function Name	Calls	**Total Time**	Self Time*	Total Time Plot (dark band = self time)
M_ff	1	20.840 s	0.006 s	
fcmthresh	2	20.497 s	0.139 s	
fcm	2	20.354 s	0.018 s	
stepfcm	54	20.167 s	18.570 s	
distfcm	54	1.597 s	1.597 s	

Fig. 9 Time of computation for FCM for sample Fig. 7b

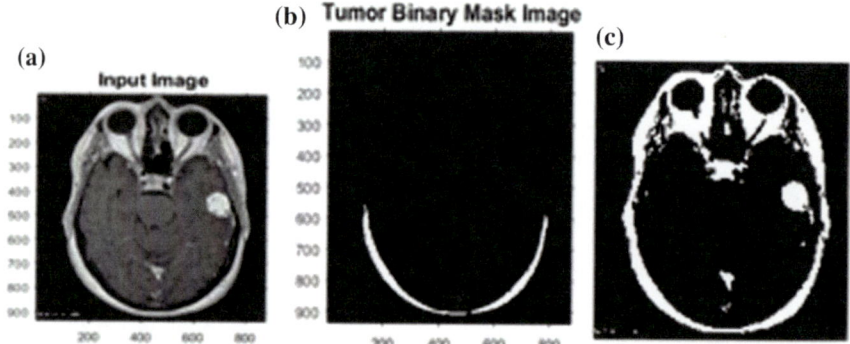

Fig. 10 **a** Original image, **b** clustered and thresholded image by k-means, **c** clustered and thresholded image by IFCM

5 Conclusion

In this paper, the performance metrics of our proposed algorithm IFCM is analyzed thoroughly in comparison with the preexisting algorithms. The IFCM proposed in our paper uses frequency in place of data points to provide evidently improved performance. A large number of images in dataset was computed in a faster and precise manner as possible. We believe that this conceptual proposal could serve for various image segmentation purposes, apart from serving the purpose of brain tumor recognition and analysis. Various high-resolution images were segmented in a very less time possible. Other major existing methods were also analyzed and drawbacks were also briefly discussed. Preexisting work's pros and cons were discussed and further chances of the hybrid models possible were also mentioned.

References

1. Hamerly G, Elkan C (2002) Alternatives to the k-means algorithm that find better clusterings. In: Proceedings on the 11th international conference on information and knowledge management (CIKM)
2. Lu Y, Chen W (2015) Automatic multimodal brain-tumor segmentation. In: IEEE fifth international conference on instrumentation and measurement, computer, communication and control
3. Pereira S, Pinto A, Alves V, Silva CA (2016) Brain tumor segmentation using convolutional neural networks in MRI images. IEEE Trans Med Imaging 35(5):1240–1251
4. Moon N, Bullitt E, van Leemput K, Gerig G (2002) Model-based brain and tumor segmentation. IEEE
5. Telrandhe SR, Pimpalkar A, Kendhe A (2016) Detection of brain tumor from MRI images by using segmentation and SVM. In: World conference on futuristic trends in research and innovation for social welfare (WCFTR-16)
6. Ghassabeh YA, Forghani N, Forouzanfar M, Teshnehlab M (2007) MRI fuzzy segmentation of brain tissue using IFCM algorithm with genetic algorithm optimization. IEEE
7. Forghani N, Forouzanfar M, Forouzanfar E (2007) MRI fuzzy segmentation of brain tissue using IFCM with particle swarm optimization. IEEE
8. Bezdek JC (1981). Pattern recognition with fuzzy objective function algorithms. ISBN 0-306-40671-3
9. www.medinfo.cs.ucy.ac.cy

An Off the Shelf CNN Features Based Approach for Vehicle Classification Using Acoustics

Anam Bansal, Naveen Aggarwal, Dinesh Vij and Akashdeep Sharma

Abstract Vehicle classification is a trending area of research in Intelligent Transport System. Vehicle Recognition can help traffic policy makers, public safety organizations, insurance companies, etc. It can assist in various applications like automatic toll collection, emissions/pollution estimation, traffic modelling, etc. Many methods, both infrastructure-based and infrastructureless have been proposed for vehicle classification but they have certain disadvantages. In this paper, we have explored the possibility to use off the shelf Convolutional Neural Network (CNN) features for commuter vehicle classification using acoustics. To extract features from acoustic recordings taken from the vehicle, a simple CNN is designed. These features are used to classify vehicles in five main categories car, bus, plane, train, and three-wheeler using Support Vector Machine (SVM). This approach is tested on dataset having 4789 recordings and gives good accuracy as compared to simple Mel Frequency Cepstral Coefficients (MFCC) feature based deep learning and machine learning approach.

1 Introduction

One of the challenging issues in Intelligent Transport System is to detect the mode of transport used by the commuters. It becomes more challenging on roads for developing countries like India where traffic is heterogeneous and chaotic. Vehicle Classification on roads can assist the traffic management policymakers. Knowing the type and the number of vehicles plying on road, various design decisions can be laid and layout of the respective city can be planned. Recognition of type of vehicle aid in the number of applications like automatic toll collection, surveillance, accident prevention, traffic congestion avoidance, etc.

A. Bansal (✉) · N. Aggarwal · D. Vij · A. Sharma
UIET, Panjab University, Chandigarh, India
e-mail: anambansal19@gmai.com

N. Aggarwal
e-mail: navagg@gmail.com

© Springer Nature Switzerland AG 2019   1163
D. Pandian et al. (eds.), *Proceedings of the International Conference on ISMAC in Computational Vision and Bio-Engineering 2018 (ISMAC-CVB)*, Lecture Notes in Computational Vision and Biomechanics 30,
https://doi.org/10.1007/978-3-030-00665-5_110

The researchers have proposed the number of sensing techniques to detect the type of vehicles plying on the roads. The proposed solutions can be broadly categorized into infrastructure-based and infrastructureless solutions. Infrastructure-based solutions include Magnetic Sensors, Inductive Loop Detectors, Video Cameras, Infrared Sensors, etc. These techniques have certain downsides. They are costly in terms of installation and maintenance. Further, the accuracy of video camera based approaches is affected by the weather conditions and occlusion. Similarly, magnetic sensor based approaches work well in homogeneous and lane driven traffic only. Even though, some researchers have proposed infrastructureless solutions, which mainly rely on smartphone sensors such as GPS, accelerometer, etc. [1]. But for vehicle classification, these approaches have very limited accuracy.

Various researchers have used acoustics for vehicle classification [2, 3]. Since each vehicle has a specific sound pattern, so they proposed to use the roadside installed microphones to capture the vehicular acoustics and extract their features for classification using SVM. Since it is difficult to install and maintain the roadside installed microphones, we propose to use acoustics captured from commuter's smartphone. Acoustics recorded are then sent to the server where CNN is used to extract features and SVM classifier is used to determine the type of the vehicle. Along with the commuter's acoustics, only current location is sent and no identity data such as phone number or ID are sent to preserve the privacy of commuters. Main contributions of this paper are:

1. Determining the efficacy of human-driven features and off the shelf CNN features using SVM classifier.
2. Tuning of various hyperparameters of SVM classifier.
3. Improving the vehicle classification accuracy using off the shelf CNN based approach as compared to human-driven features based approaches.

Remaining paper is structured as follows: In Sect. 2, survey related to vehicle classification is demonstrated. In Sect. 3, the system overview of the proposed approach is described. In Sect. 4, various experiments and their results are listed. In Sect. 5, conclusion and future work are described.

2 Related Work

Since vehicle classification is important in developing countries, various solutions have been proposed by researchers. Infrastructure-based solutions include Inductive Loop Detectors, Magnetic Sensors, Video Cameras, Infrared Sensors, Piezoelectric Sensors, Microwave Radars, etc.

Inductive Loop Detectors [4] and Wireless magnetic sensors [5] were used for classification of vehicles. But they are sensitive to traffic and temperature. Magnetic sensors need orderly and homogeneous traffic to perform accurately. Video cameras were employed for classifying vehicles [6]. But cameras, though can monitor several lanes but are affected by occlusion [7] and adverse weather conditions. Microwave radars were used for classifying vehicles into five classes [8]. But other electronic devices interfere with electromagnetic signals of radar. Overall infrastructure-based solutions are costly in terms of maintenance and installation. Then there are off road solutions for classifying vehicles such as GPS, GSM, and accelerometers [1, 9]. They are installed either in smartphones [10] or vehicles.

Sounds from the vehicles can be used for vehicle classification. Microphones capture the sounds emitted by the vehicles which can be used for determining various traffic parameters like presence, passage, and class of vehicles. Determining these parameters aid in applications like traffic congestion detection [11], accident detection [12], vehicle classification [13]. Traffic congestion state was detected by capturing sounds through microphones installed along the roads [14, 15]. Microphones in smartphones also give considerable accuracy for traffic state congestion detection [16, 17]. Microphones are not affected by lighting conditions and visual occlusions[11], work 24/7 and have low power requirements [16].

Microphones acquired the acoustic signals and classified vehicles into four classes—horns, medium, light, and heavy vehicles using Artificial Neural Network [3]. Vehicles' sounds were acquired by roadside installed microphones and detection and classification of vehicles were performed. MFCCs features were used and classification accuracies for Artificial Neural Network (ANN) and K-Nearest Neighbour (KNN) were compared [18]. Accuracies of Gaussian Mixture Model (GMM), Hidden Markov Model (HMM), and Bayesian Subspace Methods were compared for vehicle classification. Vehicles were classified into nine classes and MFCCs were extracted. Bayesian subspace method gave 50% higher accuracy as compared to others [2]. Probability Neural Network was used as the classifier for classifying vehicles' types and vehicles' positions [13]. Considerable accuracy was obtained. kandpal et al. [19] proposed feature extraction methods that helped in classification of ground vehicles on road in three categories—car, bike, and truck using Multilayer Neural Network. Military vehicles were classified into four classes by acquiring sounds from them. Accuracies of Multilayer Perceptron (MLP) and Probability Neural Network (PNN) based on Gaussian Mixture were compared. PNN gave higher accuracy [20].

We have proposed a new approach of inferring vehicle category through acoustic recordings of vehicles. CNN is employed for feature extraction and SVM for classification of sound recordings. The proposed approach surpasses the human-driven features based approaches for vehicle classification.

3 System Overview

The acoustic data from vehicles can be used to predict the type of vehicle. The vehicles on the roads generate various sounds such as friction between tires and pavement, gears, vibrations in the engine, rotational parts, wind effect, fans, etc. [2]. These sounds act as the signature for each vehicle. We have used off the shelf CNN features based approach. In this, CNN is used for feature extraction and SVM for vehicle classification. This approach is compared with the two approaches. In the

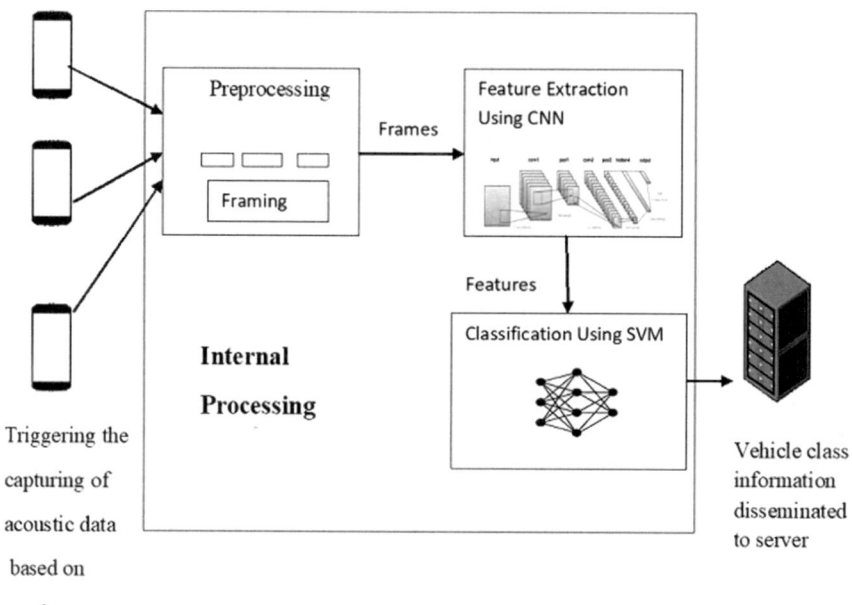

Fig. 1 System overview

first approach, MFCCs, extracted from audio signals are given as input to CNN and in the second approach, MFCCs are given as input to SVM. CNN and SVM classify the audio signals into one of the vehicles' categories.

The system overview of the proposed approach is shown in Fig. 1.

3.1 Preprocessing

In preprocessing phase, the audio signal undergoes framing. Framing is required so as to assume continuously varying signal quasi-stationary. Each audio signal is framed with the frame length of 8192. These frames undergo Discrete Cosine Transformation (DCT) and these coefficients are inputted to the CNN for feature extraction.

3.2 Feature Extraction and Classification

Features extracted from audio signals are the characteristics of vehicles that describe vehicle types. For off the shelf CNN features based approach, CNN is used for feature extraction.

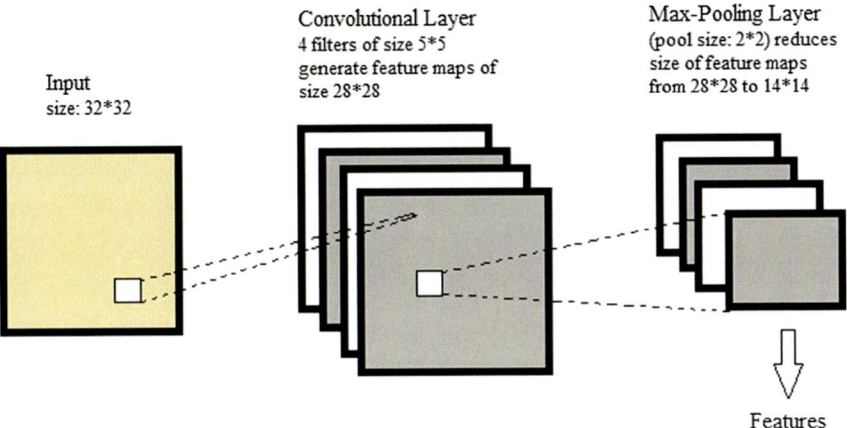

Fig. 2 Feature extraction using CNN

CNN has mainly four layers—Convolutional layer, Relu layer, Pooling layer, and Fully Connected layer. The last layer predicts the class of the vehicles. Before it, the layers extract features which act as the signature for respective class of vehicles. The convolutional layer has a number of filters. Each filter convolves across the audio signal. Each filter is trained to identify specific feature present in the audio signal. The result of filter's convolution is feature map (also called activation map). Set of activation maps become input to the next layer. Each successive layer identifies higher level features. We have used three layers of CNN—convolutional layer, Relu, and max-pooling layer. Features from max-pooling layer are given as input to SVM for classification. Figure 2 illustrates the feature extraction using CNN.

4 Experimental Setup and Results

The size of frame effects the accuracy of classification [16]. The frame size of 8192 is considered the best for traffic management applications. We have compared the accuracy of CNN, SVM, and the proposed off the shelf CNN features based approach. The number of experiments is conducted on the dataset consisting of 4789 recordings each of about 30 s [21]. Audio signals are sampled at 16 kHz sampling frequency. Each audio is classified into one of the five classes of vehicles—car, bus, train, plane, and three-wheeler.

The number of recordings for each class and the class labels are shown in Table 1. Each audio file undergoes framing with the frame length of 8192 and hop length of 2048. We experimented with training set size of 0.8, validation set size of 0.1, test set size of 0.1. The hyperparameters C and gamma are tuned to get the validation accuracy same as training set accuracy. C = 100 and gamma = 0.01 are the finally tuned parameters. Comparative analysis of three approaches is demonstrated in following subsections.

Table 1 Details of dataset

Vehicle category	No. of recordings	Class label
Car	850	0
Bus	1104	1
Plane	542	2
Train	1223	3
Three-wheeler	1070	4

4.1 MFCCs and SVM

Each audio signal undergoes framing and for each audio frame, 13 MFCCs features are extracted. These features are used to train the machine learning classifier, i.e., Support Vector Machine. We experimented with different kernel functions—Radial Basis function (RBF), Linear, and Sigmoid. RBF, Linear, and Sigmoid functions give accuracy of 72.65, 62, and 54.07%, respectively.

4.2 MFCCs and CNN

Each audio signal undergoes framing and for each audio frame, 13 MFCCs features are extracted and are inputted to Convolutional Neural Network. CNN is trained using categorical cross-entropy as the loss function. The size of mini-batch in each epoch is 32. The weights are updated using Adam after the mini-batch is processed. The first layer contains 256 neurons and input size is 13. Then there is another layer of 256 neurons and finally a fully connected layer. As an activation function, Relu is applied in between. There are 5 softmax nodes in the last layer corresponding to the number of classes. The CNN gives an accuracy of 79.97% on the test set.

4.3 Proposed Approach (CNN + SVM)

In this approach, raw audio signals undergo framing. Each frame of audio signal undergoes Discrete Cosine Transformation (DCT). The frame length of 8192 gives 8192 DCT coefficients. These 8192 coefficients are reshaped to the size of $48 * 48$, interval wise. Though the other sizes of $36 * 36$, $40 * 40$, $44 * 44$, and $52 * 52$ were also tried but $48 * 48$ yielded the best results. These are given as input to the convolutional layer of CNN. Input is convolved with filters in the convolutional layer. The dimensions of the output of convolutional layer are reduced by one max-pooling layer. The output from the max-pooling layer is flattened and given as input to SVM for vehicle classification. Experiments are performed with the different number of

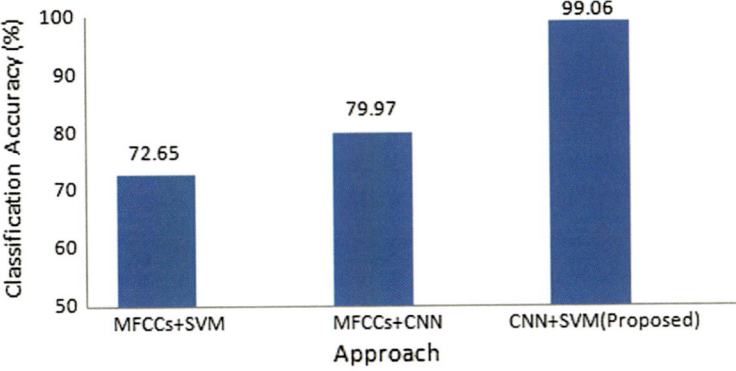

Fig. 3 Comparative analysis of three approaches

filters (4, 8, 16, 32), stride (1, 2, 4), and dimensions of filters (5 * 5, 7 * 7, 9 * 9), different pooling sizes(5 * 5, 7 * 7, 9 * 9). Figure 2 demonstrates the set of parameters for one of the experiments performed using this approach. Different kernel functions of SVM—Radial Basis Function (RBF), Linear function and Sigmoid function are used for experiments.

The best results are obtained with eight filters with filter dimension of 9 * 9, the stride of 2 and the pooling size of 9 * 9. RBF, Linear, and Sigmoid kernel functions give the accuracy of 94.89, 99.06, and 99.06% respectively.

Figure 3 illustrates the comparative analysis of three approaches.

5 Conclusion

In this paper, a new approach for vehicle classification using acoustics has been proposed. We have presented off the shelf CNN features based approach for classifying vehicles into five categories—car, bus, train, plane, and three-wheeler. The accuracy of the proposed approach is compared with SVM and CNN. It gave the accuracy of 99.06% which is quite higher than the approach using MFCCs and SVM (72.65%) and the approach using MFCCs and CNN (79.97%). We anticipate that the new approach can perform well in other acoustic domains as well by using transfer learning methodology and tuning various hyperparameters.

Acknowledgements This work is undertaken as a part of the project 'CARTS—Communication Assisted Road Transportation System' funded by ITRA, Media Lab Asia; and Design Innovation Center, Panjab University, Chandigarh, India.

References

1. Sun Z, Ban XJ (2013) Vehicle classification using GPS data. Transp Res Part C: Emerg Technol 37:102–117
2. Munich ME (2004) Bayesian subspace methods for acoustic signature recognition of vehicles. In: 2004 12th European signal processing conference, pp 2107-2110
3. George J, Cyril A, Koshy BI, Mary L (2013) Exploring sound signature for vehicle detection and classification using ANN. Int J Soft Comput 4:29
4. Gajda J, Sroka R, Stencel M, Wajda A, Zeglen T (2001) A vehicle classification based on inductive loop detectors. In: Proceedings of the 18th IEEE instrumentation and measurement technology conference, IMTC 2001, vol 1, pp 460–464
5. Kaewkamnerd S, Pongthornseri R, Chinrungrueng J, Silawan T (2009) Automatic vehicle classification using wireless magnetic sensor. IEEE international workshop on intelligent data acquisition and advanced computing systems: technology and applications, IDAACS 2009:420–424
6. Lai JC, Huang SS, Tseng CC (2010) Image-based vehicle tracking and classification on the highway. In: 2010 international conference on green circuits and systems (ICGCS), pp 666–670
7. Zhang W, Wu QJ, Yang X, Fang X (2008) Multilevel framework to detect and handle vehicle occlusion. IEEE Trans Intell Transp Syst 9:161–174
8. Urazghildiiev I et al (2002) A vehicle classification system based on microwave radar measurement of height profiles
9. Bhoraskar R, Vankadhara N, Raman B, Kulkarni P (2012) Wolverine: traffic and road condition estimation using smartphone sensors. In: 2012 fourth international conference on communications systems and networks (COMSNETS), pp 1–6
10. Singh P, Juneja N, Kapoor S (2013) Using mobile phone sensors to detect driving behavior. In: Proceedings of the 3rd ACM symposium on computing for development, p 53
11. Tyagi V, Kalyanaraman S, Krishnapuram R (2012) Vehicular traffic density state estimation based on cumulative road acoustics. IEEE Trans Intell Transp Syst 13:1156–1166
12. Ali HM, Alwan ZS (2017) Car accident detection and notification system using smartphone. LAP LAMBERT Academic Publishing, Saarbrucken
13. Paulraj M, Adom AH, Sundararaj S, Rahim NBA (2013) Moving vehicle recognition and classification based on time domain approach. Procedia Eng 53:405–410
14. Sen R, Raman B, Sharma P (2010) Horn-ok-please. In: Proceedings of the 8th international conference on mobile systems, applications, and services, pp 137–150
15. Sen R, Siriah P, Raman B (2011) Roadsoundsense: acoustic sensing based road congestion monitoring in developing regions. In: 2011 8th annual IEEE communications society conference on sensor, mesh and ad hoc communications and networks (SECON), pp 125–133
16. Kaur A, Sood N, Aggarwal N, Vij D, Sachdeva B (2017) Traffic state detection using smartphone based acoustic sensing. J Intell Fuzzy Syst 32:3159–3166
17. Vij D, Aggarwal N (2018) Smartphone based traffic state detection using acoustic analysis and crowdsourcing. Appl Acoust 138:80–91
18. George J, Mary L, Riyas K (2013) Vehicle detection and classification from acoustic signal using ANN and KNN. In: 2013 international conference on control communication and computing (ICCC), pp 436–439
19. Kandpal M, Kakar VK, Verma G (2013) Classification of ground vehicles using acoustic signal processing and neural network classifier. In: 2013 international conference on signal processing and communication (ICSC), pp 512–518
20. Maciejewski H, Mazurkiewicz J, Skowron K, Walkowiak T (1997) Neural networks for vehicle recognition. In: Proceeding of the 6th international conference on microelectronics for neural networks, evolutionary and fuzzy systems, p 5
21. Aggarwal N, Vij D, Soni S (2017) Acoustic vehicular data. Retrieved from http://pudataset.puchd.ac.in:8080/jspui/handle/123456789/14. Accessed on 10 Apr 2018

Conjunctival Vasculature Liveness Detection Based on DCT Features

S. N. Dharwadkar, Y. H. Dandawate and A. S. Abhyankar

Abstract Iris liveness detection algorithms are developed to recognize iris images were acquired from a live person who is actually present at the time of data capture. However the quality of the acquired images will decide success rate. Systems can be spoofed by using fake Photographs, video recordings, printed contact lenses, etc. Conjunctival Vasculature can be used as a biometric trait to identify liveness, paper gives focus on generation of a novel method to extract significant portion of off-angle eye called as sclera. DCT Transform based statistical features are used to find liveliness. System is tested using Extreme learning machines.

1 Introduction

In the past decade, there are found many applications based on biometric-based personal authentication technologies. Biometrics is the study of personal identification using highly unique behavioral characteristics of human beings. Biometric systems have proved their accuracy and convenience of use of requirement for better and robust security in our interconnected world. Liveness detection is measure concern for all security systems worldwide. It is used to find data is taken from live person or not. This is especially crucial for remote authentication. Number of biometric traits has been used to check liveliness [1]. Among all biometrics Conjuctival vasculature is most recent one which can be used along with IRIS.

Conjunctival vasculature recognition is most recent ocular biometric modality. It can be captured in the visible spectrum which is possible in all camera devices.

S. N. Dharwadkar (✉) · Y. H. Dandawate (✉)
E&TC Department, VIIT, Pune 48, India
e-mail: shridharwadkar@gmail.com

Y. H. Dandawate
e-mail: yhdandawate@gmail.com

A. S. Abhyankar (✉)
E&TC Department, SPPU, Pune, India
e-mail: aditya1210@gmail.com

© Springer Nature Switzerland AG 2019
D. Pandian et al. (eds.), *Proceedings of the International Conference on ISMAC in Computational Vision and Bio-Engineering 2018 (ISMAC-CVB)*, Lecture Notes in Computational Vision and Biomechanics 30,
https://doi.org/10.1007/978-3-030-00665-5_111

1171

Simona Crihalmeanu and Arun Ross explained conjunctival vasculature can be used as a soft biometric [2]. Due to rich layer of blood vessel pattern this trait cannot be spoofed easily hence can be used as an assisted technique for robust iris recognition and checking liveliness. They provide detailed explanation for an extraction of conjunctival vasculature pattern from multispectral eye images. After minutiae extraction HU invariant moments were calculated. Identification was done on the basis of correlation. Accuracy was obtained 100% but experimentation was being carried on Limited database of 49 objects with manual process used for segmentation. Seyed Mohsen Zabihi, Hamid Reza Pourreza et al. explained use a classifier-based method for segmentation and extraction of blood vessels from conjunctival images [3]. In their proposed work, Contrast Limited Adaptive Histogram Equalization (CLAHE) method was used to distinguish vessels from background as a preprocessing task. It also explains further, extraction of feature vector by using Local Binary Patterns (LBPs) and finally Artificial Neural Fuzzy Inference System (ANFIS) classifier is used for segmentation of pixel vessels with background. Morphological operations were used as a post processing step applied to improve the results of segmentation process along with removal of noise. However algorithm was tested on 40 conjunctival images. Number of researchers makes use of Iris for liveness detection. Kang R young Park et al. introduced the novel method for fake iris detection [4]. They tracked changes in dimensions of pupil along with local iris features in presence of visible light. Also to detect the change of local iris features, multiple wavelet filters having Gabor and Daubechies bases were used. To estimate accuracy of fake iris, SVM (Support Vector Machine) were used. Virginia Ruiz-Albacete and Pedro Tome-Gonzalez explain in their excellent review discussed emergence of various ambiguities due to direct attacks which of iris-based verification. Fake database was created by printing iris images by using commercial printer and then, presenting images in front of iris sensor [5]. He also highlights different attacks like effect of light reflections or eye movement, pupil response to a sudden lighting event on recognition. Huang, Changpeng Ti, Qi-zhen Hou, et al. have done An experimental Study of Pupil Constriction for checking liveliness [6]. They made use of pupil constriction, to differentiate real and live irises. Vikas Gottemukkula, Sashi Kanth Saripalle et al., used fusion of iris and conjunctiva for recognition [7]. They design a weighted fusion scheme to combine IRIS and conjunctiva. Experiments were conducted on 50 persons indicate that such a fusion scheme gives reduction in equal error rate by of 4.5% as compared with iris recognition.

IRIS-based system can be spoofed by using direct attacks so to add robustness, conjunctival vasculature based modality can be used along with iris. As this modality works in visible spectrum along with robustness it is gaining more and more popularity. Section 2 explains methodology of Fake database generation. It also highlights novel algorithm developed for automatic Sclera extraction from database image. Section 3 provides analytical discussion of a liveliness detection system in terms of statistical features based on DCT transform. Section 4 provides concluding remarks and future scope.

2 Proposed Methodology

In a Conjunctival vasculature based liveness detection an eye image entered in the system can be differentiated as real or fake, so the extraction of image features from region of interest is most important task. Algorithm can be summarized as follows.

2.1 Image Acquisition for Database Generation and Experimentation

Liveness detection based on conjunctiva has found limited exposure in literature. So there is need for generation of an off-angle fake iris database [8]. Database of 719 off-angle IRS images is generated out of which 600 are real images. Out of that 60% are male volunteers and remaining are female volunteers with age group from 20 to 35 years. Figure 1 provides samples of Off-Angle Iris images. Off-Angle eye images

Fig. 1 Original eye image

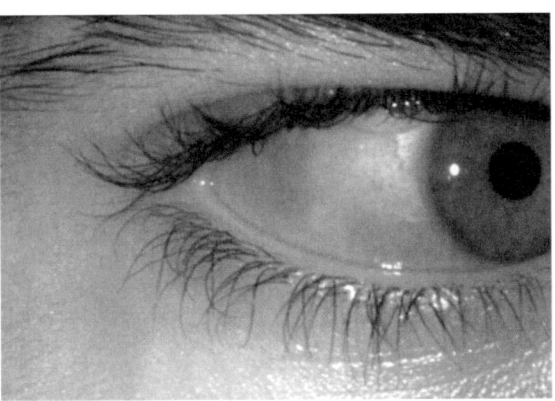

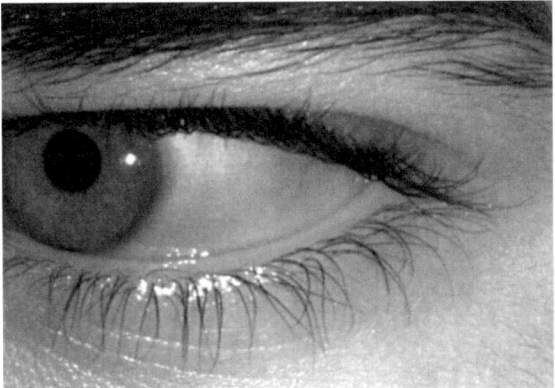

Fig. 2 Fake eye image

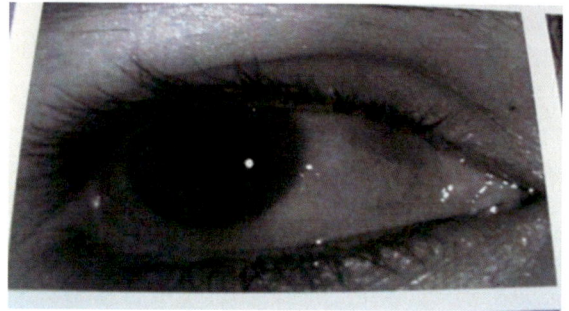

are captured by using CMITech scanner is a binoculars-type iris biometrics imaging device with Iris image pixel resolution of 640×480 pixels. As conjunctival features can be capture in visible spectrum fake database is created by taking photograph by 11 megapixel Sony digital camera. Figure 2 shows sample of fake database images.

A ROI Extraction

After image acquisition significant portion of an eye called as conjunctival vasculature needs to be separate out from as eye image. Proposed preprocessing algorithm follows following steps.

(1) Capture an eye image.
(2) Binarize Input Image by using thresholding.
(3) Apply Adaptive Histogram Equalization.
(4) Crop eye image by drawing Concentric circles with radii equal to that of size of an Iris. Integro-differential operator proposed in [9] will be used for iris localization.
(5) Find different areas of connected components in the binary Image.
(6) Find Maximum area component which extracts white portion of an eye.
(7) Prepare Binary mask.
(8) Apply it on original image to extract sclera pattern.
(9) Fit maximum square from extracted ROI.

(a) **(b)**

(c) **(d)**

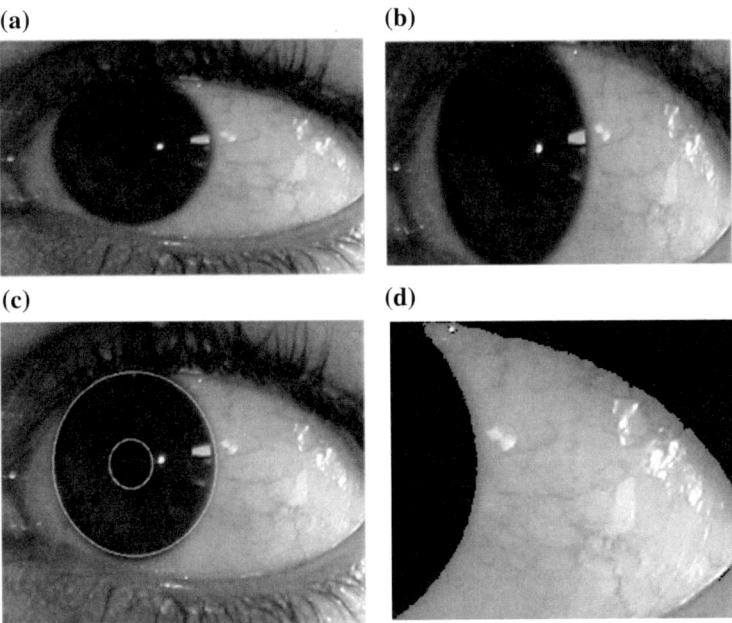

Fig. 3 **a** Original eye image, **b** iris localization figure, **c** cropping based on iris localization, **d** ROI extraction

B Feature Extraction

After extraction of Maximum Square from extracted ROI, DCT transform was applied to highlight low-frequency coefficients compare with high frequency components. DCT transform can be calculated as shown in Eqs. 1–3 respectively.

$$C_k = \frac{2}{N} w(k) \sum_{n=0}^{N-1} x_n \cos\left(\frac{2n+1}{2N}\pi k\right), 0 \le k \le N-1 \tag{1}$$

and

$$x_n = \sum_{k=0}^{N-1} w(k) C_k \cos\left(\frac{2n+1}{2N}\pi k\right), 0 \le n \le N-11 \tag{2}$$

where

$$w(k) = \sqrt{2}, k = 0 \text{ and } w(k) = 1, 1 \le k \le N-1 \tag{3}$$

The use of DCT can produce better results with less computational cost Statistical features like mean, Standard deviation, Variance and Mean were calculated from transform coefficients. First order features can be calculated as follows:

$$\text{Mean} = \frac{1}{mn} \sum_{(r,c) \in W} g(r, c),$$ (4)

where "g" is the image, "r" and "c" are the row and column coordinates respectively, within image size of "$m \times n$"

$$\text{Variance} = \frac{\sum f.(X_1 - \bar{X})^2}{N}$$ (5)

$$\text{Standard Deviation} = \sqrt{\frac{\sum f.(X_1 - \bar{X})^2}{N}},$$ (6)

where $\bar{X}$ is mean and N is size of an image

$$\text{Median} = \text{argmin}_n, \ \alpha(H(n) - |n - \alpha|$$ (7)

Figure 3 highlights different steps used for extraction of conjunctiva.

C Extreme learning Machine Approach

Extreme learning machine is used to classify liveliness of an image. ELM is Single-layer feed forward neural network which provides much faster speed than traditional feed forward network learning algorithms. Advantage lies in assigning random input weights and hidden layer biases of SLFNs [10]. After the input weights and the hidden layer biases are chosen randomly, inverse operation of the hidden layer output vector can then be used to determine the output weights (linking the hidden layer to the output layer) of SLFNs.

3 Experimental Results

Experimentation was carried on 719 images with 600 live and 119 fake images Cross-validation ($k = 2$, 5- and 10fold) is performed to calculate accuracy of the system. Mean, Standard Deviation, Variance and Median were used as a member of feature vectors for training and testing of the system. Sigmoid activation function with 10 numbers of neurons is used for training for optimum performance. Table 1 results proved that system performs better for 10fold cross-validation. Experimentation carried out on Matlab R2015a software with Intel i5 processor with 1.7 GHz speed. Table 2 shows Percentage accuracy of discrete cosine transform with all features for ($K = 2/5/10$) fold cross-validation.

Figure 4 shows calculation of equal error rate (EER) from Receiver Operating Characteristic (ROC) plot. EER is measured at intersection of $-45°$ line with the characteristics. EER with all feature vectors found to be 0.1. Area under Curve (AUC) is 0.987 provides good quality of Classifier. Table 2 shows accuracy of system for different combinations of feature vectors with 10fold cross-validation.

Table 1 Percentage accuracy of DCT transform for cross validation

Transform/method	Features	Percentage accuracy		
DCT Activation function:Sigmoid No. of neurons = 10	Mean + std. deviation + variance + median	$K = 2$	$K = 5$	$K = 10$
		93.88	94.02	94.02

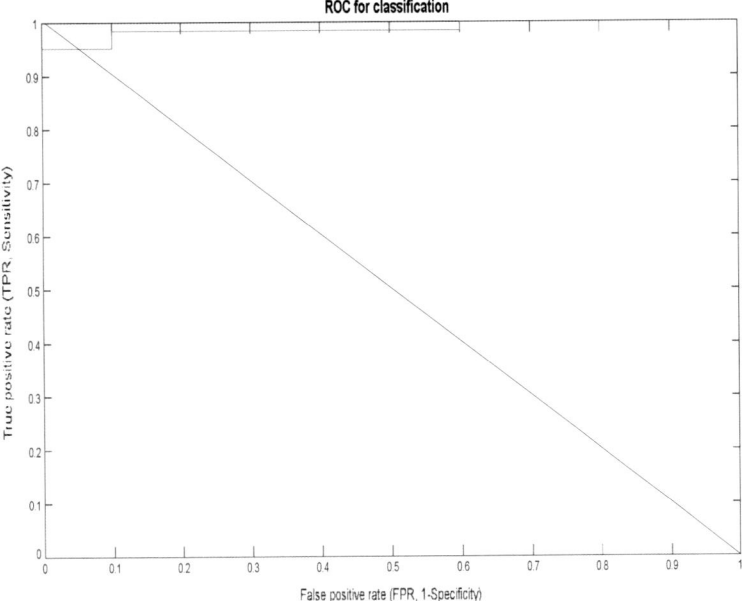

Fig. 4 ROC plot

Table 2 Percentage accuracy of DCT transform for various combination of features

Features	Percentage accuracy
Mean	90.13
Std. deviation	87.21
Variance	90.54
Median	83.59
Mean + variance	93.88
Mean + std. deviation	93.88
Variance + median	86.65
Mean + std. deviation + variance	93.88
Std. deviation + variance + median	86.23
Mean + variance + median	94.02
All features	94.02

4 Conclusion

Paper provides preliminary work based on use of Conjuctival vasculature for liveness detection. It presents novel technique for Sclera extraction from an off-angle eye image. Robustness of system demands reduction of equal error rate. Further investigation is required which transform provides maximum accuracy. Performance of the system needs investigation depending on environmental conditions at the time of database creation.

References

1. Schuckers S (2002) Spoofing and anti-spoofing measures. Inf Secur Tech Rep 7:56–62
2. Crihalmeanu S, Ross A (2011) On the use of multispectral conjunctival vasculature as a soft biometric. In: Proceedings of IEEE workshop on applications of computer vision (WACV), Kona, USA, Jan 2011, pp 204–211
3. Zabihi SM, Pourreza HR, Banaee T Vessel extraction of conjunctival images using LBPs and ANFIS. Int Sch Res Netw ISRN Mach Vis 2012(Article ID 424671)
4. Park KR, Whang MC, Lim1 JS, Cho Y (2006) Fake iris detection based on multiple wavelet filters and hierarchical SVM. Springer, ICISC 2006, LNCS 4296, pp 246–256
5. Ruiz-Albacete V, Tome-Gonzalezm P (2008) Direct attacks using fake images in iris verification. Springer, BIOID 2008, LNCS 5372, pp 181–190
6. Huang X, Ti C, Hou Q-z (2013) An experimental study of pupil constriction for liveness detection. In: IEEE workshop on applications of computer vision (WACV), pp 252–258
7. Gottemukkula V, Saripalle SK (2012) Fusing iris and conjunctival vasculature: ocular biometrics in the visible spectrum. In: IEEE conference on technologies for technologies, pp 150–155
8. Schuckers SAC, Schmid NA, Abhyankar A (2007) On techniques for angle compensation in nonideal iris recognition. IEEE Trans Syst Man Cybernetics Part B 37(5):1176–1190
9. Daugman J (1993) High confidence visual recognition of persons by a test of statistical independence. IEEE Trans Pattern Anal Mach Intell 15(11):1148–1161
10. Huang G-B, Zhu Q-Y, Siew C-K (2006) Extreme learning machine: theory and application. Neurocomputing 70:489–501

Unusual Social Event Detection by Analyzing Call Data Records

V. P. Sumathi, K. Kousalya, V. Vanitha and N. Suganthi

Abstract The availability of call data records provides an opportunity to identify unusual social events occurring in the society in an effective manner. The visitors participating in the events are the most important stakeholders for event organizers to improve their success rate. Visitors global positioning system (GPS) enabled device provides spatial data that are used to identify the visitors' presence during event time. Mobile call data records (CDR) represent a real spatial data source to detect occupied visitors, but provide less accuracy in terms of time and space resolution. Using spatial data, it is possible to detect unusual events. In this paper, the method for detecting unusual social events in various locations is proposed. The CDRs are used to detect new visitors who participated in the event and total number of visitors participated is computed for free-to-view social event. The information extracted from preprocessed CDRs are utilized to identify new visitors and to compute total visitors present in an event place effectively. The tower-wise visitor's details and CDR details provide information about the visitors' movement as well as CDRs distribution pattern during the event time. The experiment is conducted on real CDRs provided by a telecommunication service provider (TSP) servicing in a larger city. Results show that the proposed method provides accurate identification visitors involved in unusual social events compared to the state-of-the art methods.

V. P. Sumathi (✉) · V. Vanitha · N. Suganthi
Department of Computer Science and Engineering, Kumaraguru College of Technology, Coimbatore, Tamil Nadu, India
e-mail: sumathi.vp.cse@kct.ac.in

V. Vanitha
e-mail: vanitha.v.cse@kct.ac.in

N. Suganthi
e-mail: suganthi.n.it@kct.ac.in

K. Kousalya
Department of Computer Science and Engineering, Kongu Engineering College, Perundurai, Tamil Nadu, India
e-mail: kouse@kongu.ac.in

© Springer Nature Switzerland AG 2019
D. Pandian et al. (eds.), *Proceedings of the International Conference on ISMAC in Computational Vision and Bio-Engineering 2018 (ISMAC-CVB)*, Lecture Notes in Computational Vision and Biomechanics 30,
https://doi.org/10.1007/978-3-030-00665-5_112

1179

1 Introduction

In the past 5 years, mobile users and their societal-related activities have been increasing enormously. In developing countries, almost 70% of the people have smart mobile phones and have been participating in chatting and messaging all the time. A CDR is a record of a call setup and completion, and its format varies among telecom providers or programs, some of which are allowed to be configured by the user. It is also known as call detail record. Apart from mobile users' call and SMS data, the TSP creates data related to mobile network and mobile user profile. Naturally, the CDRs are in large volume and have privacy and security. Analysis of these data helps in deriving meaningful information like people's commuting behavior within a city [15, 17], inferring attractiveness towards public events and significance of social trends. It also helps in city planning [9, 14], marketing, customer profiling, food consumption patterns, disease spreading patterns, people's response to armed conflict, natural disasters, social event occurrence and its impact.

There are two kinds of events: (i) pay-to-view where the details of the event are already well known by organizer and these events are pre-planned. The number of participants, event duration and event location are fixed during the planning time, and (ii) free-to-view event where the participant's details may not be available until the day of the event, but the location and event duration are fixed. The free-to-view social event has to be analyzed to understand various factors like participant count for efficient crowd management, event planning and resource prioritization. In addition, understanding of user mobility patterns and events of interest [7, 8] are also of significance. This offers support for traffic and road management.

One of the important benefits of the social event analysis is identifying more involved visitors, who are very important for measuring the event success of an event organizer. The success can be measured using direct results like visitor count, involvement of the visitor and indirect factors like feedback from visitor. To understand the details of a visitor, it is necessary at first to find the retention of the existing visitor, the total count of the visitor, the visitors' mobility during the event time and the visitors' involvement in the event.

2 Review of Literature

CDRs contain both spatial and temporal data, facilitating the study of the mobility patterns and social behavior of individuals as well as the crowd. This rich source of data has also been taken into consideration to understand the success of an event by estimating the visitor count. Conventional methods of estimating the participant count like sampling techniques to estimate the number of participants or count the number of feedback collected may be biased. Z-score based event attendance estimation approach has been proposed [4, 10] with median error of less than 15%. It first identifies the towers associated with the event location and uses the CDRs generated by the network cells identified during the event to estimate the number of attendees. During the event, the total calls and SMS made by the mobile users are found to be directly proportional to the crowd size in that area [1, 3]. CDRs have been utilized to measure the effectiveness of billboard advertising. A system that estimates the effectiveness of an advertising based on the number of mobile phones near the billboard and infers people's preferences by using freely available event information on the internet is proposed [12].

The occurrences of social events like strikes or riots have been studied using mobile data [5]. The Bayesian location inference framework is proposed to detect unusual social gathering and to indicate the users who would have attended the event, and when and where an event has happened. Probabilistic approach has been employed to take into account the erratic antenna jumps [16]. The user mobility patterns are analyzed and the origins of people attending an event have been shown to be strongly correlated to the type of event. In addition, predictive analytics has been used to understand the users' interest patterns [2, 11]. The digital footprints created through mobile phones are utilized to rank the attractiveness of the place [6]. The social event recommendation system using K-nearest neighbor (K-NN) algorithm has been designed using mobile phone data [13]. Solving crowd estimation problem can help event managers predict the success of an event, estimate the number of participants and identify participant cluster which is engaged with mobile phone during the event time. As the CDR generated depends on the number of visitors in the event location, the problem of bias can be avoided.

2.1 Motivation

The present paper focus on analyzing the large amount of CDRs generated by telecommunication equipment, which is used to make decision to identify the unusable social event, to analyze the behavior of the visitors involved in the event, to identify the tower-wise CDR cluster and visitor count engaged mostly with mobile phone.

3 Materials and Methods

Social events are analyzed using CDR attributes. The six attributes related to social behavior analysis are total calls, number of incoming calls, number of outgoing calls, calls made to individuals physically located in the city, number of unique contacts, fraction of contacts called in the previous month and so on. The current study makes an attempt to identify unusable free-to-view events held at various locations in a city. The challenge is that the event location considered is highly distributed and the duration of the event is more than 7 h. To find out the visitor count, the amount of time spent by an individual at the event is considered. To get an accurate number of visitors, the CDRs of people who have their home or work base in that particular event location are removed.

3.1 Data Set

The CDRs are collected from a mobile service provider of Bengaluru, a city in India. The input data set contains CDRs for one month for both call and SMS. The call type attributes contains four types of values like 0, 1, 2, and 3. The values 0, 1 represent incoming and outgoing calls while 2, 3 indicate incoming and outgoing SMS. The total call duration is in the form of seconds. IMSI is used to identify the presence of other state visitors in the event while IMEI is used to identify the visitors with more than one Subscriber Identity Module (SIM). The call initiation and termination are identified using starting tower identity and ending tower identity numbers. The call status indicates success and failure of the call or SMS. Network cell details are utilized to know about the tower details like location of the tower and unique tower identity. The input tower data contain more than 10,000 tower details. All these towers are erected within Bangalore city. The latitude and longitude are used to compute the radius of an event area. The distance is calculated by using the formula as in Eq. (1).

$$
\begin{aligned}
\text{Calculated distance (in Kilo Meter)} = {} & ACOS(COS(RADIANS(90 - Lat1)), \\
& \times COS(RADIANS(90 - Lat2)) \\
& + SIN(RADIANS(90 - Lat1)) \\
& \times SIN(RADIANS(90 - Lat2)) \\
& \times COS(RADIANS(Long1 - Long2)) \times 6371
\end{aligned}
$$

$$(1)$$

where Lat1 and Lat2 denotes latitude and Long1 and Long2 denote longitude. The system is implemented using Python language, Graphlab with Turi machine learning environment.

3.2 Preprocessing of CDRs

The mobile CDRs observed by different mobile service providers do not have uniform format and structure. The values observed are large in size. Almost 50 million CDRs are generated every day. There is a need to process the CDRs generated and to convert them into a suitable form for applying data analytics. The data preprocessing step is quite important in any data mining application to achieve higher accuracy. The pre-processing involves missing value handling and outlier detection.

3.2.1   Missing Value Handling Techniques for Mobile CDR

The collected CDRs may contain missing value records due to system malfunction, network failure, mobile phone failure, power failure, and so on. For estimating visitor count, the missing value CDRs are identified and removed. This missing value removing technique is termed as reduction of data set. This technique is not suitable in all aspects of CDR analysis. The important task is to identify the unique users present in event location. The missed value record removal does not provide much effect on visitor estimation. The mobile user, making a call or SMS, is enough to identify the unique visitor in that location.

3.2.2   Outlier Detection and Removal from CDR

CDR needs to be pre-processed to remove the outliers. Outliers are mobile users who make a huge number of calls or SMS than the normal mobile users. The outlier call details are identified by counting the total number of CDRs generated by each user per day. If the total number of CDR associated with one particular user is greater than the specified threshold value, then that user will be considered as outlier. The threshold is selected by analyzing the call distribution.

3.2.3   Analyzing New Visitors' Presence in a Specific Location

Visitors' presence during the event is analyzed by first identifying the presence of regular users in the event location other than event day, and then removing these users from the CDRs generated in that event location during the event. This step is repeated for several days before and after the event day. The mobile users, who make and receive at least one call or SMS every day, are considered as regular users. It is understood that these users are living, working or commuting in the event location other than the event day and event time. The remaining CDRs are associated with the presence of new visitors in the event area during event time and the same helps in estimating the visitor count.

3.2.4   Analyzing Visitors' Presence During an Event Time

During the event day, the presence of visitors for a fraction of time at the time of the event is directly proportional to the presence of the visitor in the event. In the same way, the fraction of time in which the visitor is there outside the event time is inversely proportional to the presence of the visitor in the event. The inter-CDR time is calculated by considering time difference between the first CDR generated and second CDR generated on the same day. To identify the amount of time spent by the user in the event location, Maximum Inter-CDR Time (MICT) is computed by considering first and last call or SMS made by the visitors. MICT is the time difference between the first CDR and the last CDR generated on the same day. It is computed for every visitor in the event location on the day of the event. Its value is used for the estimation of visitor count. The threshold value is fixed at 60% based on the events like sports, conferences, cultural, social events, and so on and the event location whether it is distributed or centralized. The CDRs with MICT value less than the threshold value are considered as the normal mobile users.

3.3 Tower-Wise Visitor Clustering Algorithm

In usual events, the actual participant is known by counting the number of tickets sold on event day, which is called ground truth. However unusable events, the visitor count is not always equal to ground truth because (i) some visitors may not make any call or SMS during the event day and (ii) all the visitors may not belong to the same service provider. The visitor clustering algorithm is given in Fig. 1.

3.4 Tower-Wise CDRs Clustering Algorithm

During unusual events, the CDR count gets either increased or decreased based on the nature of the event. The CDR count always depends on the number of visitors involved in the event day in that location. The tower-wise CDR clustering is used to understand the visitors' behavior belonging to the same service provider as given in Fig. 2.

Visitor clustering algorithm (CDRs and tower location details)

Input: CDRs filtered based on date, location, duration of event and tower identity
Output: Tower-wise participant cluster.

1. Extract both call and SMS CDRs associated with each mobile user.
2. Combine both call and SMS CDRs associated with a particular mobile user.
3. Find the time of first CDR generated during a particular day.
4. Find the time of last CDR generated during a particular day.
5. Compute the time difference between first and last CDR which is equal to MICT of visitor.
6. The visitors count is computed based on MICT value of individual user.
7. The visitors are clustered based on tower usage pattern.

Fig. 1 Visitor cluster algorithm

CDR clustering algorithm (CDRs and Tower location Details)

Input: CDRs filtered based on date, location, event duration and tower identity
Output: Tower-wise CDR cluster.

1. Extract both call and SMS CDRs associated with specific location.
2. Combine both call and SMS CDRs associated with all mobile users.
3. Identify and remove all outliers as well as regular users. Consider only the new user present in the event.
4. The visitor count is computed based on MICT value of individual user.
5. The identified visitors' CDRs are filtered and considered for clustering.
6. Cluster the CDR based on date, time and tower usage details.

Fig. 2 CDRs clustering algorithm

4 Result and Discussion

By nature, CDRs are in large volume (Bid data), completely raw (not suitable for data mining), pattern extraction is complicated, real-time data and useful features must be identified. Limitations of CDRs are call frequency that depends on wealthier people, zero tower issue while non-mobile users are excluded. The users considered are educated who lives in urban areas. The pit falls of CDRs are high density of towers in cities, lower density of towers in rural area and frequent users versus new users. The users are making more than 100 calls or SMS considered as outliers. After removing the outliers, the CDR distribution pattern throughout the day is shown in Fig. 3. The data of all the days, except March 10, 2013, are obtained and found to be of similar pattern of distribution. Minimum of 99.50% of mobile users are considered as normal user. Similarly the call distribution patterns of all Sundays in March 2013 are compared in Fig. 4. In both the graphs, March 10 shows dissimilar pattern compared to all the other days. Some unusual events that occurred on that day are clearly shown in the graphs.

Towers present within 1 km (30 towers) in and around the event location are considered. This place always has more people during the night time. The event time duration from 6.00 p.m. to 12.00 p.m. is considered for counting new visitors in that location. The visitors count was more on all Sundays 2013, except March 10th as shown in Fig. 5. It clearly shows that some unusual things have occurred on that day.

Table 1 shows the new visitor count on all Sundays. The days considered are 3, 10, 17, 24, and 31 of March 2013. The users present in all the days (intersection) are considered and their CDRs are removed while those present in any one of the days (union) other than the day considered (March 10) are removed. The new visitor count is 454 (intersection) and 219 (union) on March 10. Some unusable events could occur on March 10.

After identifying the visitor count, the visitors are clustered based on their tower usage pattern during the event day and time. Figure 6 clearly shows the tower-wise

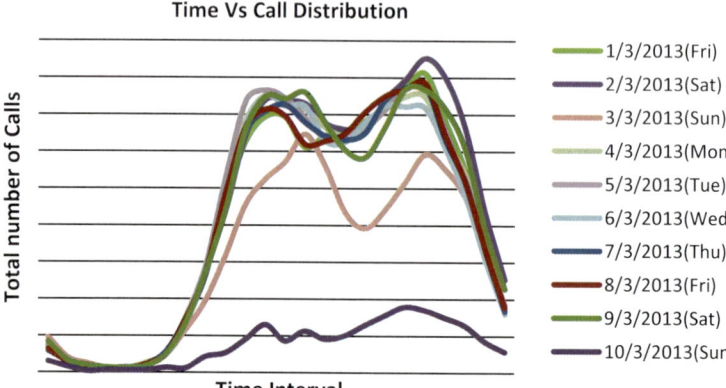

Fig. 3 Call distribution for 10 days

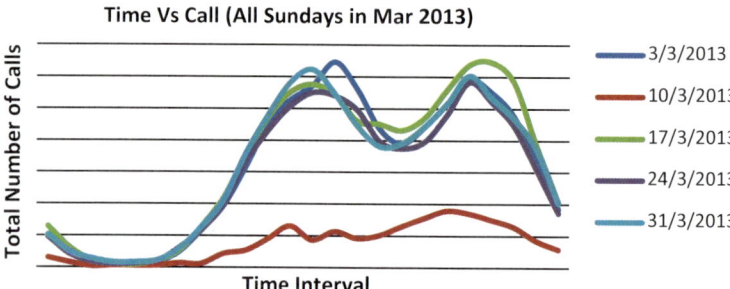

Fig. 4 Call distribution for all Sundays in May 2013

Fig. 5 Call distribution (during event time in all Sundays)

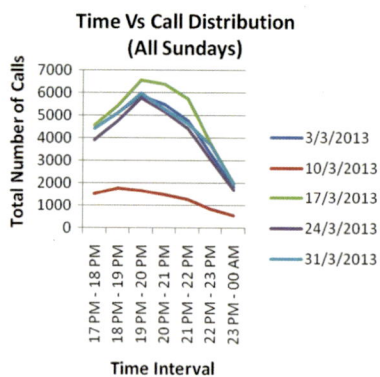

visitor count distribution. The new visitor CDRs are clustered based on tower identity. The tower-wise CDR count distribution is clearly shown in Fig. 7.

Table 1 New visitors' count using mobile CDR

New visitors' count		
Event day (intersection of incoming and outgoing CDRs)	Filtered by	
	Common user on all days (intersection)	All common users on all days (union)
3/3/2013	5704	4978
10/3/2013	454	219
17/3/2013	6318	5426
24/3/1023	5220	4354
30/3/2013	5700	4921

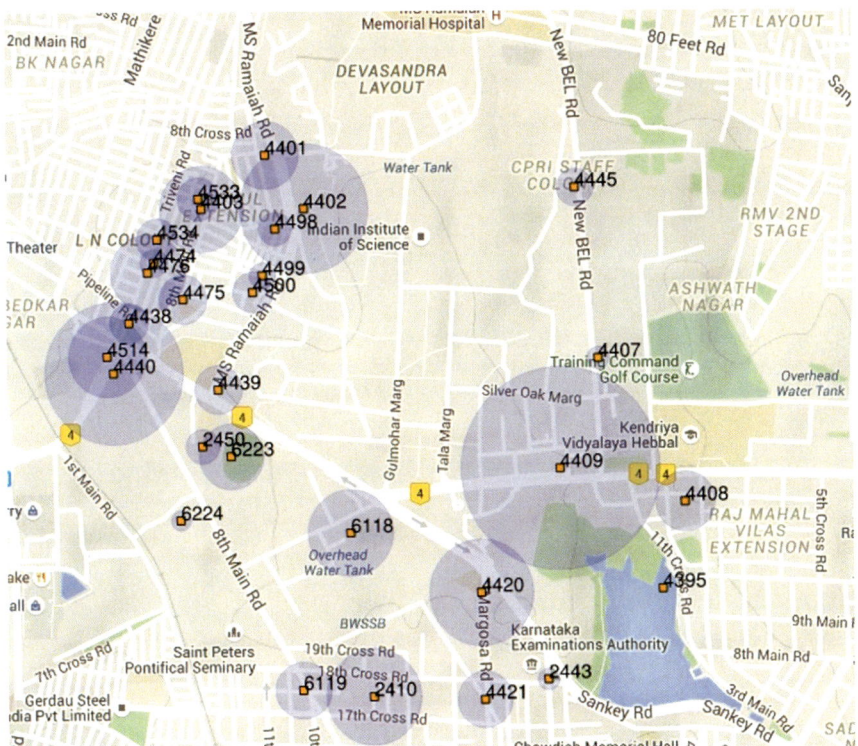

Fig. 6 Tower-wise visitor distribution

5 Conclusion

The mobile CDRs collected from single TSP are used to identify the new visitors present in an unusual event that occurred in a city. Every day the mobile CDR count is around 52 million records. The outliers, who make more number of calls in the

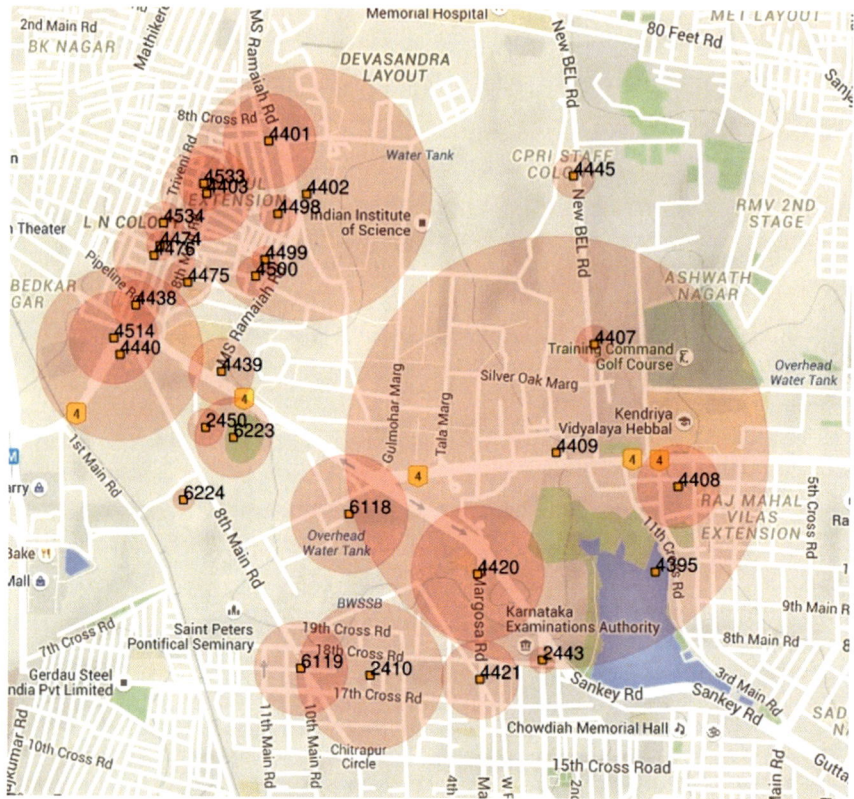

Fig. 7 Tower-wise CDRs distribution

event area, are identified and removed. New visitors present in the event are also identified and counted. The towers present in and around the event location are also identified and the CDRs are extracted from the total collected CDRs. Likewise, the duration of the visitors' stay in the event area is identified by computing MICT. The visitor count identified is less compared to all the other days. Some unusual events occurred on this day are identified. The tower-wise visitors and CDRs are analyzed and it is found that most of the visitors' associated CDRs are serviced by the towers erected near to the event location. Further research directions using CDRs are visitor level mobility between the events, visitors' behavioral analysis, population mobility changes during emergency events like earth quake and a bomb blast and classification of cities based on spatiotemporal data.

Acknowledgements I thank Indian Academy of Science (IAS) for having given me an opportunity to take up research under Summer Research Fellowship Programme at Indian Institute of Science (IISc), Bengaluru. It has provided me valuable suggestions and data set for doing the present work in Super Computer Research Centre in IISc.

References

1. Botta F, Moat HS, Preis T (2015) Quantifying crowd size with mobile phone and twitter data. R Soc Open Sci 2(5):150–162
2. Calabrese F, Di Lorenzo G, McArdle G, Pinelli F, Van Lierde E (2015) Real-time social event analytics. Netmob
3. Calabrese F, Pereira FC, Di Lorenzo G, Liu L, Ratti C (2010) The geography of taste: analyzing cell-phone mobility and social events. LNCS, pp 22–37
4. Davies L, Coleman R, Ramchandani G (2010) Measuring attendance: issues and implications for estimating the impact of free-to-view sports events. Int J Sport Mark Spons 12(1):6–18
5. Ferrari L, Mamei M, Colonna M (2014) Discovering events in the city via mobile network analysis. J Ambient Intell HumIzed Comput 5(3):265–277
6. Girardin F, Vaccari A, Gerber A, Biderman A, Ratti C (2009) Quantifying urban attractiveness from the distribution and density of digital footprints. Int J Spat Data Infrastruct Res 4:175–200
7. Hadden J, Tiwari A, Roy R, Ruta D (2007) Computer assisted customer churn management: state-of-the-art and future trends. Comput Oper Res 34(10):2902–2917
8. Lee AJT, Chen Y-A, Ip W-C (2009) Mining frequent trajectory patterns in spatial–temporal databases. Inf Sci 179(13):2218–2231
9. Louail T, Lenormand M, Ros OGC, Picornell M, Herranz R, Frias-Martinez E, Ramasco JJ, Barthelemy M (2014) From mobile phone data to the spatial structure of cities. Sci Rep 4
10. Mamei M, Colonna M (2016) Estimating attendance from cellular network data. Int J Geogr Inf Sci 1–21
11. Oliver N, Rosario B, Pentland A (1999) Graphical models for recognizing human interactions. In: Advances in neural information processing systems, pp 924–930
12. Quercia D, Di Lorenzo G, Calabrese F, Ratti C (2011) Mobile phones and outdoor advertising: measurable advertising. Institute of Electrical and Electronics Engineers
13. Quercia D, Lathia N, Calabrese F, Di Lorenzo G, Crowcroft J (2010) Recommending social events from mobile phone location data. In: IEEE 2010 data mining (ICDM), University of Technology, Sydney, Australia. IEEE, pp 971–976
14. Ratti C, Frenchman D, Pulselli RM, Williams S (2006) Mobile landscapes: using location data from cell phones for urban analysis. J Environ Plan B: Plan Design 5:727–748
15. Toole JL, Lin Y-R, Muehlegger E, Shoag D, González MC, Lazer D (2015) Tracking employment shocks using mobile phone data. J R Soc Interface 12(107):2015018
16. Traag V, Browet A, Calabrese F, Morlot F (2011) Social event detection in massive mobile phone data using probabilistic location inference. In: IEEE 2011 privacy, security, risk and trust (PASSAT) and 2011 IEEE third international conference on social computing (SocialCom) 2011 IEEE third international conference, MIT Media Lab, Boston, MA, USA. IEEE, pp 625–628
17. Williams NE, Thomas TA, Dunbar M, Eagle N, Dobra A (2015) Measures of human mobility using mobile phone records enhanced with GIS data. PloS One 10(7):e0133630

Drunk Driving and Drowsiness Detection Alert System

Vivek Nair and Nadir Charniya

Abstract Advancement of safety features to avert drunk and drowsy driving has been one of the leading technical challenges in the automobile business. Especially in this modern age where people are under serious work pressure has led to higher crash rates. To prevent such accidents this paper discusses the use of nonintrusive techniques by using visual features to determine whether driver is driving in alert state. Drowsiness detection has been implemented using HAAR Cascade for face and eye closure detection and yawn detection implemented using Template matching in visual studio 2013. For drunk state detection, an alcohol sensor (MQ-3) has been implemented to avoid drunk driving. If the driver is found to be in drunk or drowsy condition, then an alarm would be generated and the driver being alerted using a buzzer and a vibrator that can be placed in the seatbelt or under driver seat thus preventing from mishaps taking place.

1 Introduction

Scientists have been working for over a decade on designing driver inattention monitoring framework. Over the years, several improvements in driver safety have been made yet a significant number of serious accidents still occur all over the world. Driver drowsiness and drunk state being the major reasons. Each year approximately 60,000 automobile accidents occur due to sleepiness related problems. Studies indicate that 25–30% of driving accidents are related to drowsiness [1]. Major approaches developed to detect driver inattention are classified as physiological, driving-behavior-based, and visual-feature-based approaches. In pragmatic applications, visual-feature-based approaches are preferred since they are naturally nonintrusive to the driver [2]. The most common usage of yawning detection is in driver

V. Nair · N. Charniya (✉)
EXTC, V.E.S. Institute of Technology, Mumbai, India
e-mail: nadir.charniya@ves.ac.in

V. Nair
e-mail: nvivek94@gmail.com

© Springer Nature Switzerland AG 2019
D. Pandian et al. (eds.), *Proceedings of the International Conference on ISMAC in Computational Vision and Bio-Engineering 2018 (ISMAC-CVB)*, Lecture Notes in Computational Vision and Biomechanics 30,
https://doi.org/10.1007/978-3-030-00665-5_113

fatigue detection systems, where yawning is one factor among others, such as percentage eye closure, eye blink rate, etc. [3]. The U.S. national highway traffic safety administration (NHTSA) fatality analysis reporting system encyclopedia shows that there were approximately 55,926 vehicles involved in collisions in 2007 [4]. India revealed that about 40% of the road accidents have occurred under the influence of alcohol in a study done by alcohol and drug information centre (AIDC) [5].

2   Project Objective

This work aims towards saving valuable lives and prevent road accidents by developing a framework to avoid drunk and drowsy driving. The work has been developed using visual features that are nonintrusive in nature and can easily detect driver drowsiness which is monitored using HAAR Cascade files from OpenCV, yawn detection via template matching and drunk state using MQ-3 sensor. An alert is generated if driver is found to be drunk or drowsy state with help of buzzer and vibrator, thus, providing an all-round protection to driver and others present in the automobile.

3   Face and Eye Detection

Techniques using visual features have been implemented using computer vision approach for the detection of drowsiness. Exploiting visual features focuses on extracting facial features like face, eyes, and mouth [6]. Analyzing the state of eyes and mouth can provide observable cues for drowsiness detection process. This section focuses on how face and eyes have been successfully detected using the HAAR Cascade, which is based on Viola–Jones algorithm [7].

Working of Face detection is as described.

The working of Face detection using HAAR Cascade is as shown in Fig. 1. Where in first the image acquired on system start up, the images are gray-scaled and histogram equalization done for faster operation and better detection. The HAAR Cascade files were used from OpenCV library to detect the face which was marked with rectangle on detection. Here faces detected are marked with rectangles as shown in Fig. 2. If more than one face found by camera, then the next step was to find the largest face (as per the largest rectangular area found; as the driver, will be the one placed closest to the camera) and then set the largest face as Region of Interest (ROI). Driver's eyes were detected from the ROI as shown in Fig. 3 using HAAR Cascade for faster computation time.

Figures 3 and 4 shows the eyes being detected from the largest detected face. Figure 5 provides with the overall flow on how drowsiness of the driver would be detected using visual features. As described earlier once the face is detected, the largest face is set as ROI and from which the eye was detected using HAAR Cascade

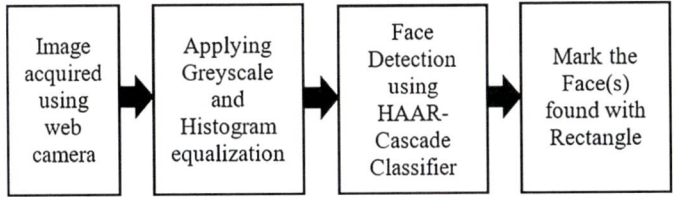

Fig. 1 Steps used to for face detection

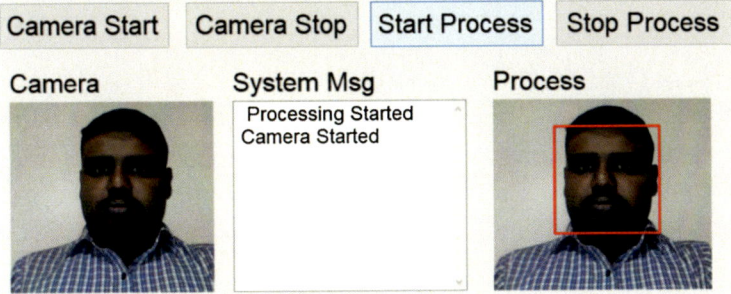

Fig. 2 Face detection using HAAR Cascade classifier

Fig. 3 Detection of eyes from face

[8]. If the eyes were found to be in closed state for a duration of three seconds or more an alert would be generated with buzzer and vibrator to alert the driver.

4 Yawn Detection

Yawn detection has been implemented using Template matching. Here the matching done between the template patch sliding over the image patch using normalized correlation coefficient. There were six methods overall to find template matching [9]. Normalized Correlation Coefficient provided with the best results as shown in Fig. 6.

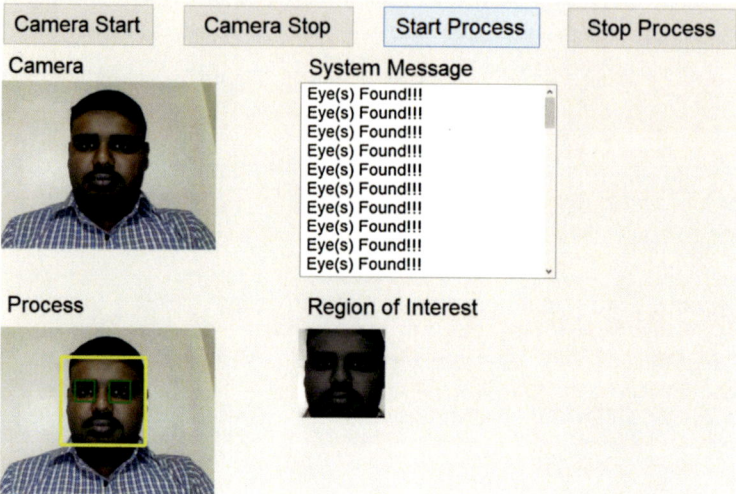

Fig. 4 Results for eye detection

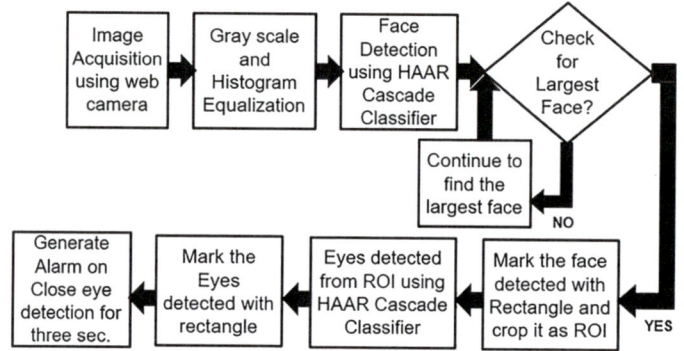

Fig. 5 Flow diagram to determine drowsiness of a driver using visual features

$$R(x, y) = \frac{\sum_{x',y'} \left(T'(x', y') \cdot I'(x + x', y + y') \right)}{\sqrt{\sum_{x',y'} T'(x', y')^2 \cdot \sum_{x'y'} I'(x + x', y + y')^2}}, \quad (1)$$

where,

$$T'(x', y') = T(x', y') - \frac{1}{(w.h). \sum_{x'',y''} T(x''y'')}. \quad (2)$$

$$I'(x + x', y + y') = I(x + x', y + y') - \frac{1}{(w.h). \sum_{x'',y''} I(x + x'', y + y'')}. \quad (3)$$

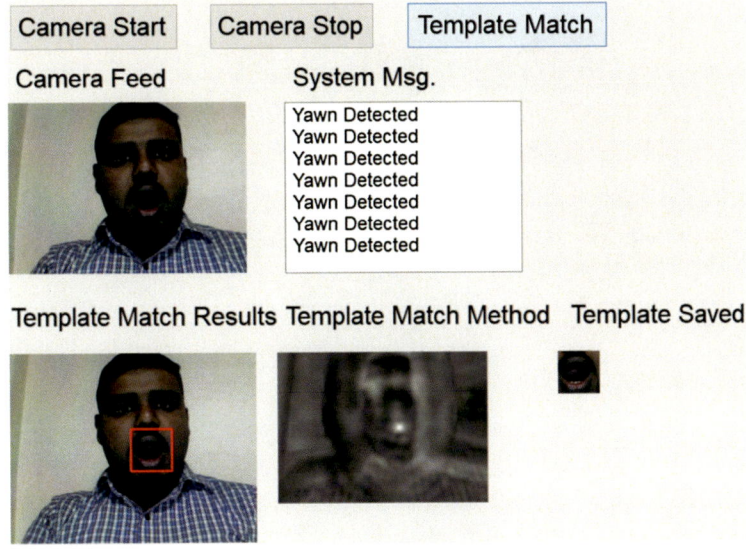

Fig. 6 Yawn detection using normalized correlation coefficient

$I_{(x,y)}$ Value of image pixel in location (x, y)

$T_{(x,y)}$ Template value of image pixel in location (x, y)

$R_{(x,y)}$ Resultant value in location (x, y) for Normalized Correlation Coefficient

Source Image W (Width).H (Height) of Source image pixels.

Template Image w (width).h (height) of Template image pixels.

Size of resulting images $W - w + 1.H - h + 1$ pixels.

Yawn detection as shown in Fig. 7 has been divided into two stages: training and testing respectively. Figure 7a shows the training stage; on image being acquired, manually select the yawn template of the driver via a mouse drag over yawn region and then resize the image and save the yawn template this concludes the training part. Figure 7b describes the testing phase where we first set a threshold value for with a match would be generated and perform template matching using the correlation coefficient normalized method. On yawn detection, a rectangle will be drawn of the area where yawn has been detected and generate an alert with help of buzzer and vibrator to alert the driver.

5   Drunk State Detection

For drunk state detection, a simple MQ-3 alcohol sensor has been used which easily detects ethanol in the air [10, 11]. This sensor has high sensitivity to alcohol and

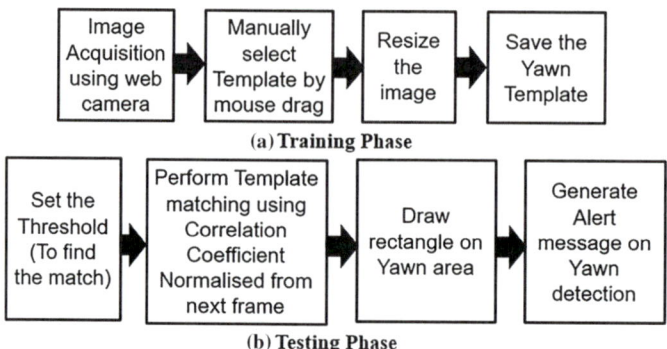

(a) **Training Phase**

(b) **Testing Phase**

Fig. 7 Flow diagram to determine drowsiness of a driver from Yawn state

small sensitivity to Benzine. Its fast response with stable and long life makes it the best choice for drunk state detection. It is commonly used in several breath analyzers or breath testers for the detection of ethanol in the human breath. The core system is the cube which is made of an Alumina tube covered by SnO_2 (tin dioxide) as shown in Fig. 8. Coil on being heated the SnO_2 ceramics act as a semi-conductor, resulting in more movable electrons, which increases current flow. When the alcohol molecules in the air meet the electrode, ethanol burns into acetic acid resulting in more current being produced. So, greater the alcohol molecules, more current generated. This change in current results in alarm generation [12].

Figure 9 describes the working of the system. When a driver enters the car and on turning "ON" the ignition the driver first gives a breath sample to which if found to be in drunk state the emergency lights glow and the engine is turned "OFF". If driver found to be sober then the web camera is used to capture driver's visual features. A yawn template is to be provided to find yawn detection. After which the system is continuously monitoring driver's behavior. On eye closure or on yawn detection an

Fig. 8 Construction of
MQ-3 Gas sensor [12]

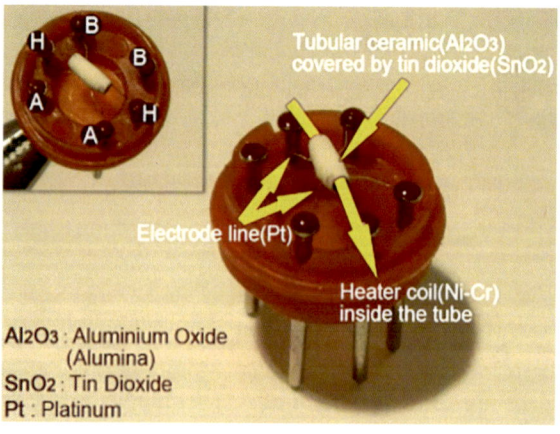

Fig. 9 The flow diagram of the entire working of system

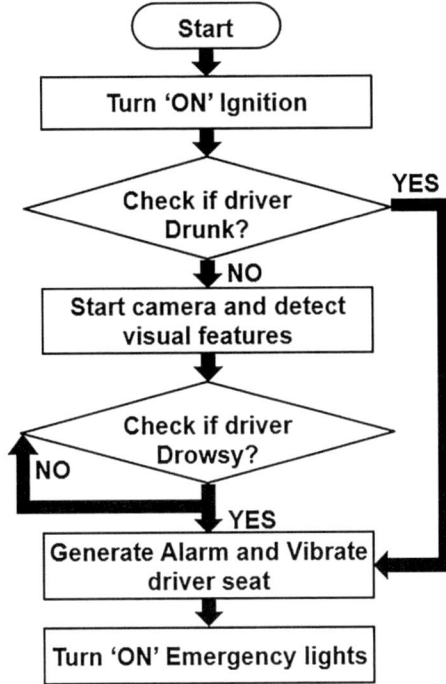

alert is generated with help of a buzzer and vibrator motor placed (under driver seat or on seat belt) which will help the driver to get back to alert state. On the outside, the emergency lights would be turned "ON" to avoid accidents. The alert will be continuous until the driver himself turns "OFF" the alarm.

The system diagram of driver drunk and drowsy state detection is as shown in Fig. 10. An iBall C8.0 web camera with 4.0 MP video resolution at 30fps was used for capturing driver's visual features in real time which would be displayed on the PC screen. The PC is connected to the LPC2148 microcontroller via an USB to TTL connection for serial communication. Once the communication is made the system becomes ready for execution. Once the driver being found to be in drunk or drowsy state the LCD displays message stating the driver is Drunk or Drowsy respectively. The MQ-3 alcohol sensor is directly attached to the microcontroller and provides an alert on drunk state detection by turn on alarm and vibrators and turning off the engine (that is the relays are turned off) so that the driver is not allowed to drive in drunk state. The buzzer is used to generate an alarm to alert the driver and the vibrator used which can be placed under the driver's seat or at driver's seatbelt to alert the driver from drowsy state. The parking LED's or the emergency lights are used to alert the other vehicles mainly coming from the rare end of the vehicle to alert the driver of other vehicle that the vehicle in front is having some issues thus avoiding accidents from happening and providing overall safety to all passengers in the vehicle.

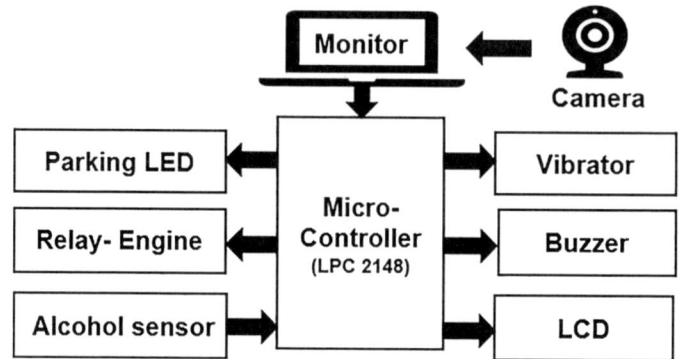

Fig. 10 System block diagram of driver drunk state and drowsiness detection with alert system

6 Results

The HAAR Cascade classifier has been implemented for face and eye detection, where in the.xml files namely haarcascade_frontalface_default and haarcascade_eye has been used to detect the face and eyes respectively in the continuous real time frame. For yawn detection template matching is used to monitor driver yawn state.

Drowsiness detection at varying lighting conditions is as follows Tables 1, 2 and 3.

Comparison between different varying conditions with 50 samples.

From the results obtained as shown in Figs. 11, 12, 13, 14, 15, 16 and 17, it is clearly visible that accuracy changes with varying lighting conditions. The best results were obtained on evenly lit facial lighting condition.

Table 1 For bright face: for 50 samples

Eye closure detected (3 s)	46
Eyes detected	46
2 eyes detected	30
False detection	04
Accuracy (%)	92

Table 2 For dark face (low light): for 50 samples

Eye closure detected (3 s)	38
Eyes detected	41
2 eyes detected	25
False detection	12
Accuracy (%)	76

Table 3 For evenly lit face: for 50 samples

Eye closure detected (3 s)	50
Eyes detected	50
2 eyes detected	42
False detection	00
Accuracy (%)	100

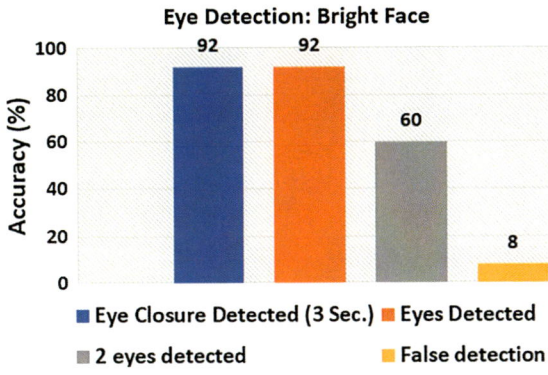

Fig. 11 Eyes detected for bright face

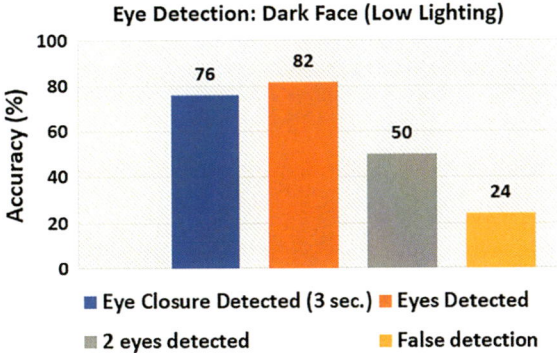

Fig. 12 Eyes detected for dark face (low lighting)

For Yawn detection: Template matching using Normalized Correlation Coefficient (C_{CN}) method has been implemented the results obtained in Tables 4, 5, 6 and 7.

From Tables 4, 5, 6 and 7 observations, it was concluded that the best threshold to be set to get the best yawn detection result for any background was 0.75.

For drunk state detection: The overall system accuracy (under ideal conditions) obtained from the best lighting condition (evenly Lit face condition) as observed in Table 3, setting a threshold of 0.75 was observed from Tables 4, 5, 6, 7 and Alcohol sensor accuracy was obtained as shown in Table 8. Table 9 shows the overall average accuracy of the system using Figs. 18, 19, 20, 21, 22, 23, 24 and 25 respectively.

Table 4 Template matching: bright background

C_{CN}	Yawn detected	Total samples	Accuracy (%)
0.95	2	20	10
0.90	14	20	70
0.85	15	20	75
0.80	19	20	95
0.75	20	20	100
0.70	20	20	100
0.65	20	20	100
0.60	19	20	95
0.55	19	20	95
0.50	0	20	0

Table 5 Template matching: black background

C_{CN}	Yawn detected	Total samples	Accuracy (%)
0.95	7	20	35
0.90	15	20	75
0.85	18	20	90
0.80	18	20	90
0.75	19	20	95
0.70	20	20	100
0.65	0	20	0
0.60	0	20	0
0.55	0	20	0
0.50	0	20	0

Table 6 Template matching: plain background

C_{CN}	Yawn detected	Total samples	Accuracy (%)
0.95	0	20	0
0.90	12	20	60
0.85	20	20	100
0.80	20	20	100
0.75	20	20	100
0.70	18	20	90
0.65	0	20	0
0.60	0	20	0
0.55	0	20	0
0.50	0	20	0

Fig. 13 Eyes detected for evenly lit condition

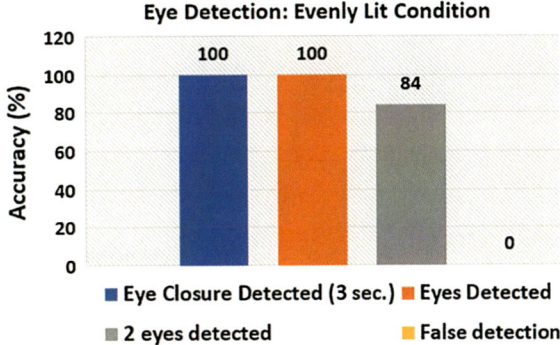

Fig. 14 Eye closure detection (3 s) for varying lighting conditions

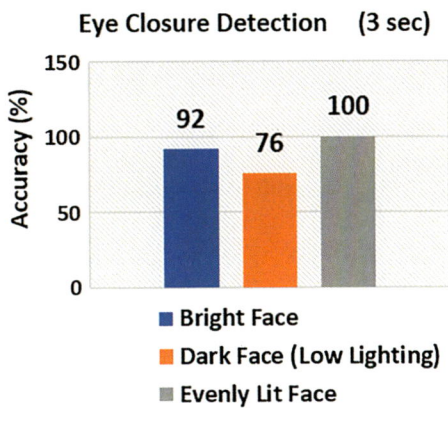

Fig. 15 Eye detected for varying lighting conditions

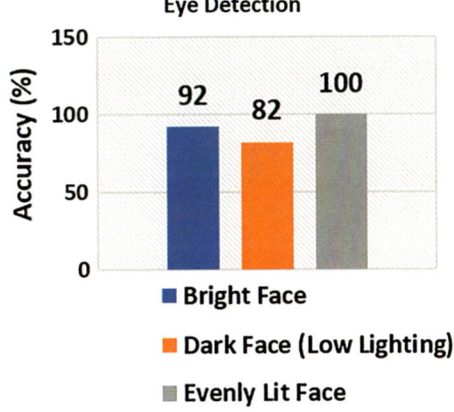

Fig. 16 Two eyes detected
for varying lighting
conditions

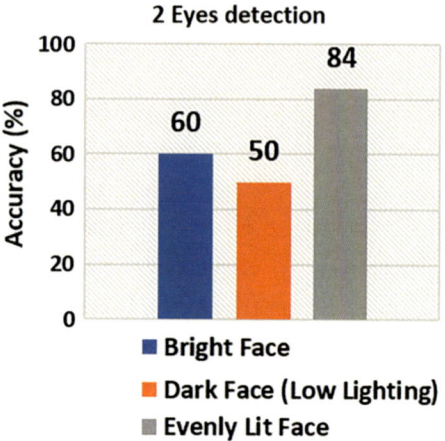

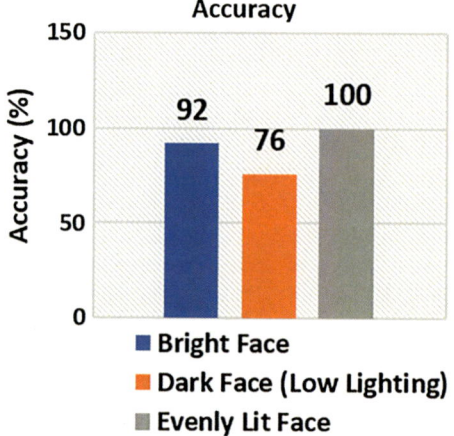

Fig. 17 Overall accuracy for varying lighting conditions

7 Conclusion

Over the decade, several drunk state and drowsiness detection techniques have been developed, even so there have been increase of accident cases due to drunk driving and driver drowsiness. This system has been implemented using nonintrusive techniques that do not bother the driver while driving thereby increasing chances to find the driver drowsy state condition and by using a simple alcohol sensor (MQ-3) to detect drunk driver condition. This framework uses the face, eye detection for drowsiness detection that has been merged with yawn detection so that in case a driver wears glasses or shades while driving, still the drowsy state can be detected using yawn results. The proposed system had very high accuracy when tested in well-lit conditions. The hardware to alert the driver during drowsy and drunk state has also been developed

Table 7 Template matching: textured background

C_{CN}	Yawn detected	Total samples	Accuracy (%)
0.95	0	20	0
0.90	2	20	10
0.85	14	20	70
0.80	19	20	95
0.75	20	20	100
0.70	20	20	100
0.65	20	20	100
0.60	12	20	60
0.55	0	20	0
0.50	0	20	0

Table 8 Alcohol sensor detection: for 30 samples

Test	Result
Detection	28
No detection	00
False detection	02
Accuracy (%)	93.33

Table 9 System accuracy for 30 samples

Readings	Results	Accuracy (%)
Eye closure detection	27	90.00
Yawn detected	29	96.67
Drunk detection	28	93.33
Overall	84	93.33

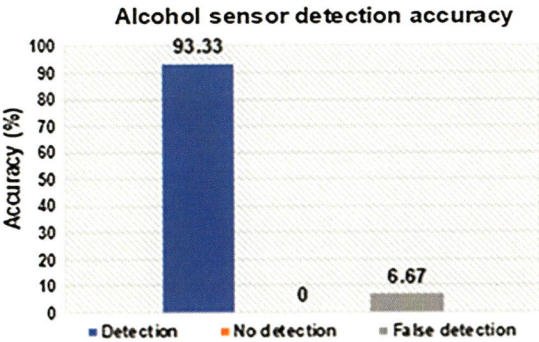

Fig. 18 Accuracy of alcohol sensor

with help of buzzer and vibrator motor along with the turning "ON" of emergency (parking) lights to alert other drivers. Here the accuracy of the entire system has

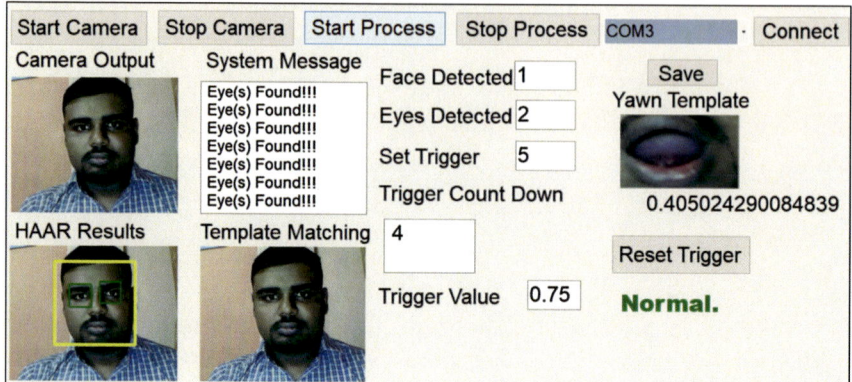

Fig. 19 Setup when driving in alert state

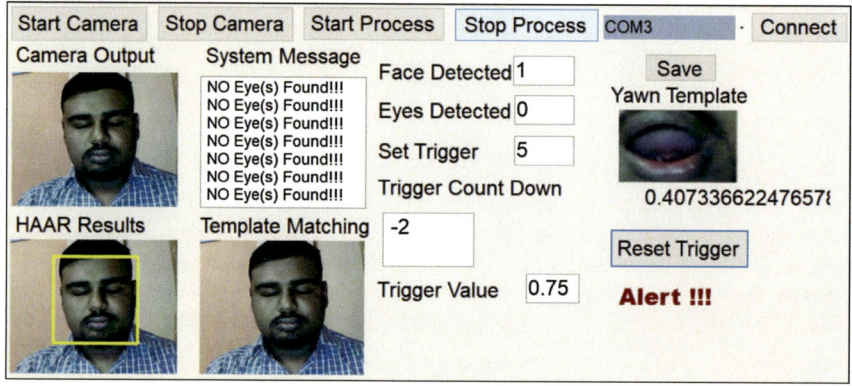

Fig. 20 Setup when driving in drowsy state (eye closure)

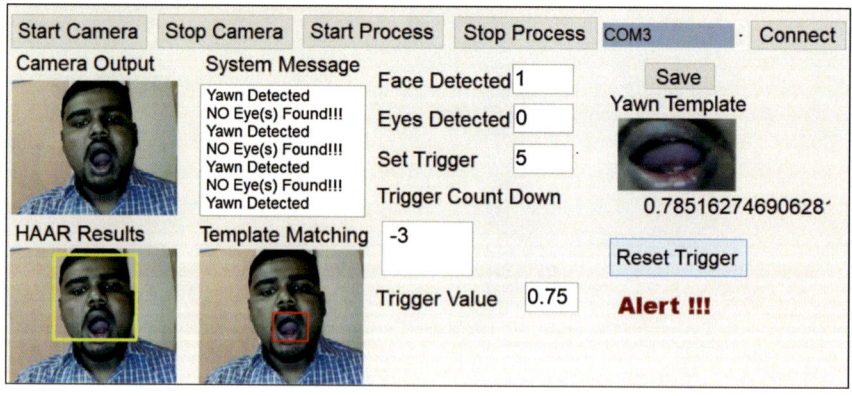

Fig. 21 Setup when driving in drowsy state (yawn detection)

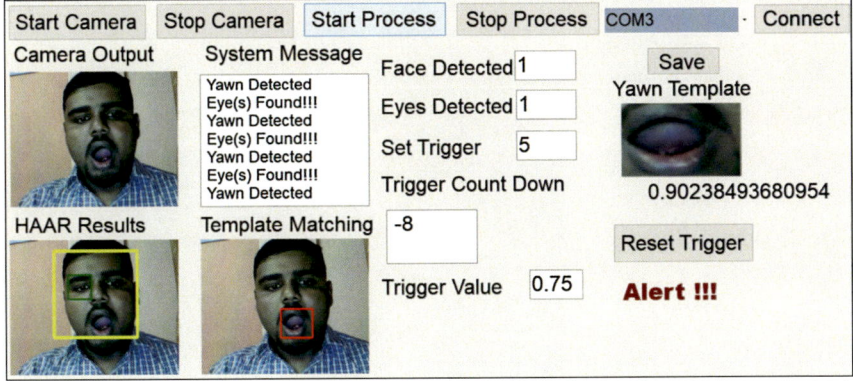

Fig. 22 Alert generation on any one condition being detection

Fig. 23 Alert message
displayed on LCD on
drowsiness detection

been calculated using varying lighting conditions for eye closure detection using HAAR Cascade classifier and yawn detection using template matching in different background conditions with varying threshold levels. From the observations, it was observed that the best detection was found on evenly lit conditions for eye closure detection. In addition to this, the best match for yawn detection was observed keeping a threshold of 0.75. Overall average accuracy of the system was found out to be 93.33%. The accuracy of the system could be increased under low lighting conditions using a better resolution camera.

Fig. 24 Alert message
displayed on LCD on drunk
state detection

Fig. 25 Overall system
accuracy of the work

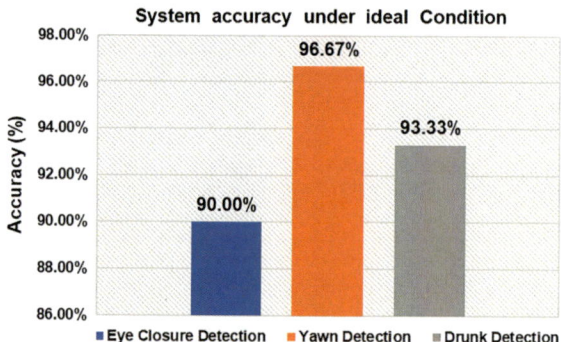

Acknowledgements The authors would like to thank the anonymous reviewers for their valuable comments and suggestions. I (Vivek Nair—Author) would like to clarify that the participant (driver) is myself and give consent to be used in the study.

References

1. Kaplan S, Guvensan MA, Yavuz AG, Karalurt Y (2015) Driver behavior analysis for safe driving: a survey. IEEE Trans Intell Transp Syst 16(6):3017–3032
2. Mbouna RO, Kong SG, Chun M-G (2013) Visual analysis of eye state and head pose for driver alertness monitoring. IEEE Trans Intell Transp Syst 14(3):1462–1469
3. Omidyeganeh M, Shirmohammadi S, Abtahi S, Khurshid A, Farhan M, Scharcanski J, Hariri B, Laroche D, Martel L (2016) Yawning detection using embedded smart cameras. IEEE Trans Instrum Meas 65(3):570–582

4. Chacon-Murguia MI, Prieto-Resendiz C (2015) Detecting driver drowsiness: a survey of system designs and technology. IEEE Consum Electron Mag 4(4):107–119
5. Sivakumar T, Krishnaraj R (2012) Road traffic accidents (Rtas) due to drunken driving in India challenges in prevention. IRACST—Int J Res Manag Technol (IJRMT) 2(4):401–406
6. Bhandari GM, Durge A, Bidwai A, Aware U (2014) Yawning analysis for driver drowsiness detection. IJRET: Int J Res Eng Technol 3(2):502–505
7. Viola P, Jones M (2001) Rapid object detection using a boosted cascade of simple features. In: Conference on computer vision and pattern recognition
8. Nair VR, Charniya NN (2017) Drunk driving and drowsiness detection. In: International conference on intelligent computing and control (I2C2), Coimbatore, India, vol 2, pp 30–35
9. Kaehler A, Bradski G. Learning OpenCV3, Computer vision in C++ with OpenCV library
10. Dong W, Cheng CQ, Kai L, Bao-hua F (2011) The automatic control system of anti-drunk-driving. In: International conference on electronics, communication and control (ICECC), Ningbo, China. IEEE, pp 523–526
11. Wu Y-C, Xia Y-Q, Xie P, Ji X-W (2009) The design of an automotive anti-drunk driving system to guarantee the uniqueness of driver. In: International conference on information engineering and computer science (ICIECS), Wuhan, China. IEEE
12. Eun Jung Park (2008) Sensor report—MQ-3 Gas sensor. http://sensorworkshop.blogspot.in/2008/04/sensor-report-mq3-gas-sensor.html

A Survey on Intelligent Face Recognition System

Riddhi Sarsavadia and Usha Patel

Abstract Face recognition system is a computer's capability which gives it a vision of performing two fundamental operations the detection and the recognition of a human face. With the advancement of machine learning algorithm and image processing techniques the accuracy of face recognition system has been significantly improved. The objective of this paper is to give a detailed survey of a few face recognition algorithm with their features and limitations. The basics of face detection and face recognition techniques along with their approaches are described in the section.

1 Introduction

The human's face is a complex because of facial components which can change after some time. Humans have a very good capacity to recognize several faces learned throughout their lifetime and can identify faces in milliseconds, even after years of separation, but the system does not have the same capacity. So, for this purpose, we need to create a face recognition system to make our system as intelligent as humans. Face recognition by systems can be significant and has a huge amount of usage for security purpose for organizations, access control for higher level authorities, national defences, etc. [1]. Face recognition incorporates fundamentally three-errand face detection, feature extraction and recognition [1].

As shown in Fig. 1 the first step of the method is face identification has its own challenges which are posture invariant, the existence or deficiency of occlusion, physical components, image orientation, facial expression, and imaging conditions [2]. After detection of a face from an image, the next step would be feature extraction in which is used to extract only features that would be used for mapping with

R. Sarsavadia (✉) · U. Patel
Nirma University, Ahmedabad 382481, India
e-mail: riddhisarsavadia9@gmail.com

U. Patel
e-mail: ushapatel@nirmauni.ac.in

© Springer Nature Switzerland AG 2019
D. Pandian et al. (eds.), *Proceedings of the International Conference on ISMAC in Computational Vision and Bio-Engineering 2018 (ISMAC-CVB)*, Lecture Notes in Computational Vision and Biomechanics 30,
https://doi.org/10.1007/978-3-030-00665-5_114

Fig. 1 Face recognition system

image database for face recognition. For face recognition step we have mainly three approaches which are the holistic approach, feature-based and hybrid approach [1].

Using Face recognition approaches, we can get recognition rate of frontal face, oriented face (up to 30°) and props (beard and eyeglasses with an existing database) up to 80–85%. To overcome these challenges, we use hybrid approaches for face recognition. When we recognize face with eyeglasses using face recognition techniques, there will be some problems due to the thickness of eyeglasses, the size of a frame of eyeglasses, reflective property of glasses etc. So, we use eyeglass removal [1] to remove eyeglasses after detecting a face. In this project work, the whole implementation would be done using OpenCV (Open Computer Vision) [3].

OpenCV (Open source Computer Vision) is a set of libraries with programming functions which has the main objective of real-time image processing [4]. OpenCV was developed by Intel's research centre by Willow Garage in 1999 at Nizhny Novgorod (Russia) [3]. It is freely available for commercial as well as non-commercial usage. The method is very helpful for developing highly efficient computer vision applications.

2 Face Detection Approaches

Some of the main face detection methods are discussed here.

(1) Some comprehensive approaches have been derived by some well-known researchers of the human faces. Their rules are robust but the fundamental hindrance in applying those ideas is the difficulty of translating human intelligence into set guidelines.

(2) Featured-based methods: Some invariable parameters of human faces are helpful in detecting face-cut, skin tone, skin texture, etc. But capabilities of determining those parameters might be severely impacted due to colour reproduction, noise, grains, etc.

(3) Template matching: Thanks to scale, pose, and shape of the human face which is given as an input, have to be matched with predefined template in the machine.

(4) Appearance-based method: In this type, the methods are learnt from the examples and references in the images. Machine learning techniques, statistical analysis approach can be applied to find relevant parameters of face from the overall image. These metrics will be used in overall analysis of the input. However, in the template matching, the templates and methods are predefined by the researchers.

A part of the primary face detection techniques is mentioned below.

(1) Defining the right face matching techniques are defined by the rules and regulations provided by the experts. The main challenge in this method is to analyse the given image and comprehend it as the human element, must be done in very limited set of rules.

(2) Featured-based techniques: Invariable aspects of appearances are monetize for identifying surface, skin shading. Be that as it may, the output from such machine calculation can be extremely underestimated because of light, focus, exposure, etc.

(3) Template coordinating: The input of this method is manipulated in the confined template. Be that as it may, the execution here suffers because of the undefined parameters like shape, scale and posture.

(4) Appearance-based technique: In these types of strategies, the agendas are predefined by experts. While, the formats in the other strategies are found from the cases in images. Machine Learning and statistical investigation are the procedures which can be manipulated to find the significant qualities of face and other.

The main purpose of face discovery is to reveal that if there are any faces in the image or not and if face is present then return an extent of each face. For face detection step of our IFRS, we need to detect face using some facial components like eyes, nose, mouth, face contour with some facial features like with beard or without beard, with eyeglasses or without eyeglasses and with an orientation of face. There are several techniques which includes these cases which are shown in Table 1.

As shown in Table 1, Haar classifier can exist for multiple face detection with facial features like with eyeglasses and without eyeglasses, with beards and without beards can be detected as well as it is faster among previous approaches and we can detect faces with more than $45°$ orientation of faces and higher detection rate among all approaches discussed earlier.

3   Feature Extraction

The previous unit gives discussion about face detection, now, in this area, well discuss how to remove valuable and minimal highlights for further recognition procedure. Figure 2 shows basic flow of feature extraction after face detection.

When input image to an algorithm is too large for the further processing then this step is necessary. Using feature extraction, we can get reduced size of an image for further procedure by which we can get lesser time for whole procedure as compared to without reducing an image. Most of the methods for feature selection and extraction are also used for later steps. Some of them are described in Table 2.

Table 1 Comparative analysis of face detection approaches

Papers	Technique	Components	Working cases	Issues
"Human face detection in a complex background", G. Yang and T. S. Huang	Hierarchical knowledge based [2]	Eyes, nose, mouth	Frontal faces in uncluttered scenes	Translating knowledge into rules, detect faces in different poses
"Finding faces in cluttered scenes using random graph matching" T. Leung, M. Burl, and P. Perona	A computational approach to edge detection [7]	Edge of face, nose, eyes	Edges detected closed to true edges	Real world conditions and scaling and thresholding of image
"An introduction to face recognition technology" S.-H. Lin	Integration of skin colour, size and shape [1]	Skin colour, edge of face, size	Detect face at different orientations up to 45° and for facial features (glasses and beards also)	Lightning conditions, only single face
"Automatic face identification system using flexible appearance models" A. Lanitis, C. J. Taylor, and T. F. Cootes	Deformable templates: active shape model (ASM) [8]	Edge of face, eyes, eyebrows, nose, mouth, lips	Single frontal face	Multi-faces, variation in pose, scale and shape
"Eigenface-based facial recognition" D. Pissarenko	Eigenfaces for recognition [9]	Eyes, nose, mouth, face contour	Only frontal faces	Lighting conditions, facial expression, orientation
"Human face detection in a complex background" G. Yang and T. S. Huang	Example based learning for view-based human face detection [2]	Eyes, nose, mouth, eyebrows, face contour, lips	Can detect faces when lightening is nearly frontal and less computational time than eigenface	Lightning conditions
"Detecting faces in images: a survey" M. H. Yang, D. J. Kriegman, and N. Ahuja	Neural network based face detection [10]	Eyes, nose, mouth, face contour, lips	Capture complex class conditional density	Requires external tuning
"Rapid object detection using a boosted cascade of simple features" P. Viola and M. Jones	Haar classifier: robust real-time face detection by Viola and Jones [11]	Eyes, nose, mouth, face contour, lips	Detect face faster than 15 times of other approaches with different facial conditions	Requires more memory

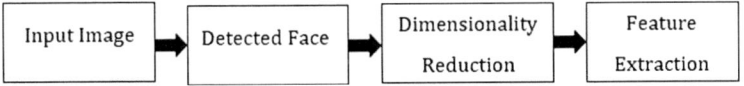

Fig. 2 Flow of feature extraction

Table 2 Comparative analysis of feature extraction

Technique	Working cases	Issues
Principle component analysis (PCA), Kernel PCA, weighted PCA [12]	Eigenfaces and dimensionality reduction	The transformed dimensions will be uncorrelated from each other, only uses second order statistics
Linear discriminant analysis (LDA), Kernel LDA [13]	Eigenvector based with supervised map; LDA with kernel method	Only small sample size problem can be solved
Independent component analysis (ICA) [14]	Transformed dimensions are independent and higher order statistics are used for dimensionality reduction	Cannot linearly separate independent sources in smaller subspace
Linear binary patterns (LBP) [4]	Uniform binary patterns will contain at most two-bitwise transactions from 0 to 1 or vice versa	Image may be blurred

4   Face Recognition Approaches

As discussed in the previous section, the images which are reduced using feature extraction technique would be used for recognition process to decrease the comparison time of an image with the database. There are various techniques for FRS [5] which can recognize human face within lesser time, using facial features like with eyeglasses or without eyeglasses, with bread or without bread, multiple face recognition at a time, orientation of face etc. [6]. IFRS techniques which incorporate these characteristics are discussed in Table 3.

5   Survey Analysis

There are various approaches for face recognition which are discussed in the previous section. In this section, complete comparative analysis of face recognition techniques with respect to real-time approach along with some facial components like facial features, props, orientation, lightning conditions, etc. are described here, as shown in Table 4, combination of different algorithms and techniques gives the accurate result, i.e. Hybrid techniques for real-time implementation of the system.

Table 3 Comparative analysis of face recognition approaches

Approach	Technique	Working cases	Issues
Holistic	Principal component analysis [12]	Requires full facial information	Facial features like with eyeglasses and with bread etc. and multiple faces at a time
Holistic	Two dimensional PCA: a new approach to appearance-based face representation and recognition [15]	Recognize single face with up to 45° orientation	Multiple faces
Holistic	Face recognition using LDA-based algorithms [13]	Recognize more than 45° oriented face	Recognize maximum 5 faces at a time
SVM	SVM: training support vector machines: an application to face recognition [16]	SVM classifies facial features using a hyper-plane	Lightning conditions and only frontal face can be recognized
Feature based	Face recognition by elastic bunch graph matching (EBGM) [17]	Faces as a graph	Recognize up to 22° oriented faces
Feature based	Face recognition by active appearance models [18]	Facial shapes, facial expressions etc.	Lightning conditions

Table 4 Survey analysis

Components	Algorithms					Hybrid techniques	
	PCA	ICA	LDA	LBP	EBGM	Fisherfaces	Proposed sys
Features							
Eyes	Yes	Yes	Yes	Yes	Yes	Yes	Yes
Nose	–	–	–	–	Yes	–	–
Ears	–	–	–	–	Yes	–	–
Face contour	–	–	–	–	–	–	–
Mouth	Yes	Yes	Yes	Yes	Yes	Yes	Yes
Props							
Eyeglasses							
With existing dataset	Yes	Yes	Yes	–	–	Yes	Yes
Without existing dataset	–	–	–	–	–	–	Yes
Beards	–	–	–	–	–	–	–
Orientation							
Up to 45°	–	–	–	–	–	–	Yes

6 Conclusion

The major approaches for face recognition are discussed in this survey paper. The complete comparative survey analysis of face recognition techniques with respect to real-time approach shows that implementing hybrid techniques, i.e. combination of various techniques gives better and accurate results than implementing single technique for building intelligent face recognition system.

References

1. Lin S-H (2000) An introduction to face recognition technology. Inf Sci 3(1):1–8
2. Yang G, Huang TS (1994) Human face detection in a complex background. Pattern Recogn 27(1):53–63
3. https://en.wikipedia.org/wiki/Computervision, "Computer vision."
4. https://www.learnopencv.com, "Opencv."
5. Jain AK, Li SZ (2011) Handbook of face recognition. Springer
6. Craw I, Tock D, Bennett A (1992) Finding face features. In: European conference on computer vision, pp 92– 96. Springer
7. Leung T, Burl M, Perona P (1995) Finding faces in cluttered scenes using random graph matching. In: Proceeding of IEEE PAMI conference, pp 637–644
8. Lanitis A, Taylor CJ, Cootes TF (1995) Automatic face identification system using flexible appearance models. Image Vis Comput 13(5):393–401
9. Pissarenko D (2002) Eigenface-based facial recognition, vol 1, pp 4–9
10. Yang M-H, Kriegman DJ, Ahuja N (2002) Detecting faces in images: a survey. IEEE Trans Pattern Anal Mach Intell 24(1):34–58
11. Viola P, Jones M (2001) Rapid object detection using a boosted cascade of simple features. In: Proceedings of the 2001 IEEE computer society conference on computer vision and pattern recognition, 2001. CVPR 2001, vol 1. IEEE, pp I–511
12. Moon H, Phillips PJ (2001) Computational and performance aspects of pca-based face-recognition algorithms. Perception 30(3):303–321
13. Lu J, Plataniotis KN, Venetsanopoulos AN (2003) Face recognition using lda based algorithms. IEEE Trans Neural Netw 14(1):195–200
14. Bartlett MS, Movellan JR, Sejnowski TJ (2002) Face recognition by independent component analysis. IEEE Trans Neural Netw 13(6):1450–1464
15. Yang J, Zhang D, Frangi AF, Yang J-Y (2004) Two-dimensional pca: a new approach to appearance-based face representation and recognition. IEEE Trans Pattern Anal Mach Intell 26(1):131–137
16. Hearst MA, Dumais ST, Osuna E, Platt J, Scholkopf B (1998) Support vector machines. IEEE Intell Syst Appl 13(4):18–28
17. Wiskott L, Fellous J-M, Kuiger N, Von Der Malsburg C (1997) Face recognition by elastic bunch graph matching. IEEE Trans Pattern Anal Mach Intell 19(7):775–779
18. Edwards GJ, Cootes TF, Taylor CJ (1998) Face recognition using active appearance models. In: European conference on computer vision. Springer, pp 581–595

Real-Time Health Monitoring System Implemented on a Bicycle

Rohith S. Prabhu, O. P. Neeraj Vasudev, V. Nandu, J. Lokesh and J. Anudev

Abstract It is evident that many systems have been developed and are being developed these days related to health monitoring. The main requirement is the continuous real-time health monitoring where the data, i.e. the body vitals/health parameters, can be easily understood by the user through an application interface, and this would also be shared with the corresponding physician who can be aware of the patient's vitals at all times. This is a study of methods where a real-time health monitoring system can become an integral part of the society whereby its intention mainly aimed at gaining awareness of each ones health and its implementation leading to focus on a health record which is linked to web also information reaching the doctors in time for continuous monitoring. The obese people are able to control their weight. The project being 'Real Time Health Monitoring System Implemented On A Bicycle' aimed at designing and assembly of a non-invasive health monitoring system. The conversion of cycle energy will also be taken into account to supply for the health monitoring devices and charging of the display.

R. S. Prabhu · O. P. Neeraj Vasudev · V. Nandu · J. Lokesh · J. Anudev (✉)
Department of Electrical and Electronics Engineering, Amrita School of Engineering, Amrita
Vishwa Vidyapeetham, Amritapuri, India
e-mail: anudevj@am.amrita.edu

R. S. Prabhu
e-mail: rohithsprabhu96@gmail.com

O. P. Neeraj Vasudev
e-mail: neerajpaderi@gmail.com

V. Nandu
e-mail: nandhuvijay009@gmail.com

J. Lokesh
e-mail: lokeshkj006@gmail.com

© Springer Nature Switzerland AG 2019
D. Pandian et al. (eds.), *Proceedings of the International Conference on ISMAC in Computational Vision and Bio-Engineering 2018 (ISMAC-CVB)*, Lecture Notes in Computational Vision and Biomechanics 30,
https://doi.org/10.1007/978-3-030-00665-5_115

1 Introduction

- At least 200,000 deaths from heart disease and strokes every year could be prevented by monitoring vital signals regularly [1, 2]. Nowadays, real-time health monitoring systems are required not only for patients to monitor their own health parameters but also for their doctors to be able to continuously assess the patient [3].
- Relating to the system presented here the patient's heart rate, calories burned, blood pressure and body temperature are mainly observed. A methodology where doctors can have full supervision and be able to understand the daily routine will help in better treatment of the patient.
- People can manage their daily routine check-up at home. In addition, this is important to provide people with continuous monitoring in non-clinical environments. However, such health management only can be achieved if the computer-based portable monitoring devices with smart sensor technologies are available [4].

2 System Architecture

See Fig. 1.

2.1 Health Monitoring

The real-time patient health monitoring system is one which incorporates various physical parameters of a human body like heart rate, blood pressure and body tem-

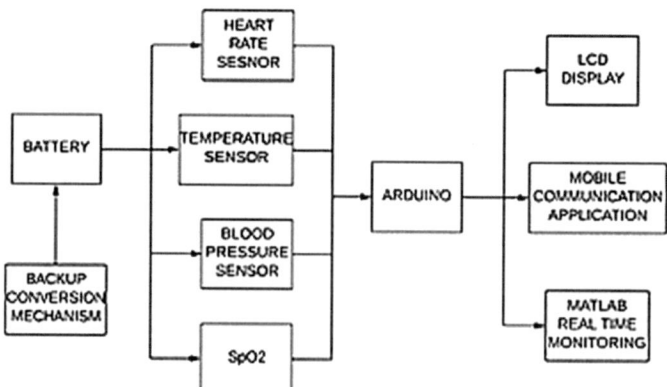

Fig. 1 Overview block diagram of the system

perature [5, 6]. This model is set to be implemented to acquire the pulse signal received from our body and is further used to measure heartbeat of a patient. Calories burned will be displayed in addition to this analysis of the ECG waveform is also considered. A microcontroller board can be used for analysing inputs from the patient. The display will be given to the mobile application, LCD display for the miscellaneous readings and continuous monitoring of the user carried out by the doctor.

2.2 Energy Conversion

The conversion part is considered to give the supply for the displays, phone and reuse of energy is seen. The dynamo set-up is used to convert the kinetic energy to electrical which after getting sufficient energy for recharging the rechargeable battery. We can step down to the required voltages of the system in which the supply voltage is required. Charging of the phone is also evident in this system.

3 Health Parameters

Here below are the methods proposed for the calculation of heart rate, temperature, blood pressure and calories. The following seem to be easy to incorporate into the system and ability to get the desired routine output is considered.

3.1 Heart Rate

This is a module designed to easily connect to the Arduino to be able to harness the heart rate of the person. It poses to be a very efficient design as the implementation and assembly onto the bicycle becomes easy. It is seen that the beats per minute are displayed in the LCD display hereby it shows the value that is seen in the serial monitor of the Arduino where usually from this which will be taken as the input directly into the system for the patient monitoring (Figs. 2 and 3).

The whole system integrated together as shown in Fig. 4 brings about heart rate monitoring system. The alert messaging, cloud storing is illustrated, and the possibility for the real-time monitoring considered makes this system one of the useful even in remote areas as the patients can have up to data assessment from their physicians.

Fig. 2 Prototype displaying BPM

3.2 Calories

In this system, the method where input is heart rate monitored and are theoretically able to get the calories burnt of the person who will be using the system. The required contents needed from the formulae will be given as input by the user.

$$\text{Male:} \frac{((-55.0969) + (0.6309 \times \text{HR}) + (0.1988 \times W) + (0.2017 \times A))}{(4.184) \times 60 \times T} \tag{1}$$

$$\text{Female:} \frac{((-20.4022) + (0.4472 \times \text{HR}) - (0.1263 \times W) + (0.074 \times A))}{(4.184) \times 60 \times T} \tag{2}$$

where

HR   Heart rate (in beats/minute)
W Weight (in kilograms)
A Age (in years)
T Exercise duration time (in hours)

With this, it is great for obese patients as they can set the required target and be able to achieve it with the supervision of their doctors.

The heart rate is efficiently taken in a short time and with less expense without using time-consuming and expensive clinical pulse detection systems then calculates the calories burned during an exercise from measured heartbeat [7].

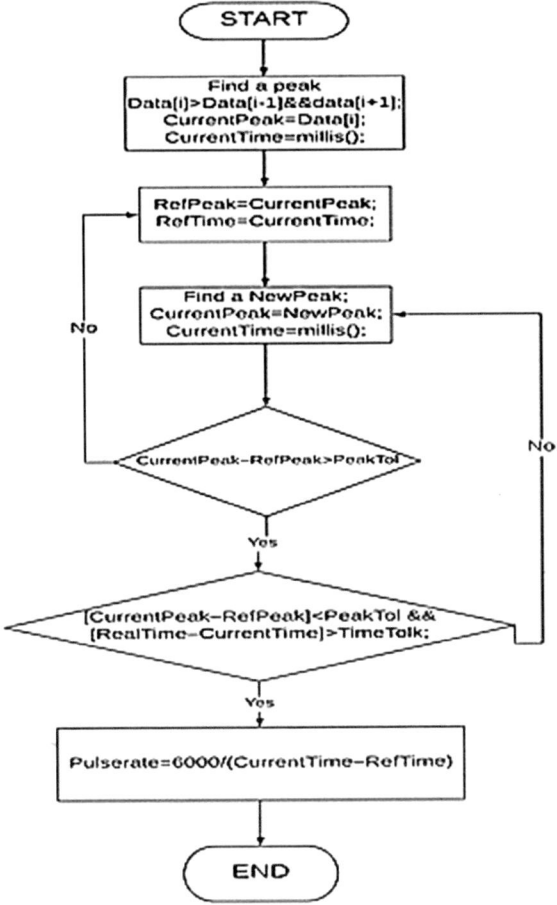

Fig. 3 Flowchart of the algorithm for getting BPM

3.3 Blood Pressure

The blood pressure is considered as the good indicator of the status of cardiovascular system. Doctor's prefer sphygmomanometer, being the most accurate device, to measure the blood pressure presently. A sphygmomanometer consists of occlusive cuff which is wrapped around the arm and a column of mercury to find the systolic and diastolic pressure. When the doctor exerts pressure on the occlusive cuff, an audible sound is produced and this sound can be heard by a stethoscope. The audible sound is known as 'korotkoff'.

The readings from the sphygmomanometer at a point when korotkoff sound start is a measure of systolic pressure and at a point where korotkoff sound stops on

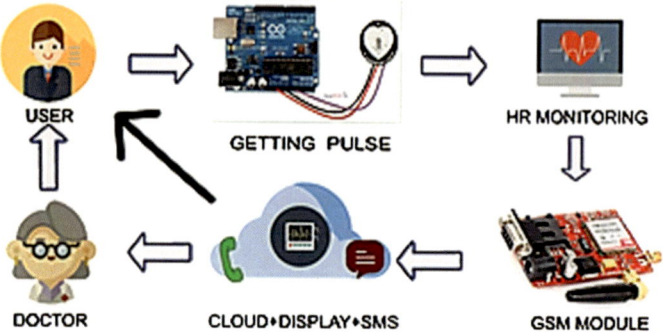

Fig. 4 Visualization of the heart rate monitoring system

lowering occlusive pressure is a measure of diastolic pressure. The blood pressure is represented in mm hg [8].

Flowchart shown in Fig. 5 gives a better understanding of how blood pressure is measured.

The pressure occurred in the arteries when the ventricles are squeezing out blood under high pressure is known as systolic blood pressure and the pressure occurred when the ventricles are filling up with the blood is known as diastolic blood pressure. This diastolic blood pressure will be low pressure compared to systolic. The net amount of blood pump out from the heart with respect to time will be the cardiac output of a person and the net amount of blood returned to the heart with respect to time is termed as venous return. For a normal healthy person, the cardiac output is equal to the venous return; this will be approximately equal to 5 L/min. The whole system integrated together as shown in Fig. 6 brings about BP monitoring system. The physician will have full control over the workout and health variations in each scenario as the block diagram below represents [9].

3.4 Temperature

LM35 is a temperature sensor which can measure the temperature more accurately compared to a thermistor. The sensor circuit is sealed and not subjected to oxidation. LM35 generates a higher output voltage than thermocouples and may not require the output voltage to be amplified. LM35 output voltage is proportional to Celsius temperature. Its scale factor is 0.1 V/°C. It does not require any external calibration or trimming and it maintains an accuracy of ±0.4 °C at room temperature and ±0.8 °C for range of 0–100 °C. It has low self-heating capability. LM35 Circuit Diagram draws only 60 μA from its supply [10]. Basically, it is operated under 4–30 V. Differential amplifiers are used. There are two inputs like non-inverting (+) and inverting (−) for the temperature sensor and an output pin (Fig. 7).

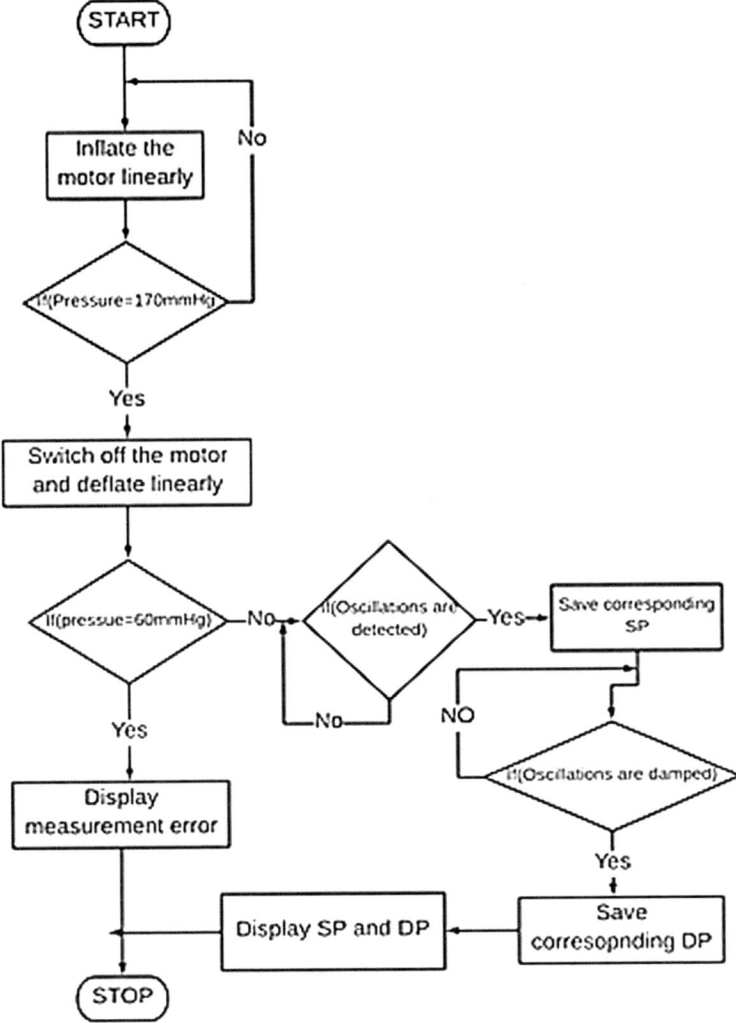

Fig. 5 Flowchart of the blood pressure monitoring system

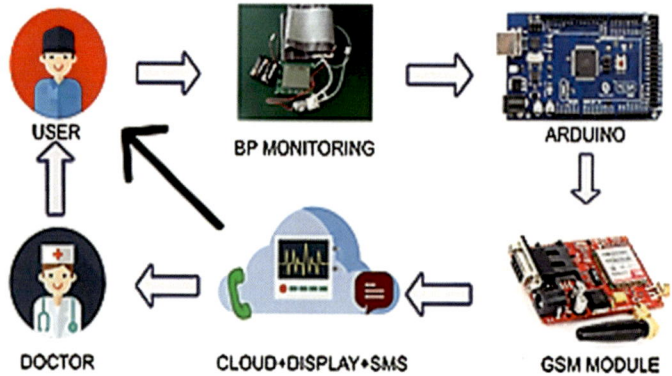

Fig. 6 Visualization of the blood pressure monitoring system

Fig. 7 Flowchart of
temperature monitoring
system

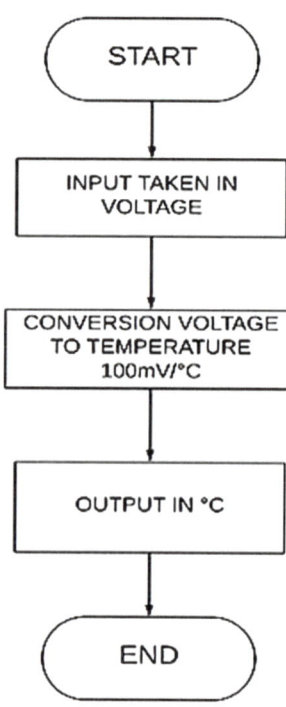

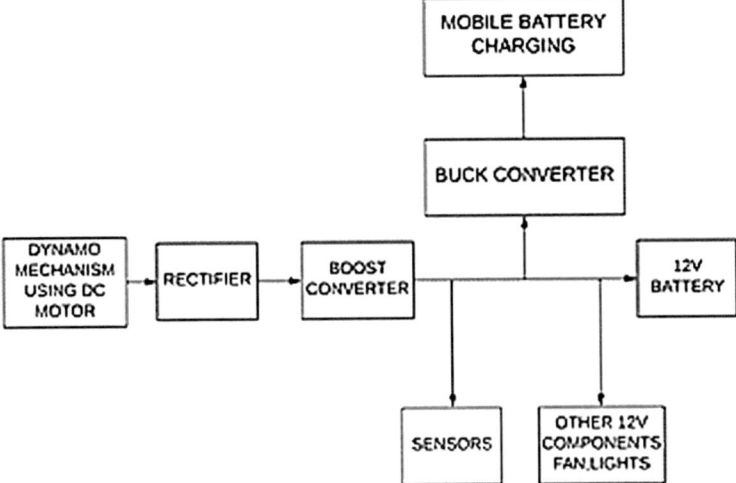

Fig. 8 Energy conversion block diagram

4 Bicycle Conversion

The conversion part plays an integral part where all the charging of the displays and system supply is provided by regeneration using the dynamo technique (Fig. 8).

4.1 Cycle Dynamo

Cycle dynamo is an energy harvester that can convert the mechanical energy produced while riding bicycles into electrical energy and thus can be utilized to power electronic gadgets requiring low power. Dynamos are available in many ranges but most of these are rated such that to deliver up to 1 A of current. This value of current is insufficient to charge the 12 V battery of the proposed system. Hence, an alternative mechanism of current generation has been implemented using a DC geared motor. The rating of the motor is such that it is having a nominal voltage of 12 V at 1000 RPM. The full load current can go up to 7 A. A current of 2 A can be delivered if the output power of the motor is 25 W. A wheel will be fixed on the shaft of the motor and is forced to touch the cycle wheel using spring mechanism. The radius of the wheel attached to the shaft can be determined using these calculations:

Normally, cycle wheel diameter varies from 45 to 60 cm.
So we take $D = 45$ cm (minimum value).
So circumference of the wheel $= [pi * D] = 140$ cm [approx.].
Assume that the cycle rotates at 300 rpm (particular case).
Let the rpm of motor be 1000 rpm.

So the ratio of rpm of cycle wheel to the rpm of the wheel attached to the shaft = 1:3 (approx.).
So that the circumference of the wheel attached to the shaft = 140/3 = 46.66 cm.
So radius = 46.66/(2 * pi) = 7.43 cm.
Hence, we can use motor shaft wheel of radius greater than 7 cm for this system.

4.2 Rectifier

Depending upon the polarity of rotation of the shaft, output from the dynamo can be varying. To achieve a steady and smooth direct current output, a full wave rectifier circuit is employed.

4.3 Converters

4.3.1 Boost Converter

To change the default, constant output voltage greater than the input voltage can be obtained using boost converters or step-up converter. Boost converter employs a voltage feedback technique to step up the input voltage of variable or fluctuating nature to a higher constant DC output. Its main advantages are low losses high efficiency. The function of boost converter is enabled by the tendency of inductor to resist changes in current. The inductor plays the role of a load on the occasion of charging and during discharging, it takes the role of an energy source. The rate of change of current is the dependent factor by which the voltage produced during discharge phase is related (Fig. 9).

The input DC voltage ranging from 2 to 12 V, coming from the rectifier output, is to be stepped up to 14.5 V using the prescribed boost converter delivering up to 1.8 A.

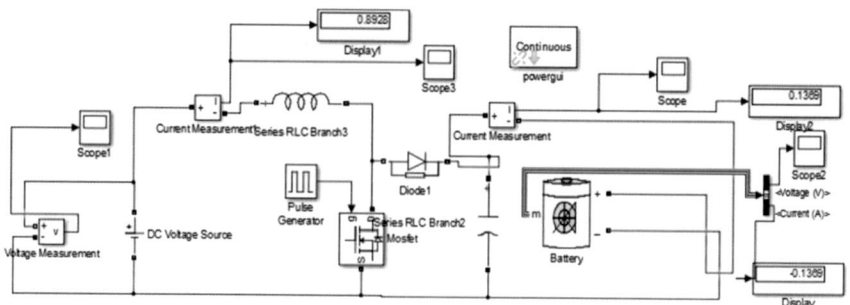

Fig. 9 Simulated circuit of boost converter in MATLAB

4.3.2 Buck Converter

The voltage supplied through the input is getting stepped down by using IC7805. This IC is suitable for low power applications. It steps downs from input of 12 V to output 5 V. Using this convertor with a heat sink to absorb the excess heat produced. The proposed system can be powered by a rechargeable lead-acid battery of 7 Ah and 12 V. Since the sum of currents of all the sensors, converters, microcontroller, GSM module and LCD displays is around 5 A, the prescribed capacity of the battery can power the system efficiently. This genre of battery can deliver high currents and has a remarkable tolerance to overcharging. Cost is low compared to other batteries having same rating.

5 Future Scope

Each and every user will have their own login ID and password where the account will be linked with a hospital; hence, they can cut the costs and develop a better relationship with the doctors. The system not only aimed to be implemented on bicycles but can be installed anywhere like wheelchairs, vehicles and even used as it is. Blood can be reached to a person in need on time as the system will be linked to all users being able to notify those with the same blood group. It becomes easy to locate people in critical condition; hence, reaching hospital in time for treatment cause system is linked with net and alert messaging also becomes useful.

6 Conclusion

This system poses to be eco-friendly, economic, less power and socially oriented mainly targeting the common people of all age groups. Aimed at keeping each one health-conscious aware and helping those in need to receive their health parameters before it is too late. The advantage of being able to instal this system not only in a cycle brings about taking into consideration the comfort of the user. The health of patients can be monitored day to day. Proposed method is an easy way of health monitoring in real time and communication between the doctor and the patient. Energy is fully utilized as to the cycle implementation being for patients who are obese to be able to lose weight and keep track of their health. All the above after being monitored physicians can access this through the web and give the required instructions then needed to be followed. Travel and time are reduced with this system.

References

1. Centers for Disease Control and Prevention (CDC) (2013) Vital signs: avoidable deaths from heart disease, stroke, and hypertensive disease-United States, 2001-2010. MMWR. Morbidity and mortality weekly report 62.35, pp 721
2. Aboughaly AA, Iqbal D, Abd El Ghany MA, Hofmann K (2017) NICBPM: non-invasive cuff-less blood pressure monitor. In: 2017 29th international conference on microelectronics (ICM)
3. Ray I, Alangot B, Nair S, Achuthan K (2017) Using attribute-based access control for remote healthcare monitoring. In: 2017 fourth international conference on software defined systems, pp 137–142
4. Wang L-H, Hsiao Y-M, Xie X-Q, Lee S-Y (2016) An outdoor intelligent healthcare monitoring device for the elderly. IEEE Trans Consum Electron 62(2):128–135
5. Anudev J, Jacob Raglend I (2012) Analytical study of GIC based current source model. In: IEEE international conference on advances in engineering, science and management, pp 219–222
6. Anudev J, Jacob Raglend I (2012) Analytical study of Howland current source model. In: 2012 international conference on computing, electronics and electrical technologies, pp 314–318
7. Karthik Reddy G, Lokesh Achari K (2015) A non invasive method for calculating calories burned during exercise using heartbeat. In: IEEE sponsored 9th international conference on intelligent systems and control (ISCO)
8. Czerwinski D, Wojcicki P, Zientarski T. On time blood pressure prediction with the use of PPG signals
9. Rajevenceltha J, Santhosh Kumar C, Anand Kumar A (2016) Improving the performance of multi-parameter patient monitors using feature mapping and decision fusion. In: 2016 IEEE region 10 conference (TENCON), pp 1515–1518
10. Sali S, Parvathi CS (2017) Integrated wireless instrument for heart rate and body temperature measurement. In: 2017 2nd international conference for convergence in technology (12CT)

A Fuzzy Rule-Based Diagnosis of Parkinson's Disease

D. Karunanithi and Paul Rodrigues

Abstract Neurodegenerative brain disorder is the root cause of Parkinson's Disease (PD). Neurodegenerative is the process of impairment of brain cells. PD is diagnosed through clinical methods. Hope this research work helps to identify the intensity of the PD. Fuzzy Inference System is used to identify the PD and its intensity. Fuzzy rules, Mamdani Fuzzy Inference, Membership Functions, and Defuzzification are the process used to obtain accurate results. Oxford Parkinson's Disease Detection Dataset is used for this research work. Among 23 fields in the dataset, only four fields FoH, DFA, Spread1, and Spread2 are chosen for analyzing the PD diagnosis and intensity. These four fields values are categorized into three sets: one is PD affected subjects, the second set is common values for both PD affected subjects and healthy subjects, and the third set is completely healthy subjects. Intensity values are measured from low to maximum as 0–100. Eighty-one rules are framed to calculate the PD intensity. We hope this FIS model is a novel method for identifying the PD intensity and helps doctors to diagnose and treat the patients in an effective way.

1 Introduction

Parkinson's Disease is a neurodegenerative brain disorder. When the brain reduces the secretion of the chemical called as neurotransmitter, neurotransmitter it directly impacts the body movements. This neurotransmitter is called as dopamine. The place where this chemical secretes in the brain is referred to as substantia nigra [1]. When these brain cells in the part of substantia nigra reduces or dies then it is said to be neurodegeneration [2].

D. Karunanithi (✉)
Computer Science and Engineering, Manonmaniam Sundaranar University,
Tirunelveli, Tamil Nadu, India
e-mail: karunanithid@gmail.com

P. Rodrigues
Computer Science and Engineering, King Khalid University, Abha, Saudi Arabia
e-mail: drpaulprof@gmail.com

© Springer Nature Switzerland AG 2019   1229
D. Pandian et al. (eds.), *Proceedings of the International Conference on ISMAC in Computational Vision and Bio-Engineering 2018 (ISMAC-CVB)*, Lecture Notes in Computational Vision and Biomechanics 30,
https://doi.org/10.1007/978-3-030-00665-5_116

In the eighteenth century, James Parkinson found the symptoms of this disease and wrote an article as "Shaking Palsy" [3]. PD affects the people after middle age and the ration is greater for male. PD is diagnosed by five stages from stage 1 to 5 as the severeness increases [4]. Electrocardiography [5] and voice signals are used for PD diagnosing [6–9]. Majority of the people with PD get affected with their voice [10–12]. So the speech recording dataset is taken for the processing and diagnosing the PD using Fuzzy Inference System.

2 Materials

Max Little from the University of Oxford created a dataset. This dataset consists of various voice recording measures of 31 people among 23 were PD affected and 8 were healthy. He created this dataset in collaboration with National Centre for Voice and Speech, Colorado. It consists of 23 columns of different voice measures. Each subject's voice recordings are made five or six times and recorded the values in the respective column [13].

3 Methodology

3.1 Fuzzy Inference System (FIS)

Fuzzy Inference is a process of formulating the mapping from given input(s) to output(s) using fuzzy logic [14]. A fuzzy set is one in which the elements do not belong completely to only one set, but do belong to that sets to a certain extent [15]. An FIS was built on the three main components: fuzzifier, inference engine, and defuzzifier.

Figure 1 shows the detailed structure of the FIS. Normally, the database contains definitions such as information on fuzzy sets parameter with a function that has been defined for every existing linguistic variable. The basic rules can be constructed either from a human or automatic generation, where the searching rules using input–output data numerically.

There are several types of FIS, namely, Takagi–Sugeno, Mamdani, and Tsukamoto. An FIS of Mamdani is most commonly widely used methodology. It was proposed by Ebrahim Mamdani [Mam75].

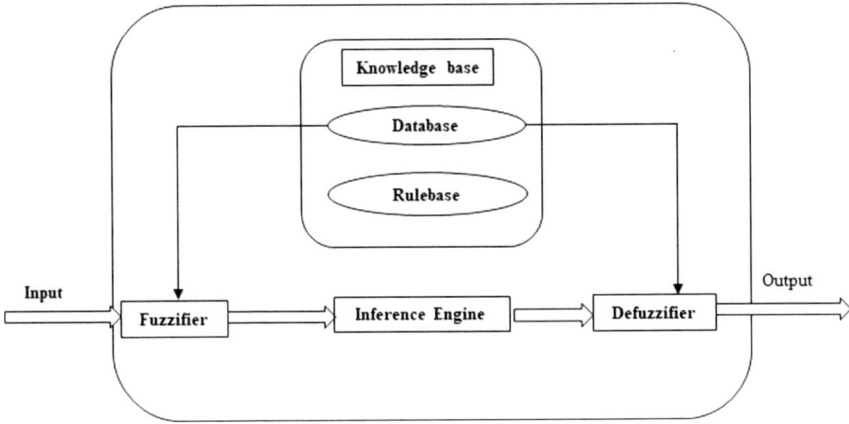

Fig. 1 Architecture of fuzzy inference system

3.2 Rules for Fuzzy Inference System to Diagnose PD

Sample fuzzy rules for predicting the status of PD are mentioned in Table 1. Data are taken from the Oxford Parkinson's Disease Dataset to frame these rules. Out of 23 fields in the PD dataset, only four fields are chosen for diagnosing PD using fuzzy logic. The four fields are FoH, DFA, Spread1, and Spread2. In the value analysis of the four fields, a clear difference between healthy and PD persons is observed. Also, these four fields correlate with my previous analysis of the same dataset using artificial intelligence. Three ranges had been identified in each field, which are a range of low values that completely fall with the PD affected persons, and mid range of values which belongs to both PD affected subjects and healthy subjects. And max. range of values completely belongs to healthy subjects. The middle range of values which belongs to both categories falls the fuzziness. Eighty-one rules are framed with these four fields to calculate the Parkinson's Disease Intensity (PDI) using fuzzy rules. PDI is calculated by nine ranges of values starting from 0, 12.5, 25, 37.5, 50, 67.5, 75, 87.5 to 100.

3.3 Fuzzification and Defuzzification

FIS designed for diagnosing Parkinson's Disease uses Mamdani Inference System. The defuzzifier uses the famous Center of Gravity (COG) method [16]. The formula for COG is

$$x* = \frac{\int \mu A(x).x \mathrm{d}x}{\int \mu A(x).\mathrm{d}x} \tag{1}$$

Table 1 Fuzzy rules for diagnosing PD

Rules for fuzzy inference system for diagnosis of Parkinson's disease
If (FoAvg is High) and (DFA is Low) and (Spread1 is Low) and (Spread2 is Low), then (PDI is Healthy)
If (FoAvg is High) and (DFA is Low) and (Spread1 is Low) and (Spread2 is Medium), then (PDI is VeryLow)
If (FoAvg is High) and (DFA is Low) and (Spread1 is Low) and (Spread2 is High), then (PDI is Low)
If (FoAvg is Medium) and (DFA is Medium) and (Spread1 is High) and (Spread2 is Low), then (PDI is Avg)
If (FoAvg is Medium) and (DFA is Medium) and (Spread1 is High) and (Spread2 is Medium) then (PDI is BAvg)
If (FoAvg is Medium) and (DFA is Medium) and (Spread1 is High) and (Spread2 is High), then (PDI is High)
If (FoAvg is Low) and (DFA is High) and (Spread1 is High) and (Spread2 is Low) then (PDI is High)
If (FoAvg is Low) and (DFA is High) and (Spread1 is High) and (Spread2 is Medium), then (PDI is VHigh)
If (FoAvg is Low) and (DFA is High) and (Spread1 is High) and (Spread2 is High), then (PDI is Max)

4 Experiment Results

4.1 Membership Functions to Diagnose PD

Mamdani FIS is used to diagnose the Parkinson's Disease. Four input member functions FrequencyAverage, DFA, Spread1, and Spread2 and one output member function Parkinson's Disease Intensity (PDI) are used in this experiment.

4.1.1 Membership Function for Frequency Average

From the analysis of the field Frequency Average (FoH), there are six recordings for each subject. The PD subjects and healthy are separated into two columns, and each subject's values are averaged and sorted in ascending order. After sorting, the three ranges of values are categorized as Frequency Average High (FoH), Frequency Average Medium (FoM), and Frequency Average Low (FoL). The range of values identified by the comparison of the healthy and PD subjects frequency average is FoH: 223.66 to 243.81, FoM: 114.30 to 203.90, and FoL: 97.94 to 113.01 (Fig. 2).

In this, FoH range of values belongs to healthy subjects. FoM range of values belongs both to healthy and PD affected subjects. Fuzziness observed in these ranges of values and another category FoL range of values belong completely to PD affected subjects. Three member functions were created to represent these values.

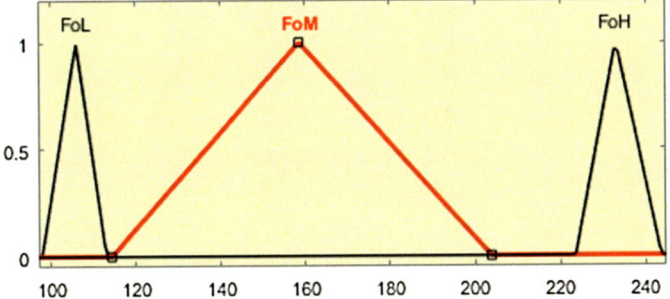

Fig. 2 Membership function for frequency average

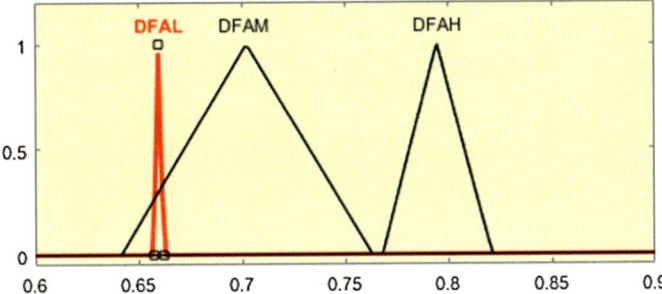

Fig. 3 Membership function for DFA

4.1.2   Membership Function for DFA

DFA field values are analyzed in the similar way FoH and categorized as DFA Low (DFAL), DFA Medium (DFAM), and DFA High (DFAH). The observed range of values for DFAL: 0.6383 to 0.6416 belongs to healthy subjects, DFAM: 0.6417 to 0.7634 belongs to both healthy and PD affected subjects where fuzziness occurs, and DFAH: 0.7686 to 0.8213 belongs to PD subjects. The membership functions for the above three are made and shown in Fig. 3.

4.1.3   Membership Function for Spread1

Spread1 field is analyzed in a similar pattern and categorized as Spread1 Low (SP1L), Spread1 Medium (SP1M), and Spread1 High (SP1H).

The range of values observed in SP1L: −7.59 to −6.70 belongs to healthy subjects, SP1M: −6.88 to −5.99 belongs to healthy and PD subjects, and SP1H: −5.63 to −3.66 belongs to PD subjects. The membership functions for Spread1 categories are made through FIS and shown in Fig. 4.

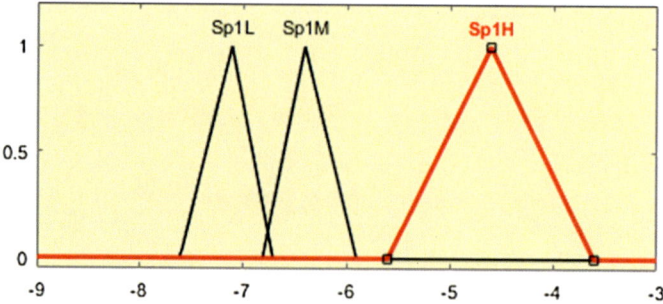

Fig. 4 Membership function for Spread1

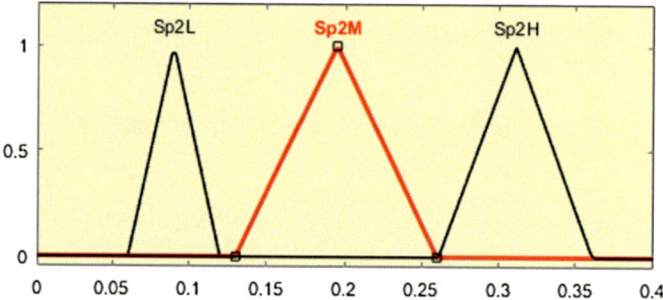

Fig. 5 Membership function for Spread2

4.1.4 Membership Function for Spread2

Spread2 field is analyzed in a similar pattern and categorized as Spread2 Low (SP2L), Spread2 Medium (SP2M), and Spread2 High (SP2H). The range of values observed in SP2L: 0.06 to 0.12 belongs to healthy subjects, SP2M: 0.13 to 0.26 belongs to healthy and PD subjects, and SP2H: 0.26 to 0.36 belongs to PD subjects. The membership functions for Spread2 categories are made through FIS and shown in Fig. 5.

4.1.5 Membership Function for Parkinson's Disease Intensity (PID)

From the four inputs FoH, DFA, Spread1, and Spread2, 81 rules are created to calculate the PDI with eight categories. The eight categories are mapped with the member functions Healthy (HTY), Parkinson's Disease Intensity Very Low (PDVIL), Parkinson's Disease Intensity Low (PDIL), Parkinson's Disease Intensity Below Average (PDIBAvg), Parkinson's Disease Intensity Average (PDIAvg), Parkinson's Disease Intensity Above Average (PDIAAvg), Parkinson's Disease Intensity High (PDIH), and Parkinson's Disease Intensity Very High (PDIVH).

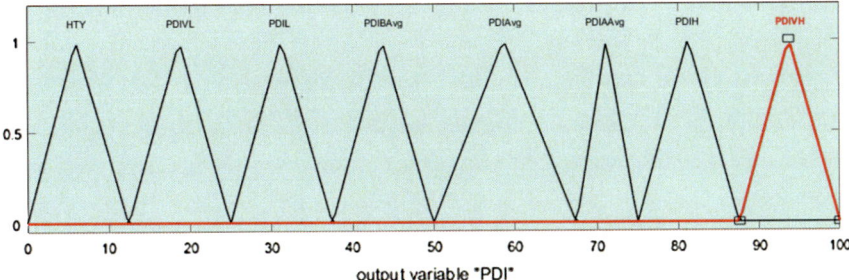

Fig. 6 Membership function for PDI

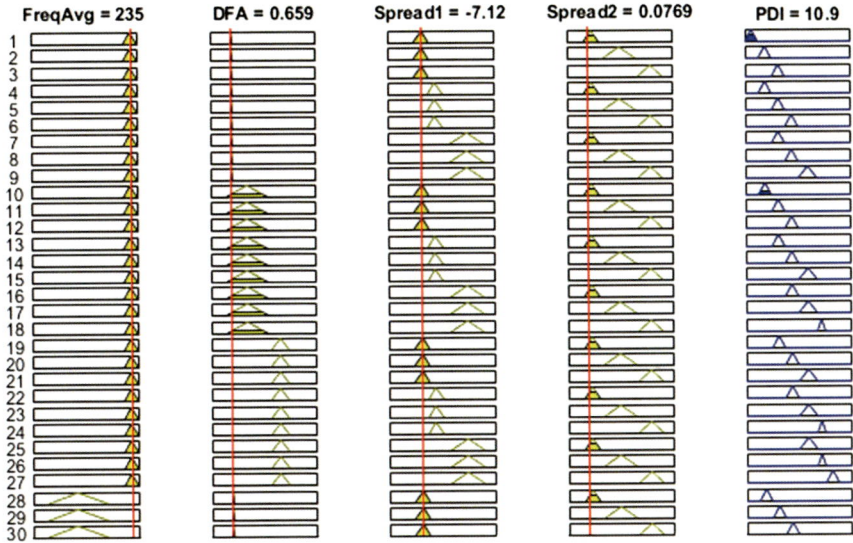

Fig. 7 Representation of fuzzy inference system for HTY

The measurements fixed for the eight member functions are HTY: 0 to 12.4, PDIVL: 12.5 to 24.9, PDIL: 25 to 37.4, PDIBAvg: 37.5 to 49.9, PDIAvg: 50 to 67.4, PDIAAvg: 67.5–74.9, PDIH: 75–87.4, and PDIVH: 87.5–100. Through the four inputs, various levels of intensity of the disease are calculated from the voice recordings of healthy and PD subjects. This fuzzy is a new method for identifying the intensity of the disease. The results will be more accurate and much useful for the doctors to treat the patients beyond the stages identification. The output member function is shown in Figs. 6, 7, 8 and 9.

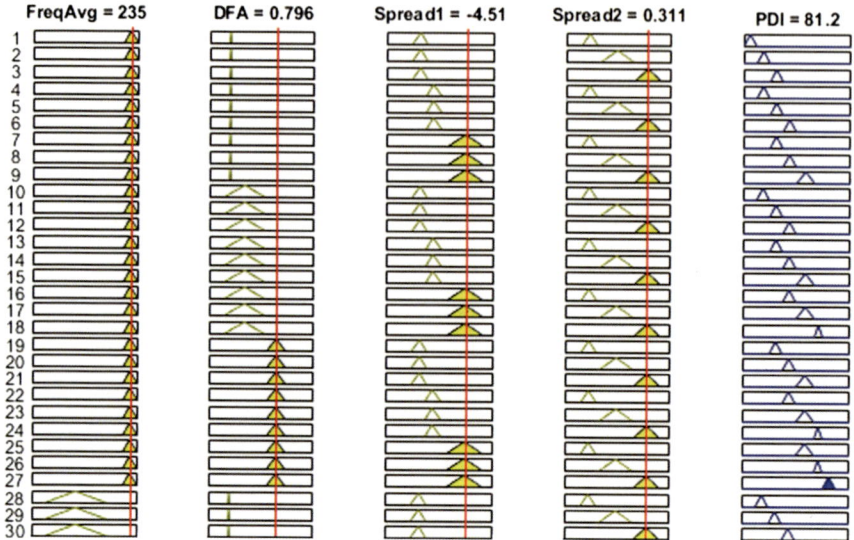

Fig. 8 Representation of fuzzy inference system for PDH

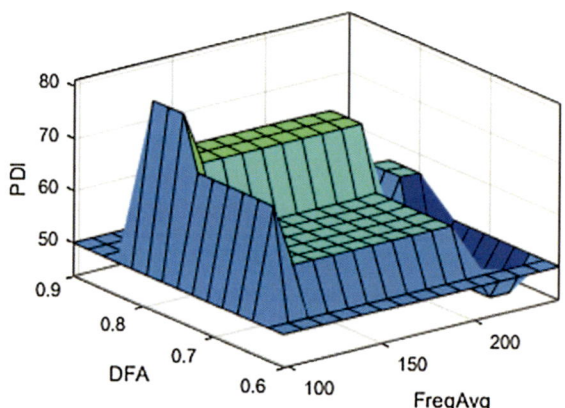

Fig. 9 Surface graph PDI with respect to DFA and FreqAvg

5 Conclusion

PD normally diagnosed by identifying the physical symptoms. Five stages of PD were identified by the severeness of the symptoms. In this research paper, the intensity of the PD was identified with eight levels from PD Oxford Dataset. In the value analysis of 23 fields, the chosen four fields, FoH, DFA, Spread1, and Spread2, values clearly show the difference between PD affected subject and healthy subjects. Also, few values are common for both subjects. Fuzziness occurs in these middle values of both subjects. Using fuzzy inference system, the intensity of the PD was able to be identified clearly. Eight levels of PD intensity were predictable through various possible combinations of four column values. With the four fields, tree structure was made with all possible combinations of four fields with the categories healthy, PD affected, and middle common values of between healthy and PD subjects. The intensity is calculated with the values 12.5 for fuzzy values of healthy and PD subjects and named as PPDV and 25 for complete PD affected values category and named as PDV. If any of the four fields values completely falls under the PD affected values, then the value of 25 is assigned. If any of the four field values falls under the middle category values that is the values which are common for both healthy and PD affected subjects 12.5 value is assigned. Eighty-one combinations were made with tree structure, and 81 rules are framed for each path of the tree. Using this PDV and PPDV, the final intensity was calculated from the range of 0–100. Eight categories of intensities: HTY, PDVIL, PDIL, PDIBAvg, PDIAvg, PDIAAvg, PDIH, and PDIVH were able to be obtained accurately through this FIS. This methodology might be useful for doctors to know the intensity of PD and treat the people accordingly.

References

1. Woo Y, Lee J, Hwang S, Hong CP (2013) Use of an adaptive-neuro fuzzy inference system to obtain the correspondance among balance, gait, and depression for Parkinson's disease. J Korean Phys Soc 62(6):959–965
2. http://www.parkinson.org/understanding-parkinsons/what-is-parkinsons. [updated in 2018]
3. Gelb D, Oliver E, Gilman S (1999) Diagnostic criteria for Parkinson disease. Arch Neurol 56(1):33–39. https://doi.org/10.1001/archneur.56.1.33.PMID9923759
4. http://www.parkinson.org/understanding-parkinsons/what-is-parkinsons/The-Stages-of-Parkinsons-Disease
5. Pezard L, Jech R, RuÊzĬicĬka E (2001) Investigation of non-linear properties of multichannel EEG in the early stages of Parkinson's disease. Clin Neurophysiol 122:38–45
6. Ene M (2008) Neural network-based approach to discriminate healthy people from those with Parkinson's disease. Math Comput Sci Ser 35:112–116
7. Caglar MF, Cetisli B, Toprak IB (2010) Automatic recognition of parkinson's disease from sustained phonation tests using ANN and adaptive neuro-fuzzy classifier. J Eng Sci Des 1:59–64
8. Gil D, Johnson M (2009) Diagnosing Parkinson by using artificial neural networks and support vector machines. Glob J Comput Sci Technol 9:63–71
9. Duffy RJ (2005) Motor speech disorders: substrates, differential diagnosis and management, 2nd edn. Elsevier Mosby, St. Louis
10. Ho AK, Iansek R, Marigliani C, Bradshaw JL, Gates S (1998) Speech impairment in a large sample of patients with Parkinson's disease. Behav Neurol 11:131–137. https://doi.org/10.1155/1999/327643
11. Logemann JA, Fisher HB, Boshses B, Blonsky ER (1978) Frequency and co-occurrence of vocal-tract dysfunctions in speech of a large sample of Parkinson patients. J Speech Hear Disord 43:47–57
12. Sapir S, Spielman JL, Ramig LO, Story BH, Fox C (2007) Effects of intensive voice treatment (the Lee Silverman Voice Treatment [LSVT]) on vowel articulation in dysarthric individuals with idiopathic Parkinson disease: acoustic and perceptual findings. J Speech Lang Hear Res 50:899–912. https://doi.org/10.1044/1092-4388(2007/064)

13. Center for Machine Learning and Intelligent Systems (2008). http://archive.ics.uci.edu/ml/datasets/Parkinsons
14. Khezri R, Hosseini R, Mazinani M (2014) A fuzzy rule-based expert system for the prognosis of the risk of development of the breast cancer. Int J Eng (IJE) Trans Basics 27(10):1557–1564
15. Camara C, Warwick K, Bruña R, Aziz T, del Pozo F, Maestú F (2015) A fuzzy inference system for closed-loop deep brain stimulation in Parkinson's disease. J Med Syst 39(11)
16. Hamidzadeh J, Javadzadeh R, Najafzadeh A (2015) Fuzzy rule based diagnostic system for detecting the lung cancer disease. J Renew Nat Resour Bhutan 3(1):147–157

A Comprehensive Study of Retinal Vessel Classification Methods in Fundus Images for Detection of Hypertensive Retinopathy and Cardiovascular Diseases

J. Anitha Gnanaselvi and G. Maria Kalavathy

Abstract Quantitative studies for classification of retinal vessels using new computer-assisted retinal fundus imaging system have allowed the researchers to understand the influence of systemic on retinal vascular caliber. These retinal vascular caliber changes reflect the cumulative response to cardiovascular risk factor. Hypertensive retinopathy can be detected in earlier stage by analyzing the retinal image. Nowadays, it is obvious that there is a relationship between changes in the retinal vessel structure and the most common diseases such as hypertension, stroke, cardiovascular diseases, those can be detected by noninvasive retinal fundus image. The proposed approach of applying an image processing technique, the aforementioned disease can be diagnosed earlier by retinal fundus image. To achieve the precise measurement of the retinal image parameters, the classification of blood vessels such as arteries and veins is necessary. These classifications of arteries and veins can be achieved through the retinal fundus image. The retinal vessel classification is based on visual and geometric features from these classified images into arteries and veins for the detection of hypertensive retinopathy, stroke, and cardiovascular risk factor. This classification of retinal fundus image is essential for early diagnosis of aforementioned diseases. The retinal arteriolar caliber which is narrower and smaller, that is associated with older age, will predict the incidence of diabetic retinopathy and cardiovascular risk factor. Similarly, retinal venular caliber which is wider, that is associated with younger age, will predict the incidence of risks of stroke and coronary heart diseases. This could suggest the possibility of using this model of fundus image in classification approaches. Finally, the selected attributes of classification are applied through the genetic algorithm with radial basis function neural network for diagnosis of the disease in order to improve the classification accuracy with less computational cost time.

J. A. Gnanaselvi (✉)
Research Scholar, Faculty of Information and Communication Engineering, Anna University, Chennai, India
e-mail: anithagnanaselvi@gmail.com

G. M. Kalavathy
Department of CSE, St.Joseph's College of Engineering, Chennai, India
e-mail: maria_kalavathy@yahoo.co.in

© Springer Nature Switzerland AG 2019
D. Pandian et al. (eds.), *Proceedings of the International Conference on ISMAC in Computational Vision and Bio-Engineering 2018 (ISMAC-CVB)*, Lecture Notes in Computational Vision and Biomechanics 30,
https://doi.org/10.1007/978-3-030-00665-5_117

1 Introduction

Diabetic retinopathy (DR) means common problem of diabetes which damages the
retinal vascular area. Mostly, it affects the blood vessels in retina. One of the major
issues in diabetic retinopathy is visual impairment. This is because of the new blood
vessel growth in retina in proliferative retinopathy. There are several categories to
predict the early diagnosis of diseases. One of the best methods is to predict the DR
by using the fundus image, this fundus image is considered to be the initial and basic
screening process in diabetic retinopathy prediction.

Among the several blood vessels, the retinal blood vessel network is visible and
is suitable for noninvasive imaging method in our body. So, the retinal blood vessel
is considered to be the reliable tool for early diagnosis of DR. In order to perform
the analysis, the classification of blood vessel is done based on geometric features
of vessel network. In retinal image analysis, the accurate measurement of retinal
vessel parameters is considered to be an important problem in eye research. There
are several parameters measured from the retinal vessel. This includes the thickness
and curvature of the vessels and also the measurement of the arteriolar–venular ratio.

The arteriolar–venular ratio is considered as the essential parameter for early
prediction and diagnosis of diseases which includes the hypertension, stroke, cardiovascular disease in youngsters, and retinopathy in child. Though there are several sets
of rules that have been defined for measuring the ratio of arteriolar–venular. These
include the distance from the optic disk margin; in Japan, the ratio is measured and
calculated as 0.25–1 of the optic disk diameter. In the U.S., it is stated as 0.5–1 for
measuring the arteriolar–venular ratio calculation.

This arteriolar–venular ratio calculation measurement includes other areas also
such as localization of optic disk, retinal vessel diameter, measuring accurately,
retinal vessel network image analyzing, and also for the classification of blood vessels
which includes arteries and veins. This classification of blood vessel is considered
as the basic step for calculating the arteriolar–venular ratio. This arteriolar–venular
ratio is important for calculating the classification of blood vessels in an efficient and
effective way.

The quantitative calculation of retinal vascular caliber is highly influenced in the
clinical significant association outcomes such as detection of stroke and coronary
heart diseases. This can be achieved by the retinal fundus image.

2 Related Work

In the modern world, by the early diagnosis many diseases can be controlled that
includes the life-threatening diseases. Though there are several approaches that are
existing already for the early diagnosing diseases, this deals with the two categories.
First category used the machine learning popular algorithm. Genetic algorithm which
is used for the feature selection of the attributes is included in the second category

and it uses the radial basis functional neural network for the classification purpose in the attributes. The diabetic retinopathy diagnosed by the Pima Indian Diabetes Dataset is used for the classification purpose. The results show that it will minimize the cost of computation time and is also better in classification [1, 2].

The diabetic retinopathy (DR) mainly affects the retinal vessels in the macular region which is located in the center portion of the retina in the fundus image. These studies are based on the four main methods. It contains the preprocessing of the image and then enhancement of the fundus image. Third is the segmentation process which will segment the vessels, and finally it includes the proposed foveal avascular zone segmentation approach which is used to diagnose the diseases. This shows the results of an average in performance metrics such as specificity, sensitivity and accuracy from theses diseases are diagonsed earlier [3, 4].

There are various approaches to quantify the width and tortuosity of retinal vessels to detect cardiovascular diseases. The artery-to-venous ratio will predict the narrow arterial and venous dilatation for the detection of stroke and cardiovascular heart diseases. Here, the artery-to-venous classification method is used which provides the accurate region around the optic disk [5].

The segmentation of retinal vascular blood vessels into veins and arteries is used for quantifying the ratio of artery-to-venous diameter. It is developed to predict the cardiovascular heart disease and stroke in children. The features are extracted from the databases and used for the training and testing phase algorithm for all the vessels and it will point out whether it is vein or artery. The artery-to-vein classification approach using the receiver operator characteristic shows better results and has detected the artery and vein in the fundus image [6].

In order to achieve the accurate calculation and measurement of parameter in diagnostic features, the arteries and veins are essential. The classification parts of blood vessels in the fundus image are evaluated by different databases in different criteria. These approaches focus on the geometric features and statistical model in spatial and transform domain [7, 8].

3 Methodology and Measurement

The common method of the blood vessel classification is by utilizing the fundus image. First step is segmentation of vessels, second step is selecting the region of interest for classifications of vessels. Third step is extracting the features from various parts of the vessels. Fourth step is to classify the vectors of features. At last, final is the combination of results to determine the final label of vessel. This is considered as the existing way to classify the vessels in the fundus image.

The proposal is based on the classification of vessels into two major criteria. They are represented as automatic and semiautomatic methods.

(a)   **(b)**

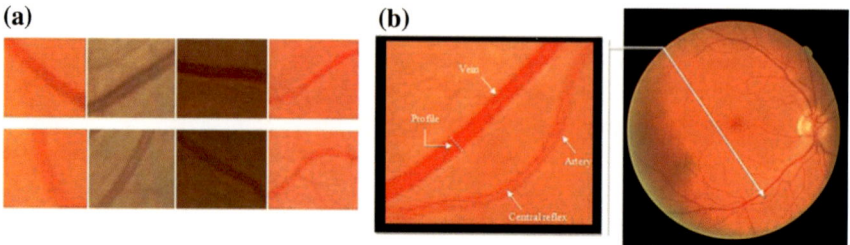

Fig. 1 **a** Sample images of vein represented in first row and artery represented in second row in fundus images, **b** Specifying central reflex and profile in a piece of a fundus image

3.1 Semiautomatic Method

The semiautomatic method for retinal vessel analyzes the artery and venous individually and it calculates geometrical and topological pixel features of each segment. This method uses the anatomical properties of veins and arteries. The semiautomatic method is a hybrid of graph-theoretic method with domain-specific knowledge and is capable of analyzing the entire vasculature (Fig. 1).

3.2 Automatic Methods

In this automatic approach, position around the optic disk is within 0.5–2 diameter of disk from its center portion and is segmented into four zones, in which each one contains one of the major arches. In these, the red channel and hue channel in each vessel segment are represented as the most discriminative features for classification. As the results, in two adjacent vessels, the darker reddish vessel is considered as vein, the vessel that has more color uniformity is considered as vein (Fig. 2).

After feature extraction, the classification criterion of vessels is based on fuzzy clustering algorithm. The Euclidean distance of each pixel from the mean value of features is calculated. After vessel classification, the data is available by vessel identifying the variation in pixels to discriminate the arteries and veins. Twenty-six fundus images have been taken in this study for assessment, in which ten fundus images were used to develop the algorithm, and 16 images were used for validation. Reported results on 15 validation images show the overall error of 12.4%. This classification procedure is performed around the optic disk and it is more preferable for optic-disk-centered images.

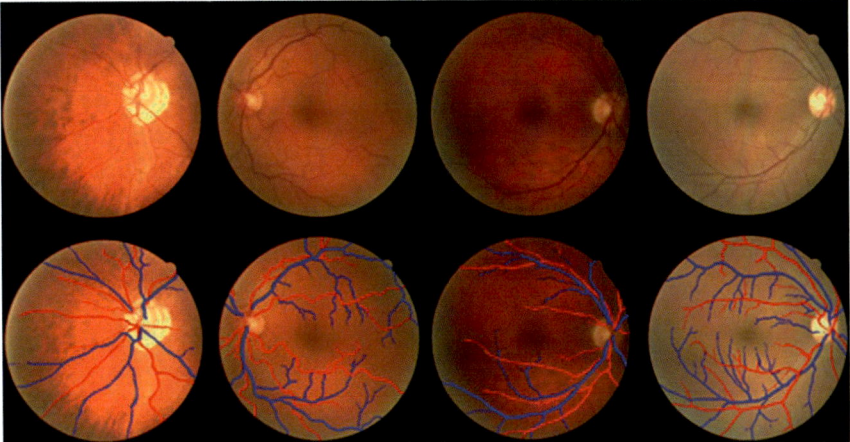

Fig. 2 First row represented some samples of fundus images from dataset and second row represented their ground truth. The arteries are represented in red lines, and the veins are represented in blue lines

3.3 Retinal Vessel Measurements

The retinal vessel calibration is based on the computer-assisted program to find out the accurate size of the fundus image. The calculation consists of actual size of the central retinal artery, the same procedure is applicable for vein, and it calculates the ratio of these abovementioned variables which is known as the artery-to-vein ratio (AVR). The result is obtained and is represented as shown in Fig. 3.

The measurement of retinal vascular caliber formula is implemented based on the theoretical and empirical methods and it can be achieved by larger quantity of retinal image with its branch point. The measurement of individual trunk vessel and branch vessel is obtained by square root mean deviation model for observing the data.

In today's population, it is difficult to find out the problem such as hypertension diabetes and ocular alignments problem such as diabetic retinopathy and glaucoma [9]. Those diseases can be measured by the retinal vessel through fundus Image. This can be measured through the refraction/axial length and retinal vascular caliber by retinal AVR (artery-to-venous) ratio calculation. However, it is very essential to identify the narrower artery and wider venous and this will result in the smaller artery-to-venous ratio. Figure 3 represents the arteriolar–venular ratio calculation measurement zones.

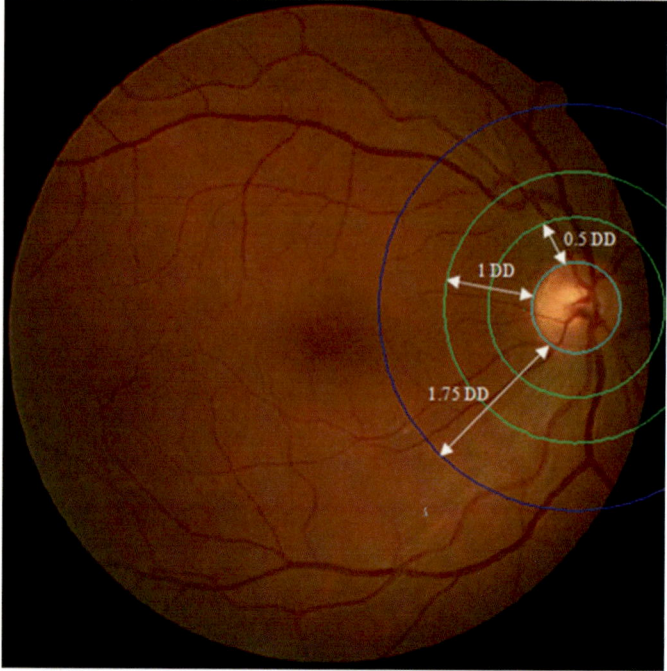

Fig. 3 Arteriolar–venular ratio (AVR) calculation measurement zones, DD = optic disk diameter

4 Results and Discussion

4.1 Proposed Methodology Using GA with RBF NN Classification of Diseases

The proposed methodology is implemented with genetic algorithm, with radial basis neural network function for classification of attributes. Here, we used Pima Indian Diabetes Dataset to perform the best diagnosis of diabetes patients in datasets. Here, the proposed method in this abovementioned algorithm is implemented to detect the diabetic patients in Pima Indian Diabetes Datasets. The genetic algorithm based on selection, crossover, and mutation operation is performed in order to obtain better attribute features and it achieves the shorter in training and testing time, that is, computational cost and time, better storage capacity, and finally increases the classification rate. Table 1 represents the proposed algorithm architecture.

The radial basis functional neural network is based on supervised learning approach. In this proposed method, the radial basis functional neural network has been used for the classification in order to diagnose the diabetes disease (Fig. 4).

The radial basis functional neural network is a supervised feedforward process which consists of one hidden layer of hidden units, which states the radial basis

Table 1 Proposed algorithm

Step1: Initialization
Step2: Store the Pima Indian Diabetes Dataset
Step3: Start the parameters of genetic algorithm
Step4: Run the genetic algorithm
Step5: (a) First generation process starts
Step6: Selection While stopping criteria not met do
(b) Crossover
(c) Mutation
(d) Selection
End
Step6: Applying the radial basis functional neural network for classification
Step7: Dataset attributes are trained
Step8: Calculation of accuracy and measuring the error
Step9: Testing the datasets
Step10: Calculation of accuracy and measuring the error
Step11: Stop the process

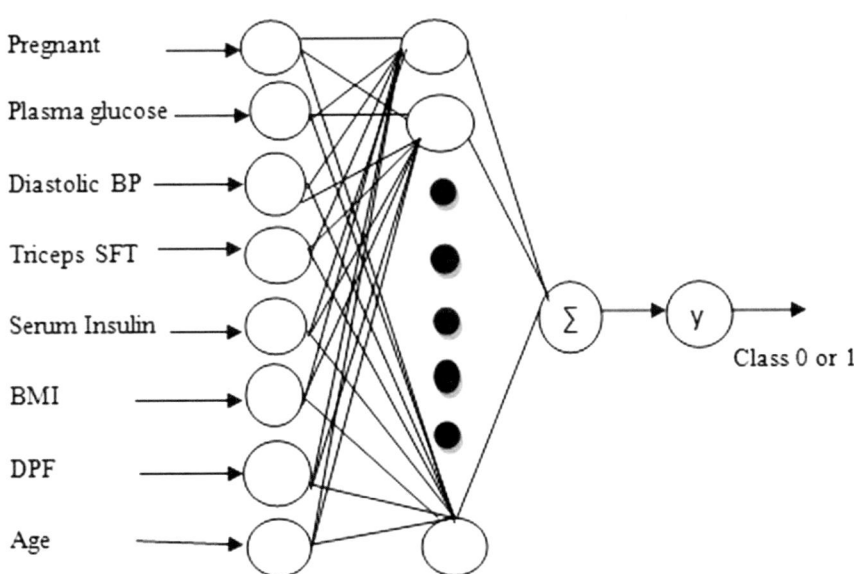

Fig. 4 Feedforward neural network model for diagnosis of diabetes

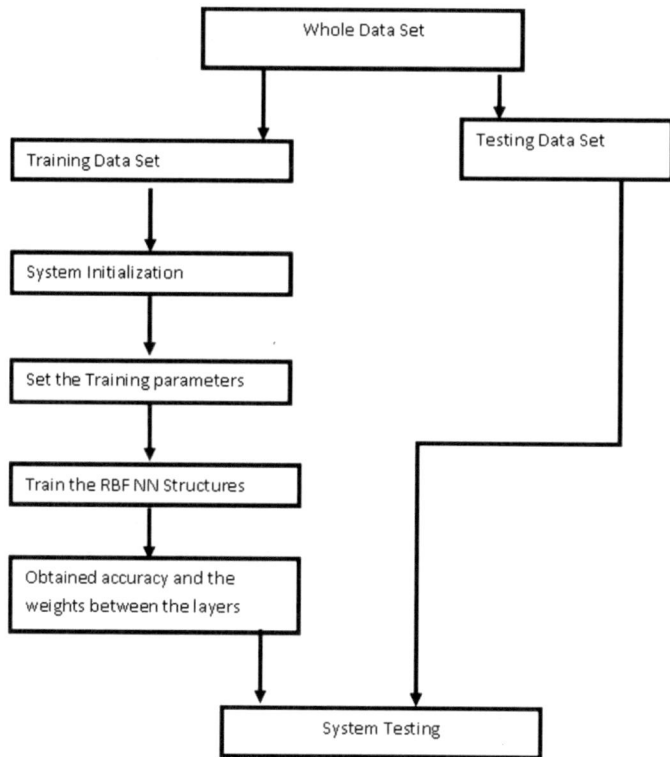

Fig. 5 The radial basis functional neural network methodology

Table 2 Genetic algorithm attributes selection

Dataset	Number of attributes	Name of the attributes	Number of instances	Number of classes
Pima Indian Diabetic Datasets with genetic algorithm	1	Plasma glucose tolerance test	768	2
	2	Serum insulin		
	3	Body mass index		
	4	Age		

functions (RBFs). These radial basis functional neural networks require a desired response to be trained in pattern classification studies. Particularly, in the present study, a training algorithm is used which normally uses a gradient descend rule for the training attributes of these networks. Figure 5 represents the radial basis functional neural network methodology (Figs. 6, 7 and Tables 2, 3).

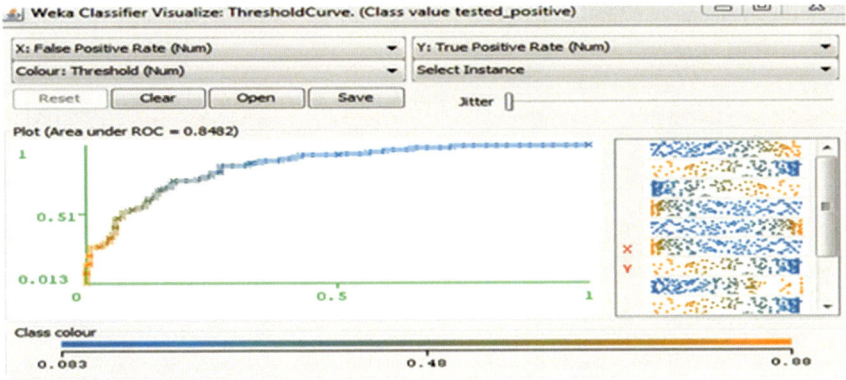

Fig. 6 ROC graph GA_RBF NN on PIDD

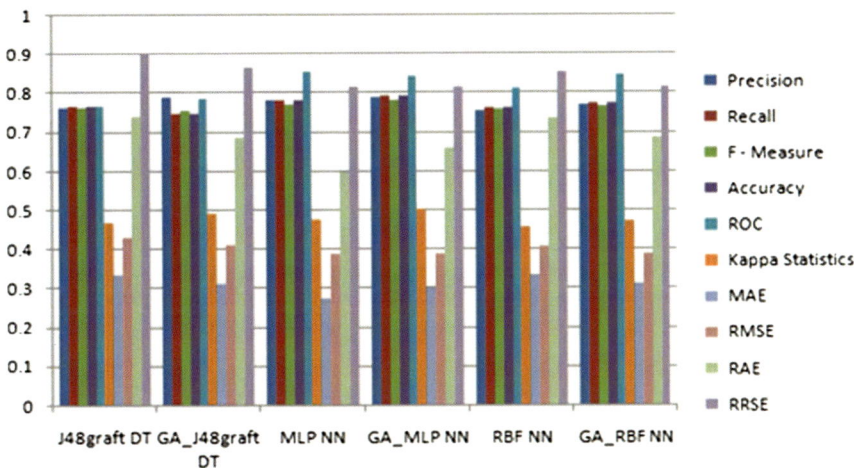

Fig. 7 Evaluation GA_RBF NN performance for PIDD

5   Conclusions and Future Directions

Diabetic blood sugar when above the desired level is considered as the world's widespread disease. The contribution of this proposed method was developed and implemented with genetic-algorithm-based radial basis functional neural network. It also determines and assesses the diabetic classification and detection of artery and vein in order to estimate the various problems including the blindness, blood pressure, kidney diseases, coronary heart diseases, nerve damages, and so on. In this proposed study, firstly, the classification has been done on Pima Indian Diabetic Datasets using radial basis functional neural network, and then using genetic algorithm for attributes selection, and thereby performing classification on the selected attribute. The results obtained are more interesting and may also happen for the exploration of the dataset.

Table 3 Results of genetic algorithm radial basis functional neural network for Pima Indian Diabetic Datasets

Measure	Training set evaluation	Testing set evaluation
Precision	0.75	0.75
Recall	0.76	0.76
F-measure	0.756	0.77
Accuracy	0.766	0.76
ROC	0.819	0.80
Mean absolute error	0.55	0.49
Root mean squared error	0.41	0.42
Relative absolute error	0.71	0.72
Root relative squared error	0.85	0.86
Kappa statistics	0.45	0.46

These classifications of arteries and veins can be achieved through the retinal fundus image. The retinal vessel classification is based on visual and geometric features from these classified images into arteries and veins for the detection of hypertension, stroke, and cardiovascular risk factor. From this, we tend to conclude that the proposed approach will help physicians to improve or take accurate decisions to do work speedily with less expense. In future, these proposed approaches can also be used for other kinds of diseases.

Acknowledgements All the images are taken from "Pima Indian Diabetes Dataset" which is publicly available.

References

1. Choubey DK, Paul S (2017) GA_RBF NN: a classification system for diabetes. Int J Biomed Eng and Technol 23(1):71–91
2. Yadav P, Ruhil N (2016) Blood vessel detection for diabetic retinopathy. In: IEEE international conference. 978-9-3805-4421-2/16/$31.00
3. Nugroho HA, Dharmawan DA, Litasri (2017) Automated segmentation of foveal avascular zone in colour retinal fundus images. Int J Biomed Eng and Technol 23(1):1–18
4. Sun C, Wang JJ, Mackey DA, Wong TY (2009) Retinal vascular caliber: systemic, environmental and genetic associations. Surv Ophthalmol 54(1):74–94
5. Joshi VS, Reinhardt JM, Garvin MK, Adramoff MD (2014) Automated method for identification and artery-venous classification of vessel trees in retinal vessel networks. PloS 9(2):e88061. www.plosone.org
6. Niemeijer M, van Ginneken B, Abramoff MD (2016) Automatic classification of retinal vessels into arteries and veins. In: Proceedings of SPIE-the international society for optical engineering, 7260, 72601F. https://doi.org/10.11117/12.813826
7. Miri M, Amini Z, Rabbani H, Kafieh R (2017) A comprehensive study of retinal vessel classification methods in fundus images. J Med Signals and Sens 7(2):59–70

8. Guzman JC, Melin P, Prado-Arechiga G (2017) Neuro-fuzzy hybrid model for the diagnosis of blood pressure. In Nature-inspired design of hybrid intelligent systems. Springer International Publishing, pp 573–584. https://doi.org/10.1007/978-3-319-47054-2_37
9. Agarwal A, Williams GH, Fisher NDL (2005) Genetics of human hypertension. Trends in Endocrinol and Metab Elsevier. 10.10.16/j term

Estimation of Parameters to Model a Fabric in a Way to Identify Defects

V. Subhashree and S. Padmavathi

Abstract Fabric defect detection is a quality check process which can locate and identify defects caused during the production process in the textile industry. Automated defect identification system uses computer vision and pattern recognition techniques whose performance depends majorly on the quality and quantity of the input dataset. A wide range of parameters is considered for decision process which compromises the accuracy of the system. This paper aims to estimate suitable parameters for the defect-free fabric which can be used by traditional methods to identify the defects in an efficient way. Hough-transform-based method is proposed to identify the parameters and the algorithm is experimented on various fabrics. The proposed method gives promising results when the horizontal and vertical threads are evident in the image.

1 Introduction

The defect detection at an earlier stage is an essential requirement for the fabric production industries. The quality control of the fabric affects the duration of production and is getting high productivity without the financial loss. According to paper [1], the cloth rates are scaled down by 45–65% because of the fabric defects caused during production process. To improve the quality, preparatory measures can be taken in prior before the item reaches the market [2]. The defects are caused by error patterns of the yarns or the spoils made by the machines during the production. The discontinuities of the threads on the fabric are an essential thing to be identified during manufacturing process. The identification of the minute thread defects on the fabric using man power during the production is tiresome and delayed. The manual

V. Subhashree · S. Padmavathi (✉)
Department of Computer Science and Engineering, Amrita School of Engineering, Coimbatore, Amrita Vishwa Vidyapeetham, Coimbatore, India
e-mail: s_padmavathi@cb.amrita.edu

V. Subhashree
e-mail: cb.en.p2cvi16008@cb.students.amrita.edu

© Springer Nature Switzerland AG 2019
D. Pandian et al. (eds.), *Proceedings of the International Conference on ISMAC in Computational Vision and Bio-Engineering 2018 (ISMAC-CVB)*, Lecture Notes in Computational Vision and Biomechanics 30,
https://doi.org/10.1007/978-3-030-00665-5_118

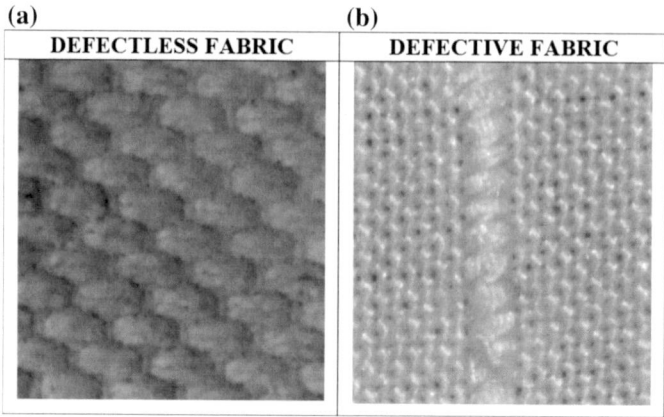

| **(a)** | **(b)** |
| DEFECTLESS FABRIC | DEFECTIVE FABRIC |

Fig. 1 **a** Defectless fabric, **b** defective fabric

method of defect identification is claimed to give recognizable results of only 70% with a limited speed. Hence, usage of automated defect identification systems helps manufacturing industries to complete the requirements within a specific time without compromising the quality.

Automated fabric defect detection is the need of the textile industries and image processing [3, 4] can be used as a solution for such problem. The automated inspection uses machine learning algorithms to identify the defects from the random patterns on the fabrics.

The algorithms use thread disorder identification in extensive range for defect detection on the fabric. The method indirectly utilizes the yarn dimensions like thickness, density, and other characteristics of the yarn. The algorithm should be able to accurately classify various types of defects on the fabric. The challenges of image-processing-based techniques include illumination difference, change in the orientation, occlusion of the yarns, etc.

It is quite common that the defects occur on different fabrics and is very less likely that all defects occur on a particular fabric. The change in the fabric becomes a factor that influences the defect detection system. When there is an increase in the input dataset, the variations on the fabric are also likely to increase. The accuracy of the defect detection system does not purely depend on the defect and in fact is much dominated by the fabric material. This phenomenon is demonstrated in Fig. 1 which shows two different fabrics with varying thread parameters. The defects occur at the thread level and the influence of these defects on the overall fabric image is very less which in turn lead to inaccuracy in classification.

There is an abundant availability of the defect-free fabric images than the defective images, which influences the dataset used for training the classifier. The above discussion demands a requirement of suitable parameters from the defect-free images which could be utilized for identifying the defects. The parameters like (i) distance between the threads, (ii) orientation of the thread, and (iii) density/thickness of the thread are few of the parameters which differ between the defective and defect-free fabric. In this paper, we attempt to model defectless fabric using the thread parameters measured from fabric images.

This paper is organized as follows: The literature survey is discussed in Sect. 2, proposed work is in Sect. 3, experimental analysis is in Sect. 4, and conclusion is in Sect. 5.

2 Literature Survey

The texture of the fabric is the most important strategy to model the defectless fabric. Texture methods used for representing defectless fabric can be majorly classified into (i) model based, (ii) spectral based, (iii) statistical, (iv) structural, and (v) hybrid and motif based [5]. GLCM and Gabor filter are most widely used applications for fabric defect detection. The relationship between the pixels is analyzed using the statistical approaches. The statistical analysis between the pixels is classified based on the first-order, second-order, and higher order methods. The gray-level co-occurrence matrix (GLCM) is categorized under the second order, which is the most predominantly used vision-based methods for defect identification on the fabrics [5]. Different features can be extracted from gray-level co-occurrence matrix as proposed by Harlick et al. that can be used to represent the texture properties as in [6] and patterns. Gabor filter is a widely used spectral method to represent the texture.

2.1 GLCM

The GLCM matrix has occurred by extracting the features which are based on second-order statistics. The features extracted by the first-order statistics are the gray-level scale distribution in the image, which does not contain the information about the different gray levels in the image. Jagdish Lal Raheja [7] proposed an effective method to extract the texture features using the distance and orientation parameters using GLCM method. The co-occurrence matrix is arranged with the corresponding frequencies of different gray levels $P\theta$, $d(I1, I2)$ to produce texture features. It measures the corresponding probability of the measured distances and the directions in a matrix [2]. The co-occurrence matrix consists of two parameters: (1) d—distance measurement between the pixels and (2) θ—orientation between the pixels.

2.2 Gabor Filter

Gabor filter is the most useful application for identifying the edge detections, documents management, and image segmentation [7].

$$g(x, y; \lambda, \theta, \varphi, \sigma, \gamma) = \exp - ((x^{\wedge'}2 + \gamma^2 y^{\wedge'}2)/(2\sigma^{\wedge}2))\cos(2\pi x^{\wedge'}/\lambda + \varphi) \quad (1)$$

Gaussian function is modulated by Gaussian kernel function. Daugman [8] proposed a method by aggregating the properties on the receptive field areas, which is the visual cortex and the sinusoidal Gabor function is shown in Eq. 1 which is for definite frequency and orientation change. Padmavathi [9] had an approach to wavelet transform using Gabor filterbank from a large set of parameters. The Gabor consists of large sets of parameters [10] like aspect ratio, bandwidth, phase offset, orientation, and wavelength. The choice of optimal parameters is always a requirement. Evolutionary algorithms [11] are recently used for optimization problems. Tsaia [12] attempts to choose optimal Gabor parameters for texture segmentation. Daugman [13] tries to capture the spatial frequencies through Visual Cortical filters.

The morphological techniques are widely used for fabric inspections [11]. The morphological operations perform better on fabric defects with hole and missing of the yarns [14]. Priya [2] was able to achieve better results by performing morphological operations on the low-contrast images. Meihong Shi [4] reduced texture erroneous and noises on the image by performing the LCD method.

The Hough transform is useful for measuring the global descriptors of the features. Mid point Hough transform is suggested by Hari in [15] to increase the speed of line detection. The fabric consists of warp and weft threads on the fabric, by finding the skew direction of warp and weft it is easy to calculate the fabric density [3]. Yildiz [1] proposed a system of the thread densities detection using the horizontal and vertical frequencies of the fabric by computing FFT. Estimation of the fabric parameters is done using the Hough transform. Ruru Pan [16] used Hough to segment out the warp and weft in both directions to find the densities between the yarns.

This paper aims to develop a model based on the thread parameters from the defectless fabric. The fabric defects depend on the thread parameters, the threads are restricted to warp and weft direction of the fabric. Hence, this paper concentrates on the parameters in these directions.

3   Proposed Work

Fabric defect detection process involving GLCM and Gabor filter usually computes the response for a large set of parameters and their average is used for defect detection. The small variations of the thread occurring in the defective fabric lose its importance when computed across wide range of parameters. When the parameters are narrow and closer to the thread parameters, the fabric defect detection accuracy can be improved. This paper estimates these parameters based on the thread parameters of a defectless fabric. The parameters estimated by the proposed method could be used to fine-tune the GLCM and Gabor filter parameters. The overall process of the proposed system is given in Fig. 2.

The input defectless image is converted into a binary image using threshold method. The binarized image is preprocessed using the morphological operations

Fig. 2 System architecture
diagram

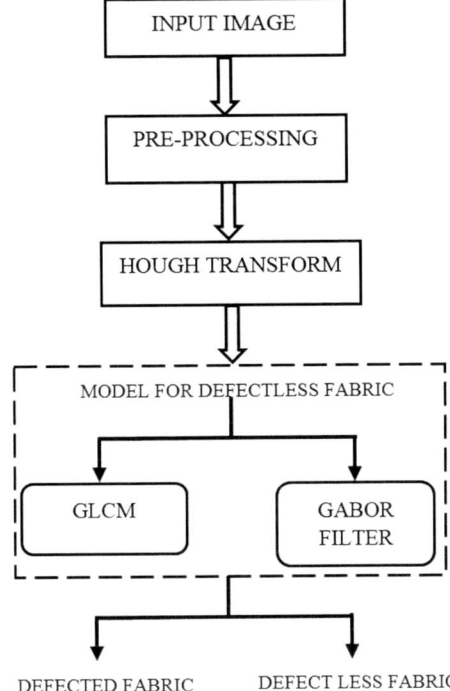

of erosion followed by double dilation to enhance the connectivity of thread parameter. The Hough transform is applied to the resulting image to detect the lines. Hough transform is used for line detection using the local measurement. The polar coordinate system is expressed as follows:

$$x \cos + y \sin = d \qquad (2)$$

where r represents the distance between the origin and the line "θ" represents the direction with reference to "r". The (d, θ) values correspond to (xi, yi), which maps to Cartesian space in the Hough space. Hough transform accumulates the count of line parameters present in the image. Dominating peaks of Hough lines are considered as thread lines. These lines are converted to spatial coordinates and the distance between them is computed. These distances are considered as thread parameters of the fabric. Finally, the refined parameters are sent to model the defectless fabric. These parameters are used to fine-tune the GLCM parameters. The features like energy, homogeneity, entropy, contrast, and correlation are calculated. The procedure is experimented on different fabrics and the results are summarized.

Table 1 GLCM features

Features	Values d = 2, Theta = 30	Values d = 2, Theta = 90	Values d = 5, Theta = 90
Energy	0.4535	0.6524	0.6333
Homogeneity	0.9335	0.7542	0.9666
Correlation	0.7826	0.8562	0.8125
Contrast	0.1332	0.1265	0.669

Fig. 3 Input image

4 Experimental Analysis

The dataset consists of 35 set of images, which includes the images at three different orientations 0,120, and 45° of camera. The dataset consists of 14 defectless fabrics, 21 defective fabrics, and four different thread thickness/thread patterns. Table 1 shows the GLCM features of the same image calculated for three different parameter sets. It could be observed that the values vary across the parameters, thus justifying the need for fine-tuning the parameters.

The input image shown in Fig. 3 is preprocessed as explained in previous section, and the resulting image is shown in Fig. 4. It is clear from the image that the morphological operations increase the visibility of the warp and weft yarn in the fabric.

Hough transform is applied to the preprocessed image and the peaks are calculated as explained in the previous section. Those dominating peaks are converted to lines and shown on the image as in Fig. 5. The distance between the lines was calculated using the dominating peak coordinates.

Fig. 4 Preprocessing

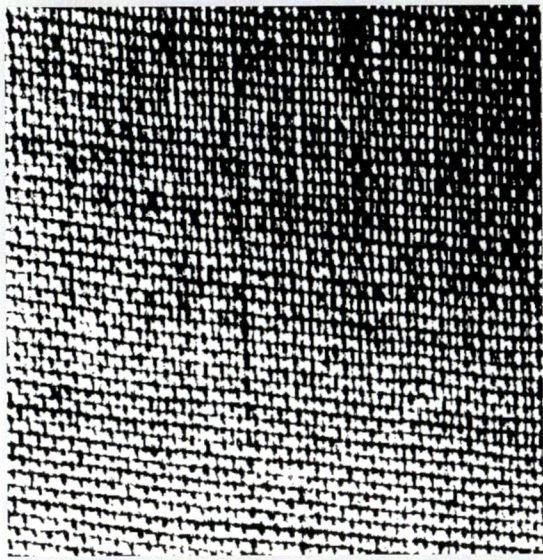

Fig. 5 Hough transform

The distance between horizontal lines is computed as 51 and distance between vertical line is computed as 38 for the test image shown. From the observed distances, the values are used for calculating the GLCM features.

From the graph, it could be observed that the calculated value is close to the observed peak. The texture correlation for the horizontal offset is shown in Fig. 6. Since the vertical thread is more enhanced than the horizontal threads in the preprocessing stage, the proposed method highlights approximately two-thread distance

Fig. 6 Horizontal offset

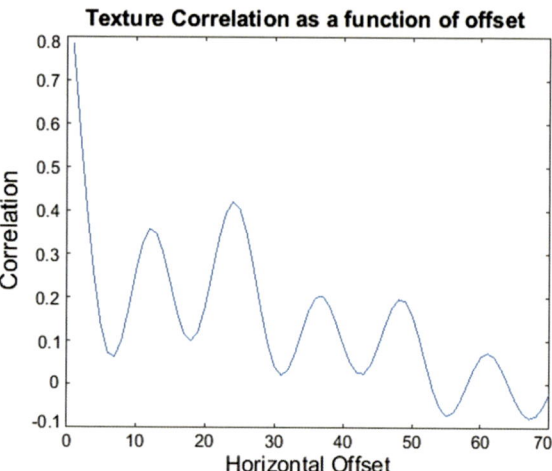

Texture Correlation as a function of offset

Table 2 GLCM feature calculation for 10 images

Image	GLCM Features				
	Homogeneity	Energy	Correlation	Entropy	Contrast
1	0.8924	0.5964	0.8841	0.8641	1.4512
2	0.8645	0.6095	0.8990	0.9745	1.3774
3	0.9245	0.5421	0.9321	0.7541	1.0121
4	0.9214	0.6041	0.9194	0.8641	1.0784
5	0.9452	0.5481	0.8974	0.6245	1.4512
6	0.9345	0.6089	0.9142	0.7541	1.5412
7	0.8945	0.5412	0.8974	0.8412	1.5427
8	0.9421	0.5601	0.9215	0.7541	1.3245
9	0.9621	0.8941	0.8614	0.5412	2.5412
10	0.9124	0.8774	0.8164	0.7845	2.6542

along the horizontal direction. Hence, the horizontal thread distance (25) is doubled (51) and given as output by the proposed method.

Table 2 shows the five major GLCM features calculated for a set of 10 images taken from different fabrics of the same thread pattern with minor variation in orientation of camera and illumination. Images 4 and 6 are taken from the same fabric, Images 1, 3, and 7 are taken from the same fabric with different translations. Images 2 and 10 are taken from the same fabric with different illuminations. From Table 2, it could be observed that the feature values for Images 4 and 6 are close to each other. It is also evident that illumination difference causes a major difference in the features extracted.

Though the proposed method gives closer results for the parameter identification, the illumination variation affects the enhancement of thread, and hence the calculation of the parameters. The accuracy of the system can be improved by choosing appropriate camera parameters for thread visibility and controlling the environmental parameters such as uniform illumination.

5 Conclusion

The parameters for GLCM and Gabor vary based on their textural patterns of the fabric. Since it is difficult to predict the parameters for a particular fabric, wide range of these parameters are used by the automated systems for defect detection. Average response of these parameters is usually considered for decision-making, and hence influencing the accuracy of identification. The selection of correct parameters according to the fabric can give profitable computation cost and better accuracy in defect identification. This paper identifies these parameters by relating them to thread parameters. The thread parameters are identified from the Hough transform.

References

1. Yildiz1 K, Senyürek1 VY, Yildiz Z (2014) A new approach to the determination of warp-weft densities in textile fabrics by using an image processing technique. J Eng Fibers Fabr 2(1)
2. Priya S, kumar TA, Paul V (2011) A novel approach to fabric defect detection using digital image processing. In; 2011 international conference on signal processing, communication, computing and networking technologies (ICSCCN). IEEE
3. Tiwari Vikrant, Sharma Gaurav (2015) Automatic fabric fault detection using morphological operations on bit plane. Int J Comput Sci Netw Secur (IJCSNS) 15(10):30
4. Shi M, Fu R, Guo Y, Bai Sh, Xu B (2011) Fabric defect detection using local contrast deviations. In: Multimedia tools and applications, vol. 52, no. I, pp. 147–157
5. Lin Chun-Cheng, Yeh Cheng-Yu (2009) Texture defect detection system with image deflection compensation. WSEAS Trans Comput 8(9):1575–1586
6. Mohanaiah P, Sathyanarayana P, GuruKumar L (2013) Image texture feature extraction using GLCM approach. Int J Sci Res Publ 3(5):1
7. Pan, R et al (2010) Automatic inspection of woven fabric density of solid colour fabric density by the Hough transform. Fibres Text East Eur 18(4): 81
8. Yildiz K, et al (2014) A new approach to the determination of warp-weft densities in textile fabrics by using an image processing technique. J Eng Fabr Fibers (JEFF) 9(1)
9. Padmavathi S, Prem P, Praveenn D (2013) Locating fabric defects using gabor filters. Int Res Eng J Sci Technol (IJSRET) 2(8):472–478
10. Bovik A, Clark M, Geisler W (1990) Multichannel texture analysis using localised spatial filters. IEEE Truns PAMI 12(1):55–72
11. Michalewicz Z (1996) Genetic Algorithms + Data Structures = Evolution Programs, AI Series, 3rd edn. Springer, New York
12. Tsaia DM, Wua SK, Chen MC (2001) Optimal Gabor filter design for texture segmentation using stochastic optimization. Image Vis Comput 19:299–316
13. Daugman JG (1985) Uncertainty relations for resolution in space, spatial frequency, and orientation optimized by two-dimensional visual cortical filters. J Opt Soc Am A 2:1160–1169
14. Sakhare K, Kulkami A, Kumbhakam M (2015) Spectral and spatial domain approach for fabric defect detection and classification. In: 2015 international conference on industrial instrumentation and control (ICIC), 28–30 May 2015

15. Hari CV et al (2009) Mid-point hough transform: a fast line detection method. In: India conference (INDICON), 2009 annual IEEE. IEEE
16. Wang Xin, Georganas Nicolas D, Petriu Emil M (2011) Fabric texture analysis using computer vision techniques. IEEE Trans Instrum Meas 60(1):44–56

Error Detection Technique for A Median Filter Using Denoising Algorithm

Maymoona Rahim and Ruksana Maitheen

Abstract This paper presents a modified design for error detection technique for a median filter, while processing the images using a digital processing system, acquisition stage capture impulsive noises along with the images. A classic filter used to eliminate. Some sliding windows of $n \times n$ matrices are created from the original image. The generated median is compared with the original pixel value at the center of the selected portion of the matrix corresponding to the image. If we found any error then the new median value is placed instead of the existing one. For storing the pixel values, a line buffer and register bank are used along with the median filter. Using the buffers the time taken for the median calculation is reduced, and as a result the total time taken for processing the image is also reduced.

1 Introduction

Image processing is a technique used to change an image into its digital form by performing some operations, to provide better visualization, image sharpening, image retrieval, and restoration. Using computers digital image can be manipulated by digital processing techniques. Image acquisition is the initial step in the image processing workflow because the process is carried out only when there is an image. Here, an image is retrieved from a hardware-based source into the system. The image which is retrieved is completely an unprocessed one.

M. Rahim (✉) · R. Maitheen
Department of ECE, Ilahia College of Engineering and Technology,
Affiliated to APJ Abdul Kalam Technological University (KTU), Muvattupuzha, India
e-mail: maymoonarahim77@gmail.com

R. Maitheen
e-mail: ruksanamaideen@gmail.com

© Springer Nature Switzerland AG 2019
D. Pandian et al. (eds.), *Proceedings of the International Conference on ISMAC in Computational Vision and Bio-Engineering 2018 (ISMAC-CVB)*, Lecture Notes in Computational Vision and Biomechanics 30,
https://doi.org/10.1007/978-3-030-00665-5_119

There are some impulsive noises added along with the image. This is also known as salt and pepper noise. This is caused by sharp disturbances in the image. The image sensors acquire unwanted noises along with the captured image. So, it is necessary to remove the errors at the acquisition stage itself to reduce the chance of noise propagation. If the unwanted noises are not removed then it will propagate to the entire system and cause low-quality image with noise.

Median filter is a nonlinear filter used to eliminate the noises at the acquisition stage. It is also named as edge-preserving filter. It has varying traits with reference to linear filters. A nonlinear filter can generate output in a nonintuitive manner. For example, if the filter was based on five values as $Y0$, $Y1$, $Y2$, $Y3$, and $Y4$ in median filter, initially, the values are sorted from lowest value to highest or vice versa. The value at the center is chosen as the median. That is, the value at position two is chosen as the median.

Field programmable gate arrays (FPGA) are semiconductor devices, which has a matrix of programmable logical blocks and are connected using programmable interconnects. FPGA can be reformed after manufacturing according to the design needs. FPGAs and targeted design technologies offer higher flexibility, quick time to market, and cheaper over all nonrecurring engineering costs for a wide range of video and imaging applications. In particular, SRAM-based FPGAs are easily and quickly reprogrammed. It provides high performance, high densities, and low cost.

The digital image sensors applicable in space applications are undergoing single event upsets (SEUs) due to the space radiations. It will damage the sensors or the entire image processing system. The cosmic rays or other energetic particles in the space collide with the endangered components of device, and may cause single event upsets. SEUs are the change of state caused by single ionizing particle striking the vulnerable part of the device. The SEUs are also called soft errors. SRAM-based FPGAs are also susceptible to the effect of radiation, so the implementation of median filter using SRAM-based FPGA can undergo SEUs which replaces the functionalities of median filter.

To protect the digital filters redundancy-based techniques such as dual modular redundancy (DMR) or reduced precision redundancy (RPR) is used. In DMR, the whole bit stream is replicated in case one should fail. Using these techniques error detection is possible in case of errors. In RPR, only the high-precision bits are replicated instead of replicating the full bits, which does not detect least significant change. This work is an enlargement of the paper error detection technique for a median filter presented in [1]. In this paper, the time taken for median calculation is reduced and as a result the time is consumed.

The paper is arranged as follows: Sect. 2 introduces the previous works related to the work. It gives the FPGA implementation of the original median filter. In Sect. 3, the proposed denoising architecture is explained. The experimental setup and simulation are reviewed in Sect. 4. Finally, Sect. 5 concludes the paper.

2 Related Works

The median filter is a nonlinear filter used to remove impulsive noises from the images. It is also used to provide care for protecting edges in an image. The filtering technique applied a two-dimensional window also called sliding window to the image and replaces the center pixel value with the median value calculated from the sliding window. Once the median replacement is done, the sliding window is shifted to another region of image and the process is continued until the whole image is processed. By these filtering techniques, it is possible to eliminate salt and pepper noises from the image and also preserve sharp edges. This is a type of intrinsic properties of nonlinear filter which is not achieved by a linear filter.

In sliding window filtering process, a 2D window is defined which is in an $N \times N$ square matrix (N is odd). N is odd value since to define exact center value for the window. N can be increased according to improve the filtering process and noise reduction. Thus, a 3×3 square window is chosen in this paper as the sliding window. Figure 1 explains the working of median filter using a 3×3 sliding window.

Initially, a sliding window of 3×3 matrixes is chosen from the original image. Now, the pixel values in the window are arranged in descending order. From the sorted values median is calculated. Since the order of the matrix is odd, it is easy to choose the mid-value. The new median value is compared with the value in the actual image. If any error occurs, then an erroneous signal is generated, and hence the actual center pixel value is displaced with the new median value calculated, i.e., the analyzed dead pixel is removed from the original image.

For the further processing of the entire image pixel, the sliding window of order 3×3 is shifted through the image. In FPGA system, this can be done with the help of a nine-pixel stream that eventually moves through the filter. Each group can be ordered using a node shown in Fig. 2 which consists of a comparator of eight bit and two 2:1 multiplexers. The two inputs are compared internally.

The classic sorting structure consists of 41 basic nodes shown in Fig. 3. Here, each box is identical exchange nodes illustrated in Fig. 2. The classic implementation has been substantiated to be far from the optimal one because the improved FPGA implementation has been developed over the years [2–5]. Using the Batcher bitonic sorter [3], a nine input sorting network can be created. This network consists of 28 nodes, which reduces the amount of FPGA used.

The architecture proposed by J. L. Smith uses the lesser number of exchange nodes. Here, the usage of FPGA is reduced since it accomplishes a nine-pixel partial sorting. Here, higher and lower pixel values are cluttered. It will not be a problem because it requires only the median value for replacement. Figure 4 illustrates this scheme.

Figure 5 depicts a fault-tolerant technique that activated an output error signal if a corrupted image pixel is detected. Then, a partial or complete reconfiguration can be performed to remove the error. Here, gray-shaded blocks are added to the original structure in Fig. 4 to create a range. Here, the time taken to create sliding window is larger. Once the sliding window is created by loading the nine-pixel values, the

11	15	10	12	14	14	16
13	0	14	10	15	13	13
10	12	14	15	14	14	15
16	16	14	13	13	14	14
17	17	16	17	17	13	13
15	18	19	14	14	15	16
15	15	16	17	18	19	16

ORIGINAL IMAGE

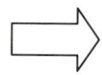

11	15	10
13	0	14
10	12	14

3X3 WINDOW

15
14
14
13

M = 12

11
10
10
0

MEDIAN SORTING

11	15	10	12	14	14	16
13	12	14	10	15	13	13
10	12	14	15	14	14	15
16	16	14	13	13	14	14
17	17	16	17	17	13	13
15	18	19	14	14	15	16
15	15	16	17	18	19	16

FILTERED IMAGE

Fig. 1 Median substitution

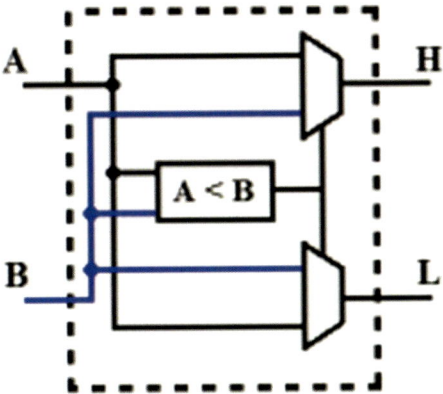

Fig. 2 Exchange node

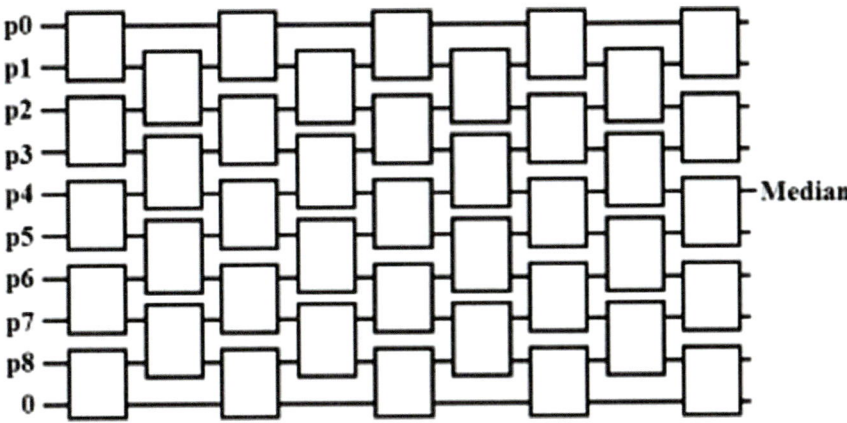

Fig. 3 Classic network

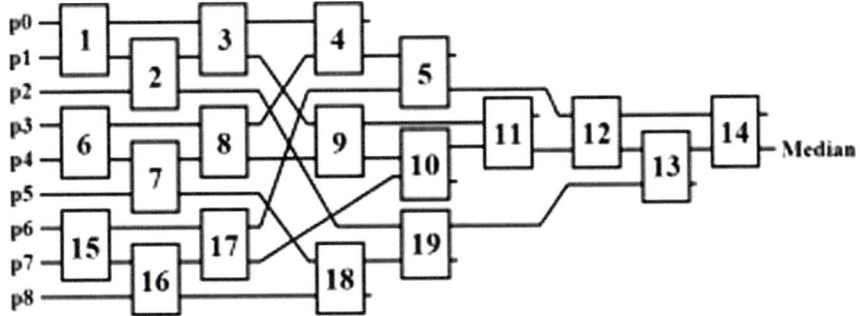

Fig. 4 Minimum exchange node network for nine pixels

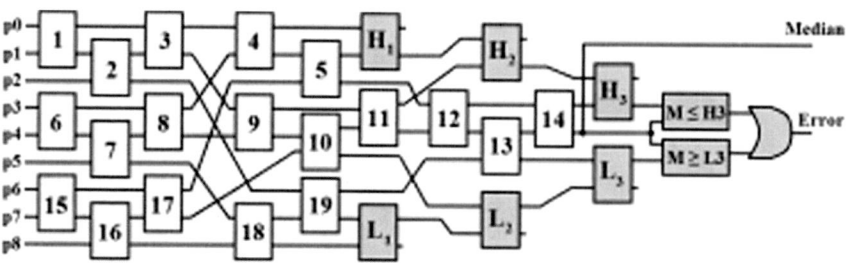

Fig. 5 Modified technique with reduced resources

median is calculated. If any error occurs, the center pixel value is reinstated with the new median value calculated. The process is performed again by shifting the sliding window and again loading the new nine-pixel values which make waste of time.

3 Proposed System

While using median filter for error detection in images the time taken to create sliding window each time is increased, and hence the entire image processing takes time. To avoid this problem, the proposed system is applicable. Instead of loading the pixel value every time, it is possible to use a line buffer for storing the pixel values, and hence the time can be consumed.

Buffers are physical memory allowed to store data temporarily and shift from one place to another if it is required. Generally, data stored in a buffer is retrieved from a hardware source. The output of buffers exactly matches the input. The system adopts a 3×3 mask, so it requires three scanning lines. Here, the concept of ping-pong arrangement is used. To recognize three scanning lines having two line buffers, it requires four crossover multiplexers. To store values in the odd rows, odd line buffers are designed and for values in the even rows even line buffers are designed, which is described in Fig. 6.

The register bank, contains nine registers, is used to collect the 3×3 pixel values of the current mask. Each register is connected serially to give the corresponding pixel values of a row and Reg4 keeps the median value of the denoised pixel value. Clearly, the denoising process for $p(i, j)$ does not start until $f(i + 1, j + 1)$ enters from the input device. The value which is stored in RB is used one by one by corresponding data detector and noise filter for the purpose of denoising.

Once the denoising process of the first set of pixel value is completed, the reconstructed value which is generated by the median filter is written back to the line buffers for storing the ith row for replacing the actual image. The denoising process is then shifted to the next set of pixel values, i.e., $p(i, j)$ to $p(i, j + 1)$, it requires only three values they are $(fi - 1, j + 2; fi, j + 2; fi + 1, j + 2)$ and are required to read into the register banks. The register banks 2, 5, and 8 are used for this purpose, and the

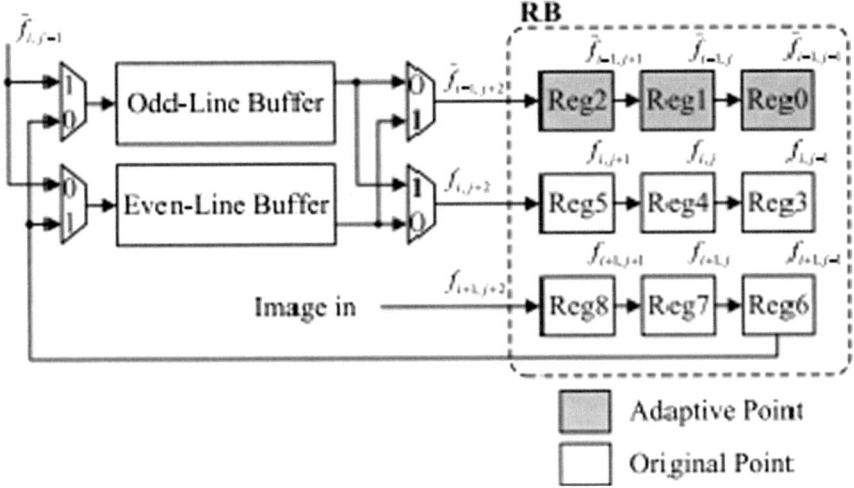

Fig. 6 Architecture of register banks

remaining six values are shifted to the corresponding register positions. Along with that the input value from the input source $f(i + 1, j + 1)$ is again written back to the buffer storing the previous rows for denoising process.

The multiplexers work in accordance with the select lines. The selection signals are set either one or zero for the purpose of denoising the odd row or even row, respectively. Figure 7 gives two examples. The figure illustrates the interconnection among line buffers and register banks. Assume, initially, we are going to denoise row 2, for that set all the four select line to zero, the samples written in row 1 and row 2 are stored in odd line buffer and even line buffer. The samples of the row are the inputs to the device, shown in Fig. 7a. Now, the previous value which is stored in register 6 and denoised output generated by the median filter are written to the odd and even line buffers. Now, the denoising process of row 2 is completed, the odd line buffer is filled with the samples of row 3. The denoised samples of row 2 are now stored in even line buffer. To denoise row 3, we have to select the select lines to 1. The previous value is stored in register 6. The denoised result from the median filter I against written back to any one of the line buffer which is demonstrated in Fig. 7b. On completion of the denoising process of row 3, even line buffer is filled with the samples of row 4 and odd line buffer is filled with the denoised samples of row 3.

4 Simulations and Experimental Results

The proposed solutions have been designed using Xilinx. In comparison with the previous technique, the sliding windows are generated with reduced time limit than the existing system which is presented in Fig. 8. Here, it takes some initial time delay

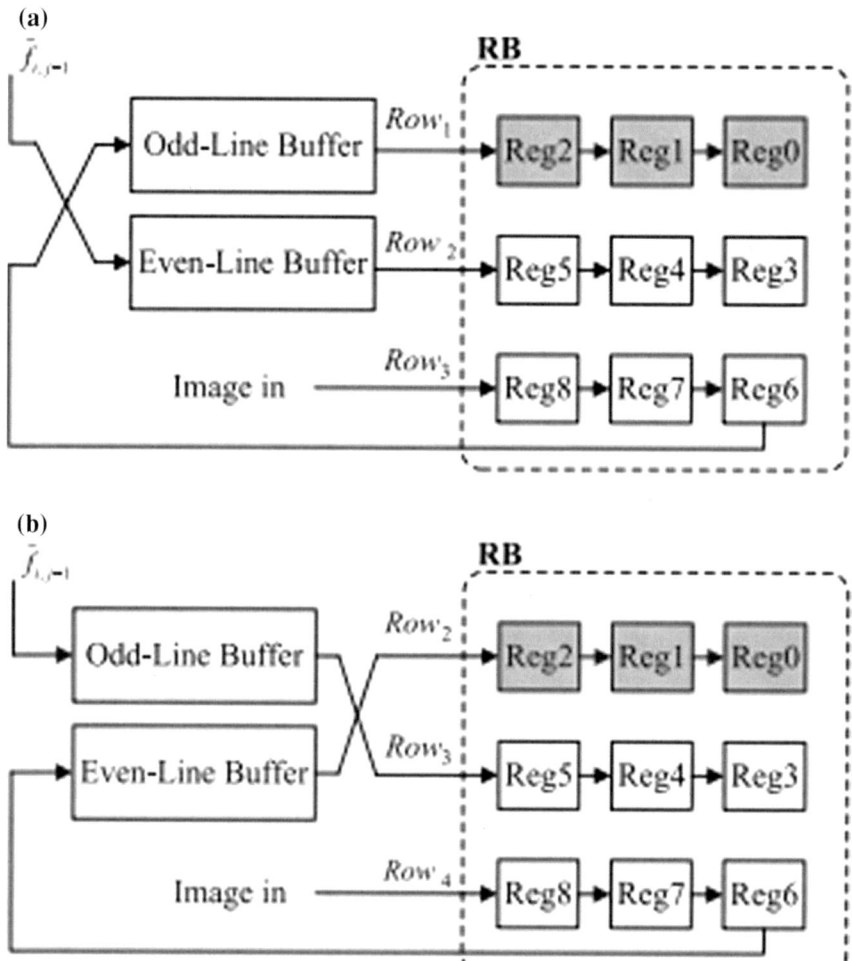

Fig. 7 Examples of the interconnections between two line buffers and RB

for loading the image into the buffer. After loading the image in each clock cycle, we are obtaining the median value from the sliding window without any delays. Hence, the image processing is carried out in a faster way than the existing techniques.

Table 1 gives the device utility during simulation. It uses lesser number of resources available which makes the system more advantageous.

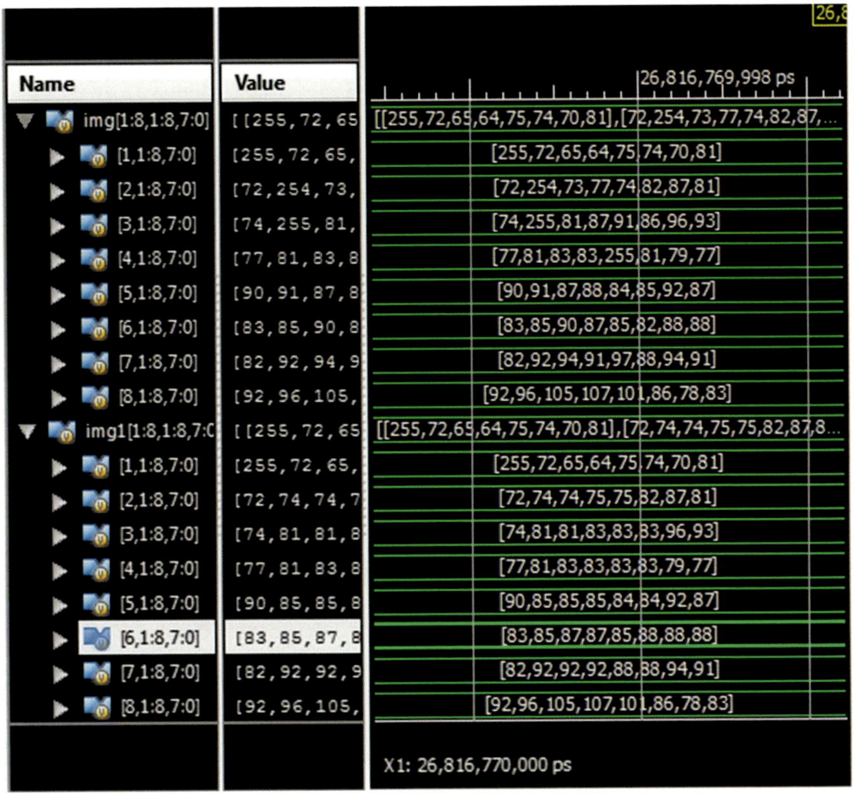

Name	Value		26,816,769,998 ps
▼ img[1:8,1:8,7:0]	[[255,72,65		[[255,72,65,64,75,74,70,81],[72,254,73,77,74,82,87,…
▶ [1,1:8,7:0]	[255,72,65,		[255,72,65,64,75,74,70,81]
▶ [2,1:8,7:0]	[72,254,73,		[72,254,73,77,74,82,87,81]
▶ [3,1:8,7:0]	[74,255,81,		[74,255,81,87,91,86,96,93]
▶ [4,1:8,7:0]	[77,81,83,8		[77,81,83,83,255,81,79,77]
▶ [5,1:8,7:0]	[90,91,87,8		[90,91,87,88,84,85,92,87]
▶ [6,1:8,7:0]	[83,85,90,8		[83,85,90,87,85,82,88,88]
▶ [7,1:8,7:0]	[82,92,94,9		[82,92,94,91,97,88,94,91]
▶ [8,1:8,7:0]	[92,96,105,		[92,96,105,107,101,86,78,83]
▼ img1[1:8,1:8,7:0	[[255,72,65		[[255,72,65,64,75,74,70,81],[72,74,74,75,75,82,87,8…
▶ [1,1:8,7:0]	[255,72,65,		[255,72,65,64,75,74,70,81]
▶ [2,1:8,7:0]	[72,74,74,7		[72,74,74,75,75,82,87,81]
▶ [3,1:8,7:0]	[74,81,81,8		[74,81,81,83,83,83,96,93]
▶ [4,1:8,7:0]	[77,81,83,8		[77,81,83,83,83,83,79,77]
▶ [5,1:8,7:0]	[90,85,85,8		[90,85,85,85,84,84,92,87]
▶ [6,1:8,7:0]	[83,85,87,8		[83,85,87,87,85,88,88,88]
▶ [7,1:8,7:0]	[82,92,92,9		[82,92,92,92,88,88,94,91]
▶ [8,1:8,7:0]	[92,96,105,		[92,96,105,107,101,86,78,83]

X1: 26,816,770,000 ps

Fig. 8 Simulated results

Table 1 Device utilization summary

Device Utilization Summary (estimated values)			[-]
Logic Utilization	Used	Available	Utilization
Number of Slice Registers	910	126800	0%
Number of Slice LUTs	3581	63400	5%
Number of fully used LUT-FF pairs	839	3652	22%
Number of bonded IOBs	18	210	8%
Number of BUFG/BUFGCTRLs	2	32	6%

5 Conclusion

Error detection during the initial stage of image processing is the most important part in image processing techniques. Median filters are helpful in removing the impulsive noises added along with the image and also these are edge-preserving filters. Along with the median filters, the use of line buffers and register banks reduces the time taken for image processing, and hence improves the system performance. The proposed system can be used in any image processing systems. This will improve the system performance by eliminating the noise at the initial stage itself.

References

1. Aranda LA, Reviriego P, Maestro JA. Error detection technique for a median filter
2. Fahmy SA, Cheung PYK, Luk W (2005) Novel FPGA-based implementation of median and weighted median filters for image processing. In: International conference on field programmable logic and applications, Tampere, Finland, pp. 142–147
3. Batcher KE (1968) Sorting networks and their applications. In: AFIPS spring joint computing conference, San Francisco, CA, pp. 307–314
4. Vasicek Z, Sekanina L (2008) Novel hardware implementation of adaptive median filters. In: 11th IEEE workshop on design and diagnostics of electronic circuits and systems (DDECS), Bratislava, Slovakia, pp. 1–6
5. Smith JL (1996) Implementing median filters in XC4000E FPGAs. Xcell 23:16

Retinal Image Processing and Classification Using Convolutional Neural Networks

Karuna Rajan and C. Sreejith

Abstract This study aims to develop a system to distinguish retinal disease from fundus images. Precise and programmed analysis of retinal images has been considered as an effective way for the determination of retinal diseases such as diabetic retinopathy, hypertension, arteriosclerosis, etc. In this work, we extracted different retinal features such as blood vessels, optic disc and lesions and then applied convolutional neural network based models for the detection of multiple retinal diseases with fundus photographs involved in structured analysis of the retina (STARE) database. Augmentation techniques like translations and rotations are done for expanding the number of images. The blood vessel extraction is done with the help of morphological operations like dilation and erosion and enhancement operations like CLAHE and AHE. The optic disc is localized by the methods such as opening, closing, Canny's edge detection and finally thresholding the image after filling the holes. The bright lesions (exudates) inside the retina are detected by the filtering operations and contrast enhancement after the removal of the optic disc. In this study, we experimented with different retinal features as input to convolutional neural networks for effective classification of retinal images.

1 Introduction

The retina is a thin layer of tissue lining the inner surface of the eyeball. The photoreceptor cells take light focused by the cornea and convert into neurosignals to the brain for visual recognition. Damages caused to the retina due to various diseases may gradually lead to loss of vision. Since the ageing of people across the world has been evolved as a major statistical trend, patients plagued by retinal diseases such as hypertensive retinopathy, age-related macular degeneration (AMD) and diabetic

K. Rajan · C. Sreejith (✉)
Calpine Labs, UVJ Technologies Pvt.Ltd, Kochi, Kerala, India
e-mail: sreejith.cherikkallil@calpinetech.com

K. Rajan
e-mail: karunarajan94@gmail.com

© Springer Nature Switzerland AG 2019
D. Pandian et al. (eds.), *Proceedings of the International Conference on ISMAC in Computational Vision and Bio-Engineering 2018 (ISMAC-CVB)*, Lecture Notes in Computational Vision and Biomechanics 30,
https://doi.org/10.1007/978-3-030-00665-5_120

retinopathy (DMR) may increase in the future. Macular degeneration is a very common eye condition and a leading cause of vision loss among people Choi et al. [1]. Diabetic retinopathy is also a reason for visual disorder due to the damages of blood vessels in patients suffering from diabetes mellitus. Retinal diseases including retinal vessel occlusion, hypertensive retinopathy, Hollenhorst emboli and retinitis are also some of the prime causes of vision impairment. Specifically, it is all because of our lifestyle changes and unhealthy practices in our day-to-day life. In many cases, visual loss can be avoided if early detection and timely treatment are provided prior to the initial stage of blindness. Thus, more precise screening programmes aided with modern approaches are needed for early treatment so as to reduce the socio-economic hardships of visual loss caused by retinal diseases. The diabetic retinopathy screening that uses fundus photographs has been widely adopted for diabetes patients Abrmoff [2]. Conducting AMD screening and DMR screening is cost-effective in a public health scenario. However, manual analysis of multiple fundus photographs for an accurate result requires efforts and expertise of the ophthalmologist. Thus, comes the relevance of the retinal image processing. It deals with the analysis of the retinal features that may help us in the diagnosis of this kind of diseases.

The entire paper is devoted to retinal image analysis and classification methods and their clinical implications. Blood vessels, optical disc, optic nerve, lesions and fovea are some of the most important structures of the human retina and are mostly used for several applications such as pathology detection inside the retina. Detection of these important structures manually is time consuming and depends upon the expertise of the user. The proposed work aims at developing an automated system that can predict the diseases by analysing these features from the retinal fundus images. The fundus (the bottom or base part of an organ) is the part of the inner eye that can be seen during an eye examination while looking through the pupil. The inner lining of the eyeball, including the retina (the light-sensitive layer), optic disc (the brightest spot inside the eye and also the head of the nerve to the eye) and the macula (the small spot in the retina where vision is the sharpest) together constitute the retinal fundus. This work focuses only on the retina; nevertheless, a brief review of the eye anatomy can be found from the following Fig. 1 [3]. In this study, we applied deep CNN for fundus photography analysis in multi-categorical disease classification. This paper explains the retinal image processing and disease diagnosis on an open retinal image database (STARE). The proposed work has two stages: feature Analysis and classification.

2 Literature Survey

The problem of retinal image analysis has already received unique attention and many works were published on this topic. Sinthanayothina et al. [4] detected features such as fovea, optic disc and blood vessels in their work. More than 100 retinal images were preprocessed by means of adaptive, contrast and local enhancement. Those regions with the largest variation in the intensity of neighbouring pixels were identified

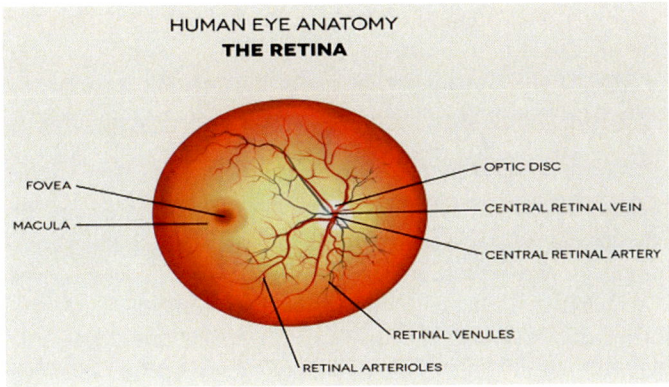

Fig. 1 Human eye anatomy

and thus optic discs were localized. Blood vessels were recognized by means of a perceptron. The foveae were identified using correlation matching. Hoover et al. [5] proposed a novel algorithm called fuzzy convergence. This algorithm identified the optic nerve as the converging point of the blood vessel network. In the absence of a strong convergence, the method identifies the optic nerve as the brightest region in the image after contrast equalization. Snchez et al. [6] proposed the Fisher linear discriminant analysis for the detection of hard exudates in the retinal images. Hard exudates were identified by making use of the colour features to define the feature space. So as to select one of these models they calculated a quantitative metric (J) to evaluate the performance of different colour spaces. Mishra et al. [7] proposed the methods like cup-to-disc ratio (CDR) and inferior superior to nasal temporal side) (ISTN) for the aim of glaucoma detection. In this work, active contour models were utilized for the identification of cup and optic disc. Veins were separated by morphological strategies as it helps in smoothing the picture.

Aslam et al. [8] proposed methods such as PCA and Hough transforms for the localization of optic nerve head. But it was not a success for fundus images having a large number of white lesions. For the fovea detection, the proposed method is the HSI transformation. The vascular segmentation was done with the help of techniques like matched filters, vessel tracking, neural net and morphological processing. Fundus photographs were used for the entire work. [9, 10] proposed a template matching method for the detection of blood vessels. For applying the entire algorithm of the matched filter, the G-plane of the image was considered. After enhancing the contrast of the image, the median filter was used to remove the noise. A bank of filters (matched filters) was applied to the image to localize and segment the veins. In this, an advanced, non-parametric tree-type classifier random forests (RF) was used for classification.

Tjandrasa et al. [11] proposed the Hough transform and active contour models that implemented the detection of the optic nerve head in retinal fundus images. The procedure begins with the image improvement utilizing homomorphic filtering for brightness correction, then continues with the expulsion of veins from the images. Blood vessels were detected by the method thresholding. Nguyen et al. [12] proposed the basic line detector which was first used as a means for vessel-background classification. Supervised methods like Kalman filter and pixel classification approaches are also used. Top hat filter, Gabor filter and matched filter were some of the techniques for the blood vessel extraction. Radha et al. [13] proposed methods like plane separation, contrast enhancement and morphological process for the blood vessel extraction. Exudates were detected and shown accurately with the help of K-means clustering and the extracted features were trained with the probabilistic neural network successfully [14, 15]. Melinsak et al. [16] suggested max-pooling layered convolutional neural networks (MPCNNs) for the retinal vessel segmentation.

Chandore and Asati [17] also proposed convolutional neural networks for the detection of diabetic retinopathy. The features discussed in this paper for the detection of diabetic retinopathy were optic disc, macula, exudates, microaneurysms, haemorrhages, etc. They simplified the labels to only two classes (without DR-class 0, with DR-class 1). On this simplified model and dataset, they got the precision of about 0.81 for class 0 and 0.88 for class 1. In the paper, Tan et al. [18] proposed some solutions not just to locate but to simultaneously segment vasculature, optic disc and fovea. They used a seven-layer convolutional neural network (CNN) to classify every pixel into one of the four classes: background, blood vessels, optic disc and fovea. They used a seven-layer CNN together with random forest algorithm to extract retinal vasculature. However, in their work, CNN was used only as a trainable feature extractor, the classification was in fact performed by an ensemble of random forest. Dasgupta et al. [19] proposed a fully convolutional neural network architecture for blood vessel segmentation.

Choi et al. [1] proposed deep learning approaches for the multi-class disease classification. In this approach, they have utilized the transfer learning along with the traditional classifiers like the random forest. They have used the VGG19-TL-RF model for the classification purpose. In this study, they obtained results with an accuracy of 30.5%, relative classifier information (RCI) of 0.052 and Cohen's kappa of 0.224 when all 10 categories were included. When the concept of transfer learning was added along with the ensemble classifiers they got an accuracy of 36.7%, 0.053 RCI and 0.225 kappa for the same 10 retinal diseases classification problem.

The purpose of this research is to develop an automated system for the analysis of retinal images, and thereby the detection of disease from 10 categories. Hence, the proposed design is a novel approach that combines the conventional analysis methods along with the modern deep learning techniques and thus leads to the detection of the retinal diseases as early as possible.

3 System Design

This section explains how the entire framework is designed and its working. The proposed system design is depicted in Fig. 2.

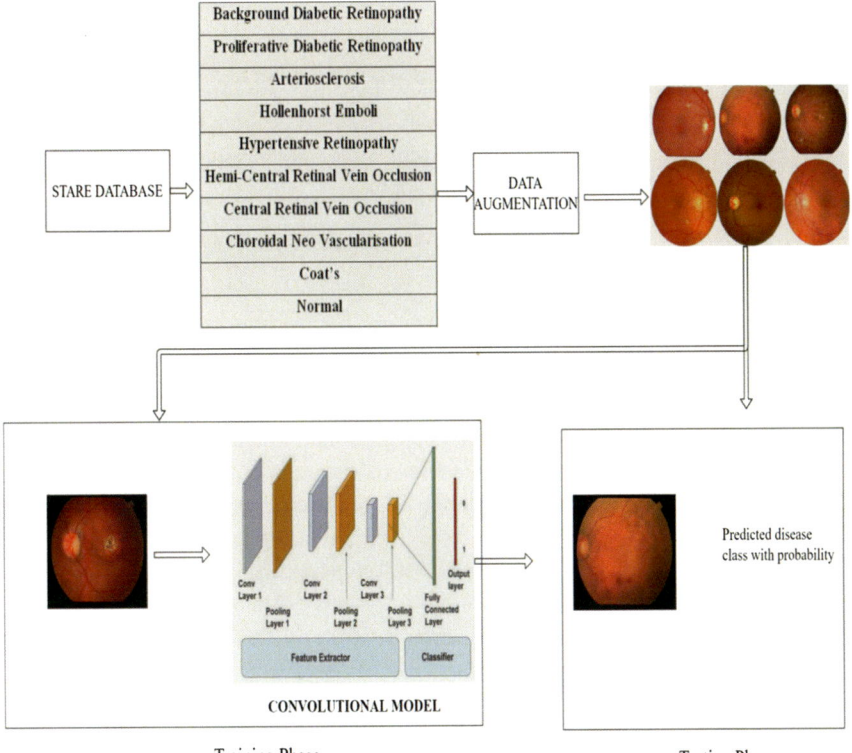

Fig. 2 Proposed system design

3.1 Dataset

For this study, the publicly accessible retinal image database for structured analysis of the retina (STARE) [20] was used, so as to develop a multi-categorical disease detection system with the help of deep learning approaches. The database contains retinal colour images captured with the help of a TRV-50 fundus camera at an angle of 35 degrees at a resolution of 605 × 700 pixels. It contains 397 images in 14 disease categories. The major problem with this dataset is unequal categorization. For example, the dataset includes only a single image of cilioretinal artery occlusion, while more than 60 images were included in the BDR. The dataset was balanced and cleaned by removing redundant data and eliminated those classes with least amount of data. Data augmentation techniques such as translation, rotation and brightness change were used for expanding the dataset. We selected 10 classes, created 100 fundus images per each disease category and eliminated redundant images. Specifically, we obtained samples with translation from the range [−10%, +10%] of the image width, with the rotation of [90, 45] and with brightness change from a range of [-10%, +10%].

3.2 Feature Extraction

Retinal feature localization and extraction are some of the important steps involved in the initial stages of image processing. Various image processing methods were used to extract/localize the prime features inside retina that may help us in further stages of disease detection. In this work, we used morphological operations for the extraction and localization of retinal features such as blood vessels, optic disc and yellow deposits. Blood vessels can be extracted with the help of a number of morphological operations such as opening (erosion followed by dilation), closing (dilation followed by erosion), contrast limited adaptive histogram equalization, adaptive histogram equalization, etc. The whole processing is done on the extracted green channel.

Yellow deposits/exudates (soft or hard) are because of the leakage of blood from broken blood vessels due to diabetes, hypertensive retinopathy, neuroretinitis and retinal vein occlusion. This kind of exudates or lesions that cause vision-threatening disorders can be detected by performing the Gaussian blurring on the retinal image and then applying binary thresholding (threshold=140) after converting it to grayscale.

Optic disc is an anatomical component of the retinal image which behaves as a recognizable proof for the retinal disease detection. It is the brightest spot inside the retina. The optic disc can be detected with the assistance of a series of morphological operations, such as closing (with a kernel size of 5×5), then erosion operation is applied with two iterations, dilation is then performed with three iterations and after that gradient operation is performed. Canny's edge detection algorithm is then applied to the image and a closing operation towards the end, to locate the optic disc.

3.3 Classification

The classification stage involves two stages, namely, training stage and testing stage. The training stage involves identifying the representative classes and developing a numerical description of the attributes of each class type through training set. Later, in the classification stage, the dataset is classified to the class it resembles the most, and finally the output stage generates the disease category of the corresponding image along with the probability. In a convolutional layer, all neurons apply the operation called convolution to the inputs, and hence they are also known as convolutional neurons. All the neurons are fully connected to the neurons in previous and post-layers. It is usually made of filters and feature maps. A pooling layer will be used just after each convolutional layer so as to decrease the spatial size (only width and height, not depth). This reduces the number of parameters, and hence computation complexity can be reduced. Also, less number of parameters avoid the overfitting. Each neuron in each layer in a fully connected layer will be connected to each other. A normal flat feedforward neural network layer is used at the end of the network after the extraction of essential features. In this work, we are using the 11-layer convolutional model with three convolutional layers, three max-pooling layers, one dropout layer, one flatten layer and three dense layers. The classification of images on the basis of disease categories are done with the help of these layers in the CNN. Different models were built on the basis of channels and extracted features

4 Experiments and Results

This section describes the techniques used in the preprocessing, the feature extraction and the detailed explanation of the models built for the purpose of classification. In this work, top 10 best classes were selected and after the data augmentation, the dataset was equalized among the classes with 100 images each. Important features needed for the disease diagnosis are extracted and localized in the image processing stage. Convolutional models are built on the basis of this data and diagnosis of the disease also occurs at this stage.

4.1 Image Preprocessing and Feature Analysis

In the feature analysis stage, important retinal features were localized and extracted needed for the disease detection. The process of vein extraction was done after all the image preprocessing tasks mentioned in the underlying stage. Morphological operations were adopted for this stage. The optic disc and lesions were also localized

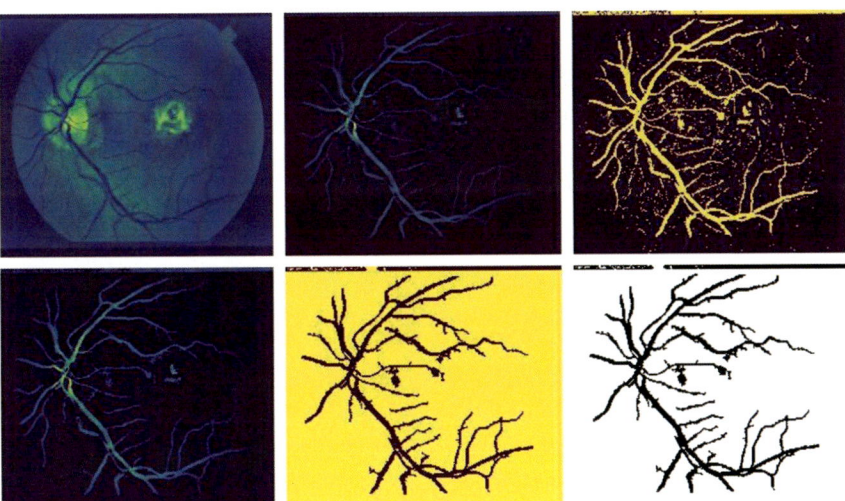

Fig. 3 Stages of blood vessel extraction

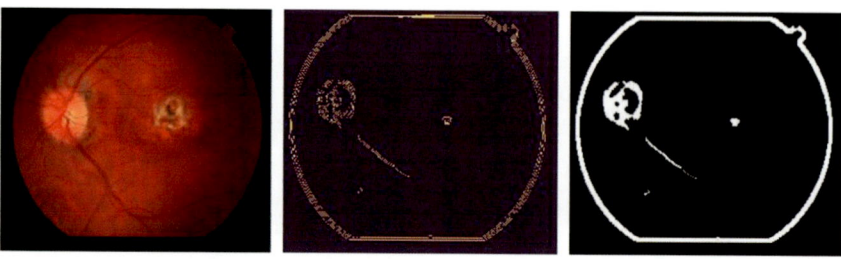

Fig. 4 Optic disc localization

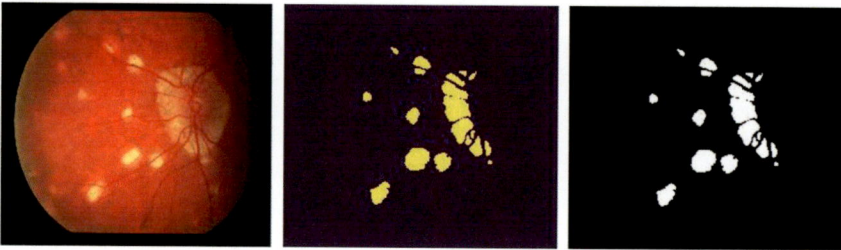

Fig. 5 Lesion localization

with the help of some enhancement and edge detection operations along with the above-mentioned morphological operations (Figs. 3, 4 and 5).

4.2 Classification

Table 1 shows the accuracy of different CNN models with different retinal features as input.

In the stage of classification, we make use of the convolutional models that are made up of three kinds of layers, convolutional (calculating the output of the connected local input neurons studying the patterns), max pooling (subsampling the inputs) and fully connected layers (predicting the class by allocating the scores). The

Table 1 Results for the classification stage

Model name	Training accuracy (%)	Test accuracy (%)
RGB model	97.8	42
Green channel	95	35
Greyscale	93	32
Blood vessel	87	38

10 class classification problem was evaluated on the basis of different models tried on the different channels and extracted features.

5 Conclusion and Future Scope

This work is a primary attempted to construct deep learning models for multi-categorical classification problems in order to detect multiple retinal diseases. The deep learning techniques in this study were promising to be applied for the diagnosis. The problems we faced are low quality and quantity of the data, variations in the shape of the structures and the ambiguous boundaries of the structures that to be extracted. As a future work, transfer learning can be adopted to improve the classification performance in order to detect multi-categorical retinal diseases. Also, this work can be modified by collecting more data in various disease categories and experimenting with deeper models that combine various retinal features.

Acknowledgements This work was fully funded by Calpine Labs, UVJ Technologies, Kochi, India. We are also immensely grateful to Mr. Bijeesh Devassy, Project Manager, UVJ Technologies and Dr. Asharaf S, Associate Professor, IIITM-K for sharing their pearls of wisdom with us during the course of this research.

References

1. Choi JY, Yoo TK, Seo JG, Kwak J, Um TT, Rim TH (2017) Multi-categorical deep learning neural network to classify retinal images: a pilot study employing small database. PLoS ONE 12(11):e0187336
2. Abràmoff MD, Garvin MK, Sonka M (2010) Retinal imaging and image analysis. IEEE Rev Biomed Eng 3:169–208
3. https://www.virginiaeyeconsultants.com/procedures/eye-conditions/retina/
4. Sinthanayothin C, Boyce JF, Cook HL, Williamson TH (1999) Automated localisation of the optic disc, fovea, and retinal blood vessels from digital colour fundus images. Br J Ophthalmol 83(8):902–910
5. Hoover A, Goldbaum M (2003) Locating the optic nerve in a retinal image using the fuzzy convergence of the blood vessels. IEEE Trans Med Imag 22(8):951–958
6. Sánchez CI, Hornero R, Lopez MI, Poza J (2004) Retinal image analysis to detect and quantify lesions associated with diabetic retinopathy. In: 26thAnnual International Conference of the IEEE engineering in medicine and biology society, IEMBS'04, vol 1. IEEE, pp 1624–1627
7. Mishra M, Nath MK, Dandapat S (2011) Glaucoma detection from color fundus images. Int J Comput Commun Technol (IJCCT) 2(6):7–10
8. Patton N, Aslam TM, MacGillivray T, Deary IJ, Dhillon B, Eikelboom RH, Yogesan K, Constable IJ (2006) Retinal image analysis: concepts, applications and potential. Prog Retinal Eye Res 25(1):99–127
9. Verma K, Deep P, Ramakrishnan A (2011) Detection and classification of diabetic retinopathy using retinal images. In: 2011 Annual IEEE India conference (INDICON), pp 1–6. IEEE
10. Priyadharsini BH Devi MR (2014) Analysis of retinal blood vessels using image processing techniques. In: 2014 international conference on intelligent computing applications (ICICA), pp. 244–248. IEEE

11. Tjandrasa H, Wijayanti A, Suciati N (2012) Optic nerve head segmentation using hough transform and active contours. Indones J Electr Eng Comput Sci 10(3):531–536
12. Nguyen UT, Bhuiyan A, Park LA, Ramamohanarao K (2013) An effective retinal blood vessel segmentation method using multi-scale line detection. Pattern Recogn 46(3):703–715
13. Radha R, Lakshman B (2013) Retinal image analysis using morphological process and clustering technique. Signal Image Process 4(6):55
14. Wang H, Hsu W, Goh KG, Lee ML (2000) An effective approach to detect lesions in color retinal images. In: 2000 Proceedings of IEEE conference on computer vision and pattern recognition, vol 2. pp. 181–186, IEEE
15. GeethaRamani R, Balasubramanian L (2016) Retinal blood vessel segmentation employing image processing and data mining techniques for computerized retinal image analysis. Biocybern Biomed Eng 36(1):102–118
16. Melinščak M, Prentašić P, Lončarić S (2015) Retinal vessel segmentation using deep neural networks. In: VISAPP
17. Chandore V, Asati S (2017) Automatic detection of diabetic retinopathy using deep convolutional neural network
18. Tan JH, Acharya UR, Bhandary SV, Chua KC, Sivaprasad S (2017) Segmentation of optic disc, fovea and retinal vasculature using a single convolutional neural network. J Comput Sci 20:70–79
19. Dasgupta A, Singh S (2017) A fully convolutional neural network based structured prediction approach towards the retinal vessel segmentation. In: 14th International symposium on biomedical imaging, pp. 248–251. IEEE
20. STARE Dataset, http://cecas.clemson.edu/~ahoover/stare/

Comparison of Thermography and 3D Mammography Screening and Classification Techniques for Breast Cancer

Sureshkumar Krithika, K. Suriya, R. Karthika and S. Priyadharshini

Abstract Breast cancer, without doubt is one of the leading reasons for fatality among women in the world after lung cancer. Awareness and accessibility to better screening and treatment protocols will have a major impact in improving the survival rates. Moving away from the traditional methods of mammography and biopsy methods, newer techniques provide faster and efficient results to ensure early start of treatment. Therefore, a comparison study has been performed to weigh the pros and cons of thermography and 3D mammography as screening methods, followed by their respective processing and classification procedures. The ease of screening, extent of radiation, percentage of false positives, efficient segmentation, clustering, and novel classification are all considered and a conclusive result is obtained determining the better of the two processes. This could potentially revolutionize the way breast cancer is diagnosed and treated for women of all ages and walks of life.

1 Introduction

In the era of automation that we live in today, it is vital to devise methods that take down the barrier of manual detection, processing, and classification of medical images. This will not only save time but also be instrumental in reducing errors that, in the case of cancer, may cause unimaginable emotional and physical strain to the patients. Minimal access to advanced medicine and poor awareness in the developments of the medical and scientific fields are the main reasons for the less survival rates in breast cancer, especially among developing countries. The staggering numbers in various countries are Gambia (12%), Algeria (38.8%), India (52%), and Brazil (58.4%) and this situation can be ameliorated by sensitizing the use of modern medical practices. In light of the issue, the various screening processes are compared and their respective advantages and disadvantages are concluded. Among the different processes, the respective processing and segmentation steps are carried

S. Krithika (✉) · K. Suriya · R. Karthika · S. Priyadharshini
Easwari Engineering College, Chennai, India
e-mail: krithi96@gmail.com

© Springer Nature Switzerland AG 2019
D. Pandian et al. (eds.), *Proceedings of the International Conference on ISMAC in Computational Vision and Bio-Engineering 2018 (ISMAC-CVB)*, Lecture Notes in Computational Vision and Biomechanics 30,
https://doi.org/10.1007/978-3-030-00665-5_121

out for thermogram and 3D mammogram [1]. Since they are of different nature, that is, thermogram is an RGB image and mammogram is a black and white image, naturally their processing steps differ and cater to their specific needs, while mammogram is filtered and processed by wavelet and discrete cosine transforms. Thermogram is processed by curvelet transform. Once the required features are extracted from the image, a database is created and used for classification in SVM which is a machine learning algorithm that could provide automated results by comparing the acquired features with the existing support vector values. The SVM classifier in place of biopsy can tremendously reduce the 2–3 weeks waiting period for results giving a clear picture of the manifestation of the unwanted growth.

2   Overview of Various Screening Methods

Breast self-exam is very instrumental in discerning a problem at the early stage. To confirm the actual presence and manifestation of the tumor, rigorous screening methods are being used around the world. Apart from mammogram and thermogram, the most prevalent ones are as follows:

- Sonogram,
- MRI imaging, and
- Molecular imaging.

As per traditional practice, these methods are followed up biopsy. During the wait time, there is a high possibility for the development of contralateral breast cancer (CBC) [2]. This is a condition where there is growth of cancer on the other breast while the original one is present or it develops post-removal of earlier cancer.

2.1   Sonogram

An ultrasonic wave is sent through the breast tissue and the presence of any abnormality will present itself as reflections from the surface that is captured as a sonogram. This is used to check whether a lump is filled with fluid (a cyst) or if it is a solid lump. This method can only be used as an auxiliary to mammogram and not as an independent procedure. The possibility of false positives is also comparatively high to that of mammogram (Fig. 1).

2.2   MRI Imaging

Breast magnetic resonance imaging (MRI) is more invasive than mammography because a contrast agent is given through an IV before the procedure. Magnetic

Fig. 1 Sonogram depicting a fluid cyst (*Source* https://www.dic-kc.com/blog/2016/mammogram-reports-and-bi-rads-categories-4-5)

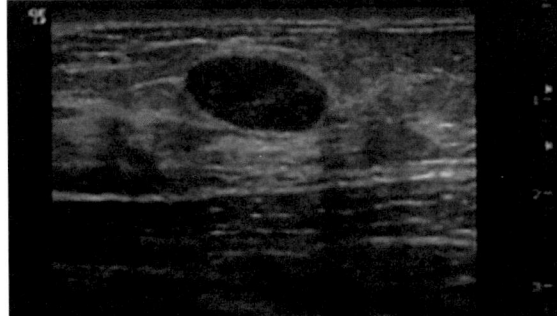

Fig. 2 MRI imaging of cancer in the right breast (*Source* https://breast-cancer.ca/mrifacts/)

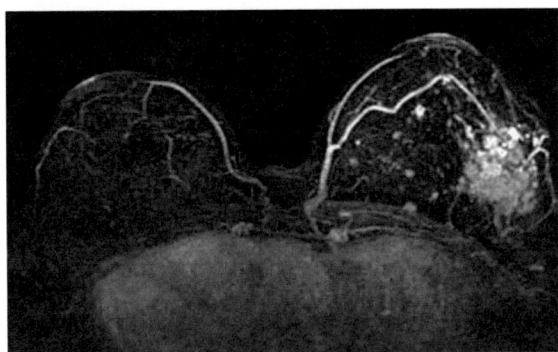

fields are used to create images of the breast which in turn aid in staging of the cancer more than detection. Though accuracy levels have improved in recent times, this method was not opted for a very long period due to its complexity and prolonged time of operation (Fig. 2).

2.3 Molecular Imaging

Molecular imaging is a new and upcoming method. Though it may give the required results with the advent of new technology and safer equipment, this procedure is still uncommon in developing countries. In addition to that, a nuclear tracer is injected into the veins to clearly illuminate the structure of the tumor [3]. This may cause long-term defects in the patient. Since this mechanism depends on the breast structure and tissue it is advisable only for older women with denser breasts (Fig. 3).

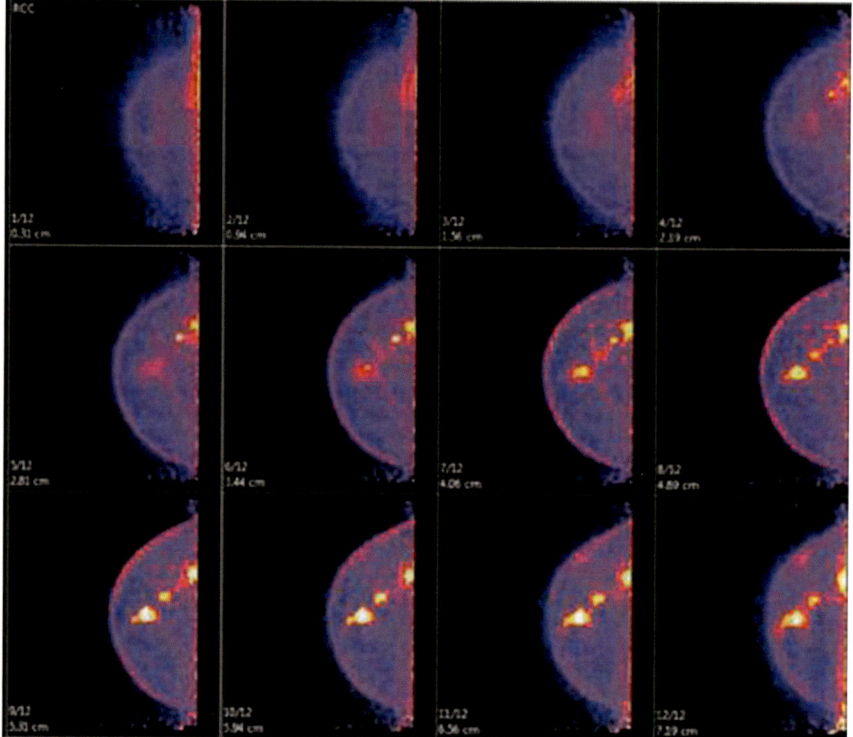

Fig. 3 Molecular imaging (*Source* HHS Author Manuscripts PMC2748346, p. 1073)

3 3D Mammography

3D mammography is similar to 2D mammography in the fact that it uses X-rays to produce images of breast tissue in order to detect lumps, tumors, or other abnormalities. Though it uses the same radiation, the advantage from traditional mammography is that it produces high-quality images, producing a crystal clear 3D construction of the breast, enabling the doctor to find out if there is anything to be concerned about. It captures multiple slices of the breast at various angles. 3D mammograms have the ability to discern a cancer cell from within layers of tissue as supposed to the 2D mammogram that only shows the growth of a mass as shown in Fig. 4.

3.1 *Advantages and Disadvantages*

Research has shown that mammograms are the reason for the decline in fatality rates in women all over the globe. Despite this feedback, definite answers can be obtained

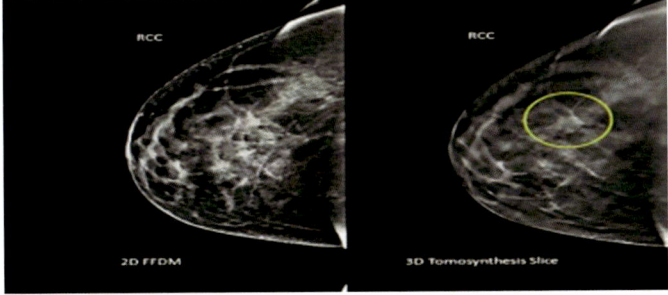

Fig. 4 Comaprison of 2D and 3D mammograms (*Source* www.google.com)

only after multiple screenings which expose the patient to constant and recurrent radiation [3]. This might set off the growth of more cancerous cells. Research shows that there is multiple recall screening sessions that are scheduled due to initial false reports. The 3D mammogram has a false-positive rate of up to 4%, though it is small number, it has the ability to cause mental and emotional trauma until the discrepancy is cleared.

3.2 Acquisition Procedure

The woman is positioned before a 3D mammography machine and her breasts are compressed by two metal plates to hold the flesh in place and still for the duration of the screening which usually lasts about 10–20 min. When ready, the machine is started and a robotic arm will move in an arc over the woman's breasts while multiple X-ray images are taken in various angles to reconstruct the 3D image of the breast as best as can be done. The dose of radiation is similar to 2D mammography but slightly higher in quantity to produce images with better contrast and structure.

3.3 Preprocessing

After the image is acquired, noise filtering is carried out using Gaussian or Wiener filter. Adaptive filters like Wiener are better for removing nonlinear noise. The image is first contrast enhanced and then salt and pepper or Gaussian noise is added and removed using adaptive filtering [4] (Fig. 5).

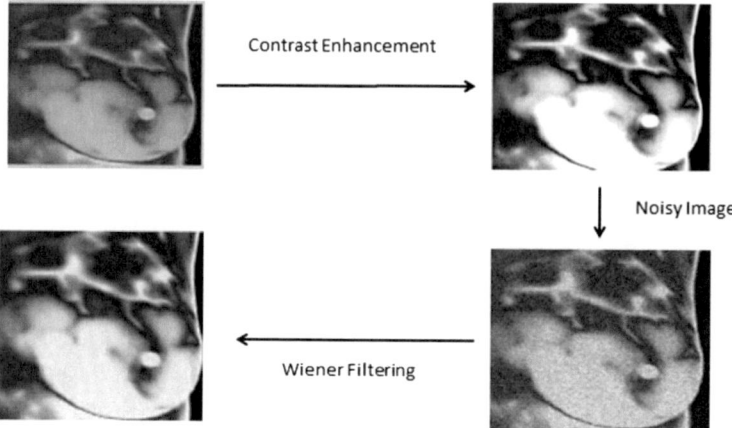

Fig. 5 Noise filtering of the 3D mammogram (*Source* Results from this study)

3.4 Feature Extraction

Extracting necessary features from the image is the main process before starting the classification process. Here, instead of discrete cosine transform, we use spherical wavelet transform. The spherical wavelet transform (SWT) proposed by Starck. It has better results as far as discerning features and evaluating them. SWT was initially used only for astronomical images but due to its focus on minute discontinuities and its inherent property to visualize the image in many parts to uncover all the necessary features, it has been deemed suitable for mammograms [5]. The main goal of this method is to reduce redundancy which results in smaller datasets and easier computation for classification. It is an isotropic transform that utilizes a wavelet pyramid to capture statistical features in an isotropic field. This helps in extracting structural features in medical images (Fig. 6).

Once the features are extracted using SWT, it is stored in a testing database ready to be compared with the training database using the SVM classifier.

4 Thermography

Thermogram is the depiction of infrared radiations from objects which exist at temperatures above absolute zero. The intensity of radiation increases with temperature and is portrayed as various temperature levels on a thermal image [6]. A thermogram is captured in a private room where the patient's body is cooled to a very low temperature causing a drop in normal blood flow. A thermo cam captures the radiations from various parts of the body, in this case the breasts. The presence of a cancerous growth is associated with the excessive formation of blood vessels and inflammation

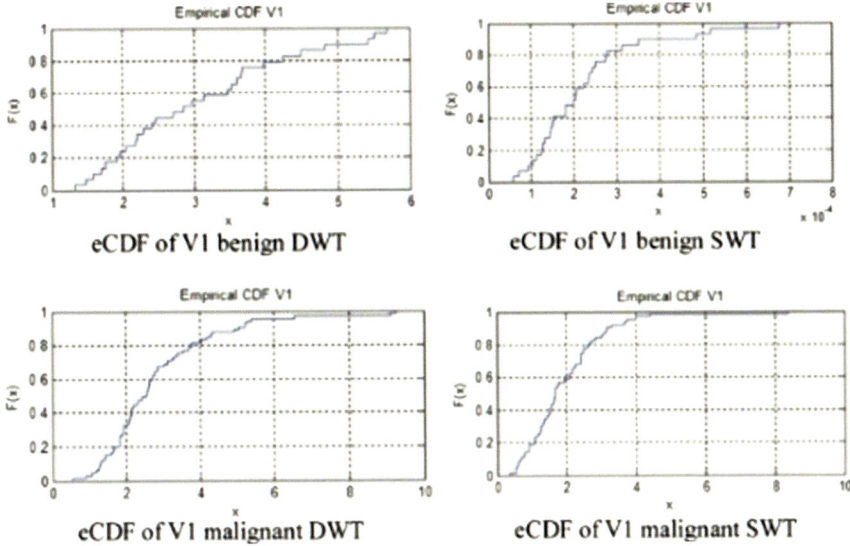

eCDF of V1 benign DWT   eCDF of V1 benign SWT

eCDF of V1 malignant DWT   eCDF of V1 malignant SWT

Fig. 6 Comparison of DWT and SWT for benign and malignant cells (*Source* Technol Cancer Res Treat, p. 507)

Fig. 7 Thermogram showing a high-temperature region with tumor growth (*Source* http://www. integratedhealthsolutions. com/medical-thermography/ breast-cancer-risk/)

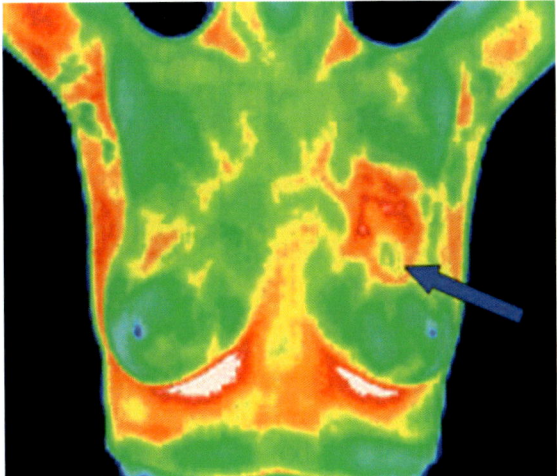

in the breast tissue. These show up on the thermal image as areas with a higher skin temperature. On further inspection of the breast, the extent of the tumor spread and the layers it has penetrated can be found out (Fig. 7).

4.1 Advantages Over Other Screening Methods

Thermography, though is a relatively new approach has a lot of advantages over the other methods, thereby making it highly sought after procedure. Researches have shown that thermography has the ability to diagnose breast cancer at least 10 years in advance [3]. Its most distinguishing benefit is that it can be used for detecting tumors in women below the age of 40 which is not the case for other procedures. It does not depend upon the structural aspects of the breast but the thermal aspect of the tissue, and hence it gives more conclusive answers for all women irrespective of their age, size of breast, tissue density, or whether they are pregnant and lactating. Since no radiation is used, no chemical inputs into the bloodstream and any other invasive technique, thermograms are an easy and affordable way to screen for abnormality. Repeat scanning is an issue in mammograms. This exposes the patients to excess radiation, whereas thermogram is a one-time procedure. Due to the high probability of real-time imaging, thermograms are used to monitor patients even after surgery for removal of infected tissue. The RGB infrared image is better suited for processing and feature extraction making the classification process simpler. Though it is effective and 100% harmless, obtaining high-contrast images with less signal-to-noise ratio is still proving to be difficult. This can be corrected later on using transforms and preprocessing adaptive filters. Despite these minor drawbacks, thermography is considered the best choice for screening.

4.2 Processing and Segmentation

After acquisition of the image, a set of processing steps that carry out de-noising and enhancement operations on the image is carried out. Other screening methods produce black and white images as supposed to the RGB of the infrared image, and therefore the processing steps are different and in various stages. Primitive noise cancellation filters are insufficient and adaptive filters like Gaussian filters are used to remove the nonlinear noise. After filtering, contrast enhancement is done on the image to sharpen the edges and intensify the difference in the background and foreground colors giving specificity to the colors in the image. Areas that exhibit as red-colored portions generally may host abnormal cells and the blue-colored regions, on the contrary may be devoid of any tumor (Fig. 8).

Preprocessing is followed by segmentation where the region of interest (ROI) is deducted. This is vital in understanding the extent of the growth and to make targeted treatment plans. For the acquired contrast-enhanced image, a segmentation process called clustering is performed. Here, the image is segregated and sequestered based on the K-means algorithm. It separates the initial image into three separate regions based on the primary colors. This aids in narrowing down the region where the tumor cells may be present. Iterations are done, specifically on each cluster to conclude the presence or absence of abnormal cell growth. K-means clustering algorithm (Hartigan

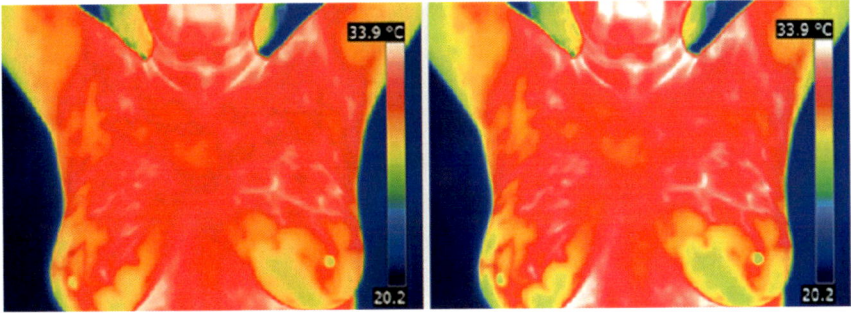

Fig. 8 Initial thermal image and contrast-enhanced image (*Source* Results from this study)

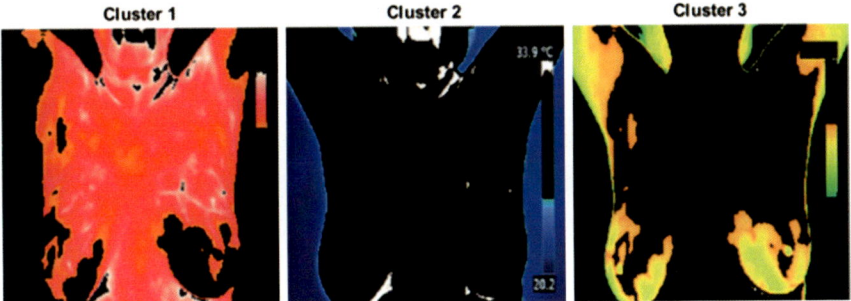

Fig. 9 Three-way clustering of thermogram using K-means algorithm based on RGB model (*Source* Results from this study)

and Wang 1979) (Lloyd 1957) (MacQueen 1967) is an investigatory algorithm in data analysis which subtracts the required area from the background and divides the image into layers, separating the hotter region from the background image [7]. Hot spot detection on thermal images becomes easier using the algorithm and it locates K-means value throughout a data highlighting its representation. It is an iterative algorithm in which K-means values are spread throughout the set in accordance with its Euclidean mean value (Fig. 9).

4.3 Feature Extraction

A thermal image has copious features that are extracted and analyzed to come to a definitive conclusion about the nature of the tumor. Features that include mean, variance, standard deviation, energy density, entropy, contrast, skewness, kurtosis, smoothness, and many others are extracted using curvelet or ridgelet transform. This transform has surpassed its predecessor in segregating the features more thoroughly and not missing any abnormal value, thereby making it a useful tool for collecting

Fig. 10 Depiction of
anisotropic decomposition
using curvelet transforms
(*Source* International Journal
of Biomedical Imaging
ID 136034, Fig. 8)

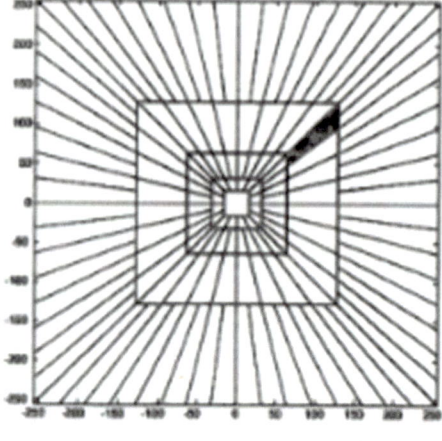

Fig. 11 Comparison of
curvelet and wavelet
de-noising capabilities
(*Source* International Journal
of Computational
Engineering Research, vol.
04)

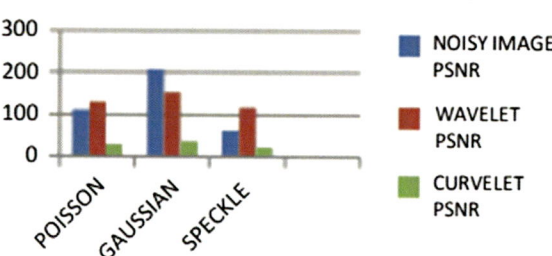

the feature database. Curvelet transform is a multiscale geometric wavelet transform, and it is more efficient in representing edges, curves, and singularities than traditional wavelet. Curvelet combines the linearity of the wavelet transform along with its own layering procedure to achieve the optimal rate of convergence by simple thresholding. Multiple decompositions slice the image into innumerable segments, thereby capturing point discontinuities in the linear structures. Curvelets in addition to a variable width have a variable length and a variable anisotropy giving it the ability to detect and measure features that are harder to notice in plain sight [6]. The image is decomposed into slices in each of the layers and the discontinuities are calculated comparing the linearity and nonlinearity of the neighboring values (Figs. 10 and 11).

The only drawback in a thermal image is that its noise level is high due to environmental factors. Also, obtaining a high-contrast image with one screening is difficult. The use of curvelet transform helps in de-noising the thermal image and the latter is resolved by the contrast enhancement. Moreover, by looking at the segmentation and layering of the curvelet transform, it gives better results in feature extraction as well as noise filtering when compared to wavelet transform [8].

The features that are finally extracted and are instrumental in the classification process are as follows:

- THERMAL DENSITY: Thermal density is the measure of difference between the textural densities in the thermal image and the complementary function.

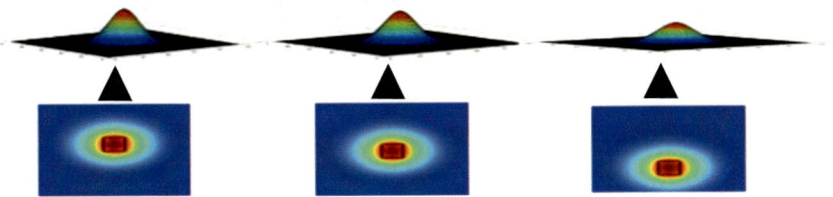

Fig. 12 Grouping of feature descriptors into high, medium, and low probability (*Source* Results from this study)

BW1 = binary thermal image,
BW2 = binary image area for the complementary function image,
Total1 = bwarea(BW1),
Total2 = bwarea(BW2), and
Thermal density = total2-total1.

- ENERGY DENSITY: The energy is expected to be high if there is a high probability of repeated pixel pairs, where $P(i, j)$ are the pixel values at the (i, j) coordinates of the image.

$$\text{Energy} = \sum_i \sum_j \{P(i, j)\}^2$$

- ENTROPY: Entropy measures the randomness of a gray-level distribution.

$$\text{Energy} = \sum_i \sum_j P(i, j) \log P(i, j)$$

- Homogeneity,
- Skewness,
- Kurtosis,
- Mean and variance, and
- Contrast.

Similar descriptors are grouped together into arrays that form the testing database which is further compared with a training database in accordance with the support vector (Fig. 12).

5 Classifications for Thermogram and 3D Mammogram

Classification techniques usually consist of taking a biopsy and analyzing it for abnormalities that suggest the presence of cancer or benign cells. The traditional method for classification is taking a biopsy of the infected tissue and analyzing it for the nature, cause, and effects of the tumor. This procedure is highly invasive and takes prolonged time to provide a conclusive answer. Therefore, the fastest kernel classification machine learning method called fast/core support vector machine is used to train the medical database with the acquired features extracted [9]. These existing values are compared and correlated with the features and values extracted from the curvelet transform and wavelet transform for thermogram and mammogram, respectively. An SVM train is added to the existing image and the computer calculates the probability of malignant cells present. An SVM vector is calculated and the probabilities are placed over the vector in regions of high, low, and reference [10]. Values that are proximal to the vector, either high or low, are considered as manifestations of malignancy.

The various classification outputs that can be derived are as follows [11]:

- True positive: This result confirms the cells as malignant and the SVM computes the affected area and accuracy of calculation by multiple iterations.
- True negative: This result confirms the tumor cells as benign and prompts the computer to analyze a different cluster in the odd chance of the cells being present in other layers and regions of tissue.
- False positive: This is an unacceptable result that points out the presence of malignant cells when in reality, they are benign.
- False negative: The SVM algorithm fails to detect the presence of cancer cells because their manifestation is not very prominent and the features do not match with the exact threshold values of the training database.

The FSVM is used to make sure that the classification provides satisfactory results with least number of iterations and high accuracy.

5.1 Thermogram

After the classification result finalizes the presence of cancer, identification of the exact location of the cells is carried out. This is performed by minimum enclosing ball (MEB) algorithm. The malignant cells may not only be present over the surface tissue layer but it may have also penetrated into the interior walls. MEB is used to detect a boundary of infected cells and sequester them in the shape of spheres [12]. This spherical area is decomposed and further layers are inspected for abnormality and a graphical output is presented depicting the exact cell locations in the entire thermogram (Fig. 13).

5.2 Mammogram

Various classifiers are used on mammograms like Parzen's classifier and linear discriminant classifier (LDC). It has been found out that best results are obtained when

Fig. 13 MEB boundary detection. Red dots represent malignant cells, blue dots are benign cells (*Source* Result from this study)

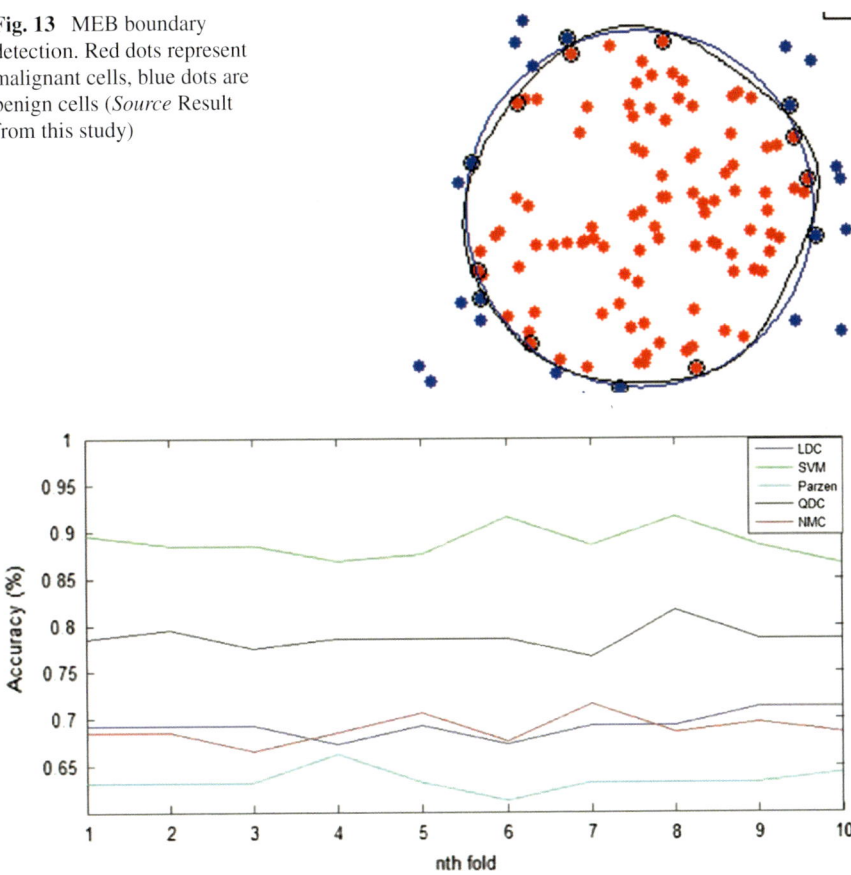

Fig. 14 Graph of accuracies for various classifiers when SWT is used for feature extraction (*Source* Technol Cancer Res Treat, p. 510)

SWT and SVM are combined [5]. After classification as benign or malignant, further boundary detection processes are not yet developed for mammograms (Fig. 14).

6 Results

After the acquisition, processing, segmentation, feature extraction, and classification techniques have been performed on both types of images, thermogram and 3D mammogram, a conclusive result with accuracy has been coined. For the mammogram, when SWT features are combined with SVM, a maximum accuracy of 88.80% is achieved while DWT features in combination with SVM yielded a maximum accuracy of 81.73%. Similarly, for an infrared image, curvelet transform features combined with SVM classification yielded an accuracy of 95.16%, sensitivity of 96.77%, and specificity of 98.38%.

7 Conclusions

With the cumulative results from both images, we can conclude that thermogram combined with SVM classifier provides a better accuracy in distinguishing the nature of the tumor, consequentially paving way for earlier detection as well as diagnosis and treatment. Though mammograms are the most sought after and reliable techniques for detection, the follow up processes do not give satisfactory results. The ultimate goal is to provide a platform for women of all ages and walks of life to be able to detect the cancer at an early stage and get the necessary help they need as soon and as effectively as possible. This procedure not only revolutionizes the field of medicine but also bridges the gap between image processing, medicine, data sciences, and machine learning.

References

1. Gonzalez RC, Woods RE. Digital image processing, 3rd edn. Pearson Education
2. Akinyemiju TF (2013) Risk of asynchronous contralateral breast cancer: multiple approaches for a complex issue. Gland Surg 2(2):110–113
3. Cowley G (2017) Mammography vs. thermography: comparing the benefits. Medical News Today
4. Geoge KMJ, Dhas DAS (2017) Preprocessing filters for mammogram images: a review. In: Emerging devices and smart systems (ICEDSS). ISBN: 978-1-5090-5555-5. IEEE Xplore 19 Oct 2017
5. Ganesan K, Acharya UR, Chua CK, Min LC, Abraham TK (2014) Automated Diagnosis of Mammogram images of breast cancer using discrete wavelet transform and spherical wavelet transform features. Technol Cancer Res Treat 13(6):605–615
6. Mahmoudzadeh E, Zekri M, Montazeri MA, Sadri S, Dabbaggh ST (2016) Directional SUSAN image boundary detection of breast thermogram. IET Image Process 552–560
7. Shmmala FA, Ashour W (2013) Color based image segmentation using different versions of K-means in two spaces. Global Adv Res J Eng Technol Innov 1(9):030–041. (ISSN: 2315-5124)
8. Shukla M, Changlani S, A Comparative study of wavelet and curvelet transform for image
9. Catanzaro BC, Sundaram N, Keutzer K (2008) Fast support vector machine training and classification on graphic processors. In: UCB/EECS
10. Phan J, Moffitt R, Dale J, Petros J, Young A, Wang M (2005) Improvement of SVM algorithm for microarray analysis using intelligent parameter selection. In: 2005 IEEE-EMBS 2005 27th Annual International Conference of the IEEE Engineering in Medicine and Biology Society, pp. 4838–4841
11. Shrivastava P, Singh P, Shrivastava G (2014) Image classification using SOM and SVM feature extraction, (IJCSIT) 5(1):264–271
12. Cervantes J, Li X, Yu W, Li K (2008) Support vector machine classification of large data sets via minimum enclosing ball clustering. Neurocomputing 71(4–6):611–619

IEFA—A Fuzzy Framework for Image Enrichment

Ankita Sheoran and Harkiran Kaur

Abstract In this work, an Image Enhancement Fuzzy Algorithm (IEFA), a technique for image enhancement has been proposed and developed. IEFA formulates the mapping from a given input to an output using fuzzy logic. IEFA improves the contrast of low-contrast images. The technique begins the process of image enrichment by modifying membership functions and designing fuzzy if–then rules that exist as a sophisticated bridge between human knowledge on one side and the numerical framework of the computers on the other side. The algorithm converts image properties into fuzzy data and further fuzzy data into crisp output through defuzzification. Further, to evaluate the performance of the proposed technique, the developed technique has been compared with "Histogram Equalization (HE) and Contrast Limited Adaptive Histogram Equalization (CLAHE)." It has been observed that PSNR and CII of the proposed algorithm (using a test image) are 25.56 and 1.13, respectively. These metrics are 0.078 and 6.603% more effective than the metrics of existing algorithms.

1 Introduction

Present-day applications require various kinds of images and pictures as sources of information for interpretation and analysis. On applying conversion of an image from one form to another form, such as digitized, scanned, transmitted, stored, and so on, some of the degradation occurs at the output state. Hence, the image has to undertake the process of image enrichment which consists of collection of techniques that seek to improve the visual appearance of an image.

Image enhancement techniques have wide number of applications in the field of medical science imaging, art studies, forensic science, and atmospheric science.

A. Sheoran · H. Kaur (✉)
Computer Science and Engineering Department, Thapar University, Patiala 147001, Punjab, India
e-mail: harkiran.kaur@thapar.edu

A. Sheoran
e-mail: ankita.shrn@gmail.com

© Springer Nature Switzerland AG 2019
D. Pandian et al. (eds.), *Proceedings of the International Conference on ISMAC in Computational Vision and Bio-Engineering 2018 (ISMAC-CVB)*, Lecture Notes in Computational Vision and Biomechanics 30,
https://doi.org/10.1007/978-3-030-00665-5_122

Image enhancement improves the quality of the image so as to prevail over the limitations of the human visual system. The incentive of image enhancement techniques includes higher visual quality, extracting the hidden details in the image, increasing the contrast of low-contrast image, enhancing image features for further processing, and many more. In image enhancement, the low-quality image is transformed into high quality with the intent of improving the look of the image [1]. Instead of increasing the inherent information of the data, enhancement upsurges the range of features.

1.1 Image Enhancement (IE)

In IE process, the visual look of image is perked up or is converted in a form that can be easily understood by human eyes or machines. Noisy image data is very difficult to enhance, which is necessary for many research and application areas. There are three main categories in image enhancement technique which are given as follows:

i. Spatial domain methods

"Spatial domain method" functions directly on pixels. In spatial domain method, according to rules, the pixel values are adapted that are dependent on the original pixel value (local or point processes). On the other hand, in many ways, the pixel values can be merged with or compared to other pixels within instant neighborhood. Consider the input image $r(j, k)$ and processed image $s(j, k)$, then the transformation $s(j, k) = Q[r(j, k)]$, where Q is an operator on r defined over some region of (j, k). The operator Q is applied at every position (j, k) to yield output s at that position. The process uses pixels in the area of image spanned by vicinity [2]. The two most popular conventional methods in spatial domain are histogram: specifications and equalization, and adaptive neighborhood histogram equalization [3].

ii. Frequency domain methods (DFT)

In frequency domain method, Fourier transformation of image is used. Transformation of image into two dimensions even with fast transformation is a very time-consuming task, thus making it less appropriate for real-time processing [3]. Sharp conversions and edges in image give extensively to high-frequency content of Fourier transformation [2]. The overall appearance of the image over smooth areas is due to low-frequency content in Fourier transformation. It is easy to study the idea of filtering in frequency domain and so improvement of image $r(j, k)$ can be completed in the frequency domain based on DFT [2]. This is chiefly beneficial in convolution if the spatial extent of the point spread sequence $h(j, k)$ is larger than the convolution theory.

$$s(j, k) = h(j, k) * r(j, k) \tag{1}$$

where $s(j, k)$ is enhanced image and $r(j, k)$ is input image.

iii. **Fuzzy domain method**

Fuzzy set theory is suitable for handling diverse uncertainties in image processing and computer vision applications. Fuzzy approaches comprise different sets that help in recognizing, characterizing, and processing of image which as a whole describes the fuzzy theory. Fuzzy logic has three processes: image fuzzification of crisp value to fuzzy values, membership function modification, and defuzzification to get back the crisp values. Fuzzy image enhancement involves mapping of gray level into membership function with the intent of generating an image of greater contrast than the input image. This is achieved by assigning a larger influence to the pixel intensity that is nearer to the average pixel intensity and to those that are beyond the average intensity of the image.

1.2 Fuzzy Logic

It is a multivalued logic in which variables can have truth values in the range 0–1 and can be any real number, whereas the truth values of variables in the classical Boolean logic can only be the "crisp" values "0 or 1" that is "completely true" or "completely false." Fuzzy logic is engaged to deal with the impression of partial truth, in which the truth value may vary between "completely true" and "completely false" in which membership function manages the degree. Process:

(a) Fuzzifying input values that are pixel data into fuzzy membership functions.
(b) Implement all applicable rules from the Rule Base to calculate the fuzzy output functions.
(c) Defuzzify the output membership functions for "crisp" output values.

Fuzzy logic systems can perform well even with imprecise and incomplete input data or even if data is not reliable, because in fuzzy logic the output is an agreement of the inputs and all the defined rules in the Rule Base [4]. By optionally adding weights to each rule, the degree up to which it affects the output can be controlled. These rule weights can also be based on outputs of other rules. The weights can be changed or can be kept static. The criterion for deciding the weights is consistency or reliability and how important the rule is.

2   Design and Implementation of Framework

In this work, the "Mamdani Fuzzy Inference System" is chosen over "Sugeno Fuzzy Inference System" because the system has to take care of three diverse ranges of pixel strength, namely, bright, dark, and gray. In Sugeno FIS, it is beyond the bounds of possibility to unite rules designed for the system as single spike is used for every designed rule.

2.1 *Workflow of the System*

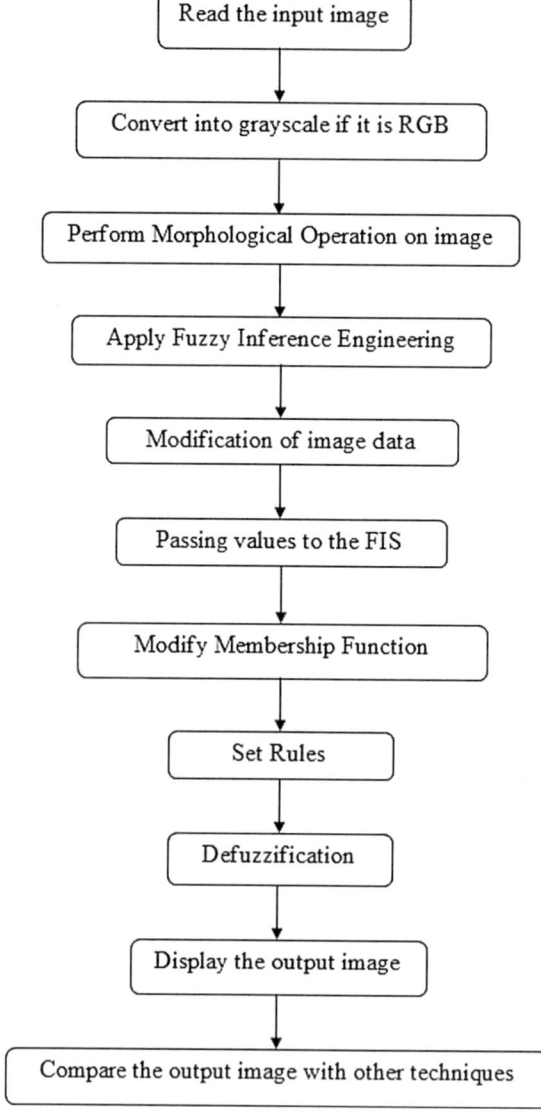

2.2 Description of Image Enhancement Fuzzy Algorithm (IEFA)

Step 1: Morphological processing which includes the following:

  i. Convert the image into grayscale if it is RGB.
 ii. Convert the image into double format.
iii. Evaluate size of the image.
 iv. Calculate minimum, maximum, and mean gray level intensity of the image.

Step 2: Transforming the image data into fuzzy domain data that is performing fuzzification.

The FIS toolbox of MATLAB comprises FIS editor, rule editor, rule viewer, and membership function editor. The FIS editor shows general information of FIS. It displays the input and output variables. Here, the input variable is "pel_in" and output variable is "pel_out." The "ALGORITHM I" converts the image pixel data into fuzzy data. The "pel_in" takes the output of ALGORITHM I, that is, fuzzified pixel value.

Logical Structure of Algorithm I
FOR grayvalue = each pixel of the image
IF (grayvalue is between 0 AND minimum)
fuzzy=0;

   ELSE IF (grayvalue is between minimum AND mean)

      fuzzy = (1/(mean - minimum)) * minimum + (1/(mean - minimum)) * grayvalue;

   ELSE IF (grayvalue between mean and maximum)

      fuzzy = (1/(maximum - mean)) * mean + (1/(maximum - mean)) * grayvalue;

   ELSE (grayvalue between maximum AND 255)
      Fuzzy = 1;
   END

END

The low-contrast input image is transformed to membership plane with membership function where its values lie in the range 0–1. The membership function adopts any value in the interval 0–1. Three triangular membership functions are taken for both input and output at three intensity levels of the image, respectively. For each membership function, range is adjusted in membership function editor toolbox of FIS.

Step 3: Membership modification.

Using Fuzzy Inference System (FIS) toolbox, the membership values are modified. FIS consists of inbuilt membership functions for both input and output variables. New membership values are defined for the pixels with respect to their input pixel intensity, and based on variation in their intensity level the membership values are assigned. The membership values are modified for the pixels considering their gray level intensity values, and hence membership values are allotted to the pixels on the basis of variation in their intensity level.

Logical Structure of Algorithm II
FOR pixel = every pixel value

   IF (pixel_value is between 0 AND 0.5)

new_pixel = 5 * (pixel ^ 3);

   ELSE

new_pixel = 1 − 5*((1 − pixel) ^ 3);

   END

Step 4: Set rules.

The rules are constructed in the graphical rule editor interface of fuzzy inference system tools. Fuzzy rules are defined for the modified membership values with respect to the pixel intensity. The ranges of both output and input membership functions have been considered in defining the fuzzy rules. The rules are added by taking into account the parameters of the selected triangular membership function for the system. The designed IEFA has been designed and implemented on the basis of the following fuzzy rules:

  i. If pel_in is Dark then pel_out is More_dark.
  ii. If pel_in is Gray then pel_out is More_gray.
 iii. If pel_in is Bright then pel_out is More_bright.

   As a whole, the input of the if–then rule is the present value of the input variable (pel_in) and the output is the whole fuzzy set (More_dark, More_gray, or More_bright).

Step 5: Convert fuzzy data into grayscale enhanced data that is performing defuzzification.

Defuzzification is the final step in Mamdani FIS. For defuzzification process, combine output fuzzy collection and the final output is a single crisp number. The image is converted to the pixel plane from fuzzy plane. The output of Algorithm III is crisp number that enhanced pixel values.

Logical Structure of Algorithm III
Set maximum intensity and minimum intensity for enhanced image as:
maxI = 255;
minI = 0;
FOR new_pixel = each pixel value

   IF (pixel <= minimum)

pixel_enhanced = 0;

   ELSE IF (pixel is between minimum AND maximum)

pixel_enhanced = ((maxI - minI) * result(x,y) + minI);

   ELSE

pixel_enhanced = 255;

   END

END

3   Experimental Setup

To assess the proposed technique, different low-contrast images have been taken. The following image enhancement techniques have been applied to the input images. The techniques are given as follows:

3.1   Histogram Equalization Technique

Figure 1 shows the original low-contrast image and the histogram representation of it.
   Figure 2 shows histogram equalized image and its histogram. Histogram equalization technique evens out the plot so that it covers the complete range of brightness with almost similar amount of pixels for every brightness range. Also, complete range of brightness value is covered and the graph is balanced.

3.2   Contrast Limited Adaptive Histogram Equalization (CLAHE)

"CLAHE" is different from the ordinary HE in respect that the adaptive technique equalizes numerous histograms, each of which corresponds to a different segment

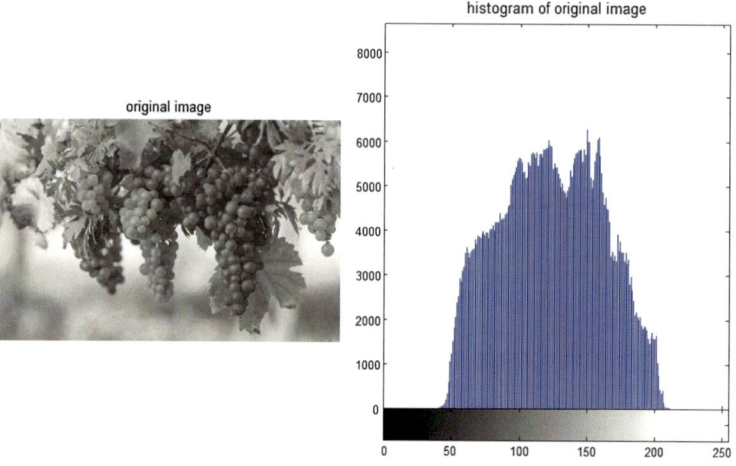

Fig. 1 Original low-contrast Grapes1.jpg and its histogram plot

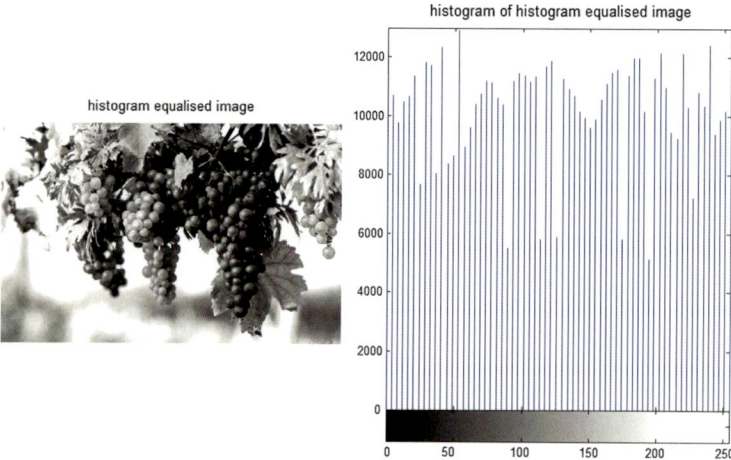

Fig. 2 Histogram equalized Grapes1.jpg and its histogram plot

of the image and redistribute the lightness values of the image using them. Figure 3 shows the CLAHE improved image.

3.3 Image Enhancement Fuzzy Algorithm (IEFA)

The proposed work aims at improving the contrast of low-contrast image by reducing the noise, without overenhancing the contrast. The method is based on intensity-level mapping onto fuzzy plane with the membership function transformation that

Fig. 3 Applying CLAHE on Grapes1.jpg

Fig. 4 Applying IEFA on Beach1.jpg

Fig. 5 Applying IEFA on Rock1.png

increases the contrast of the original low-contrast image. Figures 4, 5, 6, and 7 display the visual outcomes attained after applying IE. The radiology images in Figs. 6 and 7 were taken from radiology department of a local hospital.

4 Performance Evaluation

After applying the abovementioned image enhancement techniques on low-contrast images, the results have been obtained. The performance analysis of various tech-

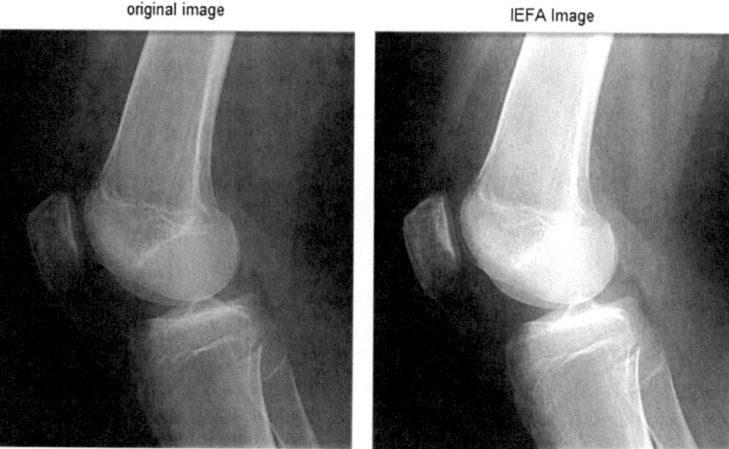

Fig. 6 Applying IEFA on Medical1.jpg

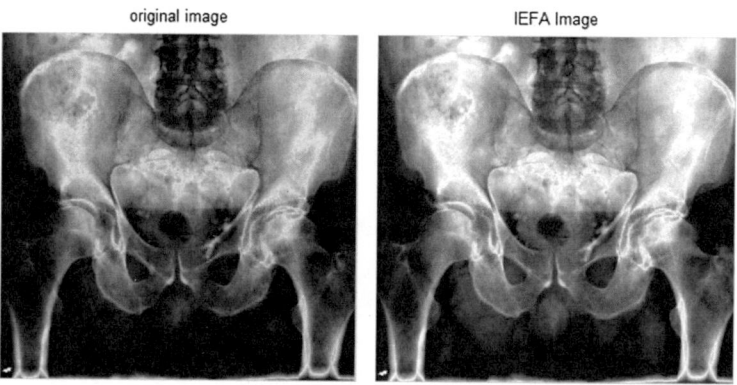

Fig. 7 Applying IEFA on Medical2.jpg

niques has been done using different parameters that is "Peak-signal-To-Noise Ratio (PSNR)" and "Contrast Improvement Index (CII)."

PSNR

"PSNR is the division of the maximum possible power of a signal to the power of corrupting noise affecting image quality." For reconstruction to be of higher quality, the value of PSNR should be high. Table 1 shows comparison of PSNR of enhancement techniques.

CII

Contrast improvement index is the division of $C_{enhanced}$ to $C_{original}$, where $C_{enhanced}$ is average contrast of new image and $C_{original}$ is the average contrast of input image. A higher value of CII is always favored. Figure 8 displays a bar representation of CII

Table 1 Comparison of PSNR of various enhancement techniques

Image name	PSNR of the original image	PSNR of histogram equalized image	PSNR of CLAHE image	PSNR of IEFA
Grapes1.jpg	25.25	25.56	25.34	25.56
Grapes2.png	25.26	25.57	25.41	25.55
Beach1.jpg	25.27	25.54	25.34	25.56
Rock1.png	25.25	25.55	25.41	25.55
Field1.png	25.27	25.54	25.35	25.58
Field2.jpg	25.25	25.55	25.33	25.55
Field3.png	25.27	25.54	25.37	25.57
Field4.jpg	25.25	25.54	25.39	25.54
Medical1.jpg	25.30	25.56	25.28	25.43
Medical2.jpg	25.34	25.52	25.45	25.41

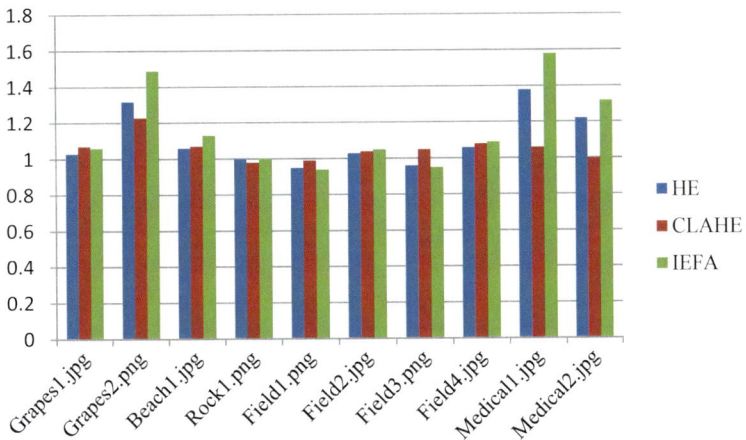

Fig. 8 CII evaluation of HE, CLAHE, and IEFA

achieved by applying the techniques on low-contrast images. From this figure, it is inferred that proposed IEFA yields the highest value of CII (Table 2).

5 Conclusion

The foremost focus of IEFA is on image enhancement of low-contrast images using fuzzy technique. For implementation, three membership functions are defined for each input and output, and rules are designed in FIS. Experiment has been performed on low-contrast images and the outcome of the proposed IEFA is compared with the

Table 2 CII analysis

Image name	HE	CLAHE	IEFA
Grapes1.jpg	1.03	1.07	1.06
Grapes2.png	1.32	1.23	1.49
Beach1.jpg	1.06	1.07	1.13
Rock1.png	1.00	0.98	1.00
Field1.png	0.95	0.99	0.94
Field2.jpg	1.03	1.04	1.05
Field3.png	0.96	1.05	0.95
Field4.jpg	1.06	1.08	1.09
Medical1.jpg	1.38	1.06	1.58
Medical2.jpg	1.22	1.00	1.32

outcome of HE and CLAHE. To contrast these algorithms, "Contrast Improvement Index (CII)", and "Peak-Signal-To-Noise Ratio (PSNR)" have been used as the performance metrics. From CII analysis, it is concluded that "Histogram Equalization (HE)" and "Contrast Limited Adaptive Histogram Equalization (CLAHE)" yield less values of CII. This is because the focus of HE is only on the global contrast of image. This also leads to the loss of the local details. In CLAHE, because of nonuniform lightning, ambiguity is introduced in image which appears as imprecise boundaries during digitization. IEFA results in better CII values as compared to CLAHE and HE by solving the problem of vagueness and imprecise boundaries through fuzzy sets and linguistic variables.

5.1 Future Scope

In future, the work can be extended to colored images that offer more real-life implementations. The algorithm can be modified to give better results for medical images. The improvised outcome for image enrichment is also used in real-time augmentation of neuroevolution.

References

1. Yaman O, Karakose M (2016) Development of image processing based methods using augmented reality in higher education. In: 15th International conference on information technology based higher education and training (ITHET). IEEE
2. Patil M (n.d) Design of novel fuzzy based method for contrast. Int J Electr Electron Data Commun. ISSN 2320-2084
3. Kaur T, Sidhu RK (2016) Optimized adaptive fuzzy based image enhancement techniques. Int J Sig Process Image Process Pattern Recogn 9.1:11–20
4. Tizhoosh HR (2000) Fuzzy image enhancement: an overview. Fuzzy techniques in image processing. Physica-Verlag HD, pp 137–171

Hidden Markov Random Field and Gaussian Mixture Model Based Hidden Markov Random Field for Contour Labelling of Exudates in Diabetic Retinopathy—A Comparative Study

E. Revathi Achan and T. R. Swapna

Abstract Diabetic Retinopathy (DR) is one of the important causes of blindness in diabetic patients. Diabetes that affects the retina is called diabetic retinopathy. Diabetic retinopathy occurs due to the damage of blood vessels in retina and increase in the level of glucose. Different pathologies are normally seen in DR such as microaneurysms, hard exudates, soft exudates, cotton wool spots and haemorrhages. We have done a comparative study of Hidden Markov Random Field (HMRF) and Gaussian Mixture Model (GMM) based HMRF for automatic segmentation of exudates and the performance analysis of both methods. The preprocessing consists of candidate extraction step using greyscale morphological operation of closing and initial labelling of exudates using K-means clustering followed by contour detection. In contour detection, we have analysed two approaches, one is GMM-based HMRF and the other is HMRF. DIARETDB1 is the dataset used.

1 Introduction

Diabetes is a metabolic disease which occurs due to a high level of sugar. The high level of sugar produces symptoms like polymia, polydipsia and polyphagia. DR is the diabetic eye disease which occurs due to damage to the retina and may cause blindness. The treatment and monitoring of retina are important to diagnose and monitor the progress of the disease as there are no warning signs for this disease. There are two types of diabetic retinopathy such as nonproliferative diabetic retinopathy and proliferative diabetic retinopathy [1]. At the stage of nonproliferative diabetic

E. Revathi Achan · T. R. Swapna (✉)
Department of Computer Science and Engineering,
Amrita School of Engineering, Amrita Vishwa Vidyapeetham, Coimbatore, India
e-mail: tr_swapna@cb.amrita.edu

E. Revathi Achan
e-mail: cb.en.p2cvi16006@cb.students.amrita.edu

© Springer Nature Switzerland AG 2019
D. Pandian et al. (eds.), *Proceedings of the International Conference on ISMAC in Computational Vision and Bio-Engineering 2018 (ISMAC-CVB)*, Lecture Notes in Computational Vision and Biomechanics 30,
https://doi.org/10.1007/978-3-030-00665-5_123

retinopathy, the leakage of blood vessels occur. In proliferative diabetic retinopathy, abnormal blood vessels are formed at the back of the eye and can burst or bleed and can cause vision loss. The pathological signs of DR are microaneurysms, haemorrhages, exudates and cotton wool spots [2]. There are different modalities used to identify diabetic retinopathy such as colour fundus photography, fundus fluorescein angiography and optical coherence photography. Fundus photography is used to examine the retina by biomicroscope with magnifying lens. Fundus angiogram is used to detect the leakage in blood vessels. Optical coherence tomography is a light-sensitive tissue to sense at the back of the eye which uses light waves to detect the abnormality in retinal images. It gives the depth information of the image. The use of light gives more axial resolution than any other imaging technique.

2   Literature Survey

Authors	Methodologies	Outcomes	Remarks
Hussain F. Jaafar et al. [3]	1. Coarse and fine segmentation 2. Split and merge technique	99.4% of accuracy	Due to distinctive performance measures, the proposed method may be successfully applied to images of variable quality
Choundhury S et al. [4]	1. Fuzzy c means based feature extraction 2. Preprocessing step includes smoothening and green channel of image	Using SVM the exudates of 97.6% accuracy is obtained	Detected only blood vessel density
Harini R., Sheela N. [5]	1. Preprocessing step techniques such as image resizing, contrast enhancement, grey and green colour extraction 2. Fuzzy c means and morphological operation	Accuracy of 99.67%, sensitivity of 100%, specificity of 95.83%	Features collected have considered for classification are area, microaneurysms, haemorrhages
Choundhury S et al. [4]	1. Machine learning techniques 2. Classifiers such as Gaussian mixture model, KNN, GMM and support vector machines	53 images contain exudates	More time complexity and future work can be directed towards detection of neovascularization

(continued)

(continued)

Authors	Methodologies	Outcomes	Remarks
R. Priya, P. Aruna [6]	1. SVM and neural network methods 2. Area, perimeter, radius of features are captured	The performance measures of SVM have an accuracy of 97.608% and probabilistic neural network has an accuracy of 89.60%	Mainly, SVM outperforms the other model
Ramon Pires [7]	1. Features are extracted and given to the input of SVM and bossa nova technique	Detection performance of 96.4% for hard exudates and 93.5% for red lesions	Bossa nova is used for mid-level feature characterization and image representation for different lesion detectors
Raju Maher et al. [8]	1. Preprocessing step consists of image contrast enhancement such as histogram equalization, morphological operators followed by binarization	That 96.9% of sensitivity, 96.1% specificity and 97.38% of accuracy	Automatic identification of image processing techniques for abnormalities in retinal images
Kusakunniran W et al. [9]	1. Four features are extracted such as colour, contrast, focus and illumination 2. Grabcut algorithm	Image segmentation based on the iterative selection and grabcut algorithm	Images are qualified on the basis of quality assessment and segmentation of diabetic retinopathy images
B. V Shilpa, T. N Nagabhushan [10]	1. Novel-based ensemble approach 2. Morphological operations compared with logical operations	89.13% is the positive predicted value 100% is the negative predicted value	Use of canny edge detection algorithm has further improved the accuracy of boundary detection, along with the removal of false connected components
Arslan Ahmed et al. [2]	1. Green channel, histogram equalization, Morphological smoothening are the preprocessing techniques 2. KNN and SVM classifiers are used	The proposed method has sensitivity of 97.39%, specificity of 98.02%, accuracy of 97.56% and AUC of 0.97%	These methods are using hybrid classifier
Mahendra Gandhi, Dr. R. Dhanasekharan [11]	1. Converting the image into greyscale and edge detection algorithm 2. Features are extracted using GLCM and given to the input of SVM	SVM classifier is used to detect exudates	This paper not only confirms the disease but also tends to measure the severity of the disease
Ganesh S, Dr. A.M. Basha [12]	1. Krisch's algorithm is used and classified	Blood vessel extraction from retinal images was done	All the features are extracted and analysed but the correct verification was not done

(continued)

(continued)

Authors	Methodologies	Outcomes	Remarks
Sindhura et al. [13]	Preprocessing step such as closing operation, RGB to HSI, adaptive histogram equalization	Percentage of exudates detected is 40.16%	Nilblack's thresholding is used and superimposed the image onto the original image for showing exudates
Shevta, Gurmeen Kaur [14]	Support vector machine and multilinear discriminant analysis	Detection of exudates was done up to 50%	The review for detection of DR using fundus images and approaches SVM and MDA
Anup. V. Deshmukh [15]	Enhanced using brightness transform function	Achieves 99% of accuracy	The local mean and entropy-based region growing technique is applied to classify exudate and non-exudate pixels
Amrita Roy Chowdury, Sreeparna Banerjee [16]	Contrast enhancement, optical disc is detected and segmented using fuzzy c means	The automatic detection can help the doctors for accurate detection of cotton wool spots and also longitudinal study of retinal image	All the spots are not detected properly
Rakshitha T. R. et al. [17]	1. Wavelet transform method 2. Techniques are examined using PSNR ratio	Calculating the peak-to-signal ratio from datasets	Best results of detection of edges compared to other techniques
Yitian Zhao et al. [18]	New saliency-based technique	Detection of leakage in fluorescein angiography	The experimental results show that it outperforms one of the latest competitors and performs leakage detection
Meenu Vijayan et al. [19]	Life assessment between patients based on questionnaire method	Evaluated the cost for different treatment modalities	Study on quality of life assessment in diabetic retinopathy among patients
Parvathy Ram, Swapna T. R. [1]	A method for profiling of hard exudates and microaneurysms	Quantification algorithms are used	Not detected automatically and used the crop operation to detect the exudates

3 Methodology

The methodology of the approach is as shown in Fig. 1. The standard database DIARETDB1 is used as an input image to detect the exudates. It consists of 89 colour fundus images. The database consists of original images and ground truth images of soft and hard exudates. The image is captured at 50° field of view using fundus camera. This dataset is defined as calibration level 1 fundus images. In the candidate extraction step, the exudate regions are extracted using greyscale morphological closing with size 3×3 structuring element. Closing operator is used to enlarge the boundaries of foreground and also preserves the background region. In contour detection step, we have used both HMRF- and GMM-based HMRF. HMRF comprises two algorithms, expectation maximization algorithm and MAP estimate.

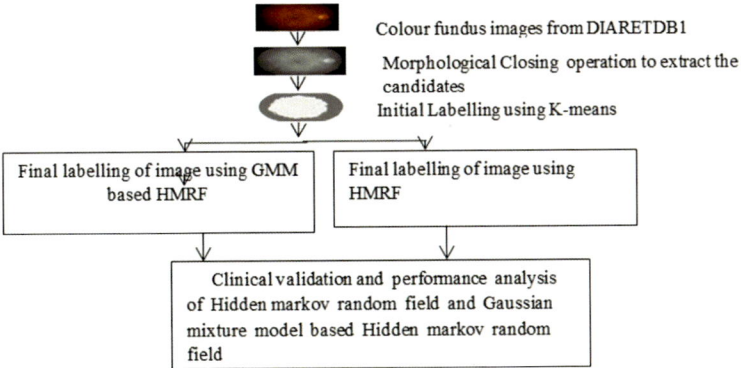

Fig. 1 Methodology

Using expectation maximization, the intensity distribution of each region is segmented using Gaussian distributions with parameters such as mean and covariance denoted by $\theta_{x_i} = (\mu_{x_i}, \sigma_{x_i})$.

3.1 Algorithm for Detection of Exudates

(a) Read an input image,
(b) Candidate extraction by grayscale morphological closing,
(c) Initialize the foreground and background pixels by HMM- and GMM-based HMM using MAP estimation and expectation maximization algorithm to find the exact boundaries.

3.2 Hidden Markov Random Field-Expectation Maximization Algorithm for Contour Detection

For the given image $\mathbf{Y} = (y_i \ldots y_n)$ and for each y_i is the intensity of a pixel, configuration of labels is defined in $\mathbf{X} = (x_1 \ldots x_n)$, where $x_i \, \varepsilon \, L$. L is a set of labels. Expectation maximization algorithm is used to estimate the parameters $\boldsymbol{\theta} = \{\theta_l | l \, \varepsilon \, L\}$.

(a) Initial parameter set $\boldsymbol{\theta}^{(0)}$ and in Expectation (E) step at tth iteration conditional expectation is

$$Q(\boldsymbol{\theta}|\boldsymbol{\theta}^t) = \sum_{x \varepsilon \chi} P\big(X|Y, \boldsymbol{\theta}^{(t)} \ln P(X, Y|\boldsymbol{\theta})\big), \tag{1}$$

where chi represents the configuration of labels.

(b) In M-step: Now maximize $Q(\theta|\theta^{(t)})$ to obtain the next estimate

$$\theta^{(t+1)} = \text{argmax } Q(\theta|\theta^{(t)}) \tag{2}$$

Then, let $\theta^{(t)} \rightarrow \theta^{(t+1)}$ and repeat from the E-step.

(c) $G(z;\theta_l)$ represents the Gaussian distribution function with parameters

$$\theta_l = (\mu_l, \sigma_l)$$

$$G(z;\theta_l) = \frac{1}{\sqrt{2\pi\sigma_l^2}} \exp(-(z-\mu_l)^2/2\sigma_l^2). \tag{3}$$

(d) Estimate the labels using current parameter set by MAP algorithm.

(e) Calculate the posterior distribution of all $l\varepsilon L$ in pixels y_i.

(g) Update the parameters of (μ_l, σ_l).

3.3 Maximum A Posteriori Algorithm of HMRF

(a) Already we have an initial estimate x^0,which we observed from EM algorithm.

(b) Provided x^k for all $1 <= i <= N$,

$$U(Y|X,\theta) = \sum [(y_i - \mu_{x_i})^2/2\sigma_{x_i}^2 + \ln \sigma_{x_i}] \tag{4}$$

is the likelihood energy Eq. (4).

(c) Clique potential is defined on pairs of neighbouring pixels

$$v_c(x_i, x_j) = 1/2(1 - I_{x_i}, x_j). \tag{5}$$

(d) Repeat step 2 until $U(Y|X,\theta) + U(X)$ converges or a maximum k is achieved.

3.4 Gaussian Mixture Model Based Hidden Markov Random Field

(a) Initial parameter set $\theta^{(0)}$ and expectation step at tth iteration conditional expectation is calculated as per Eq. (1) and used to determine which data belong to which Gaussian components.

(b) In maximization step of $Q(\theta|\theta^{(t)})$ is calculated as per Eq. (2) and used to recompute the Gaussian model parameters. GMM with g components can be represented using $\theta_l = (\mu_{l,1}, \sigma_{l,1}, w_{l,1})\ldots(\mu_{l,g}, \sigma_{l,g}, w_{l,g})$. When comparing the hidden Markov model of Gaussian distribution function of $\theta_l = (\mu_l, \sigma_l)$, this model has weighted probability:

$$G_{\min}(z, \theta_t) = \sum_{c=1}^{0} w_{l,c} G(Z; \mu_{l,c}, \sigma_{l,c}).$$ (6)

(c) Maximum a priori algorithm for labels is estimated using $x^* = \text{argmin}\{U(Y|X, \theta) + U(X)\}$, where $U(X) = \sum_{c \in C} v_c(x)$ is the clique potential. And, in the image, one pixel has almost four neighbourhood pixels. Then, clique potential is defined in Eq. (4) in the HMRF model but in the Gaussian model, this equation of constant coefficient ½ is replaced by variable coefficient as β.

$$v_c(x_i, x_j) = \beta(1 - I_{x_i}, x_j).$$ (7)

After initial labelling, the final labelling is done using HMRF- and GMM-based HMRF. Both are using expectation maximization algorithm and maximum a posteriori algorithm. But, the difference between them are parameters discussed above.

4 Result Analysis

In Fig. 2, the ground truth images are already labelled by medical experts as shown in Fig. 2(a–c). The colour fundus image in DIARETDB1 is the input as in Fig. 2 (d–f) and then preprocessing technique using morphological closing operation for candidate extraction is applied as in Fig. 2(g–i). K-means segmentation is used for finding initial labels and k value is initialized as three, such as one for foreground and other two for background and exudate/non-exudate portions. The final labelling is done using expectation maximization, MAP algorithm and initialized the number of iteration values as 10 to get the refined segmentation and to remove non-exudate portions of image as in Fig. 2(m–o).

As shown in Fig. 3, the DIARETDB1 is the database and preprocessing is done using grayscale closing operation for candidate extraction. Then, initialized the value of $k = 3$ (k-means) for foreground, background and exudate/non-exudate portion for segmentation. This method uses $g = 3$ (Gaussian components). The k-means are used for clustering/grouping a given data by minimizing the distance between them and the centroid. The Gaussian mixture model is used to determine the probability that a point 'g' is in cluster 'k' as in Fig. 3(j–l). In the expectation maximization step, we determine which data belong to which Gaussian component and in maximization step it recomputes the GMM parameters as shown in Fig. 3(m–o).

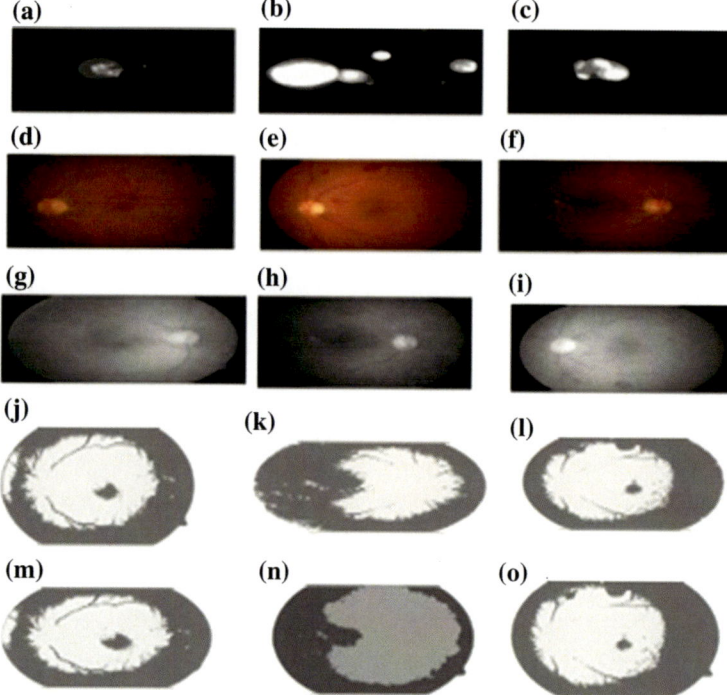

Fig. 2 **a–c** Ground truth images, **d–f** DIARETDB1, **g–i** closing operations, **j–l** initial labelling using k-means, **m–o** HMRF

As per Table 1, DIARETDB1 consists of 89 images in total with dimension 1500×1152. Ground truth images are provided for DIARETDB1, where hard and soft exudates are marked by various experts. The pixel-wise classification is done by comparing the ground truth image and using results of HMRF- and GMM-based HMRF. The accuracy of a HMRF has the slight difference when compared to GMM-based HMRF. In the case of hard exudates, the HMRF achieves sensitivity of 66% and GMM-based HMRF achieves sensitivity of 56%. In case of soft exudates, the HMRF achieves sensitivity of 71% but the GMM-based HMRF achieves sensitivity of 43%. The HMRF achieves recall of 66% and GMM-based HMRF achieves recall of 55%. According to pixel-wise classification, the two methods are used to detect exudates but HMRF is more efficient than GMM-based HMRF from the observed results. In this paper the performance analysis of HMRF and GMM based HMRF are shown below:

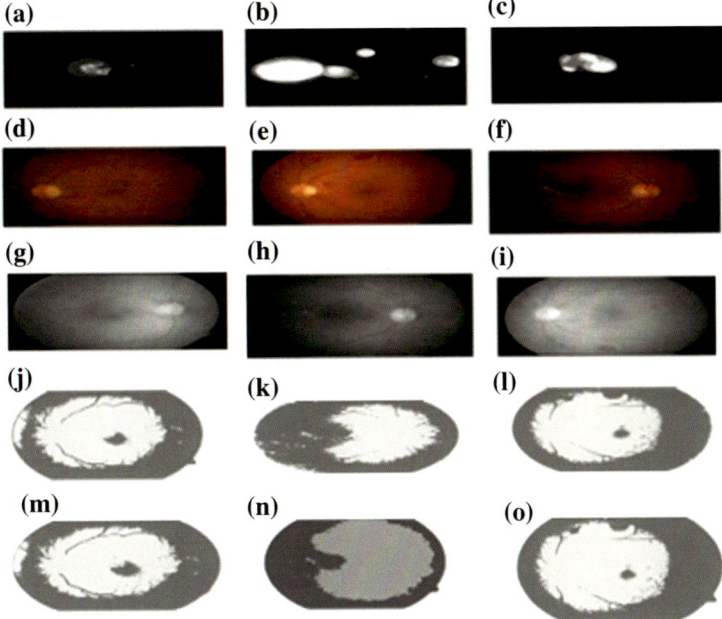

Fig. 3 **a–c** Ground truth images, **d–f** DIARETDB1, **g–i** closing operations, **j–l** initial labelling using k-means, **m–o** GMM-based HMRF

Table 1 Performance analysis of HMRF- and GMM-based HMRF

Pixel-wise classification	Hidden Markov model		Gaussian mixture model based Hidden Markov model	
	Average count of hard exudates	Average count of soft exudates	Average count of hard exudates	Average count of soft exudates
Accuracy	0.378026	0.376202	0.375152	0.373237
Sensitivity	0.664328	0.712173	0.56065	0.435036
Specificity	0.37393	0.375626	0.371404	0.393031
Precision	0.0176	0.001625	0.017016	0.001188
Recall	0.664328	0.712173	0.556559	0.435036

5   Conclusion

In this paper, we have done a comparative study of probabilistic models in colour fundus images to detect exudates. Automatic segmentation of exudates will help the clinicians to determine the disease level of diabetic retinopathy. The preprocessing technique is done using morphological operation with the standard dataset DIARETDB1. All the two methods are used to detect exudates, but HMRF is more efficient than GMM-based HMRF from the observed results.

References

1. Ram P, Swapna TR (2018) Profiling of hard exudates and microaneurysms to evaluate the progress of diabetic retinopathy. Int J Pure Appl Math 118(Special Issue 8):127–131
2. Ahmad A et al (2014) Image processing and classification in diabetic retinopathy: a review. In: 2014 5th European workshop on visual information processing (EUVIP). IEEE
3. Hussain F, Nandi Jaafar AK, Al-Nuaimy W (2010) Automated detection of exudates in retinal images using a splitand-merge algorithm. In: European signal processing conference, EUSIPCO
4. Choudhury S et al (2016) Fuzzy C means based feature extraction and classification of diabetic retinopathy using support vector machines. In: 2016 International conference on communication and signal processing (ICCSP). IEEE
5. Harini R, Sheela N (2016) Feature extraction and classification of retinal images for automated detection of diabetic retinopathy. In: 2016 Second international conference on cognitive computing and information processing (CCIP). IEEE
6. Priya R, Aruna P (2012) SVM and neural network based diagnosis of diabetic retinopathy. Int J Comput Appl 41(1)
7. Pires R et al (2014) Automatic diabetic retinopathy detection using BossaNova representation. In: 2014 36th Annual International Conference of the IEEE Engineering in medicine and biology society (EMBC). IEEE
8. Maher R, Kayte S, Bhable S, Kayte J (2015) Automated detection of diabetic retinopathy in fundus images. Inter J Emer Res Manag Tech 4(11):137–145
9. Kusakunniran W, Rattanachoosin J, Sutassananon K, Anekkitphanich P (2016) Automatic quality assessment and segmentation of diabetic retinopathy images. In: Region 10 conference (TENCON), 2016 IEEE, pp 997–1000
10. Shilpa BV, Nagabhushan TN (2015) An ensemble approach to detect exudates in digital fundus images. In: 2016 Second international conference on cognitive computing and information processing (CCIP). IEEE. Sing fuzzy C means. Int J Comput Appl 113(11)
11. Gandhi M, Dhanasekaran R (2013) Diagnosis of diabetic retinopathy using morphological process and SVM classifier. In: 2013 International conference on communications and signal processing (ICCSP). IEEE
12. Ganesh S, Basha AM (2015) Automated detection of diabetic retinopathy using retinal optical images. Int J Sci Technol Manage 4(2)
13. Sindhura A, Deva Kumar S, Ramakrishna Sajja V, Gnaneswara Rao N (2016) Identifying exudates from diabetic retinopathy images. In: 2016 International conference on advanced communication control and computing technologies (ICACCCT). IEEE, pp 132–136
14. Kaur G (2015) Review on: detection of diabetic retinopathy using SVM and MDA. Int J Comput Appl 117(20)
15. Deshmukh AV et al (2015) Features based classification of hard exudates in retinal images. In: 2015 International conference on advances in computing, communications and informatics (ICACCI). IEEE
16. Chowdhury AR, Banerjee S (2015) Detection of cotton wool spots from retinal images using fuzzy C means. Int J Comput Appl 113(11)
17. Rakshitha TR, Devaraj D, Prasanna Kumar SC (2016) Comparative study of imaging transforms on diabetic retinopathy images. In: IEEE international conference on recent trends in electronics, information and communication technology (RTEICT). IEEE, pp 118–122
18. Zhao Y et al (2017) Intensity and compactness enabled saliency estimation for leakage detection in diabetic and malarial retinopathy. IEEE Trans Med Imaging 36(1):51–63

19. Vijayan M, Jose R, Jose S, Abraham S, Joy J (2017) Study on quality of life assessment in diabetic retinopathy among patients with type 2 diabetes. Asian J Pharm Clin Res 10:116. https://doi.org/10.22159/ajpcr.2017.v10i7.18095
20. Roychowdhury, S, Koozekanani DD, Parhi KK (2014) Dream: diabetic retinopathy analysis using machine learning. IEEE J Biomed Health Inform 18(5):1717–1728
21. Kayte SN, Maher RS, Kayte CN (2015) Automated identification of hard exudates and cotton wool spots using biomedical image processing. Int J Comput Appl 131(5)

AUGEN: An Ocular Support for Visually Impaired Using Deep Learning

Reema K. Sans, Reenu Sara Joseph, Rekha Narayanan,
Vandhana M. Prasad and Jisha James

Abstract Among the wide varieties of technologies, mobile phone technology has become popular and the usage of mobile phone applications is increasing day by day. Most of the modern mobiles are able to capture photographs. This can be used by the visually impaired to capture images of their surroundings which is then used to generate sentences that can be read out to the give visually impaired people a better knowledge of their surroundings. The content of an image is described automatically to them by which they can avoid seeking help from people around them. Computer vision is a field which can be used for gaining information from images or videos. The tasks which the human visual system can do can be done using computer vision. Visually impaired people can use these technologies in order to get better understanding of their surroundings.

1 Introduction

The problem in artificial intelligence which connects computer vision and natural language processing is describing automatically the contents of an image. To describe the contents of an image in proper English sentences is a challenging task. In this project, the system will be trained with different images. During training phase, the

R. K. Sans · R. S. Joseph · R. Narayanan · V. M. Prasad (✉) · J. James
Department of Computer Science & Engineering, Muthoot Institute of Technology & Science, Ernakulam, India
e-mail: vandhanasprasad1996@gmail.com

R. K. Sans
e-mail: reemaksans1996@gmail.com

R. S. Joseph
e-mail: reenusarajoseph@gmail.com

R. Narayanan
e-mail: rekhanarayanan96@gmail.com

J. James
e-mail: jishajames@mgits.ac.in

© Springer Nature Switzerland AG 2019
D. Pandian et al. (eds.), *Proceedings of the International Conference on ISMAC in Computational Vision and Bio-Engineering 2018 (ISMAC-CVB)*, Lecture Notes in Computational Vision and Biomechanics 30,
https://doi.org/10.1007/978-3-030-00665-5_124

probability of producing the correct caption for the contents of the image should be maximized. During this training period, the system will be able to identify different objects contained in the image. How these objects relate to each other as well as their attributes and the activities they are involved in must be also known. This knowledge is then expressed in a meaningful sentence. Natural language processing is an application used for analysis and synthesis of natural language. The tasks such as automatic summarization, translation, speech recognition, etc. can be organized and structured by developers using NLP. NLP is characterized as a hard problem in computer science. The most difficult problems for artificial intelligence are processing of natural language. Many of the difficulties revolve around the issues of contexts. Neural networks are used in large areas, mostly to find the relationships between the data. They are used in traditional computer architecture to problems that the computer system cannot perform as good as the human visual system in areas like image recognition, making generalizations, that sort of thing. Researchers these days are constructing networks that are better at these problems. There are different neural networks used for training so as to perform natural language processing. It is a complex task to create a network manually, that will solve the difficult problems.

2 Literature Survey

Literature survey was done by referring papers related to the topic.

Paper [1] implemented the concept of Neural Image Caption (NIC) generator which can automatically view an image and generate a description in plain English. The key concept of NIC is Convolutional Neural Network (CNN) that encodes the image into a representation which is then followed by Recurrent Neural Network (RNN) which generates a correct caption for the image used as the input.

Paper [2] describes a system for caption generation which attempts to incorporate a form of attention with two variants, i.e., a hard attention mechanism and a soft attention mechanism. Since this system includes attention, it has the ability to visualize what the model sees. Standard backpropagation methods are used for training soft deterministic attention mechanism and hard stochastic attention mechanism trained by maximizing an approximate variation lower bound.

In paper [3], Recurrent Neural Network (RNN) is used to overcome the problem faced in spoken language understanding, which is semantic slot filling. A major task in spoken language understanding is to automatically extract semantic concepts.

In image sequence recognition, text recognition is one of the important tasks. Paper [4] describes a neural network architecture, in which feature extraction, sequence modeling, and transcription are integrated into a united framework. The proposed neural network model is named as Convolutional Recurrent Neural Network (CRNN), as it combines DCNN and RNN, and constructs an end-to-end system for sequence recognition.

Paper [5] deals with the various applications of Deep Belief Nets (DBN). In this study, DBNs to a natural language understanding problem is recognized. Natural

Language Processing (NLP) is a method by which the computers can analyze, derive, and understand meanings from the human language in a better and useful way.

Paper [6] talks about an extension of Long Short-Term Memory (LSTM) model. In particular, semantic information extracted from the image is added as extra input to each unit of the LSTM block. A Recurrent Neural Network (RNN) is a good choice to model temporal dynamics in sequences. Due to the issue of vanishing and exploding gradient, it is difficult for RNN to learn long-term dynamics.

Paper [7] depicts the mapping between the images and their corresponding description which is bidirectional. Recurrent neural network is then used to produce a visual representation of those images. The proposed system is capable of generating captions when an image is given as input as well as reconstructing visual features given in an image description.

Paper [8] uses the idea by which the humans perform task of recognizing visual sequences such as recognizing individual object, and then adding that recognized object to our internal representation of the sequence.

3 Motivation

In this era, where majority of the applications benefit the world of physically fit, there is a need to develop an application that would be a guide for the blind. Usually, they require the assistance of another individual to guide them which may create problems if the person is not trustworthy. There can be scenarios where the blind may be easily fooled. Considering all these issues into account, we decided to develop an application which favors the visually impaired. In this modern world, where technology is on its full swing in every field, there is a need for the blind to benefit from them as well. This application aims at providing the blind with a better idea of the things around them. Some of the few things that are been currently used to help visually impaired people to bear with their impairment and move on with their lives are braille, reading glasses, or a walking stick are just.

4 Proposed Method

Describing an image using properly formed sentences is a challenging task, but if successful will have a great impact. And, in this eon of technology demands a solution for this. Our project comes up with an android application that will lend a helping hand to the visually impaired people to understand their surroundings better. This application generates sentences for the images that have been captured by smartphone of the person in need, and speak out the caption being generated. Initially, the user captures the image using this application. After the image has been saved, it will be encoded into its feature set. This encoded image is then processed to generate

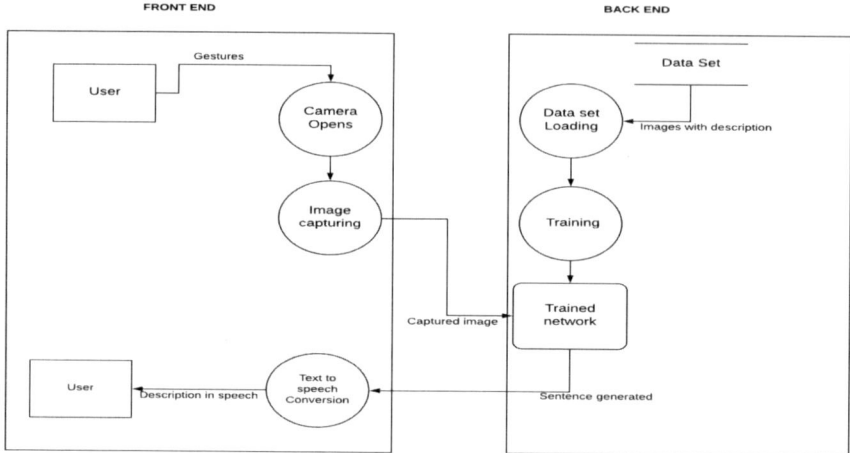

Fig. 1 System architecture

captions. The sentence so obtained is then converted to speech in order to favor the visually impaired. Therefore, the proposed system is as follows:

1. Network modeling:
   a. CNN encoding.
   b. RNN decoding.
2. Network Training:
   a. Dataset collection.
   b. Training the network.
3. Image capturing.
4. Generating captions using trained network.

Overall architecture of the proposed system (Fig. 1).

1. Network modeling

Network used is a pretrained Convolutional Neural Network (CNN)for image prepro-cessing. The image will be converted to a fixed vector representation. Then, Recurrent Neural Network (RNN) is used for generating image description. Specifically, Long Short-Term Memory (LSTM) network is used for framing description of the image. The core of the model includes an LSTM cell that processes one word at a time and which computes probabilities of the possible values for the next word in the sentence.

a. CNN encoding

In this CNN is used as an image encoder which will encode an image to its corre-sponding feature set for which it uses a pretrained. VGG16 model network with its pretrained weight is used as image encoder. Figure 2 shows architecture of VGG16 model. VGG neural network is a convolutional neural network for image classification. If an image is given as input to VGG16, it will find object name contained. It has 16 layers including convolutional layers, max-pooling layers, fully connected layers with 4096 nodes, and output layer with softmax as activation function with 1000 nodes. In this implementation, last softmax layer of VGG16 is removed and the vector of dimension (4096) is obtained from the second last layer. Figure 2 shows layers of VGG16.

b. RNN decoding

Generation of caption is achieved using an LSTM network. At the current time step, the network takes the image vector along with the partial captions and generates the word that is having the highest probability as output.

$$\theta^* = \text{argmax}_\theta \sum_{(I,S)} \log \ p(S \lor I; \theta) \tag{1}$$

Equation (1) is used to maximize the probability of generating correct description, where θ represents the parameters of the model, I represent image, and S represent sentence.

$$p(S|I) = \sum_{i=0}^{N} \log \ p \log_t(I, S_0, \ldots, S_{t-1}) \tag{2}$$

To join the probability of N sentences $S_0, S_1, \ldots, S_N$ chain rule is applied to the model as per Eq. (2).

2. *Network Training*

a. Dataset Collection

For the training and testing of network, Flickr8k dataset has been used. Flickr8k consists of 8000 images out of which 6000 are used for training, 1000 are used for testing, and remaining 1000 are used for validation.

b. Training the network

CNN and RNN are used for generating caption. For this, the network is trained using the downloaded dataset. To speed up training, each image was pre-encoded to its corresponding feature set. One hot encoding of the words is not preferred as each caption may consist of large number of unique words. Instead, an embedding model is trained that takes a word and outputs an embedding vector of dimension

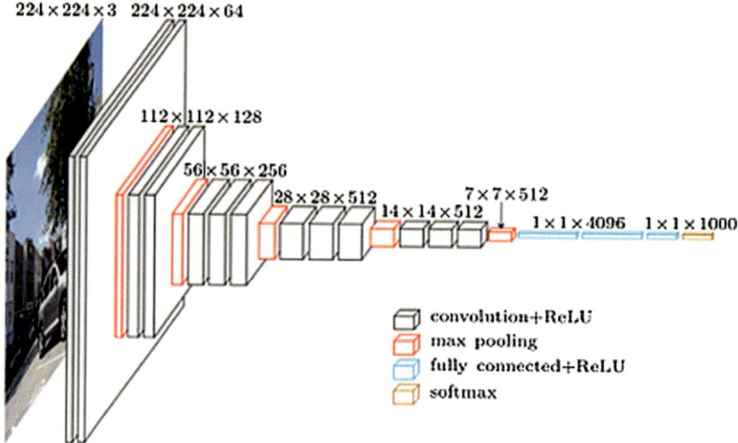

224×224×3 224×224×64

112×112×128

56×56×256

28×28×512

14×14×512

7×7×512

1×1×4096 1×1×1000

convolution+ReLU
max pooling
fully connected+ReLU
softmax

Fig. 2 VGG16 layers

(1, 128). LSTM-based network is used for generating caption. During this training phase, network is being taught how to generate caption for the image by analyzing the dataset provided. These networks were trained with 50 epochs. Once the training of the network is completed, a weight model file is created. This file will contain the learned weights of network during training. Test image which is in vector format is then given as input to the weight model for formation of caption.

3. *Image capturing*

This module is about activating the application using a gesture. Application is activated by shaking the mobile phone, as a result of which camera activity will be invoked and the user can capture the image. The images captured by the user will be sent to the server which will be stored in a file for processing. Then, captions will be generated for the same.

4. *Generating captions using trained network and converting to speech*

Caption is generated for the saved image using trained network. The image from mobile is given as input to trained network which will generate caption for the same which is saved to a file. Then, this caption is sent back to mobile phone. The caption so obtained is then converted to speech in order to favor the visually impaired.

5 Result

Our project aims to generate captions for the image given as input. For performing this, plenty of image data is required. We have access to various image datasets such as Flickr30k, MS COCO, SBU, Pascal, Flickr8k, etc. This model is trained with

Fig. 3 Input 1

Fig. 4 Input 2

Flickr8K dataset. We obtained a training accuracy of 78%. The model generates three captions for each image from which the caption with highest probability is considered.

Following figures were given as input:

The captions generated for Fig. 3 are as follows:

1. A person rows a boat over a large body of water.
2. A person riding a surfboard on a body of water.
3. A person riding on a surfboard in the water.

The captions generated for Fig. 4 are as follows:

1. A cat sits alone in dry grass.
2. Couple of kittens is standing in a field.
3. A group of chicken is standing in a field.

6 Future Work

The proposed system focuses on producing caption for images. This would work more efficiently if extended to generate the caption from a video. Some of the issues like blurred images, out of focus, etc. associated with the captured images can be resolved on using video as the input. A faster and an efficient network will help to overcome the delay required in generating captions for the image being given as input to the server.

7 Conclusion

This paper represents a developed user-friendly android application that will certainly lend a helping hand for the visionless. Most of the applications are developed to benefit the majority, i.e., the physically fit. But, there is also a need to develop an application that will make the life of the minority much easier. Usage of mobile phone applications is increasing day by day. And, nowadays there are technologies which have the potential to significantly emend the lives of visually impaired ones, which will provide them a better knowledge of the surroundings for the blind. The main inspiration for our work is the recent advancement in machine translation. AUGEN, an android application which generates meaningful sentence for the image that has been captured by the smartphone of the person in need and speak out the sentence formed to favor the visually impaired. The possibilities of using technology for making the life better each day are never ending and we are just beginning to scratch the surface.

References

1. Vinyals O, Toshev A, Bengio S, Erhan D (2015) Show and tell: a neural image caption generation. In: Computer vision and pattern recognition (CVPR), IEEE conference
2. Xu K, Ba JL, Kiros R, Cho K, Courville A, Salakhutdinov R, Zemel RS, Bengio Y (2015) Show, attend and tell: neural image caption generation with visual attention. In: International conference on machine learning, pp 2048–2057
3. Mesnil G, Dauphin Y, Yao K, Bengio Y, Deng L, Hakkani-Tur D, He X, Heck L, Tur G, Yu D, Zweig G (2015) Using recurrent neural networks for slot filling in spoken language understanding. IEEE/ACM Trans Audio Speech Lang Process 23(3):530–539
4. Shi B, Bai X, Yao C (2016) An end-to-end trainable neural network for image-based sequence recognition and its application to scene text recognition. IEEE Trans Pattern Anal Mach Intell 39:2298–2304
5. Sarikaya R, Hinton GE, Deoras A (2014) Application of deep belief networks for natural language understanding. IEEE/ACM Trans Audio Speech Lang Process 22(4):778–784
6. Leuven XJKU, Gavves E, Fernando B, Tuytelaars T (2015) Guiding the long-short term memory model for image caption generation. In: The IEEE international conference on computer vision (ICCV), pp 2407–2415

7. Chen X, Lawrence Zitnick C (2015) MindsEye: a recurrent visual representation for image caption generation. In: The IEEE conference on computer vision and pattern recognition (CVPR), pp 2422–2431
8. Ba JL, Mnih V, Kavukcuoglu K-R (2014) Multiple object recognition with visual attention. arXiv: 1412.7755 [cs.LG]

A Novel Flight Controller Interface for Vision-Guided Autonomous Drone

R. Senthilnathan, Niket Ahuja, G. Vyomkesh Bhargav, Devansh Ahuja and Adish Bagi

Abstract Aerial vehicles are rapidly exploring places in a variety of applications in the service sector such as transport assistance and other logistics involved sectors. Most applications require local perception for the aerial vehicles wherein vision system is the most powerful information. Such local perception may be designed based on landmarks on the ground below. This paper details the work where a computer vision system is aiding in landmark identification, namely, a line strip on the ground below. The vision system is implemented in a real-time controller, and a novel and simple interface solution is presented to interface the vision controller to the flight controller. The aerial vehicle under consideration is quadrotor type. The paper presents the details of the image processing algorithm along with the hardware details of the flight control interface. The vision system is developed to identify line strips which share sufficient contrast with the background under daylight condition. The image analysis is performed to continuously extract the lateral position and yaw error. Such an interface would serve as a computationally cheap and easy solution over traditional programmable flight control interface solutions such as MAVLINK.

1 Introduction

Drones are aerial vehicles which are capable of performing complex and hazardous operations, acquire data from the ground, process the data and control the flight of the drone over the desired course [1, 2]. The conventional drones are manually controlled via a remote control. An autonomous drone is the one which is controlled by a controller on-board. The flight of an autonomous drone is secured without any human intervention. The autonomy of the drone is supported by real-time image or video processing, obstacle avoidance, line following, etc., applications. The raw data is then processed and fed to the on-board flight controller. The stability of the drone in motion is generally established through the calibrated flight controller.

R. Senthilnathan (✉) · N. Ahuja · G. Vyomkesh Bhargav · D. Ahuja · A. Bagi
SRM Institute of Science and Technology, Kattankulathur, India
e-mail: senthilnathan.r@ktr.srmuniv.ac.in

© Springer Nature Switzerland AG 2019
D. Pandian et al. (eds.), *Proceedings of the International Conference on ISMAC in Computational Vision and Bio-Engineering 2018 (ISMAC-CVB)*, Lecture Notes in Computational Vision and Biomechanics 30,
https://doi.org/10.1007/978-3-030-00665-5_125

Generally, accurate calibrations such as accelerometer, speed controllers and motor testing need to be performed. Manoeuvrability in confined spaces is better achieved with the application of autonomous drones over the conventional drones which are remotely controlled. In the work presented in this paper, a quadrotor drone is utilized. Quadrotor aerial vehicle is a type of drone that consists of four rotors and two pairs of counter-rotating, fixed-pitch blades located at four corners of the body [3]. The design is simple in its mechanical structure and hence the dynamics. Furthermore, the use of four rotors allows each individual rotor to have a smaller diameter. In this way, the damage that may be caused by the rotors in case of undesirable situations is reduced [3].

Recent work reported in the literature have reported many ways of enhancement of visual perception, visual processing, redundant systems for landing and manoeuvring in structured and unstructured environments. Different algorithms have been proposed and developed to address current challenges [3–7]. A quadrotor which uses on-board processing for localization and mapping with stereo vision to manoeuvre it in indoor and outdoor GPS denied environment is presented in [5]. Although interstitial/GPS approach was used for the quadcopter to perceive the environment, it suffers from two drawbacks: necessity to receive GPS signals and lack of precision in position [7].

Vision-based aerial vehicles provide complete information of the environment in which they are located with the ability to live stream the information of their surroundings, and in the recent past, this has helped the user to manoeuvre the aerial vehicle as per their need, operated from a remote location. Vision sensor could also be used to self-estimate the attitude and altitude in correspondence with other inertial measurements [8]. The pseudo roll and pitch can be decided from the on-board video or image streams [9]. Experiments on only vision-based navigation and obstacle avoidance have been achieved on small rotary wing aerial vehicles [10]. The vision system itself faces a lot of challenges which can be due to the camera, surrounding environment or to a certain extent the aerial vehicle itself due to its body dynamics. Previous work [11] by the author involving vision guidance for drone docking involves a vision in the loop feedback apart from the motion control system realized through the flight controller. A similar approach is adopted for the current work where a cascaded control approach between the vision controller and the flight controller. The major savings is the computational cost involved in the software generation of the pulse position modulation (PPM) signals which are required for the flight controller. The current work is based on a state-of-the-art hardware PPM converter and high-speed update of PWM generation in a dedicated controller.

2 Drone Specifications

The quadrotor used for the experiment is a controlled in plus-configuration. It is controlled through a primary flight controller, namely, Pixhawk PX4. The various specifications of the components of the flight controller are listed in Table 1. The

Table 1 Specifications of the quadcopter

Specifications	Parameter
Frame	F450 Nylon fibre frame
Flight controller	Pixhawk PX4
Motors	2 × CW motors, 2 × CCW motors
ESC	30 A
Vision controller	Nvidia Jetson TK1
Camera	3 mm focal length lens
Battery	3000 mAh Li–Po
Propellers	2 × CW, 2 × CCW
Payload	1.2 kg
KV rating of BLDC motor	1000 kV
No load current	10 V: 0.5 A
Current capacity of battery	12 A/60 s
No of cells	2–3 Li–Po
Motor size	27.5 mm × 30 mm
Shaft diameter	3.175 mm
Minimum ESC specification	18 A
Type of propeller	1045 propeller
Diameter	10 in.
Pitch	4.5 in.
Propeller diameter	254 mm
Centre bore diameter	6 mm front and 9 mm reverse side
Centre seat TH	6 mm
Voltage of the motor	11.1 V
Max continuous current	30 C(90.0 A)
Input voltage	DC 6–16.8 V
Running current	30 A
Size of the motor	26 mm × 23 mm × 11 mm (L × W × H)

photograph of the fabricated quadcopter on a spherical gimbal is shown in Fig. 1. The vision algorithm is executed in a state-of-the-art embedded controller, namely, Jetson TK1 from Nvidia Corporation. The controller has an integrated Tegra processor consisting of a quad-core ARM CPU and 192 core Kepler GPU. The controller runs Linux Ubuntu version called Linux for Tegra custom tailored to suit the demands of embedded applications. The controller has GPIO which operates with logical high at 1.8 V which in some sense is a drawback since most electronics used in motion control operates on 5 V TTL logic. This requires a level translation from 1.8 to 5 V and vice versa. The proposed drone control strategy involves a standard flight controller since the robustness of commercially available flight controllers is exceptionally good. The key specifications are listed in Table 1.

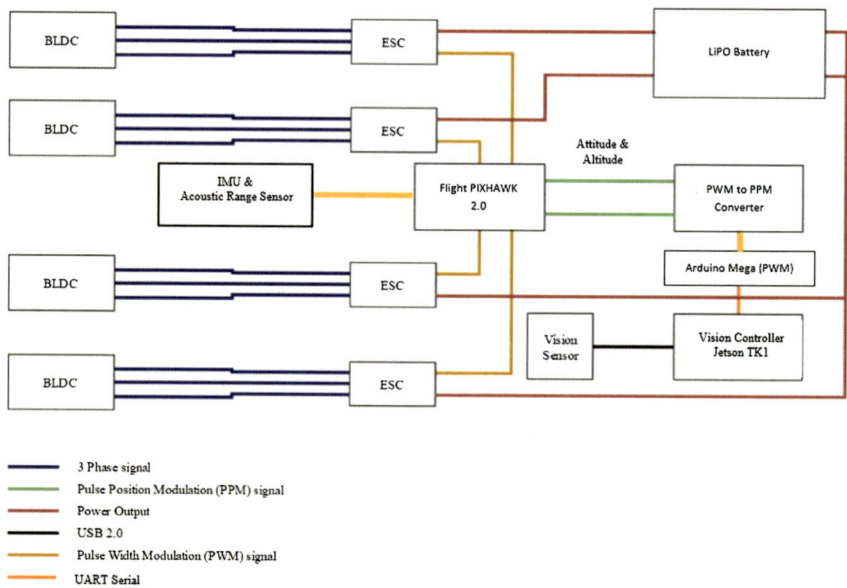

Fig. 1 Hardware Interface

In order to interface the flight controller to the vision controller, two main solutions are available. The first solution is through the MAVLINK protocol of serial commands to send and receive data from the flight controller. The second method which is adopted in the current work is to hack into the RC receiving channels of the flight controller and mimic the PPM signals for various attitude and altitude controls as it would be applied from the remote radio controller. In order to accomplish the latter method, PPM signals corresponding to pitch, roll, yaw and throttle must be generated by the vision controller which causes the flight controller to generate suitable PWM signals which get applied to the electronic speed controllers. Figure 1 illustrates the hardware interface strategy adopted in the current work.

3 Imaging Hardware

The current work uses equipment that is necessary for successful implementation of vision guidance system which is described as follows.

3.1 Scene Description

The current work is proposed for indoor environment with an area of approximately 4000 ft^2 and a height of approximately 50 ft from ground to roof. The line to be followed by the vehicle is made using a light coloured masking tape of sufficient width stuck on the ground of dark complexion. The closed enclosure ensures no wind turbulence being offered to the vehicle. It also ensures ambient daylight to enter the enclosure.

3.2 Camera

The camera helps to broadcast live image to the system. The work presented in the paper demonstrates pre-acquired images from the camera which acts as the input data to the simulated model of the quadrotor vehicle. The images are processed in order to obtain the information necessary to decide the movement of the vehicle. The camera is mounted underneath the quadcopter through a mounting, and hence, image acquisition mimics the similar acquisition pose for the camera. The camera selected is of 1.3 megapixel. The camera consists of an Aptina sensor with USB 2.0 interface. It also has a feature of day and night low light vision for environmental monitoring camera system. It has 2.1 mm wide angle lens for all kinds of camera control system. It weighs around 90 g.

3.3 Camera Modelling and Image Acquisition

The camera modelling process relates the metric real-world information to information in image space. For the current work, a scaled orthographic projection type of camera model would suit the purpose since the line strip is thoroughly planar. If a point $P(X, Y, Z)$ on the real world with reference to the camera is mapped to the point $I(x, y)$ on the image plane then the relation as per a scale orthographic projection may be defined as

$$x = s.X \; \& \; y = s.Y$$

where s is the scaling factor which is a function of the standoff distance, Z. The image acquisition adopted for acquiring the strip image sequence is asynchronous in nature. In this method, the image is first completely captured whenever a start command is read by the system. This acquisition of image takes place for a fixed time after which the capturing operation stops.

4   Vision Algorithm

The basic objective of the vision system is to obtain the error along the horizontal axis (controlled by roll motion) of the image measured from the image centre reference to the strip centre and ensure that the axis of the strip in the field of view of the camera is along the vertical axis of the image (controlled by pitch motion).

4.1   Image Processing

The current work uses on-board vision system to acquire the image of the scene which is processed by the on-board vision controller. The image captured by the camera is subjected to the vision algorithm which extracts line features from it. The first step is to convert the RGB colour image to grayscale image (Fig. 2).

The grayscale image is converted to binary image based on a predetermined threshold. The threshold intensity will make sure that the pixels having intensity less than threshold will be converted to black and the pixels having intensity greater than the threshold will be converted to white pixels. The threshold value depends upon the illumination of the scene in consideration. This makes the strip to be completely white on a black background which makes the extraction of line features conveniently. Random noise in the binarization is eliminated using area granulometry since the strip is assumed to be the largest region in the binary image. Inconsistencies in colour of the strip result in presence of holes which are suitably eliminated using morphological closing (Fig. 3).

After computing the aspect ratio and determining the orientation of the strip, its width in pixels is obtained and its mean is computed, and a skeletonized strip which is of one-pixel width is obtained.

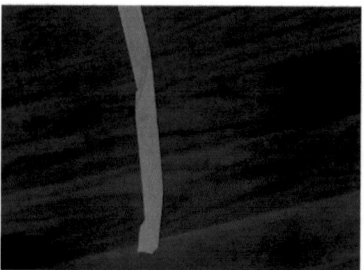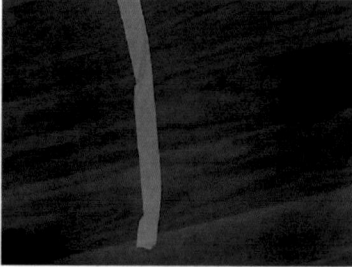

Fig. 2 RGB colour image and the corresponding monochrome version

Fig. 3 Binary image before and after noise removal and skeletonized version

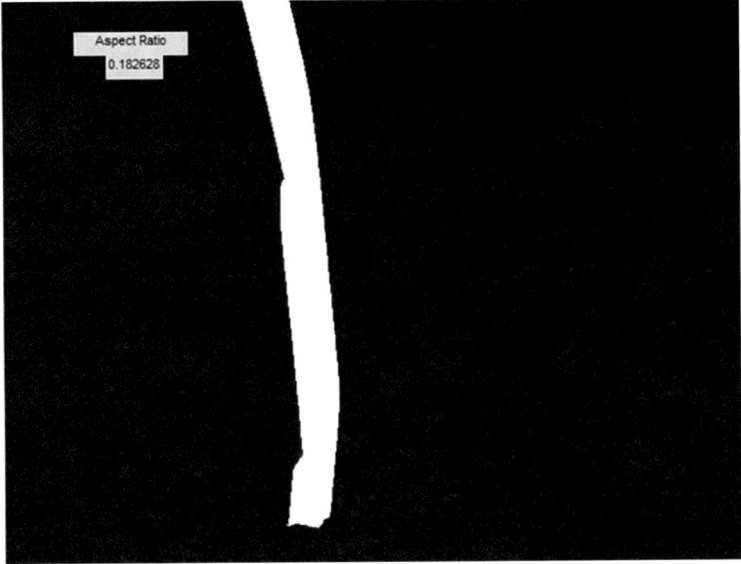

Fig. 4 Aspect ratio of the strip

4.2 Image Analysis

The vision algorithm takes input as image and carries out various operations on it to give numerical data as output at necessary stages of operation. The aspect ratio of the strip is obtained to determine the orientation of the strip in the image frame. Figure 4 illustrates the computation of the aspect ratio.

A skeleton line is obtained by processing the image. The length of this line is calculated via image analysis. The measurement of the skeletal line depends on any arbitrary orientation of the line, in the image frame captured by the camera. Either, a combination of row and column coordinates or just the row and column coordinates are taken into consideration to calculate the length of the line. After getting the total length, the centre of this line is computed by taking the mean of the corresponding coordinates. The centre of the image is obtained by taking the mean of the row and column coordinates of the complete image. A window is generated enclosing the processed image. A quarter length of the line is taken on either side from its

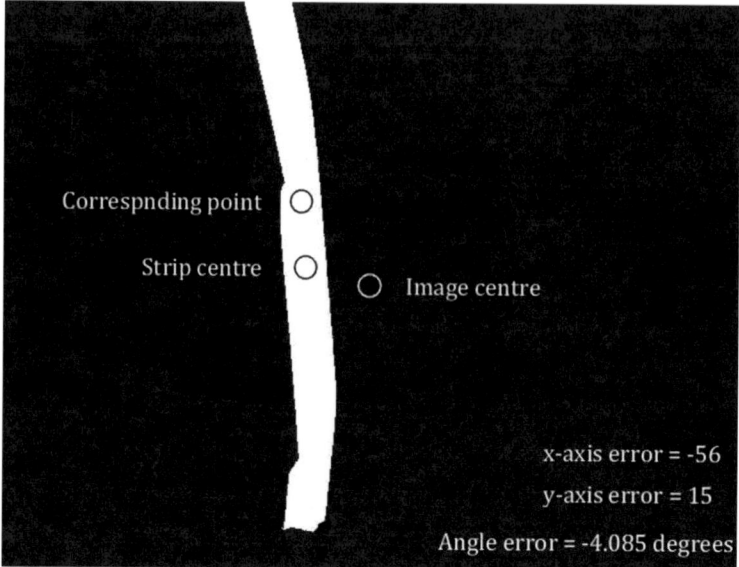

Fig. 5 Corresponding point and other results

centre, to create the window. The window intersects the line at two points. The point lying in-line with the heading direction of the quadrotor aerial vehicle is taken into consideration. The error angle between the line joining the centre point of the line and the corresponding point in the present frame and the previous frame is computed. The error between the line centre and the image centre is computed. These two errors are corrected by the flight controller to get the desired motion of the line following vehicle (Fig. 5).

5 Vision Guidance

Basically, three errors are desired to be corrected based on the information obtained from the vision system, namely, x, y, heading direction. Let dx and dy give the difference in the centroid positions which is calculated between the image centre and the strip centroid. Line following of quadrotor refers to controlling of position $P_E(x, y)$ and angular position, i.e. pitch and roll. The initial condition of the quadrotor is such that the strip is in the field of view of vision sensor. The angular positions, pitch(ϕ) and roll(θ) are controlled by the outer loop controlling positional error through vision. The altitude of the quadrotor is maintained through a separate control loop for maintaining the range from the ground below is considered with a range sensor in the loop. The quadrotor starts after ensuring the initial marker condition which in the current work is

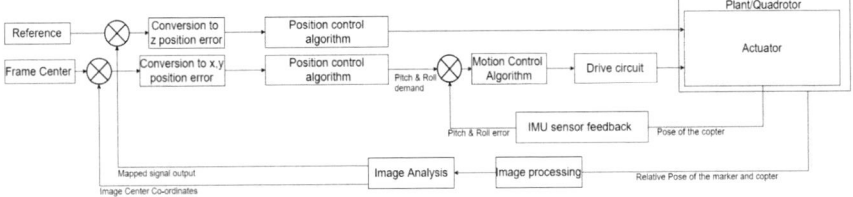

Fig. 6 Vision in the loop control block diagram

a discontinuous line strip. The same fiducial is also applicable to the ending condition. The block diagram in Fig. 6 illustrates the quadrotor control strategy adopted.

5.1 X-Y Position Control

The position control loop depends on the instantaneous position error $P_E(x, y)$ which is calculated by keeping the vision in loop so that the error signal is generated in the form of pixels. The control law is completely planned in image space. The roll and pitch angles are generated for any given error in x and y axes of the image. The position controller generates output that is further fed to control the pitch and roll of vehicle.

5.2 Attitude Control

Motion control loop is the inner loop of the system which is responsible for controlling the pitch and roll of quadrotor. As per Fig. 6, the pose-related information is given as the feed to the loop through inertial motion unit (IMU). The pitch and roll signal demand signals generated through the outer X, Y position control loop is compared to the feed signals from the IMU. The generated error signal is fed to the motion controller. The output of motion controller is then converted to motor signals through drive circuit and is used for actuation of rotors in the system.

5.3 Altitude Control

Z position control loop is required to maintain the altitude of the quadcopter. The stable altitude would ensure no magnification changes during the manoeuvring of the quadcopter. The flight controller has a default altitude measurement device which is based on atmospheric pressure. This sensor performs poorly in terms of the resolution indoor conditions; a redundant acoustic ranging sensor compatible with flight controller is used. The sensor is suitably configured in the flight controller planning software to act as replacement to the barometric sensor.

Thus, the paper presents a novel interface solution for vision-guided aerial vehicles which can be very simple in terms of the construction and principle and yet can be an effective alternative to complex but the current benchmark for flight controller communication, namely, the MAVLINK protocol.

References

1. Puri A (2005) A survey of unmanned aerial vehicles (UAV) for traffic surveillance. Department of Computer Science and Engineering, University of South Florida
2. How Jonathan P (2008) Real-time indoor autonomous vehicle test environment. Control Syst IEEE 28(2):51–64
3. Bi Y, Duan H (2013) Implementation of autonomous visual tracking and landing for a low-cost quadrotor. Optik-Int J Light Electron Opt 124(18):3296–3300
4. Meier L, Tanskanen P, Heng L, Lee GH, Fraundorfer F, Pollefeys M (2011) PIXHAWK: a system for autonomous flight using onboard computer vision. In: 2011 IEEE international conference on Robotics and automation (ICRA). IEEE
5. Chae H, Park J, Song H, Kim Y, Jeong H (2015) The IoT based automate landing system of a drone for the round-the-clock surveillance solution. In: IEEE international conference on advanced intelligent mechatronics (AIM). IEEE
6. Pattersona T, McCleana S, Morrowa P, Parra G, Luob C (2014) Timely autonomous identification of UAV safe landing zones. Image Vis Comput 32(9):568–578
7. Blösch M (2010) Vision based MAV navigation in unknown and unstructured environments. In: 2010 IEEE international conference on robotics and automation (ICRA). IEEE
8. Roberts PJ, Walker RA, O'Shea P (2005) Fixed wing UAV navigation and control through integrated GNSS and vision. In: AIAA guidance, navigation, and control conference and exhibit
9. Dusha D, Boles WW, Walker R (2007) Fixed-wing attitude estimation using computer vision based horizon detection, pp 1–19
10. Calise AJ (2003) Applications of adaptive neural-network control to unmanned aerial vehicles. In: AIAA/ICAS international air and space symposium and exposition: the next, vol 100
11. Senthilnathan R, Nikhil S, Sasikala R, Maruvada A (2017) Model based design of vision guided vertical landing system for quadrotor. Int J Control Theor Appl 10(9):803–819

Sliding Discrete Fourier Transform for 2D Signal Processing

Anita Kuchan, D. J. Tuptewar, Sayed Shoaib Anwar
and Sachin P. Bandewar

Abstract Discrete Fourier Transform (DFT) is the most frequently used method to determine the frequency contents of the digital signals. As DFT will take more time to implement, this paper gives the algorithm for the fast implementation of the DFT on the Two-Dimensional (2D) sliding windows. To fast implement DFT on the 2D sliding window, a 2D DFT (here 2D SDFT) algorithm is stated. The algorithm of the proposed 2D SDFT tries to compute current window's DFT bins directly. It makes use of precalculated bins of earlier window. For a 2D input signal, sliding transform is being accelerated with the help of the proposed algorithm. The computational requirement of the said algorithm is found to be lowest among the existing ones. The output of discrete Fourier transform and sliding discrete Fourier transform algorithm at all pixel positions is observed to be mathematically equivalent

1 Introduction

In several applications of image processing, frequency domain offers an improvement over performing a similar job in the time domain. At times the improvement is just simpler or more hypothetical algorithm. Often the largest obscurity in working in frequency domain is concern with calculation of Fast Fourier Transform [1]. If the frequency-domain data must be reorganized constantly in a real-time application, the difficulty and latency of the FFT can become a significant obstruction to achieve system goals and keeping cost and power consumption low.

A. Kuchan (✉) · D. J. Tuptewar · S. S. Anwar · S. P. Bandewar
Mahatma Gandhi Missions College of Engineering, Nanded, India
e-mail: anitakuchan@gmail.com

D. J. Tuptewar
e-mail: tuptewar_dj@mgmcen.ac.in

S. S. Anwar
e-mail: sayed_shoaib@mgmcen.ac.in

S. P. Bandewar
e-mail: bandewar_sp@mgmcen.ac.in

© Springer Nature Switzerland AG 2019
D. Pandian et al. (eds.), *Proceedings of the International Conference on ISMAC in Computational Vision and Bio-Engineering 2018 (ISMAC-CVB)*, Lecture Notes in Computational Vision and Biomechanics 30,
https://doi.org/10.1007/978-3-030-00665-5_126

For detection and processing signals, there are a number of real-time applications, such as touchscreen sensing, medical imaging and communication systems, and radar uses frequency-domain algorithms. There are many such applications where complexity should be low while reducing the latency. Sliding DFT method [2–4] gives frequency-domain information on the sample by sample basis with considerably less computations than the FFT for each update [5]. Various transform codings constitute an important module of existing image processing applications. In the sliding transform, the window is shifted by one sample ahead at a time; this procedure is repeated throughout the image. There are various algorithms like DFT for moving window [6], Walsh Hadamard Transform (WHT) [7], Running Walsh Hadamard Transform (RWHT) [8], Gray-Code Filter Kernels (GCK) [9], WHT on sliding windows [10, 11], Hopping DFT (HDFT) [12] and Sliding WHT (SWHT) [13] which works on sliding windows. For the implementation of sliding DFT, the discrete orthogonal transform becomes more complex. Various algorithms are developed in the last couple of years for the rapid calculations of the sliding orthogonal transforms. Here, sliding DFT is introduced.

The rest of the paper includes the necessary background and the key features of the proposed scheme. Also, the performance evaluation experiments are presented in the subsequent phases of the script. Remaining part presents the results and finally the conclusion.

2 Proposed Method

Considering a signal $y(n)$, which is represented as below:

$$\ldots, y(-2), y(-1), y(0), y(1), y(2), \ldots y(n), y(n+1), y(n+2) \ldots.$$

In various applications of signal processing, one might desire to study the signal with the help of sliding window of fixed size. Assuming at the time instant n, a windowed series $y(n)$ containing N number of samples is given by Eq. (1)

$$y(n) = \{y(n - N + 1), y(n - N + 2), \ldots, y(n - 2), y(n - 1), y(n)\} \quad (1)$$

after taking DFT of given sequence $y(n)$, we get $Y(n)$ which is represented as

$$Y(n) = \{Y_0, Y_1, Y_2, Y_3 \ldots Y_{N-3}, Y_{N-2}, Y_{N-1}\} \quad (2)$$

That means, to compute DFT sequence $Y(n)$, we have to perform DFT at each and every time instant n. If it is needed, then one may use FFT which is fast and computationally efficient. But this is not the scenario always, in some applications, the requirement is of only kth value of the DFT, consider Y_{k-1}, then Fast Fourier Transform (FFT) is not computationally efficient. Here with the help of FFT, we are considering complete DFT sequence and to ignore redundant values, is there any solution? The answer for this is, there we have a computationally efficient algorithm, which is known as Goertzel algorithm. This Goertzel algorithm [14] helps to get one-off Y. Similarly, when we are taking into consideration a sliding window, Sliding DFT (SDFT) is most efficient. The concept of SDFT introduced from the two succeeding time instants [11]. Consider two succeeding time instants as $n - 1$, n and $y(n - 1)$, $y(n)$ as windowed sequences which contains basically equal elements.

For example, let the window length of $N = 8$, and then at $n = 7$, the sequence is given as

$$y(7) = \{y(0), y(1), \ldots, y(6), y(7)\} \tag{3}$$

$Y(7)$ is calculated with the help of 8-DFT. And at $n = 8$, we get

$$y(8) = \{y(1), y(2), \ldots, y(7), y(8)\} \tag{4}$$

here we observe remarkable similarities between $y(n - 1)$ and $y(n)$ in the SDFT for efficient calculations.

Prior to introducing the Sliding DFT, we memorize one of the basic important properties of DFT which is circular shift property. If N-DFT of a sequence $y[n]$ is given by $Y[k]$, then we can write as

$$y[((n - m))N] \leftrightarrow W_N^{km} Y[k] \tag{5}$$

where $W_N = e^{\frac{-j2\pi}{N}}$; thus, the above equation gives circularly shifted series of $y[n]$.

Specifically, if we use circular shift property, then the sequence is circularly moved by one time instant (in the left direction), and then, the DFT value Y_k will be

$$Y_k \rightarrow Y_k e^{j2\pi k/N} \tag{6}$$

apart from importance of N, one can get $y(n)$ from $y(n-1)$ with the help of following steps:

1. Restore the foremost element of sequence $y(n - 1)$ with the very last element of $y(n)$, e.g. initiate with (3) and exchange y_0 by y_8.
2. Apply circular shift property (which will shift sequence to the left).

These two steps show the way to easy things in the DFT domain, which is represented as

$$Y_k = \sum_{n=0}^{N-1} y(n) W_N^{nk} = y(0) + y(1) W_N^k + \cdots + y(N - 1) W_N^{(N-1)k} \tag{7}$$

From Eq. (7), one can examine that the first component $y(0)$ is remained unchanged in the DFT formula, and now, apply step1 to have an in-between sequence

Fig. 1 Window of size 8 × 8

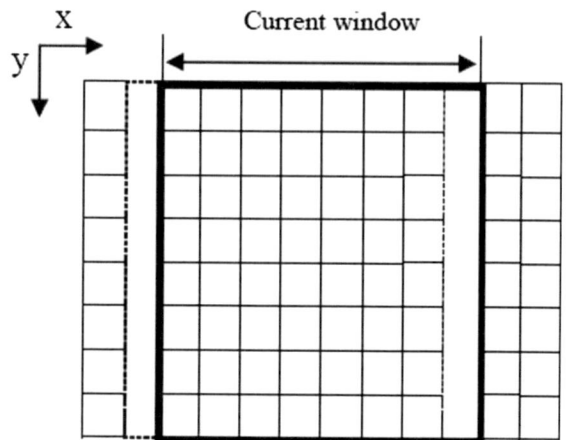

$$\widehat{y(7)} = \{y(8), y(1), \ldots, y(6), y(7)\} \tag{8}$$

we can write from (7)

$$\widehat{Y(7)} = Y_k(7) - y(0) + y(8) \tag{9}$$

now the DFT of $y(8)$ is denoted by $Y_k(8)$ then by applying property we get

$$Y_k(8) = \widehat{Y_k(7)}\, \mathrm{e}^{j2\pi k/N} \tag{10}$$

Hence, the relation between DFTs of two consecutive windowed sequences $y(n-1)$ and $y(n)$, having length N, is represented as

$$Y_k(n) = [Y_k(n-1) - y(n-N) + y(n)]\mathrm{e}^{j2\pi k/N} \tag{11}$$

Equation (11) represents the basic concept of Sliding Discrete Fourier Transform (SDFT). Only one complex multiplication and two real additions per output sample are required to implement SDFT. However, because of the recursive nature of SDFT, output must be calculated for each and every new input sample. We are applying this algorithm to 2D signals that is image. Figure 1 shows window of size 8 × 8 graphically.

Considered R_M signifies the numbers of real multiplications and R_A represents real additions. Then, the total analysis of the 2D SDFT method is summarized as

$$R_M = 4(M^2 + M) \tag{12}$$

and

$$R_A = 4M^2 + 3M + 2 \tag{13}$$

Table 1 Computational requirement of DFT, FFT and SDFT

Algorithm	Computational requirement	Window size $M \times M$
DFT	R_M	$8M^3$
	R_A	$8M^3 - 4M^2$
FFT	R_M	$4M^2 \log_2 M$
	R_A	$6M^2 \log_2 M$
2D-SDFT	R_M	$4(M^2 + M)$
	R_A	$4M^2 + 3M + 2$

Table 2 Computational requirement of DFT bins for different window sizes

Algorithm	Operation	Number of computed DFT bins			
		4×4	8×8	12×12	16×16
DFT	R_M	512	4096	13,824	32,768
	R_A	448	3840	13,248	31,744
FFT	R_M	128	768	2065	4096
	R_A	192	1152	3098	6144
2D-SDFT	R_M	80	288	624	1088
	R_A	78	282	614	1074

A computational evaluation is summarized in Table 1 where the computational analysis of 2D SDFT, FFT and DFT is shown. After comparing all these algorithms, it is found that 2D sliding DFT method is having less computational complexity in comparison with available methods. Consequently, for some applications, we suggest the use of the 2D sliding DFT method where the computational complexity is a significant concern.

This 2D SDFT algorithm calculates DFT points separately that is one sample after another. This indicates, whenever one wants to calculate DFT at few pixels, the algorithm has to implement the DFT at all time instants. In many image processing applications, all DFT bins need not to be calculated. We summarize the computational requirements of the algorithms when only some part of the DFT bins are calculated, and it is shown in Table 2.

Table 2 shows that as the number of estimated DFT bins reduces, the computational requirements of all algorithms decrease. From Table 2, we can say that the performance of the SDFT method gives better improvement when the number of DFT bins to be calculated is decreased.

The computational analysis is done for the window sizes 4×4, 8×8, 12×12 and 16×16. Graphically, it is represented in Fig. 2.

Fig. 2 DFT bins of window
size 4 × 4, 8 × 8, 12 × 12
and 16 × 16

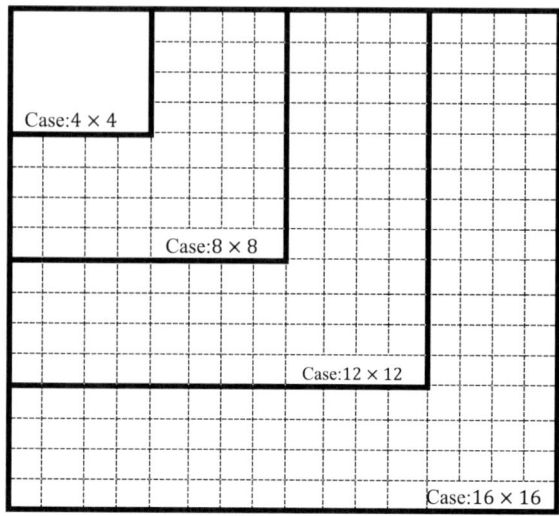

3 Experimental Results

We executed this algorithm using several CIF, 4CIF and QCIF videos with the rate of
30 frames per second. Sliding transform was applied considering the first 100 frames
of each video sequence.

Mathematically if we compare algorithms like DFT, FFT and SDFT, all these
algorithms give the same transform results. The difference is there in computational
analysis. Initially, we compute the time required for every algorithm when DFT bins
are to be calculated. The experimental results observed by giving various CIF and
4CIF sequences as input are presented in Tables 3 and 4, respectively.

Here, we denote $\Delta T\%$ the average processing time reduction of the existing 2D
SDFT as compared to SDFT. This average processing time reduction is calculated
by using the formula given below:

Table 3 Experimental results with comparison for the CIF videos

Sequences	DFT	FFT	Existing 2D SDFT	SDFT	$\Delta T(\%)$
	Time (ms)	Time (ms)	Time (ms)	Time (ms)	
City	534,940.06	69,685.00	19,875.65	14,962.73	24.71829
Soccer	534,823.75	69,685.29	19,973.48	19,401.124	2.86558
Harbour	534,716.63	69,689.14	19,836.59	18,944.464	4.497376
Average	534,826.8133	69,686.47,667	19,895.24	17,769.43,933	10.69375

Table 4 Experimental results with comparison for the 4CIF videos

Sequences	DFT	FFT	2-D SDFT	SDFT	$\Delta T(\%)$
	Time (ms)	Time (ms)	Time (ms)	Time (ms)	
City	2,256,673.25	353,330.72	83,643.72	51,640.475	38.26138
Soccer	2,256,662.25	354,086.63	83,701.13	52,137.789	37.70958
Harbour	2,258,794.18	353,240.59	83,578.79	53,547.582	35.93161
Average	2,257,376.56	353,552.6467	83,641.21333	52,441.94867	37.30086

Table 5 Experimental results for the QCIF videos

Sequences	SDFT
	Time (ms)
City	12,594.692
Soccer	11,767.736
Harbour	12,432.937
Average	12,265.121

$$\Delta T(\%) = \frac{T_{\text{algorithm}} - T_{\text{Proposed}}}{T_{\text{algorithm}}} \times 100$$

where T_{Proposed} indicates the calculated processing time of the SDFT and $T_{\text{algorithm}}$ indicates the calculated time of 2D SDFT.

The SDFT algorithm may get considerable time reduction in comparison with DFT, FFT and 2D SDFT algorithms. The various CIF video sequences performance comparison is given in Table 3. For all the video sequences, processing time of the SDFT algorithm is 10.69% less than the 2D SDFT. The experimental results show that this SDFT algorithm always gives outperforming results than the other algorithms.

We have determined the processing time by giving the 4CIF video as input, and the experimental results are shown in Table 4. This shows the SDFT method outperforms by 37.30% in comparison with the 2D SDFT algorithm.

We have also determined the processing time by giving the QCIF video as input and the experimental results are shown in Table 5.

4 Conclusion

After comparing discrete Fourier transform and fast Fourier transform algorithms, it is found that this method gives better results in higher resolution videos. The 2D sliding DFT method cannot be functional on the top or left blocks of image because of the requirement to access the foregoing pixel arrangement. As we increase the resolution of the image ratio of blocks on image borders will reduce. We can expect that this algorithm will perform better for high-resolution videos.

References

1. Farhang-Boroujeny B, Gazor S (1994) Generalized sliding FFT and its application to implementation of block LMS adaptive filters. IEEE Trans Sig Process 42(3):532–538
2. Park C-S (2015) 2D Discrete Fourier transform on sliding windows. IEEE Trans Image Process 24(3):901–907
3. Jacobsen E, Lyons R (2003) The sliding DFT. IEEE Sig Process Mag 20(2):74–80
4. Duda K (2010) Accurate, guaranteed stable, sliding discrete Fourier transform [DSP tips and tricks]. IEEE Sig Process Mag 27(6):124–127
5. Farhang-Boroujeny HB, Lim V (1992) A comment on the computational complexity of sliding FFT. IEEE Trans Circ Syst-II Analog Digital Sig Process 39(12):875–876
6. Rosendo Maias JA, Exposito AG (1998) Efficient moving-window DFT algorithms. IEEE Trans Circ Syst -II: Analog Digital Sig Process 45(2):256–260
7. Deng G, Ling A (1996) A running Walsh-Hadamard transform algorithm and its application to isotropic quadratic filter implementation. In: Proceedings of European signal processing conference (EUSIPCO)
8. Mozafari B, Savoji MH (2007) An efficient recursive algorithm and an explicit formula for calculating update vectors of running Walsh Hadamard transform. In: Proceedings of IEEE 9th international symposium on signal processing and its applications (ISSPA)
9. Ben-Artzi G, Hel-Or H, Hel-Or Y (2007) The gray-code filter kernels. IEEE Trans Pattern Anal Mach Intell 29(3):382393
10. Ouyang W, Cham W-K (2010) Fast algorithm for Walsh Hadamard transform on sliding windows. IEEE Trans Pattern Anal Mach Intell 32(1):165–171
11. Jacobsen E, Lyons R (2004) An update to the sliding DFT. IEEE Signal Process Mag 21(1):110111
12. Park C, Ko S (2014) The hopping discrete Fourier transform [sp tips and tricks]. IEEE Sig Process Mag 31(2):135–139
13. Park C-S (2014) Recursive algorithm for sliding Walsh Hadamard transform. IEEE Trans Sig Process 62(11):28272836
14. Banks K (2002) The Goertzel algorithm. Embed Syst Program Mag 15(9):3442

Automated Lung Nodules and Ground Glass Opacity Nodules Detection and Classification from Computed Tomography Images

Vijayalaxmi Mekali and H. A. Girijamma

Abstract Lung cancer health care community depends on lung cancer Computer Aided Detection system to draw useful lung cancer details from Computed Tomography lung images. Nodules growth rate indicates the severity of the disease, which can be periodically radiologist analyzed by nodule segmentation and classification. Main challenges in analyzing nodules growth rate are lung nodules of different type requires special methods for segmentation, their irregular shape, and boundary. In this paper, automatic three-phase framework for lung nodules and nodules of ground glass opacity detection followed by classification is proposed. In this work, nodule segmentation framework uses proposed automatic region growing algorithm that selects set of black pixels as seed points automatically from output binary image for lung parenchyma segmentation followed by artifacts removal to reduce disease search space. Nodules are segmented based on nodule candidates center pixels identification and intensity feature of lung nodule candidates. Segmented nodules are classified using SVM classifier and classification results are compared with other considered classifiers KNN, boosting and decision tree. In the evaluation step, it was found that SVM classifier's performance is outstanding compared to other considered classifiers in this work. Complete automation in nodule detection within very less time is the key feature of the proposed method. CT images are taken from Lung Image Database Consortium and Image Database Resource Initiative (LIDC/IDRI) public database to evaluate the performance of proposed work. An accuracy of 98% (45/46) with less computational time is achieved. The experimental results demonstrated that the proposed method achieve efficient and accurate segmentation of lung nodules and ground glass opacity nodules with less computation time.

V. Mekali (✉)
Computer Science and Engineering Department, KSIT, Bangalore, India
e-mail: duruth.viju@gmail.com

H. A. Girijamma (✉)
Computer Science and Engineering Department, RNSIT, Bangalore, India
e-mail: girijakasal@gmail.com

© Springer Nature Switzerland AG 2019
D. Pandian et al. (eds.), *Proceedings of the International Conference on ISMAC in Computational Vision and Bio-Engineering 2018 (ISMAC-CVB)*, Lecture Notes in Computational Vision and Biomechanics 30,
https://doi.org/10.1007/978-3-030-00665-5_127

1 Introduction

Lung cancer is carcinoma disease with high death rate compared to other cancers all over the world in human beings. Lung cancer killing rate is increasing day by day, approximately 1.6 millions of people globally per year. The main target for lung cancer is an elderly group of age 65 or older. Lung cancer detection study from various researchers showed that high curable rate is possible only by means of its early stage detection [1–3]. Thus, today also lung cancer detection research takes highest prominence in research and health care community. Two classes of lung carcinoma are Non-Small Cell Lung Cancer (NSCLC) and Small Cell Lung Cancer (SCLC). Most of the detected lung cancer (approximately 80%) is of type NSCLC and it is SCLC in less cases (approximately 20%). Assignment of stages to detect lung cancer is very important to determine its severity. TNM (Tumor, Nodule, Metastasis) staging is internationally accepted staging tool to stage different types of cancers [2].

Among all the medical image modalities, Low Dose Computed Tomography (LDCT) is an effective and proven GOLD STANDARD medical imaging modality that provides useful information about lung cancer by generating huge volume of image slices. With latest advanced technologies LDCT is more reliable and faster to detect very small nodules on lung area [4]. Lung nodules appear as high-intensity circular opacity with the smooth or irregular boundary of diameter 2–30 mm on LDCT [1]. Lung nodules are of different types of well-circumscribed nodules (with smooth boundary), irregular shape nodules, and nodules attached to other parts (juxta-vascular nodules attached to blood vessels and juxta-pleural nodules attached to lung pleura), solid, partly solid, nonsolid nodules, and Ground Glass Opacity (GGO). Presence of GGO on lung for long time indicates cancer. Figure 1 shows different types of nodules in CT images.

In one single scan, LDCT generates a huge amount of pathology's image slices of the area and those huge slices data set complicates radiologist work to perform manual analysis for detection of nodule pathology. Advanced Computer-Aided Detection/Diagnosis (CAD) systems based on various image processing, artificial intelligence, and machine learning algorithms are playing a prominent role in health care

(a1)　　　　　**(a2)**　　　　　**(a3)**　　　　　**(a4)**

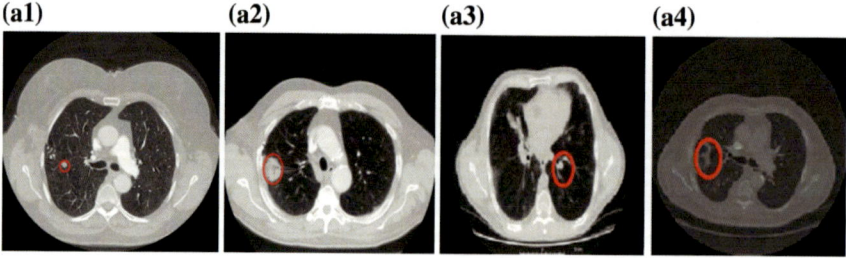

Fig. 1 Different types of lung nodules. **a**1 Circumscribed lung nodules. **a**2 Juxta-pleural nodule. **a**3 Juxta-vascular nodule. **a**4 GGO nodule

community to assist the radiologist in lung nodules detection of lung CT (Computed Tomography) images.

The major steps of lung cancer CAD system are lung parenchyma segmentation, lung nodule detection and classification of nodules as benign or malignant. Lung parenchyma segmentation is a difficult task. The main reasons are lung lobes shape is not same in all patients, high variation in geometric aspect of lung in different patients, existence of juxta-pleural and juxta-vascular nodules, overlap of chest wall, connected lung lobes. Lung region segmentation accuracy determines CAD system's nodule detection accuracy.

The rest of the paper is organized as follows.Section 1 is about lung cancer, LIDC dataset, usage of CAD system in lung cancer detection. Section 2 gives previous related work, Sect. 3 describes proposed methodology, and Sect. 4 discusses results and conclusion.

2   Related Work

Number of researchers has developed CAD systems for automatic detection of lung nodules using CT lung images and some are presented below.

Wang et al. [5] proposed lung nodules segmentation method based on 3D-extended Dynamic Programming (DP). In this work, non-nodule segmentation error rate was decreased by multidirection fusion techniques. First and second dataset set from LIDC were used and obtained true positive fraction of 75% and 71% for first and second dataset, respectively. Dehmeshki et al. [6] proposed method for automated lung nodules detection based on shape-based genetic algorithm. Spherical-oriented convolution-based filter was used in preprocessing step for input image enhancement. Seventy CT scan images were used to evaluate the proposed method, and method achieved about 90% nodule detection rate. In [7], author proposed CAD system based on fuzzy thresholding for parenchyma segmentation. For nodule segmentation and detection author used local Gaussian and mean curvatures based shape indexed map and local intensity dispersion information. Resultant CAD system was evaluated on 108 Clinical CT dataset and 90.2 detection rates were achieved.

Liu et al. [8] proposed CAD system was based on new algorithms for lung parenchyma segmentation and ROI extraction using circle shape descriptor from CT images. In this work, lung nodule candidates were predicted using ADE-Co-Forest algorithm. The resultant CAD system sensitivity was very high with low false positive. Novel CAD system for the detection of nodules attached to blood vessels from CT images was proposed by Tan et al. [9]. The author applied enhancement filters on nodule and vessel parts for nodule segmentation. To separate the attached blood vessels from nodules genetic algorithm and artificial neural network were used. Results are evaluated on CT images from LIDC.

In [10], Messay et al. the author presented a threshold and morphological operation based new fully automated CAD system for nodule detection. From each segmented nodules, 245 features were extracted to compare classifiers Fisher Linear

Discriminant classifier (FLD) and quadratic classifier. Classification results showed that FLD performance was better. Resultant CAD system used CT scans from LIDC and obtained 80.4% sensitivity. Adaptive thresholding based automatic Solitary Pulmonary Nodule (SPN) segmentation algorithm was proposed in [11]. In this work, extraction of nodule candidates method was achieved with compactness feature along with histogram analysis. True SPN from input CT images were classified using SVM classifier algorithm. In [12], the author presented fast lung nodules detection method based on thresholding for lung parenchyma segmentation, blood vessel filling by morphological operations, cylindrical shape filter for nodule enhancement, and SVM for false positive nodules reduction. Campos et al. [13] proposed novel supervised segmentation framework based on three features sets for detection of both solid and GGO lung nodules. The method used region growing algorithm for lung parenchyma segmentation and three features sets to segment lung nodules detection. The author used shape features for first type nodules segmentation, convergence matrix for second type nodules segmentation, and feature set region regression method for third type nodules segmentation. The output of three primary nodule segmentation is input for ANN to obtain accurate segmentation. For GGO nodules test, data set of CT images were taken from Nelson's trail database and CT images with solid nodules from hospital Sao Joao in Oporto, Portugal. Result was satisfactory with Jaccard coefficient 83% and dice coefficient of 93%. In [14], the author presented used region growing algorithm and Hassian algorithm for multisegmentation of lung region and lungs nodule connected vascular tree segmentation, respectively. Combined classifiers were used to classify the extracted nodule and achieved 98% of classification accuracy. Beigelman-Aubry et al. [15] proposed a solid nodule detection CAD system from Multidetector CT (MDCT) scan lung images. Along with the detection of solid nodules, other key features of this CAD were evaluation of tracking and reading time.

In this work, nodule segmentation framework uses proposed automatic region growing algorithm that selects set of black pixels as seed points automatically from output binary image for lung parenchyma segmentation followed by artifacts removal to reduce disease search space. Nodules are segmented based on nodule candidates center pixels identification and intensity feature of lung nodule candidates. Segmented nodules are classified using SVM classifier and classification results are compared with other considered classifiers KNN, boosting and decision tree.

3   Materials and Methods

The proposed CAD system consists of stages such as (1) Lung lobes segmentation (2) Lung lobes boundary refine (3) Nodules detection followed by classification (4) Comparison of classifiers performance.

3.1 LIDC Data Set

CT lung images are taken from available public Lung Image Database Consortium-Image Database Resource Initiative (LIDC-IDRI). It is internationally accepted CT lung images database. National Cancer Institute (NCI) founded LIDC in 2004 in collaboration with multiple hospitals. LIDC contains lung CT slices of size 512 * 512 with thickness 1.25–2.5 mm and pixel size in range 0.48–0.72 mm. All LIDC CT images are in Digital Imaging and Communications in Medicine (DICOM) format. The aim of this database is to promote researchers from all over the world to develop and evaluate CAD system for lung nodule detection [4, 16].

3.2 Lung Nodules Detection Methodology

Proposed methodology consist of lung parenchyma segmentation and separation of lung lobes if they are attached, the next step is artifacts removal such as mediastinum, trachea, and CT examination bed, third step is to detect lung nodules, and the last step is a classification of nodules as benign or malignant. Entire CAD system accuracy depends on nodule classification accuracy which in turn depends on the accuracy of lung segmentation. Figure 2 shows the block diagram of the proposed work.

3.3 Lung Segmentation

Objects in the input CT image are separated from image background by applying iterative thresholding algorithm given in Algorithm 1. The output generated is the binary image with objects and background information. Figure 3 explains the original input CT image, histogram of input image, and output of iterative thresholding. Binary image contains artifacts such as thorax region, CT examining bed.

Algorithm 1: Iterative Thresholding algorithm

Input: CT lung image
Output: Threshold image
Step 1: Let T_h is highest grey level in input image T_m is lowest gray level in input
    image T_h and T_m are obtained from histogram of input image.
    $T = (T_h + T_m) / 2$ (1)
Step 2: Segment the image using T
    $CT(x,y) > T$ 0   Foreground F_g
    $CT(x,y) < T$ 1   Background B_g
Step3: $T_i = (F_g + B_g)/2$ (2)
Step 4: Repeat steps 2 and 3 until Ti-Ti+1 small enough that further
    thresholding generates output with no further changes.

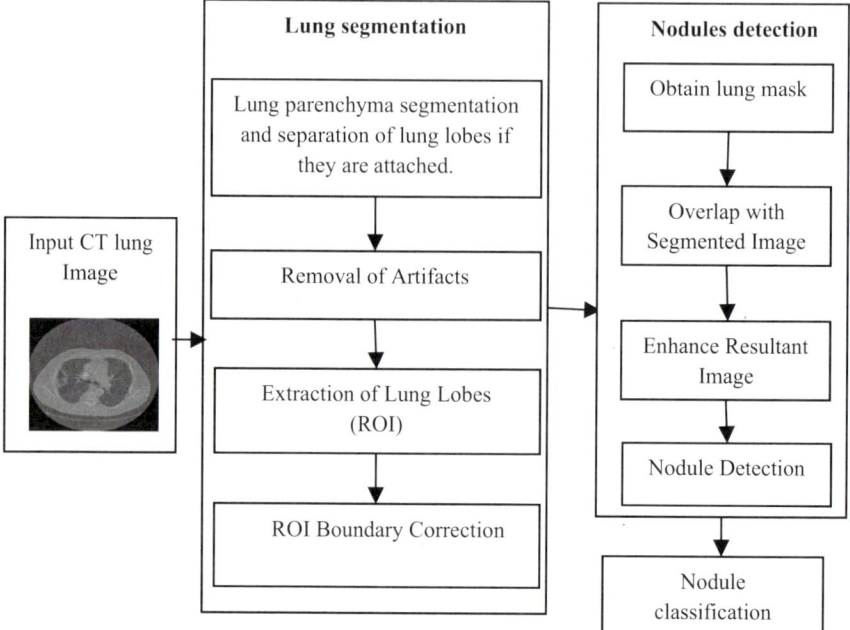

Fig. 2 Proposed CAD system to detect and classify lung nodules

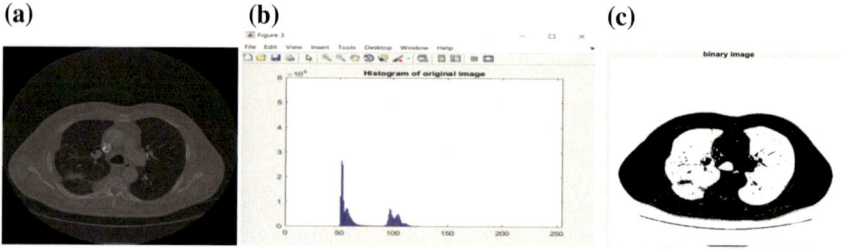

Fig. 3 a Original input CT lung image. **b** Histogram of input image. c Output binary image

Output of Algorithm 1 contains artifacts. To reduce search space for nodule detection these artifacts are cleared by first clearing the boundary of binary image followed by automated seed selection region growing algorithm given in Algorithm 2. Algorithm 2 does not require user interaction to select seed points and it automatically calculates seed points to output possible regions in binary input image. Border and artifacts cleared binary lung output image is shown in Fig. 4.

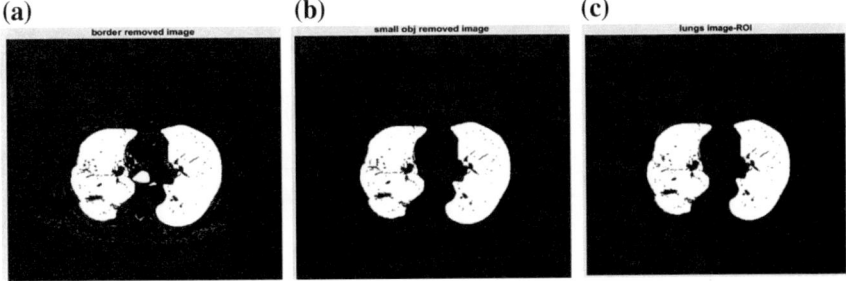

Fig. 4 **a** Binary image without border. **b** Output of automatic region growing method without artifacts and with removed small objects. **c** Extracted lung ROI

Algorithm 2: Automatic seed selection region growing algorithm

---

Input: Output Threhold image of Algorithm1
Output: Segmented image

Step 1: Sequences of black color pixels are identified by scanning binary input image in horizontal direction starting with leftmost pixel to right most pixel and vertical direction starting top most pixel to bottom most pixel. First black pixel in the set is considered as seed point for region growing.

Step 2: Let UA set of is all unallocated pixels and R_i possible seed regions,
$$UA = \{x \notin \cup_{i=1}^{n} Ri: P(x) \cap \cup_{i=1}^{n} Ai \neq \emptyset\}, \qquad (3)$$
Where I = (1,2,……..n) and P(x) is either 4 connected or 8 connected neighbors of pixel x

Step 3: If P(x) and R_j intersect, then calculate $\delta(x)$ ie similarity or difference measure between x and Intersection region R_j and $\delta(x)$ is defined as
$$\delta(x) = |gv(x) - \text{mean}\{gv(z)\}| \qquad (4)$$
where gray level at pixel x is gv(x) and mean gray level of all pixels z in the region R_j is gv(z). and j=(1,2,3…….n). Eq. 4 is calculated for all the intersected regions R_j with P(x) and minimum $\delta(x)$ is considered for inclusion of x to that region.

Step 4: Repeat Step 3 for all x ϵ UA, add a pixel y that satisfies the Eq. 3

Lung Nodules Segmentation

Lung nodule detection process consists of two steps first identify the center pixel of nodule candidates and second detect the nodules based on intensity. In segmented image, lung lobes and nodule candidates are with black edges. Noncancerous (<3 mm, benign) nodules are smaller in size compared to cancerous nodules (>3 mm, malignant). Thus, nodule size calculation is also a key feature of our nodules detection method. Black pixels collection in Fig. 4 shows the possibility of nodule candidates. Algorithm 3 first determines the center pixel of nodule candidates. In the second step using center pixels approximate elliptical boundary for each nodule candidate is identified to calculate the average intensity of pixels within the boundary. As in CT lung image, nodules appears as white spot with higher intensity value. Algorithm 3 uses this intensity feature to locate the exact nodules in the segmented image.

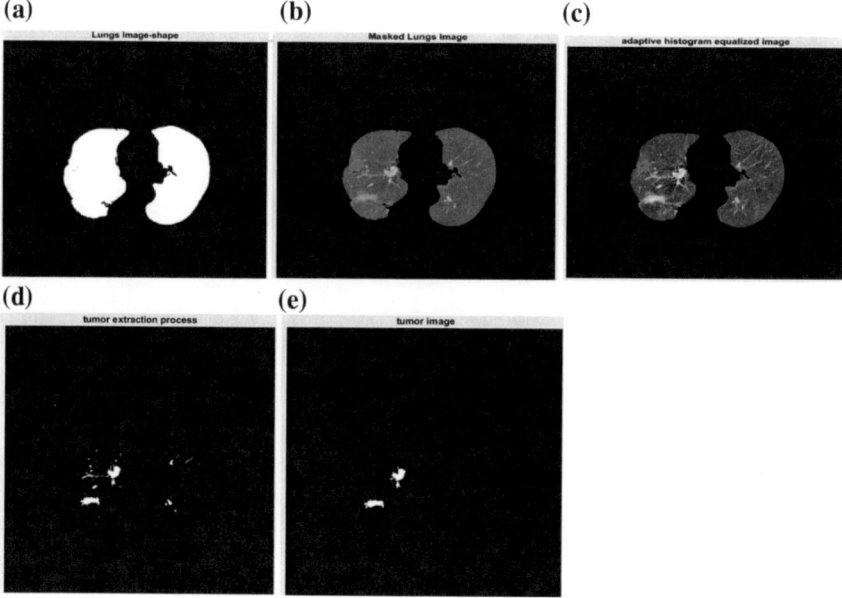

Fig. 5 **a** Lungs shape of input image. **b** Masked lung image. **c** Enhanced image. **d** Lung nodule candidates image. **e** Image with extracted nodules

Nodule candidates with average pixel intensity greater than a specified threshold are extracted and are correctly identified in lung nodules. Figure 5a–d shows nodules segmentation process of our method, Fig. 5e is output image with extracted nodules.

Algorithm 3: Nodule Segmentation algorithm

Input: Edge detected image
Output: Image with segmented nodules
Step 1: Begin with black edge pixels,
   for each considered black edge pixels
     Determine 8 – connected neighborhood black pixels and add to center pixels
   end /*Results in image CN */

Step 2: Obtain the mask image M of original input image.
Step 3: Superimpose CN with image M to obtain superimposed image S.
Step 4: Enhance the resultant image S of Step 3 to obtain enhanced image E.
Step 5: For nodule candidate i with center pixels Px_i in image E
    a) Obtain elliptical boundary region.
    b) Calculate the average of pixels intensity AP within the nodule candidate.
      if AP> NT, where NT is specified intensity to segment the nodules
        nodule candidate is correctly identified as lung nodule
        and set its center pixels as Cx_j.
      end if
   end for

3.4   Nodules Classification

The last stage of our work is a binary classification of extracted nodules. To classify the nodules, two intensity-based features sets with totally 17 features are used. First set includes five histogram-based textural features: Mean, Variance, Entropy, Energy, and Skewness. The second set includes 12 features based on Gray Level Co-variance Matrix (GLCM). Gray level of input lung CT image is used to confirm the GLCM size. Based on pixels relationship in four direction with angular difference of 45°: 0°, 45°, 90°, 135°, and distance multiple GLCM offsets are calculated. Twelve GLCM-based textural feature set includes: Sum and difference mean correlation, Inertia, Cluster tendency and shade, Autocorrelation, Maximum Probability, Homogeneity, Standard deviation, Sum and difference entropy. These considered features are used as input to SVM and other comparing classifiers. Final output image, Fig. 6a shows lung with cancer disease and Fig. 6b gives extracted nodules condition.

3.5   Classifier Validation

K-fold validation framework along with Matthews's Correlation Coefficient (MCC) given in Eq. 5 is applied to validate binary classification of SVM-RBF kernel and other considered classifiers KNN, boosting, decision tree in our classification work. Table 1 describes parameters used for MCC calculation.

$$\text{MCC} = \frac{(Tp * Tn) - (Fp * Fn)}{\sqrt{(Tp + Fp) * (Tp + Fn) * (Tn + Fp) * (Tn + Fn)}} \tag{5}$$

(a)   **(b)**

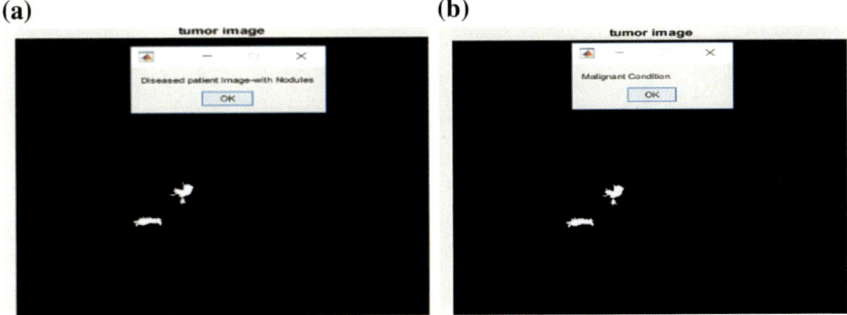

Fig. 6   a and **b** Output image with extracted nodules

Table 1 Values of Tp, Tn, Fp, and Fn		Malignant	Benign
	Malignant	Tp	Tn
	Benign	Fp	Fn

where *Tp*, *Tn*, *Fp*, and *Fn* are given in Table 1.

4   Result

Input CT images are taken from LIDC datasets and Fig. 7 shows extracted lung nodules from different input lung CT slice. In Fig. 7, upper row is the input image and lower row shows nodules segmented from input images. Totally 17 features are used to perform binary classification (benign or malignant). Receiver Operating Characteristic (ROC) curve in Fig. 8 shows SVM classifier's outstanding performance compared to other considered classifiers. Table 2 provides the accuracy and AUC values obtained using classifiers SVM-RBF kernel, boosting, decision tree, and KNN.

5   Conclusion

In this work, nodule segmentation framework uses proposed automatic region growing algorithm that selects a set of black pixels as seed points automatically from output binary image for lung parenchyma segmentation followed by artifacts removal to

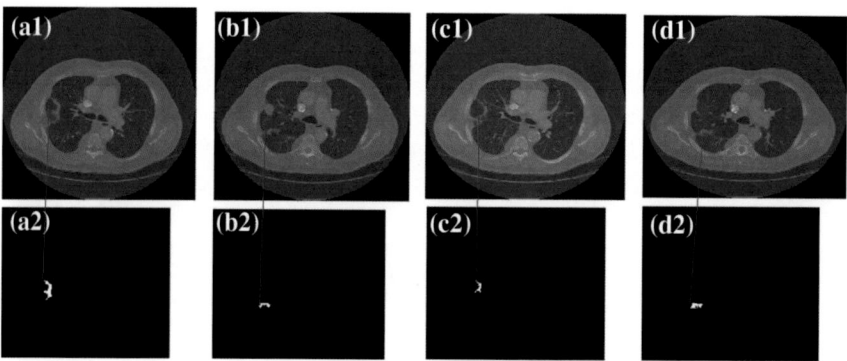

Fig. 7 **a**1 **b**1 are input lung CT images with lung nodules and **c**1 **d**1 with GGO nodules **a**2 **b**2 **c**2 **d**2 are segmented nodule images

Table 2 Comparison of classifiers performance in the proposed system

Sl. No	Classifiers	Accuracy (%)	AUC (%)
1.	SVM-RBK	98	96
2.	Boosting	90	88
3.	Decision tree	77.45	75
4.	KNN	74.13	71

Fig. 8 Receiver operating characteristic curve of classifiers

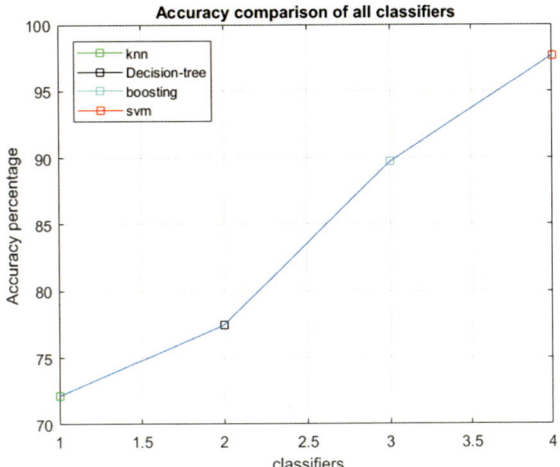

reduce disease search space. Nodules are segmented based on nodule candidates center pixels identification and intensity feature of lung nodule candidates. Segmented nodules are classified using SVM classifier and classification results are compared with other considered classifiers KNN, boosting and decision tree. In the evaluation step, it was found that SVM classifier's performance is outstanding compared to other considered classifiers in this work. Complete automation in nodule detection within very less time is the key feature of the proposed method.

Acknowledgements I thank my research guide Dr. Girijamma H A, Professor, Department of CSE, RNSIT, Bangalore, India. For supporting to complete this research article. I would like to thank public LIDC database from which images are taken to carry a reach work.

References

1. Non communicable Diseases Progress Monitor (2017) Geneva: World Health Organization; 2017. Licence: CC BY-NC-SA 3.0 IGO
2. Wender R, Fontham ETH, Barrera E (2013) American cancer society lung cancer screening guidelines, National Institute of Health, CA Cancer J Clin 63(2):107–117. https://doi.org/10.3322/caac.21172, (2013)
3. Hillary Wasserman Becky Bunn.: Lung Cancer Facts and Statistics, International Association for the Study of Lung Cancer(IACLS), WCLC, 20173999999 (2017)
4. Habib MSL (2009) A computer aided diagnosis system (CAD) for the detection of pulmonary nodules on CT scans. Systems and biomedical engineering Department, Faculty of Engineering, Cairo University, Giza, Egypt
5. Wang Q, Song E, Jin R, Han P, Wang X, Zhou Y, Zeng J (2009) Segmentation of lung nodules in computed tomography images using dynamic programming and multidirection fusion techniques. Acad Radiol 16(6):678–688
6. Ye X, Lin X, Dehmeshki J, Slabaugh G (2009) Shape-based computer aided detection of lung nodules in thoracic CT images. IEEE Trans Biomed Eng 56(7)

7. Ye J, Lin XY, Dehmeshki J, Slabaugh G, Beddoe G (2009) Shape based computer aided detection of lung nodules in thoracic CT images. IEEE Trans Biomed Eng 56:1810–1820. https://doi.org/10.1109/tbme.2009.2017027
8. Liu Y, Xing Z, Deng C, Li P, Guo M (2010) Automatically detecting lung nodules based on shape descriptor and semi-supervised learning. In: International conference on computer application and system modeling. IEEE, https://doi.org/10.1109/iccasm.20https://doi.org/10.5619447, Taiyuan, China
9. Tan M, Deklerck R, Bister BJM, Cornelis J (2011) A novel computer-aided lung nodule detection system for CT images. Med Phys 38(10)
10. Messay T, Hardie RC, Rogers SK (2010) A new computationally efficient CAD system for pulmonary nodule detection in CT imagery. Med Image Anal 14(3):390–406. https://doi.org/10.1016/j.media
11. Shao H, Cao L, Liu Y (2012) A detection approach for solitary pulmonary nodules based on CT images. In: IEEE, 2nd international conference on computer science and network technology (ICCSNT). Changchun, pp 1253–1257
12. Teramoto A, Fujita H (2013) Fast Llung nodule detection in chest CT images using cylindrical nodule-enhancement filter. Int J Comput Assist Radiol Surg 8(2):193–205. https://doi.org/10.1007/s11548-012-0767-5
13. Campos DM, Simoes A, Ramos I, Campilho A (2014) Feature-based supervised lung nodule segmentation. In: IFMBE proceedings the international conference on health informatics 42, 23 https://doi.org/10.1007/978-3-319-03005-0_7. Springer International Publishing, Switzerland
14. Elsayed O, Mahar K, Kholief M, Khater HA (2015) Automatic detection of the pulmonary nodules from CT images. IEEE. doi: https://doi.org/10.1109/intellisys.2015.7361223. SAI Intelligent Systems Conference (IntelliSys)
15. Beigelman Aubry C, Raffy P, Yang Ronald W, Castellino Philippe A, Grenier A (2007) Computer-aided detection of solid lung nodules on follow-up MDCT screening: evaluation of detection, tracking, and reading time. Chest Imaging AJR:189 https://doi.org/10.2214/ajr.07.2302
16. https://wiki.cancerimagingarchive.net/display/Public/LIDC-IDRI

A Review—Edge Detection Techniques in Dental Images

Aayushi Agrawal and Rosepreet Kaur Bhogal

Abstract Edge detection plays an important role in digital image processing applications. The main aim of edge detection is to identify the discontinuity in images, where the sharp changes in intensity take place. This research work presents the edge detection technique in dental X-ray images (panoramic radiograms), which is advantageous to separate teeth individually for better classification and identification of diseases. The objective is to study and compare the various algorithms that are Sobel, Prewitt, Canny, multiple morphological gradient (mMG), line analyzer, neural network, genetic algorithm, and infinite symmetric filter (ISF), multi-scale and multi-directional analysis with statistical thresholding (MMST), and fuzzy logic approach for edge detection in dental X-ray images. There are many difficulties in finding diseases from panoramic dental images only, and hence to overcome these difficulties edge detection is introduced. Some of the dental diseases that require edge detection for their identification are discussed. Based on capability of detecting the diseases accurately and total number of diseases detected from the dental images by the use of edge detection, comparison of results takes place.

1 Introduction

Edges can be specified as a group of immediate pixel location where a sudden variation in intensity values takes place. Edges are the boundaries among objects and background. An edge detector can be used for feature extraction, image segmentation, and object identifications [1].

In dental X-ray imaging, edge detection technology plays a very helpful role in detecting and diagnosis of the diseases. Dental radiograms are poor and complicate in some of the diseases extraction such as tooth decay, cavities [2], tooth abscess,

A. Agrawal (✉) · R. K. Bhogal
Electronics and Communication Engineering, Lovely Professional University, Phagwara, India
e-mail: aayushiagrawal20495@gmail.com

R. K. Bhogal
e-mail: rosepreetkaur12@gmail.com

© Springer Nature Switzerland AG 2019
D. Pandian et al. (eds.), *Proceedings of the International Conference on ISMAC in Computational Vision and Bio-Engineering 2018 (ISMAC-CVB)*, Lecture Notes in Computational Vision and Biomechanics 30,
https://doi.org/10.1007/978-3-030-00665-5_128

impact tooth, etc. Brightness of X-ray image is not good enough because of sequins occurring due to the presence of water on teeth.

Neural network method is used for detection of edges in dental X-ray images where inputs to the network are pixels of original image after minimizing the output error, and output pixels are replaced in the edge-detected image. Artificial neural network (ANN) is used because in this iterative training and learning takes place from input to output mapping. This algorithm is helpful in finding problems such as tooth decay, bone damage used for supporting the teeth, and dental injuries. This method may be extended for the diagnosis of major problems related to teeth [3].

Firstly, X-ray image required preprocessing for enhancement using Gaussian low-pass filter, then various edge detection techniques are applied and at last comparison of results takes place, which shows that Canny operator provides best results among Sobel and Prewitt [4]. After preprocessing of image, better edge detection techniques might be applied for detection of true edges in dental images. The algorithm of multiple morphological gradient (mMG) is applied in [5] for clear visibility of boundaries of object in panoramic radiograms. Some positions of dental caries are also defined, which says that it is divided into two groups, that are smooth surface and gap of teeth. The result of this algorithm is very useful for the identification of cavities and verification of the same is done by two dentists. Furthermore, this algorithm should be applied for the detection of more number of dental diseases.

Before edge detection in dental X-ray images the two steps that are to be followed are image enhancement and teeth segmentation so that the proper identification of every teeth occurs [6]. The problem of finding missing teeth [7] is to be solved using this procedure of edge detection. In jaw, where the teeth is missing, values of standard deviation, Euler number, and area become zero. Additionally, for proper alignment of teeth, its measurement is very important. Traditionally, gear tooth micrometer and vernier caliper are used but results are not in fraction, then minute changes in measurement lead to improper alignment of teeth. Solution of this problem is given in [8] by the use of edge and gray-based method. The results prove that it is possible to place tooth correctly at exact position. Results might be improved by soft algorithm edge detection techniques like genetic algorithms (GA) that provide flexible edge detection in very shady images [9]. Since it is also applied in dental X-ray images for identifying the diseases like tooth decay, impact tooth, and so on. Mainly GA consists of three stages; they are selection, crossover, and mutation for processing of any application. Comparison of results takes place with Roberts [10] operator and found that GA gives better results comparatively.

There is less amount of radiation in cone beam computed tomography (CBCT) as compared to panoramic radiograms because this beam is divergent and forms a cone-like structure. For the edge detection of CBCT segment, standard neural network is used, where input to this network is some parameters of image that are used and one hidden layer is also present in between input and output layer. Results are compared with Canny edge detector and it proves that NN is fast comparatively and is helpful for providing 3D reconstruction of teeth [11].

Identification of dental abscess is also major problem by the use of dental X-ray images only. For this optimal edge detection required [12] solves the same problem which states that firstly image requires removal of noise from it and after that edge detection is applied for extraction of abscess. Results of this edge-detected images are shown to the dentist and well appreciated by them. And, for root canal treatment, finding the length of root is very important parameter and that is done by the use of edge detection in dental X-ray images. The implementation of algorithm for the solution of abovementioned problem takes place by the use of MMST where Laplacian pyramid is used to decompose the image into number of levels, directional decomposition of image is done by the use directional filter banks. Statistical thresholding is done for detection of edges in an image. And, results conclude the clear and precise detection of root canal length for further diagnosis [13]. Root length might be measured more accurately by soft algorithms where output is dependent on applied rules and membership function.

The edge detection by ISEF [14, 15] is helpful in detecting dental caries in tooth decay from dental X-ray images. After that it also gives the decision for the treatment like root canal treatment or filling. Results are giving the idea where the caries-affected area is present and also verified by doctor. Since 3D reconstruction of dental images plays an important role in identification of various diseases. Some of the challenging issues are solved which are found in CBCT segments that are connectivity, noisiness, position, topological changes, and low resolution by 3D dental images [16, 17] which is obtained by doing edge detection by the use of Canny method. For the validation of method, triangulation of the sparse points is used. Moreover, this 3D reconstructed image will be used for detection of major issues related to teeth.

Sobel and Canny edge detector [18] are widely used for detection of edges as they are simply calculating the gradients of image for edge detection. Here, it is useful in detecting edges for dental age assessment. Comparison of Sobel and Canny edge detector occurs and results that Sobel is better than Canny for dental images, and then soft algorithms are used where inputs are dealing with approximation model. [19–21] shows one of them, that is, fuzzy logic approach for the detection of edges from dental X-ray segment. Fuzzy inference system (FIS) [22, 23] is used which is having three main steps that are fuzzification, knowledge base, and defuzzification. Extraction of edges depends upon the degree of whiteness and blackness of eight neighbor gray level pixels that are done by powerful rules applied to it. The results are compared with Roberts method and it is clearly visible that more precise edges are detected by fuzzy logic approach. Higher level of fuzzy logic approaches might be used for better detection of edges.

The section of this paper is prepared as follows: the basic human teeth structure is demonstrated in Sect. 2. The common dental diseases along with causes, treatment, and prevention are described in Sect. 3. Overview of edge detection in X-ray is given in Sects. 4 and 5 comprises flowcharts of each technique reviewed in this paper. In Sect. 6, results and discursion take place on the basis of outcome produced by edge detection techniques. Finally, in Sect. 7 some conclusion and future scope according to the results takes place.

2 Human Teeth Structure

Human teeth structure shown in [23] Fig. 1 mainly consists of three tissues, they are as follows:

(1) *Hard tissues*: It is also known as calcified tissues having a hard intercellular matrix and also highly mineralized. Cementum, dentine, and enamel are the hard tissues in human teeth structure.
(2) *Soft tissues*: It is present to protect and cover the root of the teeth. Tooth pulp is soft tissue found in teeth.
(3) *Supporting tissues*: It is present around the teeth and is helpful in providing the necessary support to hold teeth in use. There are there supporting tissues and they are alveolar bone, gingiva, and periodontal ligaments.

3 Dental Diseases

The major significant function of teeth is to increase appearance and individual's integration in modern society. It is also helpful to make ready the food digestive. Sometimes when humans are not aware of the importance of teeth and not brushing their teeth properly and many more causes may lead to some dental diseases [24] which are mentioned in Table 1 along with its prevention and treatments.

These dental issues also have common risk causes as noncommunicable diseases. According to the importance of dental health, WHO organized world health day in 1995 year with the name "Oral health for healthy life" [25]. A survey of oral health

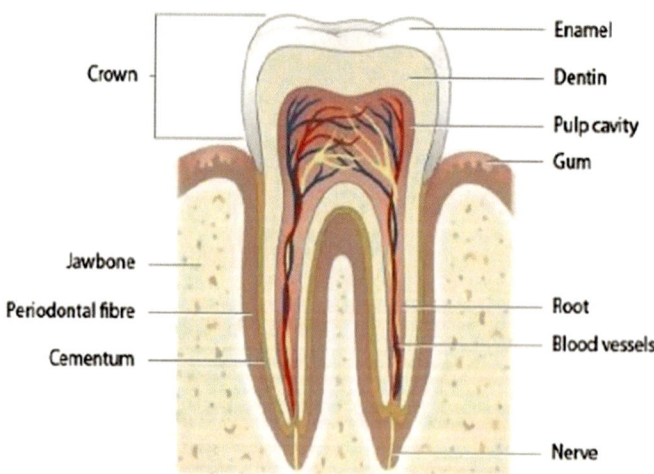

Fig. 1 Basic structure of human teeth

Table 1 Common dental diseases along its causes, prevention, possible treatment outcome, and images

S. No	Dental diseases	Causes	Prevention	Possible treatment outcome	Images
1.	Tooth decay	Improper teeth cleaning, over eating, and drinking of sugar	Washing mouth after eating, less consumption of sugar	Remineralization, fluoride treatments	
2.	Chipped tooth	Due to some accidents, cavities which make weaken the tooth	Uses of mouth guard, avoid eating the hardest food	Tooth filling, covered with crown (made up ceramics)	
3.	Stained teeth (Discoloration)	Chewing a tobacco or smoking, improper teeth brushing	Avoid chewing tobacco or smoking brush teeth twice in a day	Tooth bonding, advance whitening	
4.	Impacted teeth	Improper eruption of tooth, insufficient space in jaw	Continuously updating the dental X-ray	Tooth extraction	
5.	Sensitivity (Cold)	Due to some predefined problems and they are cavities, cracked, exposed	Avoid drinking or eating acidic food or drinks	Apply protection layer (fluoride layer), brush with fluoride gel	
6.	Hyperdontia	Actual cause is unknown but many experts believe that it is due to genetic	Not available	Surgical extraction, orthodontic treatment	

(continued)

Table 1 (continued)

S. No	Dental diseases	Causes	Prevention	Possible treatment outcome	Images
7.	Crooked teeth	Sucking of thumb, usage of baby bottle at large extent, teeth grinding	Avoid thumb sucking, avoid usage of baby bottle a lot	Mouth guard, therapy	
8.	Diastema	Different shape and size of teeth, tongue thrusting	Placement of permanent retainer	Frenectomy, dental crowns, invisalign	
9.	Periodontal diseases	Smoking, diabetes, hormonal changes in women	Avoid smoking, repeated checkup and cleaning of teeth	Flap surgery, bone and tissues grafting, deep cleaning	
10.	Bruxism	Stress, sleep disorders	Usage of bite splint at night	Biofeedback exercises, physical therapy, plastic tooth guard, relaxation techniques	
11.	Dental abscess	Smoking, diabetes, gum disorders	Brushing of teeth after every meals, avoidance of smoking	Root canal treatment, drainage of pus	
12.	Dental trauma	Hit by hard surface, rigorous sports activities	Usage of mouth guard	Tooth filling	

(continued)

Table 1 (continued)

S. No	Dental diseases	Causes	Prevention	Possible treatment outcome	Images
13.	Oral cancer	Smoking, tobacco	Avoidance of smoking and tobacco	Radiation therapy	

[26] demonstrates that it is important for every stage of life, mainly for older, adults, and seniors because of it various preventions that are needed to be followed which are discussed in next section of this paper.

4 Edge Detection in Dental X-Ray

Diagnosis of dental diseases is conventionally contained by radiographic films. Nowadays, dental X-rays came into existence for detection of several diseases such as impacted tooth, tooth decay, crooked teeth, abscess, and chipped tooth. But, the main issues with dental X-ray is that it is not very prominent. After this, edge detection techniques should be applied for better classification of dental diseases form X-ray image. Although edge detection carried out different methods as mentioned above but its main aim is to detect true edges for the proper identification of dental diseases. Several edge detection methods are discussed in the form of flowcharts in next section.

5 Experimental Analysis

Following are the methods mentioned in the form of flowcharts for detection of edges in dental X-ray images, which are discussed in Sect. 1.

Figure 2 shows line analyzer techniques for teeth, in this algorithm, image smoothing is applied for removal of noise by the use of Gaussian filter, gradients are calculated from Sobel operator. For converting blurred edge to sharp edge pixels non-maximum suppression is used, and finally double thresholding takes place to differentiate between weaker and stronger edges.

Figure 3 shows mMG method for edge detection, preprocessing of image that is cropping to extract the affected area only. Morphological operations are applied that are dilation and erosion. Dilated image is subtracted from eroded image to calculate the morphological gradient of the cropped image.

Fig. 2 Line analyzer
technique for teeth

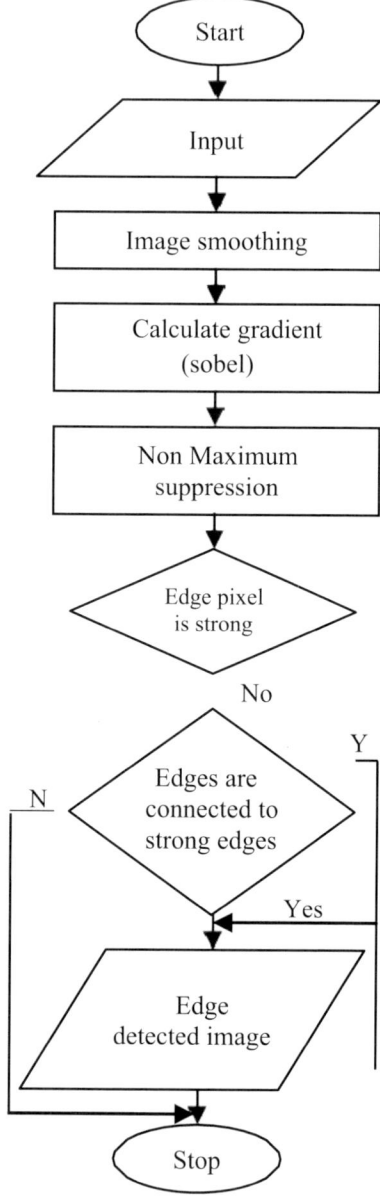

Figure 4 shows ANN training for edge detection, there were mainly two steps for detection that are pixel classification, which says value of pixels is a part of segment or not. Edge detection is done by the use of convolved mask obtained by the output of neural network.

Fig. 3 mMG method

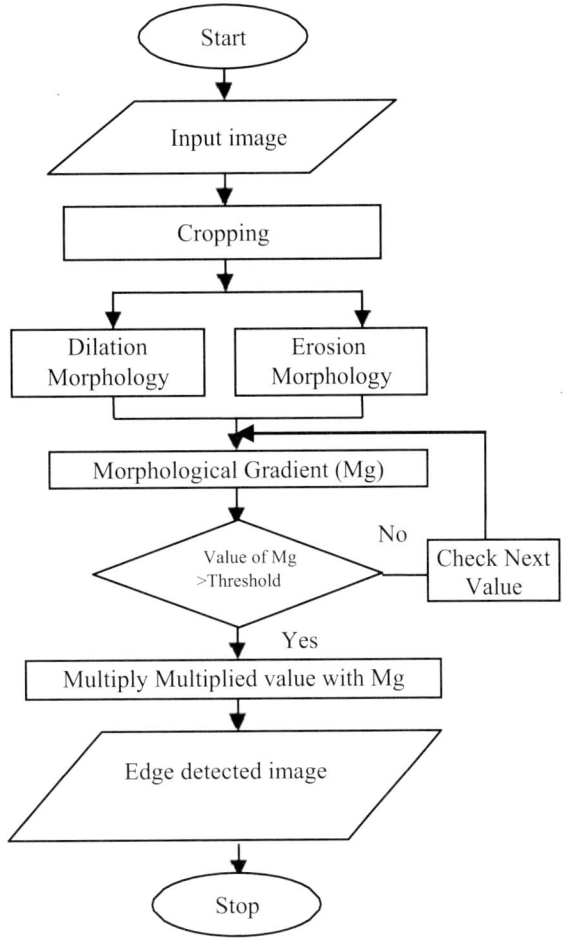

Figure 5 shows neural network and Canny algorithm for edge detection, in this both algorithms (Canny and neural network) combine to detect very accurate edge-detected image, and here output of Canny is given to input of neural network if edges are not properly detected.

Figure 6 shows ISEF method for abscess detection in dental images, and here filtering is done by the use of median filter for better enhancement of gray-scaled input image. ISEF filter is applied in both "x" and "y" direction. Binary Laplacian technique is applied by subtracting filtered image from original image. For detection for strong edges, hysteresis thresholding is used.

Figure 7 shows the method for 3D reconstruction of input image, and in this algorithm edges are detected using Canny method. Contours are detected in all direction to combine that section to form a 3D reconstructed image.

Fig. 4 Artificial neural
network method

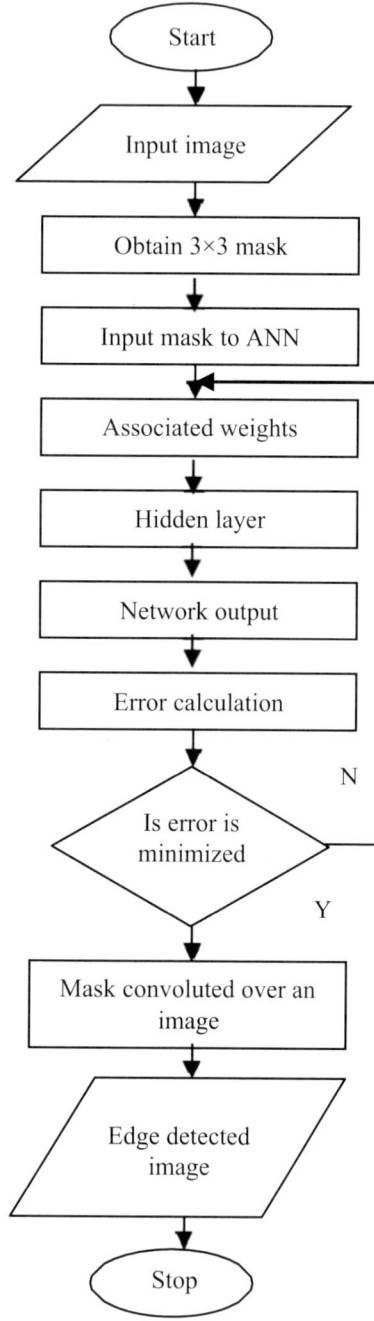

Fig. 5 Neural network
method and Canny operator

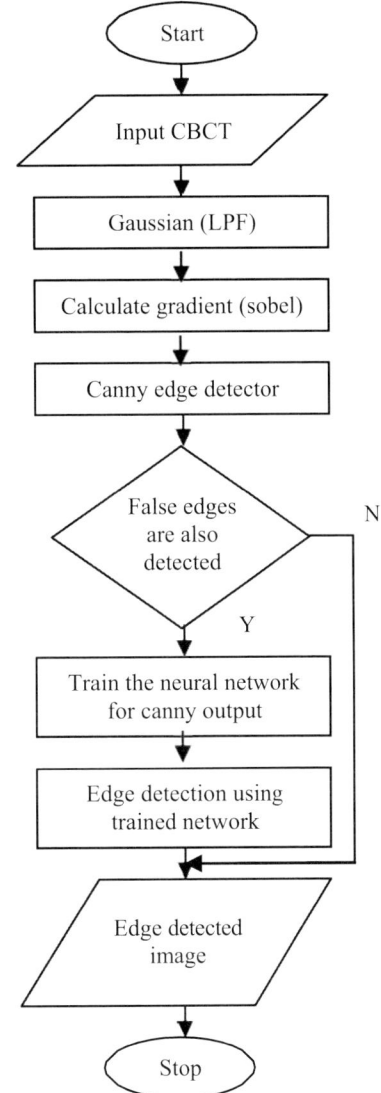

Figure 8 shows the edge detection algorithm for endodontic working length measurement, and here edge detection takes place using MMST method. In that removal of speckle noise could be done by frost filter, adaptive equalization for enhancement takes place, and Laplacian pyramid decomposition is applied to decompose the X-ray image of tooth into number of levels.

Figure 9 shows the method for detection of desired teeth, and in this histogram equalization has occurred for enhancement of image. Otsu's method is used for

Fig. 6 ISEF method

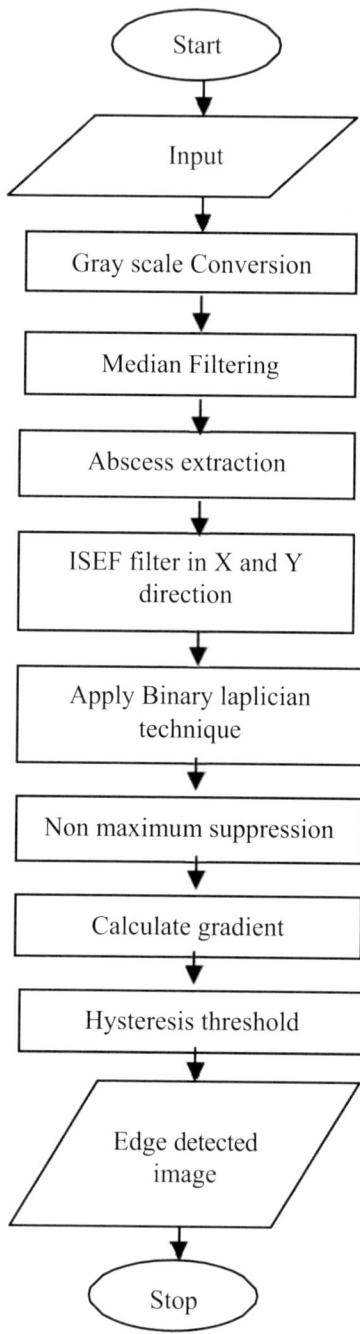

Fig. 7 Canny edge detector

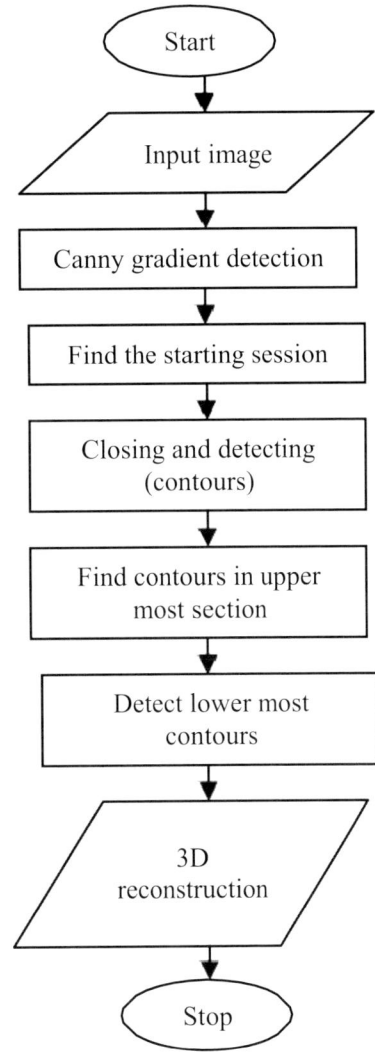

segmentation of desired area. Canny edge detection is used for detection of desired teeth.

Figure 10 shows preprocessing steps before edge detection, and here preprocessing is required for enhancement of input image, for that image resizing, grayscale conversion, and filtering by the use of Gaussian filter are done. Edges are more accurately found using Sobel or Prewitt or Canny edge detector.

Figure 11 shows fuzzy logic approach for edge detection, and here rules are dependable on eight neighbor gray level pixels, decision depends upon degree of blackness and whiteness, which decides that edges are there in the image or not.

Fig. 8 MMST method

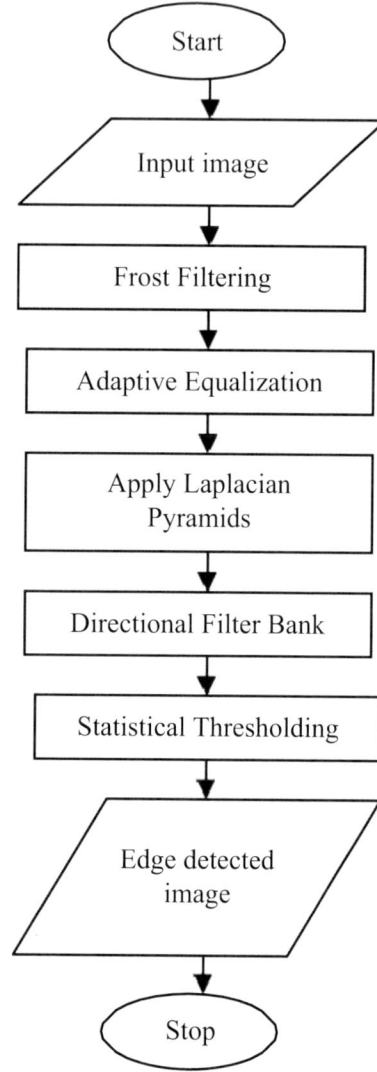

Membership function is used where predefined rules are applied for accurate detection of edges from the image.

Figure 12 shows ISEF method for caries region detection, the algorithm shows the conversion of image into gray scale for further processing. Morphological operation and filtering are required for enhancement and still the caries-affected tooth is not found, then ISEF method for edge detection is applied.

Fig. 9 Segmentation and Canny edge detector

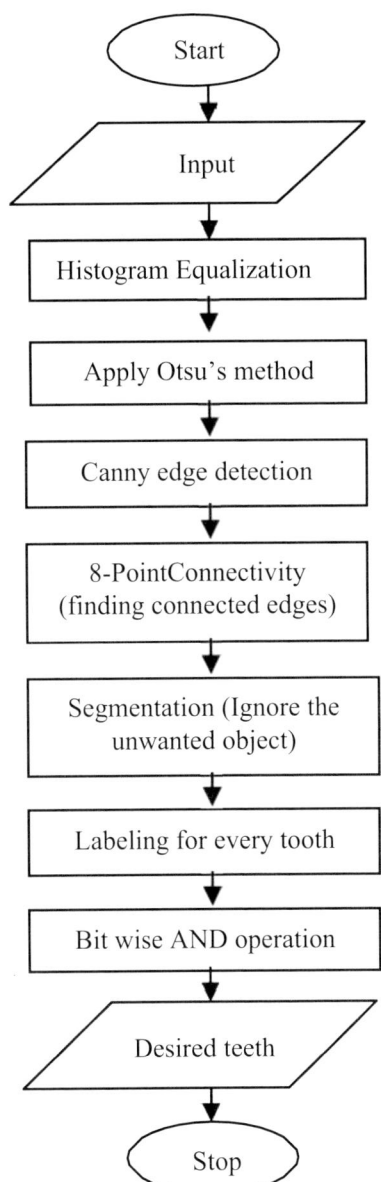

6 Results and Discussion

The results of various techniques are compared, on the basis of ability to identify the type of dental diseases or solution of issues created in dental X-ray images

Fig. 10 Sobel and Prewitt

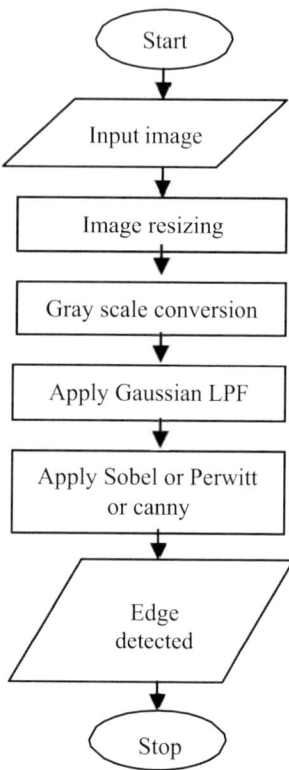

(bitewing X-ray, periapical X-ray, panoramic X-ray, and CBCT) [27] by the use of edge-detected dental images.

Table 2 shows the results of different edge detection techniques for particular dental diseases. Input to the algorithm is different type of X-ray images on the basis of total number of diseases detected, and results demonstrate that genetic algorithm, neural network, and fuzzy logic approaches give better results among all edge detection technique such as Sobel, Prewitt, Canny, mMG, and ISEF.

Fig. 11 Fuzzy method

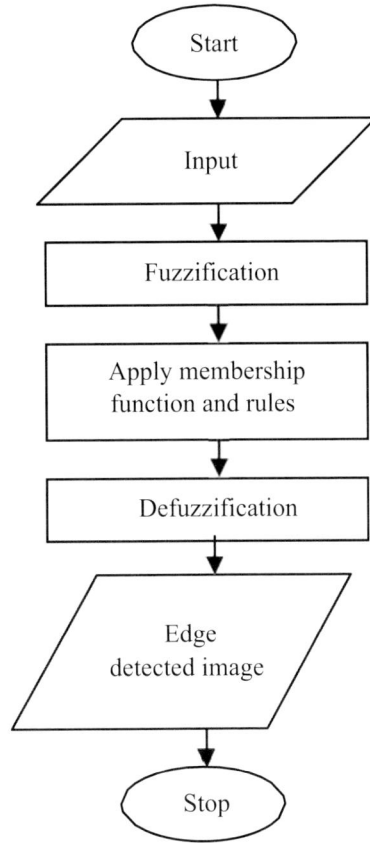

7 Conclusion

This paper is the review on several edge detection techniques applied on dental images. Each edge detectors discussed in this paper provides helpful results for identification of particular diseases such as tooth decay, dental injury, cavities, abscess. And, some edge detector gives improvement in visibilities of objects (teeth) in the dental images. Dental X-rays are not sufficient for interpretation of diseases, area of missing teeth, caries-affected areas, root length for root canal treatment, and so on. Then, edge detection of dental X-ray images takes place to overcome the disadvantages of dental X-ray images because by the detection of edges in the image easy differentiation of objects takes place, which is very helpful for doctors in better interpretation of diseases for further diagnosis. As edge detection in dental images is very helpful, it should be more accurate so that the actual diseases are interpreted by doctor for diagnosis.

Table 2 Results of various edge detection techniques

References	Type of input image	Edge-detected image outcomes
[3]	Bitewing X-ray	Identification of tooth decay, supporting bones, and dental injury
[4]	Periapical X-ray	Improvement in visibility of individual tooth
[5]	Panoramic X-ray	Identification of cavities
[6]	Bitewing X-ray	Area of missing teeth
[8]	3-D teeth image	Measurement of tooth length for proper alignment
[9]	CBCT segment	3D reconstruction of teeth
[10]	Bitewing X-ray	Improvement in identification of tooth decay, supporting bones, and dental injury
[12]	Periapical X-ray	Diagnosis of abscess
[13]	Periapical X-ray	Improvement in endodontic working length measurement
[14]	Periapical X-ray	Caries-affected areas
[16]	CBCT segment	3D reconstruction of oval cavity
[18]	Panoramic X-ray	Dental age assessment
[19]	Bitewing X-ray	Improvement in visibility of individual tooth

The results of reviewed techniques have also detected some false edges which may lead to improper diagnosis of diseases, and in that case, a precise edge detection technique is required. This precision may be achieved by the use of general type-2 fuzzy logic system (GT2FLS) [28, 29] for edge detection because this approach deals with very high level of uncertainty in the problems, by the use of footprint of uncertainty (FOU) in its membership function which allows an improved solution of real-world uncertainty.

Fig. 12 ISEF method

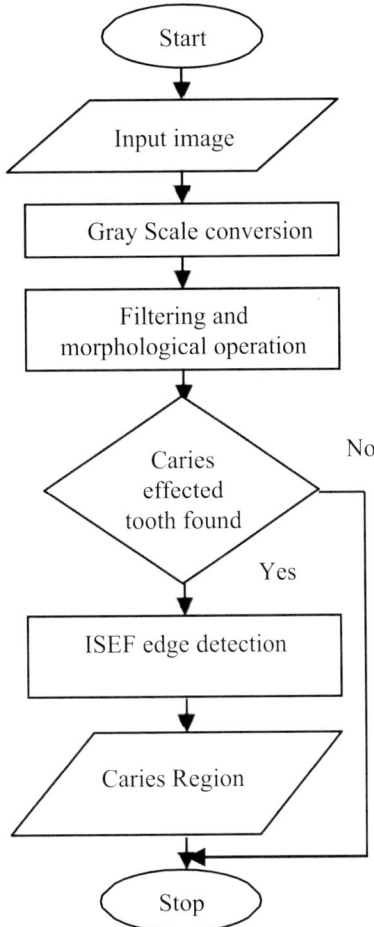

References

1. Solanki C, Godfrey WW (2016) Technique for edge detection based on interval type-2 fuzzy logic with sobel filtering. In: IEEE transactions. Doi 978-1-5090-1987-8
2. Kaushik A, Mathpal PC, Sharma V (2014) Edge detection and level set active contour model for the segmentation of cavity present in dental X-ray images. Int J Comput Appl 96(9):0975–8887
3. Senthilkumaran N (2012) Edge detection for dental X-ray image segmentation using neural network approach. Int J Comput Sci Appl (TIJCSA), 1(7)
4. Ansingkar NP, Dhopeshwarkar MG (2014) Study and analysis of edge detection techniques for segmentation using dental radiograph. Int J Eng Comput Sci 3(9)
5. Na'am J, Harlan J, Madenda S, Wibowo EP (2016) The algorithm of image edge detection on panoramic dental X-ray using multiple morphological gradient (mMG) method. Int J Adv Sci Eng Sci Technol 6
6. Croock MS, Khudhur SD, Taqi AK (2016) Edge detection and features extraction for dental X-ray. Eng Tech J 34 Part (A)(13)

7. Lin PL, Huang P–W, Cho YS, Kuo C–H (2013) An automatic and effective tooth isolation method for dental radiographs. Opto–Electron https://doi.org/10.2478/s11772-012-0051-9
8. Saoji SU, Jaini P (2014) Line analyzer techniques for teeth using edge-based method and gray-based method. In: International conference on communication systems and network technologies. https://doi.org/10.1109/csnt.2014.186
9. Senthilkumaran N (2012) Genetic algorithm approach to edge detection for dental X-ray image segmentation. Int J Adv Res Comput Sci Electron Eng (IJARCSEE) 1(7)
10. Gayathri V, Menon HP (2014) Challenges in edge extraction of dental X-ray images using image processing algorithms—a review in (IJCSIT). Int J Comput Sci Inf Technol 5(4)
11. Pavaloiu I-B, Goga N, Vasilateanu A, Marin I, Ungar A, Patrascu I, Ilie C (2015) Neural network based edge detection for CBCT segmentation. IEEE 978-1-4673-7545-0
12. Mahant PM, Desai NP, Jain KR, Mahan MG (2015) Optimal edge detection method for diagnosis of abscess in dental radiograph. IJRSI II(II)
13. Padma Vasavi K, Udaya Kumar N, Madhavi Latha M, Krihna Rao EV An edge detection scheme for endodontic working length measurement in root canal treatment for succedaneous teeth in latest trends. Circ Syst Sig Process Autom Control. ISBN: 978-960-474-374-2
14. Solanki AJ (2016) Threshold selection in ISEF based identification of dental caries in decayed tooth. Int J Electron Electr Comput Syst (IJEECS) 5(5). ISSN 2348-117X
15. Trivedi DN, Shah N, Kothari AM (2016) Dental contour extraction & matching with label contouring using ISEF algorithm on DICOM images for human identification. Int J Latest Trends Eng Technol (IJLTET) 7(2)
16. Pavaloiu I-B, Goga N, Marin I, Vasilateanu A (2015) Automatic segmentation for 3D denta reconstruction. In: ICCCNT
17. Kamencay P, Zachariasova M, Hudec R, Benco M, Radil R (2014) 3D image reconstruction from 2D CT slices 3DTV-conference: the true vision—capture, transmission and display of 3D video (3DTVCON)
18. Razali MRM, Ahmad NS, Hassan R, Zaki ZM, Ismail W (2015) Sobel and Canny edges segmentations for the dental age assessment. IEEE. DOI 10.1109
19. Bhargavi K, Jyoth S (2016) An efficient fuzzy logic based edge detection algorithm. Int J Tech Res Appl 4(3)
20. Aborisade DO (2010) Fuzzy logic based digital image edge detection global. J Comput Sci Technol 10(14) (Ver. 1.0)
21. Senthilkumaran N (2012) Fuzzy logic approach to edge detection for dental X-ray image segmentation. (IJCSIT) Int J Comput Sci Inf Technol 3(5)
22. Tangel ML, Fatichah C, Yan F, Betancourt JP, Widyanto RM, Dong F, Hirota K (2013) Dental classification for periapical radiograph based on multiple fuzzy attribute. IEEE 978-1-4799-0348-1
23. Lai YH, Lin PL (2008) Effective segmentation for dental X-ray images using texture- based fuzzy inference system. LNCS 5259:936–947
24. Moynihan P, Petersen PE (2004) Diet, nutrition and the prevention of dental diseases. Public Health Nutr. https://doi.org/10.1079/phn2003589
25. Goryawala SN, Chavda P, Udhani S, Shukla D, Pathak S, Ojha R (2015) A survey on incidence of common dental problems among patients attending dentistry OPD at a tertiary care hospital from central Gujarat. Int J Res Med
26. Harris M, Eaton KA (2011) Discussion paper, dental hyginest and dental research: a developing scene OHDM 10(4)
27. Shivpuje BV, Sable GS (2016) A review on digital dental radiographic images for disease identification and classification. Int J Eng Res Appl 6(7) (Part -5):38–42. ISSN 2248-9622
28. Melin P, Gonzalez CI, Castro JR, Mendoza O, Castillo O (2013) Edge detection method for image processing based on generalized type-2. Fuzzy Logic IEEE. https://doi.org/10.1109/tfuzz.2013.2297159
29. Melin P, Gonzalez CI, Castro JR, Mendoza O, Castillo O (2016) General type-2 Fuzzy edge detector applied on face recognition system using neural networks. IEEE 978-1-5090-0626-7

Detection of Exudates and Removal of Optic Disk in Fundus Images Using Genetic Algorithm

K. Gayathri Devi, M. Dhivya and S. Preethi

Abstract Diabetic retinopathy is one of the serious and sight-threatening complications of diabetics. The main symptom of diabetic retinopathy is the presence of exudates that results in yellow flecks due to the fluid that has seeped out of damaged capillaries. This causes the tissue in the retina to distend, resulting in hazy or unclear vision. If they are left untreated, diabetic retinopathy can cause blindness. Hence, segmentation of exudates is vital process in retinal pathologies. The proposed work involves accurate segmentation of exudates from the retinal fundus images. Initially, K-means clustering is applied on the retinal images to separate the exudates and optic disk. Genetic algorithm is used for the accurate segmentation of the exudates in which the fitness function is calculated to perform crossover between the segmented images obtained from the K-means clustering segmentation. Before performing the mutation process, the grayscale image is converted into the RGB channels. These three-segmented channels are further combined by the mutation process to obtain the genetic algorithm output. High-intensity region is determined to be the exudates and the low intensity is said to be the optic disk. The elimination of the optic disk which has the same intensity as that of the exudates is performed using watershed segmentation. Finally, the parameter validation is done after the morphological operations. This method was implemented in 10 images downloaded from CHASE and STARE database and the accuracy has been improved to 94% compared with the existing approaches.

K. Gayathri Devi (✉) · M. Dhivya · S. Preethi
Electronics and Communication Engineering, Dr. N G P Institute of Technology,
Coimbatore, India
e-mail: gayathridevik@yahoo.com

M. Dhivya
e-mail: dhivya@drngpit.ac.in

S. Preethi
e-mail: preethi.s@drngpit.ac.in

© Springer Nature Switzerland AG 2019
D. Pandian et al. (eds.), *Proceedings of the International Conference on ISMAC
in Computational Vision and Bio-Engineering 2018 (ISMAC-CVB)*, Lecture Notes
in Computational Vision and Biomechanics 30,
https://doi.org/10.1007/978-3-030-00665-5_129

1 Introduction

Diabetic retinopathy is an eye disease that causes abnormalities in the retina due to the complication of diabetes. The early identification of diabetic retinopathy is very much needed to save the vision and to provide treatment. Exudates are one of the symptoms of diabetic retinopathy. Diabetic retinopathy can be categorized into two stages such as non-proliferative stage and the proliferative stage [1]. Exudates identified in the non-proliferative stage are categorized as soft exudates and in the proliferative stage as hard exudates [2]. These exudates cause harm to the blood vessels in retina, and hence the capillaries in the vessels will leak some fluid and these will appear as shiny yellow–white dots with sharp borders called as exudates. Exudates can be identified by implementing image segmenting methods such as watershed algorithm, region-based segmentation methods, thresholding method, representation techniques, and machine learning methods [3, 4].

2 Existing Methods

Diabetic retinopathy is one of the major causes of blindness that can cause damage to blood vessels in the eye which will subsequently progress to the formation of lesion in the retina [1, 5]. Accurate exudates segmentation plays a vital role in diagnosis of diabetic retinopathy. Aqeel et al. [6] proposed an automated algorithm for the segmentation of exudates. For optimal results, texture and adaptive threshold algorithm have been incorporated for the accurate segmentation of exudates.

Pereira et al. [7] implemented a technique to extract the blood vessels in retinal fundus image for the automatic detection of diabetic retinopathy. The steps involved in this paper are preprocessing, feature extraction, ant colony optimization algorithm, and the post-processing. Two features are extracted from images and combined, in order to develop ant movement heuristics. The ant colony optimization is quite slow when compared to genetic algorithm and is efficient only in extraction of blood vessel. This proposed work does not consider the segmentation of exudates that is needed to diagnose diabetic retinopathy.

Soares et al. [8] suggested a technique to localize the optic disk (OD) and the importance of diagnosing diabetic retinopathy. The localization of the OD was done by cumulative sum field technique. The algorithm reveals to be reliable and efficient. Yu et al. [9] enhanced the detection of optic disk using directional filtering technique and level set method which requires the placement of seed point to localize the optic disk.

Radha et al. [10] developed a process to detect the exudates and to classify the retina to be normal or abnormal. The combination of multi-structure morphological process and K-means clustering technique is used effectively for retinal vessel and exudates detection. Exact detection of the condition of a retina whether it is normal or abnormal was determined successfully. For the abnormal retina, the exudates were detected by extracting the blood vessels using plane separation method. The features were extracted by applying discrete wavelet transform (DWT) and energy feature coefficients were trained with the probabilistic neural network for classifying the exudates. Morphological operations are applied on segmented image for smoothening the exudates part.

The supervised learning approach based on the multilayer perceptron (MLP) was used by Kusakunniran et al. [11] to determine the elementary seeds for the segmentation of hard exudates. In the preprocessing step, color transfer has been applied to normalize the color information to reduce the variation of color that will exist among hard exudates in different retinal images. Then, the supervised and unsupervised techniques are applied in combination to segment the exudates.

Esther et al. [12] emphasized on the segmentation of optic disk using watershed segmentation. The retinal fundus image is used by the ophthalmologists for analyzing the important features of the eye like the geometrical features of optic disk, the anterior segment of the retina, and any abnormal growth of any region in the eye. Finding the position of the optical disk in the retina plays a vital role in the diagnosis and differentiation of the exudates from the optical disk. Then, the exact contours are found using watershed transformation. The methodology for the extraction of the OD contour is mainly based on mathematical morphology along with thresholding and watershed transformation.

Vimala et al. [13] proposed an automatic and efficient technique for the identification of OD and exudates. The preprocessing of retinal images was implemented to transform the RGB image into L*A*B color space. The segmentation of OD in the preprocessed LAB image was performed using line operator and fuzzy C means clustering technique. The extraction of exudates was done by applying K-means clustering with a predefined threshold and number of clusters, and finally the classification was done using support vector machine.

Hole et al. [14] enhanced the detection of exudates by applying genetic algorithm to improve the accuracy, reduce the computational time, and for its easier diagnosis. These various approaches based on genetic algorithm were suggested to produce an image with good and natural contrast as the distribution of intensity level of the pixels in a fundus image is within a small range. Jelinek et al. [15] proposed a technique in which fusion technique was applied to obtain accurate segmentation.

Thus, based on the above literature review, the techniques that were implemented to improve the accuracy in segmentation exudates from the retinal fundus images are

- Initially, K-means clustering is applied on the retinal images to separate the exudates and optic disk.
- Genetic algorithm (GA) is applied to the clustered output in which the fitness function is calculated and then ranked to obtain the best case, average case, and worst case to improve the accuracy of segmentation.
- The elimination of the optic disk which has the same intensity as that of the exudates is performed using watershed segmentation.

3 Proposed Methods

The flow diagram of the proposed method is shown in Fig. 1, the input image was taken from the CHASE and STARE database for the segmentation of exudates. The intensity value of the pixel in the retinal images will range from 0 to 255 (for each R, G and B planes). The preprocessing was performed by converting RGB channel into the L*A*B channels and later it is converted to gray scale.

Segmentation is done using K-means clustering algorithm in which the centroids are calculated and the mean values are initialized. Genetic algorithm is used for the accurate segmentation of the exudates in which the fitness function is calculated to perform crossover and mutation in which the best case, average case, and worst case are determined. Crossover is performed between the segmented images obtained from the K-means clustering segmentation. Before performing the mutation process, the grayscale image is converted into the RGB channels. These three-segmented channels are further combined by the mutation process to obtain the genetic algorithm output.

The optic disk is eliminated using watershed segmentation which is performed only for the grayscale images. In watershed segmentation, gradient magnitude is used as a segmentation function. Background markers are used for eliminating the optic disk.

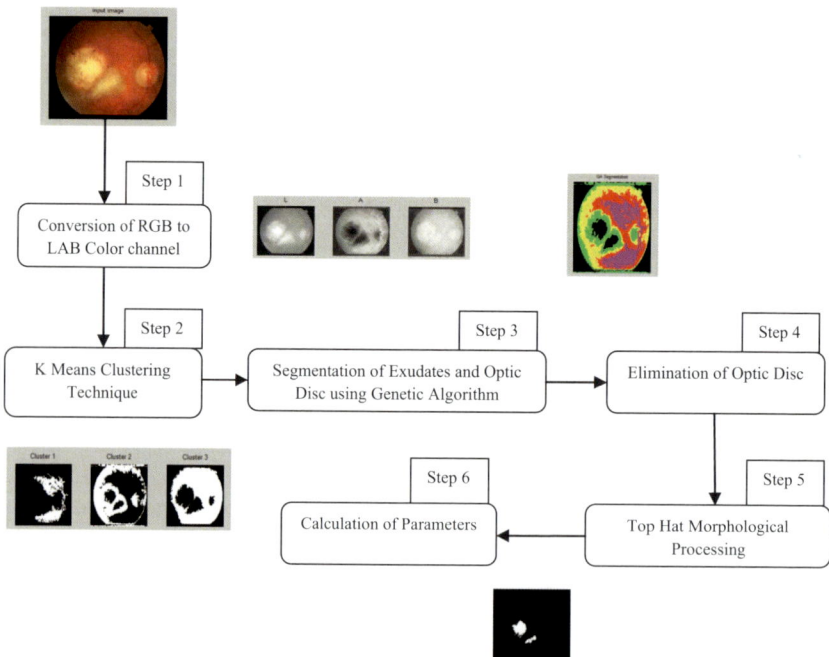

Fig. 1 Flow diagram of the proposed method

3.1 Preprocessing

The input image consisting of 700 × 605 pixels stored in TIFF format are converted into L*a*b channels using color space conversion and the preprocessed image is considered for the segmentation process. The input images are resized to focus on the region of interest and to reduce the execution time. The input image is shown in Fig. 2a and the preprocessed image, that is, resized image for the consequent process is shown in Fig. 2b.

The RGB image is converted into L*a*b* color space as it is the most exact technique for the representation of color information and can be applied to any portable device. LAB color space consists of a luminosity layer L* and two chromaticity layers a* and b* which provides information of the axis where the red–green or blue–yellow exist as the predominant color in the fundus images are red and yellow colors. The major color information of the image lies in the a* and b* layer. The flow diagram shown in Fig. 3 explains the conversion of input fundus images into L*a*b channels.

The input fundus image in RGB format is converted into grayscale format for the application of watershed segmentation for the elimination of optic disk which is shown in Fig. 4a. The LAB channel images converted into grayscale format are shown in Fig. 4b. The preprocessed image is subjected next for segmentation process, a* channel has been considered as the localized information of red which is important for further identification of optic disk and extraction of exudates.

The extracted A channel image components are grouped into a set of three clusters using K-means clustering algorithm. The main aim of using K-means clustering is to partition the image component into K clusters so that each observation will belong to one cluster to the nearest centroid that has been initialized. K-means clustering algorithm makes sure that the clusters are constructed based on comparable spatial extent and the expectation-maximization mechanism allows clusters to have different shapes.

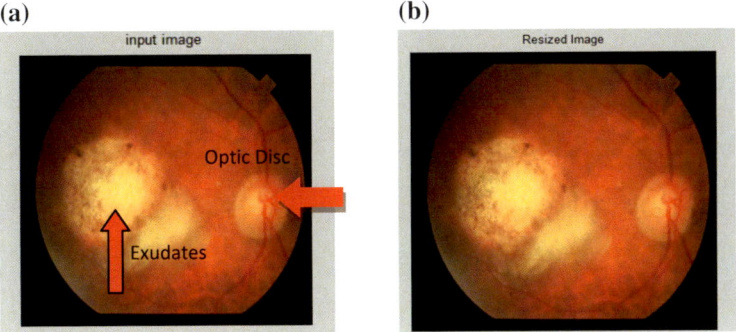

Fig. 2 **a** Input retinal image with exudates. **b** Resized image

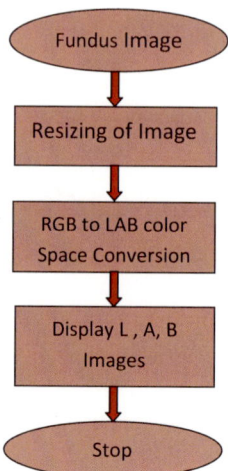

Fig. 3 Flow diagram of plane separation

(a) **(b)**

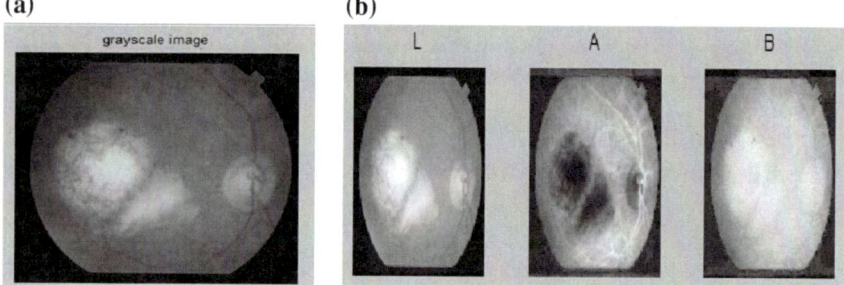

Fig. 4 **a** Grayscale image. **b** L*a*b Channels of input image

By continuous iteration of K-means algorithm with a range of K values, the cluster data sets are grouped and the results are compared. The cluster centroid and the mean distance between the data points are the two main metrics used for comparing the results across different K values. Then, the algorithm finds the relevant cluster and data set are labelled for particular pre-chosen cluster $K = 3$. The algorithm to implement the K-means clustering is composed of the following steps:

(i) Initialization process:
Initialization of K points for representing the objects to be clustered.
(ii) Grouping into clusters:
The pixels that are nearby to the initialized centroid are identified by Euclidean distance measure parameter and the grouping of pixels in the image is performed.
(iii) Optimization of centroid of clusters:

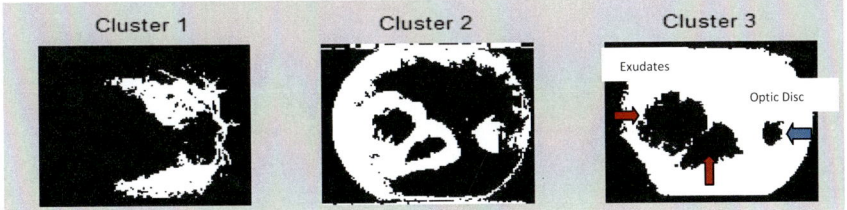

Fig. 5 Segmented output

    When all pixels have been assigned to any one cluster, the positions of the K centroids are recalculated again.

(iv) Reiteration process:

    Steps (ii) and (iii) are repeated until the centroids can no longer move. This produces the separation of the objects into clusters groups.

    The entire process can be summarized in following steps:

1. The input RGB retinal fundus image is converted into L*A*B channels and A* channel output is considered for further subsequent steps.
2. Apply K-means clustering algorithm based on Euclidean distance metric to group the pixels in the image into three clusters.
3. Every pixel in the image is labeled and it returns an index corresponding to a cluster. Every pixel in the image is labeled with its cluster index.
4. The output of K-means clustering for the three clusters segmented is shown in Fig. 5. Thus, the cluster 3 output shown in Fig. 5c containing optic disk and exudates are further considered for the segmentation of exudates.

3.2 Segmentation of Exudates and Optic Disk Using Genetic Algorithm

The clustered outputs are processed with genetic algorithm for optimization and to improve the accuracy of the segmented output. Genetic algorithm (GA) is used to obtain the accurate optimization output for the segmentation of exudates. In addition to that, this method is suitable for all retinal images with both exudates and the optic disk which is used for the diagnosis of diabetic retinopathy. Figure 6 highlights the steps that are performed in genetic algorithm. A typical genetic algorithm requires the following:

1. Representation of the clustered output solution to process in GA domain.
2. Fitness function to evaluate the solution domain.

    Each candidate solution is represented as an array of bits. Arrays of other types and structures are used in essentially the same way. To facilitate the simple crossover

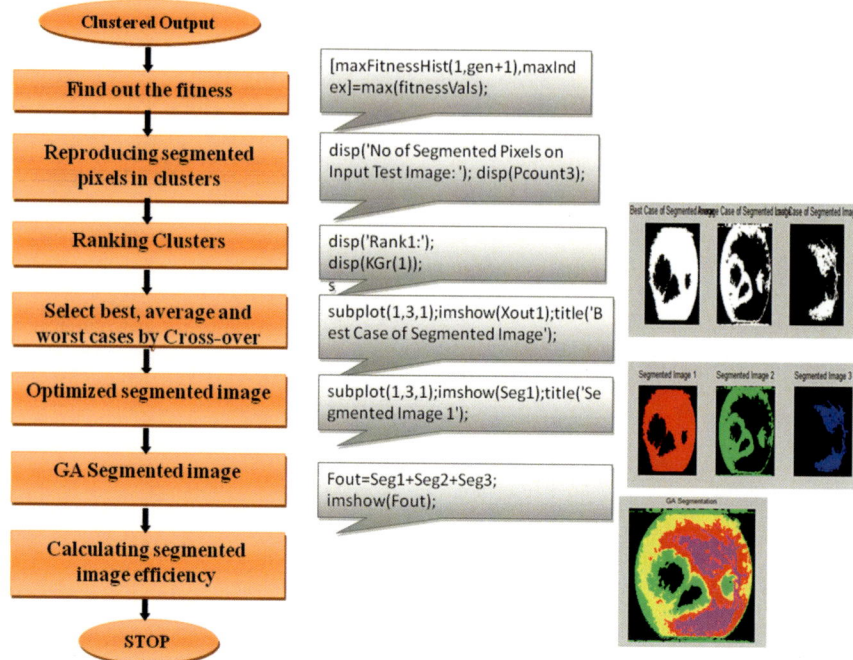

Fig. 6 Flowchart to perform genetic algorithm

operations, the parts in the images are aligned according to their fixed sizes and this main property makes the genetic representations more convenient.

The important steps that are to be performed are as follows:

Step 1 Encoding the problem in a binary string.
Step 2 Random population generation.
Step 3 Fitness calculation of each solution.
Step 4 Pairs of parent strings are selected based on fitness.
Step 5 New strings are generated with crossover and mutation process until a new population has been produced.
Step 6 Repeat steps 2–5 until satisfying solution is obtained.

Initialization Process

The genetic algorithm requires individual which should be represented by strings, here the cluster 3 output is encoded into integer strings. The region containing exudates (indicated by red arrow) and optic disk (indicated by blue arrow) are only black in color and other regions are white as shown in Fig. 5c. The population size depends on the number of pixels in exudates and optic disk region. Here, generic representation is "seeded" in these areas of interest where optimal solutions are likely to be found. Here, in our example image there are three regions ($r1$, $r2$, $r3$), two exudates and one optic disk region. A number of chromosome $\alpha = \alpha1$, $\alpha2$, $\alpha3$ are

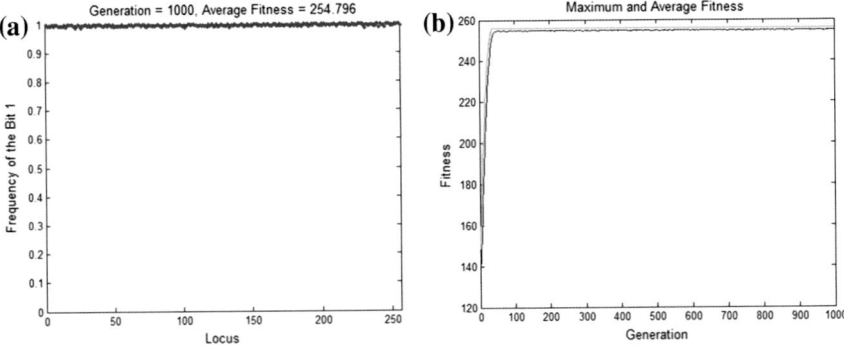

Fig. 7 **a** Fitness generation calculation. **b** Graph between generation and fitness

encoded as integer strings. The number of chromosomes will be equal to the number of segmented region of interest.

Selection Process

During each successive generation, a portion of the existing population should be selected in order to proceed further to breed a new generation. The fitness function is based on the region of selection in the cluster 3. The properties that are considered for the fitness function are size of the region and intensity of pixels which decide the similarity and the variation between regions. The fitness function can be calculated as the inverse of product of the variance and the size of each region. The best individuals are selected based on the above fitness function.

Roulette wheel selection operators are used for the selection in which the individuals with the best fitness function occupies a larger portion in the wheel and vice versa. Individual solutions are selected through a *fitness-based* process (Fig. 7a), where fitter solutions (as measured by a fitness function) are typically more likely to be selected. Graph between generation and fitness id obtained are shown in Fig. 7b.

Genetic Operators

The genetic operators like crossover and mutation are used to produce a second generation population of solutions.

After the selection of the chromosome, two points crossover operator can be applied to produce two offsprings. The length where the crossover should occur depends on the variation of intensity of pixels in the portioned region. This new solution created, typically shares many of the characteristics of its parents. Consider the representation of two parents which is shown in Fig. 8. The genetic representation of two exudates region as parent 1 and parent 2 is divided into 10 cells where each cell represents a gene. The crossover may occur between R3 and R5 of parent 1 with R3 and R5 of parent 2 and R8 and R9 of parent 1 with R8 and R9 of parent 2.

For each new child, new parents are selected and the process continues until a new population of solutions of appropriate size is generated. Although reproduction

Parent 1

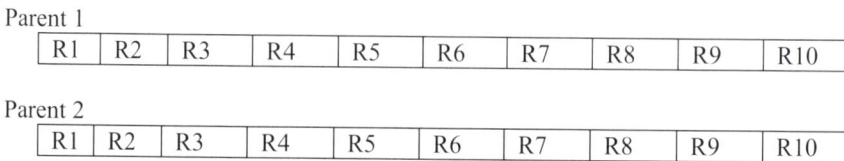

Fig. 8 Selected parents for crossover operation

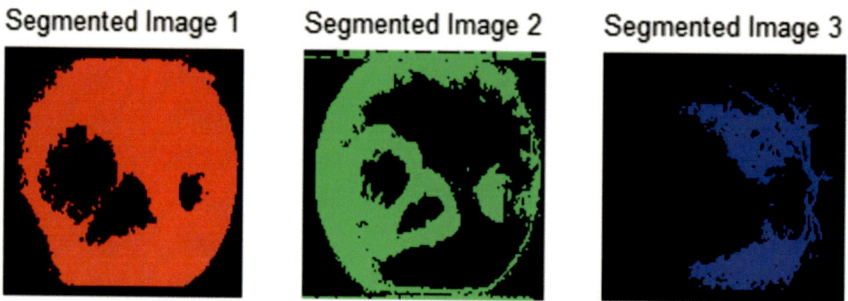

Fig. 9 Segmented images

methods that are based on the use of two parents are more "biology inspired", some research suggests that more than two "parents" generate higher quality chromosomes.

Thus, the next generation population of chromosomes that is different from the initial generation is developed. This will increase the average fitness for the population, since only the best organisms from the first generation are selected for breeding, along with a small proportion of less fit solutions. These less fit solutions ensure genetic diversity within the genetic pool of the parents, and therefore ensure the genetic diversity of the subsequent generation of children.

Termination of GA

This generational process will be repeated until a termination condition has been reached. There are some conditions that are to be considered. Common terminating conditions are as follows:

- A solution should be found that satisfies the minimum criteria.
- Fixed number of generations has been reached.
- Allocated budget (computation time/money) is reached.
- The highest ranking solution's fitness is reaching or has reached a plateau such that successive iterations no longer produce better results.
- Manual inspection.
- Combinations of the above.

Best case, worst case, and average cases are obtained and shown in Fig. 9. These segmented outputs will be considered for the termination process to find the final GA segmentation output which is shown in Fig. 10.

GA Segmentation

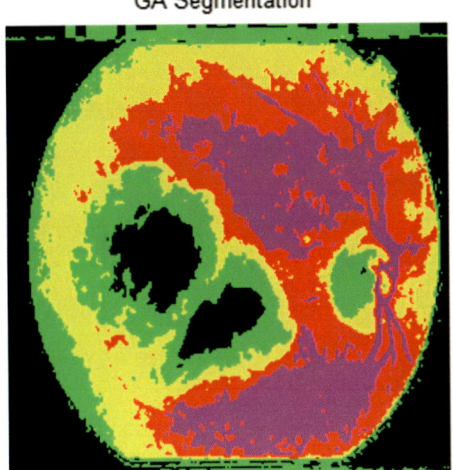

Fig. 10 GA segmentation output

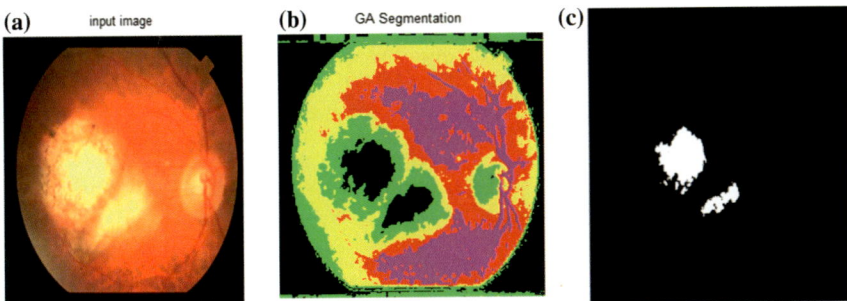

Fig. 11 **a** Input image. **b** GA segmented output containing exudates and OD. **c** Elimination of OD

3.3 Elimination of Optic Disk

Eigenvector-based OD detection is implemented to calculate the covariance matrix of the database images and the disk region is constructed from the eigenvectors. The region with minimum distance is grouped as optical disk. This work is used to distinguish the exudates and the optic disk in an image. This approach is computationally inexpensive and highly robust. Morphological operators are determined for the OD detection. Dilation, top hat transformation operators are used for the elimination of OD. The time taken by this algorithm is also very low. Watershed segmentation technique is used for average filtering and contrast enhancement as preprocessing steps. This is further processed to obtain the modified gradient vector flow algorithm for OD segmentation. Figure 11a shows the input image and 11c reveals the elimination of OD and presence of exudates.

4 Results and Discussion

The classification (or prediction) data is divided into two different classes, positives (P) and negatives (N). This classification produces four types of outcomes—true positive, true negative, false positive, and false negative.

- True positive (TP): Exudates correctly identified as exudates.
- False positive (FP): Non-exudates pixels incorrectly identified as exudates.
- True negative (TN): Non-exudates pixels correctly identified as non-exudates pixels itself.
- False negative (FN): Exudates incorrectly identified as non-exudates.

Accuracy
Accuracy is defined and calculated using Eq. 1 as the ratio of sum of TP and TN (correct identification of exudates) divided by the total number of the images.

$$ACC = \frac{TP + TN}{TP + TN + FN + FP} \tag{1}$$

Sensitivity
Sensitivity (SN) will provide us with the details of identification of exudates region. Sensitivity is calculated using Eq. 2 which is defined as the ratio of correctly predicted exudates (TP) to the total number of TP and FN.

$$SN = \frac{TP}{TP + FN} \tag{2}$$

Specificity
Specificity is also called the true negative rate (TNR) and is calculated as given in Eq. 3. Specificity gives us the correct identification of non-exudates region (TNR, expressed as a percentage).

$$SP = \frac{TN}{TN + FP} \tag{3}$$

The proposed method is compared with existing method in terms of the algorithm used and three parameters, namely, accuracy, sensitivity, and specificity and has also been utilized to evaluate the improvement in the performance of the proposed method as shown in Table 1 and the outputs for different test images are shown in Fig. 12. The proposed technique shows an improvement in all the parameters.

Table 1 Comparison of existing and proposed method parameters

Parameter	Existing technique [16]	Proposed technique
Algorithm used	Adaptive threshold algorithm (%)	Genetic algorithm (%)
Accuracy	93.69	94
Sensitivity	90.42	93
Specificity	94.60	95

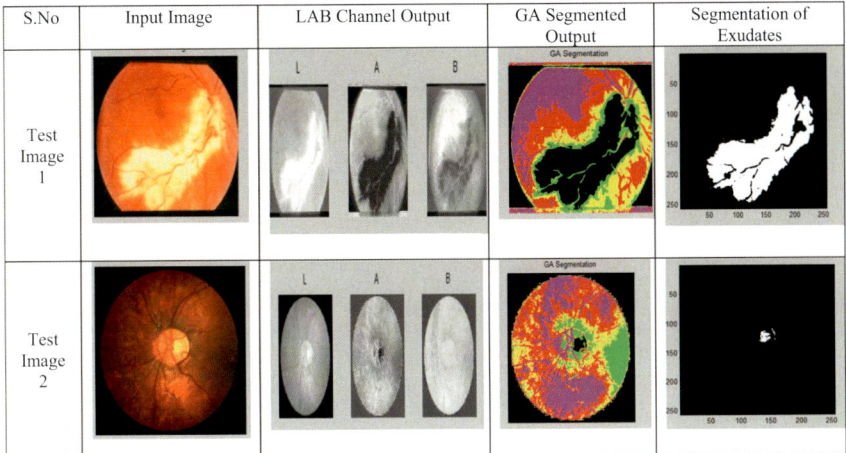

Fig. 12 Segmentation of exudates for the different test images

5 Conclusion

Diabetic retinopathy is a disease caused to a diabetic patient, it is progressive, and hence this requires continuous monitoring of the abnormalities in the eye which may lead to loss of vision. A novel method that aims to improve the accuracy in the segmentation of exudates to help the ophthalmologists for the identification and the computer-aided diagnosis of diabetic retinopathy has been proposed and analyzed in this work.

References

1. Antal B, Hajdu A (2012) An ensemble-based system for microaneurysm detection and diabetic retinopathy grading. IEEE Trans Biomed Eng 59(6):1720–1726
2. Deshmukh AV, Patil TG, Patankar SS, Kulkarni JV (2015) Features based classification of hard exudates in retinal images. In: 2015 international conference on advances in computing, communications and informatics (ICACCI), pp 1652–1655

3. Pires R, Avila S, Jelinek HF, Wainer J, Valle E, Rocha A (2014) Automatic diabetic retinopathy detection using bossa nova representation. In: 2014 36th annual international conference of the IEEE engineering in medicine and biology society (EMBC), pp 146–149

4. Roychowdhury S, Koozekanani DD, Parhi KK (2014) Dream: diabetic retinopathy analysis using machine learning. IEEE J Biomed Health Inform 18(5):1717–1728

5. Esmaeili M, Rabbani H, Dehnavi AM, Dehghani A (2012) Automatic detection of exudates and optic disk in retinal images using curvelet transform. IET Image Proc 6(7):1005–1013

6. Aqeel, AF, Ganesan S (2014) Automated algorithm for retinal image exudates and Drusens detection, segmentation, and measurement. In: 2014 IEEE international conference on electro/information technology (EIT), pp 206–215

7. Pereira C, Gonçalves L, Ferreira M (2015) Exudate segmentation in fundus images using ant colony optimization approach. Inf Sci 296:14–24

8. Soares I, Castelo-Branco M, Pinheiro AM (2016) Optic disc localization in retinal images based on cumulative sum fields. IEEE J Biomed Health Inform 20(2):574–585

9. Yu H, Barriga ES, Agurto C, Echegaray S, Pattichis MS, Bauman W, Soliz P (2012) Fast localization and segmentation of optic disk in retinal images using directional matched filtering and level sets. IEEE Trans Inf Technol Biomed 16(4):644–657

10. Radha R, Lakshman B (2013) Retinal image analysis using morphological process and clustering technique. Sig Image Process 4(6):55

11. Kusakunniran W, Wu Q, Ritthipravat P, Zhang J (2018) Hard exudates segmentation based on learned initial seeds and iterative graph cut. Comput Methods Programs Biomed 158:173–183

12. Esther JJJ, Sophia SG (2014) Detecting optic disc in digital fundus images using stochastic watershed transformation. IJREAT Int J Res Eng Adv Technol 2(1)

13. Vimala GAG, Mohideen SK (2013) Automatic detection of optic disk and exudate from retinal images using clustering algorithm. In: 2013 7th international conference on intelligent systems and control (ISCO), pp 280–284

14. Hole KR, Gulhane VS, Shellokar ND (2013) Application of genetic algorithm for image enhancement and segmentation. Int J Adv Res Comput Eng Technol (IJARCET) 2(4):1342

15. Jelinek HF, Pires R, Padilha R, Goldenstein S, Wainer J, Bossomaier T, Rocha A (2012) Data fusion for multi-lesion diabetic retinopathy detection. In: 2012 25th international symposium on computer-based medical systems (CBMS), pp 1–4

16. Wisaeng K, Hiransakolwong N, Pothiruk E (2015) Automatic detection of exudates in retinal images based on threshold moving average models. Biophysics 60(2):288–297

Analysis of Feature Ranking Methods on X-Ray Images

H. Roopa and T. Asha

Abstract Mycobacterium causes an infectious disease called tuberculosis which can be diagnosed by its various symptoms like fever, cough, etc. Tuberculosis can also be analyzed by understanding the chest X-ray of the patient which is revealed by an expert physician. The chest X-ray image contains texture and shape-based features which are extracted from X-ray image using image processing concepts. This paper presents implementation of various feature weighting methods on the extracted features of X-ray images. These feature weighting methods are analyzed using linear regression model and Linear Discriminant Analysis (LDA) model. The performance of various feature weighting methods is compared and found that the accuracy of weights by PCA using linear regression model is 98.75% which is better than other methods.

1 Introduction

Mycobacterium causes infectious disease tuberculosis (TB) [1] which usually affects lungs. When an infected person coughs, TB is transmitted from one person to another through air. TB affects lungs but can also appear in other parts of body like spine, bones, brain, and kidney. Chest X-ray findings, fever, loss of weight, and coughs are some symptoms of TB. To raise awareness about TB, world TB day is recognized on March 24 to support prevention of TB. The level of TB in all country is monitored by World Health Organization (WHO) [2]. Every year more than 9 million people develop TB and 1.5 million die from the disease.

Features represent information in an image. To understand the information present in an image, it must be analyzed and measured in various angles to get relevant information in a particular domain like medical image, satellite images, etc., so,

H. Roopa (✉) · T. Asha
Department of C. S. E, Bangalore Institute of Technology, K. R. Road, Bangalore, India
e-mail: roopatejas@gmail.com

T. Asha
e-mail: asha.masthi@gmail.com

© Springer Nature Switzerland AG 2019
D. Pandian et al. (eds.), *Proceedings of the International Conference on ISMAC in Computational Vision and Bio-Engineering 2018 (ISMAC-CVB)*, Lecture Notes in Computational Vision and Biomechanics 30,
https://doi.org/10.1007/978-3-030-00665-5_130

feature extraction plays an important role in analyzing medical image like TB image. Feature ranking is one of the feature selection methods where the features are ranked using certain defined measures.

Linear Regression (LR) model is a method to find a relationship between one dependent variable and series of changing independent variables by the straight line that represents the linear equation of the observed data. Linear Discriminant Analysis (LDA) separates two or more classes to find the best linear combination of features.

The aim of this paper is to analyze extracted features of chest X-ray images. Then, these features are ranked using different feature weighting methods like Relief, Principle Component Analysis (PCA), Support Vector Machine (SVM), etc. and examined using linear regression model and LDA model for classifying TB disease.

The paper is organized as follows: a brief overview of related work is given in Sects. 2 and 3 describes method for extracting and ranking features of X-ray image, and analyze these by applying Linear Regression (LR) model and LDA model. Section 4 gives details of experimental illustration and results of various ranking methods are explained in Sect. 5. Finally, Sect. 6 concludes the paper.

2 Related Work

For medical image analysis, Perner et al. [3] developed data mining and image processing tool using image mining concepts. Descriptions of list of attributes as given by experts are stored in a database and then a classification technique decision tree induction tree is applied to this to extract expert knowledge. This tool was used for various applications like breast MRI data, etc. Asha et al. [4] used Association Rule Mining (ARM) techniques on TB data sets to improve TB disease prediction. The symptoms of TB were considered and many descriptive rules were written and these were combined with an association classification technique used for predicting TB.

Zou et al. [5] proposed a Max-Relevance-Max-Distance (MRMD) feature extraction method, which provides stabilized feature selection method using new measure index for ranking the features that improve the efficiency of classification. Bravi et al. [6] used concave approximation of zero-norm function for ranking the features and evaluated it using SVMs with Gaussian kernel. Chung et al. [7] used feature selection and Taguchi genetic algorithm together on DNA microarray data and then the performance was evaluated using KNN with Leave-One-Out Cross-Validation (LOOCV).

Razmjoo et al. [8] developed an efficient way to rank features in an incremental manner which can be used in classification methods. Yihui et al. [9] extracted wavelet features from microarray data, then SVM was used for classification. Zyout et al. [10] extracted textual pattern from mammogram images and Particle Swarm Optimization

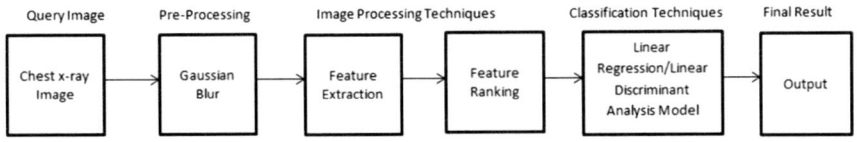

Fig. 1 Feature extraction and feature ranking of X-ray image

(PSO) was applied to select the most discriminative features, then SVM was used for classification. City block distance measure was used by Roopa et al. [11] for segmenting chest X-ray image which helps in diagnosis of TB.

3 Proposed Methodology

Features which are relevant and informative with respect to TB disease should be considered. The X-ray image must first be preprocessed and then important features are extracted from the affected region of X-ray using image processing methods. These features are ranked using various methods and evaluated using LR and LDA models. The feature extraction and ranking of an X-ray image is shown in Fig. 1.

The steps involved in extracting and ranking features from X-ray image are

1. X-ray image is taken as an input.
2. Preprocessing:
   Noise and redundant data in the X-ray image are removed using Gaussian blur filter method.
3. Feature extraction:

Geometric features and texture-based features [12] can be used to measure the characteristics of TB. Both the shape descriptors and texture descriptors were extracted from X-ray image. Extracting as many features from the region of interest of TB is one of the concerns in this work which is done by applying roipoly () function using MATLAB software.

Shape-based features were used to measure the region of interest in a TB image. The statistics like area, perimeter, coordinates of region centroid, major axis length, minor axis length, eccentricity, orientation, etc. are some of the characteristics of shape-based feature extraction which is obtained by analyzing external boundary of the X-ray image. These features are obtained using function called as regionprops ().

Texture descriptors proposed by Haralick [13] defines 14 statistics that can be calculated from the co-occurrence matrix of the image. Texture feature extraction refers to surface characteristics and appearance of an object in an image. Entropy can be found using entropy (), graycrops () function can be used to extract homogeneity, contrast, energy, correlation, numel (UL) where UL represents uniformity, mean, standard deviation can be calculated using $SD = sqrt(VR)$, where VR represents the variance and skewness using function skewness(). All these feature values were extracted from chest X-ray images.

4. Feature ranking:

    i. Weight by PCA:
       The attribute weight of TB data is obtained using component created by PCA. The attribute weight reflects the importance of attributes with respect to class attribute.

    ii. Weight by chi squared:
       For each attribute of TB data, the value of chi-squared statistic with respect to class attribute is calculated to obtain the weight of the attribute.

    iii. Weight by correlation:
       The weight of feature of TB data is calculated by computing value of correlation for each feature of TB data with respect to class attribute. The attribute weight is obtained by absolute value of correlation.

    iv. Weight by information gain:
       The attribute weight of TB data is obtained by calculating each attribute information gain with respect to class attribute.

    v. Weight by relief:
       The features of TB data values are distinguished between instances of same and different classes that are near each other to obtain weights of features.

    vi. Weight by SVM:
       The attribute weight of TB data is obtained by coefficients of hyperplane which is calculated by SVM with respect to class attribute.

5. Classification model:
   Linear regression model is applied on ranked features. Here, features are selected using M5 prime method by eliminating collinear features with a minimum tolerance of 0.05. Then, LDA model is applied to evaluate the ranked features and the results were compared with LR model.

6. Final result:
   Features obtained from X-ray images are classified as affected or normal image using LR and LDA model and the performance of feature weighting methods was evaluated using tenfold cross-validation technique.

4 Experimental Illustration

Consider TB-affected image, initially, the image is preprocessed to remove noise using Gaussian function. Then, texture- and shape-based features are extracted using MATLAB based on image processing concepts. The results of feature extraction method of TB X-ray image are illustrated in Figs. 2, 3, and 4, respectively.

The affected region of TB image is marked and then the marked region is masked to extract shape-based and texture-based feature represented in Figs. 3 and 4, respec-

Fig. 2 TB image

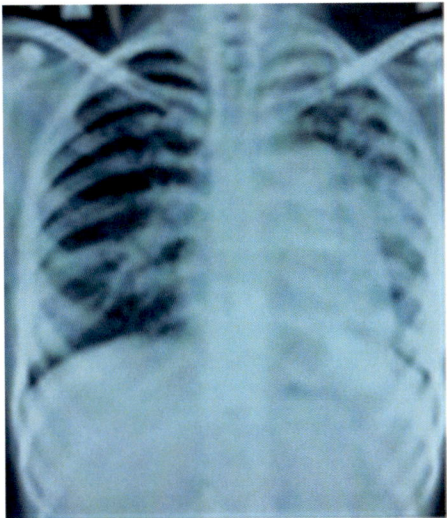

Fig. 3 Marked image

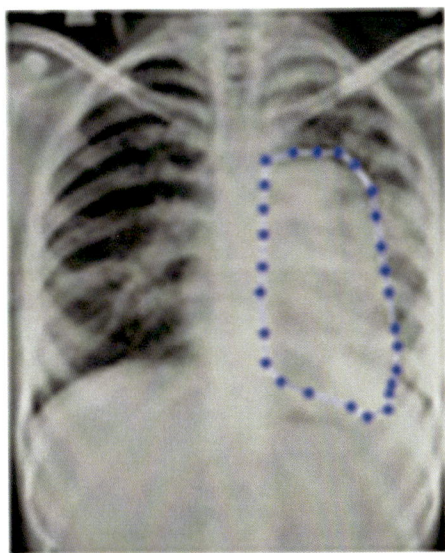

tively. Using the image processing methods, texture-based and shape-based feature values extracted from the masked image of TB are stored in a file. Table 1 gives information of extracted values of a single TB image.

For the implementation, we have considered 47 TB-affected images and 30 normal images. Features from X-ray images were extracted using MATLAB software and feature weighting and evaluation using classification model were performed using rapid miner software.

Fig. 4 Masked image

Table 1 Extracted feature values of a single TB image

Entropy	2.0954
Skewness	0.2724
SNR	6.1205
Homogeneity	0.9964
Contrast	0.16
Energy	0.4991
Correlation	0.9933
Total mean	110.4313
Variance	110.0112
Standard deviation	10.4886
Uniformity	111
Area	1744
Perimeter	172.625
Centroid	193.9060 218.2534
Major axis	65.7998
Minor axis	36.1286
Eccentricity	0.8358

All extracted feature values were used as an input for weighting the features using various feature ranking methods, and these features were evaluated using LR and LDA models separately to classify the X-ray image as affected or normal. The performance of the classification model was analyzed using tenfold cross-validation technique.

Table 2 Comparison results of feature ranking methods using LR and LDA models

Classification accuracy

Feature ranking methods	Linear regression model	Linear discriminant analysis model
Weight by PCA	98.75	93.57
Weight by chi squared	96.07	90.8
Weight by correlation	96.07	90.8
Weight by information gain	96.07	85.8
Weight by relief	96.07	82.14
Weight by SVM	94.82	78.2

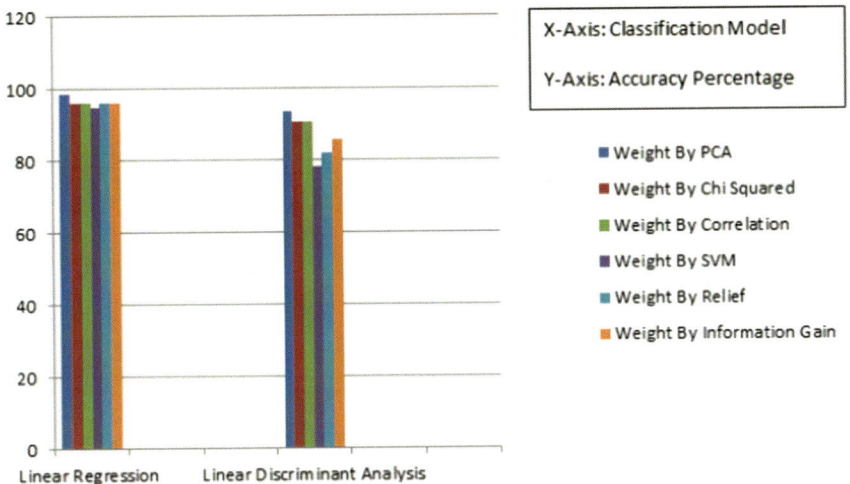

Fig. 5 Comparison of feature ranking methods using LR and LDA models

5   Results

Wrapper feature selection method is applied on extracted features of TB data. These features are ranked using various ranking methods and evaluated using LR and LDA models. The performances of these models are shown in Table 2 and Fig. 5, respectively.

The accuracy using weight by PCA of LR model is good when compared to LDA model, and also with respect to other feature weighting methods as shown in Table 2 and Fig. 5, respectively.

The implementation snapshots of feature weighting by PCA and SVM are given in Figs. 6, 7, 8, and 9, respectively.

```
PerformanceVector:
accuracy: 93.57% +/- 10.20% (mikro: 93.51%)
ConfusionMatrix:
True:    Affected         Normal
Affected:        47      4
Normal: 1        25
precision: 96.15% (positive class: Normal)
ConfusionMatrix:
True:    Affected         Normal
Affected:        47      4
Normal: 1        25
recall: 81.67% +/- 32.02% (mikro: 86.21%) (positive class: Normal)
ConfusionMatrix:
True:    Affected         Normal
Affected:        47      4
Normal: 1        25
AUC (optimistic): 1.000 +/- 0.000 (mikro: 1.000) (positive class: Normal)
AUC: 0.500 +/- 0.000 (mikro: 0.500) (positive class: Normal)
AUC (pessimistic): 0.000 +/- 0.000 (mikro: 0.000) (positive class: Normal)
```

Fig. 6 Performance of feature weighting by PCA using LDA model

```
PerformanceVector:
accuracy: 98.75% +/- 3.75% (mikro: 98.70%)
ConfusionMatrix:
True:    Affected         Normal
Affected:        47      0
Normal: 1        29
precision: 96.67% +/- 10.00% (mikro: 96.67%) (positive class: Normal)
ConfusionMatrix:
True:    Affected         Normal
Affected:        47      0
Normal: 1        29
recall: 100.00% +/- 0.00% (mikro: 100.00%) (positive class: Normal)
ConfusionMatrix:
True:    Affected         Normal
Affected:        47      0
Normal: 1        29
AUC (optimistic): 0.983 +/- 0.050 (mikro: 0.983) (positive class: Normal)
AUC: 0.983 +/- 0.050 (mikro: 0.983) (positive class: Normal)
AUC (pessimistic): 0.983 +/- 0.050 (mikro: 0.983) (positive class: Normal)
```

Fig. 7 Performance of feature weighting by PCA using LR model

PerformanceVector

```
PerformanceVector:
accuracy: 78.21% +/- 23.10% (mikro: 77.92%)
ConfusionMatrix:
True:   Affected        Normal
Affected:        41        10
Normal: 7        19
precision: 73.08% (positive class: Normal)
ConfusionMatrix:
True:   Affected        Normal
Affected:        41        10
Normal: 7        19
recall: 60.83% +/- 41.84% (mikro: 65.52%) (positive class: Normal)
ConfusionMatrix:
True:   Affected        Normal
Affected:        41        10
Normal: 7        19
AUC (optimistic): 1.000 +/- 0.000 (mikro: 1.000) (positive class: Normal)
AUC: 0.500 +/- 0.000 (mikro: 0.500) (positive class: Normal)
AUC (pessimistic): 0.000 +/- 0.000 (mikro: 0.000) (positive class: Normal)
```

Fig. 8 Performance of feature weighting by SVM using LDA model

6 Conclusion

The proposed work uses image processing concepts to extract relevant and important features from a chest X-ray image to diagnose whether a person is TB infected or not, by preprocessing data and feature extraction process. Then, wrapper feature selection methods are implemented on these extracted features and ranked using various methods. These features are classified using linear regression and LDA classification model to analyze the TB disease. The performance of feature weighting by PCA using LR model is examined and found that the accuracy is 98.75% which is better when compared to the accuracy of other feature weighting methods. Future work is to consider more images and carry out the implementation on this huge data set.

PerformanceVector

```
PerformanceVector:
accuracy: 94.82% +/- 6.36% (mikro: 94.81%)
ConfusionMatrix:
True:    Affected          Normal
Affected:        47        3
Normal: 1        26
precision: 96.30% (positive class: Normal)
ConfusionMatrix:
True:    Affected          Normal
Affected:        47        3
Normal: 1        26
recall: 84.17% +/- 30.38% (mikro: 89.66%) (positive class: Normal)
ConfusionMatrix:
True:    Affected          Normal
Affected:        47        3
Normal: 1        26
AUC (optimistic): 0.973 +/- 0.059 (mikro: 0.973) (positive class: Normal)
AUC: 0.973 +/- 0.059 (mikro: 0.973) (positive class: Normal)
AUC (pessimistic): 0.973 +/- 0.059 (mikro: 0.973) (positive class: Normal)
```

Fig. 9 Performance of feature weighting by SVM using LR model

References

1. Asha T, Murthy KNB, Natarajan S (2012) Data mining techniques in the diagnosis of tuberculosis. INTECH open access Publisher
2. Global Tuberculosis Report 2015-http://apps.who.int/iris/bitstream/10665/191102/1/9789241565059_eng.pdf
3. Perner P (2002) Image mining: issues, framework, a generic tool and its application to medical-image diagnosis. Eng Appl Artif Intell 15(2):205–216
4. Asha T, Natarajan S, Murthy KNB (2011) A study of associative classifiers with different rule evaluation measures for tuberculosis prediction. IJCA special issue on "Artificial intelligence techniques—novel approaches & practical applications" AIT
5. Zou Q, Zeng J, Cao L, Ji R (2016) A novel features ranking metric with application to scalable visual and bioinformatics data classification. Neurocomputing 173:346–54
6. Bravi L, Piccialli V, Sciandrone M (2017) An optimization-based method for feature ranking in nonlinear regression problems. IEEE Trans Neural Netw Learn Syst 28(4):1005–1010
7. Chuang L-Y et al (2011) A hybrid feature selection method for DNA microarray data. Comput Biol Med 41(4):228–237
8. Razmjoo A, Xanthopoulos P, Zheng QP (2017) Online feature importance ranking based on sensitivity analysis. Expert Syst Appl 85:397–406
9. Liu Y (2009) Wavelet feature extraction for high-dimensional microarray data. Neurocomputing 72(4):985–990
10. Zyout I, Czajkowska J, Grzegorzek M (2015) Multi-scale textural feature extraction and particle swarm optimization based model selection for false positive reduction in mammography. Comput Med Imaging Graph 46:95–107

11. Roopa H, Asha T (2016) Segmentation of X-ray image using city block distance measure. In: 2016 International conference on control, instrumentation, communication and computational technologies (ICCICCT), Kumaracoil, pp 186–189
12. Roopa H, Asha T (2018) Feature extraction of chest X-ray images and analysis using PCA and kPCA. Int J Electr Comput Eng (IJECE) 8(5)
13. Haralick RM (1979) Statistical and structural approaches to texture. Proc IEEE 67(5):786–804

Salient Object Detection for Synthetic Dataset

Aashlesha Aswar and Arati Manjaramkar

Abstract Salient object detection essentially deals with various image processing and video saliency methodologies such as object recognition, object tracking, and saliency refinement. When image contains diverse object parts with cluttered background then using background prior we perform salient object detection through which we get more accurate and robust saliency maps. This paper introduces the analysis of salient object detection using synthetic dataset which also deals with negative interference of image that contains diverse object parts with cluttered background. Earlier study uses contrast prior but nowadays researchers use mainly boundary connectivity for improving the results. So, for detecting salient object we used four stages: first, we use SLIC superpixel method for image segmentation. Second, we use boundary connectivity which distinguishes the spatial layout of image region by considering image boundaries. Third, we use background measure and for reducing the noise in both foreground and background regions. Lastly, we use optimization framework through which we acquire a clean saliency map.

1 Introduction

Attention is the process of concentrating on one aspect of the environment and ignoring other things. So, visual attention has two stages: first the attention is distributed constantly over the external visual scene and processing of data is performed sequentially, second the attention is concentrated to a specific area of the visual scene and processing is performed serially. So, the word salient means most noticeable which is different from other parts at least in some aspect like color, texture, shape, etc. and object is nothing but a shape. There are various applications of visual attention according to researchers. So, the salient object detection is motivated by various

A. Aswar · A. Manjaramkar (✉)
Department of Information Technology, S.G.G.S. I E & T, Nanded, India
e-mail: akmanjaramkar@sggs.ac.in

A. Aswar
e-mail: aashleshaaswar@gmail.com

© Springer Nature Switzerland AG 2019
D. Pandian et al. (eds.), *Proceedings of the International Conference on ISMAC in Computational Vision and Bio-Engineering 2018 (ISMAC-CVB)*, Lecture Notes in Computational Vision and Biomechanics 30,
https://doi.org/10.1007/978-3-030-00665-5_131

applications like automatic image cropping, image retargeting, and adaptive image display on small devices, image collection browsing, advertising design, image and video cropping, and object segmentation [1]. Recent studies witnessed that visual attention helps in object detection, tracking, and recognition [2].

Basically, all the methods use the properties of backgrounds and object, detect the object on the basis of contrast prior and background prior. Previously for detecting salient object almost all saliency methods uses contrast prior where the contrast between the object and their background region is high but recently background prior is used for saliency computation in which it is considered that the image boundary regions are preferably background [3].

While we using boundary prior we came across drawback because of that we consider all the image boundary is background and when the object is partially connected to the image boundary then it is not considered as salient object. For avoiding this drawback, we used boundary connectivity method. The detection procedure of salient object is categorized into four stages: first image segmentation, image boundary connectivity, salient object detection using background measure, and optimization.

Early method works on various datasets such as MSRA10K, ECSSD, DUT-OMRON, etc. Early models used human eye fixation datasets as researcher inspired by biological methodology.

Ma and Zhang [4] propose a framework which outputs three levels of attention such as attended view, areas, and points. They proposed a contrast-based saliency map and fuzzy growing algorithm which stimulates the selection of human perception, using this they can accurately locate attention areas. This model enlightens the borders so it is not used for detecting the salient object.

Basically, salient object detection algorithms main goal is to segment the salient object from image and it is evaluated using foreground mask and boundary boxes, these methods use low-level cues called contrast prior. Contrast prior uses the unique identity of the object and differentiate object and background using contrast between color, regions, etc.

Hemami et al. [5] detect the limitation and requirements of saliency maps for various previous methods. Based on the analysis they proposed a frequency-tuned approach which uses the low-level features such as color and luminance to compute saliency in images. Cheng et al. [6] propose histogram-based contrast (HC) which accomplished a faster result and high-quality saliency map is used for reducing cost of computational efficiency. Perazzi et al. [7] use a contrast and spatial distribution to obtain an accurate pixel saliency maps. They decompose image into homogeneous elements to compute saliency using Gaussian filter. Hence, the assumption is that the photographers do not crop the object from the image. So, the contrast prior is not useful. And, when the object boundary moderately touches to image boundary in that contest the boundary prior also fails or destroyed.

Wei et al. [9] propose the geodesic saliency (GS) which is used to get high accuracy and speed and which is also more robust to detect salient object. This method overcomes the limitations of background prior by connecting the image boundary regions using an edge with an appropriate boundary weight. Yang et al. [8] propose the manifold ranking on graph to detect the salient region and uses the two-stage approach with foreground and background measures or queries for ranking thorough which saliency map is generated.

Zhu et al. [9] propose a robust measure that uses both contrast prior and background prior. The minimization of optimization framework is used to get uniform salient object detection. And, the smoothness term is used to get clear object and background segmentation. Alexe et al. [10] measure the object characteristics in Bayesian framework and they propose an object measure which combines various measures. Zhang et al. [11] propose a cascaded architecture using the ranking SVM which generates an ordered set of proposals for windows that contains objects. Cheng et al. [12] propose BING, i.e., binarized version of gradient feature which generates the objectness proposals and it is tested using few atomic operations. Jiang et al. [13] propose a saliency detection algorithm which integrates three cues such as uniqueness, focusness, and objectness (UFO) and it gives top performance.

It is difficult to attribute the results to the specific algorithm properties because saliency maps are not smooth. Saliency filters show that using Gaussian filter they formulated contrast and saliency estimation in a combined way [7]. So, using boundary prior rather than obtaining background from image boundary, we focus on the object detection using objectness proposal techniques [12].

We then use a superpixel objectness measure and by thresholding this measure, we appropriately obtain background and foreground regions. Then, assign saliency values to superpixels using the proposed robust saliency measure called background connectivity. Then, we use an optimization framework rather than using weighted summation and multiplication or combining the cues. So, the optimization framework proposed by [9] in which the cost function assigns the values to superpixels and it is called foreground weights. And then, the minimization function is used to assign image background regions by taking lower values and assigning foreground regions with higher values. The smoothness function is used to uniform that obtained saliency maps in cost function.

We evaluated our method on various benchmark datasets: MSRA-1000 [5], SED1, SED2 [16], CSSD [2], and most importantly on synthetic images dataset. And proposed method also gives extraordinary results on synthetic image datasets.

Rest of the paper is organized as follows: Sect. 2 describes the methodology that contains segmentation and various methods to detect salient object. Section 3 describes the saliency optimization framework. Experimental results on various datasets are discussed in Sect. 4 and Sect. 5 concludes the paper.

2 Boundary Connectivity

2.1 Structuring Your Paper

Instead of assuming image patch as background or image boundary as background, we can easily connect the image boundary. The proposed method states that when the image patch region is deliberately connected to the image boundary then that image patch is background; otherwise, it is a salient object. This measure characterizes the spatial layout of image region with respect to image boundaries. It is stable in image content variations because it buses an instinctive geometrical interpretation and this also makes it unique.

The boundary connectivity measure detects the background at very high precision with very satisfactory recall using a single threshold. And, it also handles exclusively background images without objects and enhances a traditional contrast computation.

2.2 Definition

As compared to background regions, the object regions are much less connected to the image boundary that is why object and the background regions in general images are different in their spatial layout. This is shown in Fig. 1.

The synthetic image consists of four regions. As per the visual perception, the red region is clearly a salient object because it is big in size, compact, and only moderately connected to image boundaries. And, it is clear that the green and white regions are background because these regions are heavily connected to image boundaries. The pink region is too small in size and it is slightly connected to image boundary and that is why it is considered as a partially cropped object and it is not a salient object.

So, the boundary connectivity measure states that how heavily an image region R is connected to image boundary. Then, the boundary connectivity BondCon is as

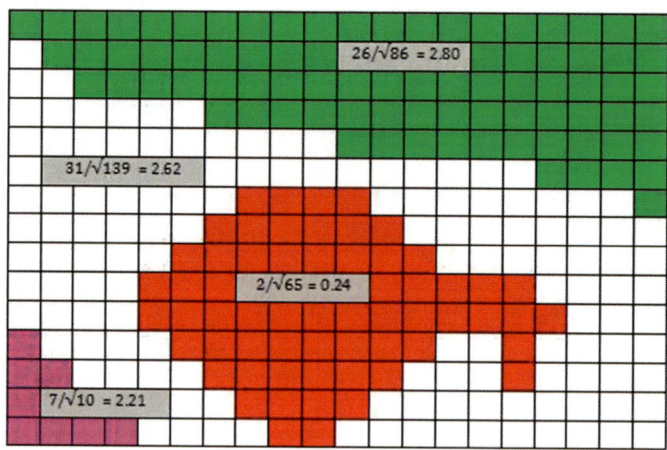

Fig. 1 Example of boundary connectivity. The synthetic image consists of four regions associated with BondCon values. BondCon is small for object region of image and more for background region of image

$$\text{Bondcon}(R) = \frac{|\{p|p \in R, p \in \text{Bond}\}|}{\sqrt{|\{p|p \in R\}|}} \tag{1}$$

where p is image patch, Bond is set of image boundary patches and it uses an instinctive geometrical interpretation. It is the ratio of regions perimeter on boundary to square root of its area or the regions overall perimeter. We used the square root of its area because the measure remains stable across different image patch resolutions. As shown in Fig. 1, the boundary connectivity is small for object regions and large for background regions.

2.3 Saliency Computation

Image segmentation is a process of portioning a digital image into multiple segments. The main goal of segmentation is to simply change the representation of an image into something which is more meaningful and easy to analyze. Figure 2 shows the flow of method, and Fig. 2b shows the output after applying SLIC superpixel segmentation.

Image segmentation itself is a challenging job; if we choose hard segmentation method, then it is very difficult for parameter and algorithm selection. So, we used a soft segmentation approach called SLIC method by [14] and it has high computational efficiency in which the image is abstracted as a set of superpixels. And, based on the experiments we find that 190–210 superpixels are sufficient for 300 * 400 resolution of image.

The SLIC superpixel method uses the CIELab color space. We connect all the adjacent superpixels (p, q) by constructing an undirected weighted graph and find the Euclidean distance between the average colors and it is denoted as $d_{\text{ap}}(p, q)$. Then, a collected weight with their shortest path on graph is called geodesic distance and is calculated between any two superpixels and is denoted as $d_{\text{ge}}(p, q)$.

Fig. 2 Flow of method: **a** Input image, **b** SLIC superpixel segmentation, **c** Probability of background using Eq. (6), **d** Background weighted contrast using Eq. (7), **e** Optimized saliency map using Eq (8)

$$d_{ge}(p, q) = \min_{p_1 = p, p_2, \ldots, p_n = q} \sum_{i=1}^{n-1} d_{ap}(p_i, p_{i+1}) \tag{2}$$

If for comfort, we define $d_{geo}(p, q) = 0$. After this, we calculate the "spanning area" of each superpixel p as

$$\text{Area}(p) = \sum_{i=1}^{N} ex\left(-\frac{d_{ge}^2(p, p_i)}{2\sigma_{clr}^2}\right) = \sum_{i=1}^{N} S(p, p_i) \tag{3}$$

where N is the number of superpixels.

As per Eq. (3), the area (p) is computed as the soft area of the region where the image patch p belongs. We note that $S(p, p_i)$ in the summation is in $(0,1)$ and distinguish how much superpixel p_i is subscribed to the p's area. We consider when superpixel p_i and p are in horizontal region, then note $d_{ge}(p, q) = 0$ and $S(p, p_i) = 1$ that ensures superpixel p_i adds a unit area to the area of p.

And when superpixel p_i and p are in different regions, then there is at least one edge that exists on their shortest path and $S(p, p_i) = 0$ that ensures superpixel p_i does not contribute to the area of p. So, by experiments, we find that the execution is strong when parameter σ is within (5–15) and then we set it to 12.

Correspondingly, we find the length along the boundary and set one for superpixels on image boundaries and set it to zero otherwise. And then, we compute the boundary connectivity of p as shown in Eq. (4).

$$\text{BondCon}(p) = \frac{\text{Len}_{bond}(p)}{\sqrt{\text{Area}(p)}} \tag{4}$$

where $\text{Len}_{bnd}(p)$ is the length of boundary for superpixel p. For calculating Bond-Con(p), the shortest path between all adjacent superpixel is calculated using Johnson's method [15].

Then, between any two superpixels add the edges. As shown in Fig. 3, the boundary connectivity values of background regions are enlarged [3]. This is useful due to closing of foreground objects and the physically connected background regions are separated.

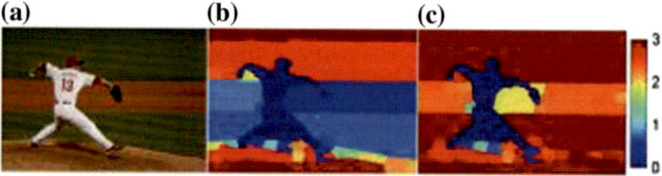

Fig. 3 Image enhancement by connecting its boundaries: **a** Input image, **b** BondCon without linking image patches, **c** BondCon is improved by linking boundary patches

Then, the weighted contrast of background of image gives an important information for salient object detection and it enhances the general contrast computation. The regional contrast is calculated by the summation of contrast's overall appearance distance to all other regions to weight by its spatial distance [7].

$$\text{Contr}(p) = \sum_{i=1}^{N} d_{\text{ap}}(p, p_i) W_{sp}(p, p_i) \tag{5}$$

where $W_{\text{sp}}(p, p_i) = ex\left(-\frac{d_{\text{sp}}^2(p,p_i)}{2\sigma_{\text{sp}}^2}\right)$ is calculated using distance weighted matrix. The distance between two superpixels is denoted by $d_{\text{sp}}^2(p, p_i)$. Figure 2c shows the result after applying Eq. (5).

Then, we present a new term called probability of background. It is calculated by mapping the superpixels boundary connectivity value and it is shown in Eq. (6).

$$W_i^{\text{bkg}} = 1 - ex\left(-\frac{\text{BondCon}^2(p_i)}{2_{\text{bondCon}}^2}\right) \tag{6}$$

When boundary connectivity value is very large, its value is near to one, otherwise, its value is near to zero. And the probability of background is denoted by W_i^{bkg}.

So, for better result, we set BondCon to be one but if we set the value of BondCon in between 0.5 and 2.5, then our result is inconsiderable because it would not affect the previous result. Then, we calculate the summation of distance appearance with spatial distance and probability of weight to enhance the contrast and it is shown in Eq. (7).

$$w\text{Contr}(p) = \sum_{i=1}^{N} d_{\text{ap}}(p, p_i) W_{sp}(p, p_i) W_i^{\text{bkg}} \tag{7}$$

As per Eq. (7) object region of image will get large probability of background from background region of image and then their contrast increases, and on the other side the background region of image will get very less probability of background from the object region of an image and then their contrast is reduced. Figure 2d shows the result after applying Eq. (7).

3 Saliency Optimization

In general, there are various saliency cues which are merged using assigning weight to summation and multiplication. Alternately, we use a framework to optimize saliency. In saliency optimization, we merge the background weight and foreground weight used in [7]. And, it is also merged with smoothness function and it is stated as

$$\sum_{i=1}^{N} W_i^{\text{fg}}(t_i - 1)^2 + \sum_{i=1}^{N} W_i^{\text{bkg}}(t_i)^2 + \sum_{i,j} W_{ij}(t_i - t_j)^2 \qquad (8)$$

where Eq. (8) shows that the ultimate value of saliency which is assigned to p_i is t_i and after the minimization of smoothness cost function W_i^{bkg} is denoted as the background weight corresponding to superpixel p_i and W_i^{fg} is denoted as the foreground weight corresponding to superpixel p_i. High foreground weight W_i^{fg} supports superpixel p_i to take values near to one and high W_i^{bkg} supports to take the value of p_i near to zero. And, W_{ij} is a smoothness term for which we used the same parameter setting which is used in [3]. Figure 2e shows the result after applying Eq. (8).

4 Experimental Results

We evaluated the proposed approach on four benchmark datasets and a synthetic image dataset. These datasets are MSRA-1000 [5], SED1, SED2 [16], where MSRA-1000 contains large images with complex background and this is the most widely used dataset for salient object detection. SED1 is a dataset which contains the images with only one object and on the other hand, the SED2 dataset which contains the images with two objects and both datasets contain the object with different sizes and datasets. Synthetic images datasets contain 40 images which contain 3D objects icons, etc. Results on all datasets are compared ground-truth mask labeled by humans. Figure 4 shows experimental results of proposed methods on different datasets.

4.1 Precision and Recall

We evaluate a method using precision–recall curves and MAE, i.e., mean absolute error. Precision is the fraction of relevant instances means pixels assigned correctly among the retrieved instances means against the total number of pixels assigned salient, whereas recall is the fraction of relevant instances that is retrieved as the pixels labeled correctly in relation over total relevant instances in the image means to the number of ground-truth pixels.

Table 1 shows that using threshold as two, the proposed measure can detect background with very decent recall and high precision. Figures 5 and 6 show the performance of boundary connectivity and geodesic saliency based on precision and recall for synthetic datasets.

Fig. 4 Experimental results of proposed methods on different datasets

Table 1 Background precision/recall for superpixels with boundary connectivity >2 on three benchmark datasets

Benchmark	Boundary connectivity		Geodesic saliency	
	Precision (%)	Recall (%)	Precision (%)	Recall (%)
MSRA	98.3	77.3	98.3	63.6
SED1	97.4	81.4	96.5	69.6
SED2	95.8	88.4	94.7	65.7

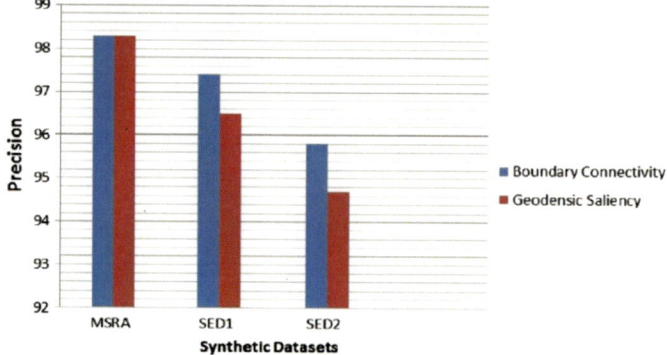

Fig. 5 Performance of boundary connectivity and geodesic saliency based on precision for synthetic datasets

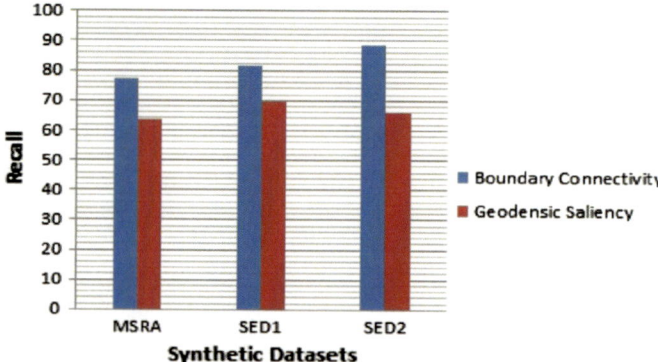

Fig. 6 Performance of boundary connectivity and geodesic saliency based on recall for synthetic datasets

4.2 Mean Absolute Error

Mean absolute error (MAE) measures the similarity between saliency map and the ground truth of images [7]. For a saliency map S, ground-truth mask is G, then MAE is defined as

$$\text{MAE} = \frac{1}{w * h} \sum_{x=1}^{h} \sum_{y=1}^{w} |S(x, y) - G(x, y)| \tag{9}$$

where h and w denote the height and weight of image. The proposed approach performs better than the state-of-the-art methods as per MAE.

5 Conclusion

In this paper, we used a background measure to detect salient object which produces very clean and precise saliency maps. Using boundary connectivity, the background region is produced instinctively. Using SLIC superpixel segmentation it is very easy to segment the image. Optimization method is used for reducing the noise from image and smoothing it. We perform experiments on various benchmark datasets such as MSRA [5], SED1, SED2 [16], synthetic datasets and achieve state-of-the-art results. In future work, we develop some advanced method to utilize background prior.

References

1. Liu T, Sun J, Zheng N, Tang X, Shum H (2007) Learning to detect a salient object. In: CVPR
2. Yan Q, Xu L, Shi J, Jia J (2013) Hierarchical saliency detection. In: CVPR
3. Zhu W, Ling S, Wei Y, Sun J (2014) Saliency optimization from robust background detection. In: CVPR
4. Ma Y-F, Zhang H-J (2003) Contrast-based image attention analysis by using fuzzy growing
5. Achanta R, Hemami S, Estrada F, Susstrunk S (2009) Frequency-tuned salient region detection, pp 1597–1604
6. Cheng M-M, Zhang G-X, Mitra NJ, Huang X, Hu S-M (2011) Global contrast based salient region detection, pp 409–416
7. Perazzi F, Krähenbühl P, Pritch Y, Hornung A (2012) Saliency filters: contrast based filtering for salient region detection, pp 733–740
8. Lu H, Ruan X, Yang C, Zhang L, Hsuan Yang M (2013) Saliency detection via graph-based manifold ranking
9. Zhu W, Wei Y, Wen F, Sun J (2012) Geodesic saliency using background priors
10. Alexe B, Deselaers T, Ferrari V (2012) Measuring the objectness of image windows 34(11)
11. Zhang Z, Warrell J, Torr PHS (2011) Proposal generation for object detection using cascaded ranking svms, pp 1497–1504
12. Cheng M-M, Zhang Z, Lin W-Y, Torr PHS (2014) BING: Binarized normed gradients for objectness estimation at 300 fps. In: IEEE CVPR
13. Jiang P, Ling H, Yu J, Peng J (2013) Salient region detection by ufo: Uniqueness, focusness and objectness. In: ICCV'13, pp 1976–1983
14. Machiras V, Decenciere E, Walter T (2015) Spatial repulsion between markers improves watershed performance. Mathematical morphology and its applications to signal and image processing. Springer International Publishing, pp 194–202
15. Johnson DB (1977) Efficient algorithms for shortest paths in sparse networks. J ACM 24(1):1–13
16. Alpert S, Galun M, Basri R, Brandt A (2007) Image segmentation by probabilistic bottom-up aggregation and cue integration. In: CVPR

Atherosclerotic Plaque Detection Using Intravascular Ultrasound (IVUS) Images

A. Hari Priya and R. Vanithamani

Abstract The atherosclerotic plaque deposition in artery is a type of cardiovascular disease and is a major factor of death. Mostly larger and high-pressure vessels such as the femoral, cerebral, renal, coronary, and carotid arteries are influenced by atherosclerosis. Hence, the characterization of plaque distribution and its liability to rupture are mandatory to judge the degree of risk and to schedule the treatment. Intravascular ultrasound (IVUS) is an ultrasound imaging modality which uses a unique catheter. The catheter will be provided with an ultrasound probe which is miniaturized and is connected to the lateral end. In this paper, segmentation of the IVUS image using fast marching method (FMM) is done and will be followed by the feature extraction. Feature extraction methods such as local binary pattern (LBP), speeded up robust feature (SURF), and histogram of oriented gradients (HOG) are used and the resulting image is classified using Euclidean distance classifier. By comparing the results of subjected feature extraction techniques, LBP method is found suitable for the detection of atherosclerotic plaque.

1 Introduction

Atherosclerotic plaque formation is a diseased condition of an artery which is characterized by the narrowing of the lumen in initial stages followed by the building up of plaques. During the initial stages, there will be no symptoms recorded. A fatty substance develops on the surface of the inner walls of arteries which is known as plaque. It causes thickening of the artery and the artery may lose its elasticity in

A. Hari Priya (✉)
Medical Electronics, Avinashilingam Institute for Home Science
and Higher Education for Women, Coimbatore, Tamil Nadu, India
e-mail: haripriyaarivukkarasu@gmail.com

R. Vanithamani
Biomedical Instrumentation Engineering, Avinashilingam Institute
for Home Science and Higher Education for Women, Coimbatore, Tamil Nadu, India
e-mail: vaniraj123@yahoo.co.in

© Springer Nature Switzerland AG 2019
D. Pandian et al. (eds.), *Proceedings of the International Conference on ISMAC in Computational Vision and Bio-Engineering 2018 (ISMAC-CVB)*, Lecture Notes in Computational Vision and Biomechanics 30,
https://doi.org/10.1007/978-3-030-00665-5_132

further stages. Further plaque deposition will lead to the narrowing of artery. It may also result in thrombosis. The artery may be blocked suddenly or embolus may be formed out of this thrombosis. This embolus (condition in which the clot travels along with the blood) may lead to stroke. Partial block of the artery by plaque may lead to increased blood pressure in that artery [1].

Many of the risk factors for atherosclerosis can be controlled and factors such as age and family history cannot be changed. Smoking, devoid of physical exercise, unhealthy diet, and over consumption of alcohol can be controlled. Another major cause for atherosclerosis is the obesity [2].

IVUS is a catheter-based imaging system that allows the technicians or healthcare professionals or interventional cardiologists to get the images of diseased blood vessels from lumen of the vessel [3]. Lumen characteristics, size of the vessel, diseased artery segments, and characteristics of plaques can be obtained from IVUS images [4, 5].

In this work, different feature extraction methods such as local binary pattern (LBP), HOG, and SURF were used to detect the atherosclerotic plaque using an IVUS image. Then, the suitable feature extraction technique to detect plaque is identified [6].

2   Materials and Methods

Processing of the IVUS image is done to detect the plaque formation inside an artery. It is a step-by-step process which includes the segmentation of IVUS image, followed by feature extraction and classification. The flow diagram of the IVUS image processing for identification of abnormal artery is given in Fig. 1

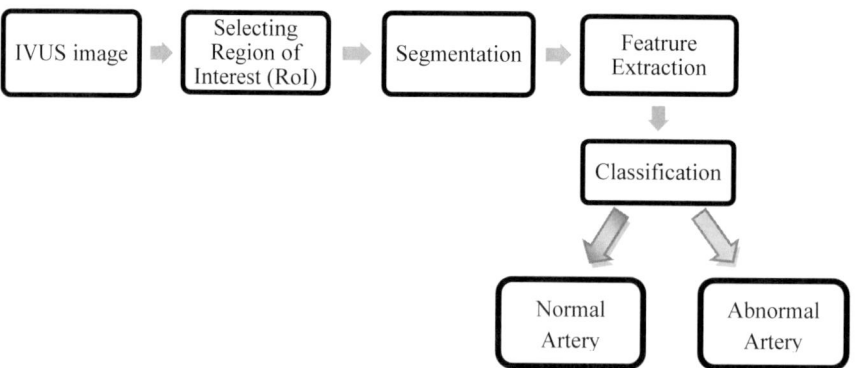

Fig. 1 Flow diagram of the IVUS image processing

2.1   Region of Interest

Input image is chosen from the dataset and then the RoI is selected manually. To exclude the lumen from the processing, the RoI is selected. In future, lumen can be automatically removed from the consideration to increase the quality of processing.

2.2   Segmentation

Partitioning the digital image into divergent regions is termed as image segmentation. The segmentation process will turn the input image into an image with more useful and meaningful information. Segmentation can locate the edges and boundaries more typically [7]. Image segmentation allocates a label to each pixel of the image and pixel with the same label is grouped which may share some similar characteristics. Fast marching method (FMM) is used to segment the IVUS image [8].

FMM It solves the boundary value problems numerically [8]. The eikonal equation is given in Eqs. 1 and 2

$$|\nabla u(x)| = 1/f(x) \quad \text{for}(x \text{ belongs to } \Omega) \tag{1}$$

$$U(x) = 0 \text{ for}(x \text{ belongs to } \delta\Omega) \tag{2}$$

For non-flat domain problems, Eq. 3 is used

$$|\Delta su(x)| = 1/f(x) \tag{3}$$

FMM algorithm The domain is partitioned into mesh and the mesh points are referred as nodes. Each node is assigned as x_i and the corresponding value is represented in Eq. 4

$$U_i = U(x_i) \approx u(x_i) \tag{4}$$

Nodes are named as far (nodes not visited yet), considered (nodes that are visited and tentatively assigned), and accepted (nodes which are visited and values are permanently assigned).

Step 1   For every node x_i, the value of U_i is given in Eq. 5

$$U_i = +\infty \tag{5}$$

Label these x_i nodes as far and for nodes $x_i \in \delta\Omega$ set $U_i = 0$ which is labeled as accepted.

Step 2   For all nodes x_i use the eikonal update formula to compute the new value for $\widehat{U}$. If $\widehat{U} < \widehat{U}_i$, then set $U_i < \widehat{U}$ and label x_i as considered.

Step 3   Assume $\dot{x}$ be the node considered with the lowest value U. $\dot{x}$ is labeled as accepted.

Step 4   For every neighbor x_i of $\dot{x}$ which is not accepted, the tentative value $\widehat{U}$ is calculated.

Step 5   If $\widehat{U} < \widehat{U}_i$ then set U_i as $\widehat{U}$. If x_i was labeled as "Far", then the label should be considered as updated.

Step 6   If there is a "Considered" node, return to step 3 or abort.

2.3   Feature Extraction

Feature refining or extraction starts from an initial set of deliberate data and builds features that are considered to be revealing and necessary which can lead to better manual explications [9]. Grouping a subgroup of the embryonic features is termed as feature selection. The determined features should have the pertinent information from the input data. Thus, the desired process can be done using diminished representation instead of using the whole data for further processing [10, 11]. In this paper, feature extraction techniques like LBP, HOG, and SURF are used to refine features from the segmented IVUS image.

LBP   LBP is a feature refining technique which uses the neighboring pixels to extract features. It is considered to be a substantial feature refining technique for texture classification [12, 13].

LBP algorithm.

Step 1   Divide the segmented image into pixels.

Step 2   Select a pixel and compare the nominated pixel to all neighboring pixels (eight neighboring pixels). It is done for all pixels starting from a corner pixel and follows the pixel along clockwise or counterclockwise circle.

Step 4   The particular bit in the binary array is assigned either to 0 or to 1. If the considered pixel's value is more than the neighboring pixel's value, then assign "0" or assign "1".

Step 5   The binary number thus obtained is converted into the decimal number. This resultant decimal number is stored in the center pixel. Thus, LBP mask is created by processing each pixel possessed by the input image.

Step 6   LBP histogram is calculated where the LBP mask value ranges from 0 to 255. And so, the obtained LBP descriptor will be of 1×256 in size.

Step 7   Concatenate the histogram values to get the features extracted.

Histogram of Oriented Gradient (HOG) HOG is used to recognize the objects. This method of feature extraction counts the incidence of gradient orientation value in every cell of the image. As it is employed on local cells, the HOG is advantageous over the other descriptors. As it acts on a local cell, HOG is stable to photometric and geometric transformations [13].

HOG Algorithm The HOG algorithm involves five steps such as

Step 1   Compute the gradient values. A point distinct derivative mask is applied in one or both of the directions (horizontal and vertical) to get the gradient values. The filter kernels used are $[-1,0,1]$ and $[-1,0,1]^{\mathrm{T}}$.

Step 2   In this step, the cell histograms are created. With the values got from the gradient calculation, each pixel (within the cell) launches a weighted value for a histogram channel (orientation based).

Step3   Locally, normalization of the gradient strengths is done to record changes in lighting and contrast. The normalization of gradient strength is processed by grouping the cell into spatially connected blocks.

Step 4   Let the non-normalized vector be v, possessing all histograms (within a given block), $\|v\|k$ be the k-norm for $k = 1, 2$ and e be a small constant. Equation 6 is used as normalizing factor.

$$f = \frac{v}{\sqrt{\|v\|_2^2 + e^2}} \tag{6}$$

Step 5   The HOG can be adapted for recognition of an object by providing them as features.

Speeded Up Robust Features (SURF) SURF is used to observe and recognize objects, registration of image, and classification of data. It employs Hessian blob detector and Haar wavelet response to refine features from the given input data. Hessian blob detector is used to observe the interest points. The total of Haar wavelet result around the considered point is the feature descriptor of the SURF method of feature extraction [14].

SURF Algorithm The surf algorithm involves the following steps:

Step 1   Square-shaped filters are used as Gaussian smoothing approximation.

Step 2   The Hessian blob detection is done. The Hessian matrix determinant is the measure of local change around the considered point. The maximal determinant points are chosen. For a point $p = (x, y)$ in an image A, the Hessian matrix $H(p, \sigma)$ is

$$H(p, \sigma) = \begin{pmatrix} Lxx(p, \sigma) & Lxy(p, \sigma) \\ Lyx(p, \sigma) & Lyy(p, \sigma) \end{pmatrix} \tag{7}$$

where the elements of the matrix are second-order derivative's convolution of Gaussian with the image $A(x, y)$ at the x point.

Step 3 Images are smoothened repeatedly using Gaussian filter and then they are subsampled to reach the higher level of the pyramid, thus consecutive stairs with various mask measures are calculated.

$$\sigma \text{ approx} = \text{current filter size} \times (\text{base filter scale}/\text{base filter size})$$

Step 4 The reproducible orientation has been fixed based on data from a region of circle around the considered point. Then, a region of square is assigned to the orientation selected and the SURF descriptors are refined from it.

Step 5 The Haar wavelet results within a circular locality in both x-direction and y-direction of radius 6 s near (around) the considered point are calculated where s is the scale with which the considered point is detected. Sum the vertical and horizontal responses within the particular window. The local orientation vector is calculated by summing the two responses. The placement of the considered point is defined by the longest of the vectors calculated.

Step 6 The interest region is further divided into smaller square regions, then the Haar wavelet results are refined at sample points which are regularly spaced. To gain more robustness, the resultant riposte is weighted with Gaussian.

2.4 Classification

Classification is the technique of gathering the testing data into a preset class based on the trained data set containing observations whose class is already known [15]. In this paper, Euclidean distance measurement is used for classification.

Euclidean Distance The straight-line space between two pixels is said to be the Euclidean distance. The space between two points in Euclidean space is the Euclidean distance [16]. In 2-D space, the distance between $R = (a_1, a_2)$ and $S = (b_1, b_2)$ is defined as

$$\sqrt{(a_1 - b_1)^2 + (a_2 - b_2)^2} \tag{8}$$

3 Results and Discussion

The suggested method is tested using the IVUS image obtained from the database available at https://www.dropbox.com/s/el674ocdp9uojro/Training_Set.zip?dl=0.

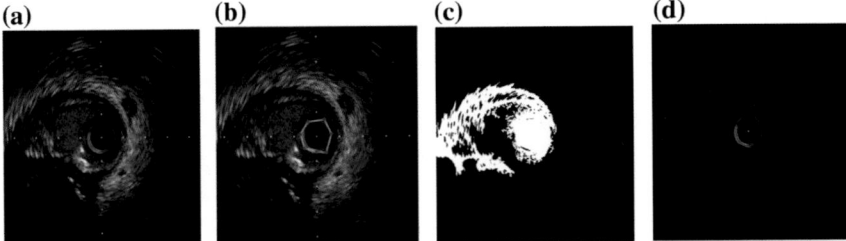

Fig. 2 Segmentation of test image-1 using FMM method: **a** Original image, **b** Selection of RoI, **c** Inverted image, **d** Segmented image

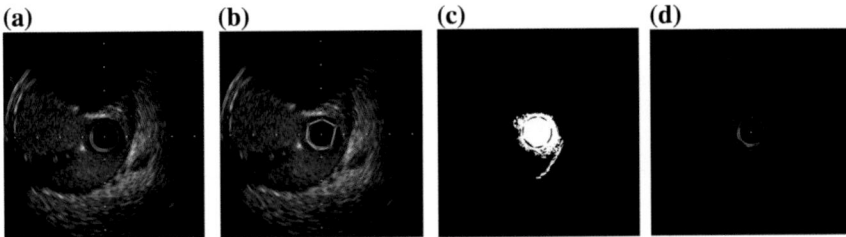

Fig. 3 Segmentation of test image-2 using FMM method: **a** Original image, **b** Selection of RoI, **c** Inverted image, **d** Segmented image

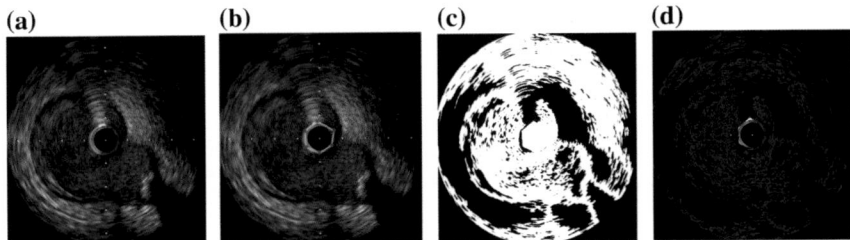

Fig. 4 Segmentation of test image-3 using FMM method: **a** Original image, **b** Selection of RoI, **c** Inverted image, **d** Segmented image

MATLAB is used to test the algorithm. Twenty-five images are taken to validate the results. The segmented results of test images 1, 2, 3, and 4 are shown in the figures from Figs. 2, 3, 4, and 5.

The results are tabulated in Table 1.

From the results obtained, the sensitivity can be found using Eq. 9

$$TP/(TP = FN) \tag{9}$$

The specificity can be found using Eq. 10

$$TN/(TN + FP) \tag{10}$$

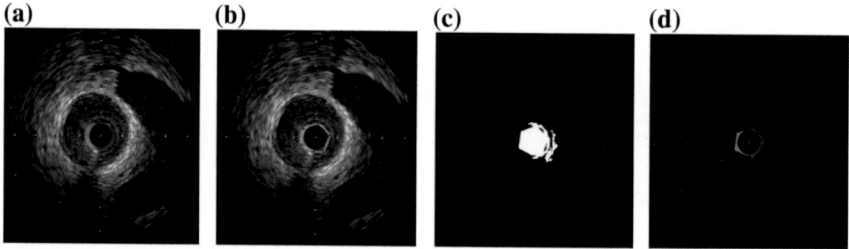

Fig. 5 Segmentation of test image-4 using FMM method: **a** Original image, **b** Selection of RoI, **c** Inverted image, **d** Segmented image

Table 1 Results of the tested artery using different feature extraction techniques

Algorithm	Classification results			
	TP	TN	FP	FN
LBP	7	1	1	16
HOG	1	7	1	16
SURF	3	5	1	15

Table 2 Sensitivity, specificity, and accuracy of detection of abnormal artery using LBP, HOG, and SURF techniques

Algorithm	Sensitivity	Specificity	Accuracy (%)
LBP	0.3043	0.5	92
HOG	0.0589	0.875	68
SURF	0.1666	0.8333	72

Equation 11 can be used to find the accuracy

$$TP = FN/(TP + FP + TN + FN) \tag{11}$$

The sensitivity, specificity, and accuracy for detecting the abnormal artery using LBP, HOG, and SURF are given in Table 2.

3.1 Conclusion

Atherosclerotic plaque can make arteries narrower, leading to reduced or blocked blood flow which in turn may cause heart failure [17]. Thus, it is very mandatory to diagnose the atherosclerotic arteries accurately [18]. In this work, the IVUS images are segmented using fast marching segmentation. Three feature extraction techniques such as LBP, HOG, and SURF are used. From the results obtained, the LBP is found to be a more precise method for detecting atherosclerotic artery compared to HOG and SURF. Thus, the LBP is a better technique to process the IVUS image than the HOG and SURF techniques for detecting atherosclerotic plaque.

References

1. Taki A, Roodaki A, Setarehdan SK, Avansari S, Unal G, Navab N (2013) An IVUS image-based approach for improvement of coronary plaque characterization. Comput Biol Med 43:268–280
2. Taki A, Roodaki A, Pauly O, Setarehdan SK, Unal G, Navab N (2009) A new method for characterization of coronary plaque composition via IVUS IMAGES. IEEE-2009, pp 787–790
3. Zhang Q, Wang Y, Ma J, Shi J (2011) Contour detection of atherosclerotic plaques in IVUS images using ellipse template matching and particle swarm optimization. In: 33rd annual international conference of the IEEE EMBS, pp 5174–5177
4. Chen F, Ma R, Liu J, Zhu M, Liao H (2018) Lumen and media-adventitia border detection in IVUS images using texture enhanced deformable model. Comput Med Imaging Graph 66:1–13
5. Zakeri FS, Setarehdan SK, Norouzi S (2017) Automatic media-adventitia IVUS image segmentation based on sparse representation framework and dynamic directional active contour model. Comput Biol Med 89:561–572
6. Shi Y, Witte RS, O'Donnell M (2005) Identification of vulnerable atherosclerotic plaque using IVUS-based thermal strain imaging. IEEE Trans Ultrason Ferroelectr Freq Control 52:844–850
7. Balocco S (2014) Standardized evaluation methodology and reference database for evaluating IVUS image segmentation. Comput Med Imaging Graph 38:70–90
8. Wien TU, Ives St (2016) Comparison of the parallel fast marching method, the fast iterative method, and the parallel semi-ordered fast iterative method. In: The international conference on computational science, vol 80, pp 2271–2275
9. Krishnan S, Athavale Y (2018) Trends in biomedical signal feature extraction. Biomed Signal Process Control 43:41–63
10. Anam S, Misawa H, Uchino E, Suetake N (2012) Parameter tuning by PSO for fuzzy inference-based coronary plaque extraction in IVUS image. IEEE-2012 2012:1426–1429
11. Roodaki A, Taki A, Setarehdan SK, Navab N (2008) Modified wavelet transform features for characterizing different plaque types in IVUS images. A feasibility study. IEEE-2008, pp 789–792
12. Mahale VH, Ali MH, Yannawar PL, Gaikwad AT (2017) Image inconsistency detection using local binary pattern (LBP). In: 7th international conference on advances in computing & communications, vol 115, pp 501–508
13. Korkmaz SA, Binol H (2018) Classification of molecular structure images by using ANN, RF, LBP, HOG, and size reduction methods for early stomach cancer detection. J Mol Struct 1156:255–263
14. Kan S-C, Cen Y-G, Cen Y, Wang Y-H, Voronin V, Mladenovic V, Zeng M (2017) SURF binarization and fast codebook construction for image retrieval. J Vis Commun Image Represent 49:104–114
15. Caballero KL, Barajas J, Pujol O, Rodriguez O, Radeva P (2007) Using reconstructed IVUS images for coronary plaque classification. In: Proceedings of the 29th annual international conference of the IEEE EMBS, pp 2167–2170
16. Deza MM, Deza E, Marie M (2009) Encyclopedia of distances. Springer, pp 94
17. Li X, Li J, Jing J, Ma T, Liang S, Zhang J, Mohar D, Raney A, Mahon S, Brenner M, Patel P, Kirk Shung K, Zhou Q, Chen Z (2014) Integrated IVUS-OCT imaging for atherosclerotic plaque characterization. IEEE J Sel Top Quantum Electron 20
18. Dehnavi SM, Babu MSP, Yazchi M, Basij M (2013) Automatic soft and hard plaque detection in IVUS images: a textural approach. In: Proceedings of 2013 IEEE conference on information and communication technologies, pp 214–219

Correlative Feature Selection for Multimodal Medical Image Fusion Through QWT

J. Krishna Chaithanya, G. A. E. Satish Kumar and T. Ramasri

Abstract A novel image fusion technique is proposed in this paper to achieve an efficient and informative image by combining multiple medical images into one. This method accomplishes quaternion wavelet transform (QWT) as a feature representation technique and correlation metric for feature selection. Here, the correlation-based feature selection is accomplished to extract the optimal feature set from the sub-bands thereby to reduce the computational time taken for fusion process. QWT decomposes the source images first and then the feature selection process obtains an optimal feature set from the low-frequency (LF) as well as high-frequency (HF) sub-bands. The obtained optimal feature sets of both LF sub-bands and HF sub-bands are fused through low-frequency fusion rule and high-frequency fusion rule and the fused LF and HF coefficients are processed through IQWT to obtain a fused image. Various models of medical image are processed in the simulation and the performance is evaluated through various performance metrics and compared with conventional approaches to show the robustness and efficiency of proposed fusion framework.

J. Krishna Chaithanya (✉) · G. A. E. S. Kumar
Department of ECE, Vardhaman College of Engineering, Hyderabad, India
e-mail: j.krishnachaitanya@vardhman.org

T. Ramasri
Department of ECE, SVUCE, SVU, Tirupati, India

© Springer Nature Switzerland AG 2019
D. Pandian et al. (eds.), *Proceedings of the International Conference on ISMAC in Computational Vision and Bio-Engineering 2018 (ISMAC-CVB)*, Lecture Notes in Computational Vision and Biomechanics 30,
https://doi.org/10.1007/978-3-030-00665-5_133

1 Introduction

Medical imaging has attained a vast research interest in recent years due to the increasing requirement of disease diagnosis and clinical investigation. Owing to various medical image modalities, every image modal can illustrate some prior information about the human body, but limited to specific purpose only. For instance, the structures of bones and hard tissues are better visualized through the computed tomography (CT) image, whereas the detailed structure of soft tissues is illustrated through magnetic resonance imaging (MRI) only. In the same manner, the details of anatomical structures are represented through MRI-T1 image model, whereas the normal tissues and pathological tissues are described through MRI-T2 only [1–4]. Further, image models including the functional magnetic resonance imaging (fMRI), positron emission tomography (PET), and single photon emission computed tomography (SPECT) provide the details about the functional information in low-resolution spatial images, useful in the detection of metabolic abnormalities and cancer-related diseases. Analysis of individual imaging model builds an extra manual burden and also results in an increased time complexity. Hence, there is a necessity to combine the multiple image models to construct a compendious figure. Medical image fusion is a compendious technique which gives a fused image as an input by fusing multiple images with different modalities. On the other hand, the fused images are more suitable to help the doctor in the treatment planning: fusion of CT and MRI images can represent the bone structures and soft tissues in order to represent the physiological and anatomical features of a human body simultaneously. A further advantage with image fusion is the reduction in the storage cost.

In past, numerous techniques are developed to achieve an efficient fused image which provides complete information about different multimodal images. The entire earlier image fusion techniques are divided into transform domain techniques and spatial domain techniques. In the case of spatial domain approaches, the fused image directly relates the pixel intensities of source image, whereas in the transform domain approaches, these are related with their transformed coefficients. Since there exists a direct relation among the pixels of input and output image, the small variations in the source image intensities result in a proportional effect in the fused image. For instance, the common drawback observed form the earlier spatial domain approaches is the reduced contrast in the fused image which represents the bad quality fused image. Due to this, the portions which need to be highlighted in the fused image would not get highlighted and the regions which need to be suppressed will get highlighted. Instead of directly operating over the pixel intensities, the transform-domain-based fusion approaches perform over transformed pixel intensities.

A novel image fusion technique is proposed in this paper to achieve an efficient and informative image by combining multiple medical images into one. This method accomplishes quaternion wavelet transform (QWT) as a feature representation technique and correlation for feature selection. The QWT is more robust to phase variations and the proposed correlation-based feature selection is robust for computational time. The source image is decomposed into sub-bands through QWT first, and then the obtained bands are processed for correlation-based feature selection and then the selected coefficients are fused into their respective modified sub-bands. Finally, the modified sub-bands are formulated as a fused mage through IQWT.

Reminder of the paper is formulated as follows: The related literature survey of the proposed work is illustrated in Sect. 2. The basic preliminaries about QWT and correlation are described in Sect. 3. The details of the proposed image fusion method are illustrated in Sect. 4. Section 5 illustrates the details of performance evaluation, and finally Sect. 6 concludes the paper.

2   Literature Survey

2.1   Spatial Domain Approaches

The approaches proposed based on principal component analysis (PCA), intensity–hue–saturation (IHS), and Brovey transform are better examples for spatial domain approaches [5–7]. To deal with sparse representation of medical image, Zhang et al. [6] proposed an image fusion technique for medical images centered to PCA and some sparse representation techniques named as SPCA. In this approach, the correlations between and within the channels are modeled effectively. Here, the sparse representation is applied over remote sensing image and the PCA is applied to achieve a reduced dimensionality to reduce the computational cost. Kaur et al. [8] proposed a PCA-based image fusion technique in the collaboration with genetic algorithm. However, all these techniques fail in the spectral degradation. Another region segmentation approach is proposed in [9] to find the regions through the morphological filtering. According to [9], the source images are initially processed for fusion through the simple averaging method. Then, the obtained fused image is segmented into different regions through the normalized cut method. Further, based on the obtained regions, the source images are subjected to region segmentation and then the segmented regions are processed for final fusion according to their spatial frequencies with respect to the regions of initial fused image. Though this method is observed to be too simple, it introduces unnecessary side effects in the fused image like reduced contrast level [10].

2.2   Transform Domain Approaches

Recently, the transform base image fusion approaches have gained a lot of research interest due to its effectiveness in the quality increment of fused images. Among those transforms, multiscale transforms (MST) is one of the most popular transforms which transforms the image in the multiscale fashion. Framelet transform [11], dis-

crete wavelet transform (DWT) [3, 12, 13], Gabor transform, non-subsampled contourlet transform (NSCT) [1, 4, 14], and contourlet transform [15] are some example
of MSTs. Compared to DWT, the CT represents the details of image features more
precisely by which the performance of fusion approach increases. A further edge perspective fusion framework is proposed by Deng et al. [16] based on the canny edge
operator and DWT. This approach mainly aims to preserve the edge features. DWT
decomposes the source image initially into the low-frequency and high-frequency
(vertical, horizontal, and diagonal) sub-bands. Further, the canny operator is applied
over vertical and diagonal sub-band images to acquire the edge information. An
effective DWT-based multi-focus image fusion approach is proposed by Yang et al.
[17] by seeing the physical meaning of wavelets. A novel coefficient selection algorithm is also described in this approach to select the optimal feature set for fusion.
Zhou et al. [18] presented a novel multiscale fusion method based on weighted gradient to solve the image misregistration problem which results in the degradation
of multi-focus images. Further, to solve the misregistration problems at the fused
images, this approach focused on the two-scale mechanism, i.e., focusing on both
large scale and small scale coefficients. Considering the limitations in the directionality of 2D separable wavelets, the WT cannot represent the edges directions more
accurately.

3 Preliminaries

3.1 Quaternion Wavelet Transform (QWT)

Considering the analytics of quaternion signal [19], axioms of the quaternion algebra
and the separability property, the QWT is derived as a natural extension from the real
and complex wavelet transforms. The QWT is a perfect transform technique for the
signals with higher dimensions. In the case of QWT, the signal with larger dimensions
can be decomposed into multiresolution levels by direct accomplishment. Since the
QWT results in phase-directed bands along with amplitude band, a more detailed
analysis can be acquired over the image with different phase variations. This is limited
in the conventional wavelets as well as in the complex wavelets. Thus, instead of
a distance similarity accomplishment in the complex wavelet pyramid, QWT uses
phase concept for top-down parameter estimation.

The QWT of an image can be defined as

$$f(x, y) = A_n^q f(x, y) + \sum_{s=1}^{n} \left[D_{s,1}^q f(x, y) + D_{s,2}^q(x, y) + D_{s,3}^q f(x, y) \right] \quad (1)$$

where A_n^q and $D_{s,p}^q, (p = 1, 2, 3)$ are the approximation sub-band images and
detailed sub-band images. The analytical extension of QWT is based on the analytical

expression of real wavelet and its two-dimensional Hilbert transform, as represented by

$$
\begin{cases}
\psi^D(x, y) = \psi_h(x)\psi_h(y) \\
\rightarrow \psi^D + i H_{i1}\psi^D + j H_{i2}\psi^D + k H_{i3}\psi^D \\
\psi^V(x, y) = \phi_h(x)\psi_h(y) \\
\rightarrow \psi^V + i H_{i1}\psi^V + j H_{i2}\psi^V + k H_{i3}\psi^V \\
\psi^H(x, y) = \psi_h(x)\phi_h(y) \\
\rightarrow \psi^H + i H_{i1}\psi^H + j H_{i2}\psi^H + k H_{i3}\psi^H
\end{cases}
\tag{2}
$$

$$
\begin{cases}
\phi(x, y) = \phi_h(x)\phi_h(y) \\
\rightarrow \phi + i H_{i1}\phi + j H_{i2}\phi + k H_{i3}\phi
\end{cases}
\tag{3}
$$

where ϕ is scaling function, ψ^D, ψ^V, ψ^H are the wavelet functions oriented to diagonal, vertical and horizontal directions, respectively. From 2D Hilbert transform, 1D Hilbert transform can be derived through the x- and y-axis as

$$
\begin{cases}
\psi_h, \psi_g = H\psi_h \\
\phi_h, \phi_g = H\phi_h
\end{cases}
\tag{4}
$$

Based on the above analytical extension, the 2D QWT can be defined as

$$
\begin{cases}
\psi^D(x, y) = \psi_h(x)\psi_h(y) + i\psi_h(x)\psi_g(y) \\
\quad + j\psi_g(x)\psi_h(y) + k\psi_g(x)\psi_g(y) \\
\psi^V(x, y) = \phi_h(x)\psi_h(y) + i\phi_g(x)\psi_h(y) \\
\quad + j\psi_g(x)\phi_h(y) + k\phi_g(x)\psi_g(y) \\
\psi^H(x, y) = \phi_h(x)\psi_h(y) + i\phi_g(x)\psi_h(y) \\
\quad + j\psi_g(x)\phi_h(y) + k\phi_g(x)\psi_g(y) \\
\phi(x, y) = \phi_h(x)\phi_h(y) + i\phi_g(x)\phi_h(y) \\
\quad + j\phi_g(x)\phi_h(y) + k\phi_g(x)\phi_g(y)
\end{cases}
\tag{5}
$$

where the representations given in first three rows of above QWT equation [20, 21] are the mathematical formulae to extract the high-frequency coefficients from the image along diagonal, vertical, and horizontal direction, respectively. Further, the last row is for the extraction of QWT low-frequency coefficients from the image.

3.2 Correlation Measure

Once both low approximation sub-bands and detailed sub-bands are obtained through QWT, they are needed to be fused to obtain a single approximation and detailed sub-band image. One important point to be noticed is that the obtained bands consist of most redundant information, especially in the high-frequency bands. The QWT extract totally four bands, one is amplitude and three are phase-oriented bands. Further, all the three phase bands are oriented in three directions such as horizontal, vertical, and diagonal at the detailed band extraction. If all these bands are processed for fusion process, it results in a more computational burden over the system and also consumes too much time to obtain a fused image. Hence, this work proposed a new feature selective coding based on the correlations of features in all the bands.

Further, the fusion needs to be applied over the coefficients which are more informative and also should not affect its visual quality. To achieve a better fusion performance, this work accomplished directional filters by which the variations in all directions are predicted much efficiently. Once the regions with fewer variations are predicted through directional filters, those regions are processed for fusion. For this purpose, the correlation is measured between all the coefficients of all the sub-bands. The coefficients with maximum correlation are only selected for fusion. This is applied over every sub-band and only a set of feature is extracted from every band. In the case of proposed QWT-based fusion framework, the QWT decomposes the image into low frequencies and high frequencies followed by magnitude and three phase bands. The proposed correlation measure evaluation is applied over every band and a new feature set with maximum correlation is extracted for both source images A and B, then the extracted new and optimal feature set of both images are processed for fusion. The simple architecture of correlative-measure-based feature selection is shown in Fig. 1.

For this purpose, the proposed correlative-measure-based coefficient selection algorithm is developed and termed as correlative feature selection (CFS). In this way, this approach of coefficient selection results in the feature set with optimal nature and is also highly informative with less burden.

4 Fusion Framework

The following step-by-step process illustrates the detailed explanation of the proposed image fusion technique. Initially, consider the input images with two different models A and B. Apply the proposed fusion mechanism over these two images to obtain a fused image as follows:

Step 1 Apply QWT over both source images A and B to decompose them into the low-frequency (Approximations, A) and high-frequency (Details, D) sub-band images. Apply orientational filter over both LF and HF sub-band images to extract the wavelet coefficients in particular orientation l. Let us

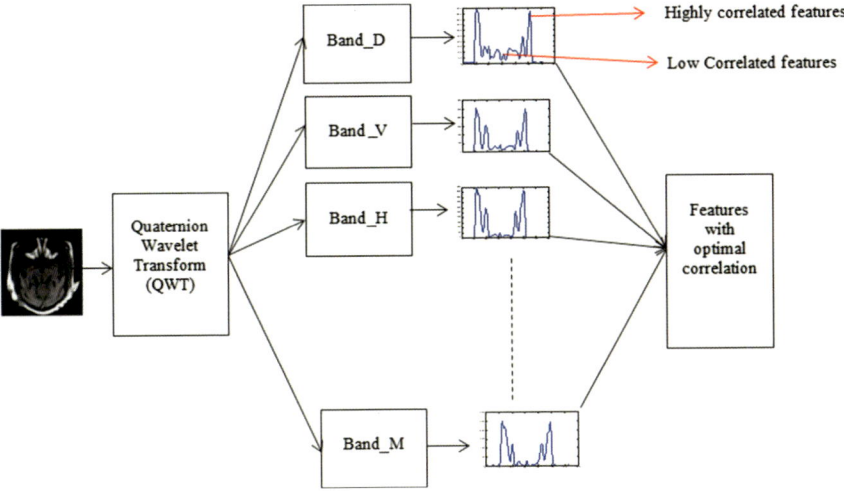

Fig. 1 Schematic of correlation-based feature selection

consider the approximate sub-band $A_{k,l}^{A}$ which is of the input image A and the approximate sub-band $A_{k,l}^{B}$ which is of the input image B at kth orientation and lth direction, respectively. Further, consider the detailed sub-band image $D^{A}(x, y)$ is of the input image A and $D^{B}(x, y)$ is of the input image B. Further, apply the CFS over all the obtained sub-band images.

Step 2 Perform the CFS over the approximation bands of both images to obtain an optimal feature set by which the extra complexity over the system reduces.

Step 3 Perform fusion of approximation sub-band images $A_{k,l}^{A}$ and $A_{k,l}^{B}$ through the proposed fusion rule based on the angular consistency [22] to obtain a fused approximation sub-band image, $A^{F}(x, y)$.

In general, the approximate sub-bands fusion is carried out through the most popular "averaging rule." However, due to an issue of reduced contrast level in the fused image, this rule is violated and a novel fusion rule is derived considering the angular consistencies of approximation sub-bands. The proposed angular consistency rule for the fusion of approximation sub-bands is formulated as

$$A^{F}(x, y) = \begin{cases} A^{A}(x, y) & \text{if } A_{c}^{A\theta}(x, y) > A_{c}^{B\theta}(x, y) \\ A^{B}(x, y) & \text{if } A_{c}^{A\theta}(x, y) < A_{c}^{B\theta}(x, y) \\ \frac{1}{2}(A^{A}(x, y) + A^{B}(x, y)) & \text{if } A_{c}^{A\theta}(x, y) = A_{c}^{B\theta}(x, y) \end{cases} \quad (6)$$

where $A_{c}^{A\theta}(x, y)$, $A_{c}^{B\theta}(x, y)$ are the angular consistencies of approximations of input images A and B, correspondingly.

Step 4 Perform the CFS over the approximation bands of both images to obtain an optimal feature set by which the extra complexity over the system reduces.

Step 5 Perform fusion of detailed sub-band images $D^A(x, y)$ and $D^B(x, y)$ based on new fusion rule proposed through the spatiofrequency energies of detailed sub-bands [23] to obtain a fused detailed or high-frequency sub-band image, $D^F(x, y)$.

In general, the detailed sub-band image fusion is carried out through the most popular "Larger Absolute Selection Rule." However, due to the loss of vast complementary information in the fused image, this rule is not used here and the new fusion is derived based on the spatiofrequency energies of detailed sub-bands. Particularly, in the case of detailed sub-bands, preservation of edges, lines are very much important which gives a more detailed analysis about the soft tissues in medical images. Further, discrimination between noise and detailed coefficients is more important by which the fusion process may mislead to consider a noise as a high-frequency coefficient and considers it as a fusing coefficient. Hence, there is a need to provide a perfect discrimination between noises and high-frequency coefficients, the proposed spatiofrequency energies take this responsibility. The proposed spatiofrequency energies-based fusion rule is formulated as

$$D^F(x, y) = \begin{cases} D^A(x, y) \text{ if } E^A(x, y) \geq E^B(x, y) \\ D^B(x, y) \text{ if } E^A(x, y) < E^B(x, y) \end{cases} \tag{7}$$

where $E^A(x, y)$ and $E^B(x, y)$ represent the "Spatiofrequency energies" of detailed sub-bands of input images A and B, respectively.

Step 6 Recreate the fused image by applying inverse QWT over the fused approximation sub-band $A^F(x, y)$ and detailed sub-band $D^F(x, y)$.

5 Simulation Results

This section illustrates the details of performance evaluation of the proposed framework quantitatively and qualitatively. To verify the performance of the proposed approach, an extensive simulation is carried out over various types of images like medical images, natural images, etc. Here, the source images of size 256 * 256 are considered and the simulation is carried through MATLAB software. The obtained fused images are shown in Figs. 2, 3, 4, 5, and 6.

Sources of the test images are as follows:

1. The Whole Brain Atlas (Harvard): http://www.med.harvard.edu/aanlib/
2. BrainWeb: Simulated Brain Database: http://brainweb.bic.mni.mcgill.ca/brainweb/

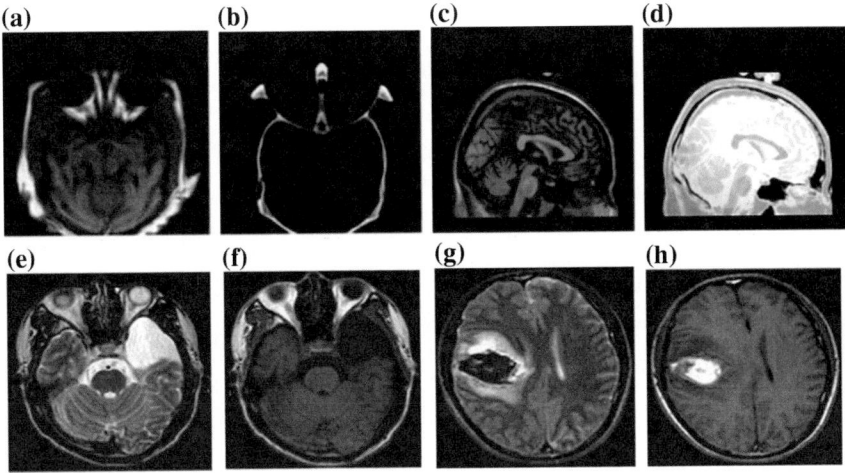

Fig. 2 Test images: **a, c** Magnetic resonance imaging (MRI), **b, d** Computed tomography (CT) images, **e, g** Magnetic resonance-T2 relaxation (MR-T2) images, and **f, h** Magnetic resonance-T1 (MR-T1) relaxation images

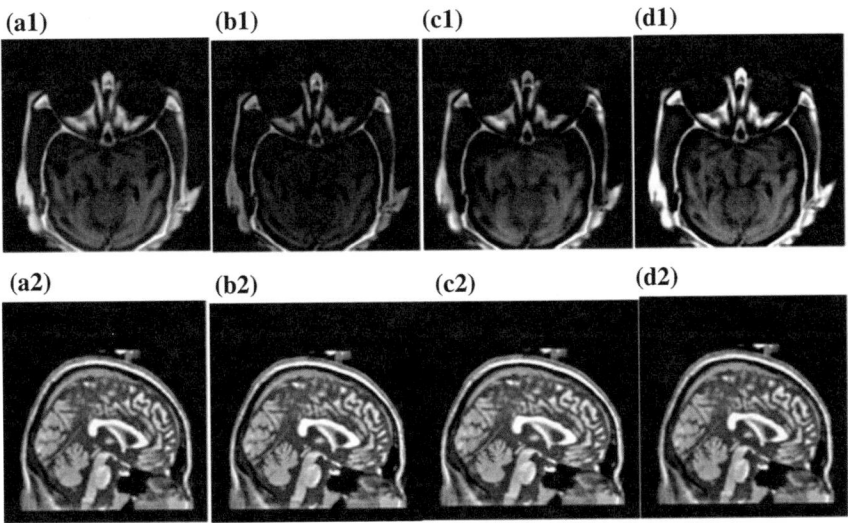

Fig. 3 Obtained results for CT and MRI images: **a**1–**a**2 Fused images through DWT, **b**1–**b**2 Fused images through CT, **c**1–**c**2 Fused images through NSCT, and **d**1–**d**2 Fused images through proposed

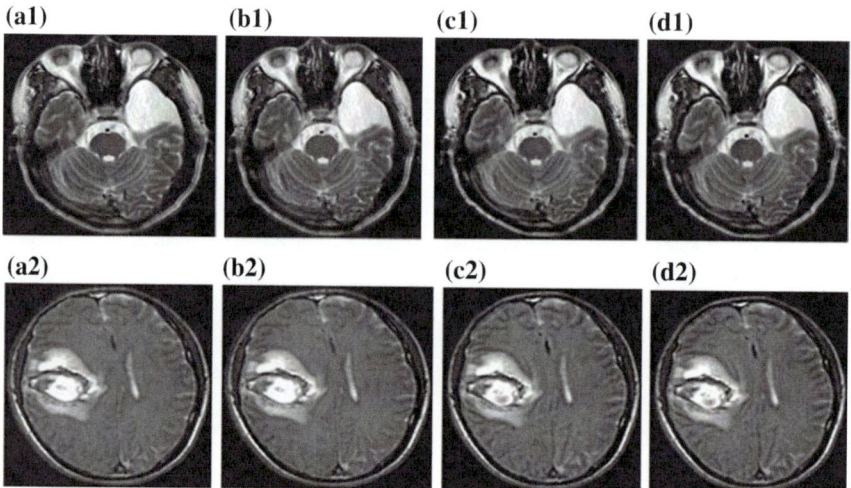

Fig. 4 Obtained results for MR-T1 and MR-T2 images: **a**1–**a**2 Fused images through DWT, **b**1–**b**2 Fused images through CT, **c**1–**c**2 Fused images through NSCT, and **d**1–**d**2 Fused images through proposed

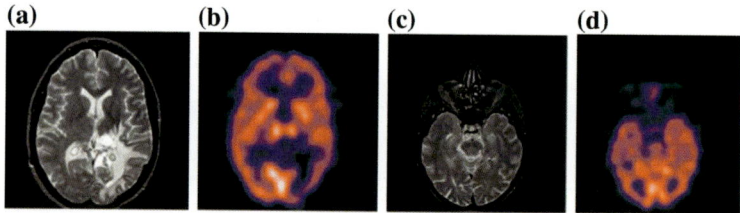

Fig. 5 Source images: **a**, **c** MRI images, **b**, **d** SPECT images

Under the performance evaluation, five objective evaluation measurement parameters are adopted to evaluate the fusion performance. There are weighted quality fusion index (WQFI) [24], local quality index (LQI) [24], edge-dependent fusion quality index (EFQI) [24], and $_/$ [25] which measures the transmission of edge features and visual features from input images to fused images, and mutual information (MI) [26] which measures the amount of information transferred from input images to output fused images. The range of Q_0, Q_w, Q_E, and $Q_{AB/F}$ lies between 0 and 1 and the range of MI is above 1 and it varies from image to image. These performance metrics are evaluated for all the above test image sets and are formulated in Figs. 7, 8, 9, 10, and 11.

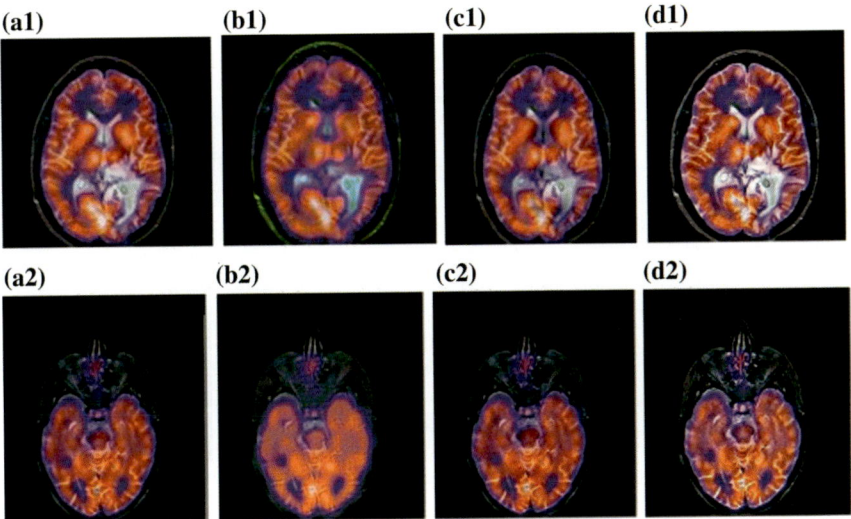

Fig. 6 Obtained results for MR-T1 and MR-T2 images: **a**1–**a**2 Fused images through DWT, **b**1–**b**2 Fused images through CT, **c**1–**c**2 Fused images through NSCT, and **d**1–**d**2 Fused images through proposed

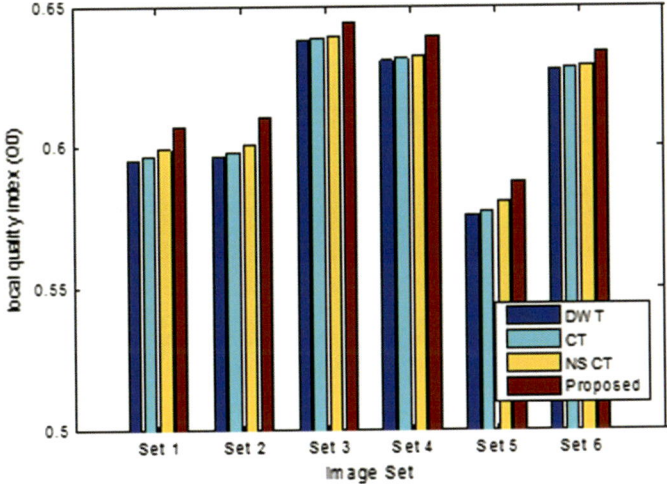

Fig. 7 Local quality index comparison

6   Conclusion

A novel image fusion technique is proposed in this paper to achieve an efficient and informative image by combining multiple medical images into one. This method accomplishes QWT as a feature representation technique and correlation for feature

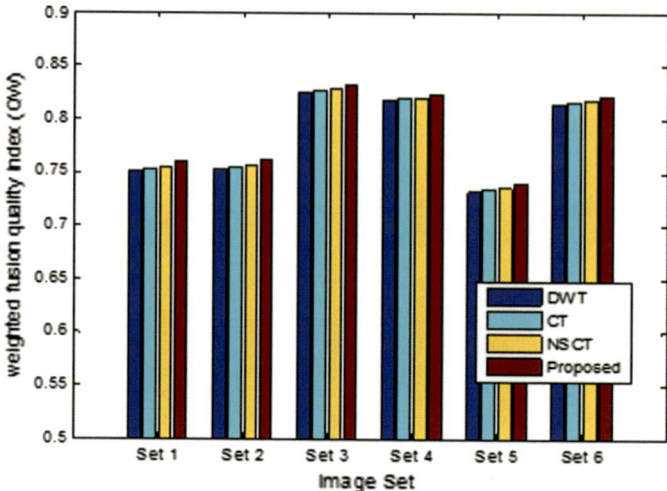

Fig. 8 Weighted fusion quality index comparison

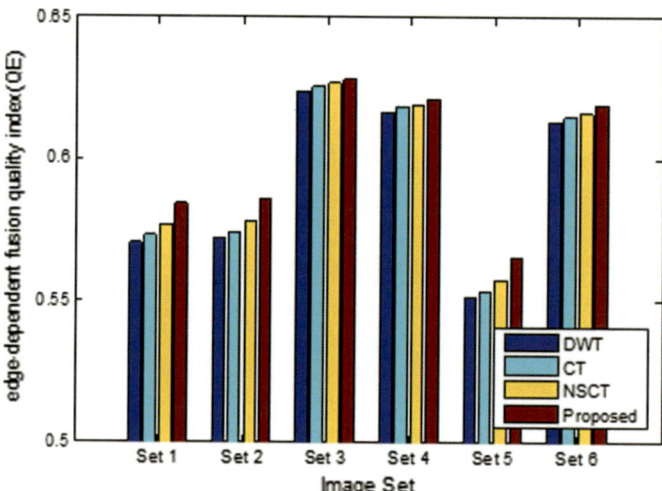

Fig. 9 Edge-dependent fusion quality index

selection. Compared to the conventional multiscale transform techniques such as DWT, CT, and NSCT, the QWT decomposes the image into the phase-deviated bands also by which more information will be revealed about the frequency characteristics of image. Further, the proposed correlation-based feature selection optimizes the feature set to be fused, reduces the extra computational time. The obtained simulation result shows an outstanding performance.

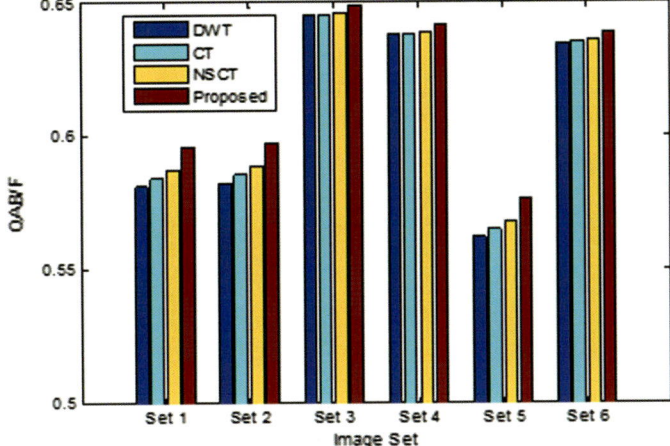

Fig. 10 QAB/F comparison

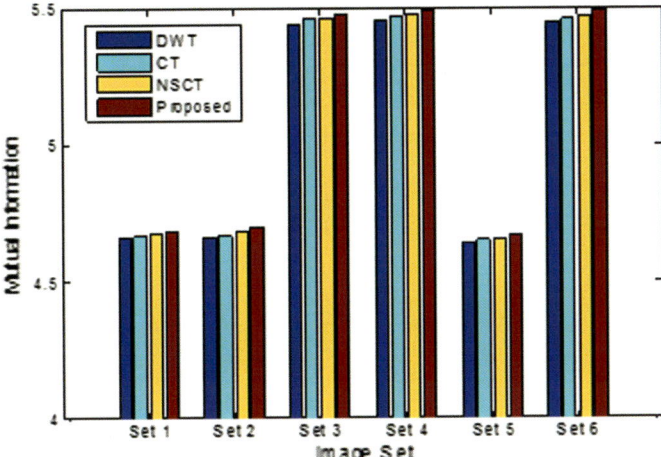

Fig. 11 Mutual information comparison

References

1. Gomathi PS, Kalaavathi B (2016) Multimodal medical image fusion in non-subsampled con-tourlet transform domain. Circ Syst 7(8):1598–1610
2. Valdés Hernández MDC, Ferguson KJ, Chappell FM, Wardlaw JM (2010) New multispectral MRI data fusion technique for white matter lesion segmentation: method and comparison with thresholding in FLAIR images. Eur Radiol 20(7):1684–1691
3. Liu Z, Yin H, Chai Y, Yang SX (2014) A novel approach for multimodal medical image fusion. Expert Syst Appl 41(16):7425–7435
4. Yang Y, Tong S, Huang S, Lin P (2014) Log-Gabor energy based multimodal medical image fusion in NSCT domain. Comput Math Methods Med 2014:1–12

5. Saravanan V, Babu G, Sivakumar R, Monie EC (2013) Medical image fusion by PCA method and implementation on FPGA. Int J Emerg Trends Sci Technol 03(05):800–806
6. Zhang X (2016) Sparse representation and PCA method for image fusion in remote sensing. In: 2016 2nd international conference on control, automation and robotics (ICCAR)
7. Salem YB (2016) Image fusion models and techniques at pixel level. In: Image processing, applications and systems (IPAS)
8. Kaur R (2016) An approach for image fusion using PCA and genetic algorithm. Int J Comput Appl 145(6):(0975–8887)
9. Li S, Kang X, Hu J, Yang B (2013) Image matting for fusion of multi-focus images in dynamic scenes. Inf Fus 14(2):147–162
10. Liu Y, Jin J, Wang Q, Shen Y, Dong X (2014) Region level based multi-focus image fusion using quaternion wavelet and normalized cut. Sig Process 97:9–30
11. Bhatnagar G, Wu QMJ, Liu Z (2013) Human visual system inspired multi-modal medical image fusion framework. Expert Syst Appl 40(5):1708–1720
12. Shen R, Cheng I, Basu A (2013) Cross-scale coefficient selection for volumetric medical image fusion. IEEE Trans Biomed Eng 60(4):1069–1079
13. Li S, Yang B (2008) Multi-focus image fusion using region segmentation and spatial frequency. Image Vis Comput 26:971–979
14. Ganasala P, Kumar V (2014) CT and MR image fusion scheme in non-subsampled contourlet transform domain. J Digit Imaging 27(3):407–418
15. Yang L, Guo BL, Ni W (2008) Multimodality medical image fusion based on multiscale geometric analysis of contourlet transform. Neurocomputing 72(1–3):203–211
16. Deng Ai (2011) An image fusion algorithm based on DWT and canny operator. Adv Res Comput Educ Simul Model 175:32–38
17. Yang Y, Huang SY, Gao J, Qian Z (2014) Multi-focus image fusion using an effective discrete wavelet transform based algorithm. Meas Sci Rev 14(2):102–108
18. Zhou Z, Li S, Wang B (2014) Multi-scale weighted gradient-based fusion for multi-focus images. Inf Fusion 20:60–67
19. Chan WL, Choi H, Baraniuk R (2004) Quaternion wavelets for image analysis and processing. In: Proceedings of international conference image processing, pp 3057–3060
20. Yin M, Liu W, Shui J, Wu J (2012) Quaternion wavelet analysis and application in image denoising. Math Prob Eng 2012, Art. no. 493976
21. Gai S, Wang L, Yang G, Yang P (2016) Sparse representation based on vector extension of reduced quaternion matrix for multiscale image denoising. IET Image Process 10(8):598–607
22. Yang Y, Tong S, Huang S, Lin P (2014) Log-Gabor energy based multimodal medical image fusion in NSCT domain. Comput Math Methods Med 2014:1–12
23. Yang Y, Tong S, Huang S, Lin P (2015) Multi-focus image fusion based on NSCT and focused area detection. IEEE Sens J 15(5):2824–2838
24. Piella G, Heijmans H (2003) A new quality metric for image fusion. In: Proceedings of the international conference on image processing (ICIP '03), Catalonia, Spain, pp 173–176, September 2003
25. Xydeas CS, Petrović V (2000) Objective image fusion performance measure. Electron Lett 36(4):308–309
26. Qu G, Zhang D, Yan P (2002) Information measure for performance of image fusion. Electron Lett 38(7):313–315

Early Detection of Proliferative Diabetic Retinopathy in Neovascularization at the Disc by Observing Retinal Vascular Structure

Nilanjana Dutta Roy and Arindam Biswas

Abstract Proliferative Diabetic Retinopathy (PDR) is the advanced stage of Diabetic Retinopathy (DR) with high risk of severe visual impairment. Neovascularization is a common scenario at this stage where abnormal vessels proliferate. This paper describes a semi-automated method for early detection of PDR around few diameters of Optic Disc (OD) in retinal images. Center of OD detection from segmented images is essentially important here because the approach focuses on Neovascularization at the Disc (NVD). Around OD center on few pixel distance window boundary, the width of major vessels are measured and counted. Finally, the major vessels are identified by distinct colors. The sensitivity and specificity results on STARE dataset of 25 images are 0.86 and 0.87, respectively. The approach shows the average accuracy as 0.88.

1 Introduction

Diabetic Retinopathy (DR) is one of the leading causes of blindness for patients who are suffering from diabetes for many years. Out of four major progressive stages of DR, Proliferative Diabetic Retinopathy (PDR) is the advanced stage where lack of oxygen supply in blood vessels triggers the proliferation of thin and fragile blood vessels. These new fragile vessels, which are most likely to leak and bleed, when they grow around certain diameter of Optic Disc (OD) region in human eye are classified as Neovascularization at the Disc (NVD). When they grow at anywhere else in the vasculature, are termed as neovascularization elsewhere (NVE). Both the categories are responsible for abnormal proliferation and can cause vitreous hemorrhage,

N. D. Roy (✉)
Department of Computer Science and Engineering,
Institute of Engineering and Management, Kolkata, India
e-mail: nilanjanaduttaroy@gmail.com

A. Biswas
Department of Information Technology,
Indian Institute of Engineering Science and Technology, Shibpur, Howrah, India
e-mail: barindam@gmail.com

© Springer Nature Switzerland AG 2019
D. Pandian et al. (eds.), *Proceedings of the International Conference on ISMAC in Computational Vision and Bio-Engineering 2018 (ISMAC-CVB)*, Lecture Notes in Computational Vision and Biomechanics 30,
https://doi.org/10.1007/978-3-030-00665-5_134

which further leads to a high risk of vision loss. Although a lot of algorithms have been developed to detect PDR in retinal images, the number is significantly less in comparison to non Proliferative Diabetic Retinopathy (NPDR) detection. A recent approach in [1] shows the extraction of vessel patterns and OD region for NVD/NVE classification by multilayered threshold. This method is dependent on a few structural and intensity based features for classification. Monitoring the openness of the major temporal arcade was proposed by the authors of [2] using single and dual parabolic models in normal and PDR images. To classify the stages of neovascularization, morphological transformations and Gaussian filtering were used in [3]. But all these methods described above, are prone to showing high false positive rates due to lack of analysis on the classification of features for NVD/NVE separately. The detection of NPDR lesions was successfully done using Hessian-based filtering approaches proposed by the authors of [4], but the value of such filters for PDR detection has not been significantly analyzed yet.

In this work, we present preprocessing and segmentation, followed by the center of OD detection, count and measurement of the width of major vessels, and coloring them in the Methodology section. Validation stage is to monitor the images regularly. Experimental evaluation section deals with the results, performance measurement, and databases we have used. An overall conclusion is stated in the Conclusion section.

This paper makes a contribution on detecting early signs of NVD PDR based on retinal vessel structure. This could further help the ophthalmologists to screen the disease and start early treatment.

An image of normal eye and image of PDR effected eye where the presence of fragile vessels are clearly seen around OD are shown in Fig. 1.

(a)   **(b)**

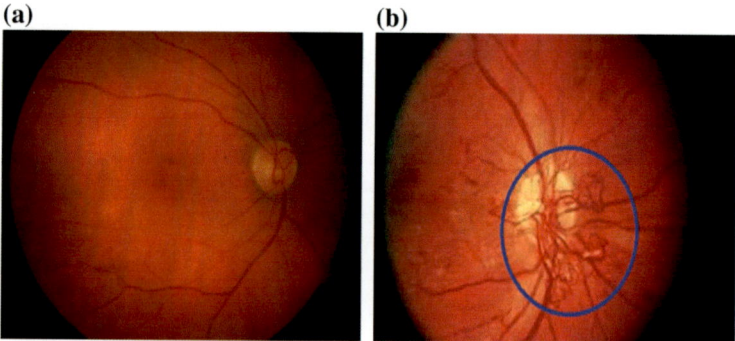

Fig. 1 Fundus images **a** normal eye, **b** PDR effected eye with NVD

2 Methodology

The task of PDR detection is accomplished based on the segmentation process followed by OD center detection, count on major vessels around OD, width measurement, and vessel coloring techniques. Initial steps include preprocessing to enhance the vasculature of the image and to ensure the image's quality against noise. Thereafter, the need for segmentation comes to validate the presence of abnormality in the image. Regular analysis of the eye of a diabetic patient, within few diameter areas around OD is indicative of the unusual changes about to happen. As the purpose of this approach is to detect neovascularization at OD (NVD), we are using the center of OD as the prime reference point from where all the vessels emanate. Repetitive count on major vessels around OD and measuring the width of them also help as a significant indication toward proliferation of fragile vessels. Conflict in vessel count raises an issue of abnormal situation whereas latter help in catching newly created tiny and fragile vessels. Coloring each vessel is used as an additional feature for validation applied over the resultant image along with the local features. The framework produces a final decision by comparing changes in number, width, and color of the tiny vessels with the stored image of the same person in the database.

2.1 Image Preprocessing and Segmentation

The fundus images from DRIVE [5] database, captured by Canon CR5 non-mydriatic 3 CCD cameras with a 45° FOV for medical imaging, have to undergo few image operations. Some common image processing steps are applied here to make the image ready for further processing. Grayscale conversion, sharpening using multiple passes of illumination distribution by CLAHE, followed by Otsu thresholding helped us to give the image a good shape at its initial stage. The grayscaled image is then passed through 2D median filtering for de-noising and finally, a smooth textured image in the binary platform is hence presented. Figure 2 shows the process for segmentation.

2.2 Center of Optic Disc

Regular monitoring on OD and its neighborhood region quicken the process of initial PDR detection as NVD originates at the optic disc region. So to track the progress, the center of OD detection plays an important role here as a reference point. In a 3×3 sized scanning window over the segmented binary image, a white pixel is detected as terminal point whose more than seven neighbors are black. Starting from the terminal points, all the straight lines are removed from the vasculature by making them as black till the junction points arrive. The resultant image is a structural 'ring' with all the closed polygons and with no single straight line. Later, the junction points are

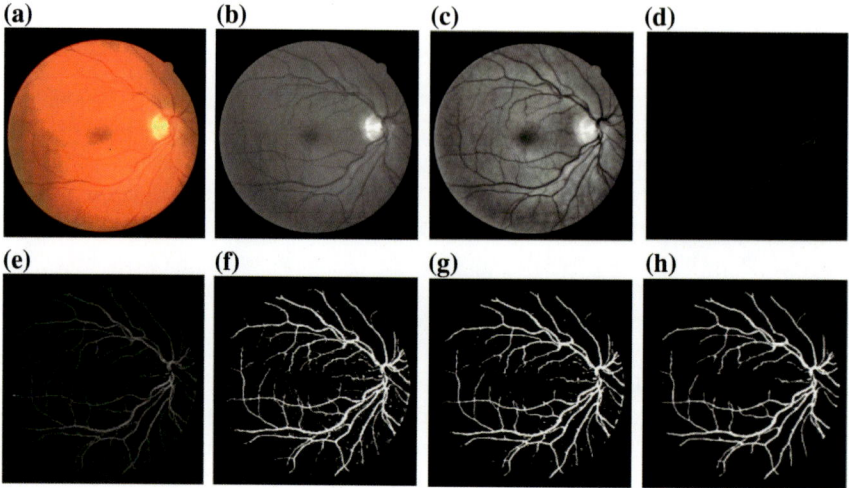

Fig. 2 a Original RGB image, **b** green channeled image, **c** AHE filtered image, **d** image after bottom hat, **e** contrast enhanced image, **f** extracted blood vessels by Otsu's thresholding, **g** median filtered image, **h** segmented image after noise removal

detected from the "ring". It is observed that the number of junction points increase at around few diameters of OD. Here, a cluster is identified with maximum number of junction points and a convex hull [6] is formed out of it. The centroid of the resultant convex hull, shown in Eqs. 1 and 2, is detected as the reference center of OD. A flow of this section is shown in Fig. 4. A result of center of optic disc detection is shown in Fig. 3a.

$$C_x = \frac{1}{6A} \sum_{i=1}^{N-1} (x_i + x_{i+1})(x_i y_{i+1} - x_{i+1} y_i) \tag{1}$$

and

$$C_y = \frac{1}{6A} \sum_{i=1}^{N-1} (y_i + y_{i+1})(x_i y_{i+1} - x_{i+1} y_i) \tag{2}$$

Hence, the point (C_x, C_y) is then detected as the center of OD in the proposed method.

2.3 Major Vessels by Measuring Width

Leaving a certain pixel distance from the center of OD detected from the previous stage, a rectangular window boundary is formed surrounding it. Scanning the window boundary clockwise determines the width of the major vessels, shown in Fig. 3b.

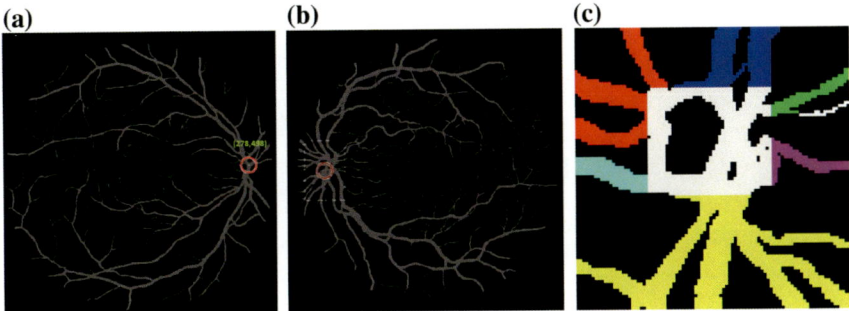

Fig. 3 Sequence of the major vessel width measurement around OD is shown **a** center of OD detected, **b** window boundary around center of OD is drawn, **c** major vessels are identified

Number of white pixels within a black-to-white and white-to-black transition pair shows the width of the major vessels. A structural analysis on the number of widths is done and an assumption on the average width is also observed.

2.4 Vessel Coloring

The same window boundary method described earlier is followed here to color the major vessels. At the outer boundary of the window, clockwise, white pixels are colored until white-to-black transition occurs. The process continues till there is no white pixel left at the outer boundary of the window [7]. Keeping detected center of optic disc at the middle, major vessels with specific width are identified by different colors, see Fig. 3c.

3 Validation Stage

Regular monitoring of the diabetic patients helps to determine the early signs of PDR. Here, two-stage validation starts by passing the fundus image of the patient through the preprocessing and segmentation phases again. Abrupt changes in the count of vessels first, asserts the abnormal presence of newly generated minor vessels. The sudden emanation of tiny vessels is caught by calculating the width of them at the second stage of validation. Finally, the presence of tiny vessels below an empirically derived threshold value is not given colors as they have failed to satisfy the coloring criteria. So, early detection of PDR by two-stage validation method helps in preventing the disease before it aggravates.

The images in Fig. 5c, d show the unusual changes in the region around optic disc center which is indicative of PDR. The proposed method can identify the generated

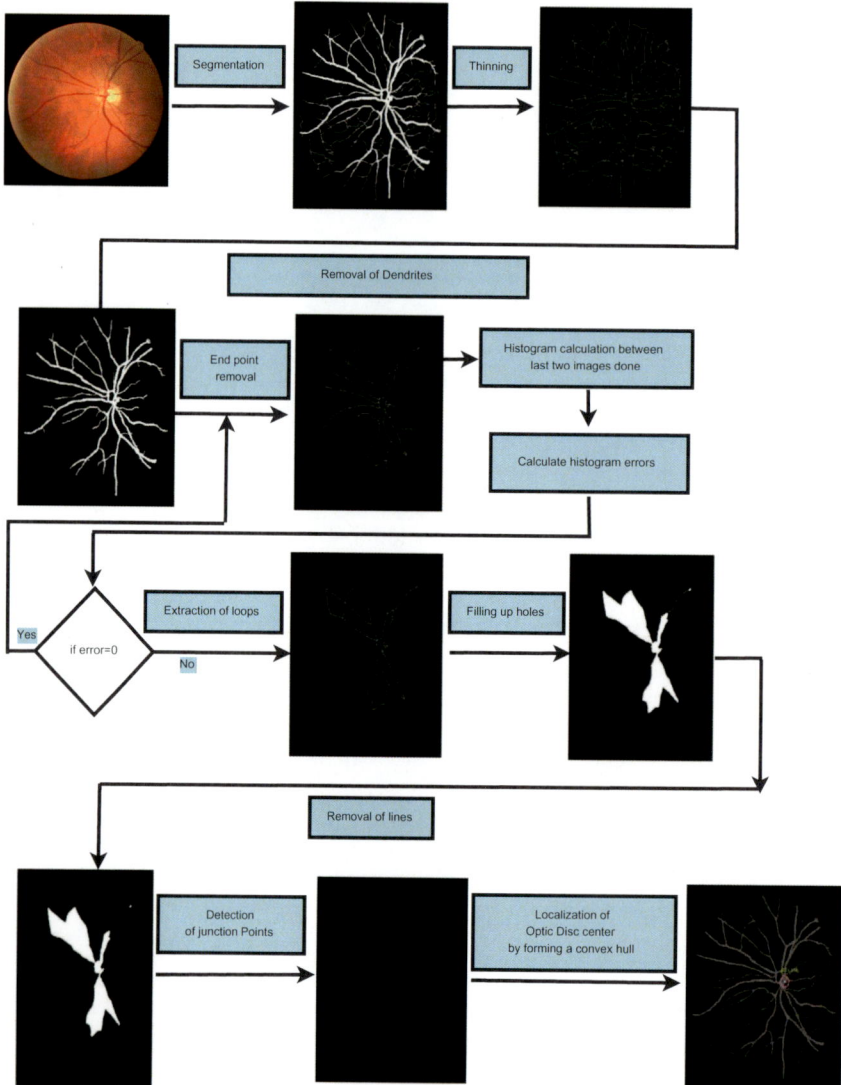

Fig. 4 Flow for optic disc center detection

tiny vessels by measuring their width and not coloring them if they failed the coloring criteria, for not being a part of major vessel. Also, the method triggers an alarming situation by finding sudden mismatch in vessel count around OD. Abrupt changes in OD area proposed in this work thus detects the early signs of PDR.

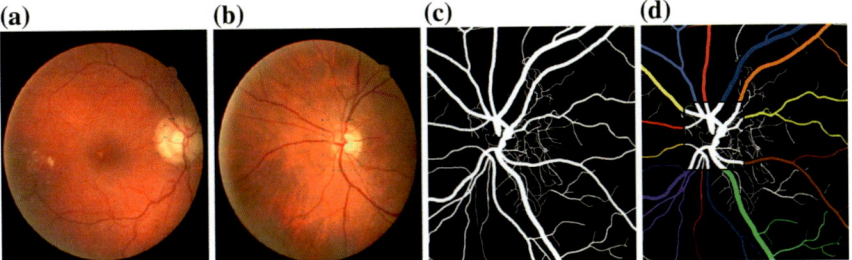

Fig. 5 Images **a** and **b** from DRIVE database are found to be normal here, but images **c** and **d** from STARE database show proliferation of fragile vessels around optic disc

4   Experimental Evaluation

4.1   Materials

Due to low prevalence of PDR images, the proposed method was evaluated on the images collected from DRIVE [5] and STARE [8] databases. The collected RGB images are passed through the several stages of preprocessing to make the images ready for the experiment. Segmentation plays a crucial role here which is done on every image by following the method described in image preprocessing and segmentation section.

4.2   Performance Measures

It is a common practice to measure the performance of an algorithm based on its sensitivity, specificity, and accuracy calculation. Sensitivity, defined in formula 3, measures the proportion of actual positives which are correctly identified. Specificity is the proportion of negatives which are correctly identified and defined in formula 4.

$$\text{Sensitivity} = \frac{TN}{TN+TP} \tag{3}$$

$$\text{Specificity} = \frac{TP}{TP+TN} \tag{4}$$

Accuracy is calculated as

$$\text{Accuracy} = \frac{(TP+TN)}{(TP+FP+TN+FN)} \tag{5}$$

where TP is true positive, TN is true negative, FP is false positive, and FN is false negative.

Table 1 Results of databases

Database	No. of images	Accuracy	Sensitivity	Specificity
STARE	25	0.88	0.86	0.87

Table 2 Comparison with other PDR detection methods

Methods	Accuracy	Sensitivity	Specificity	Level
Agurto [9]	0.94	0.96	0.83	Image
Goatman [10]	0.91	0.84	0.85	Image
Hassan [3]	0.70	0.63	0.89	Pixel
Jelinek [11]	0.90	0.94	0.82	Image
Welikala [12]	0.97	1	0.90	Image
Proposed method	0.88	0.86	0.87	Image

4.3 Results

We observe that the proposed method accomplishes the task of early signs of PDR detection in the accuracy range between 0.82 and 0.91 and achieves low false positive rates. Table 1 shows the overall result in various databases in average case and Table 2 compares it with many other existing methods. Figure 5a, b shows the experimental results on images from DRIVE [5] dataset. On both the images, the width of major vessels is computed and shown as green lines on the RGB images. At the same time, count on major vessels is also done. Since there is no abnormality found on the images, the images are declared as normal eye images. On the other hand, we found an abnormality on the images shown in Fig. 5c, d in vessels count, the presence of some tiny vessels and moreover, few vessels are not colored as they failed in the coloring criteria (Fig. 5d). We, therefore, declare them as PDR images where proliferation has already started.

5 Conclusion

PDR is the advanced stage of DR where new abnormal blood vessels form in different regions of the retina. In this work, we are focussing on the NVD where fragile vessels grow around the optic disc. DR cannot be cured completely. Photocoagulation (laser analysis) is very effective if it is done before the disease adversely harms retina. Surgical elimination of vitreous gel (vitrectomy) helps in improvement of vision, provided that the massive damage in retina has not been done. An anti-inflammatory medicine or antivascular endothelial growth factor medication injection is helpful in new blood vessel contraction process. This study helps in detecting the early signs of PDR. Also, an initiative has been taken to find the way out to detect the disease at its early stages to prevent from permanent vision loss. This work will be useful for the technical persons and an automated approach of this study will save many diabetic patients from blindness due to diabetic retinopathy. Since no symptom is seen in PDR until the disease turns into the stern, regular observation on diabetic patient's eye is essential as a preventive measure against vision loss.

References

1. Akram MU, Khalid S, Tariq A, Javed MY (2013) Detection of neovascularization in retinal images using multivariate m-mediods based classifier. Comput Med Imaging Graph 37(5):346–357
2. Oloumi F, Rangayyan RM, Ells AL (2012) Computer-aided diagnosis of proliferative diabetic retinopathy. International conference of the IEEE engineering in medicine and biology society (EMBC) 2012:1438–1441
3. Hassan SSA, Bong DB, Premsenthil M (2012) Detection of neovascularization in diabetic retinopathy. J Digit Imaging 25(3):437–444
4. Srivastava R, Wong DW, Duan L, Liu J, Wong TY (2015) Red lesion detection in retinal fundus images using frangi-based filters. In: IEEE engineering in medicine and biology society (EMBC), pp 5663–5666
5. The DRIVE database, Image sciences institute, university medical center Utrecht, The Netherlands. http://www.isi.uu.nl/Research/Databases/DRIVE/. Last accessed on 7th July 2007
6. Gonzalez RC, Eugene Woods R Digital image processing book, 3rd edn. Paperback Publishers
7. Dutta Roy N, Someswar M, Dalmia H, Biswas A (2014) Identification of distinct nerves in retinal fundus images. compImage'14, P.A. Pittsburgh, USA
8. Boyd J (1996) STARE software documentation: diskOptic disk locator. Vis Comput Lab Dept Elect Comput Eng Univ California
9. Agurto C, Honggang Y, Murray V, Pattichis MS, Barriga S, Bauman W et al (2012) Detection of neovascularization in the optic disc using an AMFM representation, granulometry, and vessel segmentation. In: Annual international conference of the IEEE, engineering in medicine and biology society (EMBC), 2012, pp 4946–4949
10. Goatman KA, Fleming AD, Philip S, Williams GJ, Olson JA, Sharp PF (2011) Detection of new vessels on the optic discusing retinal photographs. IEEE Trans Med Imaging 30:972
11. Jelinek HF, Cree MJ, Leandro JJ, Soares JV, Cesar RM Jr, Luckie A (2007) Automated segmentation of retinal blood vessels and identification of proliferative diabetic retinopathy. J Opt Soc Am Opt Image Sci Vis 24:1448–1456
12. Welikala R, Dehmeshki J et al (2014) Automated detection of proliferative diabetic retinopathy using a modified line operator and dual classification. Comput Methods Programs Biomed 114(3):247261

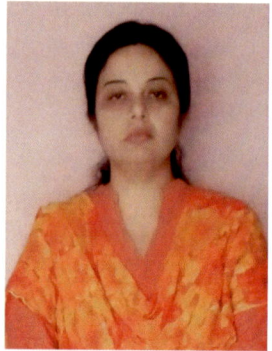

Nilanjana Dutta Roy is a researcher at Indian Institute of Engineering Science and Technology, Shibpur, with specialisation in the field of Image processing and is an Assistant Professor at Institute of Engineering & Management, Kolkata.

Arindam Biswas graduated from Jadavpur University, Kolkata, India, and received his masters and doctorate degree both from the Indian Statistical Institute, Kolkata, India. He is currently Associate Professor in the Department of Information Technology, Indian Institute of Engineering Science and Technology, Shibpur, India. His research interests include digital geometry, image processing, approximate shape matching and analysis, medical image analysis, natural language processing, and biometrics. He has published over 75 research papers in international journals, edited volumes, and refereed conference proceedings and holds one US patent.

Finding Center of Optic Disc from Fundus Images for Image Characterization and Analysis

Nilanjana Dutta Roy and Arindam Biswas

Abstract An automated method for center reference point extraction from retinal fundus images is essentially required for an untroubled image mapping in medical image analysis, image registration, and verification. This paper proposes a spadework, revealing a distinct reference point within optic disc in blood vessel structure of the human eye, analysis on which would serve as an efficient preventive measure for any ocular disease and would strengthen the image verification method along with other extracted features of the human eye at low cost. The proposed method includes segmentation from colored fundus images followed by removal of thin and tiny blood vessels which carry very less information. Removal process comes up with a few bounded polygonal structures, named ring near optic disc. From the named structures near optic disc, a cluster of junction points have been found with maximum members and we made a convex hull out of them. Finally, calculating the centroid of the formed convex hull unveils the center of the optic disc. Experiments are done on some publicly available databases called DRIVE, STARE, and VARIA. Experimental results compared to other standard methods are available in the literature.

1 Introduction

Digital image analysis on retinal images offers huge potential benefits by automated analysis process. In a research setting, an automated method exhibits its ability to examine a large number of images within marginal time and cost and diagnoses deformities more precisely than traditional observation driven techniques. Unnatural behavior in retinal reference features may be associated with the occurrence of

N. D. Roy (✉)
Department of Computer Science and Engineering,
Institute of Engineering and Management, Kolkata, India
e-mail: nilanjanaduttaroy@gmail.com

A. Biswas
Department of Information Technology,
Indian Institute of Engineering Science and Technology, Shibpur, Howrah, India
e-mail: barindam@gmail.com

© Springer Nature Switzerland AG 2019
D. Pandian et al. (eds.), *Proceedings of the International Conference on ISMAC in Computational Vision and Bio-Engineering 2018 (ISMAC-CVB)*, Lecture Notes in Computational Vision and Biomechanics 30,
https://doi.org/10.1007/978-3-030-00665-5_135

1451

retinopathies or cardiovascular diseases. To observe any abnormality in the human retina as a preventive measure, study and analysis on optic disc (OD) are essentially important. On the other hand, due to the overwhelming demand of biometric authentication against vulnerable threats in traditional security systems, strengthening it from every aspect at marginal cost has become an essential work. Optic disc is a key reference for many recognition algorithms [1, 2]. It has been asserted by expert clinicians and scientists that the structure of its blood vessels are distinct in every individual, even for the identical twins and it remains same during his life. The statement which, although literally true, is nevertheless a challenging task to prove technically. OD is the brightest region of any retinal fundus image from where all the blood vessels emanate. It is often considered as a landmark and reference for the other features of any retinal fundus image. An automatic and efficient detection of it plays a significant role for various reasons, like to locate other anatomical components in retinal images, for vessel tracking to diagnose many eye diseases even related to diabetes and for registering the image for personal authentication. Moreover, to find abnormal structures in a retinal image successfully, it is often necessary to mask out the normal anatomy from the analysis. Also, the position of OD can be used as a reference length for measuring distances in retinal images, especially from the location of macula. Localization of OD is hence an essential and challenging work. OD extraction is quite difficult since its brightness, color, and contrast of OD is similar to other components such as cotton wools and exudates [3]. Most of the algorithms to localize OD fail when pathological regions exist in retina images [4, 5]. Some other algorithms are quite expensive and suffer from high computational cost [6–8]. We have taken an initiative in support to the undeniable truth by extracting a key feature from the vascular structure, along with other invariable features present in it which would be further helpful in image registration and verification process. Hence, it is a well-suited identification characteristic for biometrics also. Therefore, identifying the right parameter plays a significant role in the process to provide a stringent and unambiguous authentication in marginal time. So keeping these situations in mind, we are trying to explore a unique reference key, say, center reference key of OD which would be robust enough to work in any situation at normal cost and thus will help in biometric template matching in future.

2 Proposed Method

The proposed method is fundamentally based on two major phases. Phase I narrates the preprocessing phase which includes segmentation from the colored retinal images. Phase II describes a way to locate the center reference point of OD. Removal of thin and tiny blood vessels called dendrites initiates the Phase II process followed by finding the centroid of a cluster points with maximum neighboring points. The below-mentioned steps further help us to define the proposal more precisely.

2.1 Vessel Segmentation and Image Preprocessing

The process starts with a segmentation of the images using a previously developed approach that uses supervised pixel classification with a Gaussian filter set and classification by a k-nearest neighbor classifier [9]. The resulting image represents the likelihood of each pixel belonging to a vessel. In order to trace the vessel path and obtain structural mapping, a connected binary vessel image is required which may be obtained using a vessel reconnection algorithm based on a graph search [10]. The binary vessel image is generated from the vessel probability image using Otsus thresholding method [11]. The Otsu threshold minimizes the intra-class variance for the foreground and the background classes which is basically vessels and non-vessel region. Then, the skeleton of blood vessels is obtained by applying mathematical morphology which reduces the vessel to a center line of single pixel width [12].

While scanning the segmented image from its top-left to bottom-right corner, the width [13] of each blood vessel have been calculated on each point upon a threshold value 5 pixels. If the width is found to be lesser than the threshold, that vessel is eliminated from the original image considering it as a dendrite.

2.2 Localization of OD Center

Keeping the importance of localizing OD in mind, we have taken an initiative to locate an exact point, the center of it within the white bright region called OD. This robust method will successfully work with low-resolution images, in presence of pathological regions and even in infected images. As the algorithm proceeds, the center of OD region has been found following the below-mentioned method, see Fig. 1 to have a concise view of the process. Figure 2b exhibits the ultimate outcome of this section from a segmented image. It has been statistically found from the binary images, that there is a normal tendency of the major blood vessels near optic disc to bifurcate and cross each other which further form a few polygonal loop-like structures, called rings. We are emphasizing on the rings here by considering them as a trademark to locate the optic disc in this algorithm. But overlapping on dendrites may also appear as same which carries redundant information. So, removing dendrites is an essential task to keep the loops near OD intact.

A morphological thinning operation is applied on the segmented fundus image to get a thinned, skeleton view of it initially, shown in Fig. 3b. Next, we tried to reveal all the rings present in segmented fundus image. Then, to remove the tiny vessels, a scan at each point of the segmented input image is being done to measure the width of all existing vessels. For any vessel, having its width as less than a threshold value of 5 pixels, is considered as dendrite and the same is removed from the original segmented image, see Fig. 3c. Here, we tried to remove all the end points or tails of blood vessels which would end up at any bounded region or polygonal loops. A white pixel whose seven neighbors are black in a 3 × 3 scanning window has

Fig. 1 Overview of the method

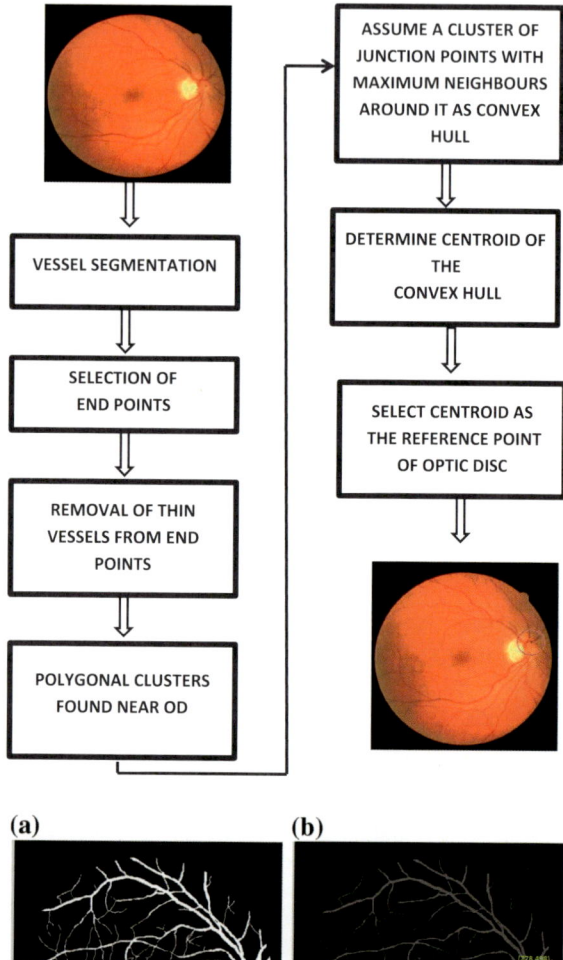

VESSEL SEGMENTATION

SELECTION OF
END POINTS

REMOVAL OF THIN
VESSELS FROM END
POINTS

POLYGONAL CLUSTERS
FOUND NEAR OD

ASSUME A CLUSTER OF
JUNCTION POINTS WITH
MAXIMUM NEIGHBOURS
AROUND IT AS CONVEX
HULL

DETERMINE CENTROID OF
THE
CONVEX HULL

SELECT CENTROID AS
THE REFERENCE POINT
OF OPTIC DISC

Fig. 2 a Segmented fundus image. **b** Optic disc located

(a)   **(b)**

been assumed as end point in this method. This comes within an iteration of few steps. At each step, end points are removed from the thinned image by making them same as background and it is being compared with the histogram of thinned image obtained in the previous iteration. Comparison is made following the histogram error calculation between both the images at every iteration. Error zero indicates that both the images are exactly same by anatomy, whereas some error shows that there are further scopes exist yet for removal of its tails. This process stops when no end points

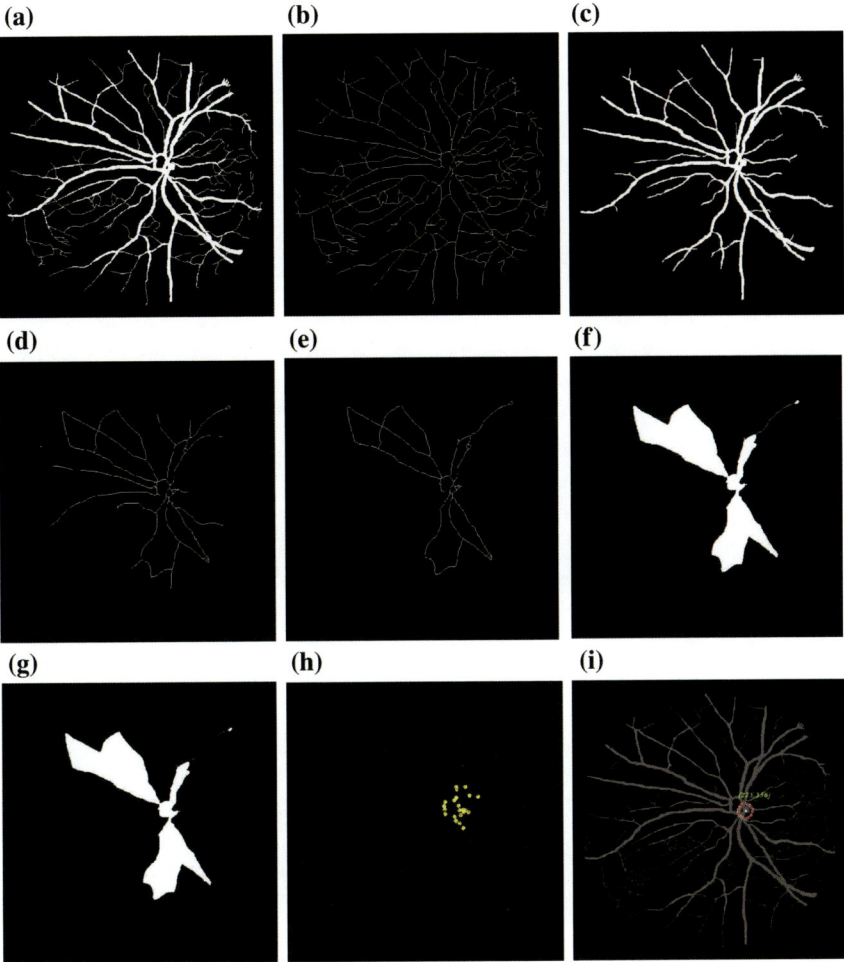

Fig. 3 **a** Segmented fundus image. **b** Thinned image obtained after applying morphological function. **c** Image after removing the dendrites. **d** Image after a few iteration. **e** After extracting the loops in the image. **f** After applying area fill function to bridge the holes. **g** Further removal of lines connecting the loops. **h** All junction points detected. **i** Optic disc located

can be further removed and the loops have been found. Following this way, all the end points are removed from the thinned images by making every eight neighborhood pixels as black, refer Fig. 3e. Histogram error 0 says both the images are found to be identical. On the flip side, the histogram error not equal to zero indicating not identical images, signifies that the end points could further be removed from the images and the process continues till the next match comes. Single line removal process stops at every junction and checks for its neighborhood regions.

2.3 Determining Convex Hull Out of Junction Points Near OD

After all the single line pixels have been removed, it turns into a smallest possible structure with all the polygonal rings, see Fig. 3e. We have used an algorithm for region filling which performs on background pixels of the binary image, starting from any seed pixel within the bounded region and covers all the regions within the bounded loops. After the polygonal holes have been filled up [14], see Fig. 3f, a scanning window of size 3 × 3 moves around horizontal direction to determine a polygon with maximum number of single lines or neighbors connected with it. Neighbor selection is again done in anticlockwise direction and by connected component analysis around the bounded region. All single lines attached with the region are now removed by making them same as background pixel. Finally, we are now left with a small colored region and the junction points [15] within that bounded region has been found. We are concerned about those selected junction points in this region as they would help us to accomplish the goal of determining the center of OD. We now try to convert those scattered points into a convex hull or minimal convex polygonal (MCP) shape to locate the centroid of the bounded area.

Converting a few points into a convex hull starts by finding the bottom most point, say P in this case. Remaining points are sorted in ascending order with respect to their angles formed with the bottom most point P. Traversing counterclockwise with left turn for the next node is safe, whereas taking right turn may discard the next point to be chosen from the remaining list saying the initial and final points are out of boundary line. The process ends after meeting the initial point from where it has been started. The ultimate junction points near OD that we have generated from the earlier steps are now being converted into a convex hull or MCP by following the Graham Scan method [16].

2.4 Centroid of MCP as Center of OD

Once the focused region has been found, it becomes an effortless job to calculate the centroid of the MCP or the convex hull. Let us assume the bounded region is a polygon with N vertices and $N - 1$ line segments; (x_i, y_i), where $i = 0$ to $N - 1$. The last vertex (x_N, y_N) is assumed to be the same as initial vertex (x_i, y_i), marking it as a closed polygon. The area of this region is then decided by

$$A = \frac{1}{2} \sum_{i=1}^{N-1} (x_i y_{i+1} - x_{i+1} y_i) \tag{1}$$

Area further helps to calculate the centroid of the bounded region where (x_N, y_N) is assumed to be same as (x_0, y_0).

$$C_x = \frac{1}{6A} \sum_{i=1}^{N-1} (x_i + x_{i+1})(x_i y_{i+1} - x_{i+1} y_i) \tag{2}$$

and

$$C_y = \frac{1}{6A} \sum_{i=1}^{N-1} (y_i + y_{i+1})(x_i y_{i+1} - x_{i+1} y_i) \tag{3}$$

Hence, the point (C_x, C_y) is then considered as the center of OD in our proposed method.

3 Results and Discussion

The purpose of this spadework was to identify the center pixel of OD in the human retina. Finding center of OD plays a significant role in many aspects as discussed earlier. Additionally, it could act as a root of a retinal vascular tree generated from the retinal vascular network. Any geometric transformation would successfully be applied here for future applications. This major extracted feature could again be used as an important biometric parameters for preparing retinal template for image registration by matching digital images. Besides, any deformities, like retinal dislocation etc., could easily be detected for medical image analysis. This section presents ultimate outcomes of both Phase I and Phase II within a single block (Table 1).

In literature, there are contributions from many other authors also in the similar area. We tried to reveal a comparative study which proves our proposed method is comparable with some of the other existing works, please see Table 2. Figure 4 shows the detected OD center on different images of DRIVE [17], VARIA [18] and

Table 1 Experimental results showing location of OD center on DRIVE database

DRIVE database							
Image	Optic disc	Image	Optic disc	Image	Optic disc	Image	Optic disc
1	(248, 84)	11	(254, 63)	21	(254, 69)	31	(263, 377)
2	(248, 484)	12	(255, 76)	22	(283, 475)	32	(266, 515)
3	(298, 71)	13	(275, 497)	23	(251, 417)	33	(257, 512)
4	(282, 346)	14	(283, 487)	24	(287, 481)	34	(281, 468)
5	(279, 78)	15	(249, 229)	25	(298, 466)	35	(267, 77)
6	(255, 470)	16	(272, 491)	26	(248, 86)	36	(279, 482)
7	(252, 483)	17	(280, 470)	27	(283, 500)	37	(274, 496)
8	(282, 491)	18	(249, 501)	28	(269, 486)	38	(290, 499)
9	(268, 89)	19	(276, 490)	29	(289, 505)	39	(282, 89)
10	(281, 466)	20	(264, 501)	30	(263, 469)	40	(282, 486)

Table 2 Comparison between different methods for OD localization

Results of other methods to detect OD

Algorithms	Result (%)	Data set	Running time (s)	Distance (pixels)	System configuration
Youssif	100	DRIVE	210	17	Intel Core 2 Duo 1.7 GHz and 512 Mb RAM
Sekhar	85	DRIVE	NA	NA	NA
Rangayyan	100	DRIVE	2294	23.2	Intel Core 2 Duo 2.5 GHz and 1.96 GB RAM
Zhu et al.	90	DRIVE	NA	18	NA
Park et al.	90.25	DRIVE	4	NA	Intel 1 GHz and 1 GB RAM
Dehghani	100	DRIVE	27.6	15.9	Intel Core 2 Duo 2.67 GHz and 3.24 GB RAM
Proposed method	92	DRIVE	20	15	Intel Core 2 Duo 2.67 GHz and 2 GB RAM

(a) **(b)** **(c)**

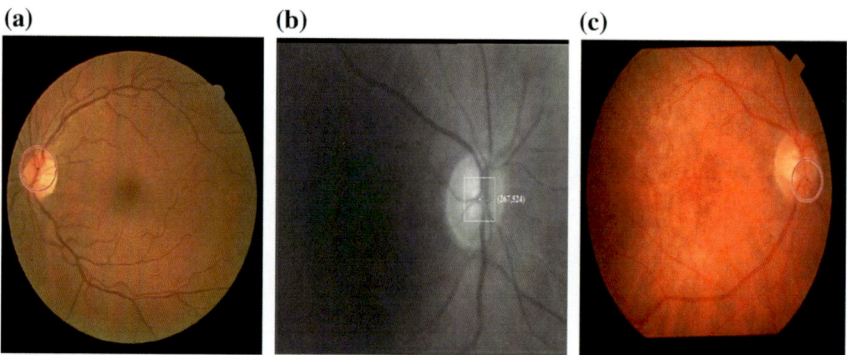

Fig. 4 Results of the proposed method on various databases. **a** Detected center of OD on DRIVE image. **b** Detected center of OD on VARIA image. **c** Detected center of OD on STARE image

STARE databases. The images of DRIVE have been captured using Canon CR5 non-mydriatic 3 CCD cameras with a 45° FOV for medical imaging.[1]

4 Conclusion

Characterization of several features from the human retina is immensely appreciable work. Diagnosis of any ocular diseases need a thorough analysis on retinal fundus images, especially on OD and bifurcation points. Also, with the rapid growth of biometric authentication, searching an efficient feature is essentially important and

---

[1]The results may vary a bit due to different clarity of the images published by other databases.

challenging task for faster computation. So keeping these primary objectives in mind, i.e., analysis of specific features for disease detection, for characterization, and to search new features for preparing a retinal template for biometric authentication, we have executed the proposed method on 32 bit MATLAB 2013a version on 32 bit Windows XP running on Intel Core-2 dual processor with 2 GB of RAM. Locating center of OD is essentially important for finding the accurate distance between Macula and OD. The experimental results are clearly showing the detected OD locations on segmented fundus images from various databases.

References

1. Farzin H, Abrishami Moghaddam H, Moin MS (2008) A novel retinal identification system. EURASIP J Adv Sig Process 2008. Article ID 280635. https://doi.org/10.1155/2008/280635
2. Ortega M, Penedo MG, Rouco J, Barreira N (2009) Retinal verification using a feature-point based biometric pattern. EURASIP J Adv Sig Process, Article ID 235746. https://doi.org/10.1155/2009/235746
3. Sumathy B, Poornachandra S (2013) Feature extraction in retinal fundus images. In: IEEE Xplore proceedings of information communication and embedded systems (ICICES), Chennai
4. Osareh A (2004) Automated identification of diabetic retinal exudates and the optic disc. Ph.D. dissertation. Department of Computer Science, Faculty of Engineering, University of Bristol, Bristol, UK
5. Li H, Chutatape O (2001) Automatic location of optic disc in retinal images. Proc Int Conf Image Process (ICIP) 2:837–840
6. Youssif AA, Ghalwash AZ, Ghoneim AS (2008) Optic disc detection from normalized digital fundus images by means of a vessels direction matched filter. IEEE Trans Med Imag 27:11–18
7. Rangayyan RM, Zhu X, Ayres FJ, Ells AL (2010) Detection of the optic nerve head in fundus images of the retina with Gabor filters and phase portrait analysis. J Digit Imag 23(4):438–453. https://doi.org/10.1007/s10278-009-9261-1
8. Zhu X, Rangayyan RM, Ells AL (2010) Detection of the optic nerve head in fundus images of the retina using the hough transform for circles. J Digit Imag 23(3):332–341. https://doi.org/10.1007/s10278-009-9189-5
9. Niemeijer M, Staal J, Ginneken B, Loog M, Abramoff M (2004) Comparative study of retinal vessel segmentation methods on a new publicly available database. Proc SPIE 5370:648–656
10. Joshi V, Garvin M, Reinhardt J, Abramoff M, Identification and reconnection of interrupted vessels in retinal vessel segmentation. In: IEEE, ISBI, image segmentation methods, vol FR-PS3a.7, pp 1416–1420
11. Otsu N (1979) A threshold selection method from gray-level histograms. IEEE Trans Sys Man Cyber 9(1):62–66
12. Rockett P (2005) An improved rotation-invariant thinning algorithm. IEEE Trans Pattern Anal Mach Intell 27:1671–1674
13. Lowell J, Hunter A, Steel D, Basu A, Ryder R, RL Kennedy (2004) Measurement of retinal vessel widths from fundus images based on 2-D modeling. IEEE Trans Med Imaging 23(10)
14. Hearn D, Pauline Baker M Computer graphics, 3rd edn. Paperback Publishers
15. Saha S, Dutta Roy N (2013) Automatic detection of bifurcation points in retinal fundus images. Int J Latest Res Sci Technol 2(2):105–108. ISSN (Online):2278-5299. http://www.mnkjournals.com/ijlrst.htm
16. Graham RL (1972) An efficient algorithm for determining the convex hull of a finite planar set. Inf Process Lett 1:132–133
17. The DRIVE database, Image sciences institute, university medical center utrecht. The Netherlands. http://www.isi.uu.nl/Research/Databases/DRIVE/. Last accessed on 7th July 2007
18. VARIA Database, Department of Computer Science of the Faculty of Informatics of the University of Corua, http://www.varpa.es/varia.html

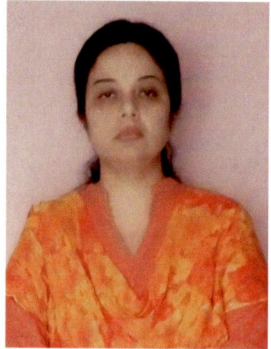

Nilanjana Dutta Roy is a researcher at Indian Institute of Engineering Science and Technology, Shibpur, with specialisation in the field of Image processing and is an Assistant Professor at Institute of Engineering & Management, Kolkata.

Arindam Biswas graduated from Jadavpur University, Kolkata, India, and received his masters and doctorate degree both from the Indian Statistical Institute, Kolkata, India. He is currently Associate Professor in the Department of Information Technology, Indian Institute of Engineering Science and Technology, Shibpur, India. His research interests include digital geometry, image processing, approximate shape matching and analysis, medical image analysis, natural language processing, and biometrics. He has published over 75 research papers in international journals, edited volumes, and refereed conference proceedings and holds one US patent.

A Robust Method for Image Copy-Move Passive Forgery Detection with Enhanced Speed

Asif Hassan and V. K. Sharma

Abstract Forgery detection of images is presently one of the fascinated research fields. Copy-move forgery is the most commonly used methods for image forgery. A novel method is proposed in this paper, which is an effective and advanced method for detecting copy-move forgery. The proposed method is a block matching technique with reduced computational speed and less computational complexities. The efficiency of outcome is also improved. The image is segmented into fixed dimensions of overlying blocks and then discrete cosine transform (DCT) is applied to each block to extract its features. Then, the mean of each block is obtained. The mean of each block is compared with other blocks to find the similarity between the blocks. The computational outcomes are shown that indicates the proposed method is robust to detect copy-move forgery efficiently with enhanced speed.

1 Introduction

An image is an artifact that represents visual insight of an event. We live in a digital world where digital images are used as a means of communication. Images are everywhere, on the Internet, social media, newspapers, etc. The authenticity of images is in question because it is very easy to edit any image using easily available image manipulating tools [1–4]. Hence, digital image forgery detection is important to authenticate the images. Image authentication methods can be classified into two modules such as active methods and passive methods. Active method is the technique in which prior information about the original image such as watermarking or signature [5–7] which is embedded inside the image is known for forgery detection. It is a drawback because in various situations prior information about the image is not available [8–11].

A. Hassan (✉) · V. K. Sharma
Bhagwant University, Ajmer, Rajasthan, India
e-mail: asif.43hassan@gmail.com

V. K. Sharma
e-mail: viren_krec@yahoo.com

© Springer Nature Switzerland AG 2019
D. Pandian et al. (eds.), *Proceedings of the International Conference on ISMAC in Computational Vision and Bio-Engineering 2018 (ISMAC-CVB)*, Lecture Notes in Computational Vision and Biomechanics 30,
https://doi.org/10.1007/978-3-030-00665-5_136

(a)   **(b)**

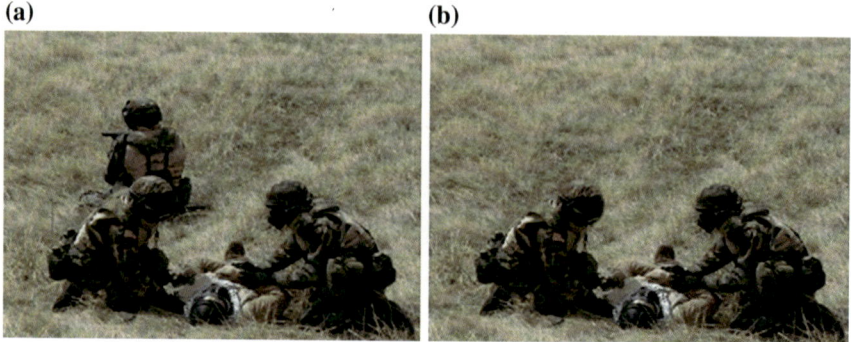

Fig. 1 Copy-move forgery: **a** Original image, **b** Copy-move forged image

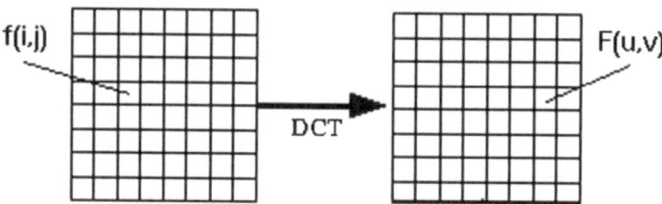

Fig. 2 Discrete cosine transforms

The Passive method is also known as the *blind technique* in which information regarding the image is not available [12, 13]. Hence, this method authenticates the image without the existence of the original information.

One of the most commonly used methods of image altering is hiding a region in the image and distorting the image information. Copy-move forgery is the most common image forgery used to hide the portion of the image. In copy-move forgery, some portions of an image are copied and posted on another area of the same image. An example of copy-move forgery is shown in Fig. 1.

The most frequently used method for image copy-move detection is block matching in which the image is divided into the same size overlying blocks then feature of each block is extracted and then each block is compared with other blocks in the same image. At last, the outcome of forgery detection is decided based on matched block features. During the feature extraction process, the essential features are chosen from the blocks using a discrete cosine transform (DCT)[1, 3]. These essential features are used to compare the blocks. DCT helps to segment the image or spectral sub-bands of differing importance corresponding to the image visual quality. The DCT is like the discrete Fourier transform which transforms the image into the frequency domain from the spatial domain as shown in Fig. 2.

Fadl and Semary (accelerated method) [1] proposed a method using K-means classifier and images of 128×128-pixel grayscale images. The proposed method works on direct computation, without using the classifier, for an image size of any pixel with reduced computational steps.

The paper is organized as follows: Sect. 2 presents the proposed system in details. Section 3 presents the test and results and Sect. 4 is the conclusion.

2 Proposed Method

The core of the proposed method is to examine whether the input image contains copied regions or not and to identify the region from where it is copied from the same image.

The proposed method is explained in the following steps:

Step 1 The color image or gray image of any pixel value is taken as input for which copy-move detection must be performed.

Step 2 The Color image is converted from RGB to gray. Gray image is retained as it is.

Step 3 Images are segmented into equal size overlaying blocks. The total number of blocks depends on the pixel of the image. It is calculated using the following equations:

$$\textbf{Block size} = \textbf{2t} \tag{1}$$

where $\textbf{t} = \textbf{log2 (M} \times \textbf{N)} - \textbf{12}$, "M" is the number of rows, and "N" is the number of columns.

Therefore,

$$\textbf{the total number of blocks} = ((\textbf{M} * \textbf{N}))/(\textbf{Block size}) \tag{2}$$

The minimum value of $t = 2$

Step 4 Calculate the DCT for each block.

The general for DCT is

$$F(u, v) = \left(\frac{2}{N}\right)^{\frac{1}{2}} \left(\frac{2}{M}\right)^{\frac{1}{2}} \sum_{i=0}^{N-1} \sum_{j=0}^{M-1} \Lambda(i) \cdot \Lambda(j)$$

$$\cdot \cos\left[\frac{\pi.u}{2.N}(2i + 1)\right] \cos\left[\frac{\pi.v}{2.M}(2j + 1)\right] \cdot f(i, j) \tag{3}$$

For the input image N by M, $f(i, j)$ is the intensity of the pixel in row i and column j; $F(u, v)$ is the DCT coefficient in row $k1$ and column $k2$ of the DCT matrix.

Step 5 Calculate the mean of obtained DCT of each block.

Step 6 Then, the mean of each block is matched with all other blocks in the same image.

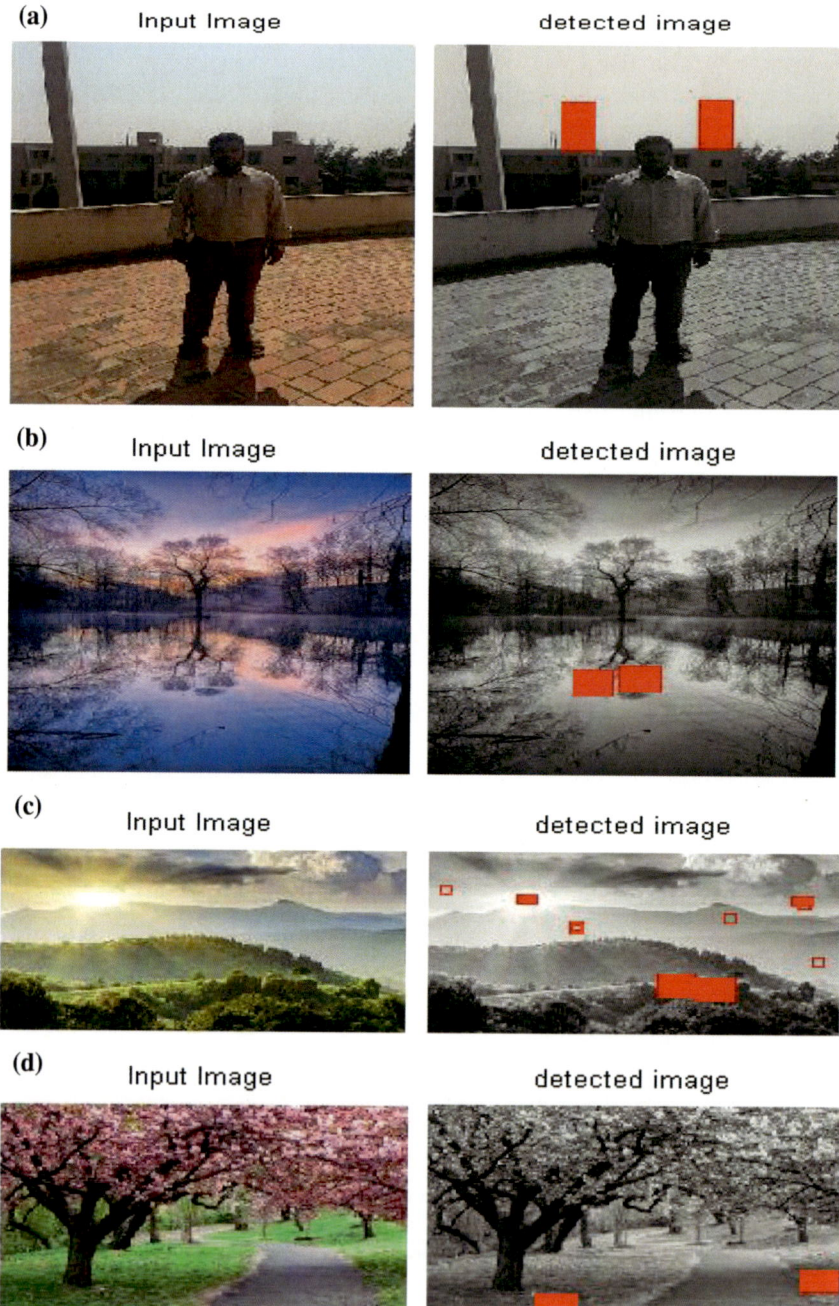

Fig. 3 **a–g** Random rectangular region detected

Fig. 3 (continued)

Step 7   If the similarity is found then the block is highlighted (considered as a copied block).

3   Simulation and Results

The experimental results are discussed in this section. The tests were carried out on the MATLAB R2013a, RAM 2 GB, and processor 2.90 GHz, the images with the different pixel values, saved in BMP format, are tested to check the computational speed and robustness of the proposed algorithm.

Figure 3 shows two images: The input image and detected the image. The random rectangular region is copied and pasted onto the same image. The highlighted region represents the copied region and pasted region. The test conducted without having prior knowledge of the images with different pixel values.

Table 1 shows the computational time of forgery detection of random rectangular regions applied on images using the proposed method and method proposed by Fadl and Semary [1] (the accelerated method). As per the table, the proposed method is improved with enhanced speed compared to the accelerated method.

Figure 4 shows that the performance of the accelerated method, which produces noise while detecting higher pixel value images, and hence its performance is reduced compared to the proposed method. As the pixel values are more, the accelerated method produces noise. The proposed method is more accurate even for higher pixel value. This is highlighted in the first row and last row of Table 1.

Table 1 Comparison of computational time between the proposed method and the accelerated method

Figure	Image pixels	Proposed method (s)	Accelerated method (s)
3a	304×408	44.99	78.70 (with noise)
3b	183×275	13.07	23.76
3c	300×168	14.09	21.09
3d	284×177	13.09	22.09
3e	284×177	13.95	22.60
3f	276×183	14.17	22.03
3g	400×300	42.23	84.96 (with noise)

(a) **(b)**

Fig. 4 a, b Noise during detection in the accelerated method

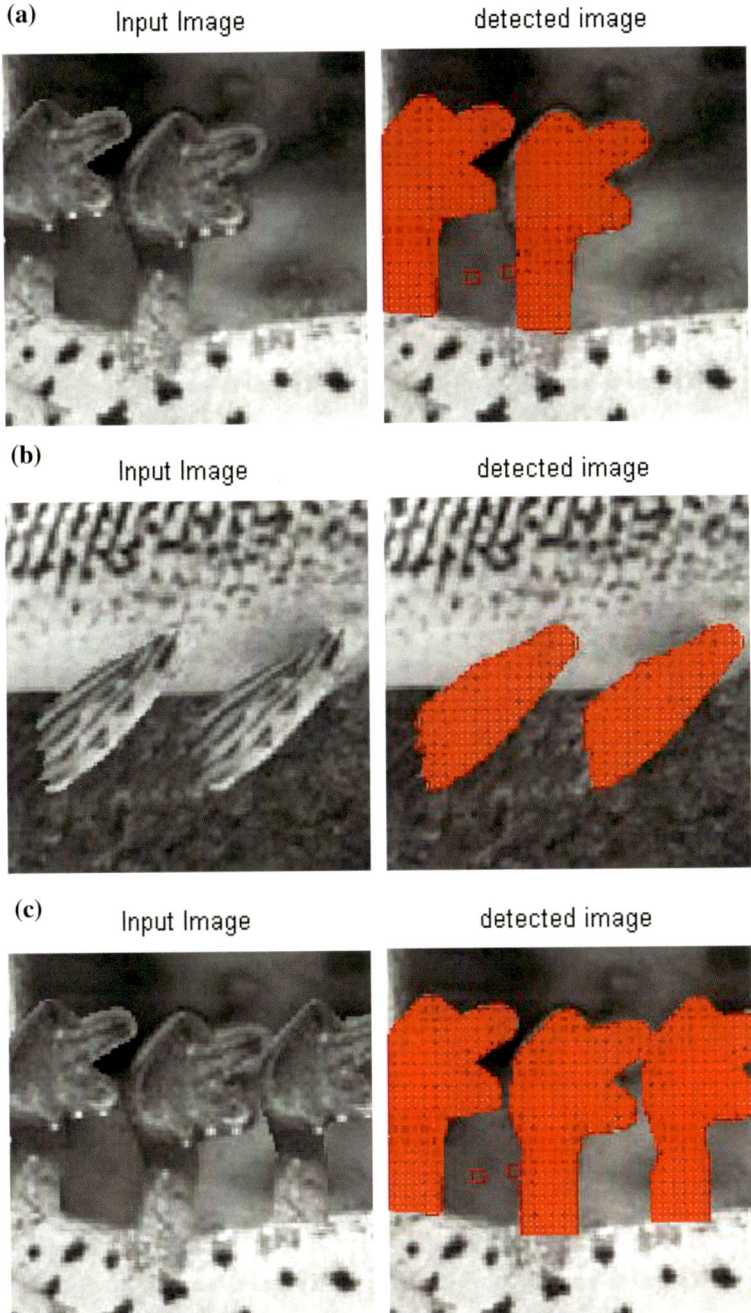

Fig. 5 a–c Copy-move detection for irregular regions

Table 2 Comparison of computational time between the proposed method and the accelerated method

Figure	Image pixels	Proposed method (s)	Accelerated method (s)
5a	128×128	3.17	9.46
5b	128×128	3.00	6.42
5c	128×128	3.34	6.86

Figure 5 shows an irregular region that is copied and pasted onto the image. The test shows the detection for the single region and multi-regions. All the images are 128×128 pixel values.

Table 2 shows the computational time of forgery detection of random rectangular regions applied on images using the proposed method and method proposed by Fadl and Semary [1] (the accelerated method). As per the table, the proposed method is improved with enhanced speed compared to the accelerated method.

4 Conclusion

This work is robust to detect copy-move forgery efficiently, by means of DCT features and taking the mean of DCT of blocks and comparing with other blocks of the same image. It works without any prior information about the image. The proposed work is fast and more effective for any pixel values compared to existing methods.

References

1. Fadl SM, Semary NA (2014) A proposed accelerated image copy-move forgery detection. In: Visual communications and image processing conference, IEEE
2. Lynch G, Shih FY, Liao HYM (2013) An efficient expanding block algorithm for image copy-move forgery detection. Inf Sci 239:253–265
3. Huang Y, Lu W, Sun W, Long D (2011) Improved DCT-based detection of copy-move forgery in images. Forensic Sci Int 206(1):178–184
4. Lin HJ, Wang CW, Kao YT (2009) Fast copy-move forgery detection. WSEAS Trans Sig Process (World Sci Eng Acad Soc) 5(5):188–197
5. Bhargava N, Sharma MM, Garhwal AS (2012) An improved image authentication technique using random-sequence based secret-sharing scheme. In: 2012 International conference on radar, communication and computing (ICRCC), 21–22 Dec. 2012. IEEE 07 Feb 2013
6. Katzenbeisser S, Petitcols FAP (2000) Information techniques for stenography and digital watermarking. Norwood, A, Artec House
7. Alam S, Jamil A, Saldhi A (2015) Digital image authentication and encryption using digital signature. In: 2015 international conference on advances in computer engineering and applications (ICACEA), 19–20 Mar 2015. IEEE 23 July 2015
8. Singh VK, Tripathi RC (2011) Fast and efficient region duplication detection in digital images using sub-blocking method. Int J Adv Sci Technol 35:93–102

 9. Liu G, Wang J, Lian S, Wang Z (2010) A passive image authentication scheme for detecting region duplication forgery with rotation. J Netw Comput Appl 34(5):1557–1565
10. Sebe N, Liu Y, Zhuang Y, Huang T, Chang S-F (2007) Blind passive media forensics: motivation and opportunity. In: Multimedia content analysis and mining. Springer, Berlin, pp 57–59
11. Chen C-H, Tang Y-L, Hsieh W-S (2014) Color image authentication and recovery via adaptive encoding. In: 2014 international symposium on computer, consumer and control (IS3C), 10–12 June 2014. IEEE 30 June 2014
12. Zhang Z, Ren Y, Ping XJ, He ZY, Zhang SZ (2008) A survey on passive-blind image forgery by doctor method detection. In: Proceedings of seventh international conference on machine learning and cybernetics, pp 3463–3467
13. Kou G, Ma Y (2015) Color image authentication method based on triple-channel spiking cortical model. In: 2015 10th international conference on broadband and wireless computing, communication and applications (BWCCA), 4–6 Nov 201. IEEE 03 Mar 2016
14. Cox IJ, Miller ML, Bloom JA (2002) Digital watermarking San Francisco. Morgan Kaufmann, Burlington

Feature Extraction and Classification of Epileptic EEG Signals Using Wavelet Transforms and Artificial Neural Networks

Upasana Chakraborty and R. Mary Lourde

Abstract Marked by unpredictable seizures, epilepsy is the fourth most prevailing chronic neural disorder. This neurodegenerative disorder can attack individuals belonging to any category or age group. Also, the resulting seizures can be of any type. There is always a possibility of misjudging the symptoms with psychogenic nonepileptic events. Thus, in addition to the common methods like functional magnetic resonance imaging (fMRI) and positron emission tomography (PET), electroencephalography (EEG) is a useful tool to differentiate epilepsy from other neurodegenerative disorders. However, EEG measures brain activity directly unlike the other two techniques that measure changes in blood flow to a certain part of the brain. Hence, EEG is most widely used. The paper focuses on conversion of time domain brain signals into time–frequency domain using wavelet transforms followed by extraction of various statistical and nonlinear features. These features are then fed to the neurons of an artificial neural network (ANN) which indicates the presence of epilepsy in an individual.

1 Introduction

Epilepsy is the fourth most common neurodegenerative disorder as per the Epilepsy Foundation. It can be characterised by frequent seizures which are unpredictable. Statistically, in the US itself, 1.3–2.8 million people are still fighting with this disorder with small kids and elderly people being the most prone [1]. Since not many people respond to the medication as expected, there is a need to identify the root causes through further research. It can help to identify why such type of seizures arise and if it is possible to prevent or reduce its impact if realised before its onset. Though

U. Chakraborty (✉) · R. Mary Lourde
Department of Electrical and Electronics Engineering, Birla Institute of Technology and Science, Pilani, Dubai Campus, Dubai, United Arab Emirates
e-mail: upasana.chakraborty96@gmail.com

R. Mary Lourde
e-mail: marylr@dubai.bits-pilani.ac.in

© Springer Nature Switzerland AG 2019
D. Pandian et al. (eds.), *Proceedings of the International Conference on ISMAC in Computational Vision and Bio-Engineering 2018 (ISMAC-CVB)*, Lecture Notes in Computational Vision and Biomechanics 30,
https://doi.org/10.1007/978-3-030-00665-5_137

there are several other ways to record the brain activity like fMRI and PET, EEG is the only one that is noninvasive, and hence do not have any harmful effects on the subject. Major part of the EEG signal is concentrated in the frequency range of 0–80 Hz. This range is subdivided into five frequency bands—delta, δ (0–3.5 Hz); theta, θ (4–7.5 Hz); alpha, α (8–13 Hz) and beta, β (14–30 Hz). Above 30 Hz lies the gamma band, γ. The amplitude of brain signals varies from around 10 to 100 μV. EEG signal amplitude of an epileptic person is greater compared to that of a normal person [2]. But, since brain activity changes with age, it can be very difficult to conclude if a person has epilepsy by looking at the amplitude of brain signals in time domain, thus making feature extraction necessary.

In this work, authors used discrete wavelet transforms (DWT) to extract the wavelet coefficients that can help to analyse EEG data in time–frequency domain. This approach unravels various features that remain hidden in only time or frequency domain analysis. Use of wavelet coefficients also helps in reducing the number of terms to deal with the huge time domain data that can be represented with few wavelet coefficients. These coefficients are further used to extract various features. Some are statistical parameters like mean, standard deviation and others are nonlinear features like fractal dimensions [3]. These features are finally fed to the input neurons in an ANN-a tool used for classification. Using a simple algorithm, it can classify whether an individual is suffering from epilepsy or not. Though several attempts have been made to do the same, there is always scope for improvement. All the simulations are performed in MATLAB.

2 Literature Review

In [3], support vector machines (SVM) is used as a classification tool. SVM is trained using certain statistical and nonlinear features. Different SVM classifiers are used for different sub-bands and based on the maximum score, final result is obtained. Using various kernel functions, seizure and seizure-free EEG were classified with a very good accuracy. Only drawback is that SVM fitting and prediction takes longer time compared to ANN. This makes the system complex. Also, kernelised SVMs are nonparametric. Support vectors are selected from among the training set with some bias value so the number of outputs is less compared to the number of inputs but at times, the number of support vectors is equal to the number of training datasets. In such situations, using ANN is a better option as SVM works best when there is only one output.

Guo et al. [4] used relative wavelet energy of frequency bands as elements for classification. EEG signal was divided into several frequency bands out of which relative wavelet energy of two bands—low and middle frequency bands were extracted. This was fed to a three-layered neural network which was used for classification with an accuracy of nearly 95%.

3 Methodology

3.1 Dataset Used

The dataset used was taken from the Centre of Epilepsy in Bonn, Germany. Data acquisition was performed by Andrzejak [5]. It consists of five sets A–E which include EEG from a normal person and from those suffering from epilepsy. Three sets were used for our analysis—A, D and E. Set A consists of EEG data from normal people with their eyes closed. Electrodes were placed on the epileptic zone of the brain for Set D and in Set E, data was recorded from individuals who had epilepsy and were having seizures at the time of experiment. The motive behind choosing these sets is that one gives idea about a normal person's EEG when his/her eyes are closed that means there is not much eye activity so EOC/visual artefacts would not be noising the EEG, whereas the other two sets inform about those suffering from epilepsy with or without seizure. Each of these sets consists of 100 trials from a single channel. Data is acquired from A to D convertor with a sampling rate of 173.61 Hz and recorded for a time span of 23.6 s. The 10–20 International Standard of electrode placement is used for data acquisition.

3.2 Wavelet Transforms

Wavelet transforms are very useful for EEG signal analysis because of their nonstationary nature which is similar to that of the signals under study. As stated above wavelet transforms help to decompose the whole signal into wavelet coefficients which then are used for feature extraction. Equation 1 shows the general expression of wavelet transforms, where $x(t)$ is the wavelet, τ stands for the translational motion of the wavelet and s is the scale parameter. $\varphi(t)$ is the transforming function also known as the mother wavelet. Different mother wavelets were used to decompose and reconstruct the signals to check which one gives a waveform closest to the original one. This led the authors conclude that out of all the other wavelet functions like Haar, Symlets, Coiflets, Daubechies (dB) was the most efficient and commonly used one [6]. All these mother wavelets use filters to decompose a signal. Daubechies function varies from dB1 to dB20 based on the order of filter used. As the order increases from 1 to 20, the number of samples chosen also increases which improves the efficiency of reconstruction. Therefore, in this study, Daubechies 20 is used to calculate wavelet coefficients. Figure 1 shows the reconstructed waveform of the delta frequency range using Haar wavelet function, whereas delta, alpha and theta band waveforms are compared in Figs. 2, 3 and 4, respectively. These are reconstructed using dB2, dB8, dB14 and dB20. Undoubtedly, it is evident that Daubechies function gives better results.

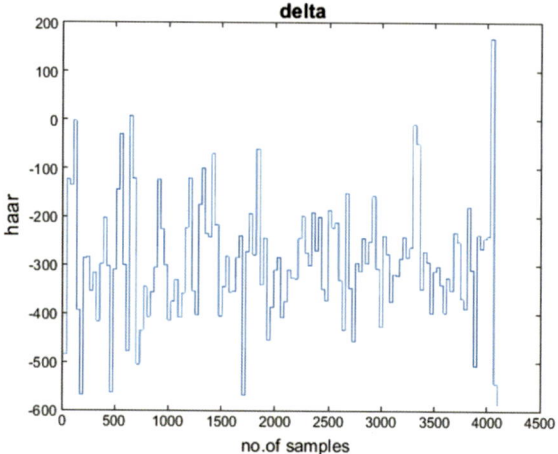

Fig. 1 Delta band plot using Haar wavelets

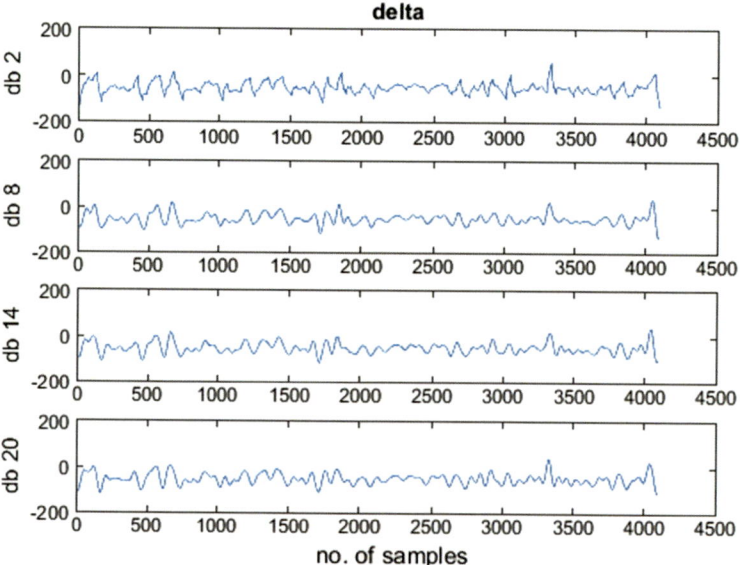

Fig. 2 Delta band plot using different Daubechies family wavelets

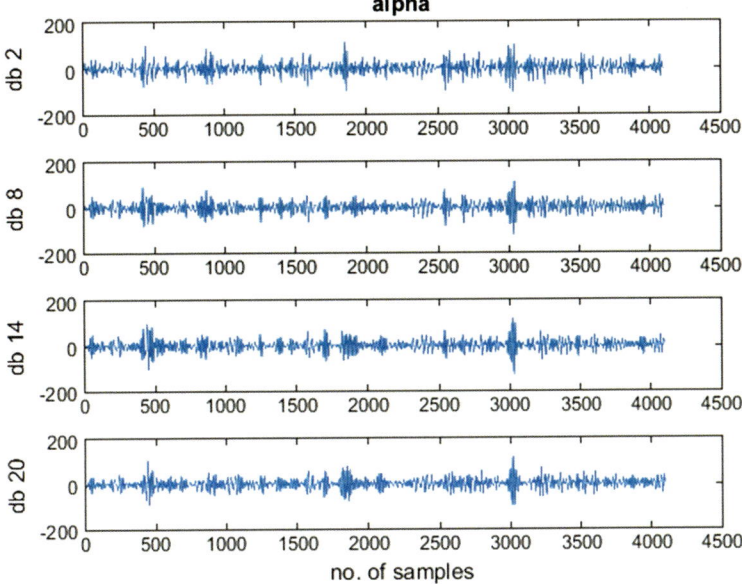

Fig. 3 Alpha band plot using different Daubechies family wavelets

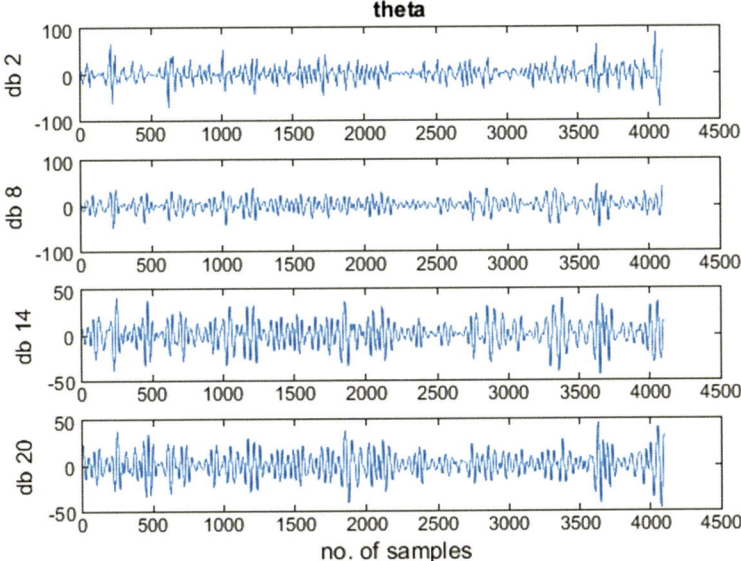

Fig. 4 Theta band plot using different Daubechies family wavelets

$$\varphi(\tau, s) = \frac{1}{\sqrt{|s|}} \int x(t)\varphi^* \left(\frac{t - \tau}{s} \right) dt \tag{1}$$

4   Feature Extraction

The levels of decomposition of an EEG signal from a channel using wavelet transforms is determined by the number of major frequency components that are of interest. Since EEG has its maximum part concentrated from 0 to 30 Hz and for the dataset
used, sampling rate is 173.6 Hz, it needs to be divided into five sub-bands. Wavelet
decomposition decomposes the signal first into two halves, one is the high-frequency
component (using high-pass filter) and other is low-frequency component (using low-
pass filter). High-pass filter here follows the wavelet function and low-pass filter is
termed as the scaling function [7]. This process needs to be repeated with the lower
frequency region till zero is reached. According to the Nyquist criterion, maximum
frequency content of the signal is $f_s/2$. To begin with, signal is halved to obtain a
lower frequency range from 0 to 86 Hz, half the sampling rate. This is termed as D1
(detailed coefficient level 1). The other half is termed as A1 (approximate coefficient
level 1).

Table 1 shows the various frequency ranges in which the signal is divided based on
wavelet coefficients. The number of levels of decomposition is calculated based on
the sampling rate of the A–D convertor. The frequency range in each level $\left[\frac{f_m}{2} : f_m \right]$
is related to the sampling frequency and the level of decomposition by the formula
mentioned in Eq. 2.

$$f_m = \frac{f_s}{2^{l*l}} \tag{2}$$

where l represents the level of decomposition [8].

Figure 5 shows how a signal is decomposed into three levels using wavelet transforms. Sets A and E were specifically used for wavelet decomposition and feature

Table 1 Wavelet coefficients with corresponding frequency bands and their ranges

Coefficient	Frequency range	Decomposition level/frequency band
D1	43.40–86.80	(1) Noise
D2	21.70–43.40	(2) Gamma
D3	10.85–21.70	(3) Beta
D4	5.42–10.85	(4) Alpha
D5	2.71–5.42	(5) Theta
A5	0.00–2.71	(5) Delta

Fig. 5 Wavelet decomposition of a signal into three levels [9]

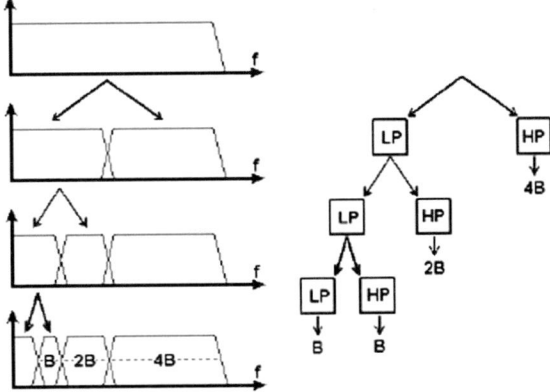

extraction. The wavelet coefficients from D1 to A5 were used to extract various features as listed below:

1. Maximum of wavelet coefficients in each sub-band.
2. Minimum of wavelet coefficients in each sub-band.
3. Mean of the wavelet coefficients in each sub-band by the formula given as

$$\mu_i = \frac{1}{N} \sum_{i=1}^{N} D_{ij}, \; i=1,2,\ldots,l.$$ (3)

4. Standard deviation of coefficients in each sub-band is given as

$$\sigma = \sqrt{\left(\frac{1}{N-1} \sum_{i=1}^{N} (D_{ij} - \mu)^2 \right)}.$$ (4)

5. Variance of the coefficients in each band is σ^2.
6. Median of the coefficients of each sub-band is given as

$$\text{Median} = \begin{cases} D^{\frac{(N+1)}{2}}, & \text{if } N \text{ is odd} \\ \frac{1}{2}D\left(\frac{N}{2}\right) + D\left(\frac{N}{2} + 1\right), & \text{if } N \text{ is even} \end{cases}.$$ (5)

7. Energy of the coefficients of each sub-band: The energy of alpha, beta, etc., bands within an EEG signal were calculated using the formula given below:

$$\text{Energy}(E_i) = \sum_{j=1}^{N} |D_{ij}|^2, \quad i = 1, 2, \ldots l$$ (6)

8. Relative wavelet energy in each sub-band: For this, the energy is first calculated for different wavelet coefficients individually (E_i). Then, the following equation gives the relative wavelet energy of the ith reconstructed coefficient:

$$\rho_i = \frac{E_i}{E_{\text{total}}}. \tag{7}$$

9. Mean square error (MSE): This feature is used to calculate the MSE given as

$$\text{MSE} = \sum_i^N \frac{(D_i - \mu)^2}{N} \tag{8}$$

10. Fractal dimension: It gives idea about the complexity of a signal, how complex are the samples that constitute a signal. If d is the distance between the first sample value and the one which is farthest, fractal dimension is calculated as follows:

$$\text{FD} = \frac{\log_{10} L}{\log_{10} d}. \tag{9}$$

L is the sum of the distance between successive samples. In all the above features, D_{ij} stands for definite wavelet coefficients, N is the total number of wavelet samples and μ is the mean of a particular set of wavelet coefficients. Wavelet coefficients for a normal and epileptic person are plotted in Figs. 6 and 7, respectively. It is evident from the two figures that the amplitude of EEG signals of patients suffering from epilepsy is more than those of normal subjects.

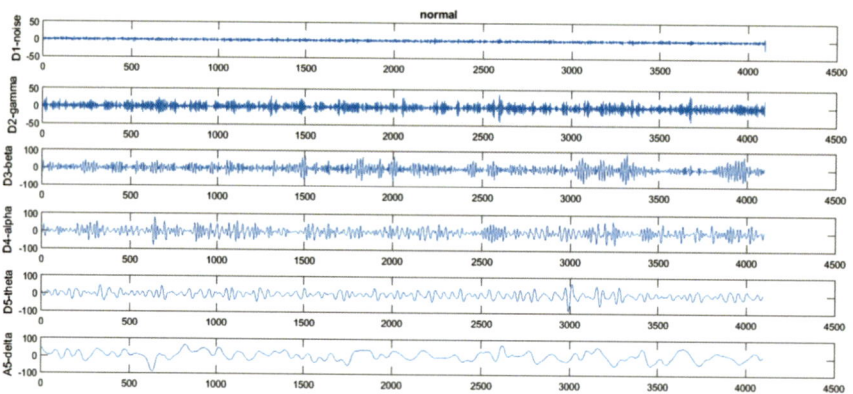

Fig. 6 Wavelet coefficients' plot for a normal person

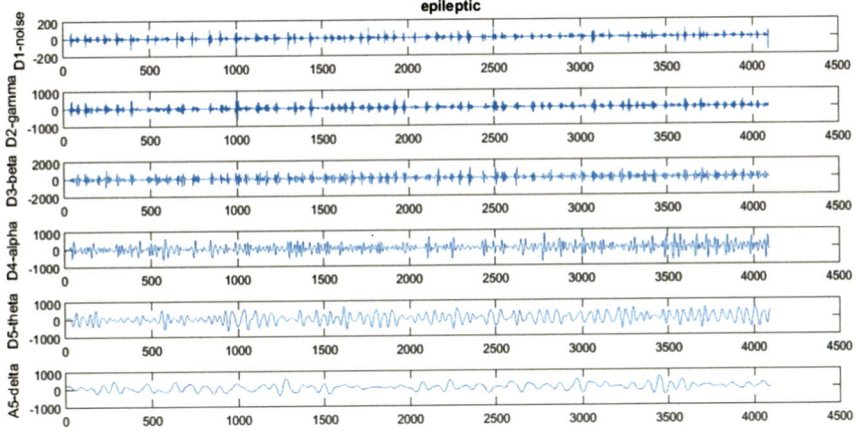

Fig. 7 Wavelet coefficients' plot for an epileptic person

5 Feature Classification

5.1 Artificial Neural Networks (ANN)

Classification can be performed using various tools like support vector machines (SVM), hidden Markov model (HMM), etc. Another efficient tool is the ANN. ANN can be either single layer with no hidden layers or multiple layers where hidden layers are introduced to increase the accuracy of the results obtained. Each neuron in the input layer is connected to all the other neurons in its next layer using some weights or so-called 'biases' [10]. These biases are the result of nonlinear activation functions applied to the input neurons. Number of hidden layers that need to be added and the number of neurons in each layer is determined experimentally and varies according to the application. Most common ANN algorithm in use is the backpropagation algorithm.

5.2 Classification Using ANN

ANN-a multilayer perceptron is used in the paper. These features were fed into a neural network where each input node represents a feature. The model used for this work consists of ten input neurons each representing a feature and two output neurons representing the two classes of subjects—normal and epileptic. In each of these stages, an activation function is used to calculate the final result which can be passed onto the next stage. For the first set of hidden layers, 'tansig' activation function is used, whereas 'purelin' is used for the last set. Features extracted using both normal and epileptic data were randomly distributed to form a dataset. This set

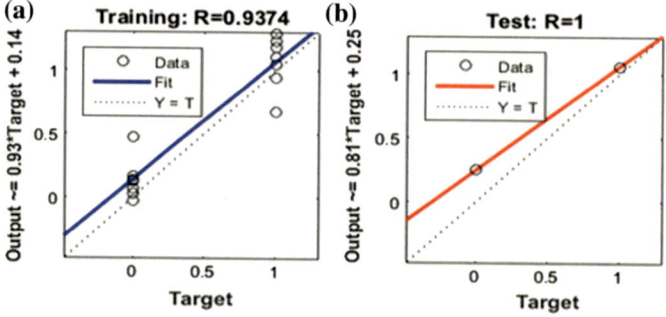

Fig. 8 Regression plot for **a** training data and **b** testing data

was randomly divided into three categories, namely, training, validation and testing. For better analysis, normal subject's dataset was assigned the binary value of 01, whereas that of epileptic patients was assigned 10. After several attempts of training the neural network, the testing dataset was used to test the accuracy of classification. Difference between actual output and that predicted by neural network can be used to calculate MSE. In order to get rid of the redundancies if any, the input dataset used for feature extraction was normalised. To check the maximum accuracy obtained, the network was trained several times using different combinations of hidden layers and neurons in each layer. Regression plots for training and testing data are shown in Fig. 8a, b, respectively. It shows how close our data is compared to the ideal 45° line. Performance of such a system can be evaluated using the validation performance plot shown in Fig. 9a. Performance of the neural network is plotted in terms of MSE. If the MSE converges and is less than one, as is the case here (less than 10^{-2}), it is an indication that the neural network is being trained properly. Network works best at epoch 62 with a gradient of 0.0047866. An epoch gives information about the number of times the training dataset is used to update the weights. Figure 9b shows the confusion matrix obtained using the 15 sets of ten features each. It is evident that out of 15 people only 1 person was erroneously classified. Hence, the overall efficiency obtained is 96.7%. Neural network toolbox in MATLAB is used for the classification.

6 Conclusion

As the number of cases of such neurodegenerative disorders is increasing day by day, urgent attention on research related to their diagnosis is demanded. Early detection can help to lessen if not completely eradicate the symptoms and can make one aware of how to act when onset of seizure takes place. Ten features as stated above are fed to the neural network and after several rounds of training and using hit and trial, the performance was optimised. Minimum MSE obtained had a value less than

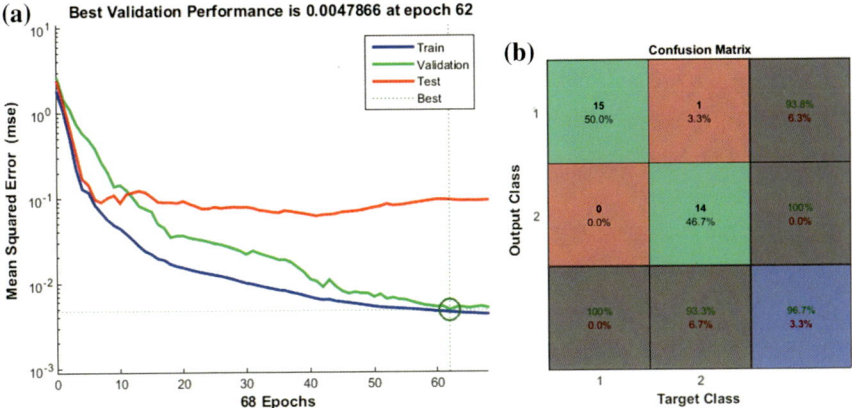

Fig. 9 **a** Performance plot and **b** confusion matrix

10^{-2}. Performance plot and confusion matrix were plotted which gave an accuracy of 96.7%. This work can be extended to develop an application for the detection of epilepsy. The EEG raw data needs to be fed to the application which then can show if the person is suffering from epilepsy or is normal.

Acknowledgements Authors would like to thank the Centre of Epilepsy in Bonn, Germany for the data provided.

References

1. Epilepsy Foundation, http://www.epilepsy.com
2. Hamad A, Houssein EH, Hassanien AE, Fahmy AA (2016) Feature extraction of epilepsy EEG using discrete wavelet transform. In: 12th international computer engineering conference (ICENCO), pp 190–195
3. Kolekar MH (2014) Machine learning approach for epileptic seizure detection using wavelet analysis of EEG signals. In: International conference on medical imaging, m-health and emerging communication systems (MedCom), pp 412–416
4. Guo L, Rivero D, Seoane JA, Pazos A (2009) Classification of EEG signals using relative wavelet energy and artificial neural networks. In: Proceedings of the first ACM/SIGEVO summit on genetic and evolutionary computation, pp 177–184
5. (Dataset) Andrzejak RG, Lehnertz K, Rieke C, Mormann F, David P, Elger CE (2001) Indications of nonlinear deterministic and finite dimensional structures in time series of brain electrical activity: dependence on recording region and brain state. Phys Rev E 64:061907, http://epileptologie-bonn.de/cms/
6. http://wavelets.pybytes.com/
7. Guler I, Ubeyli ED (2007) Multiclass support vector machines for EEG-signals classification. IEEE Trans Inf Technol Biomed xi:117–126

8. Omerhodzic I, Avdakovic S, Nuhanovic A, Dizdarevic Z (2010) Energy distribution of EEG signals: EEG signal wavelet-neural network classifier. World Acad Sci Eng Technol 61:1190–1195
9. Cheong LC, Sudirman R, Hussin SS (2015) Feature extraction of EEG signal using wavelet transform for autism classification. ARPN J Eng Appl Sci x (19):8533–8540
10. Anand SV, Selvakumari RS (2013) Detection of epileptic activity in the human EEG-based wavelet transforms and SVM. Int J Eng Res Technol (IJERT) ii(1):1–6

3D Printed Surgical Guides in Orthognathic Surgery—A Pathway to Positive Surgical Outcomes

Chitra Chakravarthy, Sanjay Sunder, Santosh Kumar Malyala and Ammara Tahmeen

Abstract Rapid advancements in robotics and computer-aided surgeries have revolutionized the field of medicine and surgery. Maxillofacial surgery has benefited from these technological advances as significant contributions have been made in the management of complex soft tissue and bony pathologies. These technologies are especially useful in patients with post-traumatic defects and also cases with esthetic facial deformities. Defects in the craniofacial skeleton are of either congenital, developmental, traumatic, or pathological ethology. The primary purpose of correcting facial anomalies is for functional rehabilitation. The esthetic rehabilitation of a patient is very challenging. It is important to achieve excellent postoperative form and function and also to minimize operative and postoperative morbidity. Rapid prototyping biomodels that are being used in recent times are playing a very significant role not only for patient education, diagnosis of defects but also in surgical planning. Their use in surgical planning reduces anesthesia time, reduces operating time, and provides better esthetic and functional results. The integration of 3D imaging and computerized surgery continues to bring about newer and better changes in the conventional surgeries making the outcome much more beneficial to the patient [1]. We present a case of facial deformity in the form of maxillary excess and retrogenia corrected using orthognathic surgery supported by the use of surgical guides fabricated using additive manufacturing.

C. Chakravarthy · S. Sunder · A. Tahmeen
Department of Oral and Maxillofacial Surgery, Navodaya Dental
College and Hospital, Raichur 584103, Karnataka, India

S. K. Malyala (✉)
Department of Mechanical Engineering, National Institute
of Technology Warangal, Warangal 506004, Telangana, India
e-mail: msantoshpdd@gmail.com

© Springer Nature Switzerland AG 2019
D. Pandian et al. (eds.), *Proceedings of the International Conference on ISMAC in Computational Vision and Bio-Engineering 2018 (ISMAC-CVB)*, Lecture Notes in Computational Vision and Biomechanics 30,
https://doi.org/10.1007/978-3-030-00665-5_138

1 Introduction

Anterior maxillary osteotomy and genioplasty are some of the most commonly performed orthognathic surgical procedures. Anterior maxillary osteotomy is indicated in patients with vertical or horizontal excess or deficiencies of the anterior maxilla. Genioplasty is a surgical procedure to correct deformity of the chin. Presurgical planning in orthognathic surgery plays an important role in the outcome of the procedure [2]. Conventionally, radiographic analysis and prediction tracings are used to give the surgeon an insight into the area of anomaly and also gave a measure of the extent of the defect in the jaws. Additionally, now 3D facial skeletal analysis has the added advantage of a three-dimensional assessment [3].

The CT scan images are reconstructed into a three-dimensional format on which virtual surgery can be performed and the osteotomy lines accurately planned are taking care of vital structures and getting a clear visualization of the osteotomy on the lingual/palatal aspect which cannot be visualized clinically [3]. The CT scan data can also be converted into a printable format on which a mock surgery can be performed and the osteotomized segment is moved to its new position and the desired result is obtained. This also allows the surgeon to adapt the plates according to the new position of the bone getting the exact contour as will be seen clinically during the surgery. This is extremely advantageous as it saves a lot of time and effort intraoperatively and gives an accurate fit of the fixation plates which is also critical to the outcome of the surgery [4, 5]. The use of CAD/CAM technology can further enhance the predictability of the surgery thereby minimizing the percentage of error. Using the surface data, it is possible to generate various surgical templates or guides which can enable ease of surgery and improved efficiency [6].

The importance of achieving a symmetrical osteotomy cut cannot be overemphasized. Being an esthetic procedure, it is critical to the outcome of the procedure. Using a 3D printed surgical template/guide for an anterior maxillary osteotomy and an advancement genioplasty help us get an accurate osteotomy and reduced surgical time to a large extent. The surgical guide plays an important role in the precise transfer of the 3D surgical planning to the patient in the operating room. This customized printed guide was used to first perform the mock surgery and adapt the plates and then the same was used to guide the osteotomy.

2 Materials and Methods

A 20-year-old female patient presented with a prognathic anterior maxilla and retrogenia. Radiographic analysis and planning were done, presurgical orthodontics was performed for about 8 months prior to surgery after which an orthognathic of anterior maxillary advancement and intrusion and advancement genioplasty were planned.

2.1 Construction of the Surgical Guides

The patient was scanned using a 128 slice SOMATOM perspective CT scanner (Siemens Somata Perspective model number 76970). The scan data was obtained in digital imaging and communications in medicine (DICOM) format which was in two and a half D format. This DICOM data was processed using MIMICS 18.0. materialized medical software to generate a 3D CAD model of the patient's scanned data. This data completely included the hard and soft tissue contents. By adjusting the threshold value, the bone data was isolated. Using MIMICS software a threshold value of 226–3071 Hounsfield units can be applied to get accurate bone detail.

The next step was to separate the region of interest from the complete scan. The mandible was first separated and a 3D CAD model was created. Similarly, the maxilla up to the infraorbital margins was isolated creating a separate 3D CAD image. These were then converted into stereolithography (STL) file formats. This is a neutral file format for all AM machine preprocessing software. The 3D medical models of maxilla and mandible were fabricated using material extrusion technique.

Virtually, a surgical simulation of the surgery was performed on the maxilla and mandible. The genioplasty cut was simulated taking care of the mental nerves, the root apices and also inclining the cut as required from the buccal to lingual cortex. The osteotomized segment was also moved virtually to present an ideal three-dimensional picture of the outcome of the surgery. The maxillary anterior segmental osteotomy was also simulated virtually keeping the cuts exactly 5 mm apart as planned.

Using the surface data generated on the 3D CAD model, a surgical template was designed (Fig. 1a) and various designs of genioplasty guides have been reported till date [7]. This genioplasty template was designed to extend from the mandibular anterior teeth all the way to the lower border of the mandible taking care to isolate the mental nerves bilaterally. The purpose of the lower border extension was to improve the accuracy of the fit of the template and also to ensure the osteotomy cuts extend into the lower border. A slot was created exactly along the virtual osteotomy line. The edges of the slot were angulated at a 30° chamfer to enable the drill or saw to follow that angulation (Fig. 1b). The maxillary template was extended along the lateral border of the piriform aperture till the second premolar on both sides. A 5 mm slot was created to guide the saw for the two cuts (Fig. 1c). The maxillary surgical guide followed the border of the piriform aperture. These surgical templates followed the surface data accurately to get a perfect adaptation. These surgical templates were then printed using a biocompatible material.

2.2 Mock Surgery

Using the templates and the 3D printed jaws, the mock surgery was performed presurgically and the virtual cuts were replicated on the models using the template. The surgical guide was then removed and the osteotomized segments were separated

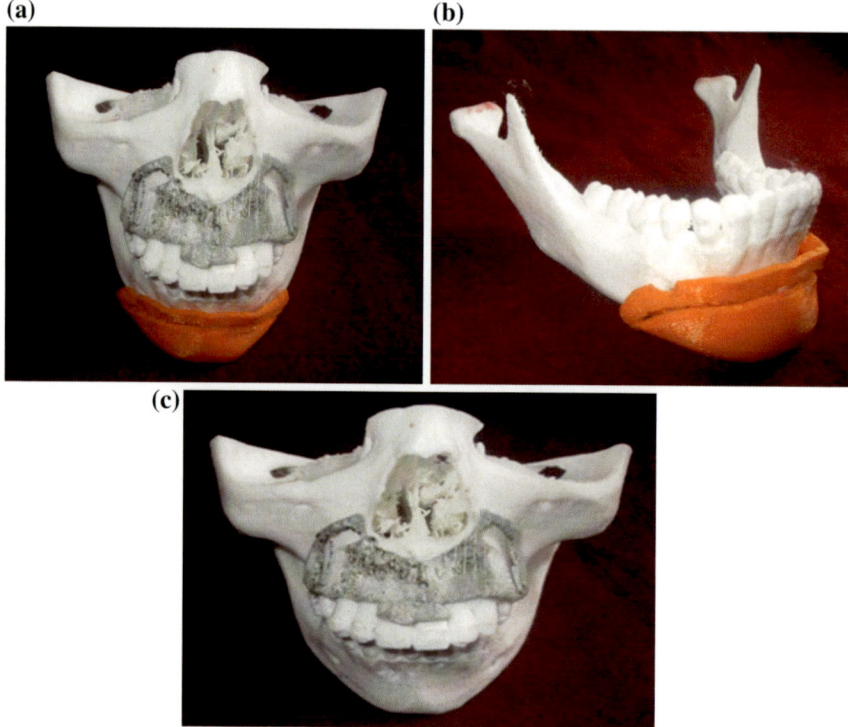

Fig. 1 a Mandible and maxillary templates, **b** Slot extensions on the genioplasty template, **c** Slot extension on the maxillary template

and repositioned as planned in the virtual surgery (Fig. 2a, b). Titanium plates were then meticulously adapted to follow the bony contours of the maxilla in its new position. The genial segment was also advanced and titanium plates were adapted to the advanced segment (Fig. 3a, b). These same pre-bent/adapted plates were then to be used during surgery after being sterilized. The surgical guides were also sterilized and used during surgery.

2.3 Surgical Procedure

Following the mock surgery, an informed consent was obtained from the patient. Under general anesthesia, the anterior mandible was exposed to the lower border. The mental nerves were identified and isolated bilaterally. The template was placed and fixed using a single monocortical screw (Fig. 4a). A surgical saw was used to perform the osteotomy as was planned following the chamfer line. The cut was extended to the lingual cortex and to the lower border. With the use of the template, the

(a) **(b)**

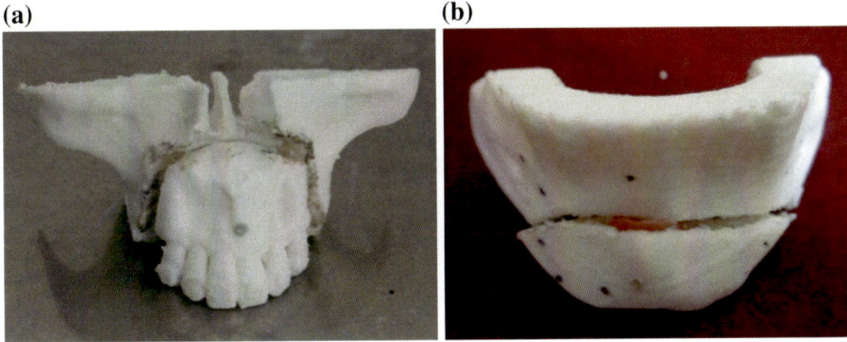

Fig. 2 **a** Osteotomized segments of maxilla, **b** Osteotomized segments of mandible

(a) **(b)**

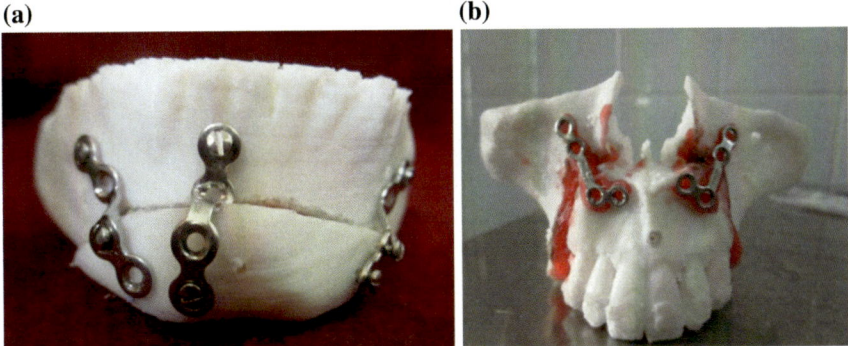

Fig. 3 **a** Adaptation of plates on 3D model of mandible, **b** Adaptation of plates on 3D model of maxilla

symmetry and position of the cut were accurate. The time taken for the surgery was also significantly reduced. Similarly, the anterior maxillary osteotomy was carried out using the templates and setback was done (Fig. 4b). The templates were removed and the osteotomized segments were mobilized to their desired positions. The positions were determined by the adaptation of the plates. Since the plates were already pre-contoured, the surgical time was further reduced. The plates were fixed as planned in the mock surgery.

3 Results and Discussion

Using the 3D printed surgical guides during osteotomy significantly reduced the time of surgery and also the osteotomy cuts could be placed more precisely without any risk of overextension or asymmetrical cuts. Postoperative results were satisfactory and the symmetry could be appreciated. Orthognathic surgery is an elective cosmetic

(a)   (b)

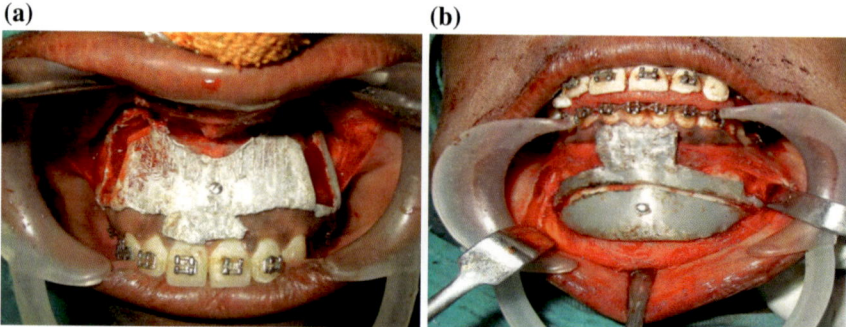

Fig. 4 **a** 3D template placed on the maxilla using a monocortical screw, **b** 3D template placed on the mandible using a monocortical screw

procedure. The outcomes of all the surgery depend to a large extent on the accuracy of the osteotomies. Symmetrical osteotomy and position of the cuts make the surgery successful. The use of the surgical template enables us to visualize the surgery virtually to make changes if necessary and also to replicate the same osteotomy line in the patient intraoperatively. The use of customized wafers and pre-bent plates has been gaining importance in orthognathic surgery [8].

Fabrication of the surgical guide has to be done with great planning keeping in mind certain key points in the design. The contour of the template has to accurately match and fit the contour of the jaw. This is of utmost importance because we do not want it to move during the osteotomy. The thickness of the template should be adequate to ensure that the template does not bend or deform during the surgery. If it is designed too thick or overextended, it will interfere with vital structures in the area. Bulky templates are difficult to use especially in the maxillofacial region where the access is limited. We should also make sure that the template is not cut during the surgical procedure which may possibly put traces of the template material into the surgical site. In this particular genioplasty template, 30° chamfer was very useful as it allows the advancement genioplasty osteotomy to be performed accurately. With the help of the 3D printed models, the patient and the clinician benefit in different ways. The models help in an accurate diagnosis and give an assessment of the extent of the facial deformity. This can be demonstrated to the patient for easy understanding of the defect. It also improves patient compliance as they are able to visualize the problem area and understand the way it needs to be treated. Surgical models help the practitioner in performing the mock surgery and visualizing the desired outcome of the surgery before it is actually replicated on the patient. This helps in improving the confidence of the surgeon. Fabrication of surgical templates is one of the major advantages of 3D printing. Once the mock surgery is done, a precise transfer of the same surgical planning can be done to the operating table with the help of these surgical guides. It helps us to replicate the mock surgery.

Adaptation of the titanium plates on the 3D model has several advantages. A lot of time and effort goes into adaptation of the plates during the surgery. Adding to the difficulty is also the presence of blood and soft tissue obstructing the view of the bony contours. Preadapted plates significantly reduce the surgical time, the exposure time to general anesthesia. It also decreases the total amount of blood loss and the time that the wound is exposed to the external environment.

The preadapted plates serve another important function. They serve as a guide for the mobilization of the osteotomized segment. This is especially useful in cases of advancement genioplasty where the segment has to be advanced arbitrarily. The bent plates serve as a guide to the amount of advancement required. The plates can be fixed accurately without wasting much operating time. Conventionally, the maxillary osteotomy takes about 90 min in our center of which significant time is taken for assessing the symmetry and for contouring the plates both of which were done easily. The surgical time was reduced to an hour for the maxillary procedure. The genioplasty also usually takes about an hour which was completed using the surgical guides in about 40 min. Fabricating of a 3D printed model or guide is now a relatively simple process for a clinician. An accurate CT scan with 1 mm cuts helps to replicate an accurate image. A helical scanner with a zero-degree gantry tilt scanning parallel to the occlusal plane is preferable to reproduce the jaws accurately. The DICOM images are converted into STL images in the laboratory. Using the MIMICS software, specific virtual cuts can be made on the bone and the osteotomized segments can be moved as desired. Some of the challenges we face using this technology need to be overcome. Since this is a relatively new field, changes are still in progress to achieve maximum benefit.

The accuracy of the models, especially in areas where the bone is thin may not be perfect. A major part of maxillofacial surgeries such as trauma and orthognathic are dependent on achieving an ideal occlusion, replication of the precise contours of the dental surface is of prime importance. This is not always possible with even a high-resolution CT scan. Sometimes a part of the data is lost due to the presence of metal artifacts such as crown and bridges, and orthodontic brackets. To overcome this issue, dental casts of the patient are scanned separately using a high-resolution scanner and the teeth are superimposed to recreate the delicate details of the teeth in the jaw. These minor drawbacks do not pose as an obstacle in the path of this evolving field as the advantages far outweigh the drawbacks.

4 Conclusion

Orthognathic surgery is a procedure that is done to improve facial esthetics. Symmetrical and accurate osteotomies and precise mobilization of the segments is of prime importance to the outcome of the surgery. The use of modern day technology in the form of additive manufacturing has added precision and ease to this challenging procedure. The use of a surgical guide allows the surgeon to easily perform the surgery as planned in the mock surgery and in the virtual surgery. It simplifies the surgery and allows improved accuracy and superior surgical outcomes. Adaptation of this technology into other surgical aspects of maxillofacial surgery will definitely give an added benefit to the surgeon in the form of reduced laboratory time used for planning [9], reduced surgical time, ease of surgery, and improved clinical outcomes.

Acknowledgements The authors declare that they have no conflict of interest.

References

1. Xiao Y, Sun X, Wang L, Zhang Y, Chen K, Guomin W (2017) The application of 3D printing technology for simultaneous orthognathic surgery and mandibular contour osteoplasty in the treatment of craniofacial deformities. Aesthetic Plast Surg 41(6):1413–1424
2. Haas OL Jr, Becker OE, de Oliveira RB (2014) Computer aided planning in orthognathis surgery—systematic review. Int J Oral Maxillofac Surg, pii: S0901-5027(14)00430-5
3. Wang L, Tian D, Sun X, Xiao Y, Chen L, Wu G (2017) The precise repositioning instrument for genioplasty and a three-dimensional printing technique for treatment of complex facial asymmetry. Aesthic Plast Surg 41:919–929
4. Rubio-Palau J, Prieto-Gundin A, Cazalla AA, Serrano MB, Fructuoso GG, Ferrandis FP et al (2016) Three-dimensional planning in craniomaxillofacial surgery. Ann Maxillofac Surg 6:281–286
5. Raphaël O (2013) Three-dimensional rapid prototyping models in cranio-maxillofacial surgery: systematic review and new clinical applications. P Belg Roy Acad Med 2:43–77
6. Lim S, Kim M, Kang S (2015) Genioplasty using a simple CAD/CAM (computer-aided design and computer aided manufacturing) surgical guide. Maxillofac Plast Reconstr Surg 37:44
7. Choi JW, Namkug K (2015) Clinical application of three-Dimensional printing technology in craniofacial plastic surgery. Arch Plast Surg 42:267–277
8. Suojanen Juho, Leikola Junnu, Stoor Patricia (2016) The use of patient = specific implants in orthognathic surgery: a series of 32 maxillary osteotomy patients. J Cranio-Maxillofac Surg 44:1913–1916
9. Wrzosek MK, Peacock ZS, Laviv A, Goldwaser BR, Ortiz R, Resnick CM, Troulis MJ, Kaban LB (2016) Comparison of time required for traditional versus virtual orthognathic surgery treatment planning. Int J Oral Maxillofac Surg 45(9):1065–1069

Image-Based Method for Analysis of Root Canal Geometry

Ankit Nayak, Prashant K. Jain, P. K. Kankar and Niharika Jain

Abstract Image-based techniques of observation and measurement are getting popular due to its accuracy and accessibility. In dentistry, endodontic files are used for root canal shaping and cleaning. Manufacturers design and fabricate different endodontic files. Endodontists examine the performance of different endodontic files. Endodontists used artificial root canal for comparison and examination of the file system. After shaping of artificial root canals, endodontic blocks are compared by qualitative observation. This article presents a methodology based on image analysis of artificial root canal for examination and comparing the shaping ability of endodontic files. In the proposed method, the images of the artificial root canal were processed in a custom-made MATLAB® program to quantify the root canal deviation and root canal transportation. Calculation of root canal curvature quantified root canal transportation. Performance of 20/0.02 NiTi hand file and 20/0.04 NiTi rotary file for glide path preparation was compared. Twenty endodontic blocks were divided into two experimental groups for canal preparation. Pre- and post-treatment root canals were compared regarding canal curvature modification and deviation for the experimental group of 20/0.02 NiTi hand file with self-adjusting file (SAF) and 20/0.04 NiTi rotary file with SAF using t-test at 95% confidence interval. Pre- and post-treatment images are processed in the custom-made MATLAB® program to find out the canal

A. Nayak (✉) · P. K. Jain · P. K. Kankar
CAD/CAM Lab, Mechanical Engineering Discipline,
PDPM Indian Institute of Information Technology, Design and Manufacturing,
Jabalpur 482005, Madhya Pradesh, India
e-mail: ankitnayak@iiitdmj.ac.in

P. K. Jain
e-mail: pkjain@iiitdmj.ac.in

P. K. Kankar
e-mail: kankar@iiitdmj.ac.in

N. Jain
Departments of Conservative Dentistry and Endodontics,
Triveni Institute of Dental Science, Hospital and Research Centre,
Bilaspur 495001, Chhattisgarh, India
e-mail: niharika.dr@gmail.com

© Springer Nature Switzerland AG 2019   1491
D. Pandian et al. (eds.), *Proceedings of the International Conference on ISMAC in Computational Vision and Bio-Engineering 2018 (ISMAC-CVB)*, Lecture Notes in Computational Vision and Biomechanics 30,
https://doi.org/10.1007/978-3-030-00665-5_139

transportation and deviation. A significant difference in an experimental group of hand file and the rotary file was found. It has been found that canal transportation in case of 20/0.02 NiTi hand file was significantly greater than 20/0.04 NiTi rotary file.

1 Introduction

Root canal shaping is the subject of excellent care and attentiveness. Root canal treatment (RCT) is the favored treatment process for the infected tooth. Endodontists used endodontic files to remove the infected tissues from the root canal followed by root canal filling and sealing. Root canal shaping is an essential step of RCT, which should be performed with great care, the success of RCT depends on the root canal filling. First, root canal shaping was done with the help of watch spring by Edwin Maynard of Washington, D.C. in 1838 [1]. After that several types of endodontic instruments were designed and fabricated by manufacturers. These instruments are based on different mechanics of root canal shaping, shaping kinematics, and manufacturer's protocols. Endodontic instrument helps dentists up to the great extent to achieve accuracy in root canal shaping. Axis deviation or root canal transportation [2–6] is the cause of root perforation [7], external tooth resorption, and tooth fracture [8, 9]. For examination and comparison of the performance of endodontic files, endodontists are using different image-based methods like X-ray, photography, or computed tomography (CT) scanning [10, 11].

CT scanning is a costly means to know the geometry of root canal, while in case of radiography, there is not any established method to extract the information from the radiograph of teeth. Some researchers have used photography for evaluation of canal curvature of endodontic blocks (Endo Training Bloc-J, Dentsply Maillefer, Ballaigues, Switzerland) [12]. Endodontic blocks are getting famous for evaluation and comparisons of different endodontic file systems [13, 14]. Since material properties and architecture of root canal are uniform for each endodontic block, this helps to maintain the uniformity in the sample population, while it is not easy to keep in case of extracted teeth. Moreover, extracted teeth of the same anatomy and uniform architecture of root canal are not natural to get for experimentation.

In this article, endodontic training blocks are used for experimentation to compare the shaping ability of the SAF [12], while it used with 20/0.02 (hand) and 20/0.04 (rotary) files for glide path preparation. After the glide path preparation, SAF is used for shaping the root canal. Prepared root canals were examined using custom-made MATLAB program for image analysis of post- and pre-treated images of endodontic blocks to find out the deviation in root canal axis.

2 Material and Method

Twenty J-shaped endodontic blocks are used for ex vitro study of endodontic files. The study is divided into three steps pre-processing, root canal shaping, and post-processing. Pre-processing of endodontic block involves cleaning of endodontic blocks in order to remove dirt and stain and this helps to get error-free images of root canal; after that the root canals are filled with red ink so that it can be identified in pre-processing images. In root canal shaping, endodontic files are used for shaping of simulated root canal as per the file manufacturer protocol for the respective endodontic file. After root canal shaping, the simulated root canals are filled with black ink and scanned again for post-processing images. Pre- and post-instrumented images were processed in MATLAB to remove noise and image artifact. Median filter by taking 2×2 sliding window was applied to remove the salt pepper noise from images. An optimum contrast value for each image was used to eliminate the image artifact. Pre-processed images were sent to the MATLAB code for analysis. Image analysis algorithm has been explained in the following flowchart.

Centering ability of root canal can be analyzed by measuring the deviation between the axis of pre- and post-operated root canals. Pre- and post-treatment images of root canal have been taken to compare and collect the information of trajectory of root canal shaping. Pre- and post-treatment images are converted into binary images followed by thinning of root canal up to one-pixel width to get the central axis of root canals. Geometric analysis of root canal axis can collect information regarding canal modification. Canal transportation and curvature modification were analyzed using custom-made MATLAB programs (Fig. 1).

Coordinates of root canal axis were calculated from the binary image of root canal axis. The axis of root canal has been converted into the series of point coordinates. Then, a circle was fitted into the curvature of root canal using Eq. 1.

$$(x - x_c)^2 + (y - y_c)^2 = r^2 \tag{1}$$

where $x_c y_c$ are the coordinates of the center of circle and r is the radius of the circle.

Radius and the center of curvature for root canal were calculated. The curvature of original and modified root canal has been compared to know about the changes in root canal geometry (Fig. 2).

Images, showing the center axis of the root canal, are superimposed and the region of interest (ROI) is cropped. Superimposed image shows the axes of pre- and post-operated root canals. After the Boolean operation between cropped and superimposed image, the flood-fill algorithm was applied. Flood-fill algorithm helps to remove the noise that occurs after the skeletization of the root canal image. This image will contain the boundary of white pixels which are not the part of deviation. To eliminate these pixels, the superimposed image was subtracted from the image of deviation. After that the pixels were counted to get the deviation in any particular case of the instrumented root canal (Fig. 3).

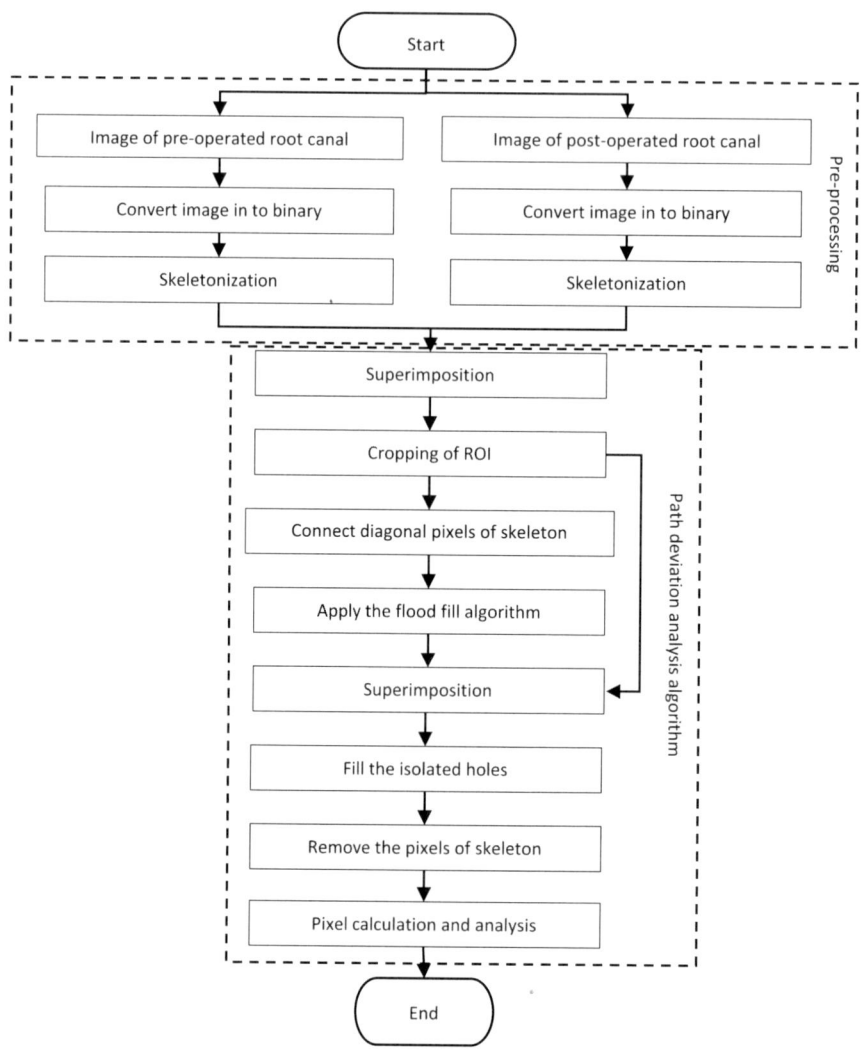

Fig. 1 Flowchart of the image analysis algorithm

3 Results

Post- and pre-treated images for hand file and mechanized file are processed through the proposed algorithm to analyze canal modification in terms of root canal transportation. The results of canal curvature modification and deviation for the experimental group of 20/0.02 NiTi hand file with SAF ($n = 10$) and 20/0.04 NiTi rotary file with SAF ($n = 10$) are compared using t-test at 95% confidence interval. Results of t-test are shown in Table 1.

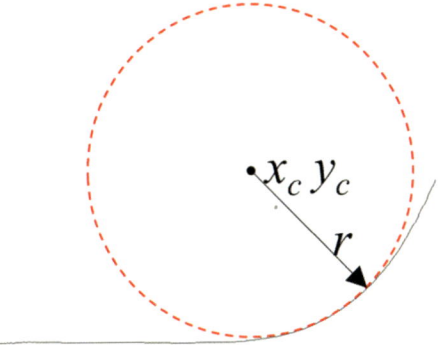

Fig. 2 Fitted circle in the root canal curvature

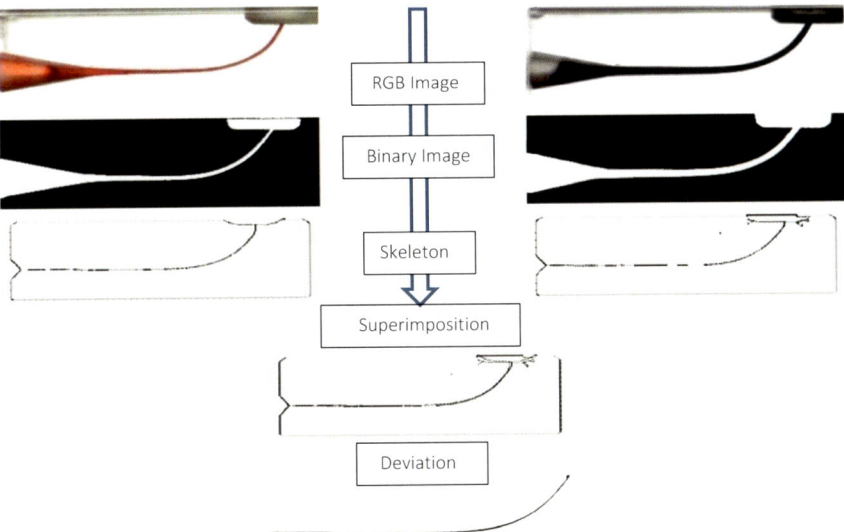

Fig. 3 Major steps of image processing

Table 1 Deviation in root canal curvature of the experimental group (Mean ± Sandard deviation)

Parameter	Group A	Group B	P value
Root canal deviation	14,783 ± 1182	7548 ± 648	$P < 0.05$
Root canal curvature modification	343 ± 24	295 ± 20	$P < 0.05$

P value for two-sample t-test at 95% CI is less than 0.05, which means that there is a significant difference in root canal curvature modification and root canal deviation between group A and group B.

4 Discussion

The image-analysis-based algorithm is developed to compare the performance of the SAF in endo training blocks, while glide path was prepared using 20/0.02 (hand) and 20/0.04 (rotary), two different files. Glide path preparation, coronal enlargement is essential for the good results of RCT. Endodontists have used a different kind of file for glide path and root canal preparation [14, 15]. Different root canal and glide path preparation techniques and protocols are introduced and tested by the endodontists. These techniques are unable to compare or examine the approaches quantitatively [3, 16, 17]. The suggested method of image analysis can be used for comparison of various processes, endodontic files, and protocols of RCT.

Computed tomography (CT) scan gives three-dimensional details of root canal architecture. CT scan is used for detailed analysis of prepared root canal [11, 18]. CT process is much costly than image scanning, and post-processing of CT data is also complicated to collect the information of root canal geometry. Instead of CT scanning, suggested methodology of image analysis is less costly to examine the endodontic block. However, it is less accurate than CT analysis but it is under the limit of tolerance zone.

Acknowledgements Endodontic blocks which are also known as J-shaped resin blocks manufactured by Dentsply Maillefer, Ballaigues, Switzerland have been used for experimentation.

References

1. Grossman LI (1982) A brief history of endodontics. J Endod 8:2–5
2. Goldberg F, Massone EJ (2002) Patency file and apical transportation: an in vitro study. J Endod 28:510–511
3. Bürklein S, Jäger PG, Schäfer E (2016) Apical transportation and canal straightening with different continuously tapered rotary file systems in severely curved root canals: F6 SkyTaper and OneShape versus Mtwo. Int Endod J 50:983–990
4. Zanesco C, Só MVR, Schmidt S, Fontanella VRC, Grazziotin-Soares R, Barletta FB (2017) Apical transportation, centering ratio, and volume increase after manual, rotary, and reciprocating instrumentation in curved root canals: analysis by micro-computed tomographic and digital subtraction radiography. J Endod 43:486–490
5. González Sánchez JA, Duran-Sindreu F, de Noé S, Mercadé M, Roig M (2012) Centring ability and apical transportation after over instrumentation with ProTaper universal and ProFile Vortex instruments. Int Endod J 45:542–551
6. Nayak A, Jain PK, Kankar PK, Jain N (2018) Computer-aided design–based guided endodontic: a novel approach for root canal access cavity preparation. Proc IMechE Part H: J Engg Medicine 232:787–795
7. Tsesis I, Fuss Z (2006) Diagnosis and treatment of accidental root perforations. Endod Top 13:95–107
8. Cheng R, Zhou X-D, Liu Z, Hu T (2007) Development of a finite element analysis model with curved canal and stress analysis. J Endod 33:727–731
9. Uddanwadiker RV, Padole PM, Arya H (2007) Effect of variation of root post in different layers of tooth: linear vs nonlinear finite element stress analysis. J Biosci Bioeng 104:363–370

10. Gergi R, Rjeily JA, Sader J, Naaman A (2010) Comparison of canal transportation and centering ability of twisted files, pathfile-ProTaper system, and stainless steel hand K-files by using computed tomography. J Endod 36:904–907
11. Gergi R, Osta N, Bourbouze G, Zgheib C, Arbab-Chirani R, Naaman A (2015) Effects of three nickel titanium instrument systems on root canal geometry assessed by micro-computed tomography. Int Endod J 48:162–170
12. Nayak A, Kankar PK, Jain N, Jain PK (2018) Force and vibration correlation analysis in the self-adjusting file during root canal shaping: an in-vitro study. J Dent Sci 13:184–189
13. Schrader C, Peters OA (2005) Analysis of torque and force with differently tapered rotary endodontic instruments in vitro. J Endod 31:120–123
14. Choi DM, Kim JW, Park SH, Cho KM, Kwak SW, Kim HC (2017) Vibrations generated by several nickel-titanium endodontic file systems during canal shaping in an ex vivo model. J Endod 1–4
15. Hof R, Perevalov V, Eltanani M, Zary R, Metzger Z (2010) The self-adjusting file (SAF). Part 2: mechanical analysis. J Endod 36:691–696
16. Jain N, Pawar AM, Ukey PD, Jain PK, Thakur B, Gupta A (2017) Preservation of root canal anatomy using self-adjusting file instrumentation with glide path files, pp 51–55
17. Capar ID, Ertas H, Ok E, Arslan H, Ertas ET (2014) Comparative study of different novel nickel-titanium rotary systems for root canal preparation in severely curved root canals. J Endod 40:852–856
18. Benyo Szilagyi L, Haidegger T, Kovacs L, Nagy-Dobo CB (2009) Detection of the root canal's centerline from dental micro-CT records. In: 31st annual international conference IEEE EMBS, pp 3517–3520

Analysis of Explicit Parallelism of Image Preprocessing Algorithms—A Case Study

S. Raguvir and D. Radha

Abstract The need for the image processing algorithm is inevitable in the present era as every field involves the use of images and videos. The performance of such algorithms can be improved using parallelizing the tasks in the algorithm. There are different ways of parallelizing the algorithm like explicit parallelism, implicit parallelism, and distributed parallelism. The proposed work shows the analysis of the performance of the explicit parallelism of an image enhancement algorithm named median filtering in a multicore system. The implementation of explicit parallelism is done using MATLAB. The performance analysis is based on primary measures like speedup time and efficiency.

1 Introduction

Parallel computing is a type of computation in which many calculations are carried out simultaneously, operating on the principle that large problems can often be divided into smaller ones, which are then solved at the same time. Parallel computations can be carried out in different ways like bit-level computations, instruction-level computations, and data- and task-level computations. Even though parallelism has been thought and employed for many years, it could not be used widely because of physical constraints; however, parallel computing has become the main requirement in computer architecture, mainly in the form of multicore processors.

Parallelism of a task in computers depends on the hardware having multicores and multiprocessors with multiple processing elements within a machine. This is different from performing the same task in multiple computers as in grids, multiple passive arrays (MPP), and clusters. There are some specialized parallel computer architectures that are used beside traditional processors, for speeding up the specific tasks.

S. Raguvir · D. Radha (✉)
Department of Computer Science & Engineering,
Amrita School of Engineering, Amrita Vishwa Vidyapeetham, Bangalore, India
e-mail: dm_radha@yahoo.com

© Springer Nature Switzerland AG 2019   1499
D. Pandian et al. (eds.), *Proceedings of the International Conference on ISMAC in Computational Vision and Bio-Engineering 2018 (ISMAC-CVB)*, Lecture Notes in Computational Vision and Biomechanics 30,
https://doi.org/10.1007/978-3-030-00665-5_140

Applications are often classified depending on the way subtasks are required to get communicated or synchronized with each other. An application needs complex parallelism if the divided tasks have to communicate with each other multiple times per unit time. In another way, applications show simple parallelism if there are subtle communications between the subtasks. It exhibits embarrassing parallelism if subtasks rarely or never have to communicate.

When implicit parallelism is employed in MATLAB, but there are some restrictions on the code where we cannot add parallelism into it. Even though there are different ways of doing the parallelism, there are some overheads behind them to get the synchronized output.

2 Literature Survey

The parallelism can be thought of parallel computing and distributed computing. In parallel computing, the parallelism is done in one/more CPUs in the same system, whereas in distributed computing it happens with multiple processors connected by communication links. The most important reason to prefer parallel computing over traditional sequential computing is the ease in applying algorithms, reduced complexity, and increased standardization [1].

Parallel computing can be in the form of implicit and explicit parallelism. Certain programming languages use automatic parallelism available in it. They are named as implicit parallelism. Explicit parallelism can be achieved by parallel computations using special-purpose directives or function calls available in the programming languages. Most of the applications require subtasks like synchronization, communication, or task partitioning to achieve parallelism. Apart from contributing to the intended parallelism of the program, there are overheads which are required to maintain the synchronization.

Various applications or fields need parallel computing for balancing the requirement of the current world. A massive task with fast execution is the requirement of the competitive world. Medical imaging, network-based processing, routing, bioinformatics, and data analytics are fields to name a few. Image processing is vital in computer vision, medical imaging, meteorology, astronomy, remote sensing, and other related fields [2]. Images of any size consume more time to do any processing. Parallel computing can be an apt way to utilize the essence of image processing applications. Distributing visualization algorithms has shown increased efficiency and increased speedup with respect to the processors used [3].

Image processing involves large quantities of data and certain applications use the same kind of operations that are performed on every part of the image. Parallelism applied to such applications is an advantage. It can be applied in various ways like data parallel, task parallel, and pipeline parallel. In data-parallel approach, each computing unit will get a piece of data and perform the operations on receiving data. The primary challenge of this approach is in making the parallelism efficiently in decomposing of data and composing the data back to its form. The next challenge is to balance the load. The image data should be distributed in such a way that the computing units get approximately the same load. This avoids unnecessary communication between computing units. In task-parallel approach [4], certain operations are grouped in tasks and will be allotted to computing units. The approach also requires efficient planning of selecting and distributing the tasks to various computing units. In pipeline-parallel approach, parallelism can be implemented if image processing application requires multiple images so that images are processed in different stages at the same time.

Parallel computing using MATLAB is one of the most widely used and easy to learn platforms. It holds many toolboxes for various applications which made it easy to use [5] and as it follows most of the basic programming languages. It is a platform for learning mathematical computing environments in technical computing. In the paper [6], authors have discussed the use of MATLAB in parallel computations of various research fields such as signal and image processing. It also says about the parallel computing toolbox (PCT) used for parallel computations.

MATLAB can utilize the multicore processors using workers that run locally [7]. The work discussed the fast and efficient computation in multicore processors. There are different ways of achieving parallelism in MATLAB. MATLABMPI, bcMPI, pMATLAB, Star-P, and PCT are to name a few. And, also, the future of MATLAB for parallel computations was discussed.

The advantages of the PCT are that it supports high-level constructs, for instance, parallel for loops and distributed arrays, it has various math functions which let the users utilize existing MATLAB. The simplest way of doing parallelism in MATLAB is using parfor, spmd-end constructs. Parfor and spmd-end assert that all the iterations of the loop are independent so that it can be done in any order or in parallel. It is easy to combine serial and parallel codes in parfor and spmd. There will be better speedup with little effort. Identification of independence in the algorithm is a trade-off for the speedup with the least effort. Independence can be achieved through interprocess communication [8]. But, such conditions can be avoided in explicit parallelism and lead to distributing the load among different processors/cores/threads by labindex of the MATLAB. Each labindex is associated with a core which improves the performance and efficiency of the system.

These parallel algorithms can work with various numbers of threads, to take all the benefits of the upcoming processors having any number of cores.

Parallelization is an add-on to the computing environment. The images are processed and analyzed to find the accuracy and reliability of the results of parallelism. The processing time is also to be considered along with its accuracy and reliability. More the processing time in parallelization, degrades the use of parallelism in the applications. Like the other stages of image processing such as preprocessing used for clarity and visibility of the image, parallelization involves in the optimization of the speed at which the image is processed.

The rapid requirement of parallelism increased the attention to use polyhedral frameworks for optimizing explicit parallelism [9].

The proposed work explores current multicore architectures to parallelize the image processing algorithms which enhance the image. Comparisons on sequential and parallelization through the usage of different number of cores available are discussed.

These parallel algorithms can work with a different number of threads and their processors with different number of cores. Load distribution among processors can be of different types like horizontal, vertical, and equal distribution (cubical).

Image processing algorithms can be parallelized using the simplest way by distributing the load to different cores than considering the limitations of many other aspects of different ways of parallelizing the algorithms.

Parallel computing is having very important significance in several image processing techniques like edge detection, histogram equalization, noise removal, image registration, image segmentation, feature extraction, different optimization techniques, and many more [7].

A different approach of parallel programming from the programmer's point of view is discussed which is known as explicit parallelism [2]. Comparison of sequential and parallel version and the performance of them are discussed.

Even though there is no full acceptance on explicit parallelism.

3   Proposed Model and Implementation

Parallelism in image processing is essential in every field for the current world. There are different ways of parallelizing the algorithms for an image. Every method is having its own advantages and disadvantages.

To avoid the limitations of various parallel programming constructs, the images are divided and sent to different cores to process in the same way. This is a kind of explicit parallelism where the cores are assigned to the same tasks on different portions of the image. This is depicted in Fig. 1.

The different tasks that can be performed using explicit parallelism are contrast stretching, histogram equalization, noise smoothing, filtering, sharpening, etc. The three algorithms considered are histogram equalization, median filtering, and sharpening.

Most of the enhancement algorithms can be improved in performance by explicit parallelism. Different parts of the image are distributed among the cores on the machine and are parallely executed. The dependency of the operations need not be checked as the image is divided into sub-images to do the same task on them.

The method is depicted in Fig. 1. An image is divided into sub-images as per the number of cores on the system. If parallelism is not added into the algorithm, it is executed by the single core and becomes the sequential algorithm. If the number of cores is two, the image can be divided into two different ways like vertical split and horizontal split. In the same way, if the number of cores is four, it can be divided in two different ways as shown in Fig. 1.

Once the algorithm is chosen, and a number of cores of the system are known, it can be run in MATLAB by assigning each of the sub-images to different cores. It can be visualized in pmode where each part is executed by labindex as shown in Fig. 2.

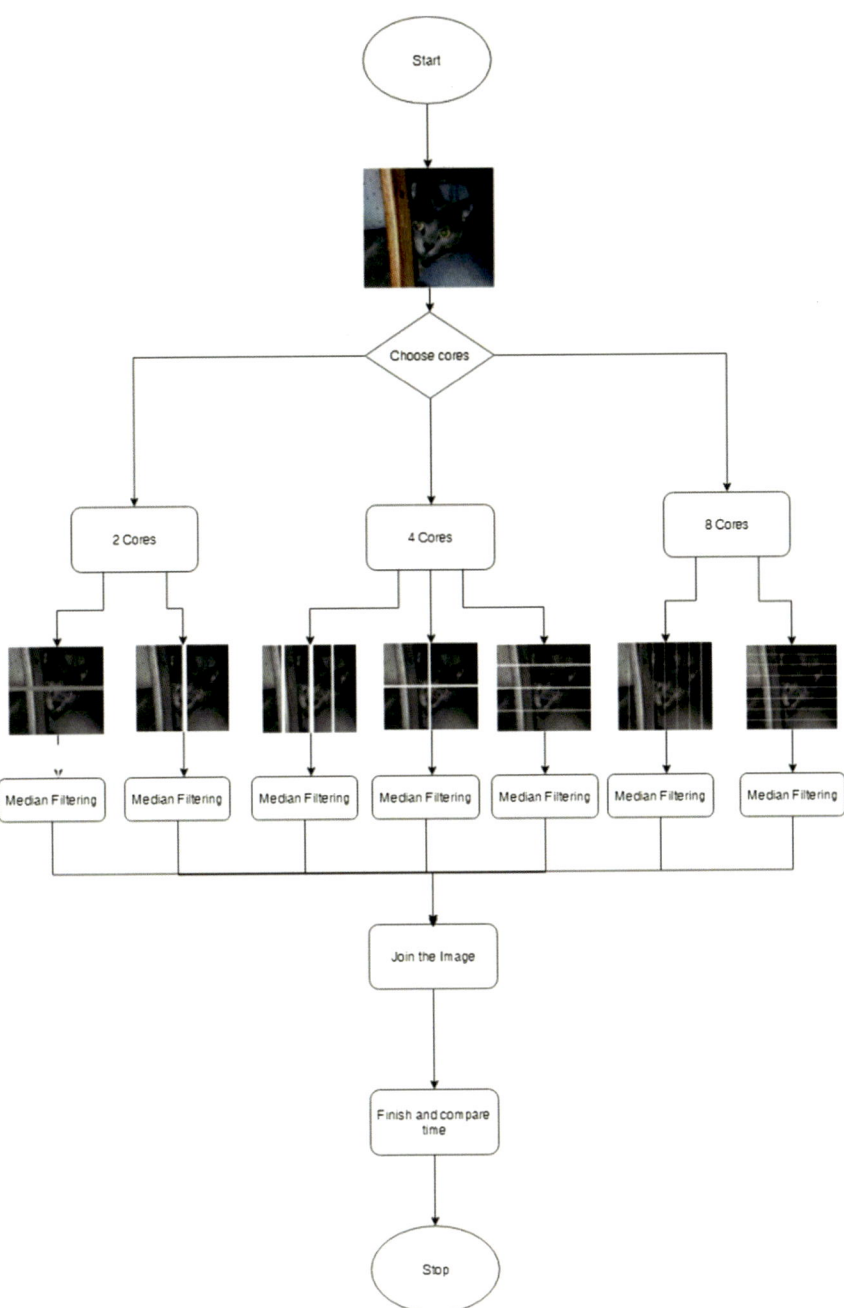

Fig. 1 System model

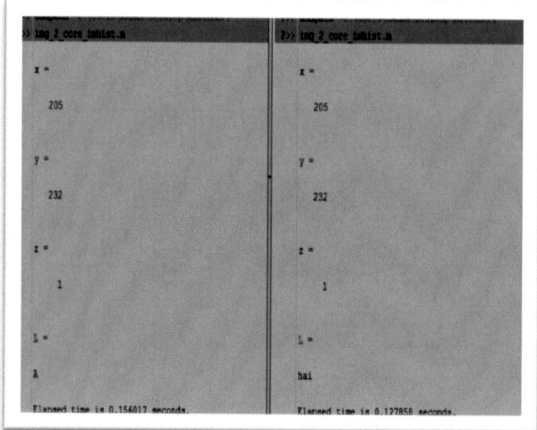

Fig. 2 Sample program running in interactive mode—pmode with two workers/labindex

A labindex is an instance of MATLAB using PCT. When a code is executed in parallel, it is run by the workers/lab indexes than by the main MATLAB. Each labindex is associated with a core of the system. Typically, MATLAB is asked to start one worker for each of the cores in the machine. Each of those workers will be run as a process by the OS, and will end up being run one worker per core in parallel.

Median filters are statistical nonlinear filters that are often described in the spatial domain and are used to remove random noise [10].

Enhancing the image is by finding the median value in a 3×3 neighborhood around the corresponding pixel in the input image. It does the filtering of the image in two dimensions. The corresponding pixels on the edges are found with the median values which are substituted with zeros for the missing neighborhood pixel values. As the median is calculated with zero values, the enhanced image may look distorted at the edges.

The histogram equalization and median filtering are the algorithms used for analyzing the explicit parallelism of the proposed model.

The algorithm is executed in a different number of labindex which is chosen according to the number of cores on the machine associated with it. The algorithm is executed on a single labindex with the original image as shown in Fig. 3. This is a sequential execution. The same algorithm is executed in parallel on two parts of the image by two labindex, four parts of the image by four labindex, and eight parts of the image by eight labindex.

The model is tested in the system with following specifications:

CPU—i7-4770 (Quad core) 3.4 GHz,
RAM—4 GB ddr3.

4   Results and Analysis

The low-resolution image considered is an image with the size 255×204, as shown in Fig. 3.

Table 1 shows the time taken by the computer to execute the algorithm using a single core (sequential), two cores, four cores, and eight cores for performing image enhancement on this image. The image is partitioned according to the number of cores used for parallelism.

For medium resolution image, the considered image is of dimensions 512×512, as shown in Fig. 4. Table 2 contains time taken by the computer to perform the algorithm with this image.

The high-resolution image used for applying this algorithm is of dimensions 1419×1001, as shown in Fig. 5. The data computed for this image is given in Table 3.

Fig. 3 Low-resolution image

Table 1 Image analysis for low-resolution image

No. of cores	Time taken	Speedup time	Efficiency (%)
1 (Sequential)	0.00404		
2 (Horizontal)	0.00593	0.68213	34.107
2 (Vertical)	0.00417	0.97048	48.524
4 (Horizontal)	0.00900	0.44947	11.237
4 (Vertical)	0.00693	0.58324	14.581
4 (Box)	0.005289	0.76485	19.121
8 (Horizontal)	0.00498	0.81266	10.158
8 (Vertical)	0.00817	0.49498	6.187

Fig. 4 Medium-resolution image

Table 2 Image analysis for medium-resolution image

No. of cores	Time taken	Speedup time	Efficiency (%)
1 (Sequential)	0.00548		
2 (Horizontal)	0.00582	0.94141	47.070
2 (Vertical)	0.01578	0.34723	17.362
4 (Horizontal)	0.00988	0.55450	13.862
4 (Vertical)	0.00964	0.63129	15.782
4 (Box)	0.009642	0.56824	14.206
8 (Horizontal)	0.00657	0.83445	10.431
8 (Vertical)	0.00641	0.85529	10.691

The speedup time from the observed values is given by

$$\text{speedup} = \frac{T_s}{T_p}$$

To find the efficiency, we use the following formula:

$$\text{efficiency} = \frac{T_s}{pT_p},$$

where p is the number of processors used,

T_s and T_p are the time taken to perform the operation in sequential and parallel methods.

Fig. 5 High-resolution image

Table 3 Image analysis for high-resolution image

No. of cores	Time taken	Speedup time	Efficiency (%)
1 (Sequential)	0.00548		
2 (Horizontal)	0.01071	0.51167	25.584
2 (Vertical)	0.01227	0.44672	22.336
4 (Horizontal)	0.01564	0.35023	8.756
4 (Vertical)	0.01808	0.30299	7.575
4 (Box)	0.01542	0.35527	8.882
8 (Horizontal)	0.01803	0.30387	3.7998
8 (Vertical)	0.01828	0.29973	3.747

5 Conclusion

Explicit parallelism is tested on image preprocessing algorithms like median filtering in three scenarios with different number of cores. The primary analysis is done on the basis of speedup time and efficiency. It has been observed that the number of partitions required for splitting the job to many cores may degrade the efficiency. The parallelization of the above algorithm has shown that the efficiency is highest when two cores are used. It is observed that the resolution does not show much variation in speedup and efficiency. The same can be tested for various algorithms with more parameters to understand the trade-off between parallelism and the overheads.

Intermediate representation can improve the efficiency of parallelism [11]. The analysis has to be extended with more parameters such as overhead time and fork time to understand the maximum advantage of explicit parallelism in image enhancement algorithms using MATLAB.

References

1. Liu F, Seinstra F, Plaza A (2011) Parallel hyperspectral image processing on distributed multicluster systems. J Appl Remote Sens 5
2. Saxena S, Sharma N, Sharma S (2013) Image processing tasks using parallel computing in multi core architecture and its applications in medical imaging. Int J Adv Res Comput Commun Eng 2(4)
3. Squyres J, Lumsdaine AB, McCandless A, Stevenson R (1996) Parallel and distributed algorithms for high speed image processing
4. Kessler C, Keller J (2007) Models of parallel computing: review and perspectives. PARS-Mitteilungen 24:13–29. ISSN 0177-0454,GI/ITG PARS
5. Alyasseri Z (2014) Survey of parallel computing with MATLAB. Cornell University Library, arXiv:1407.6878, Article
6. Kaur P (2015) Implementation of image processing algorithms on the parallel platform using MATLAB. Int J Comput Sci Eng Technol (IJCSET)
7. Saxena S, Sharma S, Sharma N (2016) Parallel image processing techniques, benefits and limitations. Res J Appl Sci Eng Technol 12(2):223–238
8. Ranjith R, Shanmughasundaram R (2015) Simulation of safety critical applications for automotive using multicore scheduling. In: International conference on control, instrumentation, communication and computational technologies (ICCICCT)
9. Chatarasi P, Shirako J, Sarkar V (2015) Polyhedral optimization of explicitly parallel programs. In: International conference on parallel architecture and compilation (PACT), pp 213–226
10. Saxena C, Kourav D (2014) Noises and image denoising techniques: a brief survey. Int J Emerg Technol Adv Eng 4(3)
11. Belwal M, Sudarshan TSB (2015) Intermediate representation for heterogeneous multi-core: a survey. In: International conference on VLSI systems, architecture, technology and applications (VLSI-SATA)

A Comprehensive Study on Character Segmentation

Sourabh Sagar and Sunanda Dixit

Abstract In identifying the characters from a given image, character segmentation plays an important role. In a given line of text, first, we have to segment the words. Then, in each word there will be a character-by-character segmentation. There have been some rapid developments in this area. Many algorithms have been implemented to increase the accuracy range and decrease the word error rate. This paper aims to provide a review of some of the developments that have happened in this domain.

1 Introduction

Lot of research work has been going on, for some time now. Results of this research have increased the accuracy range from research to further. But still there is no method which produces 100% perfect segmentation. Because the cursive handwriting does not have any correct format. People have their own style of writing. So, it will be a very difficult task to compare the input image with its database. This paper provides a review on some of the methods which are used for character segmentation.

2 Literature Survey

Manjunath and Sharath [1] provide an application that can run on any android device because they are portable. This paper uses Kohonen's algorithm. It will extract the text from the image and finds the meaning on the Internet. Zhu et al. [2] use feature coordinates as unary features. For binary features, this method uses the differences

S. Sagar (✉) · S. Dixit
Department of Information Science and Engineering,
Dayananda Sagar College of Engineering, Bangalore, India
e-mail: sourabhacademic@gmail.com

S. Dixit
e-mail: sunanda.bms@gmail.com

© Springer Nature Switzerland AG 2019
D. Pandian et al. (eds.), *Proceedings of the International Conference on ISMAC in Computational Vision and Bio-Engineering 2018 (ISMAC-CVB)*, Lecture Notes in Computational Vision and Biomechanics 30,
https://doi.org/10.1007/978-3-030-00665-5_141

in coordinates between the neighboring feature points. Kozielski et al. [3] proposed the conversion of digit recognition to handwritten recognition. Independent normalization procedure is applied for the image slices. Additional algorithm used here is a novel moment-based algorithm. Wu et al. [4] maintained a codebook containing SIFT descriptors (SDs). Here, shift descriptors signature (SDS) and scale and orientation histogram (SOH) are used to identify the writing. Six public datasets are used for this experiment. Tan et al. [5] proposed a method where strokes are generated for each line by factorizing to get the similarity matrix. To get labels for strokes, nonlinear clustering methods are applied to the similarity matrix.

Ghosh and Valveny [6] use a simplified version of an attribute model to get an integral image representation of the input document. It increases the efficiency. Performance will be increased by the introduction of re-ranking step. Naveena and Manjunath Aradhya [7] proposed a method where mixture of Gaussians is learnt by applying the expectation maximization (EM) algorithm. For segmenting the characters, branch points are used as reference points. Given results are satisfactory. Haji et al. [8] proposed an algorithm which can be complemented with any existing segmentation algorithm. More accuracy in segmentation rate is achieved. Parui et al. [9] proposed an algorithm. Script and particular intelligent segmentation strategy are independent of this algorithm. Based on different scripts, different segmentation algorithms are explored. HMM algorithm is evaluated. Chen et al. [10] mainly focus on concatenating the small segments in order to overcome the over segmentation.

Lawgali et al. [11] provide huge database of 6600 shapes of 50 writers. The name of database is handwritten Arabic character database (HACDB) which will be useful for the recognition of characters after segmentation. Sharma et al. [12] provided the method which increases the recognition accuracy in minimum amount of time. Using Euler number, the search space will be reduced. Saba et al. [13] focus on unconstrained handwritten words. It figures out the characters and compares it with the database. Artificial neural network is employed. Marti et al. [14] provide a database for English language named Lancaster–Oslo/Bergen (LOB). It will be helpful for the recognition of characters after segmentation.

Prasanna et al. [15] approach help the blind or visually impaired person. This approach will identify the characters in the taken image and gives output in the form of voice. So, a blind person can get the information from that output through hearing. Sharma and Sharma [16] proposed that identifying characters in an image is an easy task for humans but not for machines. Humans use their previous knowledge and compare the characters but the machine cannot do. Sandhya and Krishanan [17] have provided algorithm for reconstruction of the broken characters and recognition. In historical documents, the characters will be broken and blurred. In those cases, it will be used. Pardeshi et al. [18] used the wavelet-like discrete cosine transform (WDCT) to separate the handwritten words and printed words. Banumathi and Jagadeesh Chandra [19] proposed projection profile technique for character segmentation comprising preprocessing, binarization, and many other processes. Patil and Hanni [20] proposed the optical character recognition (OCR) which segments the whole document into text lines, then words, and then characters. Those characters are used for recognition.

The input for character recognition is the output of line segmentation, while doing a survey the following techniques are observed on text line segmentation.

Dixit et al. [21] proposed a new cost function which considers the line spacing and skew of each line. For baseline correction, a new algorithm is used which improves the efficiency. Dixit and Suresh [22, 23] proposed a method where adaptive histogram equalization and sliding window techniques play a very important role. This is compared with conventional text line segmentation technique which leads to an improvement. Dixit et al. [24] proposed a method for line segmentation, where based on hough lines, text lines are identified. Noise will be removed after the segmentation result. Dixit and Narayan [25] discuss a novel technique for line segmentation. Preprocessing like skew correction is done.

Survey Table

Year	Title	Methodologies used	Pros	Cons	Language
2016	Character recognition using image processing [16]	Optical character recognition (OCR)	Provides an upper edge	Accuracy depends on material	English
2016	Broken Kannada character recognition [17]	End-point algorithm	Achieves 98.9% accuracy		Kannada
2016	Line and word segmentation of Kannada handwritten text documents using projection profile technique [19]	Projection profile technique	90% successful segmentation	Word segmentation is not satisfactory	Kannada
2016	Handwritten and machine printed text separation from Kannada document images [18]	Wavelet-like discrete cosine transform (WDCT), k-NN classifier	Achieves 99.50% accuracy		Kannada
2016	Handwritten Kannada document image processing using optical character recognition [20]	Optical character recognition (OCR)	Measurable improvement	Skew detection and noise removal techniques can still be improved	Kannada
2015	Script independent online handwriting recognition [9]	Hidden Markov model (HMM)	Three segmentation strategies		English, Arabic, Bangla, Bengali
2015	Query by string word spotting based on character bigram indexing [6]	Pyramidal histogram of characters (PHOC)	Segmentation free	Retrieval method needs improvement	English

(continued)

(continued)

Year	Title	Methodologies used	Pros	Cons	Language
2014	Offline text-independent writer identification based on scale-invariant feature transform [4]	Scale-invariant feature transform (SIFT)	Writer identification	Segmentation is not easy	English, Chinese
2014	Segmentation of Kannada handwritten text line through computation of variance [25]	Computation of variance			
2014	Kannada text line extraction based on energy minimization and skew correction [21]	Skew correction, background cleaning	Segment the lines of text		Kannada
2014	Text line segmentation of handwritten documents in Hindi and English [24]	Hough line	Non-skewed also considered		Hindi, English
2014	Sliding window technique for handwritten text line segmentation [23]	Sliding window, OCR, adaptive histogram equalization		Not preferable for images with complex characteristics	Kannada
2013	South Indian Tamil language handwritten document text line segmentation technique with aid of sliding window and skewing operations [22]	Sliding window, adaptive histogram equalization	Better method than the conventional		Tamil
2013	Implementing Kannada optical character recognition on the Android operating system for Kannada signboards [1]	Kannada OCR, Kohonen's algorithm, Hilditch algorithm	Portable, easy to use, recognize, translation	Less accurate, do not support for more languages	Kannada
2013	Online handwritten cursive word recognition using segmentation-free MRF in combination with P2DBMN-MQDF [2]	Markov random field (MRF), Pseudo 2D bi-moment normalization (P2DBMN), Modified quadratic discriminant function (MQDF)	Restricts the search space	More time is needed	Chinese, Japanese

(continued)

(continued)

Year	Title	Methodologies used	Pros	Cons	Language
2013	Keyword spotting in unconstrained handwritten Chinese documents using contextual word model [10]	Contextual word model	Higher recall rate		Chinese
2013	Handwritten Arabic character database (HACDB) for automatic character recognition [11]	Building their own database	Large database		Arabic
2012	Moment-based image normalization for handwritten text recognition [3]	Normalization process, a novel moment-based algorithm	Achieved word error rate to 13.4% from 16.7%	Language model's perplexity is 258.7	French, English
2012	A new handwritten character segmentation method based on nonlinear clustering [5]	Spectral clustering based on normalized cut (Ncut), Kernel clustering based on conscience online learning (COLL)	Segment nonlinearly separable characters	Took few databases	English, Chinese
2012	Handwritten character segmentation for Kannada scripts [7]	Expectation maximization (EM) algorithm, cluster mean points	Satisfactory result	Not good for overlapping components	Kannada
2012	Statistical hypothesis testing for handwritten word segmentation algorithms [8]	Hidden Markov model (HMM)	Can be used with any segmentation algorithm	Segmentation is not perfect	English
2012	An improved zone-based hybrid feature extraction model for handwritten alphabets recognition using Euler number [12]	Euler number	Increases speed and accuracy	Not a complete mechanism	English
2011	Cursive script segmentation with neural confidence [13]	ANN, CEDAR database	Accuracy is increased		

(continued)

(continued)

Year	Title	Methodologies used	Pros	Cons	Language
2011	Kannada text extraction from images and videos for vision-impaired persons [15]	Text extraction, text recognition, speech synthesis	Blind people can get the information		Kannada
1998	A full English sentence database for offline handwriting recognition [14]	Database	A basis for character recognition		English

3 Conclusion

There has been a lot of advancement in character segmentation domain. Yet, no perfect segmentation technique is found. But, there are chances of getting perfect segmentation which shows 100% accuracy. Through this survey, we have come to realize that there are several segmentation techniques existing. Applying these different techniques, we can try to achieve good accuracy level in this domain.

References

1. Manjunath AE, Sharath B (2013) Implementing Kannada optical character recognition on the android operating system for Kannada sign boards. IJARCCE (Int J Adv Res Comput Commun Eng) 2(1)
2. Zhu B, Shivram A, Setlur S, Govindaraju V, Nakagawa M (2013) Online handwritten cursive word recognition using segmentation-free MRF in combination with P2DBMN-MQDF. IEEE, pp 349–353
3. Kozielski M, Forster J, Ney H (2012) Moment-based image normalization for handwritten text recognition. IEEE, pp 256–261
4. Wu X, Tang Y, Bu W (2014) Offline text-independent writer identification based on scale invariant feature transform. IEEE 9(3)
5. Tan J, Lai JH, Wang CD, Wang WX, Zuo XX (2012) A new handwritten character segmentation method based on nonlinear clustering. Elsevier, pp 213–219
6. Ghosh SK, Valveny E (2015) Query by string word spotting based on character Bi-Gram indexing. IEEE, pp 881–885
7. Naveena C, Manjunath Aradhya VN (2012) Handwritten character segmentation for Kannada scripts. IEEE, pp 144–149
8. Haji M, Sahoo KA, Bui TD, Suen CY, Ponson D (2012) Statistical hypothesis testing for handwritten word segmentation algorithms. IEEE, pp 114–119
9. Samanta O, Roy A, Bhattacharya U, Parui SK (2015) Script independent online handwriting recognition. IEEE, pp 1251–1255
10. Huang L, Yin F, Chen Q-H, Liu C-L (2013) Keyword spotting in unconstrained handwritten Chinese documents using contextual word model. Image Vis Comput 31:958–968
11. Lawgali A, Angelova M, Bouridane A (2013) HACDB: handwritten Arabic characters database for automatic character recognition. IEEE

12. Sharma OP, Ghose MK, Shah KB (2012) An improved zone based hybrid feature extraction model for handwritten alphabets recognition using euler number. IJSCE (Int J Soft Comput Eng) 2(2). ISSN: 2231-2307

13. Saba T, Rehman A, Sulong G (2011) Cursive script segmentation with neural confidence. IJIC (Int J Innov Comput) 7(8)

14. Marti UV, Bunke H (1998) A full English sentence database for off-line handwriting recognition. IEEE

15. Prasanna K, Ramakhanth Kumar P, Thungamani M, Koli M (2011) Kannada text extraction from images and videos for vision impaired persons. IJAET (Int J Adv Eng Technol). ISSN: 2231-1963

16. Sharma S, Sharma R (2016) Character recognition using image processing. IJAETMAS (Int J Adv Eng Technol Manage Appl Sci) 03(09):115–122. ISSN:2349-3224

17. Sandhya N, Krishanan R (2016) Broken Kannada character recognition—a neural network based approach. In: ICEEOT (International conference on electrical, electronics and optimization techniques)

18. Pardeshi R, Hangarge M, Doddamani S, Santosh KC (2016) Handwritten and machine printed text separation from Kannada document images. IEEE

19. Banumathi KL, Jagadeesh Chandra AP (2016) Line and word segmentation of Kannada handwritten text documents using projection profile technique. In: ICEECCOT (International conference on electrical, electronics, communication, computer and optimization techniques)

20. Patil MM, Hanni AR (2016) Handwritten Kannada document image processing using optical character recognition. IOSR-JCE (IOSR J Comput Eng) 18(4), Ver. VI:39–47. e-ISSN:2278-0661, p-ISSN:2278-8727

21. Dixit S, Hosahalli Narayan S, Belur M (2014) Kannada text line extraction based on energy minimization and skew correction. In: IEEE IACC (International advance computing conference)

22. Dixit S, Suresh HN (2013) South Indian tamil language handwritten document text line segmentation technique with aid of sliding window and skewing operations. JATiT (J Theoret Appl Inf Technol) 58(2). ISSN: 1992-8645, E-ISSN: 1817-3195

23. Dixit S, Suresh HN (2014) Sliding window technique for handwritten text line segmentation. IJRCEE (Int J Res Comput Eng Electron) 3(04). ISSN: 2319-376X

24. Dixit S, Sneha NU, Suresh HN (2014) Text line segmentation of handwritten documents in Hindi and English. IJRITCC (Int J Recent Innov Trends Comput Commun) 02(04). ISSN: 2321-8169

25. Dixit S, Narayan SH (2014) Segmentation of Kannada handwritten text line through computation of variance. IJCSIS (Int J Comput Sci Inf Sec) 12(02). ISSN: 1947-5500

EZW, SPIHT and WDR Methods for CT Scan and X-ray Images Compression Applications

S. Saradha Rani, G. Sasibhushana Rao and B. Prabhakara Rao

Abstract Scanning rate of medical image tools has been significantly improved owing to the arrival of CT, MRI and PET. For medical imagery, storing in less area and not losing its details are vital. So, an efficient technique is necessary for storing in a cost-effective way. In this paper, wavelet is employed to perform decomposition, and image is compressed using Embedded Zero-Tree Wavelet (EZW), Set Partitioning in Hierarchical Trees (SPIHT) and Wavelet Difference Reduction (WDR) algorithms. These algorithms are applied to compress X-ray and CT images, and compared using performance metrics. From results, it is seen that compression ratio is better in WDR for all the wavelets than SPHIT and EZW. High compression ratio, 82.47, is obtained with Haar and WDR combination for CT scan, whereas this is 32.89 for Biorthogonal and WDR combination for X-ray. The main objective of this paper is to find the optimal combination of wavelets and image compression techniques.

1 Introduction

These days, the image compression algorithms for decreasing the dimensions play a significant task in an image processing predominantly in communications, visualizations, classifications of image and storage of high-dimensional information. The majority of the mentioned applications are impracticable with no compression [1].

S. Saradha Rani (✉)
Department of Electronics and Communication Engineering, GITAM,
Visakhapatnam 530045, Andhra Pradesh, India
e-mail: ssaradarani@gmail.com

G. Sasibhushana Rao
Department of Electronics and Communication Engineering,
AU College of Engineering, Andhra University, Visakhapatnam 530003, India
e-mail: sasigps@gmail.com

B. Prabhakara Rao
Department of Electronics and Communication Engineering,
JNTUK, Kakinada 533003, Andhra Pradesh, India
e-mail: drbprjntuk@gmail.com

© Springer Nature Switzerland AG 2019
D. Pandian et al. (eds.), *Proceedings of the International Conference on ISMAC in Computational Vision and Bio-Engineering 2018 (ISMAC-CVB)*, Lecture Notes in Computational Vision and Biomechanics 30,
https://doi.org/10.1007/978-3-030-00665-5_142

1517

Fig. 1 Process of three-level
wavelet decomposition

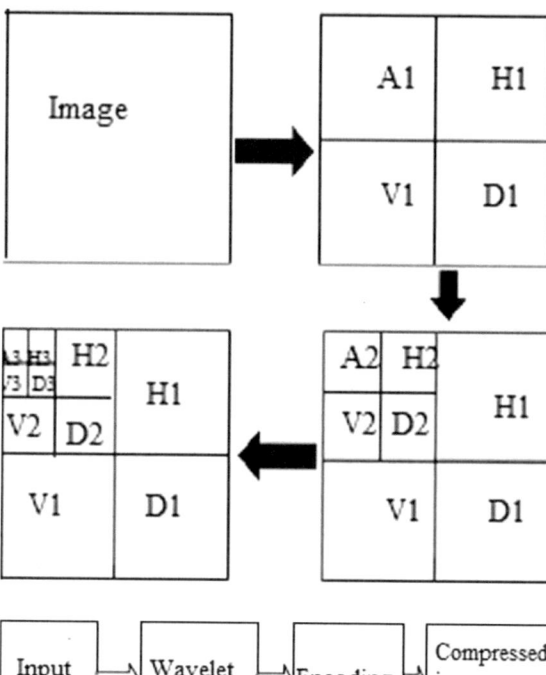

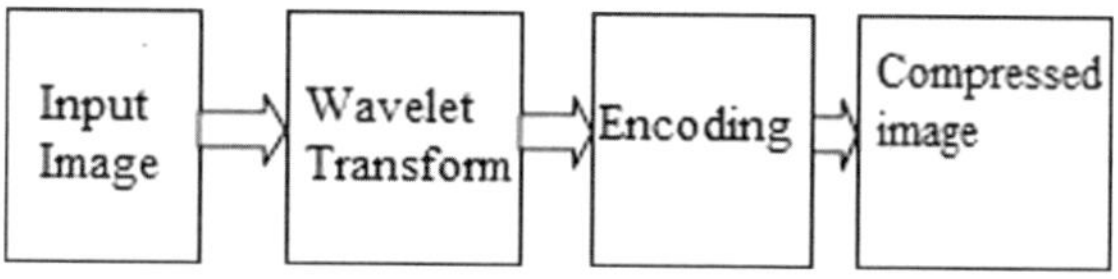

Fig. 2 Block diagram of
image compression

Compression is a procedure of lessening the quantity of required bits to signify the
data or image, without diminishing the nature of the image. By compressing the data,
it is possible to accumulate storage space capability; reducing expenditure for network bandwidth and storage equipment can boost up the transfer of file. Compression
is a means of describing the data in a conservative form fairly than its original form.
The process of wavelet decomposition involves low-pass and high-pass filtering. The
wavelet transform partitions the image into numerous sub-bands, namely, HH, HL,
LL and LH; among these sub-bands, merely LL sub-band is further decomposed,
as it has low-frequency component and noise compared to the remaining sub-band
levels [2].

The process of wavelet decomposition is shown in Fig. 1.

The compression is performed in two levels. First, the wavelet transform is implemented to decompose the image; second, encoding is applied for compressing the
image and is illustrated with block diagram shown in Fig. 2 [3]. In this paper, Haar,
Daubechies and Biorthogonal wavelets have been used for decomposition, and EZW,
SPIHT and WDR algorithms have been applied for encoding.

2 Embedded Zero-Tree Wavelet (EZW)

In wavelet-based compression, EZW method was one in all the initial and dominant algorithms. Further algorithms were shaped relying on the elementary concept of EZW [4]. The heart of EZW compression is the development of self-resemblance across the distinct image wavelet transform scales [5]. Alternatively, in wavelet decomposed image, EZW approximates the high occurrence coefficients. For the reason that the wavelet coefficients have data concerning each frequency and spatial content of a picture, removal of a high-frequency coefficient results in a minor degradation of image in a specific position of the restored image instead of that across complete picture. In this, the threshold is employed to compute a significance map of significant and insignificant coefficients. The significance map can be represented with zero trees efficiently [5]. The following are the main steps in EZW:

(a) Initializing threshold: Threshold T is initialized to the lowest power of 2 in which 'T' exceeds $|C_{\max(k,1)}|/2$, where $C_{k,1}$ is wavelet coefficients.

(b) Coding of significance map: when $|C_{k,1}| > T$, output a symbol by scanning the coefficients in a predefined manner. When this symbol inputs the decoder, it assigns $C_{k,1} = \pm 1.5T$.

(c) Refinement: in this step, every significant coefficient is refined by transmitting the next binary bit. Decoder augments the present value by $\pm 0.25T$, as soon as it receives [5].

(d) Put $T = T/2$, as well as repeat from (b) if further iterations are desired.

3 Set Partitioning in Hierarchical Trees (SPIHT)

The SPIHT encoding algorithm is developed edition of EZW. For different types of images, high values of PSNR can be achieved using SPIHT technique. It offers an improved assessment standard for successive techniques. Hierarchical trees pertain to quadtrees, described in EZW [6]. Set partitioning means the manner how quadtrees separate or split, and the wavelet transforms at a certain threshold. The coding technique uses three lists, known as List of Insignificant Sets (LIS), List of Significant Pixels (LSP) and List of Insignificant Pixels (LIP) [6]. The subsequent process illustrates the encoding using SPIHT:

(a) Initialization: set Output $n = [\log_2{}^{C\max(k,1)}]$, LSP = { }, LIP = {(k, l)} and LIS = {D(k, l)} [5].

(b) Sorting pass: considering every access in LIP, if it is significant to produce an output as '1' by removing it from LIP and adding to LSP, otherwise output as '0'. Output the significance, i.e. output its sign if significant, for every access in the LIS. Carry out set partitioning, relying upon whether it is the D(k, l) or the L(k, l) set. And update the LSP, LIP and LIS accordingly [7].

(c) Refinement pass: on considering every access in LSP, apart from those that are brought all through the sorting pass with similar 'n', encode the nth MSB [8].

Fig. 3 SPIHT algorithm
flow chart

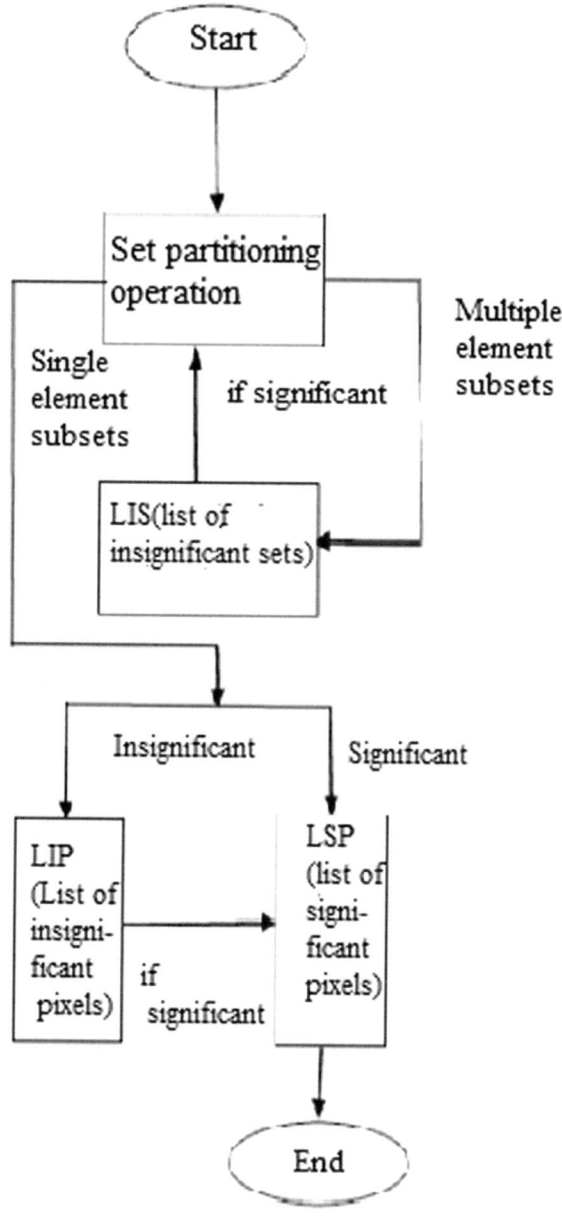

(d) In present pass, n is decremented by '1' and the steps (b) and (c) are repetitive
till 'n' reduces to zero.

The SPIHT algorithm is illustrated in flow chart shown in Fig. 3.

4 Wavelet Difference Reduction (WDR)

The imperfection in SPIHT is that it solely finds the location pertaining to significant coefficients, making hard to carry out the operations like Region of Interest (ROI) on compressed data, which rely on significant transform values precise location. That means a portion that requires increased resolution is to be selected in compressed image. WDR technique is completely easy procedure. First, an image is transformed using wavelet. Second, for the wavelet coefficients, bit-plane-based WDR encoding that is based on bit-plane coding is applied. The main five steps that constitute WDR are as follows:

(a) Initialization: select initial value of threshold 'T_0' so that 'T_0' is superior to all transformed values and at least one transform value has a level of $T_0/2$ [9].
(b) Threshold updation: assign $T_n = T_{n-1}/2$.
(c) Significance pass: for the deemed significant values, i.e. transform values that are equal or more than the threshold value, index values will be encoded by difference reduction method [10]. The difference reduction technique basically encodes the range of steps required to shift from previous significant value index to the present significant value index [10]. Significance pass output includes bit sequence produced by difference reduction and significant values signs, describing the accurate positions of significant values [10].
(d) Refinement pass: similar to the process in SPIHT method, here it generates refined bits by using standard bit- plane quantization procedure [10]. These refined bits are the improved approximation of the exact transform bits.
(e) Perform (b)–(d).

5 Performance Measures

5.1 Compression Ratio, CR

It is the quantitative relation between amount of bits required for representation of true picture to the amount of bits in compressed image [9]. The expression is

$$CR = \frac{n_1}{n_2} \tag{1}$$

where

n_1 amount of bits in true l image and
n_2 amount of bits in compressed image.

5.2 Mean Squared Error, MSE

MSE is the error metric accustomed to measure up to the excellence of different image compression methods [9].

$$\text{MSE} = \frac{1}{mn} \sum_{k=0}^{m-1} \sum_{l=0}^{n-1} (O(k, l) - C(k, l)) \tag{2}$$

where $m \text{ X } n$ is dimension of the image, $O(k, l)$ is true image and $C(k, l)$ is the approximated or compressed image.

5.3 Peak Signal-to-Noise Ratio, PSNR

The other error metric is PSNR that is used to compute up to the subjective faithfulness of the uncompressed image [9]. PSNR is the measure of quality of the compressed image.

$$\text{PSNR} = 10 \log_{10} \left(\frac{m \text{X} n}{\text{MSE}} \right) \tag{3}$$

6 Experimental Results

The results have been analysed on test image and two medical images. The wavelets Haar, Daubechies and Biorthogonal have been used to analyse the methods EZW, SPIHT and WDR, and various performance measures PSNR, MSE, CR and BPP are calculated for each of the methods for the following images.

Figure 4 illustrates the original image and compressed images of test image, and Figs. 5 and 6 depict the compressed images of CT scan of lower abdomen and X-ray of shoulder.

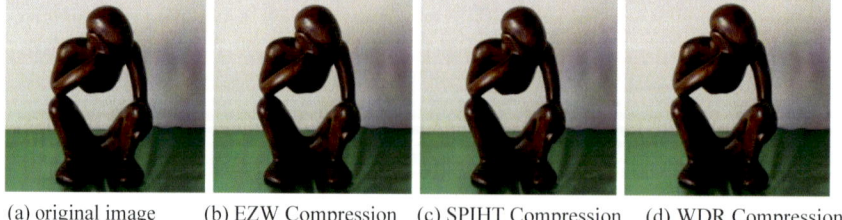

(a) original image   (b) EZW Compression   (c) SPIHT Compression   (d) WDR Compression

Fig. 4 Compression of African sculpture test image using various image compression methods

Table 1 Performance analysis of various algorithms for different wavelet and medical image configurations

Wavelet	Algorithm	CT scan of lower abdomen				X-ray of shoulder			
		C.R	PSNR	MSE	BPP	C.R	PSNR	MSE	BPP
HAAR	EZW	69.95	46.0833	0.0447	16.7876	26.21	54.66	0.2222	2.0972
	SPIHT	40.08	45.15	1.987	9.6201	10.75	45.78	1.718	0.8599
	WDR	82.47	46.08	1.602	19.7919	31.6	49.77	0.6858	2.5278
DAUBECHIES	EZW	67.79	61.46	0.0464	16.2706	19.92	54.5	0.2306	1.5934
	SPIHT	37.1	44.98	2.067	8.9034	7.88	46.52	1.448	0.6306
	WDR	79.1	46	1.634	18.1843	23.23	50.02	0.6478	1.8585
BIORTHOGONAL	EZW	66.1	60.45	0.0586	15.8832	27.44	54.51	0.2302	2.1948
	SPIHT	35.22	45.54	1.818	8.452	11.39	45.42	1.866	0.9112
	WDR	77.22	45.77	1.357	18.5356	32.89	48.22	0.979	2.631

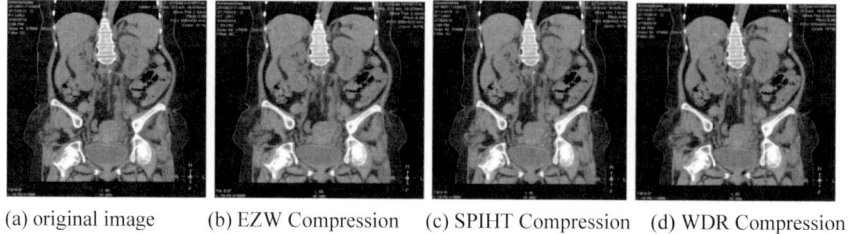

(a) original image   (b) EZW Compression   (c) SPIHT Compression   (d) WDR Compression

Fig. 5 Compression of CT scanned lower abdomen image using various image compression methods

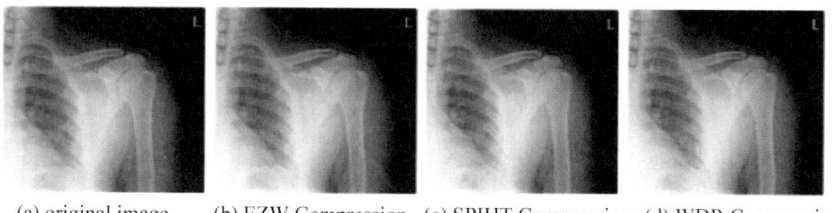

(a) original image   (b) EZW Compression   (c) SPIHT Compression   (d) WDR Compression

Fig. 6 Compression of X-ray of shoulder image using various image compression methods

Table 1 shows the performance measures of EZW, SPIHT and WDR algorithms. From the table, it is seen that the compression ratio is better for WDR algorithm for all the wavelets. For CT scanned lower abdomen image, the best measures are C.R = 82.47 and PSNR = 46.08 for Haar wavelet and for X-Ray of shoulder, C.R = 32.89 for Biorthogonal wavelet and PSNR = 49.77 obtained for Haar wavelet.

7   Conclusion

In this paper, an assessment of the discussed algorithms is carried out on CT and X-ray images. Here, the image is decomposed using Haar wavelet. After transforming, it is encoded using different algorithms. The same procedure is repeated using Daubechies and Biorthogonal wavelets. The capability of the algorithms to compress an image is analysed in terms of CR, MSE, BPP and PSNR. Quality of image is calculated by means of Mean Squared Error (MSE) and Peak Signal-to-Noise Ratio (PSNR), while compression measure is analysed via compression ratio and bits per pixel.

Acknowledgements The author would like to thank GIMSR, Visakhapatnam, for providing lower abdomen CT scan and shoulder X-ray images to carry out the research work.

References

1. Kumari L, Pandian R, Aran Glenn J (2016) Analysis of multi scale features of compressed medical images. In: International conference on electrical, electronics, and optimization techniques (ICEEOT)
2. Negahban Farnoosh, Shafieian Mohammad Ali (2013) Various novel wavelet based image compression algorithms using a neural network as a predictor. Basic Appl Sci Res 3(6):280–287
3. Gonzalez RC, Woods RE (2009) Digital image processing. Pearson Education, 2nd edn
4. Shapiro JM (1997) Embedded image coding using zero trees of wavelet coefficients, pp 108–121
5. Raja SP, Suraliandi A (2010) Performance evaluation on EZW & WDR image compression techniques. In: International conference on communication control and computing technologies
6. Raja SP (2009) Wavelet based image compression: a comparative study. In: 2009 international conference on advances in computing control and telecommunication technologies
7. Said A, Pearlman WA (1993) Image compression using the spatial-orientation tree. In: IEEE international symposium on circuits and systems, Chicago
8. Said A, Pearlman WA (1996) A new, fast, and efficient image codec based on set partitioning in hierarchical trees. IEEE Trans Circ Syst Video Technol 6(3)
9. Vaish A, Kumar M (2015) WDR coding based image compression technique using PCA. In: 2015 international conference on signal processing and communication
10. Singh P, Singh P (2012) Implementation of SPIHT and WDR algorithms for natural and artificial images using wavelets. In: Fourth international conference on computational intelligence and communication networks

Human Identification Based on Ear Image Contour and Its Properties

P. Ramesh Kumar, K. L. Sailaja and Shaik Mehatab Begum

Abstract Identity management is the process of authenticating individuals by means of security objects (traits) to confirm whether the subject is permitted to access any secured property. Ear biometrics is one of the best solutions to access any secured property, which may be private/public. In the current security surveillance, the subject is identified passively without the knowledge. Ear recognition is a better passive system where the human ear is captured to verify whether he is authorized or not. This system can possibly suit for crowd management like bus stations, railway stations, temples, cinema theatres, etc. An ear biometric system based on 2D ear image contours and its properties was proposed. In this article, three types of databases are taken as input, i.e. IIT Delhi Database, AMI Database and VR Students Sample Database, and enrolment and verification process is done with these databases based on the contour features and its properties—bounding rectangle, aspect ratio, extent, equivalent diameter, contour area, contour perimeter, checking convexity, convex hull and solidity. This approach takes less time to execute, and the obtained FAR and FRR performance parameter values are nominal when compared to other traditional mechanisms.

1 Introduction

Biometrics is a method of utilizing the physical parts of a person as a stable durable secret key. Like the fingerprints which are unique for each individual, the face, eyes, ears, hands and voice are also additionally one of a kind. Innovation has progressed to the point where PC frameworks can trace and identify the patterns, hand shapes, ear contours and other physical attributes. Utilizing biometrics gadgets are empowered

P. Ramesh Kumar (✉) · K. L. Sailaja · S. Mehatab Begum
Department of Computer Science & Engineering, VR Siddhartha Engineering College,
Vijayawada, India
e-mail: send2rameshkumar@gmail.com

K. L. Sailaja
e-mail: sailaja0905@gmail.com

© Springer Nature Switzerland AG 2019   1527
D. Pandian et al. (eds.), *Proceedings of the International Conference on ISMAC
in Computational Vision and Bio-Engineering 2018 (ISMAC-CVB)*, Lecture Notes
in Computational Vision and Biomechanics 30,
https://doi.org/10.1007/978-3-030-00665-5_143

with the capacity to check personality instantly and deny access to every other person. Utilizing biometrics for distinguishing and confirming people offers extraordinary points of interest over general customary techniques like smart cards, magnetic stripe cards, etc.

Passwords in the form of text can be overlooked or inadvertently seen by somebody lacks secrecy which makes an issue for clients. Since the biometric confirmation depends on the recognizable proof of a physical part of a person, the individual himself can act as a password in a biometric system. While choosing a biometric type, it is important to give due thought to acceptability. In this sense, the ear is a standout amongst the most appropriate contender to be utilized for biometrics. It does not change with emotions, states of mind, sadness, fear or cosmetic changes. The ear is effectively caught from a distance through a camera, regardless of whether the subject is not completely agreeable.

This makes ear-based identification [1–3] more suitable for passive intelligent security system and for crime investigation. Also, it might be understood that most of the existing biometrics system requires a large hardware set-up in a restricted background. Most of the biometrics systems require automatic recognition algorithm to identify the people based on physical, behavioural or chemical character. The researchers working on the stable biometric attributes and algorithms [3–9] have been explored broadly in recent times. Our assessment considers the deployment of ear as a biometric for human identification. In addition to these, other criteria are added such as feature extraction of ear image and its contours properties are calculated. In this paper, nine features: bounding rectangle, aspect ratio, extend, equivalent diameter, contour area, contour perimeter, checking convexity, convex hull and solidity are extracted. Human identification is done with the support of these contour features.

2   Feature Extraction

The proposed approach recognizes the human identity based on ear image contours and reduces the False Rejection Rate (FRR) and False Acceptance Rate (FAR). Here, only ear images are considered, and these images are characterized by the contour features [bounding rectangle, aspect ratio, equivalent diameter, contour area, contour perimeter, checking convexity, convex hull and solidity]. Feature values are calculated, and based on these values, the individuals are easily identified and verified.

2.1   Feature Template

The feature extraction of ear image contours is defined based on some properties, and the following features are extracted [10]:

Fig. 1 Bounding rectangle

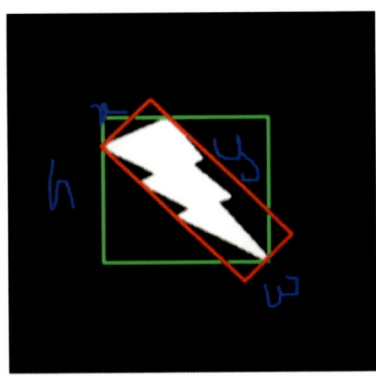

1. Bounding rectangle,
2. Aspect ratio,
3. Extent,
4. Equivalent diameter,
5. Contour area,
6. Contour perimeter,
7. Checking convexity,
8. Convex hull and
9. Solidity.

2.1.1 Bounding Rectangle

The Minimum Bounding Rectangle (MBR) is an expression of the maximum extents of a 2-D object like point, line, polygon, etc., or set of objects within its 2-D coordinate system.

The bounding rectangles are classified as follows:

(i) Straight bounding rectangle which never considers the rotation of the thing.
(ii) Rotated rectangle which considers the rotation of the thing (Fig. 1).

2.1.2 Aspect Ratio

It is the ratio between the width and height of an image.

$$Aspect\ Ratio = Width/Height$$

Width: rectangle width, and Height: rectangle height.

2.1.3 Extent

The area covered by something is called Extent.

$$\text{Extent} = \text{Object Area}/\text{Bounding Rectangle Area}$$

Bounding Rectangle Area: Area of the rectangle (W * h), and
Object Area: The area covered by the ear image.

2.1.4 Equivalent Diameter

Here, we can find the diameter of irregular-shaped objects by applying some pressure.

$$\text{Equivalent Diameter} = \sqrt{4 * \text{Contour Area}/\Pi}$$

Contour Area: Bounding area of the ear image.

Contour

It is an outline representing or bounding the shape or form of something. It moulds into a specific shape, especially one designed to fit into something else.

2.1.5 Contour Area

An outline representing or bounding the shape or form of something is called contour, and the regional area is called contour area.

$$\text{Area} = \text{contourArea}()$$

2.1.6 Contour Perimeter

The contour perimeter or arc length is the continuous line forming the boundary of a closed figure.

$$\text{Perimeter} = \text{arcLength (contour)}$$

2.1.7 Checking Convexity

This property is to check whether the curve is convex or not.

$$\text{Checking Convexity} = \text{is Contour Convex (contour)}$$

2.1.8 Convex Hull

convexHull() is used to check the curve for convexity defects and correct it.

2.1.9 Solidity

The quality or state of being firm or strong in structure. Example: wrapping a rubber band around the region.

$$\text{Solidity} = \text{Contour Area}/\text{Convex Hull Area}$$

3 Methodology

The architecture of this methodology consists of five steps (Fig. 2). Initially, raw image is taken from image acquisition device. The second step is preprocessing, where the ear image is preprocessed manually or with the help of cropping tools such as Matlab, snipping tool, etc. The third stage is image binarization. In this stage, binarized image will be obtained. In the fourth and final stage, the contour features are extracted, and the individual can be easily identified and verified by comparing with the created feature database.

1. Input image acquisition,
2. Preprocessing for ear image,
3. Image binarization,
4. Contour feature extraction and
5. Comparison.

Input Image Acquisition

The input images are taken from three databases:

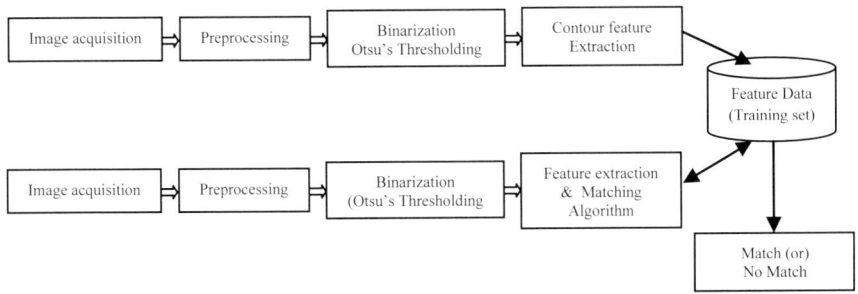

Fig. 2 Enrolment and verification process

1. IIT Delhi Database.
2. AMI Database.
3. VR Students Sample Database.

Preprocessing for Ear Image

In the preprocessing stage, the ear image which is extracted from the side view of the face gets preprocessed by removing shadows, hair, ornaments, etc., around the ear image. The ear shape is extracted by applying a series of grayscale morphological operations.

Image Binarization

The preprocessed ear image is used to develop a binarized mask which covers the required region for feature extraction using Otsu's threshold. The obtained image produces masks with different dimensions. The thresholding limit is adjusted to make the mask area less than the predefined limit.

Feature Extraction

The contour of an interested region in an image helps the image analyst to extract the shape of the object. The contours are good features for object detection and recognition. The best suitable image for contour processing is binary image, and any edge detection or threshold method can be easily applied on it to find the contours. The nine different contour features and their properties are extracted here.

Comparison

In this step, first perform image acquisition and in the preprocessing stage raw images are cropped and preprocessed images are used in the feature extraction state. Now comparison is performed and if the compared feature is existed in the database, then the feature number is displayed; otherwise, miss match function is called.

4 Performance

Different numerical measurements can be analysed to measure the performance of a biometric system. The predominant performance attributes are FAR and FRR. The system begins the operation by individual enrolment, where the unique feature data depends on the kind of biometrics collection and storage in a feature database as template along with the user identification number. Then, the process of verification/validation to authenticate the biometric feature is done by the matching algorithm to decide either acceptance (or) rejection.

False Acceptance Rate The FAR is the possibility that the algorithm mistakenly permits a non-certified person, due to error template matching [11].

False Rejection Rate The FRR is the possibility that the application erroneously discards admission to a certified individual due to template mismatch [11].

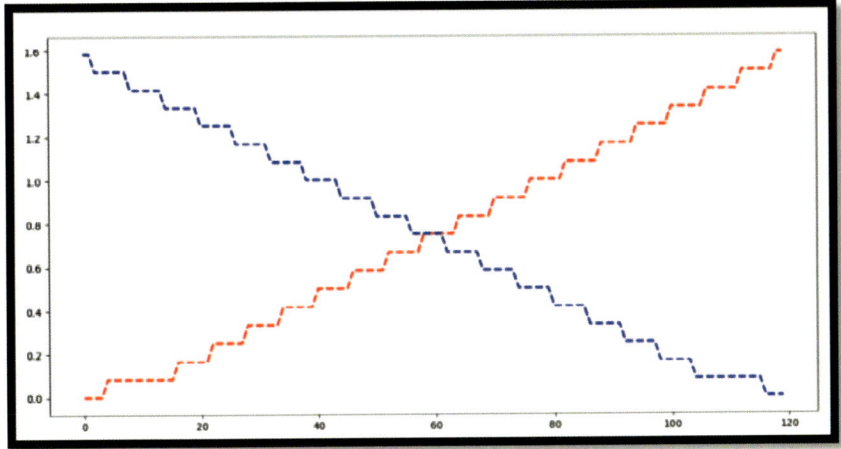

Fig. 3 Representation of FAR, FRR and CER

CER—Crossover Error Rate The CER is the representation of FAR and FRR curve, and it is the measurements of FAR and FRR intersection point.

FER—Failure to Enrol Rate The proportion of individuals who were unsuccessful to complete the registration. The reason for failure enrolment may be because of not familiar to the system, illumination (light) and surrounding environment.

Speed The time the biometric application device and verification/validation algorithm takes to enrol a sample and the time taken to verify/validate the individual subject.

Number of Templates In a general application, the highest quantity of patterns that can be stored is a concern. Most of the biometric tools are microcontroller-based systems where internal memory is limited. Therefore, the particulars of enrolment should also be limited.

Below is the performance calculation of biometric system with respect to FAR, FRR and CER. FAR and FRR values can be differentiated to maintain some threshold limit. If the value is below the threshold limit, then it is called FRR. If the threshold limit exceeds the imposters, then it is called FAR. A graph is plotted against FRR and FAR and the transaction point is determined as CER. Here both the FAR and FRR values are equal. The graphical representation of FRR and FAR with CER cross section point is shown. The red line represents FRR and the blue one represents FAR curve (Fig. 3) and Table 1 shows the extracted contour features.

Here, we compared three ear image database and calculated the performance measures (FAR, FRR, CER, speed). The databases are IIT Delhi, AMI and VR Students Sample databases with a total of 793, 699, and 100 ear images. The average time of execution is 0.33 s, and the average rate of FAR, FFR and CER is 0.8, 0.5 and 0.4%, respectively. These values are shown in Table 2.

Table 1 Extracted contour feature database

Image index	Aspect ratio	Perimeter	Area 1	k	x	y	w	h	Solidity	Extent	Equi_Diameter
0	1	5.6568542	2	True	17	6	3	3	0.5	0.22222222	1.595769122
1	2	11.656854	8	True	2	2	6	3	2	0.4444444	3.19158243
2	1	5.6568542	2	True	16	7	3	3	0.5	0.22222222	1.595769122
3	1	12.485281	10	False	16	6	5	5	2.5	0.4	3.568248232
4	1	5.6568542	2	True	13	6	3	3	0.5	0.22222222	1.595769122
5	1	11.899495	8.5	True	13	6	5	5	2.125	0.34	3.289762321
6	0.85	102.91169	177	False	1	1	29	34	44.25	0.17951319	15.01210843
7	2.8	40.142135	16	False	1	1	14	5	4	0.22857143	4.513516668
8	1	3.4142135	0.5	True	4	2	2	2	0.125	0.125	0.797884561
9	1.25	11.071068	8.5	True	6	4	5	4	2.125	0.425	3.289762321
10	1	5.6568542	2	True	28	14	3	3	0.5	0.22222222	1.595769122
11	0.75	7.6568542	4	True	29	15	3	4	1	0.33333333	2.256758334
12	1	5.6568542	2	True	28	14	3	3	0.5	0.22222222	1.595769122
13	0.75	7.6568542	4	True	38	32	3	4	1	0.33333333	2.256758334
14	1.85	33.213203	16.5	False	1	9	13	7	4.125	0.18131868	4.583497844
15	1.4	16.485281	9	False	1	8	7	5	2.25	0.25714286	3.385137501
16	2.5	33.899495	28.5	False	1	7	15	6	7.125	0.31666667	6.023896333
17	2	27.899495	18.5	False	1	8	12	6	4.625	0.25694444	4.85334231
18	2	2	0	False	9	15	2	1	0	0	0
19	1	6.2426406	1.5	True	9	13	3	3	0.375	0.16666667	1.381976598
20	1	6.2426406	1.5	True	9	13	3	3	0.375	0.16666667	1.381976598
21	0.5	2	0	False	18	88	1	2	0	0	0
22	1.66	9.6568542	1	False	14	88	5	3	0.25	0.06666667	1.128379167
23	1	6.8284271	2	False	12	89	3	3	0.5	0.22222222	1.595769122
24	2.2	50.627417	105	False	2	23	22	10	26.25	0.47727273	11.56244577
25	1	5.6568542	2	True	1	94	3	3	0.5	0.22222222	1.595769122

Table 2 Performance comparison of the three ear databases

Database name	# images in DB	Time of execution	FRR%	FAR%	CER%
IIT Delhi DB	793	0.339900	0.5	1.039	0.37
AMI DB	699	0.03500	0.5	1	0.455
VR Sample DB	100	0.05999	0.56	0.5	0.4

5 Conclusion

This work mainly focused on the contours to extract the features, such as bounding rectangle, aspect ratio, extent, equivalent diameter, checking the convexity, solidity, etc., to extract the contour properties of an ear image, which can be used to authenticate the individual on identity management systems. Here, IIT Delhi, AMI and VR Sample Ear Databases consisting of 793, 699 and 100 images, respectively, are used for experimentation. The experimental results show that this approach gives better results when compared to the traditional methods. This algorithm takes less amount of time to execute, and FAR, FRR are very low when compared to previous traditional mechanisms discussed in literature survey [12–17].

References

1. Attarchi S, Faez K, Rafiei A (2008) A New Segmentation Approach for Ear Recognition. In: Blanc-Talon J, Bourennane S, Philips W, Popescu D, Scheunders P. (eds) Advanced Concepts for Intelligent Vision Systems. ACIVS 2008. Lecture Notes in Computer Science, vol 5259. Springer, Berlin, Heidelberg
2. El-Bakry HM, Mastorakis N (2009) Ear recognition by using neural networks. In: Proceedings of the 11 th International Conference on Mathematical methods and computational techniques in Electrical engineering (pp 770–804)
3. Omara I, Li F, Zhang H, Zuo W (2016) A novel geometric feature extraction method for ear recognition. Expert Syst Appl 65:127–135
4. Kumar PR, Dhenakaran SS (2017) Structural (Shape) Feature Extraction for Ear Biometric System. In: Lobiyal D, Mohapatra D, Nagar A, Sahoo M (eds) Proceedings of the International Conference on Signal, Networks, Computing, and Systems. Lecture Notes in Electrical Engineering, vol 395. Springer, New Delhi
5. Yan P, Bowyer KW (2007) Biometric recognition using 3D ear shape IEEE Transactions on pattern analysis and machine intelligence 29(8):1297–1308
6. Kumar VN, Srinivasan B (2012) Ear biometrics in human identification system. Int J Inf Technol Comput Sci 4:41–47
7. Yuan L, Mu Z, & Xu Z (2005) Using ear biometrics for personal recognition. In: Advances in Biometric Person Authentication, Springer, Berlin, Heidelberg pp 221–228
8. Marti-Puig P, Rodríguez S, De Paz JF, Reig-Bolaño R, Rubio MP, & Bajo J (2012). Stereo video surveillance multi-agent system: new solutions for human motion analysis. Journal of Mathematical Imaging and Vision 42(2–3):176–195
9. Hurley DJ, Nixon MS, Carter JN (2000) Automatic ear recognition by force field transformations. In: IEE colloquium on vision biometrics (Ref. No. 2000/018). IET

10. Contour properties and features available in Opencv: http://docs.opencv.org/3.2.0/d3/d05/tutorial_py_table_of_contents_contours.html

11. Performance of biometrics: http://www.biometric-solutions.com/performance-of-biometrics.html

12. Pflug A, & Busch C (2012) Ear biometrics: a survey of detection, feature extraction and recognition methods. IET biometrics 1(2):114–129

13. Abaza A, Ross A, Hebert C, Harrison, MAF, Nixon MS (2013) A survey on ear biometrics. ACM computing surveys (CSUR), 45(2):22

14. Castrillón-Santana M, Lorenzo-Navarro J, Hernández-Sosa D (2011) An study on ear detection and its applications to face detection. In Conference of the Spanish Association for Artificial Intelligence, Springer, Berlin, Heidelberg pp 313–322

15. Lammi H-K (2004) Ear biometrics. Department of Information Technology, Lappeenranta University of Technology, Laboratory Information Processing, Lappeenranta, Finland

16. Choras M (2007). Image feature extraction methods for ear biometrics--a survey. In: 6th International Conference on Computer Information Systems and Industrial Management Applications (CISIM'07) IEEE. pp 261–265

17. Hurley DJ, Arbab-Zavar B, Nixon MS (2007) The ear as a biometric. In: Jain A, Flynn P, Ross A (eds) Handbook of biometrics, Chapter 7, Springer US, pp 131–150

Defocus Map-Based Segmentation of Automotive Vehicles

Senthil Kumar Thangavel, Nirmala Rajendran and Karthikeyan Vaiapury

Abstract Defocus estimation plays a vital role in segmentation and computer vision applications. Most of the existing work uses defocus map for segmentation, matting, decolorization and salient region detection. In this paper, we propose to use both defocus map and grabcut using wavelet for reliable segmentation of the image. The result shows the comparative analysis between the bi-orthogonal and Haar function using wavelet, grabcut and defocus map. Experimental results show promising results, and hence, this algorithm can be used to obtain the defocus map of the scene.

1 Introduction

Defocus estimation plays a vital role in computer vision applications including depth estimation, image quality assessment, blurring the image and image refocusing, etc. [1]. It can also be used for segmentation, matting, decolorization and salient region detection, depth estimation [2–5]. Practically, extracting a depth from a single image is very difficult because to estimate the depth of a point local features alone are not sufficient, global features should also be considered [6] defocus map provides some clues for depth estimation [7] (Fig. 1).

S. K. Thangavel (✉)
Department of Computer Science and Engineering, Amrita School of Engineering, Amrita
Vishwa Vidyapeetham, Coimbatore, India
e-mail: t_senthilkumar@cb.amrita.edu

N. Rajendran · K. Vaiapury
TCS Research and Innovation, Tata Consultancy Services, Chennai, India
e-mail: rajendran.nirmala@tcs.com

K. Vaiapury
e-mail: karthikeyan.vaiapury@tcs.com

© Springer Nature Switzerland AG 2019
D. Pandian et al. (eds.), *Proceedings of the International Conference on ISMAC
in Computational Vision and Bio-Engineering 2018 (ISMAC-CVB)*, Lecture Notes
in Computational Vision and Biomechanics 30,
https://doi.org/10.1007/978-3-030-00665-5_144

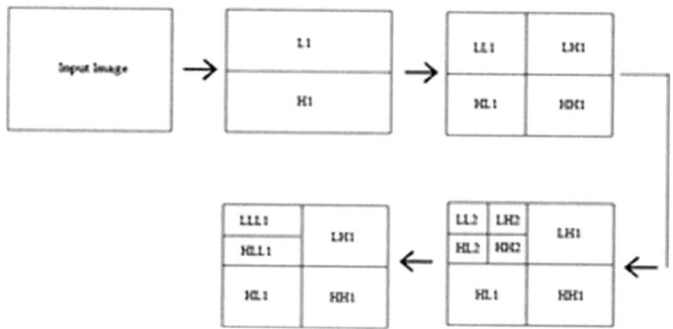

Fig. 1 Two levels of DWT decomposition

2D discrete wavelet transform (DWT) decomposition of the image can be represented [8, 9] as

$$C = X \, I \, Y \tag{1}$$

where C is the wavelet coefficients final matrix, I represents the input image and X and Y are the row filter matrix and column filter matrix, respectively. When DWT is applied on an image decomposition of the image takes place which decomposes the image into four parts in first level, namely, approximation coefficient consists of low-low frequencies, horizontal coefficient consists of low-high frequencies, vertical coefficient consists of high-low frequencies and diagonal coefficient consists of high-high frequencies. The approximation coefficient computed in the first level is used for the next level of decomposition.

To reconstruct the image, inverse discrete wavelet transform (IDWT) is used which can be represented as

$$I = X^{-1} \, C \, Y^{-1} \tag{2}$$

for the matrices which are orthogonal Eq. (2) can be modified [8] as

$$I = X^T \, C \, Y^T \tag{3}$$

There are different types of wavelet families, namely, Haar wavelet, bio-orthogonal wavelet, Daubechies wavelet and coiflets wavelets [9].

Although considerable work has been done by the research community on defocus map, we propose to use both defocus map and grabcut using wavelet for segmenting the image. We also provide a comparative analysis between the bi-orthogonal and Haar function using wavelet, grabcut and defocus map.

In this work, we have used both defocus map and grabcut using wavelet for segmenting the image. The wavelet families used in these works are Haar and bi-orthogonal function.

2 Related Works

Defocus map is used to highlight the most important subpart of the object. There are various approaches that are used for defocus estimation like active illumination method (AIM), coded aperture method (CAM), interpolation method (IM), deconvolution, etc. [10, 11]. The defocus map can also be obtained from multiple images, but the major problems that can be faced are occlusions.

Although considerable work has been done by the research community on defocus map, we propose to use both defocus map and grabcut using wavelet for segmenting the image. We also provide a comparative analysis between the bi-orthogonal and Haar function using wavelet, grabcut and defocus map.

In [7], Zhuo et al. proposed a method for finding the depth of a single image. The blur estimation method using Gaussian gradient ratio (GGR) is used to compute the depth. Initially, the defocus blur at the edges in an input image is calculated using step function. Then, the point spread function which is also known as Gaussian kernel function is convoluted with the edge pixels. The output obtained with the isotropic Gaussian kernel is re-blurred. GGR is computed between the original blur image and the re-blur image to get the maximum edge location. This results in the sparse defocus map. Due to the inaccurate blur at the edge locations, joint bilateral filtering applied to the sparse defocus map to improve the efficiency. The sparse defocus map is applied in the entire image to get the depth map; this can be achieved by using defocus blur interpolation. Matting Laplacian is used for defocus blur interpolation and the depth map is obtained. This approach is robust to noise, inaccurate blur and the adjacent neighbour edges. The limitation of this approach is that it cannot provide any information, whether the blur at the edges are obtained due to blur or defocus in the input image.

In [2], Jiang et al. proposed an algorithm for salient region detection, which uses three cues such as uniqueness, focus and objectness. In their work, the focus/defocus estimation is done by using scale-space analysis; objectness estimation is computed using window overlapping-based approach. The uniqueness is estimated using pixel level and regional level uniqueness. In [4], Namboodiri et al. proposed a method to estimate the depth from a single image using reverse heat equation and depth from defocus (DFD) using graph cuts. In their proposed method, experiments are performed on various types of images, and the results show that relative layers of depth can be estimated.

In [8], Petrova et al. proposed a method for detecting the edges in medical images using wavelets. In their proposed work they have done detailed experiments on different types of edge detection methods that can be done using wavelets like modification of approximation replaced by zero, modification of approximation using edge detectors and wavelet transform modulus maxima method. In [9], Brannock et al. proposed a method for detecting the edges using a modification of approximation replaced by zero. In their work, the compared the different types of wavelet families like Haar wavelet, bi-orthogonal wavelet, Daubechies wavelet and coiflets wavelets for edge detection on noisy images.

Grabcut segmentation is used to segment the 2D image. By drawing a rectangle in an image, the region of interest is marked. The user needs to segment the image as foreground and background regions with respect to the region of interest.

Gaussian mixture models are used to modal the foreground and the background. Grabcut is the combination of statistical modal and graph cuts of the background and foreground region.

The paper is organized as follows: In the first section, we have introduced defocus map and discrete wavelet transforms. Section 2 provides the related works. In Sect. 3, the proposed solution and description are given. Section 4 provides the defocus map estimation, and Sect. 5 gives a detail description of the results and conclusion. Finally, the conclusion and future work are given in Sect. 6.

3   Proposed Solution

The given input image is converted to monochrome image which consists of black and white image. Discrete wavelet transform is computed on the monochrome image which decomposes the image into four parts such as approximation coefficient, horizontal coefficient, vertical coefficient and diagonal coefficient. The approximation coefficient is alone used for further processing. This work explains various types of segmentation using grabcut and approximation coefficients. The experiments are done for two wavelet families, namely, Haar and bi-orthogonal function.

A. **Modification of approximation replaced by zero**

Approximation coefficient obtained by DWT is replaced with zero, which removes all the low-level frequencies and the results obtained is superimposed with the monochrome image and the edges are extracted (Figs. 2 and 3).

B. **Modification of approximation replaced by zero using grabcut segmentation**

Grabcut segmentation is used to segment the monochrome image as foreground and background regions. The foreground region consists of objects of interest and DWT is computed on the resultant output. The approximation coefficient obtained is replaced with the zeros, and the edges are extracted by superimposing zero approximation coefficient with the monochrome image.

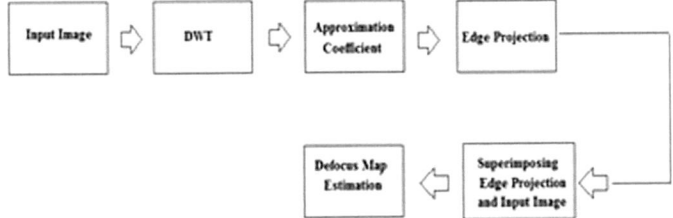

Fig. 2 Block diagram for defocus map using modification of approximation replaced by zero

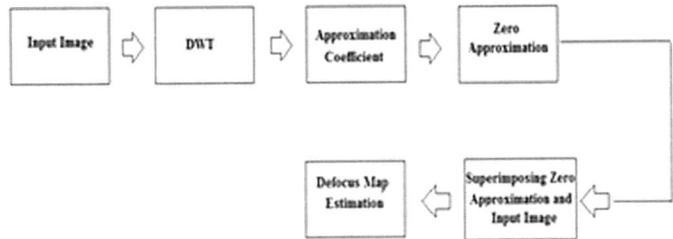

Fig. 3 Block diagram for defocus map using modification of approximation by edge projection

C. Modification of approximation by edge detection

By using approximation coefficient obtained by DWT, the edge pixels are detected using edge detectors. Then, the resultant image is superimposed with the monochrome image.

D. Modification of approximation by projection using grabcut segmentation

The foreground and the background regions are segmented using grabcut segmentation to get the region of interest. DWT is computed on the foreground region and the approximation coefficient is obtained. The edge pixel is detected on the approximation coefficient. The resultant image is superimposed with the monochrome image.

4 Defocus Map Estimation

This approach has been divided into two phases (Fig. 4):

1. Sparse defocus blur estimation and
2. Defocus blur propagation.

Initially, the gradient-based method is used to compute the defocus blur on the edges. Then, the sparse defocus map is propagated to the whole image. In sparse defocus map when the input image $I(m, n)$ is given, the edges are extracted using edge detection algorithms. Then, the defocus blur is calculated by assuming the edge

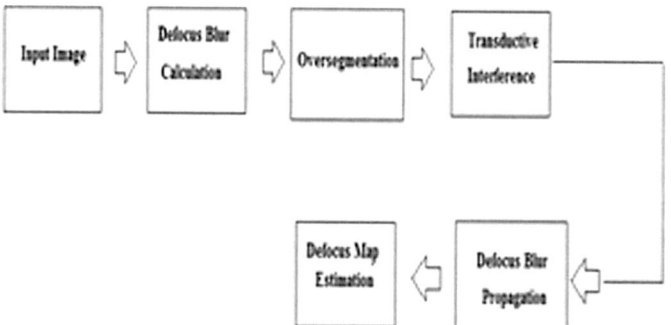

Fig. 4 Block diagram for defocus map estimation

function as $x(p) = af(p) + b$, where a and b are the amplitude and offset of the edge for image $I(m, n)$. $x(.)$ indicates the step function and p be the pixel location of the image $I(m, n)$. The edge function $x(p)$ and Gaussian kernel $G_k(p, s)$ are convoluted to obtain the defocus blur and s is the unknown standard deviation and the degree of blurness present on the edges. The defocus blur can be written as

$$D_f(x(p)) = x(p) * G_k(p, s) \tag{4}$$

Another Gaussian kernel is used to re-blur the

$$\nabla\left(D_f(x(p)) * G_k(p, s_1)\right) = \nabla x(p) * G_k(p, s) * G_k(p, s_1)$$

$$= \frac{a}{\sqrt{2\pi\left(s^2 + s_1^2\right)}} \exp -\left(\frac{p}{s^2 + s_1^2}\right) \tag{5}$$

where s_1 is the standard deviation of re-blur Gaussian kernel. Gradient magnitude ratio (G_M) between the original blur and the re-blur gives the maximum value at the edges, which is denoted as

$$G_M = \frac{D_f(x(p))}{(D_f(x(p)) * G_{k1}(p, s_1))} \tag{6}$$

$$G_M = \sqrt{\frac{s^2 + s_1^2}{s^2}} \tag{7}$$

$$s = \frac{s_1}{\sqrt{G_m^2 - 1}} \tag{8}$$

using gradient magnitude ratio the unknown standard deviation are calculated. The over-segmentation is used to find the super-pixels in the input image which can be used to find the similarity between the adjacent pixels. Simple linear iterative clustering (SLIC) is used to find the super-pixels in the given input image using k means clustering [6]. The super-pixels can be represented as $M = \{M_1, M_2 \ldots M_i\}$. In the super-pixels set M, the weighted is added to the graph $W_{cg} = (M, X, \alpha)$ where α varies from [0, 1], M is the super-pixel set and X is the edge set which contains the all two adjacent super-pixel. The weight matrix can be defined as

$$W = \left[\alpha_{ij}\right]_{n \times n} \tag{9}$$

Using over-segmentation, similarity between the adjacent pixels alone can be calculated. To calculate the similarity between any super-pixels which is not just limited to the adjacent pixel, transductive matrix is used. This is obtained by using the affine matrix which can be written as

$$T_i = (D - \varphi\alpha)^{-1} I$$
$$\varphi = (0, 1) \tag{10}$$

The initial blur function P_{mj} for each super-pixel is can be represented as

$$P_{mj} = \underset{p \in X_j}{\mathrm{med}} \, T(P) \tag{11}$$

The defocus blur propagation using affine information can be written as

$$\widetilde{P_{mj}} = \widetilde{T_i}[P_{m1}, P_{m2} \ldots P_{mn}] \tag{12}$$

5   Results and Discussion

Figures 5, 6, 7, 8, 9, 10, 11 and 12 represent the outputs obtained using edge pixel projection using bi-orthogonal and Haar wavelet functions. The comparison has been made by using bi-orthogonal and Haar functions using discrete wavelet transform with or without grabcut segmentation. Figure 5a represents the results for superimposed modification of approximation replaced by zero with original image by using bi-orthogonal function, and Fig. 5b represents the defocus map obtained for Fig. 5a; the edges are extracted but are not much clear, and the outliers are also extracted.

Fig. 5 a Superimposed
original image and zero
approximation for
bi-orthogonal wavelet
function. **b** Defocus map for
(**a**)

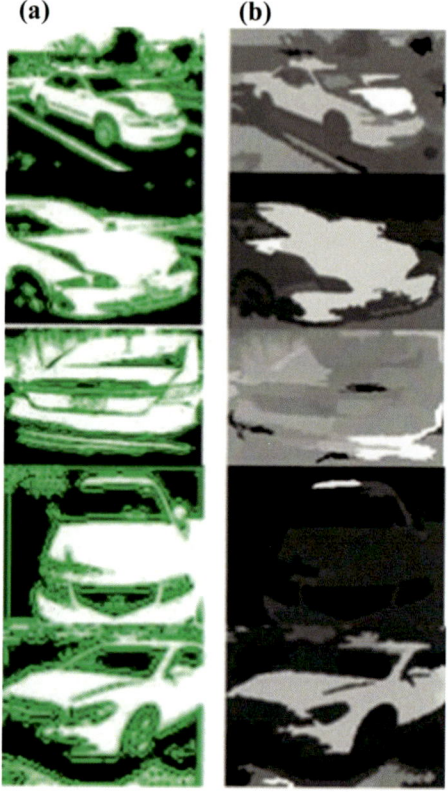

(a)   **(b)**

Figure 6a shows the results by using superimposed modification for the approximation by zero and original image with grabcut segmentation for using bi-orthogonal function from which the edges are extracted. Figure 6b shows the defocus map results for Fig. 6a which gives the better results compared to modification of approximation method by zeros using bi-orthogonal wavelet function.

Figure 7a shows the results for superimposed modification of approximation replaced by edge projection with original image using bi-orthogonal wavelet function and Fig. 7a defocus map results obtained for Fig. 7a.

In Fig. 8a, the results obtained for superimposed modification of approximation using edge with original image and grabcut segmentation using bi-orthogonal wavelet function is provided. Figure 8b shows the results for defocus map of Fig. 8a. Comparing the results obtained using bi-orthogonal function, Fig. 8 gives better segmentation result since outliers are reduced in a significant manner. In Fig. 9a, the results for superimposed modification of approximation replaced by zero with original image by using Haar wavelet function are provided. Figure 9b represents the defocus map obtained for Fig. 9a. The edges are extracted and the defocus map obtained is not clear. Figure 10a represents the results for superimposed modification of approxima-

Fig. 6 a Superimposing of original image and zero approximation using grab cut for bi-orthogonal wavelet function. **b** Defocus map for (**a**)

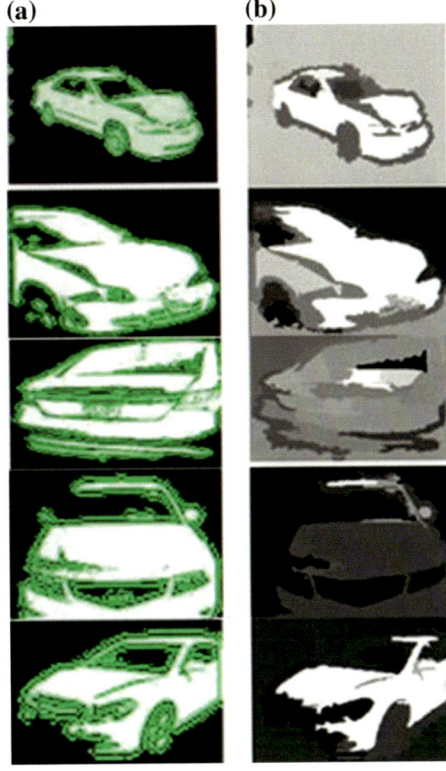

(a)   **(b)**

tion replaced by zero with original image and grabcut segmentation by using Haar wavelet function and Fig. 10b represents the defocus map obtained for Fig. 10a. The edges are extracted and defocus map obtained is better compared to modification of approximation replaced by zero with original image using Haar wavelet function.

Figure 11a represents the results for superimposed modification of approximation by edge projection with the original image by using Haar wavelet function and Fig. 11b represents the defocus map obtained for Fig. 11a.

In Fig. 12a, the results for superimposed modification of approximation replaced by edge projection with original image and grabcut segmentation by using Haar wavelet function is provided. Figure 12b represents the defocus map obtained for Fig. 12a. The edges are extracted and the defocus map obtained is better compared to modification of approximation replaced by zero with original image using Haar wavelet function. By comparing the results obtained using both bi-orthogonal wavelet function and Haar wavelet function, Haar wavelet using DWT in Fig. 12 gives better results.

(a)   **(b)**

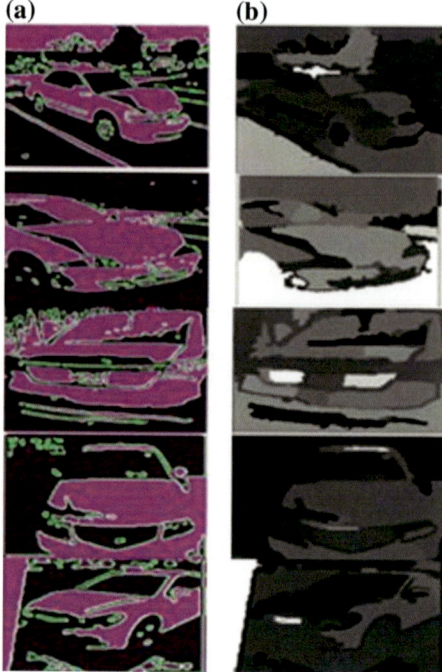

Fig. 7 **a** Superimposing of original image and edge projection for bi-orthogonal wavelet function.
b Defocus map for (**a**)

Fig. 8 **a** Superimposing of original image and edge projection using grabcut for bi-orthogonal wavelet function. **b** Defocus map for (**a**)

(a)   **(b)**

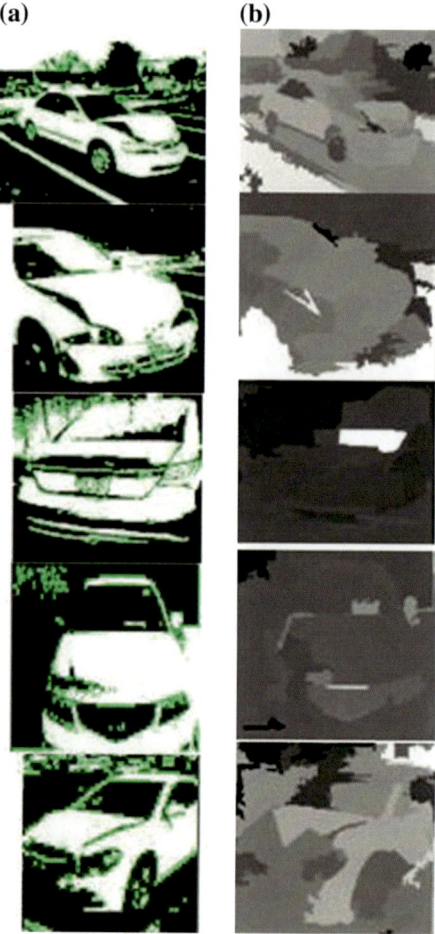

Fig. 9 **a** Superimposing of original image and zero approximation for Haar wavelet function. **b** Defocus map for (**a**)

(a) **(b)**

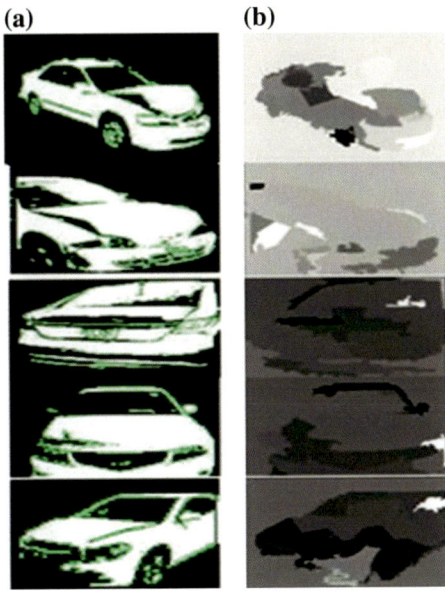

Fig. 10 **a** Superimposing of original image and zero approximation using grabcut for Haar wavelet function. **b** Defocus map for (**a**)

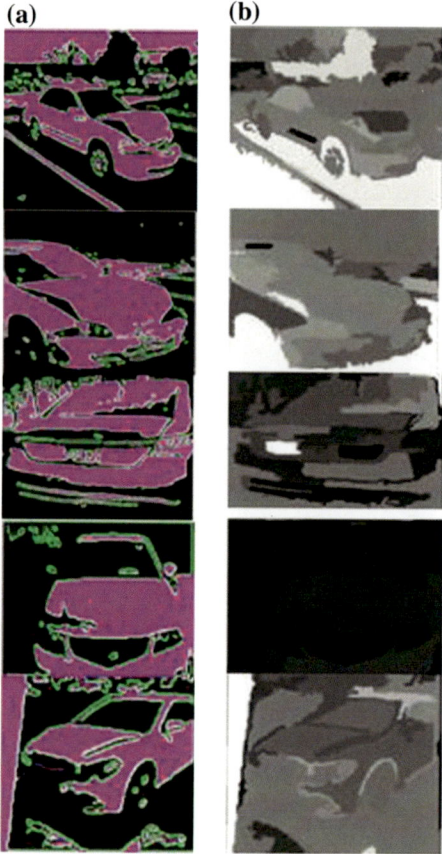

Fig. 11 **a** Superimposing of original image and edge projection for Haar wavelet function. **b** Defocus map for (**a**)

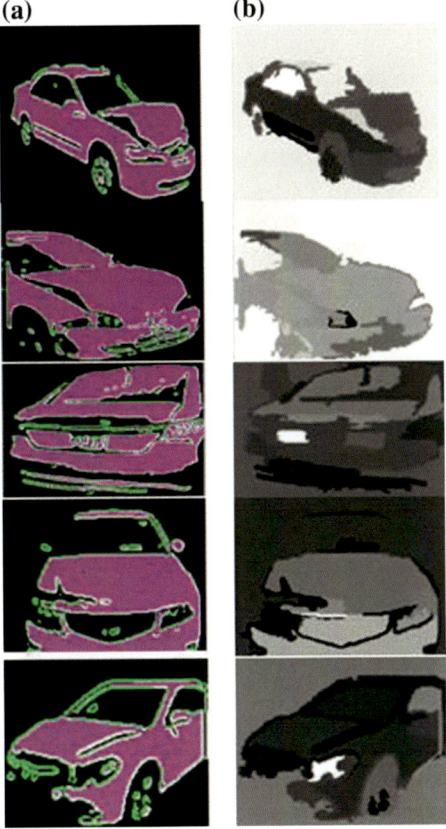

Fig. 12 **a** Superimposing of original image and edge projection using grabcut for Haar wavelet function. **b** Defocus map for (**a**)

6 Conclusion and Future Work

In this work, we have proposed a solution to use both defocus map and grabcut using wavelet for segmenting the image. The comparative analyses have been made between the bi-orthogonal and Haar function using wavelet, grabcut and defocus map.

Acknowledgements This work is performed as a part of internship with TCS Innovation Lab.

References

1. Chen D-J, Chen H-T, Chang L-W (2016) Fast defocus map estimation. In: 2016 IEEE international conference on image processing (ICIP). IEEE
2. Jiang P, Ling H, Yu J, Peng J (2013) Salient region detection by UFO: uniqueness, focusness and objectness. In: IEEE international conference on computer vision, ICCV 2013, Sydney, Australia, 1–8 Dec 2013, pp 1976–1983
3. Lin HT, Tai Y, Brown MS (2011) Motion regularization for matting motion blurred objects. IEEE Trans Pattern Anal Mach Intell 33(11):2329–2336
4. Namboodiri VP, Chaudhuri S (2008) Recovery of relative depth from a single observation using an uncalibrated (real-aperture) camera. In: 2008 IEEE computer society conference on computer vision and pattern recognition (CVPR 2008), 24–26 June 2008, Anchorage, Alaska, USA
5. Peng Y, Zhao X, Cosman PC (2015) Single underwater image enhancement using depth estimation based on blurriness. In: 2015 IEEE international conference on image processing, ICIP 2015, Quebec City, QC, Canada, 27–30 Sept 2015, pp 4952–4956
6. Ashutosh S, Chung SH, Ng AY (2005) Learning depth from single monocular images. NIPS 18:1–8
7. Zhuo S, Sim T (2011) Defocus map estimation from a single image. Pattern Recogn 44(9):1852–1858
8. Petrova J, Hostalkova E (2011) Edge detection in medical image using the wavelet transform. Report of research, Department of Computing and Control Engineering, Czech Public
9. Brannock, E, Weeks M (2006) Edge detection using wavelets. In: Proceedings of the 44th annual Southeast regional conference, pp 649–654. ACM
10. Moreno-Noguer F, Belhumeur PN, Nayar SK (2007) Active refocusing of images and videos. ACM Trans Graph 26(3):67–75
11. Levin A, Fergus R, Durand F, Freeman WT (2007) Image and depth from a conventional camera with a coded aperture. ACM Trans Graph 26(3):70–78

Multi-insight Monocular Vision System Using a Refractive Projection Model

J. Mohamed Asharudeen and Senthil Kumar Thangavel

Abstract The depth information of a scene, imaged from the inside of a patient's body, is a difficult task using a monocular vision system. A multi-perception vision system idea has been proposed as a solution in this work. The vision system of the camera has been altered with the refractive projection model. The developed lens model recognises the scene with multiple perceptions. The motion parallax is observed under the different lenses for the single shot, captured through the monocular vision system. The presence of multiple lenses refracts the light in the scene at the different angles. Eventually, the appearance of the object dimension is augmented with more spatial cues that help in capturing 3D information in a single shot. The affine transformations between the lenses have been estimated to calibrate the multi-insight monocular vision system. The geometrical model of the refractive projection is proposed. The multi-insight lens plays a significant role in spatial user interaction.

1 Introduction

A multi-insight vision is built to observe the world with a single camera system. The stereo vision system is used to observe the shape, size and localization of the target from two or more images taken at different views. The proposed work aims to retrieve the target information with the single camera system. It handles distinct perceptions with a single shot. The new lens model is constructed with five distinct lenses. The parent lens of the model holds the normal view. Its neighbourhood lens holds the information hidden in the parent lens. The designed model of the lens is termed as augmenting the refraction (ATR lens). From here, augmenting the refraction lens has

J. Mohamed Asharudeen · S. K. Thangavel (✉)
Department of Computer Science and Engineering, Amrita School of Engineering, Amrita
Vishwa Vidyapeetham, Coimbatore, India
e-mail: t_senthilkumar@cb.amrita.edu

J. Mohamed Asharudeen
e-mail: cb.en.p2cvi16002@cb.students.amrita.edu

© Springer Nature Switzerland AG 2019 1553
D. Pandian et al. (eds.), *Proceedings of the International Conference on ISMAC
in Computational Vision and Bio-Engineering 2018 (ISMAC-CVB)*, Lecture Notes
in Computational Vision and Biomechanics 30,
https://doi.org/10.1007/978-3-030-00665-5_145

been termed as ATR lens. The motion parallax is the key to the proposed lens model observed in the neighbourhood lenses.

The proposed method takes advantage of the refraction provided by the lens. The image captured using the ATR lens holds the refracted ray of the neighbourhood lenses.

The depth estimation can be done using visual cues from the images. The popular algorithm used for recovering the depth information such as stereopsis, structure from motion uses multiple images. The features extracted from a single image are not sufficient enough to gather the depth information. Global features, as well as local features, are considered for extracting the depth information [1]. The multi-insight image that contains visual cues of the same scene at the different angles gives the required depth information. The acquisition of spatial cues with ATR is more prominent. The existing work perceives the monocular cues from a single image. The proposed work acquires the visual cues from a single image. Here, the image is a multi-insight image and thus holds more spatial cues information. As an initial prototype lens model, it is made bigger in size.

This paper is arranged as follows: the background works made in calibration and a few works about the projection model are discussed. The consecutive section holds the proposed ATR lens model and then the extended pinhole model of the camera. The geometric model of the light rays traced in the ATR lens is interpreted. The estimation of the target using the proposed geometry model is observed in the later section.

2 Related Works

Cui et al. [2] presented a perspective projection model for the stereovision system. It considers the prism as a single lens. It establishes an affine transformations relation between the image points and the object points. Beardsley and Murray [3] presented the calibration method by making use of the vanishing line and vanishing point information.

Zhang [4] proposed a flexible new technique for calibrating a camera. This requires a planar pattern. The results are better without using any kind of orthogonal plates or the distinct world points. Maximum likelihood estimation is used for the refinement. Teixeira et al. [5] implemented the algorithm proposed by [4]. This gives the understanding of Zhang's method. It estimates the intrinsic, extrinsic and radial distortion parameters. Burger [6] gives a detailed report about the steps involved in Zhang's calibration method. Cipolla et al. [7] presented a method for recovering the projection matrices from uncalibrated images to reconstruct the 3D models. The geometrical intuitive method has been interpreted using parallelism and orthogonality of the architectural scenes.

Tsai [8] introduced a new 3D camera calibration technique for machine vision technology with the advantage in terms of accuracy, speed rather than the existing state of the art. Jurjevic and Gasparovic [9] presented a two-step calibration technique and 3D point cloud estimation with open source technologies. This gives the general introduction of epipolar geometry. The distortion coefficients are also discussed in detail. Heikkila and Silven [10] presented a simple technique of calibrating a camera with four steps. The linear estimation and nonlinear estimation of projection matrix are also discussed.

Barath et al. [11] developed a new method of estimating the focal length and the fundamental matrix with the help of two affine correspondences from the uncalibrated cameras. Meng and Hu [12] presented a new calibration pattern for calibrating a camera. The calibration technique is closely related to the existing Zhang's method. Feng et al. [13] presented a new transparent calibration board for calibrating cameras at multiple angles. They introduced two more new parameters such as depth and refractive index of the glass. Still, it uses the traditional calibration method. Drennan [14] gives a general idea of the camera calibration algorithm. The calibration of a camera using vanishing points and vanishing lines are also discussed in [15–18]. Michels et al. [1] used a hierarchical multiscale Markov random field to recover the depth information. The depth model is estimated using global and local features present in a single image. The visual cues are discussed in comparison with the binocular cues.

3 Proposed Works

In this section, the ATR model is elaborated. The refractive projection model has been constructed using five distinct lenses. The lenses are arranged in a fashion as shown in Fig. 1. These lenses perceive the environment at different angles as shown in Fig. 2 to form the images. These images are responsible for observing the motion parallax, it is termed as the multi-insight image. A single camera vision system is responsible for the novel vision.

Fig. 1 Lens arrangement

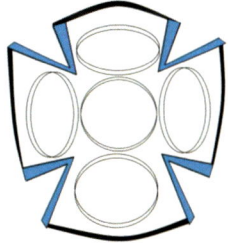

Fig. 2 ATR lens prototype

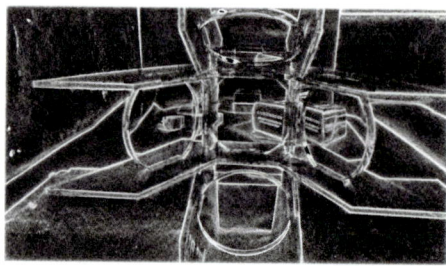

Fig. 3 Pinhole projection model

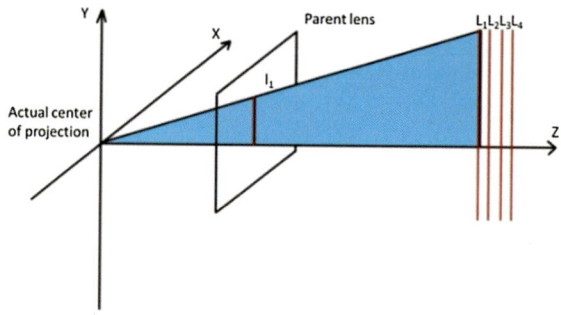

3.1 Camera Model

Image formation made by the ATR lens is described in this section. The image formation of a pinhole camera model shown in Fig. 3 is altered for the ATR lens model. The projection of a world point in space onto the image plane follows the similar triangle principle [19–21]. The centre of projection [22] in the Euclidean coordinate system is represented as COP. Let the world point be X and its image point be x. A new virtual centre of projection (VCOP) is introduced shown in Fig. 4 from which the refracted rays meet the actual COP. The VCOP is assumed for calibrating the individual lens. The motion parallax plays a significant role in ATR lens. Objects that are nearer moves faster than the faraway object which has been distinguished with its distinct lens. The occluded surfaces from the parent lens are less obscure in the neighbourhood lens.

Every individual lens holds the similar triangle property for mapping the world coordinates $(X, Y, Z)^T$ to the image coordinates. The proposed system claims that the camera coordinate frame has been replaced with lens coordinate frame as the system involves more than one lens. The neighbourhood lens still follows the similar triangle rule as parent lens. The VCOP of the neighbourhood is shifted to the actual centre of projection.

The mapping for the parent lens is from the Euclidean space R^3 to R^2. The mapping for the neighbourhood lens is also from the Euclidean space R^3 to R^2, and then from R^2 to R^2.

Fig. 4 Extended pinhole model of ATR lens

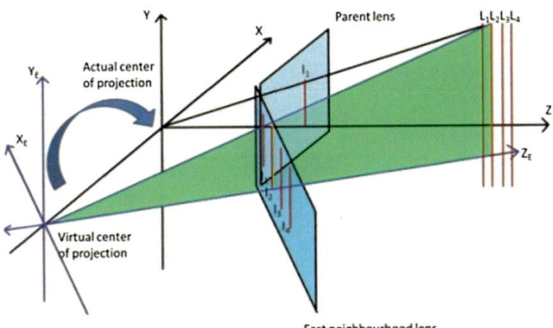

Fig. 5 Virtual centre of projection shifted to the actual centre of projection

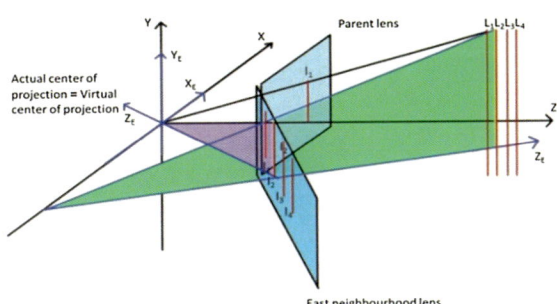

The linear mapping between the homogeneous coordinates of the world coordinate system and the image coordinate system is represented as in Eq. (1). Let x_a be the homogeneous vector of the image coordinates that correspond to the actual centre of projection, and X be the homogeneous vector of the world coordinates. Let P be the camera projection matrix.

$$x_a = PX \tag{1}$$

The neighbourhood lens follows the linear mapping similar to the parent lens between the homogeneous coordinates of the world coordinate system and the lens coordinate system as in Eq. (2), and then, it follows Eq. (3) for the linear mapping between the homogeneous coordinates of the lens coordinate system, and the image coordinate system corresponds to the actual centre of projection. Let x_L be the homogeneous vector of the lens coordinates. Let P_R be the refractive projection matrix. The refractive projection matrix represents the shifting of the neighbourhood coordinates from the VCOP to the actual centre of projection with a rotation and a translation vectors.

$$x_L = PX \tag{2}$$

$$x_a = P_R x_L \tag{3}$$

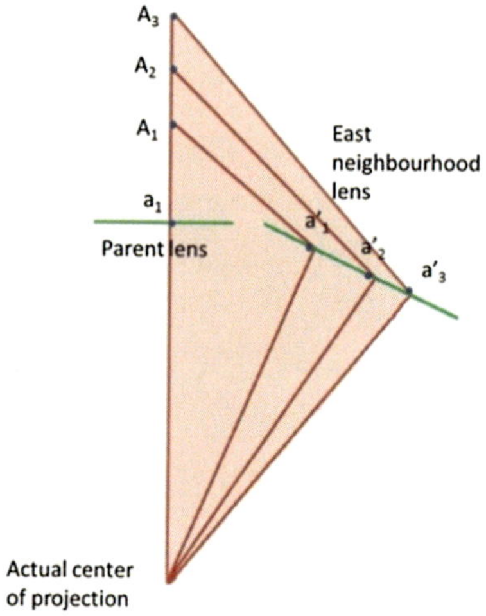

Fig. 6 Ray converges to the actual centre of projection in ATR model

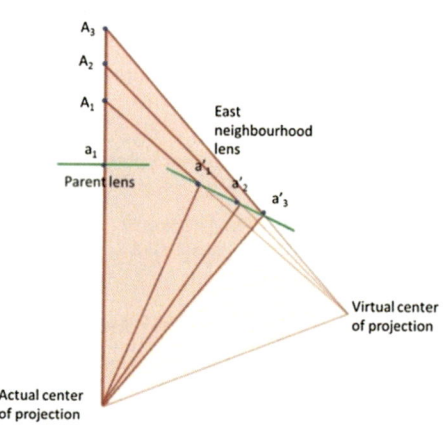

Fig. 7 Ray visualised to converge at the virtual centre of projection

The ray perception for the ATR lens is shown in Fig. 6. The parent lens sets the neighbourhood lens as an observer to retrieve the occluded information. The horizontal section of the prototype is examined instead of the entire set-up. Three lenses are placed in the set-up. The refracted rays from the neighbourhood lenses are observed in the multi-insight image. Let's consider target $A3$ is occluded by $A2$ and is occluded by $A1$. These targets are observed through neighbourhood lens. VCOP of the right neighbourhood lens is shown in Fig. 7. The VCOP is shifted to the actual centre of projection. Once it is shifted, the ray observation is traced as shown in Fig. 5.

The entire ray observation of the horizontal section is shown in Fig. 8. The occluded targets are observed by the neighbourhood lenses.

The affine transformation obeys the Thales theorem as shown in Figs. 9 and 10. The normal to the neighbourhood lens meets at a point k. The distance between the point k and the actual centre of projection of parent lens is maintained between its neighbourhood lenses to their virtual centre of projection.

According to Thales theorem, point k, centre of the lens and corresponding projection centre follow the theorem as in Eq. (4). Thus, the orientation can be estimated easily and is fixed.

$$\triangle\, ADE \text{ where } BC \| DE$$

$$\frac{AB}{BD} = \frac{AC}{CE} \tag{4}$$

The intrinsic parameter estimated from the parent lens is utilised as the intrinsic parameter for the neighbourhood lens, as the calibration parameter estimated depends on the single camera [23, 24]. The affine transformation matrix is constructed with rotation only along the y-axis for the horizontal section and along the x-axis for the vertical section of the lens. Translation is from the VCOP and the actual centre of projection. The ATR lens set-up works to find out the occluded parts in the scene with multiple perceptions. The estimation of the target follows the proposed occluded triangles relations. The model for target estimation is shown in Fig. 11. The vertex of the rays viewed as a series of the triangle with a common vertex and a common base. It has the other two vertexes that are collinear. The occluded triangles are constructed as right-angled triangles with every vertex of all triangles are collinear, and they are having neither common base nor common vertex.

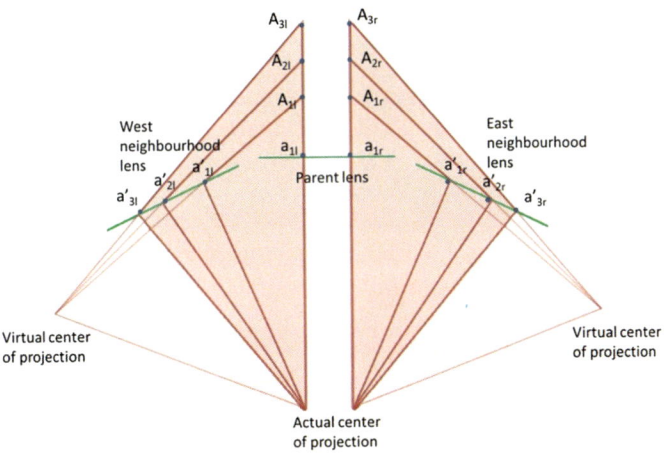

Fig. 8 Ray convergences of the entire ATR model

Fig. 9 Actual centre of projection

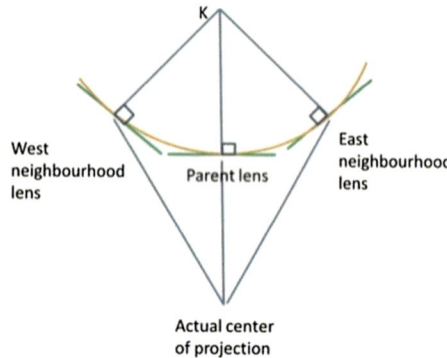

Fig. 10 Estimation of virtual centre of projection

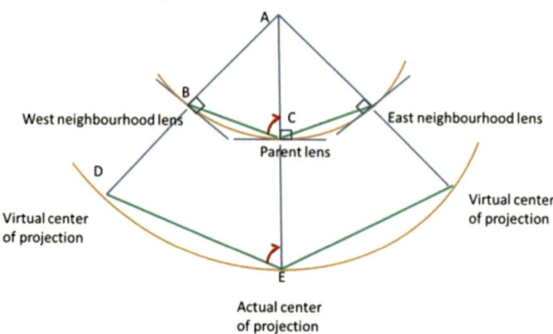

Fig. 11 Target estimation using geometrical model

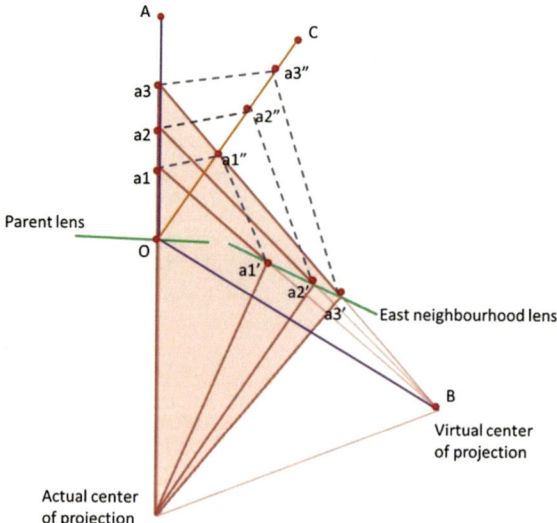

(a)   **(b)**

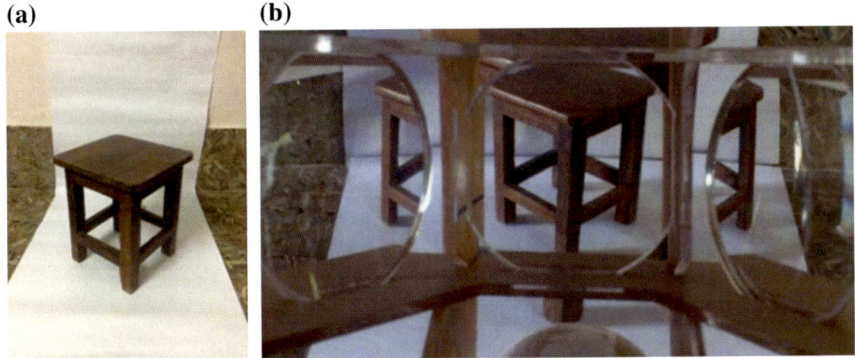

Fig. 12 **a** Image captured in a camera and **b** multi-insight image

4 Results

The ATR lens is developed with an intention to be utilised in medical imaging vision system. As it is a prototype model and needed lens calibration, this model is developed in bigger size for making the calibration easier to explore the multi-insight images. The image captured through the lens model consists of simple targets. These targets are traced using feature detectors. The image captured using a single camera is shown in Fig. 12a. The dimension of the target captured with different perceptions in multi-insight image is shown in Fig. 12b. The comparison of the image is shown in Fig. 12 to understand the visual cues are observed only at this single image about the target. The visual cues of the target are observed with different perceptions in this multi-insight image.

The image captured within the lens region gets segmented and is shown in Fig. 13a–c from west neighbourhood lens, from parent lens and east neighbourhood lens, respectively.

These segmented images are shown in Fig. 14 and have the feature mapping between them [25–28]. From Fig. 14, it is evident that the depth information from the multi-insight image can be estimated once the correspondence matching has been made. The calibration of the lens is made using Zhang's calibration method.

5 Conclusion

The refractive projective model has been modelled as an extended pinhole model. The monocular vision system developed with this model handles distinct perceptions when imaged from the inside of the patient's body. The multi-insight image obtained with this ATR lens holds more cue information about the scene. The motion parallax plays a significant role in ATR lens. The visual cues obtained from multi-insight

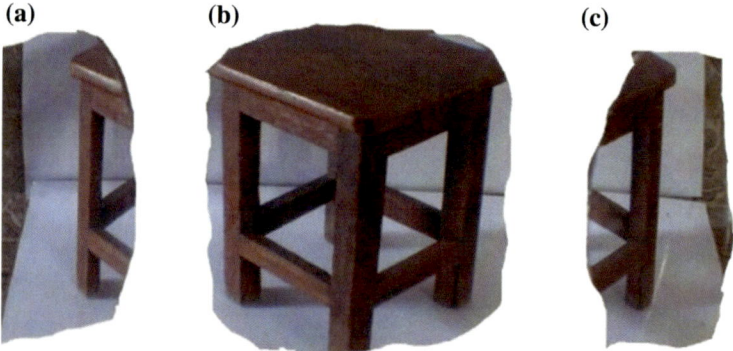

Fig. 13 Image from the lens is segmented from the multi-insight image: **a** segmented from west neighbourhood lens, **b** segmented from parent lens and **c** segmented from east neighbourhood lens

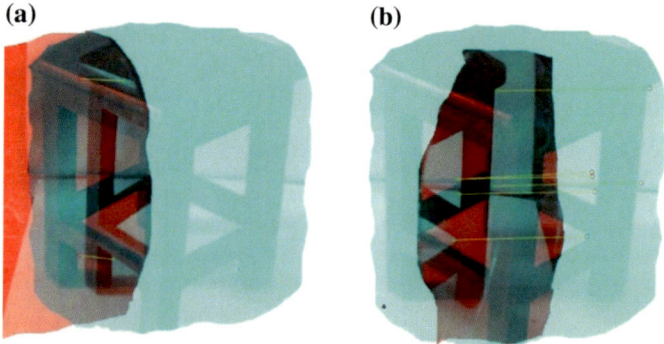

Fig. 14 Features matched with the parent lens image and the neighbourhood lens image: **a** mapping between parent lens and west neighbourhood lens image, and **b** mapping between parent lens and east neighbourhood lens image

images can be utilised for constructing the depth information. The multi-insight lens plays a significant role in spatial user interaction.

References

1. Michels J, Saxena A, Ng AY (2007) 3D depth reconstruction from a single still image. Int J Comput Vis
2. Cui X, Zhao Y, Lim K, Wu T (2015) Perspective projection model for prism based stereo vision. Opt Soc Am 23(21)
3. Beardsley P, Murray D (1992) Camera calibration using vanishing points. University of Oxford, Oxford
4. Zhang Z (1998) A flexible new technique for camera calibration. Microsoft research-technical report-98-71

5. Teixeira L, Gattass M, Fernandez M (2006) Zhang's camera calibration: step by step
6. Burger W (2016) Zhang's camera calibration algorithm: in-depth tutorial and implementation. Technical report-HGB16-05
7. Cipolla R, Drummond T, Robertson D (1999) Camera calibration from vanishing points in images of architectural scenes. University of Cambridge, BMVC99
8. Tsai RY (1987) A versatile camera calibration technique for high accuracy 3D machine vision metrology using off the shelf TV cameras and lenses. IEEE J Robot Autom RA-3(4)
9. Jurjevic L, Gasparovic M (2017) 3D data acquisition based on OpenCV for close range photogrammetry applications. Int Arch Photogram Remote Sens Spat Inf Sci XLII-1/W1
10. Heikkila J, Silven O (1997) A four step camera calibration procedure with implicit image correction. University of Oulu
11. Barath D, Toth T, Hajder L (2017) A minimal solution for two view focal length estimation using two affine correspondences. Machine Perception Research Laboratory
12. Meng X, Hu Z (2003) A new easy camera calibration technique based on circular points. J Pattern Recogn Soc: 115–1164
13. Feng H, Li Q, Feng M (2017) A novel multi camera calibration method using transparent glass calibration board. In: Proceedings of 2nd international conference on advances in materials machinery, Electronics I. American Institute of Physics
14. Drennan M (2010) An implementation of camera calibration algorithms. Clemson University
15. Orghidan R, Salvi J, Gordan M, Orza B (2012) Camera calibration using two or three vanishing points. In: IEEE proceedings of the federated conference on computer science and information systems, pp 123–130
16. Wang L-L, Tsai W-H (1990) Computing camera parameters using vanishing line information from a rectangular parallelepiped. Mach Vis Appl 129–141
17. Guillou E, Meneveaux D, Maisel E, Bouatouch K (2000) Using vanishing points for camera calibration and coarse 3D reconstruction from a single image. Vis Comput 396–410
18. Caprile B, Torre V (1990) Using vanishing points for camera calibration. Int J Comput Vis 4:127–140
19. Hartley R, Zisserman A (2005) Multiple view geometry in computer vision. Cambridge University Press, Cambridge
20. Beardsley P, Murray D, Zisserman (1992) A Camera calibration using multiple images. University of Oxford, Oxford
21. Feng M, Jia X, Wang J, Feng S, Zheng T (2017) Global calibration of multi camera based on refractive projection and ray tracing. Sensors 17:2494
22. Hartley RI, Kaucic R (2002) Sensitivity of calibration to principal point position. In: European conference on computer vision. Springer, Berlin
23. Rothwell CA, Zisserman A, Marinos CI, Forsyth DA, Mundy JL (1992) Relative motion and pose from arbitrary plane curves. University of Oxford, Oxford
24. Hansen P, Alismail H, Rander P, Browning B (2012) Online continuous stereo extrinsic parameter estimation. CVPR
25. Sreelakshmi S, Vijai A, Senthil Kumar T (2016) Detection and segmentation of cluttered objects from texture cluttered scene. In: Proceedings of the international conference on soft computing systems
26. Senthil Kumar T, Vijai A (2014) 3D reconstruction of face: a comparison of marching cube and improved marching cube algorithm. Int J Adv Image Process Tech 1:6–9
27. Nirmala R, Thangavel SK (2017) Develop, implement and evaluate a multimodal system for government organisation. In: Proceedings of 4th international conference on advanced computing and communication systems, ICACCS 2017, pp 111–117
28. Kavin D, Thangavel SK (2017) Improving capabilities for visually challenged person in library environment. In: Proceedings of 4th international conference on advanced computing and communication systems, ICACCS 2017, pp 1–10

ONESTOP: A Tool for Performing Generic Operations with Visual Support

Gowtham Ganesan, Subikshaa Senthilkumar and Senthil Kumar Thangavel

Abstract Programming has become tedious for every person these days. Learning programming languages and writing a computer program for different tasks using various programming languages is a difficult and time-consuming task. Therefore, modules are used to make programming easier and faster. Cloud computing enables applications to be accessed everywhere. The 'ONESTOP' tool will be provided as a facility to the users under the category 'Software as a Service'. The paper provides directions for enabling the same facility. It does not address the challenges for provisioning this tool on the cloud. Every module in ONESTOP consists of the operations under that category. The tool processes the input by removing fillers, identifying the operation to be performed using trie data structure and synonym mapping and displaying the result. User need not write codes or define functions. A simple sentence in English is sufficient to perform the task. The tool is easy to use and does not require any programming knowledge to use it. All the operations are performed in less time enhancing the performance of the tool. Key aspect of ONESTOP is that it does not produce any error and saves debugging time.

1 Introduction

In recent times, people assume that programming knowledge is the only need for building an application. But, for a proper structuring of a program, one has to plan the steps required to solve the problem. Then, structuring the program is important. To achieve this, modules are used. It helps programmers reuse modules thereby making

G. Ganesan · S. Senthilkumar · S. K. Thangavel (✉)
Department of Computer Science and Engineering,
Amrita School of Engineering, Amrita Vishwa Vidyapeetham, Coimbatore, India
e-mail: t_senthilkumar@cb.amrita.edu

G. Ganesan
e-mail: cb.en.u4cse14417@cb.students.amrita.edu

S. Senthilkumar
e-mail: cb.en.u4cse14451@cb.students.amrita.edu

© Springer Nature Switzerland AG 2019   1565
D. Pandian et al. (eds.), *Proceedings of the International Conference on ISMAC in Computational Vision and Bio-Engineering 2018 (ISMAC-CVB)*, Lecture Notes in Computational Vision and Biomechanics 30,
https://doi.org/10.1007/978-3-030-00665-5_146

the task easier. And, novice programmers will find it easy to use a module or function rather than writing it again.

Fidge [1] has proved in his experiment that the common programming mistakes were caused because of lack of experience in programming in a particular language, not because of the way of thinking programmatically. Hence, the time spent on learning the syntax of a computer language can instead be effectively spent to develop an algorithm or to focus more on other important aspects. ONESTOP is mainly developed to encourage people to explore and create new algorithms rather than concentrating on learning programming languages. The purpose of the tool is to provide the result of a given pseudocode so that users may spend time doing useful work. This will help in enhancing the problem-solving skills of people.

Students are exposed to various operations in a small period of time. They tend to forget it or learn its result without proper understanding. Practical knowledge always helps us remember things for a longer time than compared to theoretical knowledge. ONESTOP provides a platform for students to perform operations and view the results so that they will always remember it. Unique feature of this tool is that users need not enter the operations to be performed in the form of a code. User has to specify a query in English to perform an operation and the tool will provide the result. By doing so, it saves time and user will get a visual support of how the result will look like. And, user can select operations from the list of operations in the tool and enter input values instead of typing the whole input query.

Tokenization is an important method in identifying keywords from a sentence. It plays a major role in understanding the actual meaning of a sentence by breaking it into tokens. After splitting a sentence into tokens, tokens are analysed about their significance in the sentence. ONESTOP uses tokenization and other techniques to generate the result. The tool is developed in C++, thereby ensuring less execution time. Qt Creator is used to maintain cross-platform consistency. ONESTOP is designed meticulously that there is no possibility for an error to occur.

The paper is organized as follows: Section 2 gives the reason why the tool has been developed. In Sect. 3, various works are analysed and their limitations are discussed. Section 4 describes how the tool works. Pseudocode, screenshots of the result, comparison of similar tools and user experiments are discussed in Sect. 5. Section 6 comprises future work and Sect. 7 contains conclusion.

2 Motivation

People believe that programming skills are required to do any task. This forces people to learn various programming languages related to their task or field. They will also have to remember the syntax of every language. This can be time-consuming. Many tools have been developed in recent days which aim to reduce the need to remember syntax of various languages. ONESTOP is developed to perform operations without the need to code. Even a person without programming knowledge will be able to use this tool at ease and perform the necessary operation. ONESTOP aims to prove that programming knowledge is not always necessary to perform an operation. Even a basic sentence can be used to perform a task. Hence, people can focus on learning new algorithms or explore new topics instead of spending time on learning programming languages.

3 Related Work

Charntaweekhun and Wangsiripitak [2] proposed a system for visual programming using flowchart. In their system, when a flowchart is drawn, it is compiled and run and the output is displayed on the system environment. The main aim of this project is that novice programmers do not have to write codes. They can draw many flowcharts, link them all and when the run button is pressed, a source code in C is generated and EXE file is invoked to display the output. Similarly, ONESTOP aims to remove the burden of writing codes faced by programmers. Programmers can give commands in a form of sentence in English instead of using flowcharts to get the output. Also, giving commands in English would be much easier than drawing flowcharts.

Sarda and Jain [3] have eliminated the need to know query language to retrieve data from database. And user need not know the database schema. The system accepts any form of query which is given as a set of words, and then the words are translated into database queries with the help of metadata. The result is then provided to the user. Disadvantage of this tool is that it does not understand the purpose and usage of quotation marks in queries.

Little and Miller [4] understand the difficulty of programming using syntax. So they came up with an idea of syntax-free programming wherein a parser would generate code for the given input in the form of English words. ONESTOP provides the result and description instead of generating codes. This is because novice programmers would have a better understanding of the input provided when results are provided with description rather than providing codes.

Smith et al. [5] were concerned that all the emails could not be read or categorized since large number of emails sent by Anne Hunter, Course VI administrator are going around. So the team developed a project to categorize emails and to generate keywords and key phrases for every mail so that students will know whether the email is important or not. In this project, Tynan and others have used regular expression pattern for classification from email texts. ONESTOP also used regex patterns for classifying if input belongs to arithmetic or complex number type.

Fudaba et al. [6] have developed a tool called Pseudogen. As the name suggests, the tool generates pseudocode in English or Japanese from source code written in Python. This tool was developed since people had difficulty in understanding source codes that are written in programming languages which are not well known. People can further make use of ONESTOP to know the result and working of instructions which are provided in English.

Tahir and Ahmed [7] proposed a Tree-Combined Trie or TC-Trie which is a combination of trees and tries for fast IP address lookup. ONESTOP also uses trie data structure for prefix matching to identify the operations to be performed under a category. This trie-based search operation happens after analysing the input query.

Senthil and Ohhm [8] developed a queueing model for e-learning system. Since the tool is stored in cloud, if many people are trying to access the tool at the same time, this queueing model can be used. Sankar and others [9] conducted a survey on different machine learning algorithms. Naive Bayes algorithm was discussed and its insights are used in user experiment of this tool.

4 Methodology

In this section, the working of the tool is discussed. Thereby, the step-by-step processing of the tool will be understood.

The tool is stored in cloud so that many people can access it. There are two types of cloud users who can access the tool. (i) Guest users—only few operations like arithmetic and trigonometric operations under number operations category will be available to these users. (ii) Privileged users—these users should log in using username and password. They have to pay to use the tool. All the operations present in the tool such as the operations under number, string, matrix, complex number category can be used by privileged users.

Figure 1 explains the block diagram of the tool. Initially, the user provides a sentence in English. The sentence can be used to perform arithmetic, relational, complex, matrix, geometric and trigonometric operations. The sentence or set of words is subjected to two refining mechanisms to minimize the sentence length and aids in reducing its processing complexity. The first mechanism of input sentence reduction is to remove the articles. It is removed as it is unnecessary and does not provide any useful meaning to the sentence or for further processing. The second mechanism is to remove white spaces in the sentence. This is done as the tokens are stored in dictionary without spaces between words. White spaces are also removed in order to eliminate storing of similar tokens in the dictionary. The given sentence is converted to lower case since all the tokens are stored in lower case in the dictionary. Dictionary refers to the words used in the tool. If some characters are in lower case and some in upper case, searching those tokens in dictionary will be difficult as all possible combinations of lower and upper case words should be stored in dictionary. So, for easy searching of tokens, all the words in the given input are converted to lower case. While removing white spaces and converting the sentence to lower case, there is an exception to do these processes. The exception is when single or double quotes are used in the query. The data within the quotes are not modified and kept as such.

After processing the input, the tool checks if the input is of arithmetic or complex pattern. The pattern is a regular expression. When the input follows a definite format, regular expression is the ideal one to be used. If operations to be performed are given as a word (e.g. Add), then we find what operation it has to perform by using trie-based search. In case the operation to be performed is given in the form

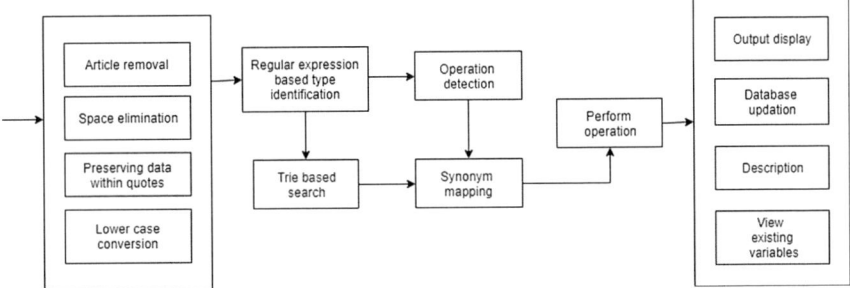

Fig. 1 Block diagram

of a symbol (e.g. +), we need not identify what operation should be performed by searching the trie. Instead, the operation will be found out with the help of regular expression patterns. In this tool, we have used two regular expression patterns: (i) [0–9] * [\+\−][1–9][0–9] * i is the pattern for complex numbers and (ii) ([0–9]|[\+\−\%\(\)]) + is the pattern used to identify arithmetic operations. In case the user specifies the wrong input, the compiler will throw an error and user can specify the right input.

If the processed input matches the arithmetic regular expression pattern, the tool identifies which arithmetic operation has to be performed and implements it. On the other hand, when the input matches the complex number pattern, tool knows that it has to perform operations in accordance to complex number category. But the tool will not know what operation should be performed (i.e.) if it has to perform addition or subtraction, etc. To help this out, ONESTOP uses trie-based search to find out the operation.

Trie belongs to tree data structure. Trie is used for searching the string efficiently. Every node in trie has index value and Boolean variable. Boolean variable is used to indicate that the right word is identified by traversing the trie until end of the word. To make the string searching efficient, trie data structure is used in this tool. By using trie, complexity of string searching has been reduced from $O(n^3)$ to $O(n^2)$. This thereby enhances the computational efficiency of the tool. When contents present in the trie is large, it gives rise to higher space complexity. This is a disadvantage of trie. In the proposed tool, we have overcome this disadvantage by inserting the possible operations of a particular category (string, arithmetic, complex, matrix, etc.) into the trie. We repeat this process of inserting the possible operations into the trie for all categories and searching is performed simultaneously. After searching the trie, we will know the categories (e.g. String, arithmetic) in which a particular operation (e.g. add) can be performed. Based on the values given in the input (for instance, either string values or numbers) to perform an operation, the tool will know which category (addition of strings in case of string and basic addition in case of numbers) will be appropriate since operations will be performed for values of same data type. If the values do not belong to a uniform data type, the tool generates an alert box and confirms with the user on basis of what data type, the operation should be performed.

Fig. 2 Trie data structure

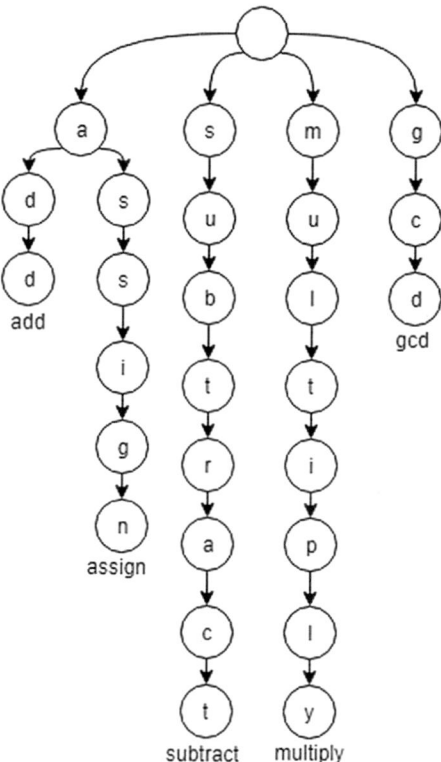

To perform search operation in trie, the tool traverses the query first. The possible operations in a particular category (say Matrix) are inserted into the trie. The tool checks if the first character in sentence matches the character in the root node of trie. If they match, next character in the input is checked with the child of the root node in trie and this process repeats for every character in the input which is checked with the subsequent child nodes until the Boolean variable in a node is encountered, search operation stops for that category and returns false. The tool then searches all other categories. In the end, we will get an array of possible categories (e.g. matrix, arithmetic, complex) which contains the operation (e.g. add) found from the input.

Figure 2 shows the example of trie data structure for the operations under the matrix category. They include add, subtract, assign, multiply and gcd. The root node is null. Trie data structure is based on prefix matching. Since prefix matching is used, common prefix of two or more words is stored only once in trie. Add, assign operations in Fig. 2 explain this. To search for multiply in the above trie, the first letter (*m*) is checked. Then, the consecutive letters are checked. On checking the last letter (*y*) in the word (multiply), the word is found and true is returned.

Map data structure is used in the proposed tool to associate similar tokens with the main token. Main tokens are assigned by the tool developer. For instance, sum, total, aggregate (similar tokens with same meaning) are mapped to add (main token). Operations are performed based on main tokens only. If suppose similar tokens are not mapped to the main token, every similar token has to be assigned the same operation individually. This will require a lot of memory space and will lead to unnecessary checking of all similar tokens. Since mapping is done in this tool, all similar tokens are collectively mapped to one main token. By doing so, the operation identification process is done in an efficient manner. Expanded words are also mapped to their corresponding ACRONYM. For instance, Total Surface Area is mapped to TSA.

The operation that has to be performed is found so far. From this, the tool will find out the minimum number of operands required to execute the operation by classifying them as unary, binary and ternary operators. Performing the operation using the operands from the input sentence comes under three ways (i) the tool will extract the values given in the input query, (ii) if previously existing variables are used in the input, the tool will check the database for the value of that variable and retrieve it and (iii) if the number of values or previously existing variables specified in the input sentence do not meet the minimum number of required values to perform the operation, prompt box will be generated and user will be asked to enter the values. From the values obtained, operations are computed, and the result is displayed to the user.

When the result is provided to the user, the result and the variables used for computation are stored in the database. The textual description provides details about how the variables are obtained. The textual description also includes how the result was obtained. Along with the result, the variables which contain the values used to perform the operation are also displayed. Thereby, user will know the value present in every variable which is used to perform the operation. User can make of these previously existing variables later by mentioning them in the input.

Another important feature of this tool is that user need not specify the operation in form of a sentence or set of words. There are inbuilt options in the tool and the user just has to select what operation he/she wants to perform. After selecting the operation to be done, user can enter the values in the prompt box or select the previously existing variables. This aims to make the tool user-friendly and reduce the time taken to enter input sentences.

5   Discussion

Pseudocode

Function: Search (input, array of words)

- Insert all the array elements received as parameter into trie data structure.
  - For i = 0 to length(input):
  (i) Search the input sub-string starting from 'i' in trie
  (ii) If match is found in trie, break the loop
    - Return the found item
(Time Complexity: O (n^2)
Space complexity: O (domain size being searched))

Steps:

1. Enter the input sentence.
2. Pre-processing the input
   - Initialise an array 'articles' containing 'a', 'an', 'the'.
- Remove the words present in the article array from input sentence.
- Eliminate spaces between words in the input(only for words outside single and double quotes)
- Convert input to lower case(only for characters outside single and double quotes)
3. If previously existing variables are used in input,

then

data types of those variables are found from the   database containing information about variables, data types and values.
4. Check the input if it matches regular expression pattern
   - if input is of complex operation pattern,

   then

   assign operation type as 'complex'.
      - else if input is of arithmetic pattern,

   then

   assign operation type as 'arithmetic'
   and perform the operation.
   - Else if the names of categories of operations are specified in the input,

then

Search (input, category names) can be invoked to identify the category by passing the category names as parameter to the trie.
5. If category of operations are identified,

then

- Assign all the operations available in that category to a temporary variable.
  - Invoke Search (input, temporary variable).
    - Obtain the returned item.
6. Else if category of operations is not found,

then

- Initialize an array 'numtypes' with four values 'number', 'string', 'complex' and 'matrix'
  - For i = 0 to length(numtypes):
  i. Assign all the operations available for numtypes[i] to a temporary variable
  ii. Invoke Search (input, temporary variable).
  iii. Add the returned operations of (ii) to an array.
    - If values required are specified in input,

then

Analyse the input values to determine the category (as string, number or complex) from (iii).
   - If category cannot be found from input value,

then

user will be asked to select the desired category from (iii).
   - Assign the operation found to 'operation' variable.
- A map variable 'otherwords' contains key value pairs where root words are given as keys.
  - Map words with same meaning to root word.
- A map variable 'operands' where keys are operation names and values are minimum number of operands required.
  - Analyse input sentence in three steps to get input values to perform the operation
  i. append the values mentioned in the input to 'input' array
  ii. values of existing variables specified in the input are added to 'input' array.
  iii. if length('input' array) < operands[operation] :
    then

    obtain the values from user.

    - Operation is performed and result is stored.

5.1 Result

Figure 3 shows input given by user, result along with description and the variables—input1, input2 and result1 are stored in the database. In this case, to calculate the total surface area of a cone, the required values are provided in the input query itself. Here, the word total surface area is mapped to TSA and value of TSA of a cone is calculated.

Figure 4 shows how the TSA of a cone is calculated using previously existing variable. Result1 contains the value got by performing the TSA calculation from input values given in Fig. 3. The value of slant height is provided by the user as 5, TSA of cone is calculated and result is provided.

In case the user just specifies what operation should be formed and does not mention any value or previously existing variable in the input query, the tool generates a prompt asking for the necessary values to perform the operation. In Fig. 5, to find TSA of a cone, user has not specified any value. So, the tool will generate a prompt box as in Fig. 6 asking for the required input (radius). Another prompt box will be generated by the tool asking the user to provide value of slant height. Based on these values, TSA of a cone is computed and result is displayed.

To perform operations using matrices, if the user provides an input query as given in Fig. 7, prompt box pops up requesting the number of matrices to be multiplied (Fig. 8). Then two more prompt boxes will pop up asking for the number of rows and columns for each matrix

User then has to enter values for each matrix as in Fig. 9 and the result, variables used and description will be displayed to the user.

If the user wants to perform complex number operations such as addition of two complex numbers as in Fig. 10, given input is checked if it matches the regular expression pattern for complex numbers and 'add' is found out to be the operation to be formed as a result of trie-based search. After this process, the real and imaginary parts of two complex numbers are added and result is displayed.

The tool is developed in a user-friendly manner wherein the user has to select what operation he/she wants to perform under a category (Fig. 11) and then enter the required values. This will reduce the time taken to enter the query by the user and make the work easy.

5.2 Similar Tools

The proposed tool has all the features which are limitations of the tools mentioned in Table 1. In ONESTOP, input sentence can be provided for complex logic also. Instead of drawing complex flowcharts as in Flowgorithm, which are difficult to understand, sentences can be specified in a sequential manner for easy understanding. Raptor enforces the user to follow predefined structure whereas ONESTOP does not expect the user to follow any structure to specify the input. To use ONESTOP, user does

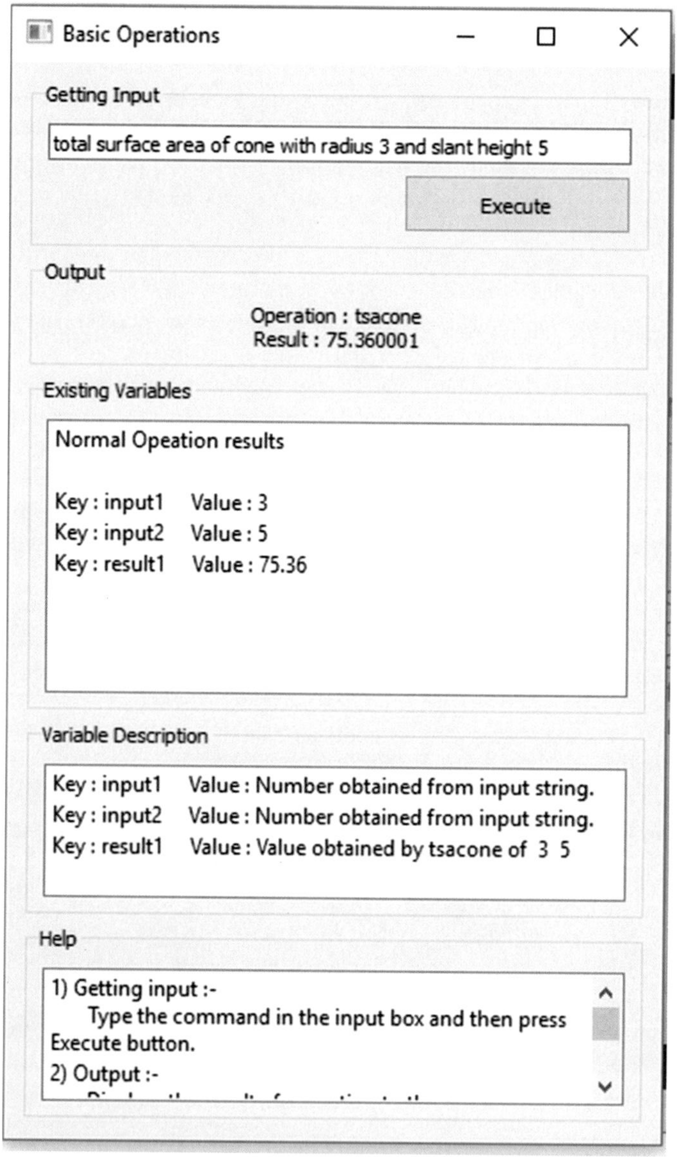

Fig. 3 Total surface area performed with all values specified

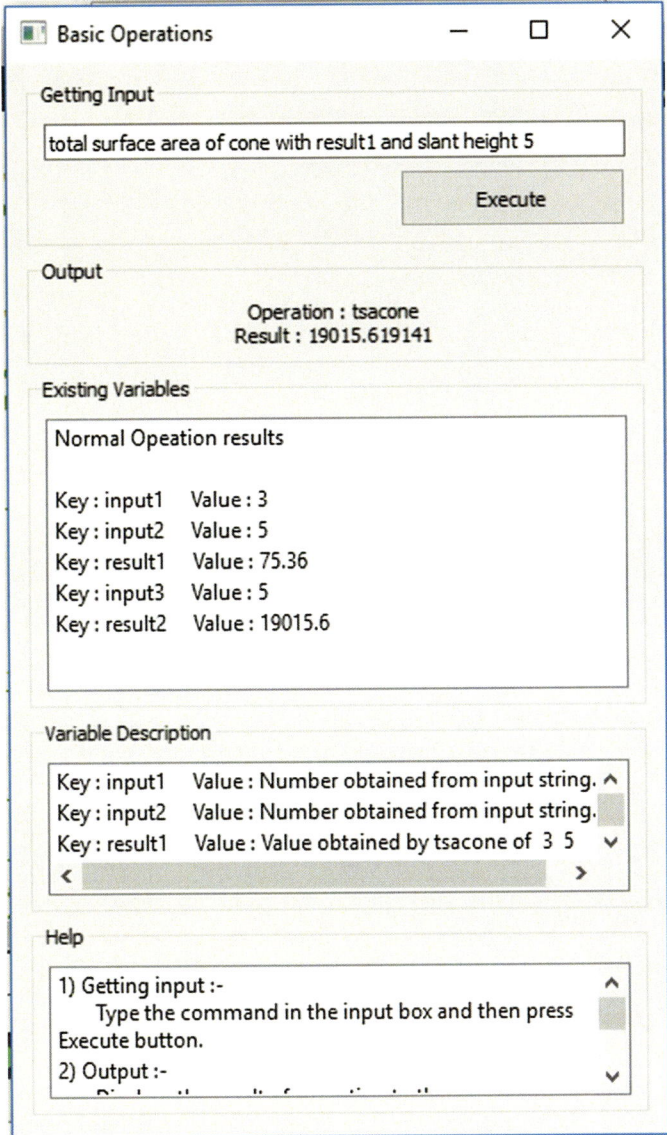

Fig. 4 Total surface area performed using existing variable

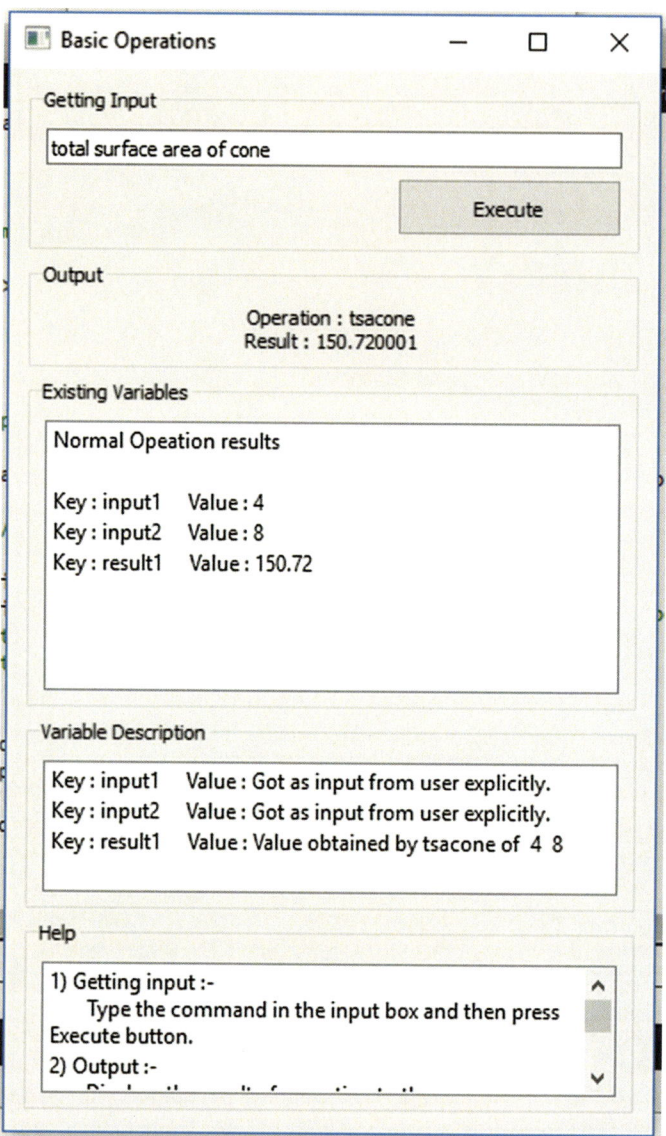

Fig. 5 Total surface area performed without values specified

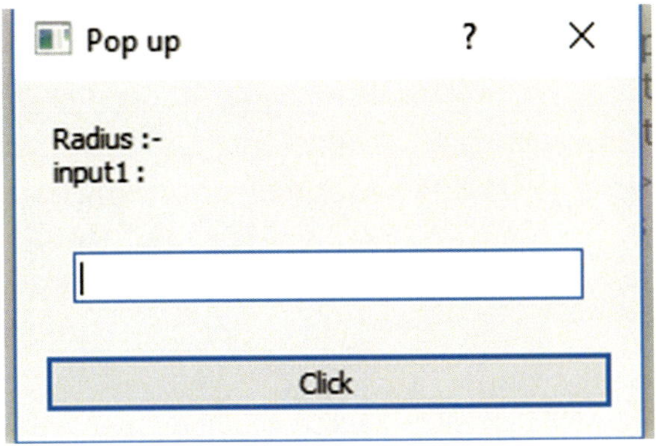

Fig. 6 Prompt box to enter value

Table 1 Characteristics and limitations of similar tools

S. No	Tool name	Input	Output	Limitations
1	Flowgortihm	Flowchart	Complete code	Altering a flowchart requires redrawing. Creating a flowchart for complex programs is difficult
2	Raptor	Flowchart with structured commands	Compiled output	Input should follow a predefined structure inside the building blocks of flowchart
3	Anchor pseudocode compiler	Code without curly braces and semicolon	Code with curly braces and semicolon	Requires programming knowledge. User has to specify complete program which abides syntactical rules with an exemption for semicolon and curly braces
4	Visual paradigm	UML diagram	C++ code	Knowledge about constructing UML diagrams is needed. The main focus is on the design
5	Codesmith generator	Code modified with template	Complete code	Searching for the function in the template is time-consuming. Function should be defined by the user

not require programming knowledge which is contradictory to Anchor Pseudocode Compiler. Visual Paradigm's primary focus is on design. On the other hand, the proposed tool focuses on completing the task rather than concentrating on design. In Codesmith Generator, function name will be present but it will not be defined in the template. However, in ONESTOP, user need not define the function as it is already defined in the tool. So, specifying the name of the operation to be done is sufficient.

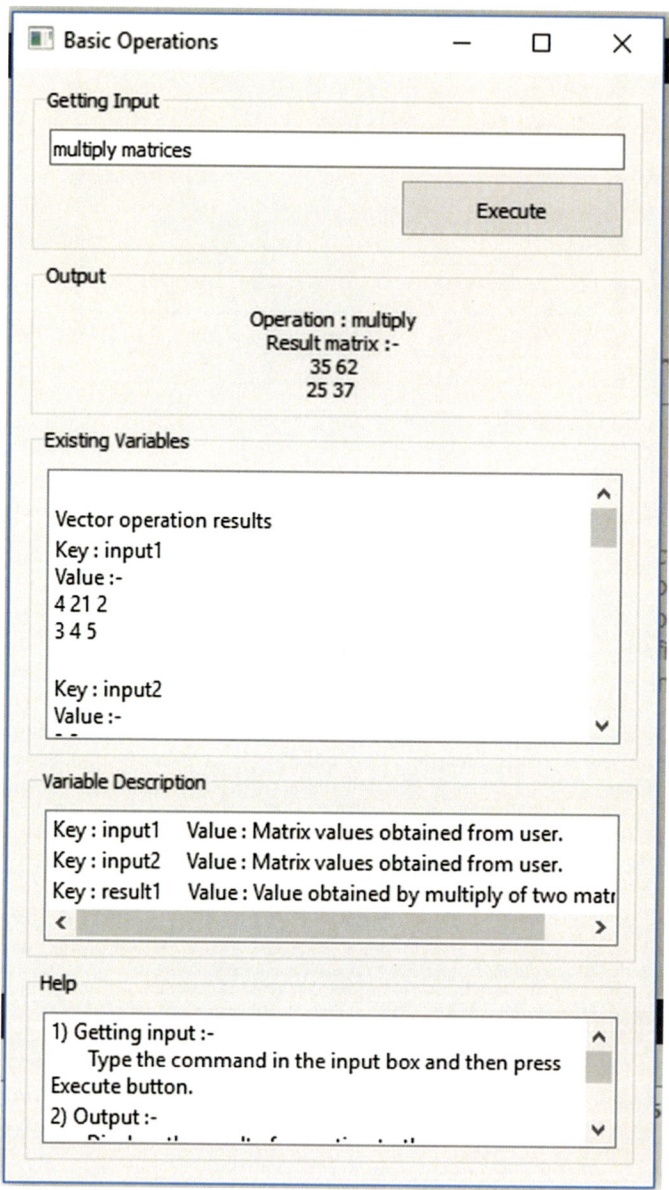

Fig. 7 Multiplication of matrices

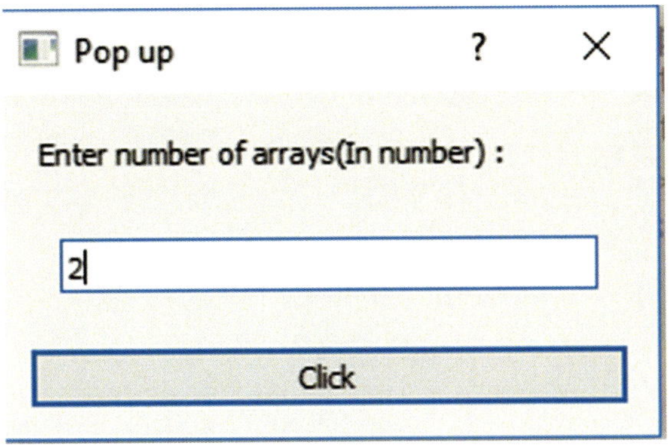

Fig. 8 Prompt box to
specify number of matrices

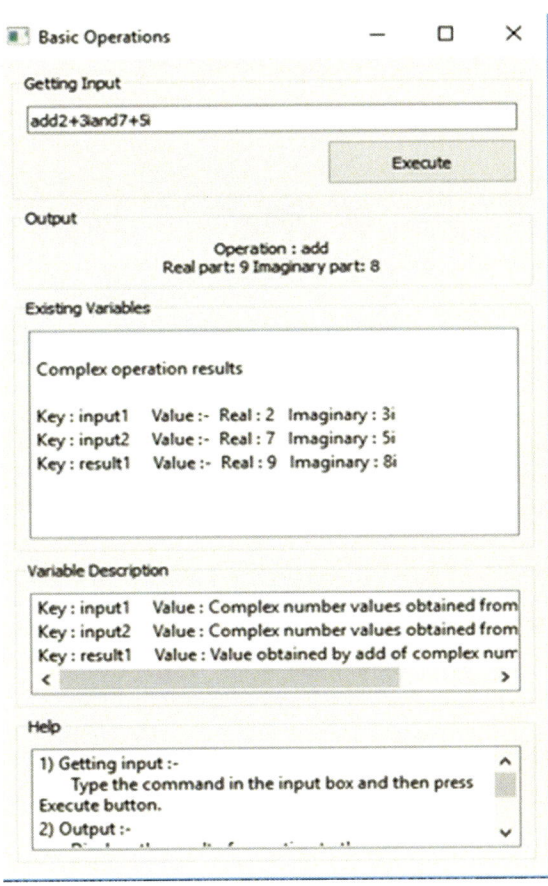

Fig. 9 Prompt box to enter
matrix values

Fig. 10 Addition of complex numbers

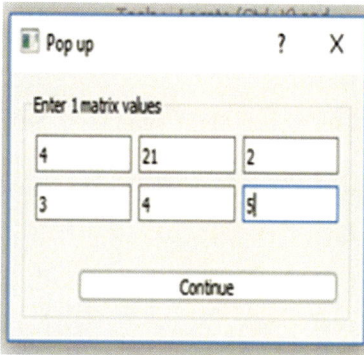

Table 2 Mean values of scores from the survey

Features	Mean value
Usability	8.13
Lack of learning	8.45
Performance	7.72
Enhancement	6.9
Result interpretation	7.54
Tool design	7.86
Scope	6.27

5.3 User Experiment

A survey was conducted among few people to rate various features of the tool. These people were of three categories: novice, intermediate and experienced programmers. They were asked to rate the features of the tool on a scale of 1–10. Table 2 shows the mean value of every feature of the tool obtained from the survey conducted among three categories of people. Usability denotes how easily the tool can be handled for performing various operations. Lack of learning indicates the amount of prerequisite knowledge required for using the tool. It also implies the efforts needed to get familiarized with the tool. Performance shows how fast the processing and execution of desired operation take place to generate the output. Enhancement signifies how much the tool can be developed further. It includes adding new features. Result interpretation is how far the user is able to understand and interpret the result. Tool design indicates how attractive the front end of the tool is. Scope defines the types and number of operations that can be currently performed by the tool.

From the results in Table 2, few conclusions can be made. Highest mean value (8.45) was for lack of learning. Usability has got the second highest mean value of 8.13 because tool is helpful for novice, intermediate and experienced programmers. Scope received the least mean value since the tool is developed for performing limited operations only. Various other types of operations can be added to the tool for further development.

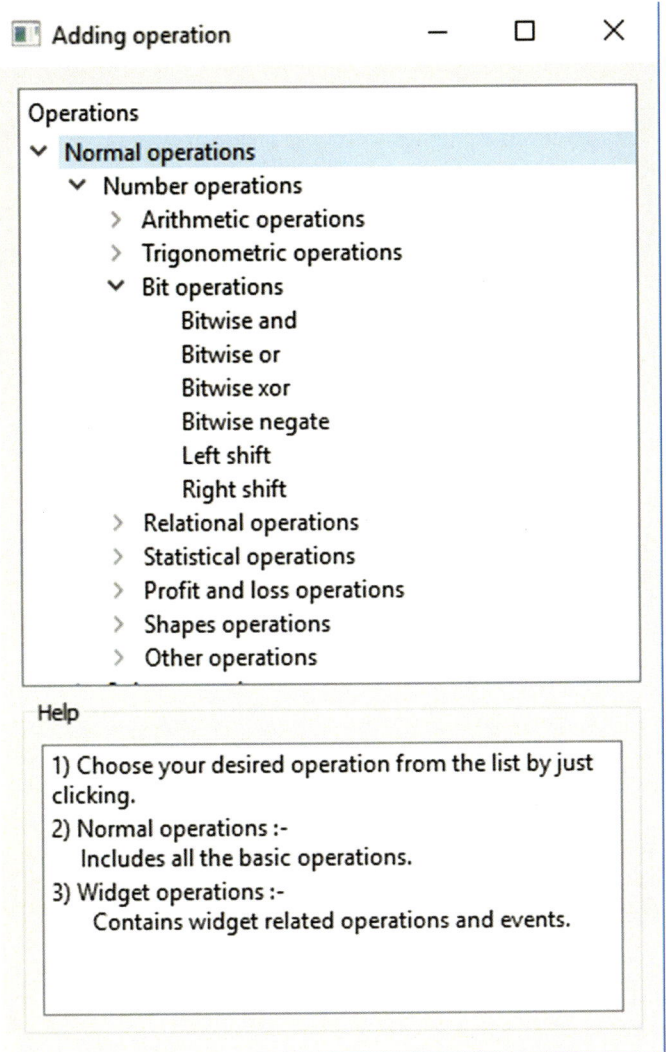

Fig. 11 Operations list

Two features which have the highest mean values are considered for further conclusion [23]. Therefore, lack of learning and usability are taken into consideration. Naive Bayes classifier is used to identify the probability of each category of people who have rated two features above 6. Posterior probability $P(c/x)$ is found by

$$P(c/x) = P(x/c) * P(c)/P(x)$$

where c is the class and x is the predictor. $P(c/x)$ is posterior probability, $P(x/c)$ is likelihood probability, $P(c)$ is class prior probability and $P(x)$ is predictor prior probability.

Let novice, intermediate and experienced programmers be the three classes. Class prior probability for each class obtained from the survey is given as

P (novice) $=0.4$.
P (intermediate) $=0.31$.
P (experienced) $=0.27$.

Likelihood probability is calculated to find the probability of having a score greater than 6 for lack of learning and usability for a particular class. Likelihood probability as calculated as follows:

P (lack of learning > 6 and usability > 6 /novice) $=0.28$.
P (lack of learning > 6 and usability > 6/intermediate) $=0.19$.
P (lack of learning > 6 and usability > 6 /experienced) $=0.15$.
Predictor prior probability is found out as
P (lack of learning > 6 and usability > 6) $=0.62$.
Posterior probabilities are calculated as follows:
P (novice/lack of learning > 6 and usability > 6) $=0.45$.
P (intermediate/lack of learning > 6 and usability > 6) $=0.30$.
P (experienced/lack of learning > 6 and usability > 6) $=0.24$.

From the above posterior probabilities, conclusions can be made that novice programmers have liked the lack of learning and usability features of the tool the most and have rated them mostly above 6 on a scale of 10. Intermediate programmers have the second highest probability of liking these two features. Experienced programmers have the least probability of rating these two features above 6 compared to the other two categories of people as they have sound knowledge in programming.

6   Future Work

ONESTOP can further be developed to perform the operations and produce the result from the commands got in many other languages since the tool developed is used to process sentences in English only. The main keyword can be in any language and the keyword identified from the input should be able to match the main keyword in order to identify the type of operation. Step-by-step processing of the input can be illustrated using animation as it will be easier for beginners to understand the working of the tool better. Graphical User Interface (GUI) is used predominantly to make a tool attractive. GUI features can be included to make designing easier. Based on the GUI requirements specified by users, ONESTOP can be enhanced to create GUI automatically. This can further be made easier for users by using drag and drop option to create GUI. Similar to Google search, already specified inputs can be stored and suggested to users while they type the sentence.

7 Conclusion

A tool which performs many operations based on the input string from user is created. The tool identifies keywords from input and performs the operation to produce the expected result. This tool is very easy to use. The user does not require specific knowledge about how to use the tool. ONESTOP was mainly developed to eradicate the fear of using or learning new programming languages and remembering their syntax. User need not learn programming languages to do various operations. A pseudocode or even a sentence can be used to perform an operation using this tool. Since the values of inputs and results are displayed in a sequential manner after performing the operation, it helps users to understand the logic of how the result was got.

References

1. Fidge C, Teague D (2009) Losing their marbles: syntax-free programming for assessing problem-solving skills. In: Conferences in research and practice in information technology series 95
2. Charntaweekhun K, Wangsiripitak S (2006) Visual programming using flowchart. In: 2006 international symposium on communications and information technologies, Bangkok, pp 1062–1065. DOI: 10.1109/ ISCIT.2006.339940
3. Sarda N, Jain A (2001) Mragyati: a system for keyword-based searching in databases. In: cs.DB/0110052
4. Little G, Miller R Syntax-free programming. URL http://up.csail.mit.edu/projects/keyword-commands/index.html
5. Smith T, Liu J, Roman E (2009) Final project: classification and keyword extraction of Anne Hunter 'jobslist' Emails
6. Fudaba H et al (2015) Pseudogen: a tool to automatically generate pseudocode from source code. In: 2015 30th IEEE/ACM international conference on automated software engineering (ASE), Lincoln, NE, pp 824–829
7. Tahir M, Ahmed S (2015) Tree-combined Trie: a compressed data structure for fast IP address lookup. Int J Adv Comput Sci Appl 6. https://doi.org/10.14569/ijacsa.2015.061223
8. Senthil Kumar T, Ohhm Prakash KI (2015) A queueing model for e-Learning system. In: Suresh L, Dash S, Panigrahi B (eds) Artificial intelligence and evolutionary algorithms in engineering systems. Advances in intelligent systems and computing, vol 325. Springer, New Delhi
9. Sankar A, Bharathi PD, Midhun M, Vijay K, Senthil Kumar T (2016) A conjectural study on machine learning algorithms. In: Suresh L, Panigrahi B (eds) Proceedings of the international conference on soft computing systems. Advances in intelligent systems and computing, vol 397. Springer, New Delhi

Performance Evaluation of DCT, DWT and SPIHT Techniques for Medical Image Compression

M. Laxmi Prasanna Rani, G. Sasibhushana Rao and B. Prabhakara Rao

Abstract Medical imaging is, visible illustration of internal of the human body in digital form. There exists a need for compression of these medical images for storage and transmission purposes with high image quality for error free diagnosis of diseases. It is necessary to develop new techniques for compression of images, resulting into reduction in cost of data storage and transmission. This paper proposes a progressive and DWT-based Set Partition in Hierarchical Tree (SPIHT) algorithm for achieving better image compression while maintaining the nature of restored image. The performance of SPIHT algorithm is evaluated and compared with simple DCT and DWT using the parameters like Peak Signal to Noise Ratio (PSNR), Mean Squared Error (MSE) and compression ratio (CR). From the results it is concluded that SPIHT algorithm produces better PSNR and MSE values compared to DCT and DWT with high image quality.

1 Introduction

Image compression is one of the important applications of digital image processing for better data transmission and storage. Generally this image data require abundant memory space for storage and higher bandwidths for transmission. To reduce the data required for digital image representation, the method of compression of image is necessary. Different applications of communications like Television, satellite and error free diagnosis of diseases in medical applications, storage and communication

M. Laxmi Prasanna Rani (✉)
Department of ECE, MVGR College of Engineering, Vizianagaram, Andhra Pradesh, India
e-mail: prassugowtham@gmail.com

G. Sasibhushana Rao
Department of ECE, AUCE(A), Andhra University, Visakhapatnam, Andhra Pradesh, India
e-mail: sasigps@gmail.com

B. Prabhakara Rao
Department of ECE, JNTUK, Kakinada, Andhra Pradesh, India
e-mail: drbprjntuk@gmail.com

© Springer Nature Switzerland AG 2019
D. Pandian et al. (eds.), *Proceedings of the International Conference on ISMAC in Computational Vision and Bio-Engineering 2018 (ISMAC-CVB)*, Lecture Notes in Computational Vision and Biomechanics 30,
https://doi.org/10.1007/978-3-030-00665-5_147

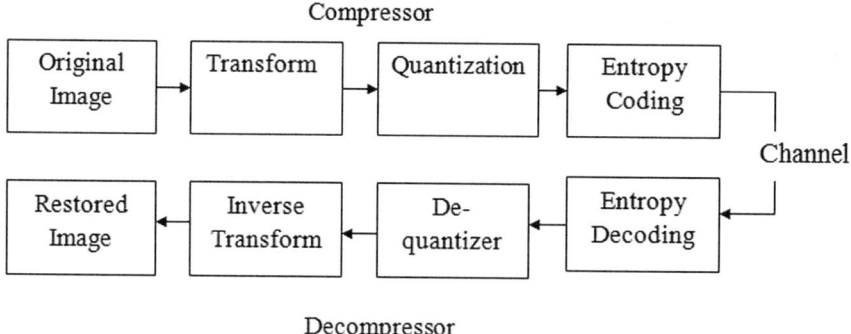

Fig. 1 Block diagram of Lossy compression

will not be feasible without compression of data [1]. The objective of an image compression technique is to reduce the repeated and irrelevant data in images such that small numbers of bits are used to represent the image in the vicinity of an "acceptable" quality of vision for the reconstructed image. By reducing the redundancy and irrelevance, the data will be stored and transmitted in effective form which is known as compression. The compression methods are also suitable for the devices have less on board memory and low battery life.

This paper recommends an image compression technique for the transmission of data in the applications of communications and wireless network. This should be maintained by balancing available bandwidth and perceived good quality of received image with minimum delay of transmission. To get better compression, the factors like algorithm complexity, compression time, computational resources and cost to be considered. These factors are essential and to be considered design or modify a compression method to make it fast and accurate to get good quality images.

The image compression techniques are of two type's i.e. lossy or lossless compression. Lossy compression techniques provide considerable decrease in data rate [2] with loss of accuracy for natural images such as photographs. Lossless compression techniques are suitable for medical images by lowering the no. of bits of the data without reducing the eminence of the image. Various image compression techniques such as transform coding and adaptive versions of techniques have developed. The block diagrams of lossy and lossless compressions are shown in Figs. 1 and 2 respectively [2].

DCT is a common compression technique for JPEG images based on block processing methods [3]. DCT represents with sum of cosine functions of an image varies at different frequencies and magnitudes. This transform has particular property of energy compaction. Image compression using the DCT is simple and get better compression ratio (CR).

The Discrete Wavelet Transform (DWT) is an efficient transform tool for compression of images due to its capability of displaying the images at different resolution also gives better compression. This wavelet transform represents with a sum

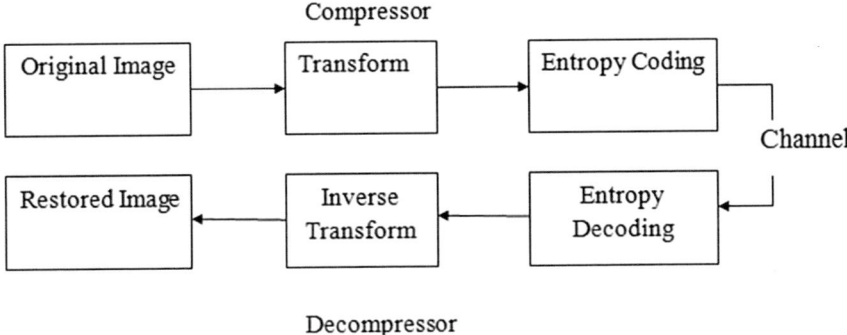

Compressor

Fig. 2 Block diagram of loss less compression

of wavelet functions at different resolution of an image due to the property of Multi Resolution Analysis. The basis function of this transform is of any function which satisfies the requirements of multi resolution analysis [2].

Set Partition in Hierarchical Tree (SPIHT) is one of the best compression algorithms which reconstruct the original image with low loss and provides good image quality. This algorithm provides excellent PSNR and improved CR and low encoding complexity than DWT and DCT [3].

This paper is organized such that introduction to discrete cosine transform, discrete wavelet transformation and SPIHT algorithm of Image is described in Sects. 1, 2 describes SPIHT algorithm and Sect. 3 illustrates the compression using SPIHT algorithm. Section 4 discusses the results and conclusions are presented in Sect. 5.

2 SPIHT Algorithm

SPIHT algorithm is award-winning compression technique has received worldwide applause and consideration since 1995. Many researchers have tested and used this algorithm which becomes the standard and advanced algorithm for compression of images. This technique is very fast and is the best among other compression techniques. It is a progressive hierarchical structure of image transmission and gives better Image eminence with high PSNR. Using wavelet transform the image is decomposed into various sub-bands as the initial step of this algorithm. The decomposition is a repetitive process until it reaches the final value. For each decomposition, there are four sub-bands i.e. approximation coefficients of low-frequency and horizontal, vertical, diagonal coefficients of high-frequency. The coefficients of different sub bands related to the same spatial position display self-similarity characteristics in the pyramid structure [4]. SPIHT establishes spatial orientation trees by defining parent children relationships between these self- similar sub bands [5]. The spatial orientation tree is an example of parent–children relationship is shown in Fig. 3.

Fig. 3 Examples of
parent-offspring
dependencies in the
spatial-orientation tree

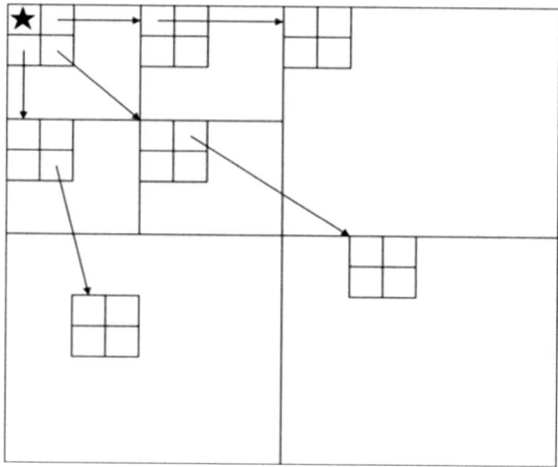

2.1 Steps in SPIHT Algorithm

There are three steps of sorting, refinement and quantization in SPIHT algorithm. Here, the image is encoded and differentiated into significant and insignificant pixels using three lists such as LIS (Least Insignificant Set), LIP (Least Insignificant Pixel), and LSP (Least Significant Pixels). The maximum number of bits is n_{max} depends on the largest pixel value in the spatial orientation tree and it can be represented by the Eq. (1).

$$n\text{max} = \left[\log_2\left(\max\left(\left|C_{i,j}\right|\right)\right)\right] \tag{1}$$

where $C_{i,j}$ represents the largest pixel value in the spatial tree.

The coefficients in the Least Insignificant Pixels, Least Insignificant Set and Least Significant Pixels are separated using the threshold value Th. The threshold value can be expressed in Eq. (2) given below.

$$\text{Threshold Value, Th} = \begin{cases} 1, & \max\left(\left|C_{i,j}\right|\right) \geq 2^n \\ 0, & \text{otherwise} \end{cases} \tag{2}$$

The steps in SPIHT algorithm are explained in the following steps using three types of passes such as sorting, refinement and quantization pass [6].

Step 1   In the sorting pass, the pixels or coefficients in the list of Insignificant Pixel (LIP) are examined to determine whether the pixels or coefficients are significant or insignificant depending on threshold value Th. If the magnitude of coefficients more than the value of threshold are found to be significant, then '1' is the output and the other bit is taken as the sign of the pixel coefficient ('1' for positive and '0' for negative). Then the significant pixel coefficient is shifted to significant pixels list. If any coefficient is less than threshold value, it becomes insignificant and that bit is taken as '0'.

Step 2   In this step, the overall wavelet coefficients in List of Insignificant Set are to be processed. The set of all offspring's of coefficients in the list of LIS are correlated with the threshold value to make a decision whether they are important or not. If the coefficients are greater than a certain threshold value then consider the coefficients as significant and move into least significant pixels; otherwise the coefficients moved into LIP as insignificant [7].

If the coefficient is insignificant, then the spatial orientation tree was a zero-tree, which is rooted by the current entry. Hence, the bit '0' is output and there is no further processing of these coefficients.

The remaining coefficients, which are coefficient descendants set except for the immediate offsprings, are shifted to the last of List of Insignificant Set. Again thresholding test is performed on all the descendants of its direct offspring's of coefficients. Once the significance test is done, the spatial tree again split into four sub-trees which are rooted by the coefficients of direct descendants and these direct offspring's are added in the end of LIS.

Step 3   In the refinement pass, all the refinement bits (nth bit) in Least Significant Pixel at current threshold are taken as output. For the next round of this algorithm, the current threshold becomes halved. This process is continuous when all the coefficients in LIS are completed and the quantization can be done to get compression.

The implementation steps of SPIHT algorithm is explained in Fig. 4.

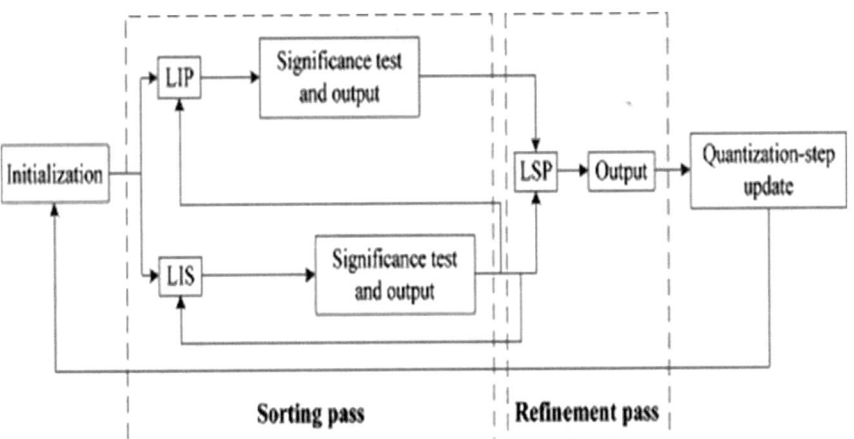

Fig. 4 SPIHT algorithm implementation steps

3 Image Compression Using SPIHT Algorithm

Compressing Images with SPIHT technique can be explained below. Initially, image is decomposed into approximation coefficients (LL) of low frequency and horizontal (LH), vertical (HL), diagonal (HH) coefficients of high frequency, by means of wavelet transform. The approximation coefficients further divided into different components depending on level of decomposition and the level equals the value of n.

The image data is encoded in SPIHT algorithm [8] after wavelet decomposition in the steps of sorting, refinement and quantization using three lists such as LIP (Least Insignificant Pixels), LIS (Least Insignificant Set) and LSP (Least Significant Pixel). When the magnitudes of coefficients less than threshold value are in the list of LIP, overall wavelet coefficients defined in the tree structure less than threshold are in LIS and finally LSP contains the set of pixel coefficients having magnitude higher than threshold are the important pixels [9].

In the sorting process, all the coefficients in the list of LIP are verified if they are significant or insignificant depending on threshold and the descendants of the coefficients in remaining lists are also tested using threshold value in the Eq. (2). The coefficients greater than thresholds are inserted into LSP and remaining coefficients are moved to LIP [5].

In the refinement process, the nth MSB of the coefficient in the LSP is taken as the final output. Next the value of n is decreased by '1'. Again sorting and refinement processes are applied to coefficients descendants until the value of n reaches '0' or the list coefficients in the LIS are completed. The quantization process is applied to the entire coefficients in list of significant pixels.

Once the encoding process is completed, the compressed image is applied to SPIHT decoding process. The true image is reconstructed with better image quality using Inverse Discrete Wavelet Transform. This complete process is explained in Fig. 5.

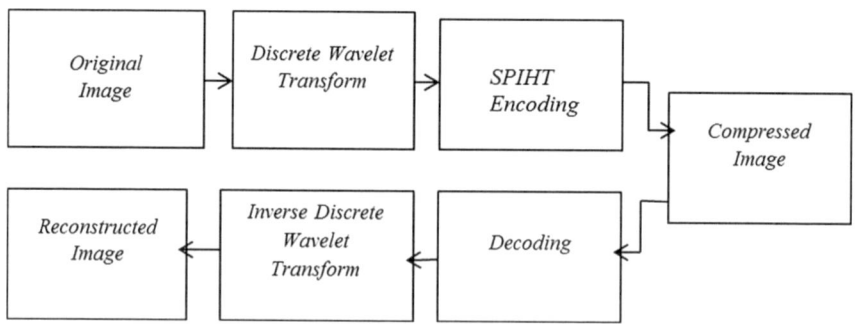

Fig. 5 Image compression using SPIHT algorithm

3.1 Performance Metrics

The performance of compressing the images using SPIHT can be obtained by calculating PSNR (Peak Signal to Noise Ratio), MSE (Mean Squared Error) and the compressed image size. PSNR is used to determine the excellence of the image. High PSNR value resembles that the reconstructed image quality is high and low PSNR value gives that reconstructed image quality is low. MSE is another parameter that measures the excellence of the image by estimating the difference between original and reconstructed images [10].

$$\text{MSE} = \frac{1}{MN} \sum_{x=0}^{M-1} \sum_{y=0}^{N-1} \left((p(x, y) - s(x, y))^2 \right) \tag{3}$$

where $p(x, y)$, $s(x, y)$ are original and reconstructed images with size of MXN.

$$\text{PSNR} = 10 \log_{10} \left(\frac{I_{\max}^2}{\text{MSE}} \right) \tag{4}$$

where $I_{\max}$ is the maximum possible pixel values of the image.

4 Results and Discussion

This paper presents a progressive compression algorithm i.e. SPIHT algorithm applied to the medical images like MRI of brain, X-ray images and a test image of lena. This algorithm is implemented in Matlab and the results are compared with present strategies of DCT and DWT. Performance of SPIHT algorithm is evaluated using the parameters of Peak Signal to Noise Ratio (PSNR), Mean Squared Error (MSE) and sizes of input and compressed image sizes.

Table 1 gives the sizes of original and compressed images using DCT, DWT and SPIHT. Table 2 shows the PSNR value obtained using MSE calculated for medical images using DCT, DWT and SPIHT algorithm respectively. Figure 6 shows the original and compressed images of the existing techniques of DCT and DWT and progressive hierarchical structure of SPIHT algorithm. From the results, it is observed that better compression obtained by DCT over DWT and better reconstruction is obtained by SPIHT algorithm with more PSNR and less MSE.

5 Conclusion

This paper progressed with SPIHT algorithm to get better image compression with progressive hierarchical structure. The values of PSNR and MSE obtained from

Table 1 Comparisons of original and compressed images sizes

Reference image	Size original image (KB)	Compressed image size(KB)		
		DCT	DWT	SPIHT
Test image (lena)	138.056	1.329	3.964	28.686
MRI (Brain1)	19.699	15.453	17.557	18.714
MRI (Brain2)	74.889	10.116	16.871	27.593
X-ray	51.462	5.077	5.016	28.342

Table 2 Comparison of MSE and PSNR values of DCT, DWT and SPIHT

Reference image	DCT		DWT		DWT Based SPIHT	
	MSE	PSNR (dB)	MSE	PSNR (dB)	MSE	PSNR (dB)
Test image (Lena)	34.2357	24.1008	29.5102	34.3105	4.2156	41.82
MRI (Brain1)	10.5186	37.9112	4.3304	43.9405	0.7534	49.36
MRI (Brain2)	1.3593	46.7975	0.2599	53.9832	0.2074	54.96
X-ray	9.2746	38.4579	0.4411	51.6851	0.0411	55.6851

| (i) | (ii) | (iii) | (iv) |

Fig. 6 (i) Original. (ii), (iii), (iv) Compressed images of DCT, DWT and SPIHT of lena

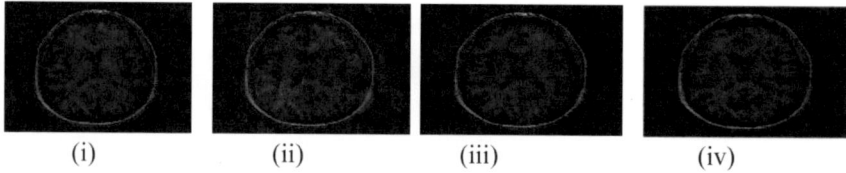

| (i) | (ii) | (iii) | (iv) |

Fig. 7 (i) Original. (ii), (iii), (iv) Compressed images of DCT, DWT and SPIHT of MRI of brain1

SPIHT, is compared with the results of DCT and DWT. From these results SPIHT algorithm exhibits good image compression and also reconstructs the original image without reducing the image quality. From the testing outcomes, it could be concluded that the SPIHT algorithm is successful and capable than prevailing strategies in phrases of PSNR, MSE (Figs. 7 and 8).

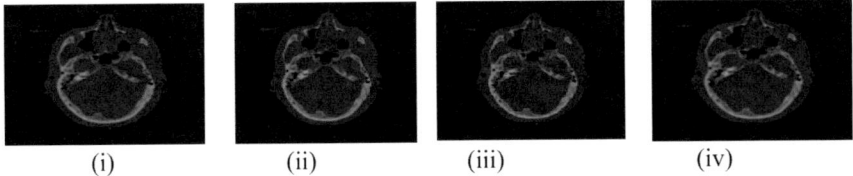

(i)   (ii)   (iii)   (iv)

Fig. 8 (i) Original. (ii), (iii), (iv) Compressed images of DCT, DWT and SPIHT of MRI of brain2

References

1. Kulkarni, AN, Gandhe ST, Dhulekar PA, Phade GM (2015) Fractal image compression using genetic algorithm with ranking select mechanism. In: International conference on communication information & computing technology (ICCICT)
2. Gonzalez RC, Woods RE (2009) Digital image processing. Pearson Education, 2nd edn
3. Telagarapu P (2011) Image compression using DCT and wavelet transformations. Int J Sig Process Image Process Pattern Recogn 4(3)
4. Shapiro JM (1997) Embedded image coding using zero trees of wavelet coefficients, pp 108–121
5. Venkataramani DY, Banu SP (2011) An efficient hybrid image compression scheme based on correlation of pixels for storage and transmission of image. Int J Comput Appl 18(3):0975–8887
6. Rehman M, Touqir I, Batool W (2015) Selection of optimal wavelet bases for image compression using SPIHT algorithm. In: Seventh international conference on machine vision (ICMV 2014)
7. Yang SH, Chang YL, Chen HC (2001) A digital watermarking scheme based on SPIHT coding. In: 2001 IEEE international conference on multimedia & expo (ICME 2001). Tokyo, Japan
8. Said A, Pearlman WA (1996) A new, fast, and efficient image codec based on set partitioning in hierarchical trees. IEEE Trans Circ Syst Video Technol 6(3)
9. Ezhilarasi P, Nirmalkumar P (2013) Efficient image compression algorithm using modified IWT and SPIHT for CMOS image sensor. Int Rev Comput Softw
10. Gu X, Ren K, Andreas JM, Hielscher H (2009) Parametric image reconstruction using the discrete cosine transform for optical tomograph. J Biomed Opt

Structural Health Monitoring—An Integrated Approach for Vibration Analysis with Wireless Sensors to Steel Structure Using Image Processing

C. Harinath Reddy, K. M. Mini and N. Radhika

Abstract Wireless sensors for structural health monitoring (SHM) with a coordinated approach to steel structure has obtained research area in the field of structural engineering due to its low cost and a wide range of applications. These systems have been used to monitor structural behaviour, and it has the potentiality to improve structural life period and also to improve public security too. In this paper, the study is done through low-cost wireless vibration sensors that are deployed on steel frame structure for vibration analysis using microcontrollers and data systems. In this project, tests were done to interface a wireless sensor to an Arduino. And this sensor data to PC via serial and visualize the data in a software called image processing in Arduino as a signal output. This paper explains the real-time deployment of SHM using sensor networks.

1 Introduction

Detecting the structural behaviour or operation of the method of implementing a property and enactment for structures is termed to as Structural Health Monitoring (SHM). Over the last few years, Wireless Sensor Networks (WSNs) have become evident and as an economical for connecting sensor networks [1]. Here, detection

C. Harinath Reddy (✉) · K. M. Mini
Department of Civil Engineering, Amrita School of Engineering, Amrita Vishwa Vidyapeetham, Coimbatore, India
e-mail: harichinnavula@gmail.com

K. M. Mini
e-mail: k_mini@cb.amrita.edu

N. Radhika
Department of Computer Science and Engineering, Amrita School of Engineering, Amrita Vishwa Vidyapeetham, Coimbatore, India
e-mail: n_radhika@cb.amrita.edu

© Springer Nature Switzerland AG 2019
D. Pandian et al. (eds.), *Proceedings of the International Conference on ISMAC in Computational Vision and Bio-Engineering 2018 (ISMAC-CVB)*, Lecture Notes in Computational Vision and Biomechanics 30,
https://doi.org/10.1007/978-3-030-00665-5_148

and characterization are interpreted as changes to geometric properties of a structure, including changes to structural functioning with properties in taking loads. Especially in conditions of loads to structures such as earthquakes or blast loading, SHM is used for model analysis testing and performance of the structure can be attained based on monitoring techniques.

Wireless sensors are used for sensing, communication and control, and provide an attractive alternative for structural mitigation. These wireless control systems have flexible installation, rapid deployment and low maintenance cost. The characterization and field test of the structure using sensors is implemented in open hardware. The system utilizes the unique features of sensors for developing low cost, compact and higher accuracy system [2]. The fundamental approach sensors monitor the seismic response or vibration analysis of building structure. These sensors have broad range of application, to monitor and control structural systems. With the designed programs and parameters, different sensors can be inputted to different software systems and to the sensing unit, for monitoring purpose [3].

The vibration analysis-based detection technique is the health monitoring parameter and is implemented using wireless sensors, with data acquisition system [4]. The term detecting technique infers a total loss of system functionality or property behaviour of structure [5, 6]. Structural property change in normal service may include stiffness of structure or earthquake vibrations [7]. Structural stiffness and strength reduction are caused due to elements of the structure when in vibrations or oscillations. Many works are already carried on the structural health monitoring of civil structures using wireless sensing technology [8–12].

In this paper, the microcontroller named Arduino UNO board is used and connected to the SW-420 vibration sensor with 5 V relay power supply unit. The working parameters are programmed in the Arduino software and compiled in the microcontroller, using the board software. With performance evaluation, results are displayed in Arduino software. In this approach, for output visualization, image processing tools with Arduino and Matlab 2013a software are shown. Since the output experimentally can be shown digitally by values in order, **image signal processing** is attained by using wireless sensors networks. It represents the set of characteristics or parameters related to the tests conducted as per the sensor deployed for monitoring.

The essential principle is to compare the frequency/time (dynamic analysis) with vibration and amplitude time simulation parameter. The results are obtained through evaluation of stiffness of the material steel frame with Matlab simulation. Both experimental and computational approaches were used for frame analysis.

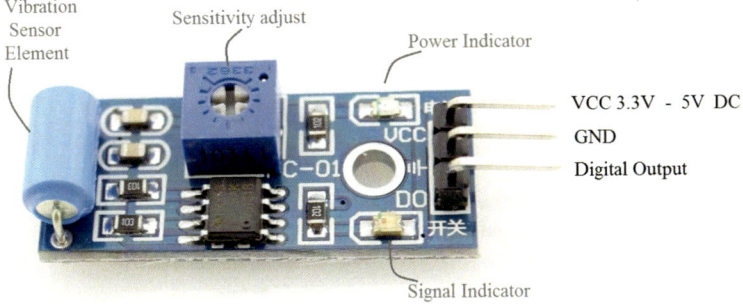

Fig. 1 SW-420 vibration sensor module

2 Materials and Parameters

Material	Sensor used	Parameters monitoring
Steel frame model-modelling on vibration motor Table. Three nos 1.5–3 mm size steel/Fe plates with 4 nos column rods of dia ϕ 10 and 12 mm	(a) Vibration SW-420 sensor	Frequency/time (dynamic analysis) (i) vibration speed versus frequency (ω) Matlab: (ii) amplitude versus time (iii) magnitude versus frequency (kHz) (iv) intensity or vibration response using FFT spectrum

2.1 Wireless Sensor System

Description:

SW-420 Vibration Sensor Arduino (Fig. 1):
The vibration sensor SW-420 is composed of jumper pin board with adjustable comparative system onboard potentiometer. This is used for sensitivity conditioning and signal indication light emitting diode (LED).

When vibration or an external force applied on the sensor module, it senses on the vibration motion, and then the output of this module gives the signal.

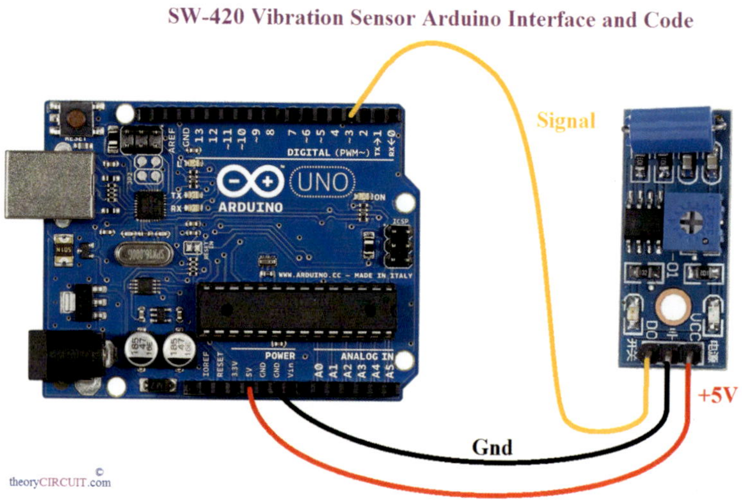

Fig. 2 Arduino circuit connection with SW-420 interface

Arduino Hookup with SW-420 (Fig. 2)

Sensor board Vcc (voltage at the common collector) pin is connected to 5 V pin; connect Gnd (ground) pin to Gnd pin of Arduino board; and connect DO (digital output) signal pin of sensor board to Arduino digital pin A3. By doing the calibration, the sensitivity threshold is adjusted.

Using USB cable connect, the entire interface of Arduino mote to the data acquisition (DAQ) system is used to visualize signals.

Experimental Investigation: Steel Frame Model test Equipment Portal frame is with three-storey (G + 3) and 1 m ht, Arduino UNO circuit board, SW-420 vibration sensor, vibration motor table and data acquisition system.

2.2 Parameter for Monitoring Vibration of Portal Frame Using Wireless Sensors

Frequency/time (dynamic analysis), vibration and amplitude, time simulation technique, 3-storey steel frame are used in a model for experimental study of structural health monitoring. Basic aspect is considering dynamic analysis which is different from static analysis. In this investigation of experiment tests, dynamic loads are applied as a function of time or frequency, and this condition of varying load application induces time- or frequency-varying response, i.e. accelerations, displacements, etc. These test characteristics make dynamic analysis more complicated and realistic than static analysis. So here in the experimental test, load application is created due to vibration motor and induces a dynamic response to steel structure. Installed

Fig. 3 Sensor position at first storey and restrained base plate in vertical direction (Y-direction) (Mechanism by concrete block)

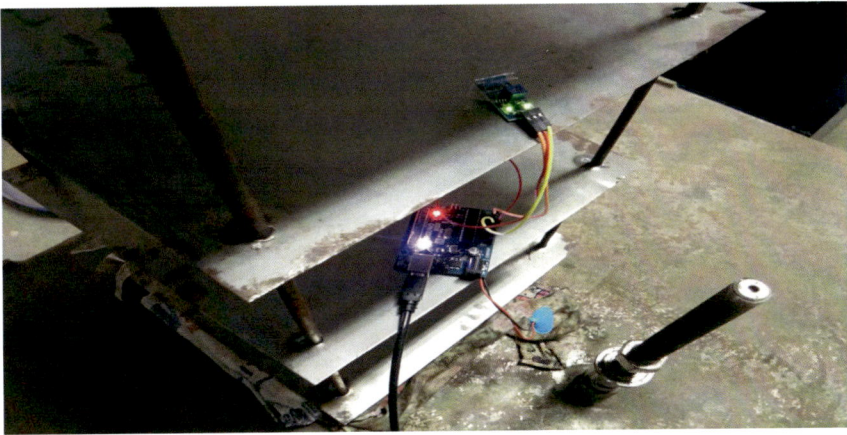

Fig. 4 Sensing interface and communication between the vibration module and Arduino with DAQ

wireless vibration sensor on three-storey frame, first storey, second storey and third storey as points (S1, S2, S3, S4...), are marked and used for testing by placing it on the vibration table (Fig. 3).

To perform and analyse the test using wireless sensing unit for the structure, active mechanism and control are implemented. These sensors kept on each floor and connected with data acquisition systems to transmit the response to the receiver (Fig. 4).

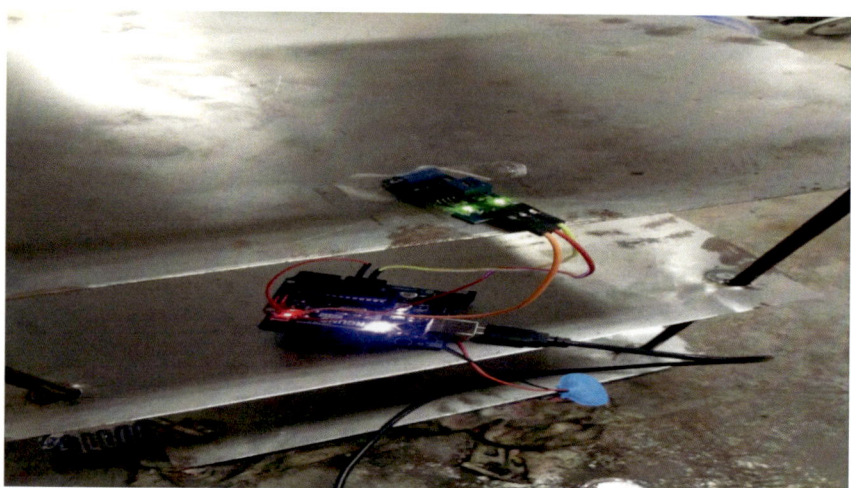

Fig. 5 Capturing of the value response of the steel structure

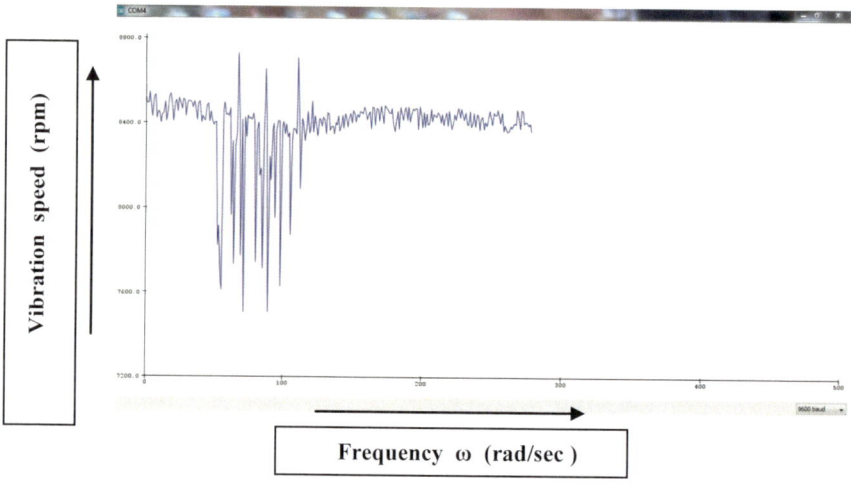

Fig. 6 Response vibration on Y-axis versus frequency (ω) on X-axis corresponding to position S1

The vibration sensor is deployed with initial arrangements and connections. The pins are connected to corresponding outputs and checked with the programmed parameters and compiled in it (Fig. 5).

Once the connections are done, sensing interface and communication are overall checked without any errors.

In Fig. 6, response signal from vibration sensor with vibration speed and frequency is attained with Arduino data acquisition system at corresponding position S1.

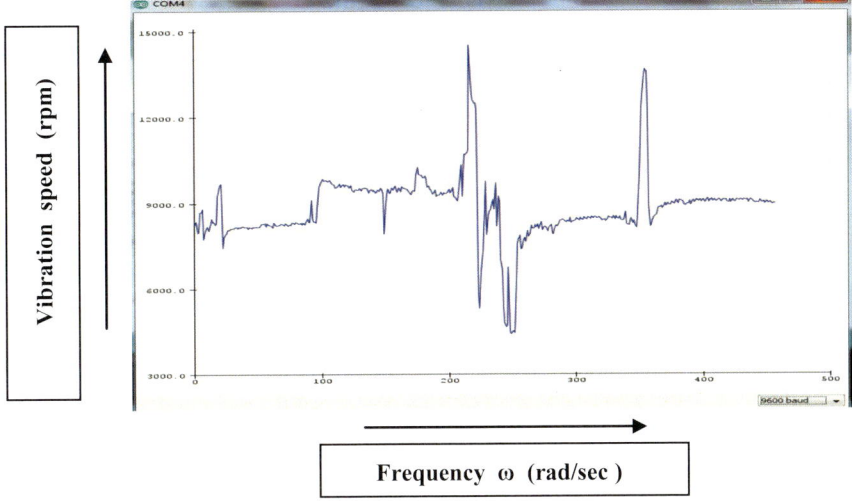

Fig. 7 Response vibration on Y-axis versus frequency (ω) on X-axis corresponding to position S2

Initial first phase of position is taken on first storey of steel frame as "S1". And the vibration motor is switched ON with certain initial speed of vibration.

At position S1, the frequency attained is captured in Arduino software from response due to sensor.

In Fig. 7, response signal from vibration sensor with vibration speed and frequency is attained with Arduino data acquisition system at corresponding position S2.

For the vibration speed of motor, at S2, considering second-phase position on second storey of steel frame, the vibration is transferred to the storey with frequency attained peak value captured in Arduino acquisition system from response due to sensor.

From this, peak value of frequency with respect to vibration speed of frame with sensors is obtained on the DAQ with Arduino and vibration sensor.

Data Obtained from Wireless Sensors to DAQ (Data Acquisition System)

Comparison between peak values are obtained from the wireless sensors measured and from those simulations using the Matlab Simulink using the measured data (Fig. 8).

From the obtained experimental test, vibration analysis for steel frame and its corresponding Matlab simulation is done by Arduino and sensor. Amplitude is taken as the parameter for initial representation of frequencies with vibration against corresponding time (Fig. 9).

For every vibration, a certain degree of magnitude is taken from an extent the steel frame attained versus corresponding frequency of motor simulation runs under which by the frame with vibration in magnitude dB (decibel) or noise (or) sound is obtained (Fig. 10).

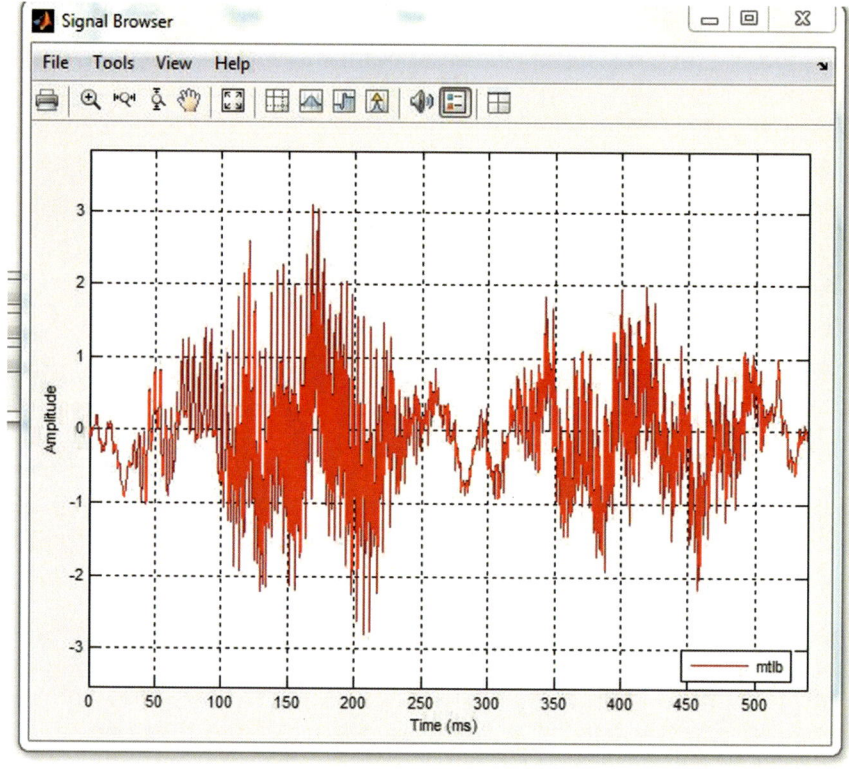

AMPLITUDE VS TIME (ms)

Fig. 8 The Matlab Simulink with corresponding amplitude vibration of the structure with time

From this, FFT spectrum graph obtained from Matlab Simulink for peak values is attained corresponding to sensor values.

From peak values obtained from experimental test using vibration sensor to steel frame, the FFT spectrum estimate is compared with corresponding frequency and magnitude of vibration.

Frame Analysis—Computational Approach

From fundamental equation $M\ddot{U}+C\dot{U}+KU = 0$ and $M\ddot{U}+C\dot{U}+KU = P(t)$ or $X(t)$, 'M' is the mass of steel frame model, 'C' is the damping condition (if damper is assumed) and 'K' is the stiffness of the model.

Both real and imaginary parts are solved and are substituted in the above equation. Mathematical analysis is done using Matlab program:

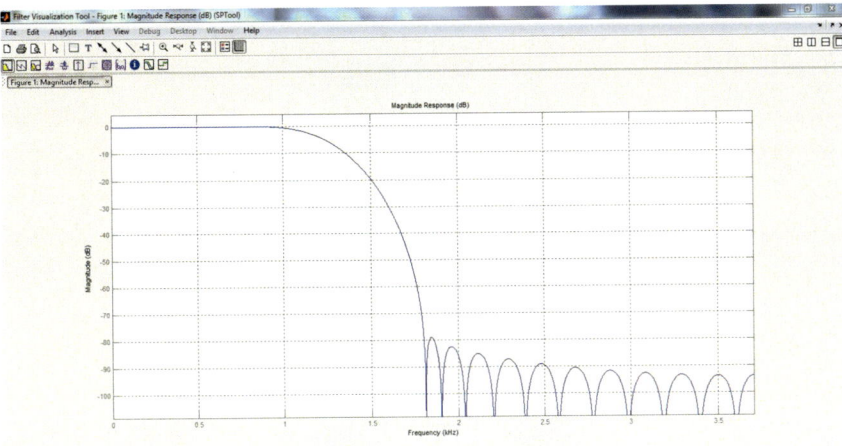

Magnitude (dB) vs frequency (kHZ)

Fig. 9 Maximum attained peak value from Matlab Simulink corresponding to wireless sensor frequency value

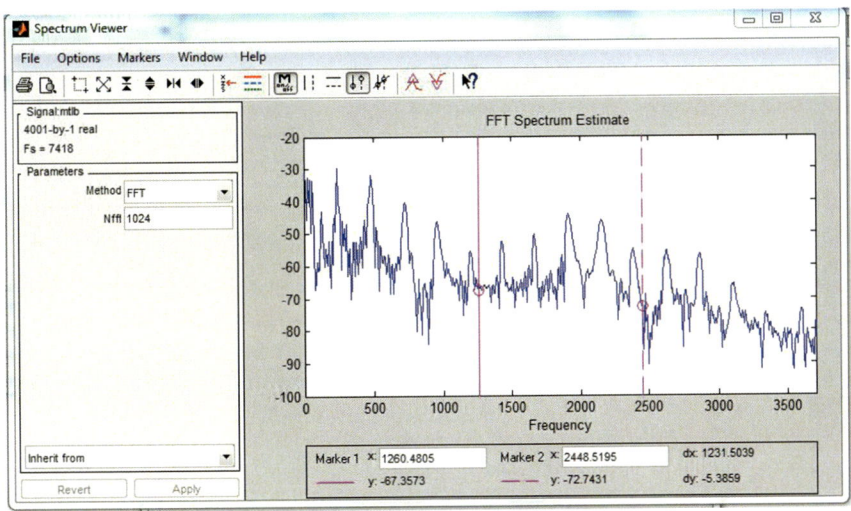

Fig. 10 Intensity or vibration response versus frequency (FFT spectrum)

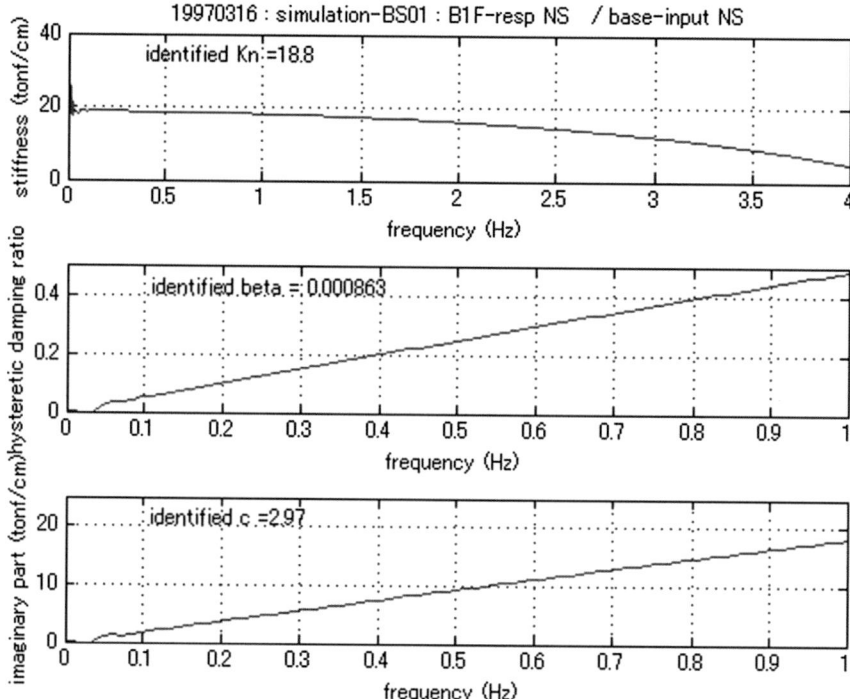

From above analysis, Kn is stiffness of structure attained by simulation analysis, 'β' (beta) is the ratio of excitation frequency to natural frequency and 'C' is the damping coefficient attained from analysis.

For the steel frame model, analysis is done in Matlab using the above parameters, and stiffness (Kn) value from experimental test is compared and analysed theoretically and mathematically using dynamic analysis as obtained.

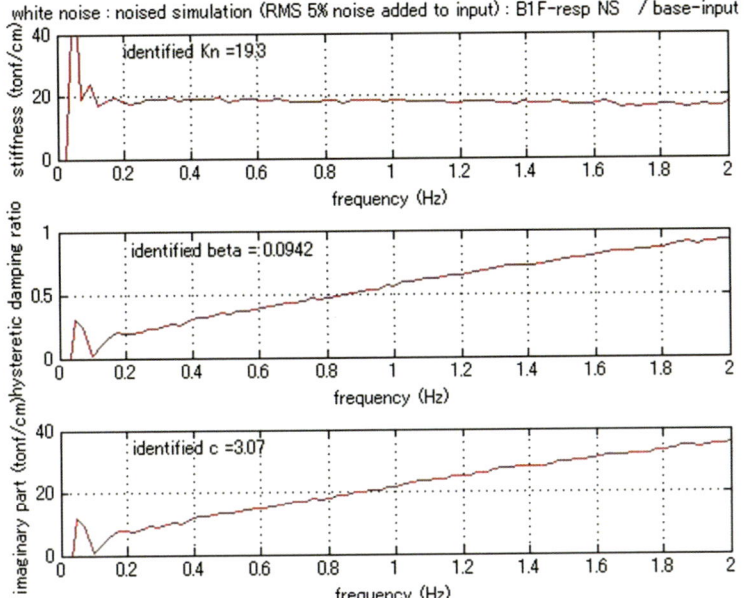

From above analysis, Kn is stiffness of structure attained by simulation analysis, 'β' (beta) is the ratio of excitation frequency to natural frequency, 'C' is the damping coefficient attained from analysis by 'NOISE' simulation approach using root mean square (RMS) method with 5% as input added (Fig. 11).

For this frame analysis, the stiffness of the steel structure (K) due to vibration analysis parameter is computationally predicted with autoregressive eXogenous (ARX) with two conditions:

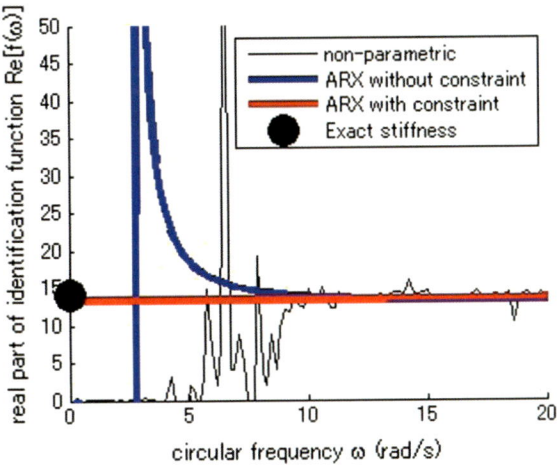

Fig. 11 Multi-loop model attained by image processing using MATLAB by auto regressive plot (ARX)

Fig. 12 Based on
autoregressive—partial least
squares (ARX) attained
multi-loop model predictive
control from Matlab

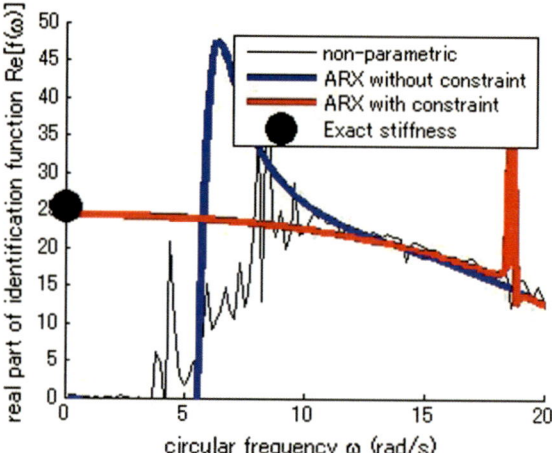

Fig. 13 Multi-loop model
predictive control attained
from MATLAB simulation
by auto regressive partial
least squares (ARX)

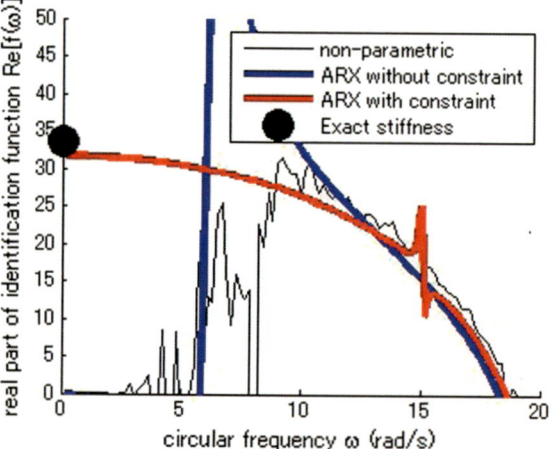

(i) Without constraint

(ii) With constraint.

The 'K' value is obtained from the mathematical dynamic equation $K = M\omega^2$, for 'M' mass steel frame and 'ω' as frequency (rad/sec).

The stiffness 'K' of steel frame mathematically obtained is compared with experimental test for both simulation and vibration analyses conducted using vibration sensor (Fig. 12).

For this frame analysis, the stiffness of the steel structure (K) due to vibration analysis parameter is computationally predicted with autoregressive exogenous (ARX) (Fig. 13).

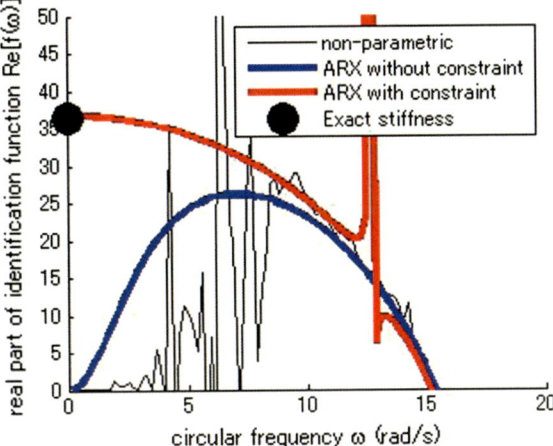

Fig. 14 Auto regressive partial least squares attained from MATLAB using multi-loop model with same frequency

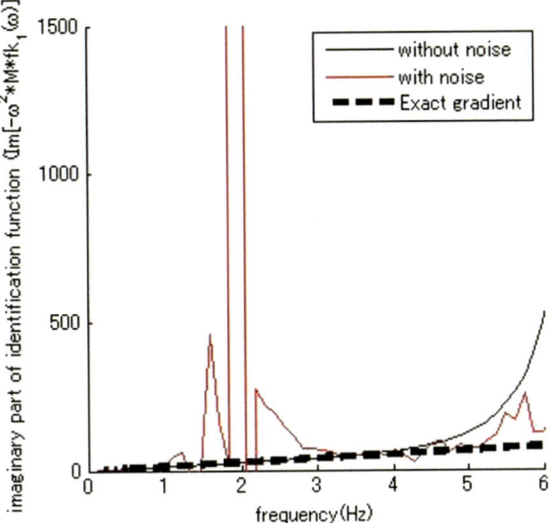

Fig. 15 Frame model vibration analysis with correlation by gradient

For this frame analysis, the stiffness of the steel structure (K) due to vibration analysis parameter is computationally predicted with autoregressive exogenous (ARX) for same frequency (Fig. 14).

For this frame analysis, the stiffness of the steel structure (K) due to vibration analysis parameter is computationally predicted with autoregressive exogenous (ARX) (Fig. 15).

For this frame analysis, exact gradient is taken for plot value as simulation with frequency (Hz) versus stiffness value at without noise condition (Fig. 16).

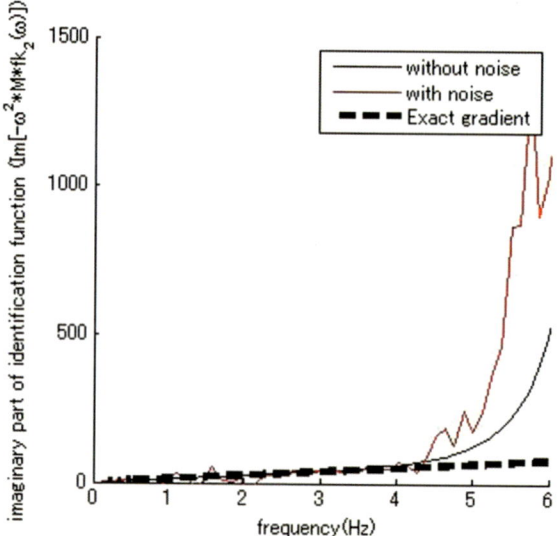

Fig. 16 Vibration analysis for frame model—with noise: real part, without noise: imaginary part correlation by exact gradient

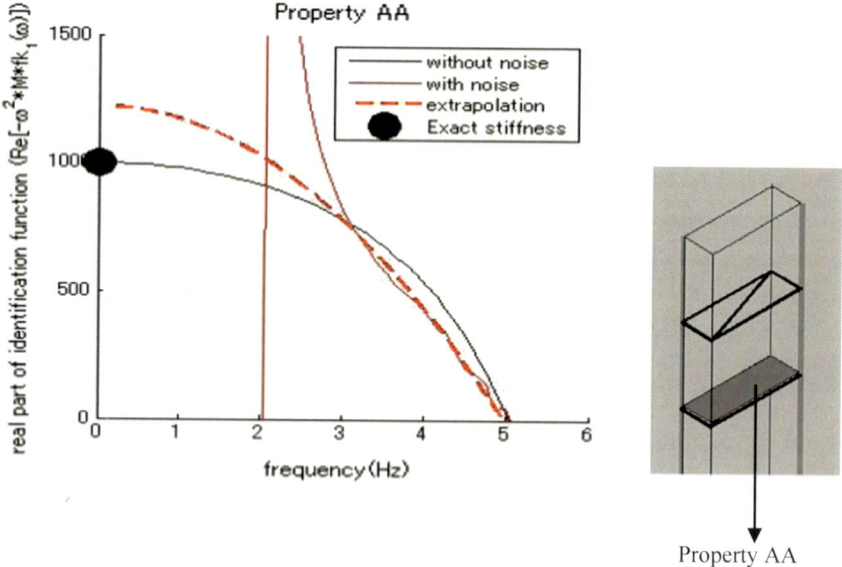

Fig. 17 Property AA—MATLAB attained graph analysis for stiffness (K) Vs Frequency (Hz)

For this frame analysis, exact gradient is taken for plot value as simulation with frequency (Hz) versus stiffness value at without noise condition. Thus, a unique discretization scheme is automatically generated (Fig. 17).

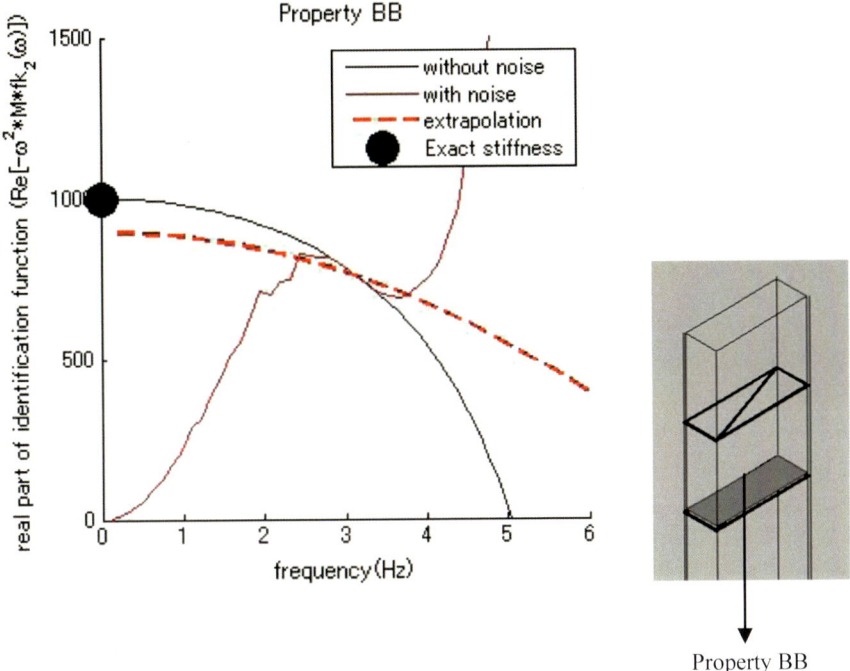

Fig. 18 Property BB—Frame analysis stiffness (K) Vs Frequency (Hz)

From the above-obtained result, the stiffness value for steel frame is same in both the cases of property position of sensors AA and BB with respect to real part of dynamic equation (Fig. 18).

From the above frame positions AA and BB, the analysis is based on the stiffness formulation without noise and with noise condition in different conditions.

From the above condition position of sensor property AA and BB, stiffness (K) of the steel frame obtained and exact stiffness value of steel frame is same in both the experimental and computational cases.

3 Conclusion

In this paper, an experimental test study of vibration analysis is performed on three-storey steel portal frame using vibration motor and computational approach is performed. Frame analysis using Matlab 2013a software and simulation technique is performed and compared with experimental approach done using wireless sensors. In this study, a new simulation algorithm is developed that generates gradient and extrapolation formulation for frame analysis. In this study, **image processing tool** is used for both programming output to Arduino and simulation for Matlab 2013 for analytical and processing approach. Significant future developments of this health monitoring of structures as a multidisciplinary research in structural dynamics, signal processing, motion and smart sensing technologies were discussed. Both experimental works in the lab and computational study were attained and discussed.

References

1. Noel AB, Abderrazak A (2017) Student member structural health monitoring using wireless sensor networks: a comprehensive survey. IEEE Commun Surv Tutorials 19(3)
2. Madhukumar N, Baiju MR (2017) MEMS based wireless sensor network for structural health monitoring. Int JA dv Res Electron Commun Eng (IJARECE) 6(8), ISSN: 2278–909X
3. Lu KC, Wang Y, Lynch JP, Lin PY, Loh C-H, Law KH (2015) Application of wireless sensors for structural health monitoring and control. In: The eighteenth KKCNN symposium on civil engineering december. Taiwan, 19–21
4. Rucka M, Wilde K (2006) Application of continuous wavelet transform in vibration based damage detection method for beams and plates. J Sound Vib 297:536–550
5. Study of SHM from an Integrated approach for structural health monitoring-2010 Prof. Rama shanker department of civil engineering Indian institute of technology Delhi
6. Rizos PF, Aspragathos N, Dimarogonas AD (1990) Identification of crack location and magnitude in a cantilever from the vibration modes. J Sound Vib 138(3):381–388
7. Reda Taha MM, Noureldin A, Lucero JL, Baca TJ ((2005)) Wavelet transform for structural health monitoring: a compendium of uses and features
8. Lynch JP, Wang Y, Sundararajan A, Law KH, Kiremidjian AS (2016) Wireless sensing for structural health monitoring of civil infrastructures
9. Shaladi R, Alasthan F, Yang C (2015) An overview on the applications of structural health monitoring using wireless sensor networks in bridge engineering. In: International conference on advances in science. In: Engineering technology and natural resources (ICASETNR-15) Aug. 27–28, 2015
10. Yan S, Ma H, Li P, Song G, Jianxin W (2017) Development and application of a structural health monitoring system based on wireless smart sensors. Shenyang Jianzhu University, School of Civil Engineering
11. Rogers CA (1990) Intelligent material systems and structures. In: Proceedings of U.S.-Japan workshop on smart/intelligent materials and systems, pp 11–33
12. Rytter A (1993) Vibration based inspection of civil engineering structures. Ph.D. Dissertation, Department of Building Technology and Structural Engineering, Aalborg University, Denmark

An Improved Image Pre-processing Method for Concrete Crack Detection

Harsh Kapadia, Ripal Patel, Yash Shah, J. B. Patel and P. V. Patel

Abstract Structure health monitoring of concrete structures has gained more attention in the recent years due to advancement in the technology. Different methods like acoustic, ultrasonic and image processing based inspection methods have been deployed to carry out an assessment of concrete structure. In this paper, work has been carried out to monitor the health of laboratory scale concrete objects using vision-based inspection. The objective is to provide a modified image pre-processing algorithms for accurate concrete crack detection. Different image processing based algorithms reviewed from existing literature were implemented and tested to detect cracks on the surface of a $15 \times 15 \times 15$ cm concrete cube. Due to random unevenness on the surface of concrete blocks, designing of an accurate and robust algorithm becomes difficult and challenging. Developed algorithm was applied to different images of concrete cubes. Receiver operating characteristics analysis and computation time analysis along with result images were discussed in the paper. In order to validate the applicability of developed algorithm, test results of crack detection on practical crack images are presented. Python was used to develop algorithm along with OpenCV library for image processing functions.

H. Kapadia (✉) · R. Patel · Y. Shah · J. B. Patel · P. V. Patel
Instrumentation and Control Engineering Department, Institute of Technology, Nirma University, Ahmedabad, Gujarat, India
e-mail: harsh.kapadia@nirmauni.ac.in

R. Patel
e-mail: 13bic037@nirmauni.ac.in

Y. Shah
e-mail: 13bic052@nirmauni.ac.in

J. B. Patel
e-mail: jbpatel@nirmauni.ac.in

P. V. Patel
e-mail: paresh.patel@nirmauni.ac.in

© Springer Nature Switzerland AG 2019
D. Pandian et al. (eds.), *Proceedings of the International Conference on ISMAC in Computational Vision and Bio-Engineering 2018 (ISMAC-CVB)*, Lecture Notes in Computational Vision and Biomechanics 30,
https://doi.org/10.1007/978-3-030-00665-5_149

1611

1 Introduction

Image processing involves processing of the digital image data to extract important information and/or features. The extracted information can either be used for autonomous machine perception or human perception or storage or transmission. To extract the information, there are numerous operations, algorithms and methods are available. Researchers around the globe are putting continuous efforts to bring improvements in them. There are numerous applications which cover a wide spectrum of engineering, science, agriculture and technology. In the majority of the cases, an image processing based product is deployed in applications to imitate the human wherein the task performed by human eye and brain is carried out by a camera and a computing device. With the advancement of technology, the prices of a good quality camera and computing devices have decreased to a greater extent. That has enabled people to explore new applications of image processing and improve the performance of the existing systems.

Majority of the inspection applications are carried out by image processing. It involves inspection of variations in a manufactured good against its specifications or may involve laboratory scale inspection of an object under test. Characteristics of the concrete structure are affected by an inferior standard of maintenance, fluctuation of temperature, an excessive amount of load. Cracks are the only external and primary indication of deformities which need to be inspected at earlier stages for strength and reliability of concrete structure [1]. Various kind of cracks is to be identified in distinct types of surfaces like bridges, building, tunnel, road, and pavement [2]. Examination of cracks in practical bases is requisite in a nuclear plant to find out any leakages to prevent a hazardous explosion. Detection and analysis of crack parts on a concrete structure are generally carried out manually in civil engineering which is time-consuming and includes human interference which may lead to high inaccuracy. Normally these inspections are carried out by experts to obtain accurate observation.

Image processing based inspection is useful due to its real-time application, precise accuracy, and cost-effectiveness and fully automated. Ultra-sonic and acoustics wave-based methods are more costly and difficult to implement in practical solution for structure monitoring [3, 4]. Imaging-based inspection approach gives a cost-effective, real-time solutions to monitor structure health [5, 6]. Various type of noises may occur in automatic crack detection such as irregular conditions in illumination, shading, and divots etc. in the images of the concrete structure. Abstruse to discriminate the difference between crack part, noise and non-crack part of the concrete surface.

Civil engineering laboratories perform a compression test, tensile test etc. on various concrete elements like a cube, column, junction etc. It is also evident that one needs to understand the concept and behaviour of different types of concrete in order to use them in practical applications. It is extremely important to assess their effectiveness by casting and testing them in the laboratory. A comparison of different image pre-processing and processing methods for crack detection have been implemented for concrete cubes of $15 \times 15 \times 15$ cm dimension. The motive behind this work is to automate and digitize the concrete cube testing process using the power of vision and computation.

2 Review on Algorithms

Researchers around the globe have explored different image processing based algorithms to address the problems of automatic crack detection. Crack detection methods using image can be based on visible, infrared, ultrasonic, laser, time of flight diffraction, etc. [4]. Different approaches like edge detection, pre-processing methods, fuzzy logic approach, neural networks, subtraction methods, various filter and thresholding methods etc. [7–10] were found in the existing literature. Applicability of these methods was tested on different surfaces of cracks like concrete structures, bridges, pipes, walls, columns, roads, pavements, tunnels etc. Most surfaces whose damage assessment and reliability are critical were considered in the literature.

Yamaguchi et al. had presented a novel approach to detect cracks in concrete surfaces [11]. They proposed an image-percolation model that identifies crack by referring to the connectivity of brightness and shape of the percolated region. They had assumed cracks were composed of thin interconnected textures. The validity of the proposed work was given by using receiver operating characteristics (ROC) analysis by means of experiments on actual concrete surface images. Fujita et al. had presented their work on crack detection [12]. They had proposed pre-processing method to suppress the effect of noises present on the concrete surfaces. An image subtraction method wherein an original image was subtracted from median filtered image to remove non-uniform illumination, shading etc. A hessian matrix based line filter was presented in the paper to emphasize on cracks. Crack part of the images was separated from background using thresholding the image with all values from 0 to 255. A comparison between 256 binary images and manually traced crack image was presented into evaluate the performance of the presented methods. Authors had done receiver operating characteristics analysis for the proposed method and included experimental results. It was concluded that the proposed methods were found effective for crack detection in practical situations. The authors did not mention the type of crack surfaces, brief literature and computation timings of the presented methods.

Atsushi et al. had proposed a crack analysis system for concrete block images. A high-resolution CCD camera was used to acquire crack images [13]. Shading correction, thresholding methods, crack tracking and labelling, feature extraction and analysis were applied to the acquired images and integrated into a tool. The authors had used sub-pixel interpolation method and a standard crack scale in order to measure physical crack area. Two calibration methods were discussed to investigate the efficiency of the proposed method. Further, improvements in the crack area measurement were suggested.

A method of crack detection for the bottom surface of concrete bridges was presented by Xu et al. Image segmentation of crack images was carried out using grey threshold iteration method and canny iteration method to achieve crack extraction [6]. Greying, histogram enhancement and gaussian filtering algorithms were used for pre-processing. Authors had presented a comparison and experimental analysis on the accuracy of two algorithms. In the conclusion, canny iteration method provided better crack extraction. Authors did not include any comparison of computation time.

An overview of different conventional and non-conventional methods of structural health monitoring was discussed by Sharma and Mehta. Variety of other non-destructive evaluation methods based on image processing were discussed along with useful outcomes, constraints, challenges etc. Applications of image processing to monitor structure health has matured and offers non-contact measurement, good amount area, mm accuracy in large structures and micron accuracy in laboratory scale structures [14].

Wang and Huang had summarized crack detection methods based on images and presented the comparison in four categories: integrated algorithm, morphological approach, percolation-based method, and practical technique. Experimental results were presented for each category of the method [10]. It was concluded that integrated algorithm is suitable for shading correction morphological approach gives better results compared to Otsu's threshold, it lacks precision do detect crack pixels. Further, percolation-based method gives best results as it operated on the image locally but the processing time is more compared to other methods. The practical technique is a semi-automatic method which requires human intervention and can give excellent performance for different types of concrete images.

3 Pre-processing Method

In recent times most of the infrastructure is composed of the concrete structure and crack detection is one of the best ways of monitoring health assessment of the structure as crack is one of the primary external, physical and visible parameter on a concrete surface to check the health of a structure. Cracks which appear on a surface are generally random, uneven, can be continuous or broken with random direction. Traditionally the inspection of a structure is carried out by trained professionals.

Vision-based methods can provide an automatic and accurate solution to the problem of automatic crack detection on a concrete surface. Application of vision-based inspection on practical structures is limited due to their large size and limited field of view of a good quality camera. In order to develop automatic crack detection system, laboratory scale concrete specimens were considered in the presented work. Cracks have characteristics like low-intensity values which matches to that of noises/dents on the surface, an area larger than noises/dents but smaller than the background. In the past several methods for crack detection have been proposed on various type of surfaces as discussed above in literature survey. A particular algorithm/method is not applicable for crack detection on all type of concrete surfaces. In this paper, different pre-processing methods were applied and tested on different cases of concrete cubes.

Figure 1 illustrates concrete cube ($15 \times 15 \times 15$ cm) image without crack and with cracks.

This paper consists discussion of pre-processing and processing methods as shown in Fig. 2. In order to remove interference objects like noises/dents present on the surface under observation, filter pre-processing was used. Gaussian, averaging, median, min-max etc. filters are widely popular. Bilateral filters can preserve the strong edges while removing the noises. Intensity difference function in bilateral filter only consider pixels with values equivalent to the central pixel value for blurring [15, 16]. Edges remain sharp due to large intensity variation but computation time is slightly higher as compared to other filters like gaussian filter. Image Subtraction is used as a pre-processing method to remove noises and further enable a better differentiation between crack and non-crack objects. Figure 3 shows sample concrete crack images considered for the study.

Fig. 1 Sample image of concrete cube with and without crack

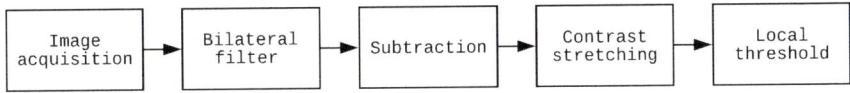

Fig. 2 Flowchart of crack detection

Fig. 3 Sample concrete crack images

3.1   Bilateral Filter Pre-processing

Bilateral filter is highly effective keeping the edges and removing unwanted noise/dents/interference objects etc. Gaussian and other filter considers nearby pixels while filtering, but does not consider the intensity of nearby pixels. Bilateral filter considers both nearby pixels and their intensity while filtering, so as to preserve edges which have large intensity variation [15]. Figure 4 shows the result of bilateral filter operation applied on sample images of Fig. 3. The result indicates that the edges were preserved and noises were removed.

3.2   Subtraction Pre-processing

A synthetic white image was subtracted from the filtered image to obtain a subtracted image. This will remove shadow present on the surface, corrects non-uniform illumination and would give a uniform illuminated image. Normally in vision based inspection system, the enclosed inspection area is provided with constant illumination conditions at the time of image capturing. Due to unsuitable conditions in the inspection, captured images were non-uniform illumination. Thresholding operation [17] was applied to a subtracted image to differentiate crack part from the background. Figure 5 illustrates result image of subtraction pre-processing method applied on sample images of Fig. 3.

Fig. 4   Result images of bilateral filter pre processing

Fig. 5   Result images of subtraction pre processing

Fig. 6 Result images of contrast stretching

3.3 Contrast Stretching Processing

Contrast stretching method was used to increase the difference between the intensity value of crack part and background. It is evident from the sample images seen in Fig. 1 that intensity values of crack part fall near to intensity values of background and noises due to greyish surface of the concrete cube. This method proves useful as it enhances the contrast of crack pixels and suppresses the non-crack pixels in the images. The range of intensity values of crack pixels was determined and enhance it while changing the pixel values of a non-crack part by multiplying pixel values with zero. This method gives more accurate results when it is applied after subtraction pre- processing method. Shading and non-uniform illumination are corrected by subtraction pre-processing method. Figure 6 shows image obtained after applying contrast stretching on subtracted image applied to sample images of Fig. 3. The result indicates shaded part and noise which are considered as cracks due to the similar intensity values.

3.4 Local Threshold Processing

In order to obtain or carry out crack detection, a threshold operation is required. It will separate crack and background part of the image which can be further utilized to perform crack analysis, measurement of crack length, width etc. and crack characterization. Traditionally, global thresholding like Otsu's thresholding is deployed in cases of bimodal (histogram) image. As already discussed and seen in Fig. 2, concrete surfaces has random noises and dents which hinder accurate crack detection. Local thresholding like niblack or adaptive mean or adaptive gaussian suits such cases [17, 18]. Here, niblack based local thresholding was used and the results of same can be seen in Fig. 7.

Fig. 7 Result images of local threshold

4 Experimental Results and Analysis

Discussed methods were tested for crack detection on more than 50 images similar
to the images seen in Figs. 1 and 3. All sample images which were captured by
DFK72AUC02 CMOS USB2.0 industrial camera by The Imaging Source GmbH,
Germany equipped with 8 mm C-mount lens by Computar, Japan. Camera has sensor
size 1/1.25" with micron CMOS sensor (MT9P031) and a rolling shutter [19]. It has
a maximum resolution of 5 MP and 6 fps. As it can be seen from above results that
most of the crack parts are identified from sample image (Fig. 3) with the combina-
tional use of two pre-processing and two processing methods. Receiver Operating
Characteristics (ROC) analysis presented in Table 1 is carried out to evaluate the
performance of methods on images of chosen 7 sample images of size 2048 × 1536
(3,145,728 pixels). Crack detection results of the chosen sample images were visually better in terms of crack detection. True positive fraction (TPF) is the ratio of no.
of crack pixels in the processed image to the no. of crack pixels in ideal image. It
signifies the ability of the method/process to be able correctly identify cracks in the
image. Process/method should also correctly to identify background as background
which is evaluated false positive fraction (FPF). It is the ratio of no. of background
pixels in processed image to the no. of background pixels in the ideal image. TPF
should be near to 1 and FPF should be near to 0 which is observed in Table 1 and chart
shown in Fig. 8. National instruments LabVIEW vision assistant tool (evaluation version) was utilized to obtain number of crack pixels and number of background pixels
in both ideal and processed image. The data mentioned in the Table 1 were obtained
using area parameter of particle analysis function available in vision assistant tool.
Computational time of methods tested on Fig. 3 is presented in Table 2. It gives a
measure of real-time implementation capability of the process/method discussed. As
a bilateral filter, contrast stretching and niblack threshold methods are both local processing methods, software code requires more computation time compared to that of
global processing methods. In Table 2 all measurement unit of time are in seconds.

Further testing of this methods are carried out on a different set of images which
consists various type of surfaces and these images are captured by smartphone. Most
of the crack parts are identified easily if there is a wide difference in the contrast
values of crack parts and background. Figure 9 row 1 shows sample image taken
from the white wall of a building, row 2 is a sample image of the crack surface taken

Table 1 ROC analysis of selected images

No. of crack pixels in ideal image	No. of crack pixels in processed image	True positive fraction (TPF)	No. of background pixels in ideal image	No. of background pixels in processed image	False positive fraction (FPF)
21,806	18,175	83.34	3,123,922	3,127,553	16.65
28,697	25,667	89.44	3,117,031	3,120,061	10.55
19,863	16,101	81.06	3,125,865	3,129,627	18.93
20,303	18,130	89.29	3,125,425	3,127,598	10.70
9766	8200	83.96	3,135,962	3,137,528	16.03
5738	4744	82.67	3,139,990	3,140,984	17.33
7543	5945	78.81	3,138,185	3,139,783	21.19

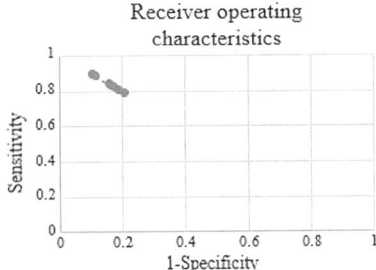

Fig. 8 ROC analysis of selected images

Table 2 Algorithm computation time for different image sizes

Method\image	Set 1 (1296 × 972)	Set 2 (2048 × 1536)	Set 3 (2592 × 1944)	Set 4(4032 × 3024)
Filter + Subtraction + Contrast stretching + Threshold	0.16	0.40	0.62	1.45

from the corridor which consists slightly uneven surface and row 3 is sample image taken from a road which has high unevenness and low contrast difference between crack and non-crack part. Figure 10 shows the threshold result of each image in the column. It is clearly seen that combination of subtraction pre-processing and contrast stretching method is effective to get more accurate crack detection results on the majority of the surfaces. The method can be very effective to find a crack if there is a significant difference of intensity value between foreground and background but for images with low contrast different like road surface shown in 3rd column of Fig. 10 accuracy is not very high.

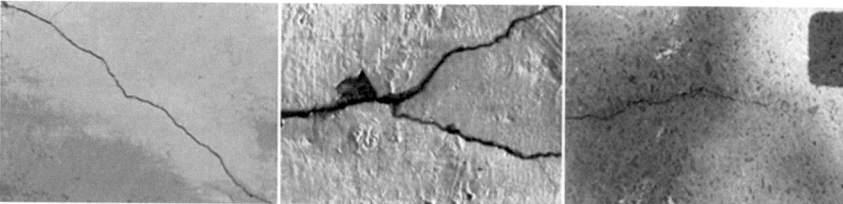

Fig. 9 Sample concrete crack images

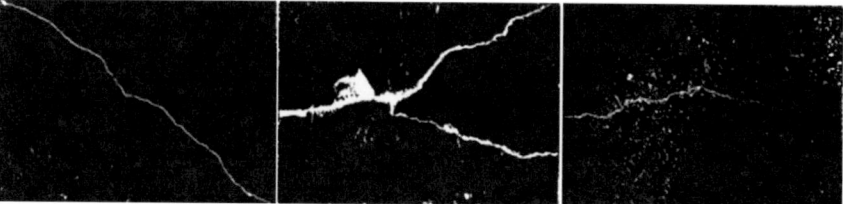

Fig. 10 Threshold images of sample images shown in Fig. 9

5 Conclusion

In this paper, two pre-processing and two processing methods are implemented for improved crack detection on different images on various crack surfaces. The methods were found effective when it is implemented on images with high contrast difference between crack part and background. Random unevenness on the surface of concrete cube presents challenges in the development of robust image processing algorithm for accurate crack detection. The proposed method detects cracks on the surface more accurately compared to the other methods suggested in the literature. Also, it is evident from the results that, a crack detection algorithm designed to identify cracks on the surface of a bridge will not provide similar accurate results when evaluated for different surfaces of concrete containing crack. The developed algorithm gives accurate results on the majority of the crack containing concrete surfaces as indicated in the Fig. 10 and presented ROC analysis. Results show that computation time increases with an increase in the size of an image. It is evident from the timings that, the algorithms can be useful for online/real-time crack detection implementation. A graphical user interface tool using python and OpenCV [20] can be developed to provide an open source crack analysis tool. It will allow engineers/students/researchers to carry on crack detection and analysis on different concrete surfaces.

References

1. Cheng C-C, Cheng T-M, Chiang C-H (2008) Defect detection of concrete structures using both infrared thermography and elastic waves. Autom Concr, pp 87–92

2. Adhikari RS, Moselhi O, Bagchi A (2014) Image-based retrieval of concrete crack properties for bridge inspection. Autom Constr 39:180–194

3. Koch Christian, Georgieva Kristina, Kasireddy Varun, Akinci Burcu, Fieguth Paul (2015) A review on computer vision based defect detection and condition assessment of concrete and asphalt civil infrastructure. Adv Eng Inform 29(2):196–210

4. Mohan A, Poobal S (2017) Crack detection using image processing: a critical review and analysis. Alexandria Eng J

5. Lee BY, Kim YY, Yi S-T, Kim J-K (2013) Automated image processing technique for detecting and analyzing concrete surface cracks. Struct Infrastruct Eng 9(6):567–577

6. Xu H, Tian Y, Lin S, Wang S (2013) Research of image segmentation algorithm applied to concrete bridge cracks. In: International conference on information science and technology (ICIST). IEEE, pp 1637–1640

7. Talab AMA, Huang Z, Xi F, HaiMing L (2016) Detection crack in image using Otsu method and multiple filtering in image processing techniques. Optik-Int J Light Electron Opt 127(3):1030–1033

8. Miyamoto A, Konno M-A, Bruhwiler E (2007) Automatic crack recognition system for concrete structures using image processing approach. Asian J Inf Technol Medwell J 6(5)

9. Choudhary GK, Dey S (2012) Crack detection in concrete surfaces using image processing, fuzzy logic and neural network. In: IEEE fifth international conference on advanced computational intelligence (ICACI). Nanjing, Jiangsu, China

10. Wang P, Huang H (2010) Comparison analysis on present image-based crack detection methods in concrete structure. In: 3rd international congress on image and signal processing (CISP). IEEE, pp 2530–2533

11. Yamaguchi T, Nakamura S, Saegusa R, Hashimoto S (2008) Image based crack detection for real concrete surfaces. IEEJ Trans Electr Electron Eng 3(1):128–135

12. Fujita Y, Mitani Y, Hamamoto Y (2006) A method for crack detection on concrete structure. In: 18th international conference on pattern recognition ICPR. IEEE, pp 901–904

13. Atsushi I, Aoki Y, Hashimoto S (2002) Accurate extraction and measurement of fine cracks from concrete block surface image. In: Proceedings of the 2002 28th annual conference of the ieee industrial electronics, pp 2202–2207

14. Sharma A, and Mehta N (2016) Structural health monitoring using image processing techniques-a review. Int J Mod Comput Sci 4(4):93–97

15. Tomasi C, Manduchi R. (1998) Bilateral filtering for gray and color images. In: Sixth international conference on computer vision, IEEE. pp 839–846

16. 2015. Smoothing Images. December 18. Accessed October 20, 2017. https://docs.opencv.org/3.1.0/d4/d13/tutorial_py_filtering.html

17. Niblack and Sauvola Thresholding. Accessed March 22, 2018. http://scikit-image.org/docs/dev/auto_examples/segmentation/plot_niblack_sauvola.html

18. Saxena LP (2017) "Niblack's binarization method and its modifications to real-time applications: a review. Artif Intell Rev 1–33

19. DFK 72AUC02 USB2.0 CMOS color industrial camera. Accessed November 2, 2017. https://www.theimagingsource.com/products/industrial-cameras/usb-2.0-color/dfk72auc02/

20. 2014. OpenCV-Python Tutorials. November 14. Accessed December 12, 2017. https://docs.opencv.org/3.0-beta/doc/py_tutorials/py_tutorials.html

Grape Crop Disease Classification Using Transfer Learning Approach

K. R. Aravind, P. Raja, R. Aniirudh, K. V. Mukesh, R. Ashiwin and G. Vikas

Abstract Grape is one of the important fruit crops which is affected by diseases. The advents of digital camera and machine learning based approaches have facilitated recognition of plant diseases. Convolution Neural Network (CNN) is one of the types of architecture used in deep learning based approach. AlexNet is a category of CNN which is used in this study for classification of three diseases along with healthy leaf images obtained from PlantVillage dataset. Transfer learning-based approach is used where the pretrained AlexNet is fed with 4063 images of above categories. The model achieved 97.62% of classification accuracy. Feature values from the different layers of the same network are extracted and applied to Multiclass Support Vector Machine (MSVM) for performance analysis. Features from Rectified Linear unit (ReLu 3) layer of AlexNet applied to MSVM achieved the best classification accuracy of 99.23%.

1 Introduction

Agricultural crops are threatened by the incidence of pests and diseases which in turn affect its production. The total loss of global production due to plant diseases is roughly estimated to be 10% [1]. Plant diseases are mainly caused by three pathogenic agents, namely, fungi, bacteria, and viruses [2]. Three diseases of grape crop, namely, black rot, black measles, and Isariopsis leaf spot, are considered in this study. Grape is one of the economically important crops with worldwide production of 7.8 million tons (2016) of which 39% are from Europe, 34% are from Asia, and 18% are from America [3]. The crop's production is affected widely by abovementioned diseases. Black rot is one of the deadly diseases caused by fungus *Guignardia bidwellii* which sometimes can cause 100% crop loss [4]. Black measles also called Esca caused by fungi are widespread in Europe which cause significant damage to the crop [5].

K. R. Aravind · P. Raja (✉) · R. Aniirudh · K. V. Mukesh · R. Ashiwin · G. Vikas
School of Mechanical Engineering, SASTRA Deemed University, Thanjavur, India
e-mail: raja@mech.sastra.edu

© Springer Nature Switzerland AG 2019
D. Pandian et al. (eds.), *Proceedings of the International Conference on ISMAC in Computational Vision and Bio-Engineering 2018 (ISMAC-CVB)*, Lecture Notes in Computational Vision and Biomechanics 30,
https://doi.org/10.1007/978-3-030-00665-5_150

Isariopsis leaf spot caused by fungi *Isariopsis clavispora* leads to serious damage to the crop under favorable climatic conditions [6].

Traditionally, plant diseases are diagnosed by experts through visual inspection which is prone to errors [2, 7]. Advances in science and technology are providing better tools and techniques for early detection, prevention, and control of pests and diseases [1]. Recent revolution in smartphones and digital cameras are paving a way to utilize these cheap image acquisition devices directly or by integration to an automation device for plant disease recognition applications [8]. Many studies have been carried out for foliage disease recognition using image processing techniques and machine learning algorithm in a variety of crops [8–21].

The recognition process involves classification of symptomatic patterns on the surface of leaves for identification of specific foliage diseases. Traditionally, disease recognition using image processing follows several steps such as preprocessing, color transformation, segmentation, feature extraction, and classification [12]. Dey et al. [13] presented a method where the initial image is cropped to reduce Central Processing Unit (CPU) processing time and storage. The image was transformed from Red Green Blue (RGB) color space to Hue Saturation Value (HSV). A thresholding operation based on Otsu method was carried out in hue channel of the HSV image for segmentation, and rotten area of the affected leaf was identified. In a study, Arivazhagan et al. [10] demonstrated a method where initially, image from Red Green Blue (RGB) space is converted to Hue Saturation Intensity (HSI) space and segmentation was carried out using thresholding operation. Textural features such as energy, homogeneity, contrast, cluster shade, etc. were extracted and provided as input to Support Vector Machine (SVM) which is used widely for classification problem. Qin et al. [11] presented a method where the pixels in the image were clustered using K median and fuzzy C-means clustering algorithms into ten classes. Based on the mean of hue component and application of classification algorithm (Naive Bayes, logistic regression, and linear discriminant analysis) using a^* and b^* values, the pixels are classified as a part of a lesion or healthy region. Among the total 129 extracted features, best features were selected using different methods (1-rule, correlation-based feature selection, and ReliefF method), and results were compared. In all the above-discussed studies, features of interest are extracted and tested for its accuracy. The accuracy of classification depends on the selected features and it is a time-consuming process.

Recent advances in the development of complex architectures using artificial neural network laid foundation for deep learning based classification. Different approaches are followed for building the network which resulted in many deep learning based algorithm, namely, autoencoder, sparse coding, Convolutional Neural Networks (CNN), and restricted Boltzmann machines. Among these algorithms, CNN is widely used for vision-based application [22]. AlexNet, GoogLeNet, Visual Geometry Group (VGG), etc. are some of the examples of CNN-based architectures. Few studies have already explored the use of CNN in plant disease classification [8, 14, 18]. In this study, three diseases of grape crops along with healthy leaves are classified using the pretrained AlexNet framework which is known as transfer learning. In addition, feature values are extracted from the different layers of a pretrained AlexNet framework and provided as input to the Multiclass Support Vector Machine (MSVM). The classification performance of the features from each layer is compared and analyzed.

2 Method

2.1 Dataset

The segmented image datasets for the three diseases, namely, black measles, black rot, and Isariopsis leaf spot (as shown in Fig. 1), along with the healthy leaves are obtained from PlantVillage dataset which consists of 54,306 images [23]. The obtained subset consisting of three diseases and healthy leaf images constitutes 4,063 segmented images.

2.2 Implementation

AlexNet is a category of CNN which consist of five convolution layers, seven rectified linear unit (ReLu), three max pooling layers, three fully connected layers, two normalization layers, two drop out layers, and one softmax layer as shown in Fig. 2. The network was pretrained with ImageNet dataset and participated in the ImageNet Large-Scale Visual Recognition challenge (ILSVR—2010) with 1.2 million images of over 1000 categories. The network ability for classification was tested and proved to be efficient than the previous benchmark. A variant of the AlexNet was implemented in ILSVR—2012 resulting in better error rates with its implementation [24].

Training of these networks with the new class of images for disease recognition requires a large dataset and time-consuming process. Hence, transfer learning tech-

(a). Black measles   (b). Black rot   (c). Isariopsis leaf spot

Fig. 1 Segmented image of the grape crop leaves from PlantVillage dataset

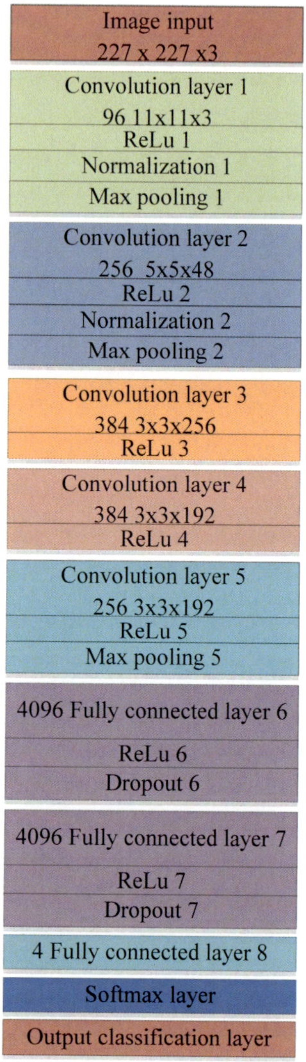

Fig. 2 AlexNet architecture
implemented for this study

Image input
227 x 227 x3

Convolution layer 1
96 11x11x3
ReLu 1
Normalization 1
Max pooling 1

Convolution layer 2
256 5x5x48
ReLu 2
Normalization 2
Max pooling 2

Convolution layer 3
384 3x3x256
ReLu 3

Convolution layer 4
384 3x3x192
ReLu 4

Convolution layer 5
256 3x3x192
ReLu 5
Max pooling 5

4096 Fully connected layer 6
ReLu 6
Dropout 6

4096 Fully connected layer 7
ReLu 7
Dropout 7

4 Fully connected layer 8
Softmax layer
Output classification layer

nique is used where the AlexNet is pretrained with ImageNet. The features or layer
weights learned from the pretraining process is preserved for its application in disease
recognition process. The size of the segmented image from the dataset is 256×256
which was resized to 227×227 as the input layer of AlexNet in Matlab can accept
image in the above-specified dimension. The convolution layers 1 and 2 is followed
by ReLu, normalization, and max pooling. ReLu is a nonlinearity-based activation
function implemented in the output of all convolution layers. Cross-channel normalization is performed to amplify the output of the excited neurons which have higher
activation and suppress the output of the local neighborhood neurons. Max pooling

is done to reduce the dimension of the output from the previous convolution layers by finding the maximum value within the kernel applied to the output of the convolutional layer. The output from the convolution layer 3 and 4 is fed to the ReLu, while output from layer 5 is provided to both ReLu and max pooling. The output from the previous layer is fed to the two 4096 fully connected layers where all the neurons are connected to each other. In dropout layers 50% of the neurons are disconnected randomly from the adjacent neurons momentarily and training is carried out using the resulting network. This method has found to lower the generalization errors. Originally, AlexNet was trained to classify 1000 categories of object; hence, it had 1000 fully connected layer. In this study, it has been replaced with four fully connected layers as only four categories are used. The softmax layers provide the probability value of the given sample image to all the class labels. The label with the highest probability is the correct class label.

The AlexNet consists of 23 layers excluding the input and the output layers. Feature values from each layer are extracted and fed as input to the MSVM. The MSVM model was implemented with "one versus all" approach where the multiclass problem was split into several binary classification problems. The result for the features from each layer is compared and discussed in Sect. 3.

2.3 Visualization of the Activation Map

In an artificial neural network with few layers and neurons, the features are extracted from the image and fed as input to the network. Contrastingly, in AlexNet, a category of CNN automatically learns to identify features. The AlexNet have five convolution layers, wherein each layer, the convolution operation is performed using a kernel matrix of specific dimension and applied over the input image. The applied matrix moves over the entire image, and an activation map or feature map is obtained as a result of the convolution operation in each convolution layer.

The number of kernel matrix and its dimension varies for each layer as shown in Fig. 2. For example, the convolution layer 1 consists of 96 kernel matrix with a dimension of $11 \times 11 \times 3$. Hence, 96 different channels of activation maps are obtained for the given input image by passing it through the first convolution layer with a total of 290,400 values. The activation map shows the specific region for which the neuron becomes active while applying the kernel over the image. The activation map can be visualized using the "montage" function in Matlab after normalization of the values. An activation map generated from the sample image of black measles is shown in Fig. 3. The activation maps are reshaped to an 8×12 grid for presenting all the 96 channels in a single image. Each channel shows a region with white, black, and gray color.

The region with white denotes positive activation, while black depicts negative activation. The region with gray is positive but not as strong as the region with white color. In Fig. 3, region with disease symptoms has high positive activation in 10th (as shown in Fig. 4) and 32nd channels, and healthy region has high negative activation,

Fig. 3 Features map obtained from convolution layer for black measles

Fig. 4 Comparison of the activation channel 10 with RGB image

whereas in 41, 45, 56, and 78th channels, it has opposite effect in activation. Figure 4 shows the comparison of activation map of 10th channel and RGB image.

The activation map obtained from convolution layer acts as the input to the second convolution layer after passing through ReLu, normalization, and max pooling layers. The convolution operation is performed using 256 kernels with the dimension of $5 \times 5 \times 48$ in the second layer. Certain mid-level features are obtained in the second layer. Subsequent pass through the other layer will result in activation maps which combines the low-level features obtained from the other layers into complex features. The features from these various layers are obtained, and its accuracy in classification using MSVM for the given set of images are analyzed which is discussed in next section (Sect. 3). Also, the accuracy resulting from AlexNet using transfer learning is discussed.

3 Results and Discussions

The system used for the study have GTX 1050 4 GB Graphics Processing Unit (GPU) with 8 GB Random Access Memory (RAM). The segmented image of dimension 227×227 from the dataset is fed to the AlexNet. 80% of the images from each class label were used for training and 20% were used for testing. Table 1 shows the number of images present in each class label.

The weight and bias learn rate of all the previous layers are not changed except the last fully connected layers. The weight learn rate is set to 10 and bias learn rate to 20. This means that the weight and bias learn rate are 10 and 20 times faster than the global learning rate, respectively. The maximum number of epoch is set as the default value 10, and a mean accuracy of 97.62% is obtained as shown in Table 2. The number of feature values obtained from each layer is shown in Table 3.

The number of feature values of ReLu layer is same as the preceding convolution layer. The dimension of features reduces as it passes through the subsequent convolution layers in the AlexNet. The accuracy of the classification using MSVM by obtaining the feature values from the five convolution layers is shown in Fig. 5. The best mean accuracy of 99.16% was obtained using the features obtained from the convolution layer 4. Feature values from the convolution layer 1 resulted in a poor accuracy of 82.61% as only simple features such as edges, color information, etc. are captured which is not sufficient for classification. The performance of the values obtained from the other layers (as shown in Table 4) was also analyzed.

Table 1 Grape leaves dataset

Class label	No. of images
Black rot	1180
Black measles	1384
Healthy	423
Isariopsis leaf spot	1076
Total	4063

Table 2 Accuracy of AlexNet in classification

Accuracy				
Black rot (%)	Black measles (%)	Isariopsis leaf spot (%)	Healthy (%)	Mean (%)
94.07	95.31	99.53	100	97.62

Table 3 Number of features from each layer

Layers	No. of feature values
Convolution layer 1	290,400
Convolution layer 2	186,624
Convolution layer 3	64,896
Convolution layer 4	64896
Convolution layer 5	43,264
Max pooling layer 1	69,984
Max pooling layer 2	43,264
Max pooling layer 5	9216
Drop 6 and 7	4096

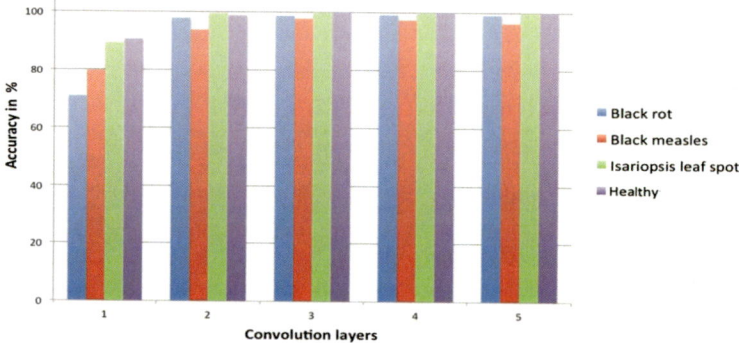

Fig. 5 Accuracy of classification for features from convolution layer 1–5

The feature values obtained from ReLu 3 resulted in the best mean accuracy of 99.23%. The mean classification accuracy has improved significantly when feature values from the different layers of the AlexNet are extracted and applied to shallow classification algorithm such as MSVM. AlexNet performance was better when compared with traditional approaches such as the study by Wang et al. [25] where two kinds of grape disease, namely, grape powdery mildew and Downy mildew with a limited dataset, were classified. The prediction accuracy was 100% using Radial Basis Function (RBF) neural network while with Backpropagation networks (BP), generalized regression networks, and Probabilistic Neural Network (PNN) yielded 94.29% of accuracy. Four shape features, 21 color features, and 25 textural features were extracted for classification, whereas in our case the features are extracted automatically. The number of dataset used in the study is relatively low of about 50 images for each category, but in our case 4063 images of standard PlantVillage dataset have been used. In another study, scab and rust of grape crop have been recognized using a total of 1478 images with Gabor filter-based features and color features. MSVM have been used for classification with an average accuracy of 86.03% which is lower

Table 4 Accuracy of classification using features from different layers

Layers	Black rot (%)	Black measles (%)	Isariopsis leaf spot (%)	Healthy (%)	Mean accuracy (%)
ReLu 1	87.29	90.97	97.21	97.65	93.28
Normalization 1	89.41	93.86	98.14	100	95.35
Max pooling 1	97.46	93.50	99.53	100	97.62
ReLu 2	97.46	94.95	100	100	98.10
Normalization 2	98.31	95.67	100	100	98.49
Max pooling 2	99.15	97.47	100	100	99.16
ReLu 3	98.73	98.19	100	100	99.23
ReLu 4	98.73	97.47	100	100	99.05
ReLu 5	97.03	97.83	100	100	98.72
Max pooling 5	97.88	98.19	100	100	99.02
4096 fully connected layer 6	96.61	97.11	100	100	98.43
ReLu 6	97.88	97.11	100	100	98.75
Dropout 6	97.88	97.11	100	100	98.75
4096 fully connected layer 7	96.61	97.11	100	100	98.43
ReLu 7	98.73	97.11	100	100	98.96
Dropout 7	98.96	97.11	100	100	98.96

than the accuracy discussed in this study [26]. Therefore, by extracting features from different layers of pretrained AlexNet and applying it to MSVM increases the accuracy significantly which is evident from the results. The results also suggest that deep learning methods are better than the methods where feature is extracted manually and applying it to the shallow learning algorithms.

4   Conclusion

Deep learning approaches require large dataset for training and results in better classification. When the number of the dataset is relatively low, a pretrained network can be used for new classification problem which is known as transfer learning. Further, the accuracy can be improved by extracting feature values from the trained AlexNet and applying it to the shallow learning based classification algorithm. The

AlexNet yielded an accuracy of 97.62%, while the classification with MSVM using feature values from different layers resulted in improved performance. Feature values from ReLu 3 resulted in best performance of 99.23% accuracy. Further evaluation can be done by using different datasets and machine learning algorithms, expanding it to other disease recognition applications.

References

1. Strange RN, Scott PR (2005) Plant disease: a threat to global food security. Phytopathology 43:83–116
2. Gavhale KR, Gawande U (2014) An overview of the research on plant leaves disease detection using image processing techniques. IOSR J Comput Eng 16(1):10–16
3. World Vitinviniculture Situation, OIV Statistical Report on World Vitinviniculture, International Organisation of Vine and Wine. www.oiv.int/public/medias/5479/oiv-en-bilan-2017.pdf
4. Sosnowski MR, Emmett RW, Wilcox WF, Wicks TJ (2012) Eradication of black rot (*Guignardia bidwellii*) from Grapevines by drastic pruning. Plant Pathol 61:1093–1102
5. Mugani L, Graniti A, Surico G (1999) Esca (Black Measles) and brown wood-streaking: two old and elusive diseases of Grapevines. Plant Dis 83:5
6. Silva CMD, Estrada KRFS, Rios CMFD, Batista BN, Pascholati SF (2014) Effect of culture filtrate of *Curvularia inaequalis* on disease control and productivity of Grape cv. Isabel. Afr J Agric Res 9(40):3001–3010
7. Garcia J, Barbed A (2016) A review on the main challenges in automatic plant disease identification based on the visible range images. Biosys Eng 144:52–60
8. Mohanty SP, Hughes DP, Salathe M (2016) Using deep learning for image-based plant disease detection. Front Plant Sci 7, Article ID 1419
9. Hiary HA, Ahmar SB, Reyalat M, Braik M, Rahamneh ZA (2011) Fast and accurate detection and classification of plant diseases. Int J Comput Appl 17(1):31–38
10. Arivazhagan S, Shebiah RN, Ananthi S, Varthini SV (2013) Detection of unhealthy region of plant leaves and classification of plant leaf diseases using textural features. Agric Eng Int CIGR J 15(1):211–217
11. Qin F, Liu D, Sun B, Ruan L, Ma Z, Wang H (2016) Identification of Alfalfa leaf diseases using image recognition technology. PLoS ONE 11(12):e0168274
12. Singh V, Misra AK (2017) Detection of plant leaf diseases using image segmentation and soft computing techniques. Inf Process Agric 4:41–49
13. Dey AK, Sharma M, Meshram MR (2016) Image processing based leaf rot disease, detection of Betel Vine (*Piper BetleL.*). Procedia Comput Sci 85:748–754
14. Sladojevic S, Arsenovic M, Anderla A, Culibrk D, Stefanovi D (2016) Deep neural network based recognition of plant diseases by leaf image classification. Comput Intell Neurosci 2016, Article ID 3289801
15. Barbedo JGA, Koenigkan LV, Santos TT (2016) Identifying multiple plant diseases using digital image processing. Biosys Eng 147:104–116
16. Mahlein AK (2016) Plant disease detection by imaging sensors—parallels and specific demands for precision agriculture and plant phenotyping. Plant Dis 241–251
17. Zhang S, Wu X, You Z, Zhang L (2017) Leaf image based cucumber disease recognition using sparse representation classification. Comput Electron Agric 134:135–141
18. Ferentinos KP (2018) Deep learning models for plant disease detection and diagnosis. Comput Electron Agric 145:311–318
19. Ali H, Lali MI, Nawaz MZ, Sharif M, Saleem BA (2017) Symptom based automated detection of citrus diseases using color histogram and textural descriptors. Comput Electron Agric 138:92–104

20. Zhang S, Wang H, Huang W, Zhuhong Y (2018) Plant disease leaf segmentation and recognition by fusion of superpixel, K-means and PHOG. Optik 157:866–872

21. Xu P, Wu G, Guo Y, Chen X, Yang H, Zhang R (2017) Automatic wheat leaf rust detection and grading diagnosis via embedded image processing system. Procedia Comput Sci 107:836–841

22. Guo Y, Liu Y, Oerlemans A, Lao S, Wu S, Lew SL (2016) Deep learning for visual understanding: a review. Neurocomputing 187:27–48

23. Hughes DP, Salathe M. An open access repository of images on plant health to enable the development of mobile disease diagnostics. arXiv:1511.08060

24. Krizhevsky A, Sutskever I, Hinton GE (2012) Imagenet classification with deep convolutional neural network. In: Advances in neural information processing systems. Lake Tahoe, pp 1097–1105

25. Wang H, Li G, Ma Z, Li X (2018) Application of neural networks to image recognition of plant diseases. In: International conference on systems and informatics. Yantai, pp 2159–2164

26. Meunkaewjinda A, Kumsawat P, Attakitmongcol K, Srikaew A (2008) Grape leaf disease detection from color imagery using hybrid intelligent system. In: International conference on electrical engineering/electronics, computer, telecommunications and information technology. Krabi, pp 513–516

Exploring Image Classification of Thyroid Ultrasound Images Using Deep Learning

K. V. Sai Sundar, Kumar T. Rajamani and S. Siva Sankara Sai

Abstract Deep learning for medical imaging has been at the forefront of its numerous applications, thanks to its versatility and robustness in deployment. In this paper, we explore various classification methodologies that are employed for datasets of relatively small in size to actually train a deep learning algorithm from scratch. Thyroid ultrasound images are classified using a small CNN from scratch, transfer learning and fine-tuning of Inception-v3, VGG-16. We present a comparison of the aforementioned methods through accuracy, sensitivity, and specificity.

1 Introduction

1.1 Background

Image classification task of a machine learning/deep learning algorithm is widely found to be useful in the medical imaging domain, wherein the algorithm is trained to classify medical images into benign or malignant in case of cancer detection and various other ailments based on the symptoms in the images. Arriving at automated diagnosis seems to be the dream of every researcher associated with computer vision coupled with biomedical imaging. Contrary to famous datasets like the ImageNet dataset, Cifar-10, and Cifar-100 which have thousands of images in each category, the availability of medical data is limited. When running a deep learning algorithm with millions of parameters, less data hurdles the training with overfitting. The model eventually tends to fail at generalization of learning giving low accuracy on the test dataset. Regularization can reduce the high variance to some extent but training a deep learning framework from scratch remains out of bounds. Data augmentation

K. V. Sai Sundar (✉) · S. Siva Sankara Sai
Sri Sathya Sai Institute of Higher Learning, Puttaparthi, India
e-mail: saisundarkandarpa@gmail.com

K. T. Rajamani
Robert Bosch, Bangalore, India

© Springer Nature Switzerland AG 2019
D. Pandian et al. (eds.), *Proceedings of the International Conference on ISMAC in Computational Vision and Bio-Engineering 2018 (ISMAC-CVB)*, Lecture Notes in Computational Vision and Biomechanics 30,
https://doi.org/10.1007/978-3-030-00665-5_151

can be employed to further boost the dataset size. Therefore, in order to arrive at accuracies which can be of deployment standard, we resort to training a smaller CNN from scratch, transfer learning—using bottleneck features from deep CNNs to train a new FC layer or a different classifier and finally fine-tuning deep architectures to classify the custom dataset.

1.2 Motivation

The thyroid ultrasound domain was chosen following consultation with local doctors who brought to our attention the high prevalence of thyroid ailments in the region. Thyroid ultrasound is predominantly used to detect thyroid nodules, classify them as benign or malignant and also to identify goiter, thyroiditis. The problem consists of binary classification initially identifying images of patients who probably require a biopsy in order to confirm malignancy of nodule and eventually multi-class classification identifying various other ailments apart from cancer. The aim of this work is to develop an automated thyroid diagnosis system that could aid the radiologist and fasten the diagnosis. In this paper, we only discuss our implementations of binary classification.

2   Dataset Description

We have used two datasets in this work. The first dataset [1] is a publically available one consisting of 298 images and their corresponding biopsy verified reports in.xml format. The TIRADS scores are given for each of the images ranging from 2 to 5 on the scale of increasing probability for malignancy. Since our task in this work dealt only with probably benign or malignant test scenario, we considered scores 2 and 3 as benign and all the scores above these as malignant. The second dataset used in this work was the local database of images from GE LOGIQ P9 which were labeled by an experienced doctor and the reports were written in word format. The various cases of cancerous nodule, thyroiditis, simple goiter, multinodular goiter, toxic goiter, and normal were present in this database. Again, we considered only the relevant images as mentioned for the previous dataset. This dataset consisted of thyroid images of 127 patients.

3   Frameworks and Hardware Used

The deep learning frameworks TensorFlow [2] and Keras [3], based in Python have been used in this work. TensorFlow, released by Google is currently one of the most widely used frameworks that come with built-in parallelism that enables usage of multi-threading to optimize learning. Keras is a high-level API that uses popular frameworks as the backend to run deep learning models. The model zoo provides the open-source implementation of the latest state-of-the-art neural networks in these frameworks. The hardware used in the training of the various methodologies discussed in the next section is as follows:

- GPUs—NVIDIA TitanX, NVIDIA Tesla K20c, and
- CPU—Intel i5-4570 CPU @ 3.20 GHz.

4 Deep Learning Architectures

Ever since the breakthrough deep learning model AlexNet won the ImagNet competition of 2012, many deep learning models have risen each bettering the previous best model. The inception [4] model was the winner of the 2014 ILSVRC 2014 image classification challenge extension of which the equally famous is Inception-v3 [5] network. The inception architecture shot into fame by making neural networks even deeper by stacking a series of inception modules. These modules consisted of parallel filters that operated on the input whose outputs were cascaded so that high-level information is not lost. The VGG-16 [6] on the other hand repeated 3×3 convolutions with three huge FC layers completing the structure. Both these architectures were designed to classify ImageNet data of 1000 classes, and trained on cluster of GPUs. Hence, fine-tuning and transfer learning use pretrained weights and modify the last few layers of these models, keeping in context the deep feature extraction by these models.

5 Classification Methodology

5.1 Training a Small CNN from Scratch

The first method constituted training a CNN from scratch using the medical data. The layer architecture is three convolutional layers with 3×3 kernels of numbers 32, 32, and 64, respectively (Fig. 1).

The features from the FC layer were classified using a regular sigmoid function into the two classes benign or malignant. The training of the CNN was done on the GPU and since the model is relatively shallow and the dataset small, the training is completed in an hour's time.

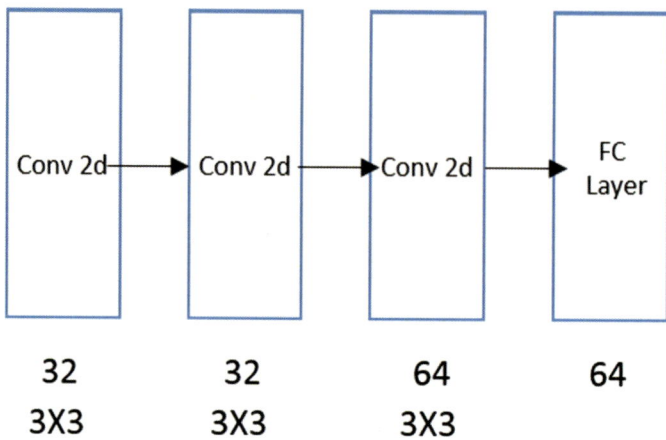

Fig. 1 CNN architecture which was trained from scratch

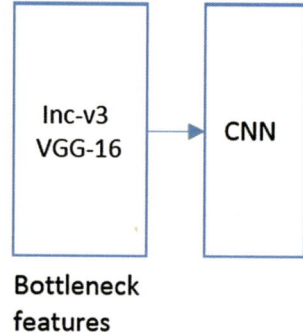

Fig. 2 Bottleneck features+CNN

5.2 Transfer Learning

Bottleneck features and the CNN: The bottleneck features from the VGG-16 and Inception-v3 were obtained and then trained on the CNN from the previous method. In VGG-16, the last three FC layers were discarded and the CNN model which we used in the first method was fed these features. The ImageNet pretrained weights were loaded into the models and after the forward pass of the image through the network bottleneck features was saved (Figs. 2 and 3).

Bottleneck features and SVM: The CNN was replaced with the popular linear classifier support vector machine which was fed the bottleneck features for classification. The simple default parameters of the SVM implementation provided by the scikit-learn were used in this method. Deep CNNs are known to be excellent feature extractors and using linear classifier to use these features proves to be an excellent way of tackling smaller datasets [7].

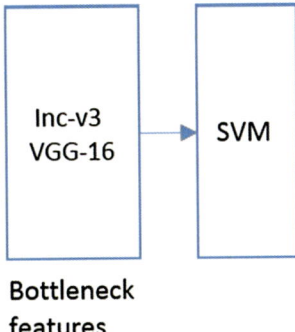

Fig. 3 Bottleneck features + SVM

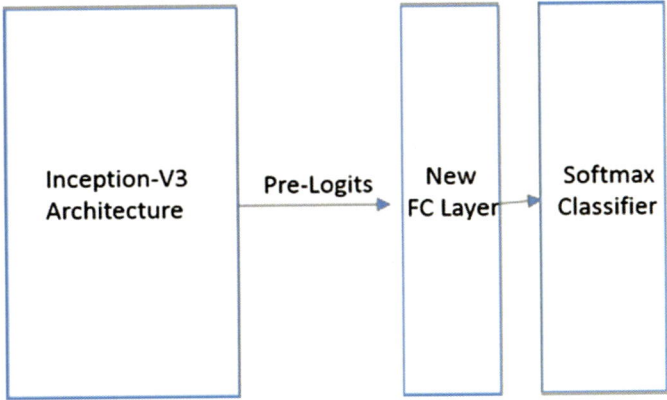

Fig. 4 Fine-tuning Inception v-3

5.3 Fine-Tuning Inception-v3 and VGG-16

The Inception-v3 model was imported with the help of tf-slim high-level API provided by TensorFlow. With the help of checkpoints provided for each of the models, pretrained models could be availed and fine-tuned. The Inception-v3 net provided in the slim API returns the list "end points" and "logits" which can be fed to a classifier to predict the class. We obtained the end points["pre-logits"] which is a layer prior to the last layer in the architecture and customized the FC layer, to give output as a binary classifier. Softmax classifier was used for the classification (Figs. 4 and 5).

The last three FC layers of the VGG-16 which contribute to huge computations were discarded, and new FC layer was attached after the "pool-5" layer. For fine-tuning, all layers above the conv 5_2 were frozen. So essentially the last three layers (excluding the FC layers) and the new custom FC layer were trained in this approach.

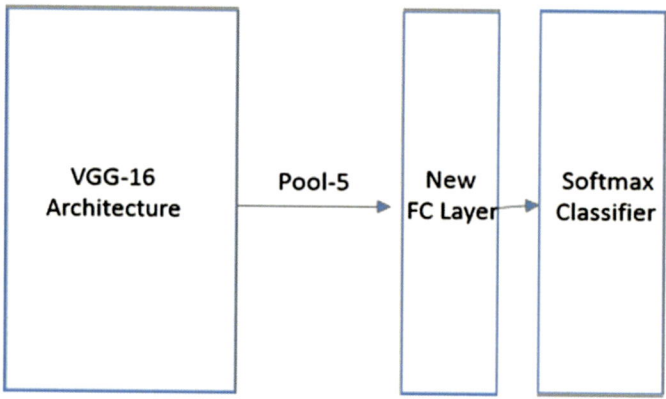

Fig. 5 Fine-tuning VGG-16

Table 1 Summary of classification results on the combined dataset

Sl No.	Model	Accuracy	Sensitivity	Specificity
1	CNN from scratch	0.82	0.82	0.89
2	VGG-16+CNN	0.79	0.84	0.74
3	Inception-v3+CNN	0.93	1.0	0.87
4	VGG-16+SVM	0.94	1.0	0.88
5	Inception-v3+SVM	0.89	0.88	0.89
6	VGG-16 fine-tuning	0.89	0.88	0.9
7	Inception-v3 fine-tuning	0.79	0.77	0.8

6 Results

The metrics used for evaluating the aforementioned methods are accuracy, i.e., ratio of the total number of correct predictions to the total number of images predicted, sensitivity, which gave an indication of true positive rate and specificity for true negative rate. The classification was done on both the datasets separately and on the combined dataset. It was observed that the first public dataset gave high sensitivity and low specificity, while the second dataset gave just the opposite. This is due to the nature of the data, wherein the first public dataset consisted of biased data with number of cancerous samples on the higher side. The local dataset on the other hand had the bias toward normal samples. This problem was handled by combing the datasets and also data augmentation achieved by flipping, rotating, and adding noise to the existing images. The table below summarizes the results obtained on the combined dataset alone. This combined dataset consisted of 2525 training samples and 613 test samples. The metrics are tabulated as percentages (Table 1).

7 Discussion

The delicate balance between the various evaluation metrics is important for performance analysis of deep learning algorithm in biomedical imaging. Owing to the fact that the dataset size is small, fine-tuning the deep architectures resulted was bettered by the other approaches. The amalgamation of linear classifier like the SVM to the architecture (especially the VGG-16) turned out to be the best model with stable metrics. Linear classifiers are considered to be useful when the deep neural networks are trained on datasets which are very different from domain in question. In order to take the classifier to a useful deployment, we are currently exploring system-based GUI and mobile-based android application of TensorFlow. Also, multi-class classification of the various thyroid ailments is being explored along with data collection and ground truth establishment in this regard.

Acknowledgements We would like to express our gratitude to our founder chancellor Bhagwan Sri Sathya Sai Baba for His inspiration. We would like to thank Department of Physics and Department of Mathematics and computer science for their support and also Dr. Narsimhan and Dr. Trimurthy from the Sri Sathya Sai Mobile hospital for their valuable insights. Consent of all participants was taken.

References

1. Pedraza L, Vargas C, Narváez F, Durán O, Muñoz E, Romero E (2015) An open access thyroid ultrasound image database
2. Abadi M, Agarwal A, Barham P, Brevdo E, Chen Z, Citro C, Corrado GS, Davis A, Dean J, Devin M et al (2017) Tensorflow: large-scale machine learning on heterogeneous distributed systems. arXiv preprint arXiv:1603.04467
3. Chollet F Keras. Github repository, https://github.com/fchollet/keras
4. Szegedy C, Liu W, Jia Y, Sermanet P, Reed S, Anguelov D, Erhan D, Vanhoucke V, Rabinovich A (2015) Going deeper with convolutions. In: The IEEE conference on computer vision and pattern recognition (CVPR)
5. Szegedy C, Vanhoucke V, Ioffe S, Shlens J, Wojna Z (2017) Rethinking the inception architecture for computer vision. In: Proceedings of the IEEE conference on computer vision and pattern recognition, pp 2818–2826
6. Simonyan K, ZissermanVery A (2015) Deep convolutional networks for large-scale image recognition, arXiv:1409.1556v6
7. Sharif Razavian A, Azizpour H, Sullivan J, Carlsson S (2014) CNN features off-the-shelf: an astounding baseline for recognition, arXiv:1403.6382
8. Chi J (2017) Thyroid nodule classification in ultrasound images by fine-tuning deep convolutional neural network. J Digit Imaging 30:477–486
9. LeCun Y, Bengio Y, Hinton G (2015) Deep learning. Nature 521(7553):436–444
10. Géron A (2017) Hands-on machine learning with Scikit-learn and tensorflow. O'Reilly Media, Inc., 1005 Gravenstein Highway North, Sebastopol
11. Codebox.: Image_augmentor, https://github.com/codebox/image_augmentor
12. Lee J-G, Jun S, Cho Y-W, Lee H, Kim GB, Seo JB, Kim N (2017) Deep learning in medical imaging: general overview. Korean J Radiol 18(4):570–584

Medical Applications of Additive Manufacturing

A. Manmadhachary, Santosh Kumar Malyala and Adityamohan Alwala

Abstract Additive manufacturing (AM) is also known as "3D printing." AM creates the physical model by construction of succeeding layers using the input material. Each succeeding layer is attached to the preceding layer to form the physical AM model. Medical industry requires the error-free exact anatomy of the patient for diagnosis, surgical planning, surgical guides, implants, etc. Since every patient has unique anatomy, AM suits as best fit for medical industry. AM medical model provides advantage of customizing each model according to the patient's specific requirement, which is not so easy in conventional way. The current work aims to explain the importance of AM medical model in the medical industry for various scenarios.

1 Introduction

Chunk Hull has got the thought of AM in the year 1986 and named it as stereolithography apparatus (SLA). The working principle behind the AM is same as inkjet printer. To generate a 3D or physical model, the input data is initially generated into form of layers using a preprocessing software [1]. AM models are to be fabricated using computer-aided design (CAD), digital imaging and communications in medicine (DICOM), and any reverse engineering (RE) data. AM provides the flexibility to change or modify the design on the models easily at any stage,

A. Manmadhachary (✉)
Department of Mechanical Engineering, Faculty of Science and Technology,
ICFAI University, Hyderabad, Telangana, India
e-mail: manmadhachary@yahoo.co.in

S. K. Malyala
Department of Mechanical Engineering, National Institute of Technology,
Warangal, Telangana, India

A. Alwala
Department of Oral and Maxillofacial Surgery, Panineeya Institute of Dental Science
and Research Centre, Hyderabad, Telangana, India

© Springer Nature Switzerland AG 2019
D. Pandian et al. (eds.), *Proceedings of the International Conference on ISMAC
in Computational Vision and Bio-Engineering 2018 (ISMAC-CVB)*, Lecture Notes
in Computational Vision and Biomechanics 30,
https://doi.org/10.1007/978-3-030-00665-5_152

without any extra efforts or wastage in time (till the fabrication stage). One of the most benefited industries using AM is medical [2]. AM provides an advantage to build multiple custom models in single batch. In a batch, each and every part can be customized as per the demand, which exactly needed in medical industry. As per recent advancements, AM has entered into the mass customization of implants for medical industry. Another major advantage with AM is any complex part that can be fabricated very easily with better accuracy. Medical industry is related to the life of the people where quality of part is most important [3]. This industry is responsible for curing of patient from suffering in a significant way to improve the quality of patient lives. Currently, AM medical models are using in dental, cosmetic surgery, hip and knee joints, bone fractures, tumor resection, etc. [4]. Apart from complex cases, the AM is also widely used in dental industry like casting models, bridges, crowns, and abutments. Most commonly used materials in medical industry are stainless steel, cobalt-chrome, titanium, and its alloys. Even though currently lots of materials are available in AM, majorly already certified milled or cast materials are easily available options for immediate use [5]. Compared to the conventional machining, AM provides superior flexibility and easy to use [6, 7]. This AM allows to reduce the number of sub-assemblies and number of parts in a product. The current chapter explains the importance of AM technology in anatomy visualization, preplanning surgery, implant manufacturing and preparation of surgical guides.

2 Procedure for Manufacturing of AM Medical Model

The development of surgical procedures using AM is shown in Fig. 1 [1]. The patient data acquired through the CT or MRI in form of DICOM will be processed through medical processing software like 3-Matic, MIMICS, 3D Doctor, 3D Slicer, etc. The data at acquisition stage is stored in the 2D format with equal thickness, which can be also called as 2½ D data. This 2½ D data can easily be converted into 3D data using medical processing software. From the complete acquired 3D data, required region of interest is separated and the 3D CAD data is generated. In process of separating required region of interest, the color, texture, and properties can be changed. The required information like bony or soft tissue is separated using Hounsfield units (HU) values from DICOM data. The required region of interest CAD model is converted into stereolithography (STL) model, which is globally accepted by all the AM machines. The STL format is surface representation of 3D CAD model in the form of triangular facets. This STL data is used as input for the preprocessing software to generate AM machine-specific code to fabricate the physical model. The AM machine reads data and lays downs or adds successive layers of liquid, powder, sheet material, or other, in a layer-upon-layer fashion to fabricate a 3D object. Few clinical cases that are solved using AM technique are explained in the later section.

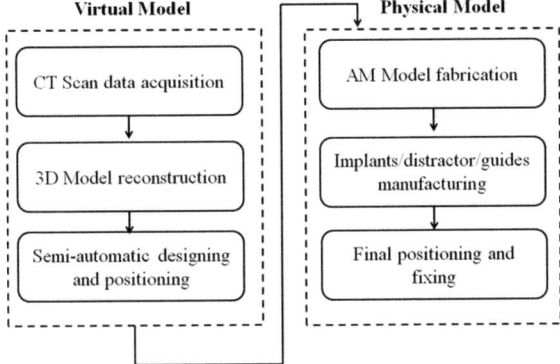

Fig. 1 Development of surgical procedures using AM medical model

3 Medical Applications

In the current chapter, all clinical case studies were conducted and patient's DICOM images procured from department of oral and maxillofacial surgery, Panineeya Institute of Dental Science & Research Center, Hyderabad.

3.1 AM in Visualization

AM model provides better understanding and visualization for the complex structures of anatomy, which is difficult to analyze using conventional 2D information. The simplest implementation of AM anatomical models into "bone resorption" is explained with a case study. The patient CBCT images were used to develop CAD model further AM model is developed. This AM model represents the patient anatomical perspective, resolution, and scale. This AM model is used to analyze the bone width in different places of maxilla and mandible, position of inferior alveolar nerve (IAN), estimation of maxilla sinus, etc. Based on this AM model, the surgeon plans implant fixing position and explain to the patient difficulties of surgery, as is shown in Fig. 2.

The below case study explains the importance of AM medical model in visualization of human anatomy, where patient (29-year male) with an accident. The complexity of this case is that patient had already met with accident in the same region, which increased complexity of the surgery. The previous surgeries metal plates present in the same region create artifacts in the CT. Due to metal artifacts, radiologist and doctor cannot identify exact location of damaged area. These CT images are processed in medical software using filters to remove artifacts and develop CAD model, as is shown in Fig. 3a. AM model was developed using CAD model; AM model is shown in Fig. 3b. Based on this AM model, surgeon identified the exact location, size, and

(a)   **(b)**   **(c)**

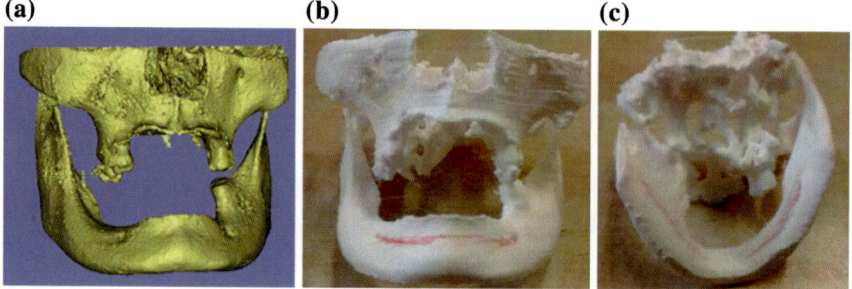

Fig. 2 **a** CAD model, **b** Front side AM model, and **c** Backside AM model

(a)   **(b)**

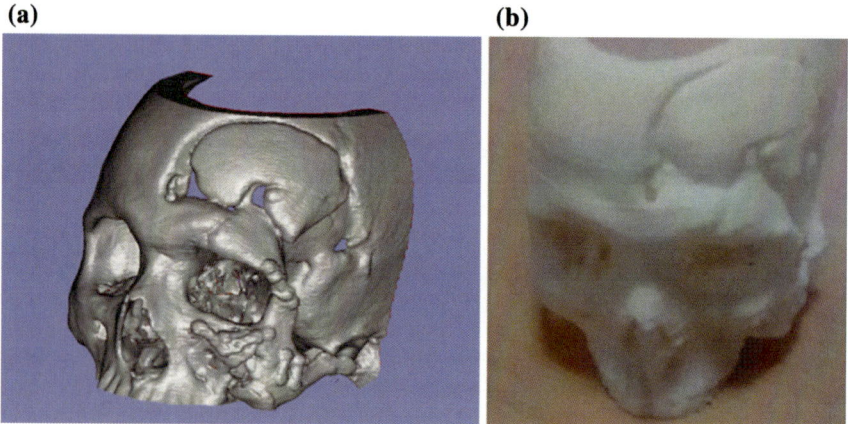

Fig. 3 **a** CAD model and **b** AM models of patient anatomy

shape of damaged area. Also, this AM model was used to bend the reconstruction plates for actual surgery, as it reduced the time for reconstruction of the frontal area.

3.2 AM in Preplanning Surgery

AM models represent exact dimensions of anatomical models of the human anatomy. AM medical models are used by surgeons to diagnose the complex structures anatomy. These AM models can be used as trial models by surgeons in case of complex surgeries. In this chapter, mandibular reconstruction and pan facial trauma surgeries were conducted using AM models. These are explained in the following sections.

This study deals with a 20-year-old male, who had a tumor in right side of mandible, as shown in Fig. 4a. He was reported with swelling and loose teeth in the right mandible. Mandibular re-establishment is challenge for the surgeon wish-

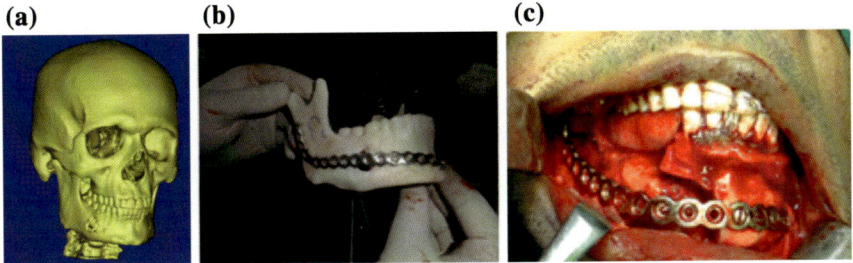

Fig. 4 **a** Tumor in mandible, **b** Bending of plate on AM model, **c** Resection of tumor and fixing of plate

ing to reconstruct its unique geometry. Reconstruction can be attained with titanium plates followed by autogenous bone grafting. Incorporation of the bone graft into the mandible provides continuity and strength required for proper esthetics. Dental implant rehabilitation at a later stage. Valuable time in the operating room is consumed in plate contouring to reconstruct the mandible. The tumor mandible CAD model was developed by using DICOM images, as shown in Fig. 4a. The CAD model is used to develop physical AM model. Mandible AM model was manufactured before the resection of tumor mandibular, which is used to prepare bridging plates before the first stage of reconstruction, as shown in Fig. 4b. Based on this AM model, the expected resection margins were designated, and a KLS Martin reconstruction plate was bent, adapted to the shape of the mandible as shown in Fig. 4c. The time needed to pre-bend the plates was minimized. The tumor was analyzed and removed based on AM model, as shown in Fig. 4c.

The current case is pan facial trauma which was treated by AM medical model. The AM medical model was used for presurgical planning [8]. Reconstruction of face can be attained with fixing of titanium plates followed by autogenous bone grafting. The patient maxilla and mandible CAD model was developed by using DICOM images, as shown in Fig. 5a. From the CAD model, we can observe the fractures of the patient maxilla, mandible and the nasal regions. The original CAD model is used to split the various fractured regions by using MIMICS software, as shown in Fig. 5b. The fractured regions of CAD models were used to develop the AM models. The various fractured regions are manufactured with different color filaments in AM process. These AM parts are used to bend and fixing of plates in fractured regions. The reconstruction plate was bent, adapted to the shape of the maxilla, mandible, and nasal regions, as shown in Fig. 5c. The time needed to pre-bend, location, and fixing the plates was minimized. These presurgical procedures and precautions are used to surgery in operation theatre, as shown in Fig. 5d and e.

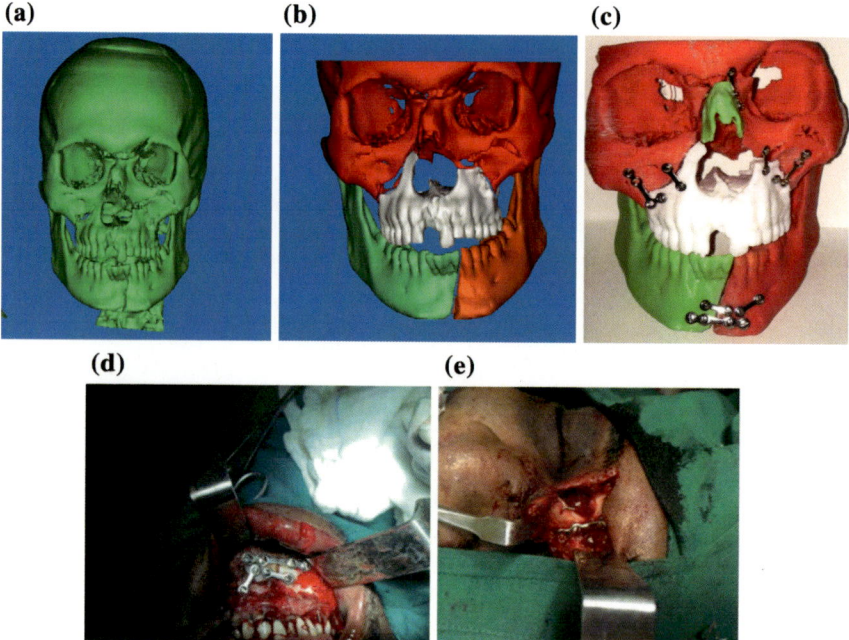

Fig. 5 a Patient CAD model, **b** Fractures in CAD model, **c** preplanning surgery on AM model, **d** fixing of plates at mandible region, **e** fixing of plates at maxilla region

3.3 AM in an Implant Design

The implant design and manufacturing with using patient-specific AM models are one of the trend in medical applications. These implants were exactly fixed to anatomical models in the human body [9]. In this chapter, design and manufacturing of distractors for cleft and tumor mandibular were developed with using AM models. These are explained in the following sections.

The current case study is regarding a 35-year male patient, the patient presented with a defect in maxillary bone which has been diagnosed as cleft alveolus based on history and clinical evaluation. It is associated with regurgitation of food and fluids into nose on consumption. On examination of CAD model, the size of the defect has been recorded as 16.6 mm based on measurement of cleft diameter in CAD model; these are shown in Fig. 6a, b at front and back side views, respectively. This CAD model was used to build a physical AM. This AM model is used for the preparation of distractor. Here, the bending, length and fixing of distractor were evaluated by using AM model. The distractor movement after bone extraction from cleft part is also planned with using AM model. The various movements of distractor are shown in Fig. 6c–e. These same preplanning surgical procedures are used at the time of actual surgery, as shown in Fig. 6f, g.

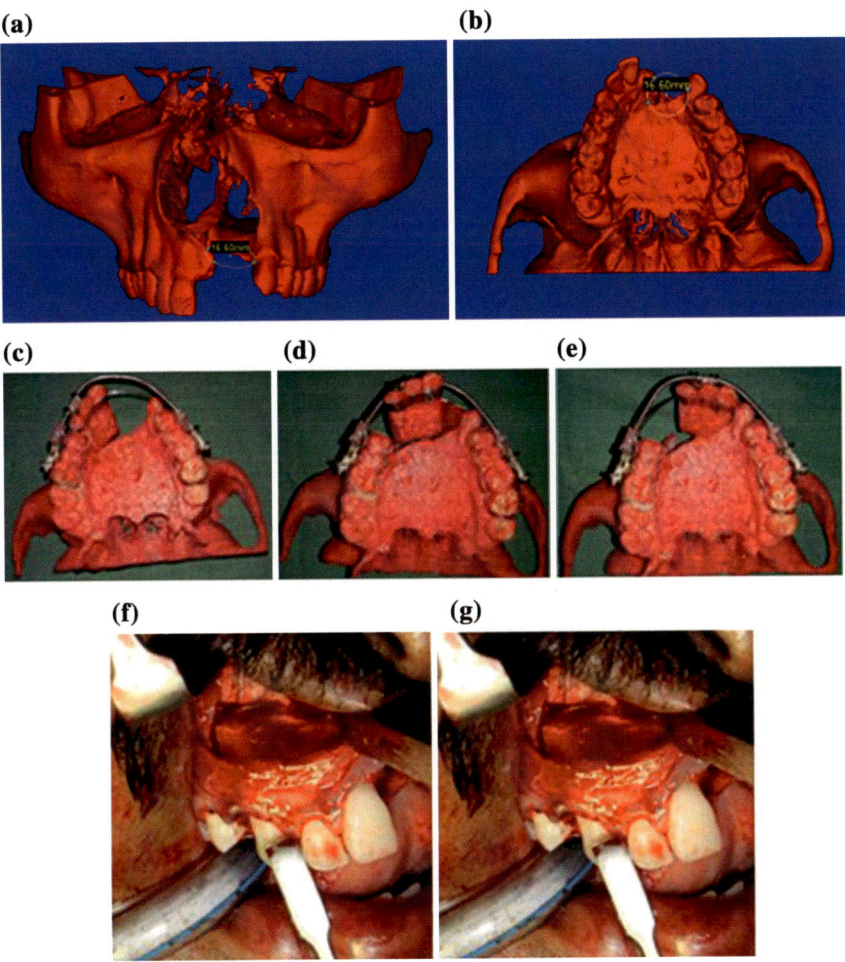

Fig. 6 Cleft in **a** front view, **b** bottom view, **c–e** surgical procedure for bone movement, **f**, **g** fixing of patient-specific distractor

The current case study is described to reconstruct a mandibular continuity defect using a technique of bifocal distraction osteogenesis, causing minimal morbidity to the patient [10]. The CT images of patient were reconstructed a CAD model, incompletely reconstructed mandibular defect, as shown in Fig. 7a. Further, AM model was developed; this defect was reconstructed using a custom-made distractor, as shown in Fig. 7b. This distractor was developed based on patient-specific dimensions such as length, width, and angle of mandible. These measurements are measured and aligned based on AM model. A mandibular segmental defect was reconstructed using the bifocal distraction technique, with the residual defect being reconstructed using an anterior iliac block graft, hence avoiding a microvascular reconstructive procedure.

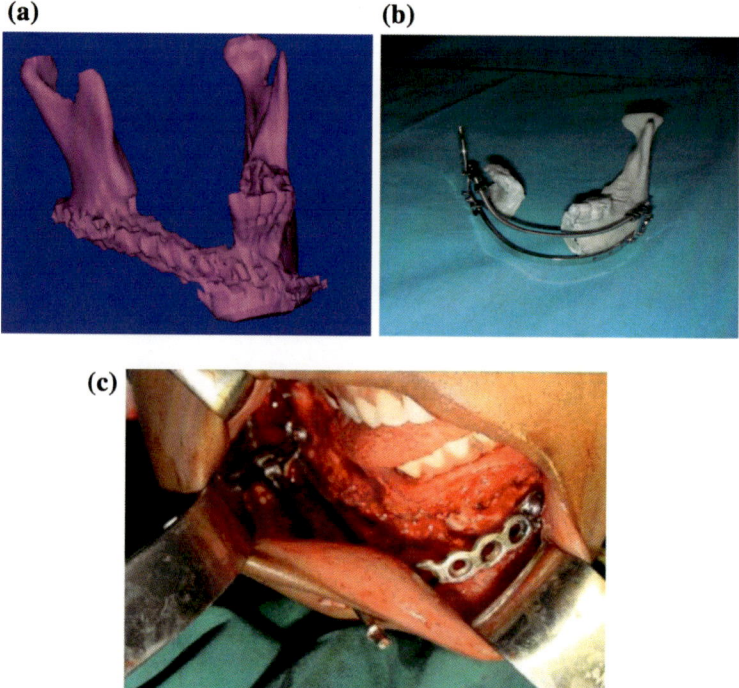

Fig. 7 **a** Mandibular defect in CAD model, **b** Patient-specific distractor and preplanning, **c** Fixing of distractor

The procedure was planned to completely avoid a bone grafting or a microvascular procedure to reconstruct the lost mandible. However, complete regeneration was reconstructed with using patient-specific distractor. Thus, patient obtained a satisfactory facial symmetry and functionally continuous mandibular bone for daily function of speech.

3.4 AM in Design and Manufacturing of Surgical Guides

AM is changing design and manufacturing of surgical guide in dentistry. Till today, the processes for dental restorations, denchers, bridges, and implants, have been limited to lost-wax technology, a method that has been used for the past 100 years. The denchers made with lost-wax method start with a gypsum model. Based on this model, the surgical guides were preparing for dental surgery. The new technologies are coming on-stream to answer that need. The FDM AM process and the material PLA are used to produce dental models [11]. The dental plastic model (mandible) models can be manufactured within 4 h depending on the size. What is more, the

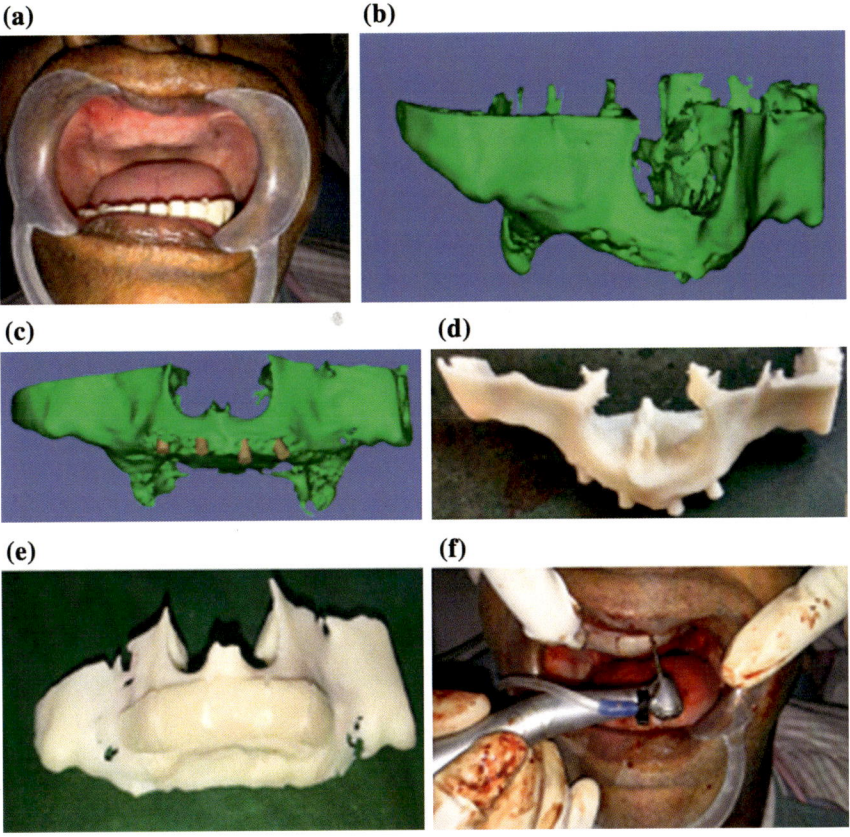

Fig. 8 **a** Patient maxilla, **b** CAD model, **c** Implants are fixed on CAD model, **d** AM model, **e** Preplanning surgery, **f** Actual surgery

plastic models are very hard-wearing, an important requirement if the crowns and tooth stumps have to be fitted and removed several times. The current case study is used for fixing of implants to set artificial teeth in maxilla, the patient maxilla as shown in Fig. 8a. 3D CAD was generated by MIMICS software from DICOM images of the patient, which is shown in Fig. 8b. Further, this CAD model was used for fixing of implants, with accurate dimensions and angels virtually, as shown in Fig. 8c. This implants designed CAD model was used to manufacture AM model, as shown in Fig. 8d. This AM model along with patient-specific implant gives surgeons to anticipate the steps in actual surgery. This AM model is used for design and manufacturing of surgical guide to fixing of implants, as shown in Fig. 8e. These AM medical model and surgical guide are using surgeons to perform a surgery for implantation with accurate dimensions and angles for the patient and also avoid the damage of superior alveolar nerve, as shown in Fig. 8f.

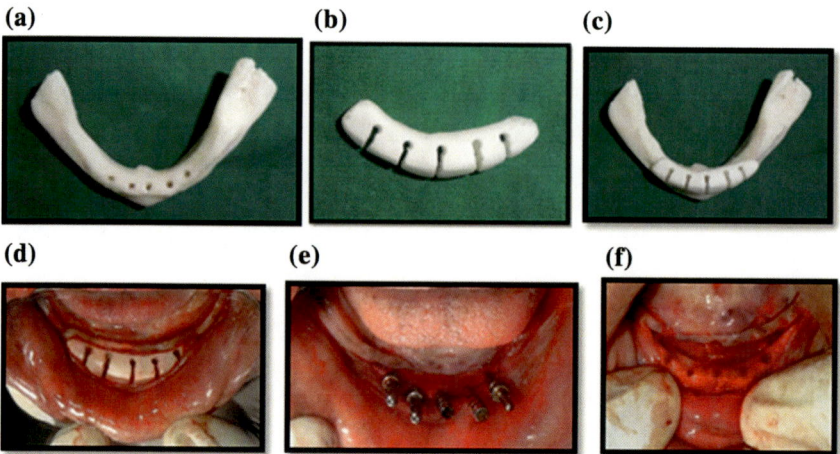

Fig. 9 **a** AM model, **b** Customized surgical drilling guide (CSDG), **c** CSDG on AM model, **d** CSDG for drilling holes on mandible, **e** Implants placement in mandible, **f** Post-operative surgery

A customized surgical drilling guide (CSDG) has proven to be beneficial to dentistry, for accurate placement of osseointegrated implants in partially mandible in flapless surgery [12]. The FDM AM process and the material PLA are used to produce a dental model, as shown in Fig. 9a. As per the need, these CSDGs have made bone or mucosa or gingival supported for the exact placement of implants, which is shown in Fig. 9b. The FDM model and CSDG model were used for preplanning of implantation for exact alignments, as shown in Fig. 9c. The surgeon's use of the CSDG allows the orthodontist and the surgeon to exact placement of implants at correct and safe position with firm and stable placement, as is shown in Fig. 9d, e and f shows placement of implants in the mandible.

4 Conclusions

The AM technology provided a precise, fast, and cheap anatomical reconstruction, which aids in shortened operation time (and therefore decreased exposure time to general anesthesia, decreased blood loss, and shorter wound exposure time) and easier surgical procedure. The application of AM in visualization was allowed to evaluate the patient anatomy conditions before and after surgery. The preplanning surgical models help medical specialists to diagnose ailments and plan for surgery. These AM models are widely used in complex surgery to ease the planning process. The implant design and manufacturing for reconstruction of the bony defect surgeries are conducted with AM medical models. Treatment planning and mock surgery were performed on the AM medical model, which reduced about 25–30% of total surgery time, thereby decreasing the complications.

Acknowledgements The authors would like to special thanks for their support, Dr. N. V. S. Shekar Reddy and Dr. U. D. Arvind, Department of Oral and Maxillofacial Surgery, Panineeya Institute of Dental Science and Research Center, Hyderabad, India for providing DICOM images and conducting surgeries in their hospital. No ethical approval was needed for these studies and all the local guidelines were followed.

References

1. Mamoru MJC, Paulo B, Dirk F, Albert JS, Kamlakar R, Naohiko S (2013) Bio-manufacturing. CIRP Ann Manuf Technol 62:585–606
2. Wohler TT (2013) Additive manufacturing and 3D printing state of the industry, Annual worldwide progress report. Wohlers Associates, Colorado, USA
3. Webb PA (2000) A review of rapid prototyping (RP) techniques in the medical and biomedical sector. J Med Eng Technol 24:149–53
4. Parthasarathy J (2014) 3D modeling, custom implants and its future perspectives incraniofacial surgery. Ann Maxillofac Surg 4:9–18
5. Farias TP, Dias FL, Sousa BA, Galvao MS, Bispo D, Past AC (2013) Prototyping: major advance in surgical planning and customizing prostheses in patients with bone tumors of the Head and neck. Int J Clin Med 4:1–7
6. Manmadhachary A, Ravi Kumar Y, Krishnanand L (2018) Finding of correction factor & dimensional error in bio-AM Model By FDM Technique. J Inst Eng (India): Ser C 99 https://doi.org/10.1007/s40032-016-0294-1
7. Manmadhachary A, Kumar MS, Ravi Kumar Y, Haranadha Reddy M, Adityamohan A (2017) Design & manufacturing of Implant for reconstructive surgery: a Case Study. KnE Eng 2:143–149
8. Alwala AM, Malyala SK, Chittaluri LR, Vasamsetty P (2016) Surgical planning in pan facial trauma using additive manufacturing medical model-a case study. Journal of Surgery Jurnalul de chirurgie. 12:125–128
9. Santosh Kumar Malyala Y, Kumar R, Alwala AM, Manmadhachary A (2017) 3D printed medical model to resolve cleft alveolus defect: a case study. KnE Eng 2:8–14
10. Alwalaa AM, Arvind UD, Malyala SK, Vasamsetty P (2018) Customization of patient specific distraction device using additive manufacture technology. Mater Today: Proc 5:4134–4137
11. Manmadhachary A, Ravi Kumar Y, Krishnanand L (2016) Improve the accuracy, surface smoothing and material adaption in STL file for RP medical models. J Manuf Process 21:46–55
12. Malyala SK, Ravi Kumar Y, Kankanala L, Vasamsetty P, Alwala A (2018) Assessment and treatment planning in maxillofacial surgery by using additive manufacturing technology. Mater Today: Proc 5:4162–4166

A Study on Comparative Analysis of Automated and Semiautomated Segmentation Techniques on Knee Osteoarthritis X-Ray Radiographs

Karthiga Nagaraj and Vijay Jeyakumar

Abstract Arthritis is a most common disease in the worldwide population targeting knee, neck, hand, hip, and almost all the joints of the human body. It is a frequently noticed problem in elder people, especially women. The severity of the disease is analyzed using the older KL grading system. Traditionally, the detection of various grades of OA (osteoarthritis) is interpreted by just a visual examination. A traditional modality, X-ray images are considered as the data for the project. The images are segmented using different segmentation techniques to extract the articular cartilage as region of interest. From the literature, eight different segmentation techniques were identified out of which seven are automated and one is semiautomated. By implementing those techniques and evaluating their performance, it is inferred that block-based segmentation, center rectangle segmentation, and the semiautomated seed point selection segmentation performs well and provides sensitivity, positive prediction value and dice Sorenson's coefficient of 100%, respectively, and specificity of 0%.

1 Introduction

Arthritis is a common disease worldwide. Osteoarthritis is the most common type of arthritis. It is also called as disk degeneration, which resembles the degeneration of the articular cartilage sandwiched between femur and tibia [18]. As the name signifies, osteoarthritis means inflammation in joints of bones. Generally, the degree of movement of any joint is from 0 to $120°$, but osteoarthritis-affected individuals' faces restriction and pain during mobility (Fig. 1).

K. Nagaraj · V. Jeyakumar (✉)
Department of Biomedical Engineering, SSN College of Engineering, Chennai, India
e-mail: vijayj@ssn.edu.in

K. Nagaraj
e-mail: karthiga16455005@bme.ssn.edu.in

© Springer Nature Switzerland AG 2019
D. Pandian et al. (eds.), *Proceedings of the International Conference on ISMAC in Computational Vision and Bio-Engineering 2018 (ISMAC-CVB)*, Lecture Notes in Computational Vision and Biomechanics 30,
https://doi.org/10.1007/978-3-030-00665-5_153

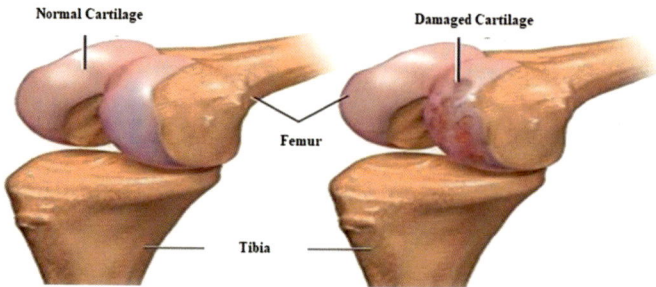

Fig. 1 Normal and OA affected knee (*Source* http://pennstatehershey.adam.com)

Table 1 Different grades of OA based on Kellgren–Lawrence system (*Source* tdmu.edu.ua)

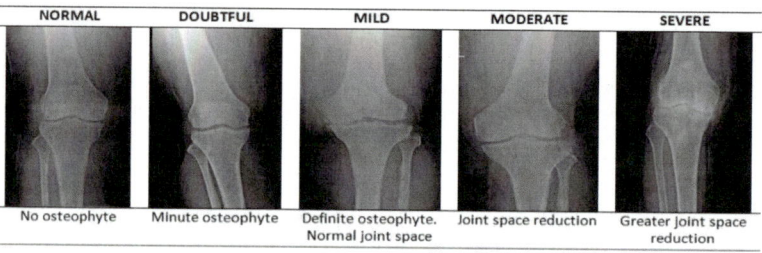

The knee disks (cartilages) are the fibrous flexible bones and they are flexible due to its high water content. These cartilages are present in all the joints and they serve as the cushion for the joints to prevent wear and tear of joints, which indirectly inhibit bone friction [6]. But, due to few other factors like, aging, obesity, over load bearing works, hereditary, hormonal imbalance, gender, trauma, etc., the cushion between the joints gets degenerated [14] and this leads to severe pain [1]. Kellgren and Lawrence (KL) system classifies osteoarthritis into five distinct grades based on disease severity. This system was proposed by Kellgren and team in 1957 and was accepted by WHO (World Health Orgaization) in 1961. The grading ranges from 0 to 4. Grade 0 means there is no characteristic symptom of OA is observed. Grade 1 means significantly doubtful symptoms of OA observed. Grade 2 means mild disk narrowing has been observed. Grade 3 means moderate narrowing of the cartilage is observed. Grade 4 means the disease is at its severe stage. Out of many systems of classification [3], this is found to be used widely by the doctors, especially in India (Table 1).

2 Materials and Methods

The real-time X-ray dataset of 300 patients in the age group of 25–80 were acquired at Manisundaram Medical Mission Hospitals, Vellore under the assistance of

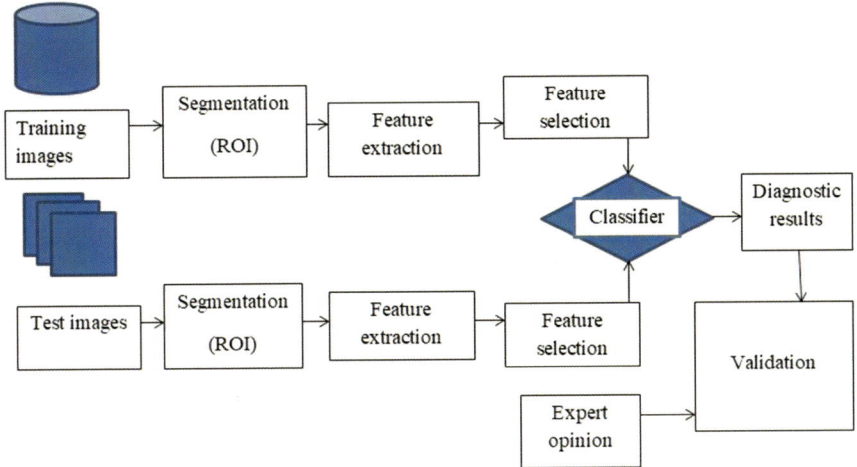

Fig. 2 Workflow of the proposed method

Dr. Manivannun K, Mr. Selva Prakash S, and Ms. Suganya S. The entire framework was performed with MATLAB 2017a (Fig. 2).

The images were acquired using X-ray equipment Wipro GE-DX 300 and pre-processed using the software Imageworks. The sample set of 25 images were taken and different kinds of segmentation techniques were applied on to them.

The proposed work used 90% of the acquired images as the training data and 10% as the test data. The images are initially segmented to impart the region of interest. Then, the segmented images proceeded with the feature extraction. From the extracted features, important features are to be selected and those features are used in training a conventional neural network [8, 10]. The obtained results are to be validated by a medical practitioner.

In this work, we have worked on segmentation using automated and semiautomated segmentation techniques. The complication in segmentation of the articular cartilage is that its gray intensity lies amidst in the gray scale and the X-ray contains various intensities of gray. Background removal fails as the cartilage and the unwanted image portion (the surrounding muscles) falls under similar gray intensities [12].

3 Segmentation

Image segmentation is a process in which information is extracted from an image. Segmentation is done on the preprocessed image to segment bones from the X-ray image, thereby to highlight the cartilage. Different segmentation techniques were [15–17]applied to find the better option [5].

The eight opted segmentation techniques are:
Automated segmentation techniques:

- Block-Based Segmentation
- Center Rectangle Segmentation
- Fuzzy C-Means Segmentation [2]
- K-Means Clustering [4]
- Marker-Based Watershed Segmentation
- Morphological Operations Based Segmentation
- Region-Based Active Contour Segmentation.

Semiautomated segmentation technique:

- Seed Point Selection Segmentation.

Each segmentation method is further described in the following sections.

3.1 Automated Segmentation

3.1.1 Block Distance Measure Segmentation

This method is a simple technique where the image is being segmented in a lossless way [13]. Initially, the image is divided into nine blocks. Out of this nine blocks, the region of interest (RoI), cartilage is extracted from the blocks 4, 5 and 6.

3.1.2 Center Rectangle Segmentation

This technique is used to mark the region other than the region of interest. The image is resized initially to avoid memory error. Then the axes of the image from both X and Y coordinates are computed. Image parameters are then defined according to the size of the region of interest. It unmarks the cartilage region and leaves the other parts of the image unmarked [20].

3.1.3 Fuzzy C-Means Segmentation

The Fuzzy C-means algorithm is an unsupervised fuzzy clustering algorithm. Clustering or cluster analysis involves assigning data points to clusters such that items in the same cluster are as similar as possible, while items belonging to different clusters are as dissimilar as possible. Let $X = (x_1, x_2,...,x_N)$ be an image with N pixels

$$J = \sum_{j=1}^{N} \sum_{i=1}^{c} u_{ij} \|x_j - v_i\|^2 \tag{1}$$

where

x_i is a multispectral data,
u_{ij} is the membership function, and
v_i is the ith cluster center.

However, most soft clustering algorithms do generate a soft partition that also forms fuzzy partition. A type of soft clustering of special interest is one that ensures membership degree of point x in all clusters adding up to one.

3.1.4   *K*-Means Segmentation

K-means is a least-squares partitioning method that divides a collection of objects into *K* groups. The initial assignment of points to clusters can be done randomly. The potential point is

$$P_n = \sum_{i=1}^{n} e^{\frac{-4x_n - x_i^2}{r_a^2}},$$

(2)

where n is data points: $X = \{x_1, x_2, x_{3\ldots} x_n\}$.

The groups obtained are such that they are geometrically as compact as possible around their respective means. Using the set of feature images, a feature vector is constructed corresponding to each pixel. Each data pixel is

$$P_n = p_n - p_1 \sum_{i=1}^{n} e^{\frac{-4x_n - x_i^2}{r_a^2}},$$

(3)

where

p_1, p_n are first and n^{th} pixel, respectively

The *K*-means can then be used to segment the image into three clusters—corresponding to two scripts and background, respectively. For each additional script, one more cluster is added. Thus, even if a pixel is wrongly clustered, it can be corrected by looking at the neighboring pixels.

3.1.5   Marker-Based Watershed Segmentation

This algorithm is based on the image morphology [7]. It considers the topological gradient of the image. The boundaries of the image are segmented. The threshold of the image T is

$$T = \left\{ \max(i) \Big| \sum_{j=0}^{i} H(j) \leq 1 - \varepsilon \right\}$$

(4)

where $H(j)$ is the histogram of image j and ε is the appropriate percentage of pixels needs pixel suppression.

It marks the foreground and the background separately. A segmentation function is computed to segment the dark regions of an image. Then, the foreground markers of each image object are computed. Segmentation function is modified, henceforth it has only the minima of both foreground and background markers [11]. For the modified segmentation function, the watershed transform is performed.

3.1.6   Morphological Operations

Step 1:   The image is subjected to median filter after being converted into gray image.
Step 2:   The edge detection algorithm used is Sobel edge detection.
Step 3:   Initially, the image is stretched and then it is dilated.
Step 4:   The holes in the image after dilation is filled.
Step 5:   The border of the image is cleared using the function.
Step 6:   Finally, the outline of the image is estimated by marking its perimeter.

3.1.7   Region-Based Active Contour Segmentation

This method is applied to segmenting images to improve the bone estimation from the X-ray images. Active contours models are used to detect objects in a given image using the techniques of curve evolution [9]. It identifies individual segments in images with multiple segments and junctions. It uses Heaviside function for smoothing

$$H\varphi(x) = \begin{cases} 1, & \varphi(x) < -\varepsilon \\ 0, & \varphi(x) > \varepsilon \\ \frac{1}{2}\left\{1 + \frac{\varphi}{\varepsilon} + \frac{1}{\pi}\sin\frac{\pi\varphi(x)}{\varepsilon}\right\} \end{cases} \tag{5}$$

Mask is a binary image that specifies the initial state of the active contour. The boundaries of the object regions (white) in mask define the initial contour position used for contour evolution to segment the image. The output image is a binary image where the foreground is white (logical true) and the background is black (logical false).

3.2 Semiautomated Segmentation

3.2.1 Seed Point Selection Segmentation

In this technique, four random seed points from the image around the region of interest is selected. By selecting the seed points, we mean the boundaries of the ROI, i.e., the rows and columns of the corresponding boundaries of the ROI [19]. The buffer for the region of interest and non-ROI is developed, respectively. The marking of the interested region and the other parts of the image is done individually and the images are obtained.

4 Results and Discussion

The real-time X-ray dataset of 300 patients was acquired and it was analyzed with MATLAB 2017a. The sample set of 25 images were taken and different kinds of segmentation techniques were applied on to them. The results are as follows (Fig. 3):

Block-based aegmentation works well for all the grades where the blocks 4–6 combinedly contain the Region of Interest (ROI). The block 5 contains the highest portion of the cartilage (Fig. 4).

In center rectangle aegmentation, the center region which contains the cartilage is alone left unmasked, whereas the non-ROI regions are masked. This method goes well with the ground truth (Fig. 5).

Fuzzy C-means segmentation segments by forming clusters with similar pixels based on its gray level, distance, connectivity, and its intensity. It works good only with the denoised grade 1 OA radiographs (Fig. 6).

K-means segmentation method separates the foreground and background into two–three clusters based on their mean value amongst the pixel intensities in the image. For X-ray images, it produces an irrelevant clustering (Fig. 7).

Morphological operations method of segmentation includes almost all basic operations like denoising, opening, stretching, dilating, filling of holes, etc. Due to the

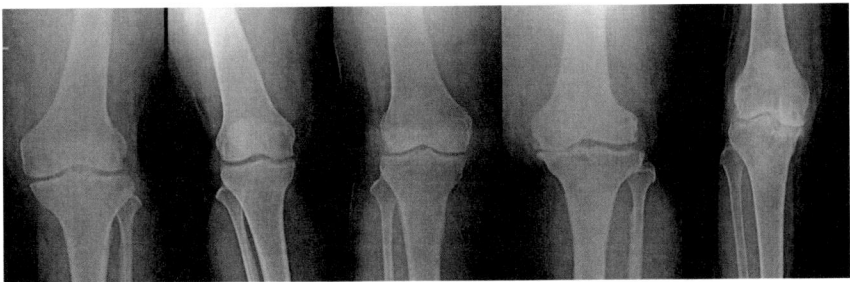

Fig. 3 Normal radiographs of grade 0–4 OA

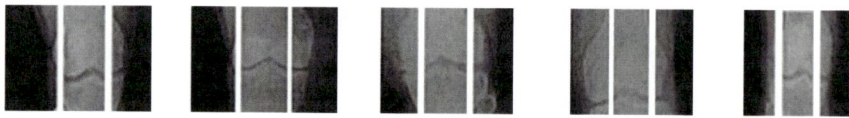

Fig. 4 Block-based segmentation for grade 0–4 OA radiographs

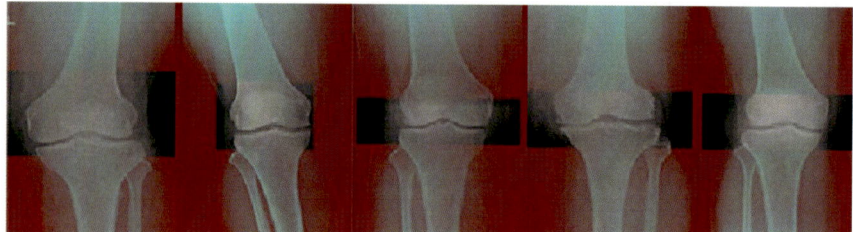

Fig. 5 Center rectangle segmentation for grade 0–4 OA radiographs

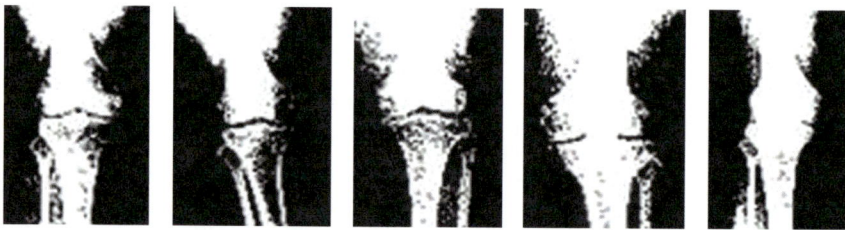

Fig. 6 Fuzzy C-means segmentation for grade 0–4 OA radiographs

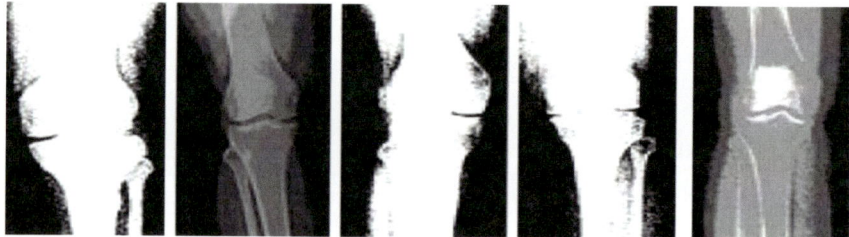

Fig. 7 K-means segmentation for grade 0–4 OA radiographs

presence of noise in the real-time acquired images, this method is unfit for X-ray images (Fig. 8).

Region-based active contour segmentation technique segments the foreground of the image from the background of the image. It shows reasonable segments in radiographs of grade 0–2 (Fig. 9).

In marker-based watershed segmentation, the background and foreground are segmented considering the topological gradient of the image. The segments are

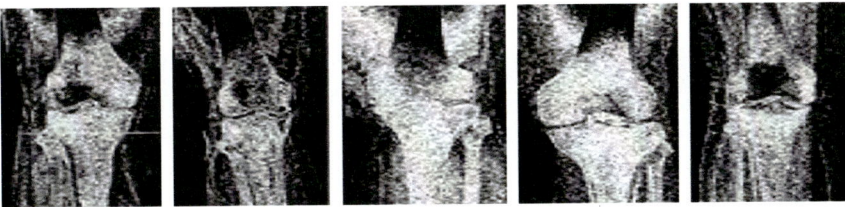

Fig. 8 Morphological operations for grade 0–4 OA radiographs

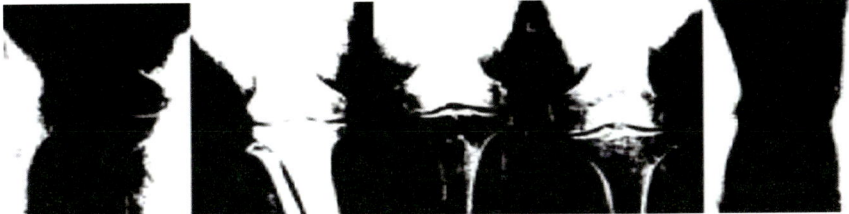

Fig. 9 Region-based active contour segmentation for grade 0–4 OA radiographs

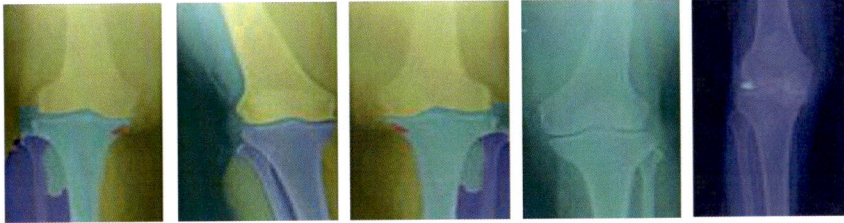

Fig. 10 Marker-based watershed segmentation for grade 0–4 OA radiographs

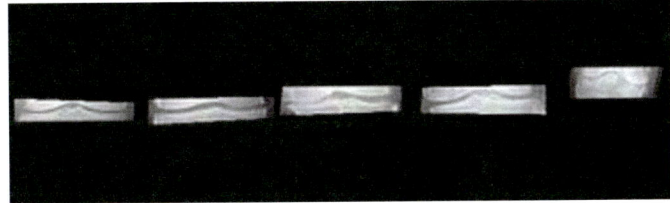

Fig. 11 Seed point selection segmentation for grade 0–4 OA radiographs

represented in the shades of RGB. This method works well for grade 1 and 2 images (Fig. 10).

Seed point selection segmentation method needs a manual selection of seed points, but it creates a defined boundary by tracing the rows and columns corresponding to the selected seed points. It goes well with radiographs of all the grades of OA (Fig. 11).

Table 2 Metrics for the performance of various segmentation techniques

Evaluation metrics (in percentage)	SE	SP	PPV	DSE
Ground truth	0	100	100	100
Segmentation techniques				
K-means clustering	0	4	4	4
Fuzzy C-means segmentation	100	75	50	5
Region-based active contour segmentation	100	0	60	10
Block-based segmentation	100	0	100	100
Marker-based watershed segmentation	100	40	40	7
Morphological operations-based segmentation	0	4	4	4
Center rectangle segmentation	100	0	100	100
Seed point selection segmentation	100	0	100	100

An evaluation metrics was computed for evaluating the effectiveness of various segmentation techniques and the parameters like sensitivity, specificity, positive prediction values, and the dice Sorenson's coefficients were used.

The confusion matrix is plotted for each and every segmentation method with its true positive, true negative, false positive, and false negative values estimated by an expert. With those confusion matrices, parameters like Sensitivity (SE), Specificity (SP), Positive Predictive Value (PPV), and Dice Sorensen Coefficient (DSC) are computed.

$$SE\ (\%) = \frac{TP}{TP + FN} * 100 \tag{6}$$

$$SP\ (\%) = \frac{TN}{TN + FP} * 100 \tag{7}$$

$$PPV\ (\%) = \frac{TP}{TP + FP} * 100 \tag{8}$$

$$DSC\ (\%) = \frac{2.TP}{2.TP + FP + FN} * 100 \tag{9}$$

On the other hand, all the above mentioned parameters are also computed for the ground truth (free-hand cropped cartilage marked by an expert). The best working segmentation techniques that go hand-in-hand with the ground truth are selected for further processing (Table 2).

5 Conclusion and Future Works

It is hence concluded that center rectangle and Block-based segmentation which are automatic segmentation techniques and the seed point selection segmentation which is a semiautomated segmentation are found to perform well. The other methods were found uncompatible to comply with X-ray images. Hence, an optimal segmentation technique is inferred for segmentation of knee images for detection of osteoarthritis. The segmented X-ray images are compatible to be used for features extraction.

This work is to conclude with a good technique that works best in segmenting the knee X-ray images which will help in automatic classification based on KL grading. The future work is to deploy the fully automatic, convolutional neural network to classify the test images based on the KL grading system.

Acknowledgements The images used for this study were obtained from Manisundaram Medical Mission Hospitals, Vellore under the supervision of Dr. Manivannun K., M.S (Ortho.), Mr. Selva Prakash S., Dip. in X-ray Technology, and Ms. Suganya S B.Sc MSW, Counselor. We duly state that my data collection does not involve patient's interference and invasive protocol.

References

1. Boniatis I, Costaridou L, Cavouras D, Kalatzis I, Panagiotopoulos E, Panayiotakis G (2006) Osteoarthritis severity of the hip by computer-aided grading of radiographic images. Med Biol Eng Compu 44(9):793
2. Chuang KS, Tzeng HL, Chen S, Wu J, Chen TJ (2006) Fuzzy C-Means clustering with spatial information for image segmentation. Comput Med Imaging Graph 30(1):9–15
3. Deokar DD, Patil CG (2015) Effective feature extraction based automatic knee osteoarthritis detection and classification using neural network. Int J Eng Techn 1(3)
4. Dhanachandra N, Manglem K, Chanu YJ (2015) Image segmentation using K-Means clustering algorithm and subtractive clustering algorithm. Procedia Comput Sci 54(2015):764–771
5. Gan HS, Sayuti KA (2017) Comparison of improved semi-automated segmentation technique with manual segmentation: data from the osteoarthritis initiative
6. Gornale SS, Patravali PU Manza RR (2016) Detection of osteoarthritis using knee X-ray image analyses: a machine vision based approach. Int J Comput Appl 145(1)
7. Hill PR, Canagarajah CN, Bull DR (2003) Image segmentation using a texture gradient based watershed transform. IEEE Trans Image Process 12(12):1618–1633
8. Kubkaddi S, Ravikumar KM (2017) Early detection of knee osteoarthritis using SVM classifier. IJSEAT 5(3):259–262
9. Lankton S, Tannenbaum A (2008) Localizing region-based active contours. IEEE Trans Image Process 17(11):2029–2039
10. LeCun Y, Bengio Y, Hinton G (2015) Deep learning. Nature 521(7553):436–444
11. Lu S, Wang S, Zhang Y (2017) A note on the marker-based watershed method for X-ray image segmentation. Comput Methods Programs Biomed 141:1–2
12. Minciullo L, Cootes T (2016) Fully automated shape analysis for detection of Osteoarthritis from lateral knee radiographs. In: 2016 23rd international conference on pattern recognition (ICPR). IEEE, pp 3787–3791
13. Navale DI, Hegadi RS, Mendgudli N (2015) Block based texture analysis approach for knee osteoarthritis identification using SVM. In: 2015 IEEE international WIE conference on electrical and computer engineering (WIECON-ECE). IEEE, pp 338–341

14. Øiestad BE, Juhl CB, Eitzen I, Thorlund JB (2015) Knee extensor muscle weakness is a risk factor for development of knee osteoarthritis. A systematic review and meta-analysis. Osteoarthritis Cartilage 23(2):171–177

15. Rahman MA, Liu S, Lin S, Wong C, Jiang G, Kwok N (2015) Image contrast enhancement for brightness preservation based on dynamic stretching. Int J Image Process (IJIP) 9(4):241

16. Roopa H, Asha T (2016). Segmentation of X-ray image using city block distance measure. In: 2016 International conference on control, instrumentation, communication and computational technologies (ICCICCT). IEEE, pp 186–189

17. Ryzhkov MD (2015) Knee cartilage segmentation algorithms: a critical literature review. Master's thesis

18. Scott D, Kowalczyk A (2007) Osteoarthritis of the knee. BMJ Clin Evid 2007:1121

19. Shamir L, Ling SM, Scott Jr, WW, Bos A, Orlov N, Macura TJ,... Goldberg IG (2009) Knee X-ray image analysis method for automated detection of Osteoarthritis. IEEE Trans Biomed Eng 56(2):407–415

20. Wahyuningrum RT, Anifah L, Purnama IKE, Purnomo MH (2016) A novel hybrid of S2DPCA and SVM for knee osteoarthritis classification. In: 2016 IEEE international conference on computational intelligence and virtual environments for measurement systems and applications (CIVEMSA). IEEE, pp 1–5

Plant Disease Detection Based on Region-Based Segmentation and KNN Classifier

Jaskaran Singh and Harpreet Kaur

Abstract The plant disease detection is the technique which can detect disease from the plant leaves. The plant disease detection has various steps which are textural feature analysis, segmentation, and classification. This research paper is based on the plant disease detection using the KNN classifier with GLCM algorithm. In the proposed method, the image is taken as input which is preprocessed, GLCM algorithm is applied for the textural feature analysis, k-means clustering is applied for the region-based segmentation, and KNN classifier is applied for the disease prediction. The proposed technique is implemented in MATLAB and simulation results show up to 97% accuracy.

1 Introduction

Agricultural area is the only area through which the food requirements of complete human race are being served. In India, around 70% of the total population relies on agriculture and grow several kinds of fruits and vegetable crops within their fields. However, there is a need for high technicality while cultivating crops that are of optimum yield and quality. It is thus important to diagnose the diseases within plants [1]. The identification of diseases is a difficult process due to which the farmers face lots of issues. A research study in which the contagious diseases of the plant are studied is known as potato plant pathology. In order to upgrade the agricultural areas, the disease detection application is applied. Accurate treatment advices are provided through disease management. A computer, digital camera as well as application software are the three components of machine vision system today. Within the application software, numerous algorithms are integrated. In biology, there are several applications that include digital image processing and image analysis technology

J. Singh
ECE Department, Chandigarh University, Gharuan, India

H. Kaur (✉)
CSE Department, Chandigarh University, Gharuan, India
e-mail: harpreet8307@gmail.com

© Springer Nature Switzerland AG 2019
D. Pandian et al. (eds.), *Proceedings of the International Conference on ISMAC in Computational Vision and Bio-Engineering 2018 (ISMAC-CVB)*, Lecture Notes in Computational Vision and Biomechanics 30,
https://doi.org/10.1007/978-3-030-00665-5_154

in order to propose advances within the microelectronics as well as computer systems [2]. The various issues that are relevant to traditional photography are also resolved here. The images that are collected from the microscopic to telescopic ranges are enhanced and analyzed with the help of this tool. Within biology, there are several applications involved. For less production of potatoes, there are numerous factors that are responsible such as early and late blights, insect damage, and roll viral diseases. Within image analysis and pattern recognition, image processing is the initial step and the most important one as well [3]. The final results of the analysis are determined through this step. The process through which an image can be partitioned into disjoint regions is known as image segmentation process.

2 Literature Review

Prakash et al. (2017) studied several image processing techniques in order to identify diseases in plants [4]. In order to identify the leaf diseases and classify them, the image analysis and classification techniques are to be implemented as an aim in this paper. There are four different parts of the proposed framework. They include image processing, segmentation of leaf using k-means clustering in order to identify the diseased areas, feature extraction, and classification of diseases. Using statistical Gray-Level Co-Occurrence Matrix (GLCM) features and classification, texture features are extracted in this approach. The Support Vector Machine (SVM) classifier is used within this proposed mechanism in order to provide better extraction of features.

Kaur et al. (2017) presented a comprehensive study related to several diseases identified within the fruits [5]. In order to minimize the time required to identify the diseases within the fruits, an automated approach is generated for identifying the diseases within fruits. The image is distorted due to the noise. Within this case, there is elaboration of denoising mechanism. As per the analysis of this study, most of the fruit crops are being affected by blight disease. The image of distorted lead is used in order to perform analysis such that the disease can be identified within these fruit crops. In order to identify the diseases within the initial stage, the singular valued analysis is used within the image processing techniques.

Dhaware et al. (2017) proposed a method which can identify the automatic leaf unhealthiness classification from leaves by using image processing [6]. Practical requisition is used here for applying the system because the images are forwarded directly as the farmers make the least efforts. Thus, the farmers that do not make much efforts are provided with advises through this technique. The plant leaves image is seized by the farmer by using the mobile camera. Without making any additional inputs, this image is forwarded to the DSS approach.

Padol et al. (2016) proposed the method that utilized SVM classification technique for identification and classification of diseases present within grape leaves [7]. Within image processing, methods such as resizing, thresholding as well as Gaussian filtering are applied. In order to execute segmentation on the leaf, k-means clustering technique is utilized in which texture and color features are utilized. Further, in order to identify the type of leaf disease, SVM classification technique is utilized. Downy mildew and Powdery mildew are the two different classes of grape leaves that were utilized within this experiment. For both of these categories, an average of 88.89% of accuracy is provided by this proposed system.

Rajan et al. (2016) proposed a mechanism that utilized image processing in order to identify pests within the crops [8]. Upon different images gathered from various sources, tests were performed using this system. At an early stage, it is possible to identify the pests within crops through this proposed method. Thus, the usage of pesticides in agricultural fields can be minimized here through which costs and environment both can be saved. As per the evaluations made, this system is termed as simple and efficient for various applications. In comparison to other manual systems, the time and accuracy level achieved through it is also enhanced.

3 Support Vector Machine

The set of relevant supervised learning mechanisms which are utilized in order to perform classification and regression are known as Support Vector Machines (SVMs). The empirical classification error can be minimized and geometric margin can be maximized simultaneously with the help of SVM. Due to this property, this is also known as Maximum Margin Classifier and it is also based on the Structural Risk Minimization (SRM). Toward the higher dimensional space in which maximal separating hyperplane is generated, the input vector is mapped by SVM. In order to partition the data, two parallel hyperplanes are generated on each side of the hyperplane. The hyperplane that maximizes the distance amongst two parallel hyperplanes is known to be the separating hyperplane [9]. It is assumed here that the larger is the margin or distance amongst these parallel hyperplanes, the better is the generalization error of the classifier. The data points that are considered here are in the form as given in Eq. (1).

$$\{(x1, y1), (x2, y2), (x3, y3), \ldots, (xn, yn)\} \tag{1}$$

Here, $yn = 1/-1$ is a constant that is used to represent the class in which point xn belongs. The numbers of samples involved are represented by n. The p-dimensional real vector is each xn. In order to guard against the variables that have larger variance, scaling is important. With the help of partitioning the hyperplane, the training data is viewed here which includes

$$w.x + b = 0 \tag{2}$$

Here, "b" is a scalar values and the p-dimensional vector is represented by "w". Perpendicular to the separating plane, the vector "w" is indicated. The margin is increased by adding the offset parameters "b". The hyperplane is forced to pass the origin in case "b" is not present due to which the solution can be restricted. There is an involvement of parallel hyperplanes here as well which can be described by the equations below

$$w.x + b = 1 \tag{3}$$

$$w.x + b = -1 \tag{4}$$

4 Proposed Methodology

The technique is proposed for the plant disease detection. The plant disease detection techniques consist of the following phases:

1. Preprocessing phase: In the first phase, the image is taken as input and the input image is converted to the gray scale image.
2. Textural feature analysis: The textural features of the input image are analyzed using GLCM algorithm. This algorithm, statistical texture analysis is performed on the observed combinations of intensities for calculating texture features. This calculation is done using the intensities at particular positions that are relevant to each other within an image. There are first-order, second-order and higher order statistics classified on the basis of the number of intensity points or pixels present within each combination. The method through which the second-order statistical texture features can be extracted is known as Gray-Level Co-occurrence Matrix (GLCM) method [10]. In numerous applications, this technique has been applied. Within an image that has number of rows and columns that are equal to the number of gray levels is known as GLCM matrix. The relative frequency through which two pixels are partitioned through a pixel distance (Δx, Δy) and their intensities are i and j for an element is defined as $P(i, j | \Delta x, \Delta y)$. Amongst the gray levels i and j at "d" displacement distance and (θ) angle, the matrix element $P(i, j | d, \theta)$ can be considered to have second-order statistical probability values. For an input image that has G gray levels from 0 to G–1 and $M \times N$ neighborhood, the intensity at sample m is considered as $f(m, n)$ for neighborhood line n. Hence,

$$P(i, j | \Delta x, \Delta y) = W Q(i, j | \Delta x, \Delta y) \tag{5}$$

Here, (Table 1)

$$W = \frac{1}{(M - \Delta x)(N - \Delta y)} \tag{6}$$

$$Q(i, j | \Delta x, \Delta y) = \sum_{n=1}^{N-\Delta y} \sum_{m=1}^{M-\Delta x} A \tag{7}$$

and

Table 1 Mathematical formulation of features [12]

Feature	Mathematical formulation
Contrast	$\sum_i \sum_j (i-j)^2 g_{ij}$
Correlation	$\frac{\sum_i \sum_j (ij) g_{ij} - u_x u_y}{\sigma_x \sigma_y}$
Energy	$\sum_i \sum_j g_{ij}^2$
Homogeneity	$\sum_i \sum_j \frac{1}{1+(i-j)^2} g_{ij}$
Mean, M	$\sum_{i=0}^{L-1} g(i) P(g(i))$
Standard Deviation, S	$\sqrt{\sum_{i=0}^{L-1} (g(i)-M)^2 P(g(i))}$
Skewness	$\frac{1}{S^3} \sum_{i=0}^{L-1} (g(i)-M)^3 P(g(i))$
Entropy	$\sum_{i=0}^{L-1} P(g(i)) \log_2 P(g(i))$
Kurtosis	$\frac{1}{S^k} \sum_{i=0}^{L-1} (g(i)-M)^k P(g(i))$
RMS	$\sqrt{\frac{1}{L*L} \sum_{i=0}^{L-1} \sum_{j=0}^{L-1} (g(i,j)-I)^2}$

$$A = \begin{cases} 1 \; if \; (m,n)=1 \text{ and } f(m+\Delta x), n+\Delta y) = j \\ 0 \quad \text{elsewhere} \end{cases} \tag{8}$$

5 Region-Based Segmentation

In this phase, the k-means clustering is applied for the region-based segmentation. The k-means clustering algorithm consists of the following phases:

1. Input data: The image which is taken as input is considered as the data on which segmentation needs to implemented.
2. Calculate arithmetic mean: The arithmetic mean of the input data is calculated which defines centroid point of the data.
3. Formation of segments: The Euclidean distance is calculated for the central point and points which have similar distance is clustered in one cluster and other in the second.
4. Selection of ROI: The segmented image is selected as ROI which is given input to next phase of classification.

6　Classification of Disease

The step of classification will classify the input image into defined disease. The ROI which is selected in the last step is taken as input. The KNN classifier is a technique that is used within the pattern recognition process to classify the objects on the basis of closest training examples present within the feature space is known as K-Nearest Neighbor algorithm (KNN). In this instance-based learning algorithm, there is only local approximation of the function and until classification, all the computation is held upon. Amongst all the machine learning algorithms, KNN is the simplest one. Through the majority vote of neighbors, an object is classified [11]. The class that is most common amongst its k-nearest neighbors is assigned that object. The object is directly assigned to the class of its nearest neighbor in the case when $k = 1$. For both classification and regression predictive issues, NN is utilized. The ease to interpret output; compute the time; and the predictive power available are the three important aspects that are considered while evaluating any technique. By default, Euclidean distance is used by $knn()$ function and the following equation is used to calculate it.

$$D(p, q) = \sqrt{(p_1 - q_1)^2 + (p_2 - q_2)^2 \cdots + (p_n - q_n)^2} \tag{9}$$

Here, the subjects that have "n" properties and are to be compared are represented as "p" and "q".

KNN is also computed by deciding the number of neighbors that can be selected here which is represented by parameter "k". The performance of KNN algorithm is affected on large scale by the choice of value of k. The variance generated due to random error can be minimized by maximizing k. The balance amongst overfitting and underfitting can be maintained on the basis of the selection of appropriate value of k. The output of the KNN classifier will give the disease name.

7　Results and Evaluation

From the publically accessible dataset "Plant Village", a database of around 40 images of potato leaves is gathered and used in order to evaluate the proposed algorithm. Image of 10 healthy leaves and 30 diseased leaves are present within the database used for conducting experiments. The complete database is partitioned into 2 different sets during the experiments. There are 24 images (60%) of images present within the training set and 16 images (40%) of images present within the testing set. The KNN that includes linear Kernel is utilized in order to perform classification. Various performance parameters such as accuracy, sensitivity, recall, and F1 score are utilized in order to evaluate the performance of this classification model. There is around 97% of accuracy achieved for classification.

Table 2 shows the performance measures applied. Comparisons are made amongst potato and other species only since there is not much work proposed on disease detection of potatoes. The performance of the KNN classification technique is presented in Table 2.

Table 2 Performance measures of classification

Class	Precision (%)	Recall (%)	F1 score (%)
0: Leaf Minar	89	94	92
1: Mosnic virus	97	93	95
2: White fly	98	98	98
Average/total	97	97	97

(a) Input Image

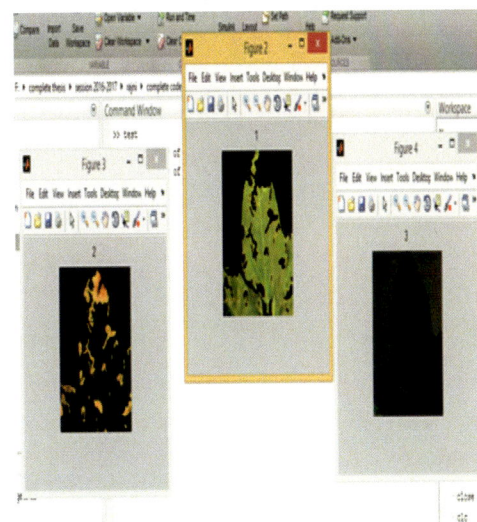

(b) Segmented Image

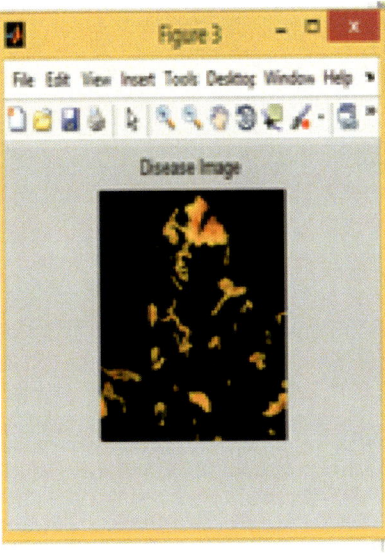

(c) Diseased Image

Fig. 1 Maximum margin hyperplanes for an SVM trained with samples from two classes

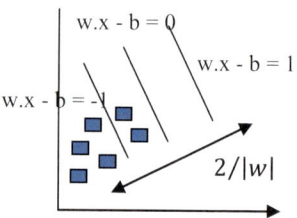

As shown in Fig. 1, a is the input image for plant disease detection. b Apply k-mean region-based segmentation. c Diseased image.

8 Conclusion

In this research paper, it is concluded that plant disease detection is the approach to detect diseases from the plants. In this work, the GLCM algorithm is applied for the textural feature analysis, k-means clustering is applied for the region-based segmentation, and KNN classifier is applied for the disease prediction. The simulation of the proposed modal is done in MATLAB and results are shown in form of figures and tables. The simulation results illustrated that accuracy is achieved up to 97% of disease prediction.

References

1. Rastogi A, Arora R, Sharma S, Leaf disease detection and grading using computer vision technology and fuzzy logic. In: 2015 2nd international conference on signal processing and integrated networks (SPIN)
2. Khirade SD, Patil AB (2015) Plant disease detection using image processing. In: 2015 international conference on computing communication control and automation
3. Waghmare H, Kokare R, Dandawate Y (2016) Detection and classification of diseases of grape plant using opposite colour local binary pattern feature and machine learning for automated decision support system. In: 2016 3rd international conference on signal processing and integrated networks (SPIN)
4. Prakash RM, Saraswathy GP, Ramalakshmi G (2017) Detection of leaf diseases and classification using digital image processing. In: 2017 international conference on innovations in information, embedded and communication systems (ICIIECS)
5. Kaur K, Marwaha C (2017) Analysis of diseases in fruits using image processing techniques. In: International conference on trends in electronics and informatics ICEI 2017
6. Dhaware CG, Wanjale KH (2017) A modern approach for plant leaf disease classification which depends on leaf image processing. In: 2017 international conference on computer communication and informatics (ICCCI—2017)
7. Padol PB, Yadav AA (2016) SVM classifier based grape leaf disease detection. In: 2016 Conference on advances in signal processing (CASP)

8. Rajan P, Radhakrishnan B, Suresh LP (2016) Detection and classification of pests from crop images using support vector machine. In: 2016 international conference on emerging technological trends (ICETT)
9. Duan X, Zhao T, Li T, Liu J, Zou L, Zhang L (2017) Method for diagnosis of on-load tap changer based on wavelet theory and support vector machine. In: 2017. In: The 6th international conference on renewable power generation (RPG)
10. Parvez A, Phadke AC (2017) Efficient implementation of GLCM based texture feature computation using CUDA platform. In: International conference on trends in electronics and informatics, ICEI
11. Zhang Zhongheng (2016) Introduction to machine learning: k-nearest neighbors. Ann Transl Med 4(11):218
12. Islam M, Dinh A, Wahid K (2017) Detection of potato diseases using image segmentation and multiclass support vector machine. In: 2017 IEEE 30th canadian conference on electrical and computer engineering (CCECE)

Analyzing e-CALLISTO Images: Sunspot Number, 10.7 cm Flux and Radio Solar Bursts

R. Sreeneebus, Z. Jannoo, N. Mamode Khan, C. Monstein
and M. Heenaye-Mamode Khan

Abstract This paper investigates on the sun's activity based on sunspot images using time series of sunspot numbers and absolute 10.7 cm flux that would also enable to establish relationship between these indices and provide possible forecasts. Moreover, these indices are also compared with the monthly number of solar bursts detected by the Mauritian Radio Telescope which forms part of the e-CALLISTO network.

1 Introduction

This study provides an insight on the sun path with respect to the different long existing indices which are the sunspot images which as the name suggests are the number of spots on the sun and the 10.7 cm radio flux which the waves detected on earth in the wavelength of 10.7 cm coming from the sun and is also referred to as $F_{10.7}$ and compare these two. Once this is done, we will study the solar burst data from a rather recent project known as the Compound Astronomical Low cost Low frequency Instrument for Spectroscopy and Transportable Observatory, e-CALLISTO for short.

R. Sreeneebus (✉) · Z. Jannoo · N. Mamode Khan
Department of Statistics and Economics, University of Mauritius, Moka, Mauritius
e-mail: ravi.sreeneebus1@umail.uom.ac.mu

Z. Jannoo
e-mail: z.jannoo@uom.ac.mu

N. Mamode Khan
e-mail: n.mamodekhan@uom.ac.mu

C. Monstein
Institute for Particle Physics and Astrophysics, ETH Zurich, Zurich, Switzerland
e-mail: monstein@astro.phys.ethz.ch

M. Heenaye-Mamode Khan
Department of Software and Information Systems, University of Mauritius, Moka, Mauritius
e-mail: m.mamodekhan@uom.ac.mu

© Springer Nature Switzerland AG 2019
D. Pandian et al. (eds.), *Proceedings of the International Conference on ISMAC in Computational Vision and Bio-Engineering 2018 (ISMAC-CVB)*, Lecture Notes in Computational Vision and Biomechanics 30,
https://doi.org/10.1007/978-3-030-00665-5_155

The e-CALLISTO aims to build solar radio spectrometers to observe solar bursts within the radio frequency and since 2009, Mauritius has been contributing to it. The aim of this paper is to use the data gathered by the Mauritian radio observatories at Bras d'Eau and compare them with the existing indices that measures the activity of the sun like the sunspot number and 10.7 cm flux in order to explore the relationship between them and perhaps even predict the number of solar bursts that will occur in the future. The e-CALLISTO has been put in place by the Institute for Astronomy, ETH Zurich, and FHNW Windisch, Switzerland spearheaded by Christian Monstein has data as early as 2002.

1.1 Sunspot Number

Wolf's sunspot number is given by the following formula [1].

$$R = k(10\,g + s) \tag{1}$$

where

R Sunspot number,
g Total number of sunspot groups visible,
s Total number of sunspots,
k A scaling factor.

The data for this index is available at the Sunspot Index and Long-term Solar Observations (SILSO) website. This data is essentially in the form of a time series and can be analysed as such. In order to do a forecasting of the data, a simple Auto Regressive-Moving Average model can be applied. In fact various such models have been proposed. The first of which was an ARMA (2, 0), then ARMA (8, 1) and ARMA (9, 0) [2].

While the sunspot number might be the oldest index for solar activity, it is not the only one. Another popular way of monitoring the sun's activity is by looking at the waves emitted by the sun in the radio frequency.

1.2 The 10.7 cm Solar Radio Flux

The 10.7 cm solar radio flux also referred to as $F_{10.7}$ is an indication of solar activity. It is the average amount of radiation emitted in the 2800 MHz frequency or 10.7 cm wavelength and while the name this index is flux, it really is a flux density [3]. The data for the $F_{10.7}$ is available on the NOAA website in the forms of daily and monthly data. These data are taken from The National Research Council Canada Dominion Radio Astrophysical Observatory in Penticton. A number of versions for the $F_{10.7}$ data is available including absolute flux, adjusted flux, and rotational averages, amongst others. For this paper, the absolute flux was used.

Since the data provided by the e-CALLISTO is in the range of radio frequency, this index was chosen to look into the relationship between the two and use this as a basis to then compare them with solar burst data observed by the Mauritian Radio Telescopes (MRT) to find a potential statistical link between them.

The organization of this paper is as follows: In the next section, we display some of the relevant findings in this field illustrating the relationship between the variables followed by a section on the proposed research on the e-Callisto Data. In Sect. 4, the fitting of the different time series models [4–6] is made. The conclusion is presented in the last section.

2 Related Work

Most if not all of them have used the adjusted flux for the relationship however in this study, the absolute flux has been used. The graph showing the difference between the two is shown below.

In Fig. 1, the green part represents the adjflux and the blue part is the absflux. Looking at the graph, the two sets of data are very similar and in fact the correlation between the two is exactly 1 with the sole difference being that the adjflux is higher than the absflux.

When it comes to a relationship between the absolute 10.7 cm flux data and monthly sunspot number was determined. When the $F_{10.7}$ is plotted against the sunspot number, a mostly linear relationship is seen between the two with a slight curve for the smaller values as shown below (Fig. 2).

So using these observation and the past studies [7] as a reference about the monthly adjusted 10.7 cm radio flux data and monthly sunspot number, equations connecting these quantities was developed. When looking at these past studies quite a few ones were suggested by different researchers including linear ones given by equation below,

$$F = 60.1 + 0.932\, N_s \tag{2}$$

where F is the 10.7 cm flux and N_s is the sunspot number.

More complex nonlinear ones which attempted to re-conciliate slight curvature for the initial portions of the graph and the mostly linear features afterwards. Some polynomials [8] are given by

$$F = 65.2 + 0.633\, N_s + 3.76 \times 10^{-3}\, N_s^2 - 1.28 \times 10^{-5}\, N_s^3 \tag{3}$$

and

$$F = 67.0 + 0.572\, N_s + 3.31 \times 10^{-3}\, N_s^2 - 9.13 \times 10^{-6}\, N_s^3 \tag{4}$$

Furthermore equations with exponentials components were suggested [7, 9] given by equations below respectively.

$$F = 0.448\,N_s\left(2 - e^{-0.027\,N_s}\right) + 66 \tag{5}$$

and

$$F = 67 + 0.97\,N_s + 17.6\left(e^{-0.035\,N_s} - 1\right) \tag{6}$$

By looking at all these equations, it formed a basis which was then adapted to the data at hand and 2 equations were developed. A linear one and a non-linear one which used exponential given by equation

$$F = 0.535\,N_s + 60.708 \tag{7}$$

and

$$F = 62 + 0.65\,N_s + 16\left(e^{-0.035\,N_s} - 1\right) \tag{8}$$

and in order to test these two, the values were compared to the different sun cycles recorded starting with cycle 18 to cycle 24 as shown in table (Table 1).

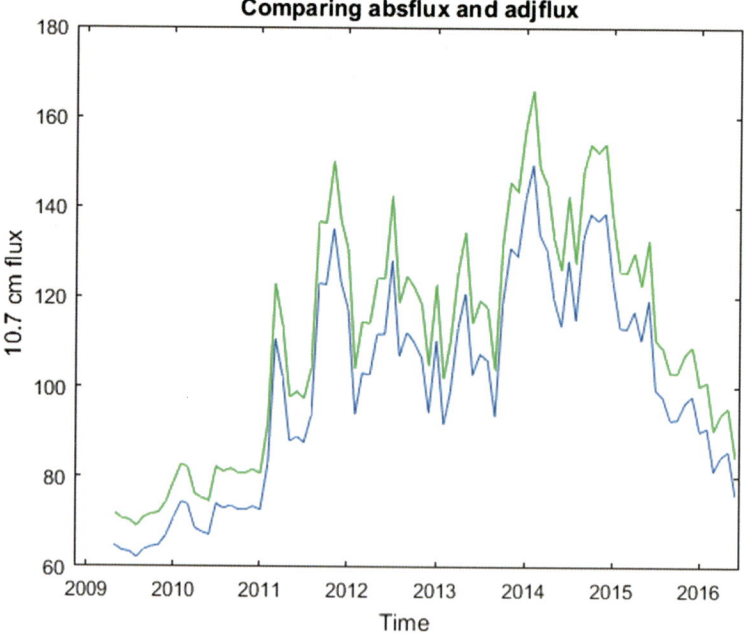

Fig. 1 Absolute flux and adjusted flux

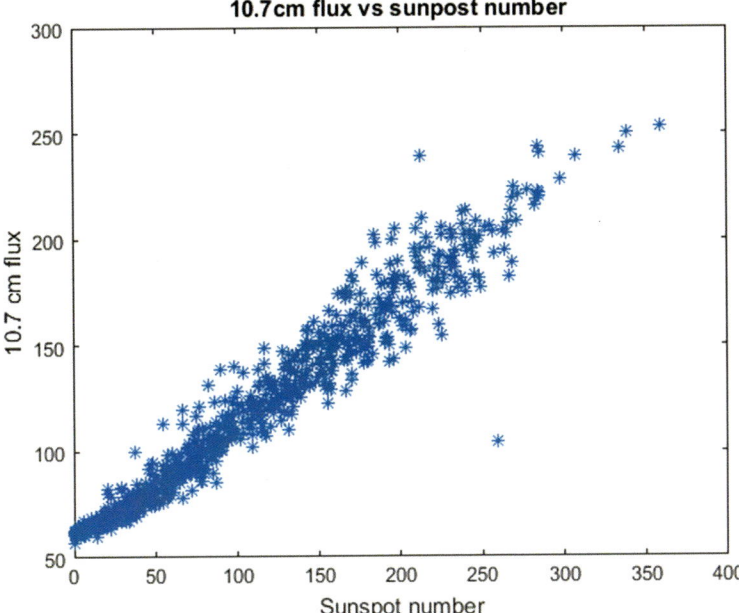

Fig. 2 $F_{10.7}$ versus sunspot number

Table 1 Errors of the 2 methods with respect to sun cycles

Cycle	Linear method		Non linear method	
	RMSE	R^2 (%)	RMSE	R^2 (%)
18	13.23	7.55	13.53	7.90
19	8.32	0.99	8.30	0.99
20	5.77	0.74	7.03	1.09
21	12.90	3.15	14.25	3.85
22	12.63	3.45	8.31	1.50
23	15.02	2.95	13.72	2.46
24	9.46	12.17	9.57	12.47

It can be seen that the error generated for both were close and sometimes the linear model was better while in others, the non-linear model had lower errors. Moreover, by adapting the equation proposed by Tapping [9] about the yearly standard deviation,

$$\sigma_F = 0.23 \, \sigma_{N_s}^{1.33} \tag{9}$$

the following equation was developed

$$\sigma_F = 0.23 \, \sigma_{N_s}^{1.27} \tag{10}$$

The errors associated with this equation is as follows:
$R^2 = 19.63\%$,
$RMSE = 3.90$.

3 Proposed Research: e-CALLISTO Data

The e-CALLISTO data is a relatively new a way to look at the sun and allows smaller countries like Mauritius with limited resources to contribute to the study of solar bursts at ground level in the radio frequency.

3.1 Data Gathering

The study of the e-CALLISTO data was first suggested by Dr. Beeharry of the Physics department, at the University of Mauritius. This study provided useful information on the set up and decisions taken by the university to get involved in the project. Furthermore, the data was collected by the MRT telescopes in the year 2009, 2010 and the first 3 months of the year 2011. The data provided was provided in the form shown in Table 2.

For the rest of the data, excel files containing time stamps of the different important activities detected were provided and to analyse these data, the database of the project's website corresponding to the MRT files and at those particular times were consulted.

These data known as spectrogram were available in two forms, namely FITS and image files in the PNG format. The most common way to analyse this type of data is to use the FITS files and with the help of a FITS viewer but due to some problems, another method was devised. A free app on google chrome known as Desmos graphing calculator https://www.desmos.com/calculator was used whereby each PNG file was uploaded and the start time and stop time of each burst was determined as well as the burst frequency range, shown in Table 2 as high and low.

The user interface of this program is seen in Fig. 5 with one of the e-CALLISTO file uploaded to it.

This method being a bit out of the ordinary presents itself with some limitations in the fact that it is not as accurate as the FITS viewers which allow to pin point

Table 2 Example of MRT data

Date code	Burst duration time		Burst description			Burst frequency range	
	Start	Stop	Type	Subtype	Intensity	High	Low
05/01/2009	6:18:50AM	6:19:09AM	III	G	2	116	45

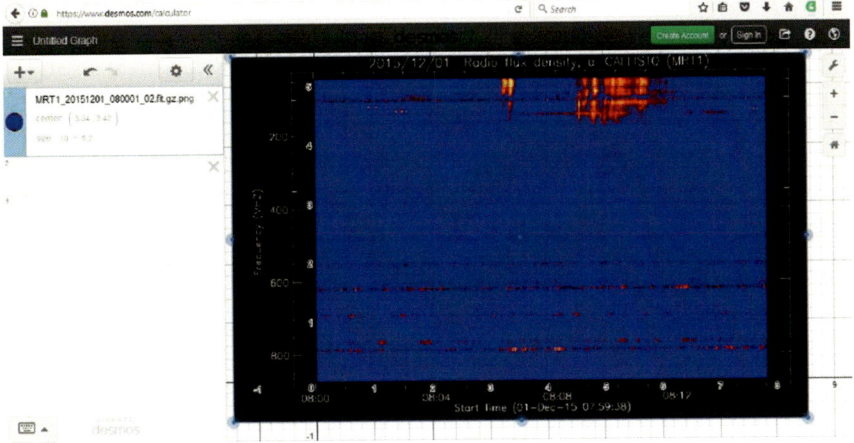

Fig. 3 User interface of Desmos graph calculator

Table 3 Compiled burst data

Date code	Burst duration time			Burst description tion	Burst frequency range		
	Start	Stop	Duration	Type	High	Low	Burst length
05/01/2009	6:18:50AM	6:19:09AM	16 s	III	116	45	71

exact locations on picture returning their exact coordinates and errors exist specially in identifying the length of the different bursts which is represented in Fig. 3 by the header of high and low and as the name would suggest are the maximum and minimum points respectively of the solar bursts. The error estimated for each of these entries would be of ± 5 MHz and that for the time one will about ± 3 s.

With the help of these entries and the burst catalog published by Monstein [10], the different bursts were identified and the Desmos calculator and the data after everything is compiled together is shown in a Table 3.

The duration was calculated by subtracting start from stop and burst length was from subtracting low from high.

3.2 Identification of the Different Types of Radio Solar Bursts

Type I

The type I bursts known as noise storm bursts are generally occur below 350 MHz with a length can be very small, with only a few MHz but others of around 30 MHz have also been detected. Their duration most often occurs in the range of 1 s to 1 min

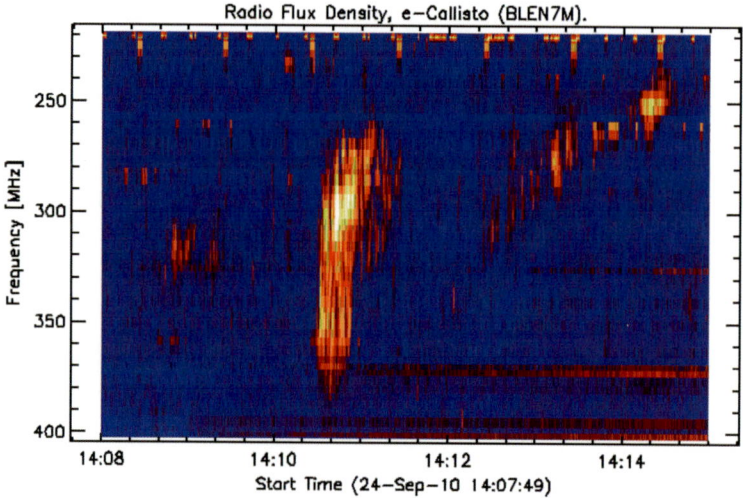

Fig. 4 Type I solar burst or noise storm

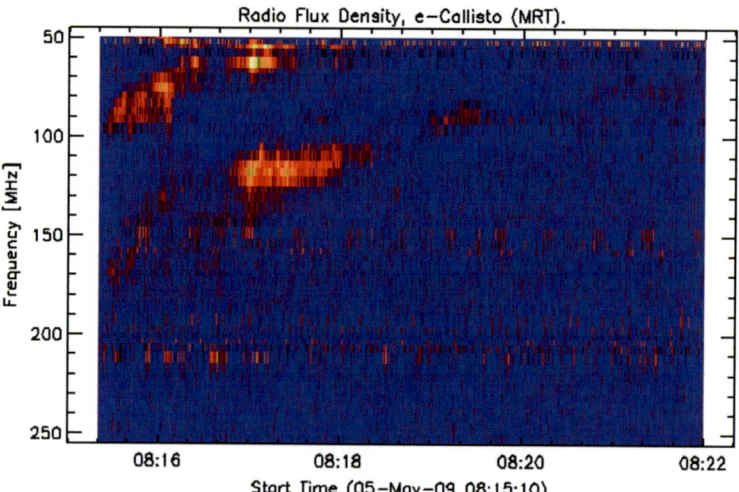

Fig. 5 Type II solar burst with fundamental and harmonic signature

[10]. An example of a type I burst can be seen in Fig. 4. The burst is as a yellow and red spot and the biggest one is the burst in question.

Type II
Slow drift bursts on the images, generally appear long and narrow [10]. Since they generally occur in massive bursts known as CMEs, their duration and length can be quite varied. In Fig. 5, the burst can be seen at the top left.

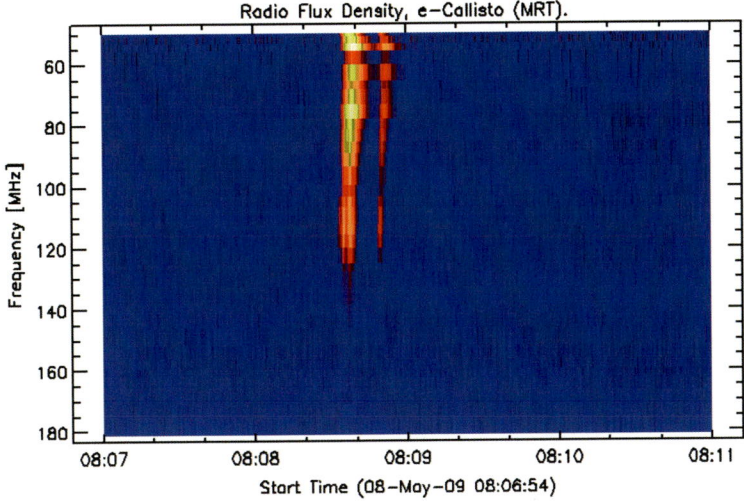

Fig. 6 Type III solar burst

Type III

Fast drift bursts are the most common bursts, and they generally occur at the top of the spectrogram [10]. They may occur individually or in groups. Most often, they last for a few seconds but large groups can go in the minute range. In Fig. 6 at the top we can see a couple of type III bursts.

Type IV

These are the biggest bursts and their duration from hours to even days [10]. Figure 7 shows a type IV burst.

Type V

The type V bursts are closely related to the type III bursts and appear as a flag [10]. Figure 8 shows a type V burst.

If we compare Figs. 6 and 8, you can notice that they are very similar, with the type V having a slightly longer duration.

When all the observations are accounted for, the total number of bursts detected by the Mauritian radio telescopes between May 2009 and June 2016 are 2390 in number and these were divided in terms of type as follows:

Type I: 154.
Type II: 104.
Type III: 2076.
Type IV: 12.
Type V: 37.
Others: 6.

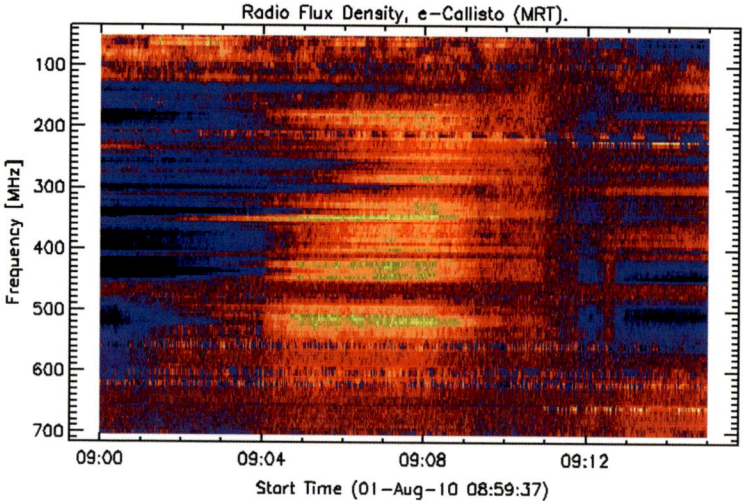

Fig. 7 Type IV solar burst

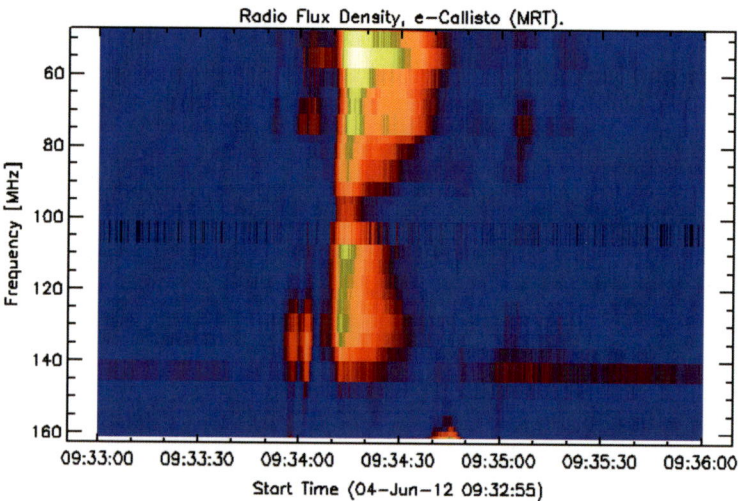

Fig. 8 Type V bursts

4   Analysis and Results: Fitting the Auto Regressive—Moving Average Modelling

4.1   Monthly Sunspot Number

4.1.1   ARMA (9, 0)

$$
\begin{aligned}
X_t = {} & 0.5753\,X_{t-1} + 0.1129\,X_{t-2} + 0.0908\,X_{t-3} + 0.0959\,X_{t-4} \\
& + 0.0288\,X_{t-5} + 0.0538\,X_{t-6} - 0.0272\,X_{t-7} - 0.0025\,X_{t-8} \\
& + 0.0362\,X_{t-9} + W_t
\end{aligned}
\tag{11}
$$

The error measures associated with this model are as follows;

AIC = 7805.61,
BIC = 7857.67,
mean error = −0.044,
root mean square error = 25.18,
mean absolute error = 18.10,
mean percentage error = Inf,
mean absolute percentage error = Inf,
mean absolute scale error = 0.93.

4.1.2   ARMA (8, 1)

$$
\begin{aligned}
X_t = {} & -0.3438\,X_{t-1} + 0.6805\,X_{t-2} + 0.2068\,X_{t-3} \\
& + 0.2394\,X_{t-4} + 0.1241\,X_{t-5} + 0.0322\,X_{t-6} \\
& - 0.0027\,X_{t-7} + 0.0038\,X_{t-8} + W_t + 0.9690\,W_{t-1}
\end{aligned}
\tag{12}
$$

The error measures associated with this model are as follows;

AIC = 7804.57,
BIC = 7856.63,
mean error = −0.31,
root mean square error = 24.97,
mean absolute error = 18.20,
mean percentage error = Inf,
mean absolute percentage error = Inf,
mean absolute scale error = 0.93 (Table 4).

Table 4 Comparing predicted values with actual values of sunspot number

Month/year	Forecast value for ARMA (9, 0)	Forecast value for ARMA (8, 1)	Actual value
01/2017	27.23	27.39	26.1
02/2017	30.80	28.27	26.4
03/2017	30.17	29.39	17.7
04/2017	30.31	28.90	32.3
05/2017	31.72	31.47	18.9
06/2017	33.08	31.70	19.2
07/2017	34.25	33.90	17.8
08/2017	34.73	33.84	32.6
09/2017	35.14	36.03	43.7
10/2017	36.04	36.05	13.2

4.1.3 Monthly 10.7 cm Radio Flux

ARMA (9, 0)

$$X_t = 0.7275\,X_{t-1} + 0.0170\,X_{t-2} + 0.1145\,X_{t-3} + 0.0732\,X_{t-4}$$
$$+ 0.0018\,X_{t-5} + 0.0546\,X_{t-6} - 0.0035\,X_{t-7}$$
$$- 0.0110\,X_{t-8} - 0.0032\,X_{t-9} + W_t \tag{13}$$

The error measures associated with this model are as follows;

AIC $= 6842.87$,
BIC $= 6894.92$,
mean error $= -0.25$,
root mean square error $= 14.08$,
mean absolute error $= 9.75$,
mean percentage error $= -1.43$,
mean absolute percentage error $= 8.01$,
mean absolute scale error $= 0.99$.

4.1.4 ARMA (8, 1)

$$X_t = -0.2417\,X_{t-1} + 0.7217\,X_{t-2} + 0.1298\,X_{t-3}$$
$$+ 0.1815\,X_{t-4} + 0.0684\,X_{t-5} + 0.0538\,X_{t-6}$$
$$+ 0.0259\,X_{t-7} + 0.0053\,X_{t-8} + W_t + 0.9761\,W_{t-1} \tag{14}$$

The error measures associated with this model are as follows;

AIC $= 6838.28$,
BIC $= 6890.34$,

Table 5 Comparing predicted values with actual values of 10.7 cm flux

Month/year	Forecast value for ARMA (9, 0)	Forecast value for ARMA (8, 1)	Actual value
01/2017	69.99	70.28	67.42
02/2017	72.01	71.90	67.59
03/2017	72.61	72.82	66.56
04/2017	73.18	73.073	73.31
05/2017	73.712	74.01	67.73
06/2017	74.20	74.29	69.47
07/2017	75.01	75.32	72.10
08/2017	75.88	75.69	72.16
09/2017	76.61	76.68	86.90
10/2017	77.28	77.01	68.31

mean error $= -0.25$,
root mean square error $= 14.04$,
mean absolute error $= 9.74$,
mean percentage error $= -1.41$,
mean absolute percentage error $= 7.80$,
mean absolute scale error $= 0.99$ (Table 5).

For both the sunspot number and the 10.7 cm flux. The predictions for the first 2–4 entries seems decent and ones afterward, are to a lesser degree.

4.2 e-CALLISTO Data

4.2.1 ARMA (9, 0)

$$X_t = 0.2476\, X_{t-1} + 0.0545\, X_{t-2} + 0.0592\, X_{t-3} + 0.1079\, X_{t-4}$$
$$- 0.0497\, X_{t-5} - 0.0892\, X_{t-6} + 0.1344\, X_{t-7}$$
$$- 0.0701\, X_{t-8} - 0.2532\, X_{t-9} + W_t \tag{15}$$

The error measures associated with this model are as follows;

AIC $= 799.28$,
BIC $= 826.27$,
mean error $= 0.83$,
root mean square error $= 22.08$,
mean absolute error $= 17.08$,
mean percentage error $=$ Inf,
mean absolute percentage error $=$ Inf,
mean absolute scale error $= 0.80$.

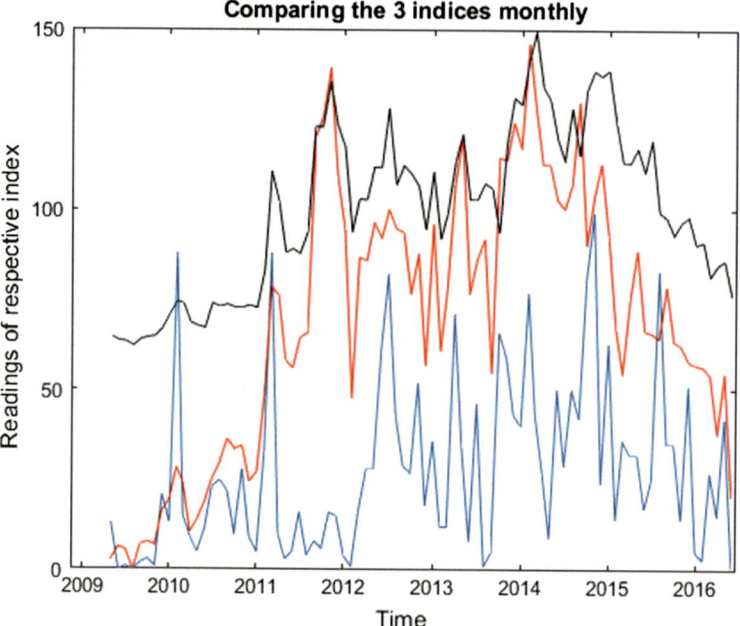

Fig. 9 Monthly time series of all 3 data sets

4.2.2   ARMA (8, 1)

$$X_t = -0.5886\,X_{t-1} + 0.2941\,X_{t-2} + 0.1138\,X_{t-3}$$
$$+ 0.138\,X_{t-4} + 0.0523\,X_{t-5} - 0.0996\,X_{t-6}$$
$$+ 0.102\,X_{t-7} + 0.00111\,X_{t-8} + W_t + 0.8964\,W_{t-1} \qquad (16)$$

The error measures associated with this model are as follows;

AIC $= 800.87$,
BIC $= 827.87$,
mean error $= 0.32$,
root mean square error $= 22.28$,
mean absolute error $= 17.39$,
mean percentage error $=$ Inf,
mean absolute percentage error $=$ Inf,
mean absolute scale error $= 0.81$.

Since the data available was not very extensive, all the data was used to fit the models and if a forecast were to be done, there would be no real data to compare with as the process of analysis of the number if bursts detected per month is quite time consuming.

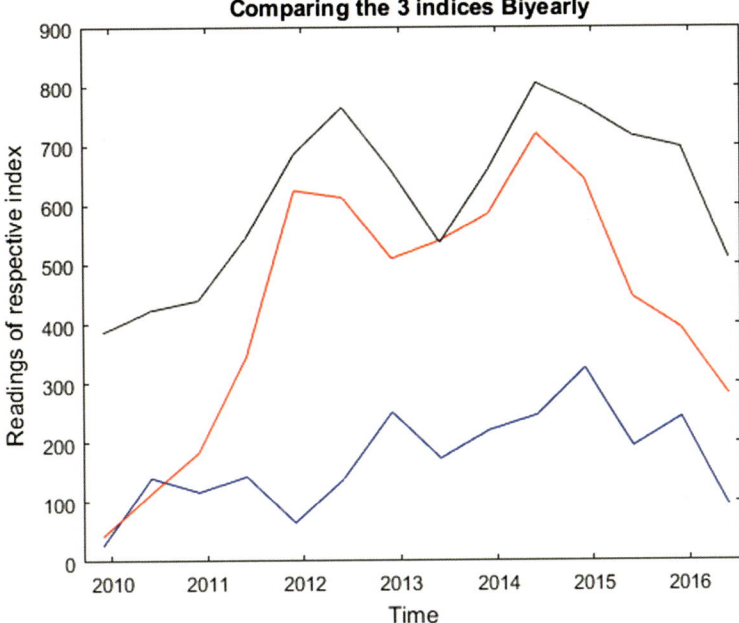

Fig. 10 6 month interval time series of all 3 data sets

4.3 Relationship Between the Solar Burst Data and the 2 Indices

When comparing the 3 data sets for the monthly period between May 2009 till June 2016 is shown below.

In Fig. 9, the red line is the sunspot number, the black line is 10.7 cm flux, and the blue line is the number of burst data detected each month.

Looking at the graph, it is tough for any clear interpretation and the correlation can help us better understand which is as follows:

Correlation between burst data and sunspot number $= 0.46$.

Correlation between burst data and F10.7 $= 0.51$.

These numbers are lower than anticipated and one explanation can be because of the data used for burst data was only the data observed in Mauritius. To remedy this, the cumulative data for longer periods that is over periods of 6 months. Therefore the data was divided from periods January to June of a particular year were summed up and that of July and December of that year were summed up and the graph of this particular data sets are shown in Fig. 10.

Similarly to Fig. 9 in Fig. 10 the red line is the sunspot number, the black line is 10.7 cm flux, and the blue line is the number of burst data. The correlation is as follows:

Correlation between burst data and sunspot number $= 0.67$.

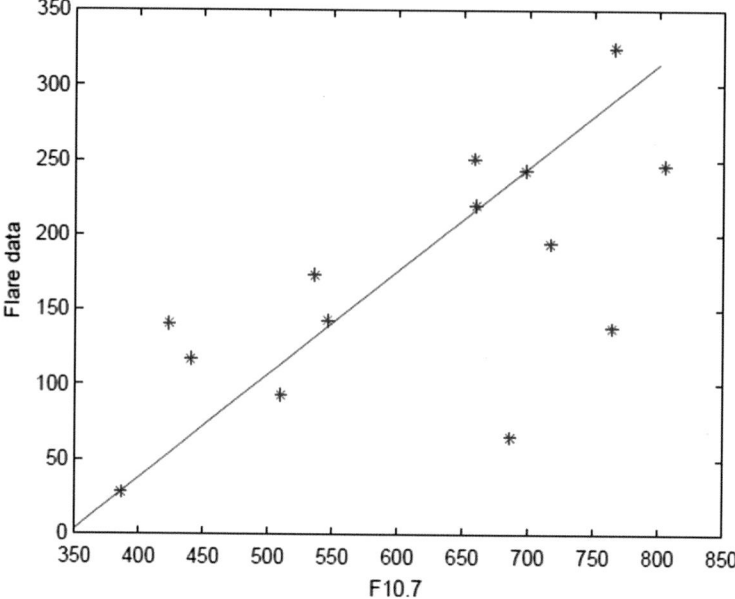

Fig. 11 Relationship between solar burst data and F10.7

Correlation between burst data and F10.7 = 0.74.

Here the correlation is higher and we can observe that it is higher for the burst and $F_{10.7}$. This can be due to the fact that both are in the Radio frequency.

The 10.7 cm flux was used because they are both in the radio frequency and the higher correlation between the two variables.

While a proper equation connecting the two variables could not be derived but a general observation that can be observed is that overall, as the flux increases so does the number of solar bursts as shown in Fig. 11 by the black line.

5 Concluding Remarks and Future Developments

The 11 year cycle was shown using the Fourier transform and time series analysis of the 2 indices that is the sunspot number and 10.7 cm flux as well as that of the solar burst data were carried out with the ARMA (9, 0) being the best model in most cases and the ARMA (8, 1) also being a decent model. The relationship between the absolute 10.7 cm flux and the sunspot number was determined as well as one between the standard deviation connecting the 2 indices. Moreover a general relationship between the sun's activity and that of the solar bursts was shown. As the sun's activity increased so did the number of solar bursts detected.

As for future works, gathering data from the e-CALLISTO using all the stations around the world and make an exhaustive database of all bursts detected to then compare it to the sunspot number and the 10.7 cm radio flux can be a good place to start. Moreover at least the data for a compete sun cycle pertaining to the radio solar bursts will also be very helpful in this endeavour to better understand the behaviour of the sun.

Acknowledgements I would like to thank Dr. Beehary of the Physics department of the University of Mauritius as he was the one who provided the data from the Mauritian Radio Telescope and all those working at MRT for gathering the data.

References

1. Box GE, Jenkins G, Reinsel GC (2008) Time series analysis: forecasting and control. Wiley
2. Werner R (2012) Sunspot number prediction by an autoregressive model. Sun Geosphere 7:75–80
3. Tapping K (2013) The 10.7 cm solar radio flux (F10.7). Space Weather 11:394–406
4. Andersen A, Weiss A (1984) Forecasting: the box-jenkins approach. In: Makridakis et al. (eds) The forecasting accuracy of major time series methods. Wiley. Chichester, England
5. Box GEP, Jenkins G (1970) Time series analysis. Forecasting and Control, Holden-Day, San Francisco
6. Box GEP, Pierce DA (1970) Distribution of the residual autocorrelations in autoregressive-integrated moving-average time series models. J Am Stat Assoc 65:1509–1526
7. Hathway D (2015) The solar cycle. Mail stop: 258–5, NASA ames research center, Moffett field, CA 94035, USA, 1–81
8. Johnson WR (2009) Power law relating 10.7 cm flux to sunspot number. Astrophysics and Space Science 1–8
9. Tapping KF, Valdes JJ (2011) Did the sun change its behaviour during the decline of cycle 23 and 24. Solar Phys 337–350
10. Monstein C (2011) Catalog of dynamic electromagnetic spectra. Phys Astron Electron Work Bench

A Homogenous Prototype Design for IED Detection Using Subsurface Ground-Penetrating RADAR

Alagarsamy Gautami and G. Naveen Balaji

Abstract Security is a prime factor of a nation's peace. Nowadays, terrorists are using Improvised Explosive Devices for their attacks. This paper presents a method to locate, characterize, and identify the IED in landmines using a subsurface technology called Ground-Penetrating RADAR with electromagnetic imaging. By this technique, the homogenous nature of soil, moisture content, dielectric constants of air, soil, and explosives are analyzed. To measure the attenuation and scattering losses of IED underground, a homogenous 2D prototype model is designed with TNT explosives as IED. By simulation, the scattering loss and attenuation are characterized over the step frequency range from 0.5 to 3 GHz. The step frequency shows that the optimal frequency to detect IED in the subsurface scan was 2 GHz. Based on the experimental results, the interference is more in dry soil than wet soil and also the dimension of landmine is directly proportional to the amount of scattering.

1 Introduction

Improvised Explosive Devices is an improvised device incorporating destructive, poisonous, pyrotechnic chemicals which are designed to destroy or to create chaos and catastrophe action. Locating, characterizing, and identifying the IED are a challenging task for any military organization to combat terrorism. IED detection can be performed by bulk detection and trace detection. Bulk detection is an image processing method to detect explosives and its computer vision characteristics, whereas trace detection method implies the chemical and physical characteristics of the explosives materials. Nowadays, wireless technology plays an important role in demining the IED in landmines [1]. IED detection method includes various sensors such as thermal infrared imagery sensors, passive magnetic anomaly sensor,

A. Gautami (✉) · G. Naveen Balaji
SNS College of Technology, Coimbatore, TamilNadu, India
e-mail: aa.gautami@gmail.com

G. Naveen Balaji
e-mail: yoursgnb@gmail.com

© Springer Nature Switzerland AG 2019
D. Pandian et al. (eds.), *Proceedings of the International Conference on ISMAC in Computational Vision and Bio-Engineering 2018 (ISMAC-CVB)*, Lecture Notes in Computational Vision and Biomechanics 30,
https://doi.org/10.1007/978-3-030-00665-5_156

chemical sniffers sensors, ground-penetrating RADAR, and the microwave sensors. To counter this problem of minesweeping and mine clearance, an efficient dual sensor technology called Ground-Penetrating RADAR (GPR) with electromagnetic imaging is addressed in this paper.

2 Background

IED in landmines are buried or implanted on the surface of ground or buried within the ground in a regular pattern or distributed irregularly. It is necessary to understand the construction of IED and its characterization [2]. The common components of IED include switch, initiator, main charge, power source, and a container. Figure 1 shows the construction of IED. There are two major aspects to understand an IED action.

(a) Tactical Characterization.
(b) Technical Characterization.

Tactical characterization is a plot planned to conduct an IED incident, whereas technical characterization includes the hierarchial design of IED with the necessary technical and forensic information of the design.

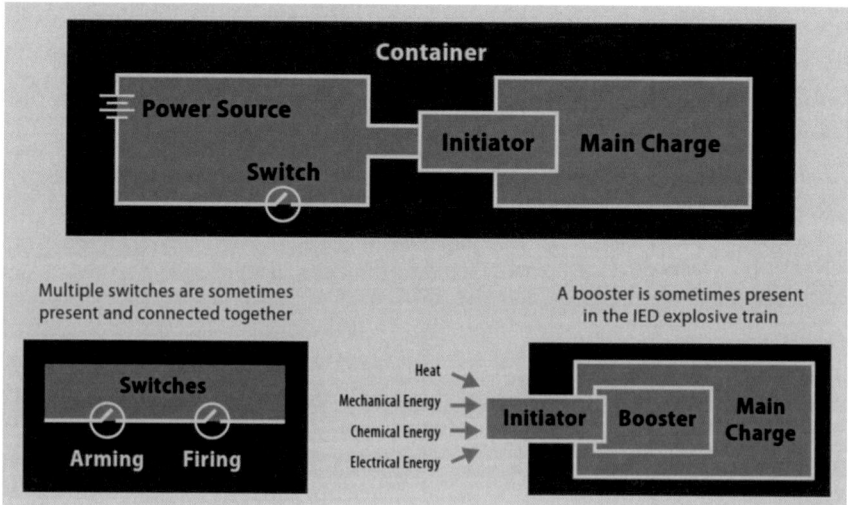

Fig. 1 Construction of IED

Fig. 2 Military vehicle to detect IED using LIBS

2.1 Novel Approaches for IED Detection

The detection of IED depends on various nonhomogenous geometric factors such as depth, size, moisture content, and incident frequency of the soil. By adapting these parameters, some of the novel approaches are listed below.

Spectroscopic approaches: Spectroscopic detection system converts the bomb fumes or vapors to ionized samples in the form of waves [1]. Some of the approaches are:

- Ion Mobility Spectroscopy.
- Mass Spectroscopy.
- Terahertz spectroscopy.
- Infrared spectroscopy.
- Laser-Induced Breakdown spectroscopy (LIBS).
- E-Noses.

Figure 2 shows the military vehicle using LIBS spectroscopic method to detect IED. The spectroscopic methods are accurate and efficient but identification of plastic IED, automated real-time detection, marking, and notification of IED are challenging factors. To overcome these problems, a noninvasive technology called subsurface sensing is modeled to recover IED from landmines [3].

3 Ground-Penetrating RADAR

GPR is an innovative approach for subsurface mapping on collecting RADAR signals from IED. The GPR consists of a transmitter and a receiver which emits electromagnetic waves on IED and detects the target by reflection [4]. Based on the velocity of the reflected signal, soil types, and nature of material, the target is characterized and

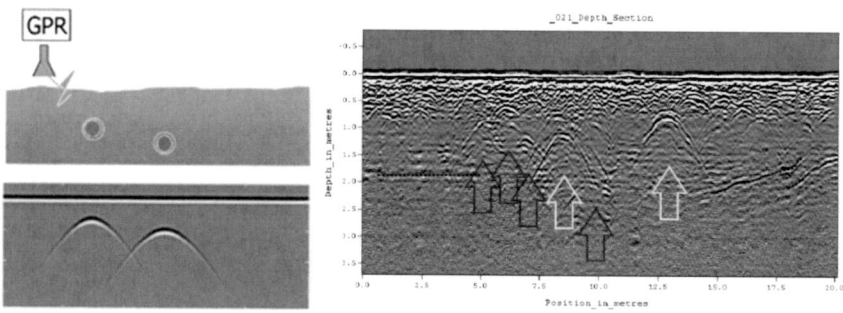

Fig. 3 Target reflection pattern of RADAR from IED

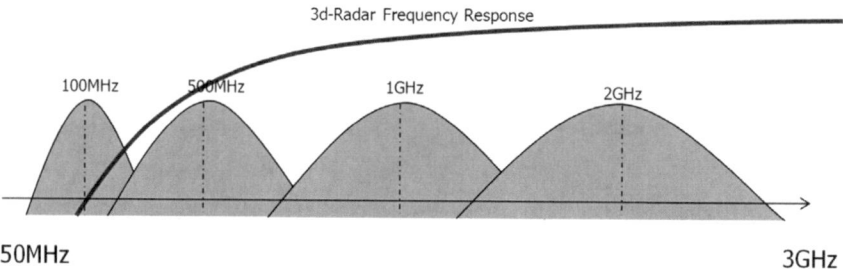

Fig. 4 3D RADAR frequency range coverage for multiple array antenna

identified. Figure 3 indicates target reflection pattern (hyperbola) due to the distance between RADAR and target.

In proper soil conditions, 3D mapping of the geographical area improves good detection than 2D mapping. To perform 3D pattern multiple parallel antennas are mounted in an array on RADAR. Figure 4 shows 3D frequency coverage for multiple array antenna design.

Some of the important features and benefits of 3D GPR are as follows:

- Target detection with real-time analysis for route clearance.
- Array of antennas at variable dimensions can be mounted.
- Detection algorithms for target can be employed.
- Demining system is efficient.

3.1 Data Acquisition Techniques

Step frequency approach is one of the techniques which are used for data collection. In this approach, an incremental frequency steps are transmitted by antenna to the target along the subsurface [5]. The choice of step frequency paves more advantages

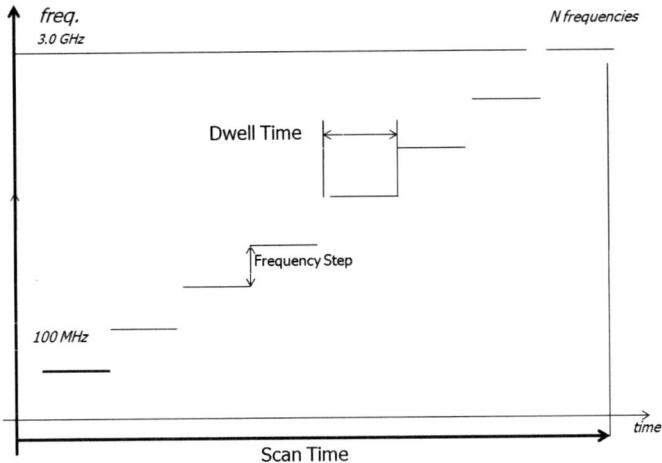

Fig. 5 Graphical representation of step frequency signal

Table 1 The physical parameters of soil and TNT explosives to measure attenuation losses

Material	Relative permittivity	Relative permeability	Conductivity
Air	439.2	1	0
Dry soil	1273 + 31i	2.9	0.004
Wet soil	1756 + 395i 4	4	0.049
TNT	2.9	1	4.8e−4

as shown in Fig. 5, such as maximum integration time, medium resolution is sufficient for greater soil attenuation, and lower interference with radio signals.

3.2 Prototype Design for IED Detection

To determine the GPR experimental results a prototype is designed using COMSOL Multiphysics software with the IED in Landmine scenario. Figure 8 is a 2D prototype model to simulate the parameters and losses to detect GPR actions to locate IED. To promote the modeling the finite element modeling is tested to extract results from the complex environment. Some of the features are adopted in the representation such as two homogenous surfaces which are perfectly matched. These layers absorb all the incident signals from transmitter. The step frequency of 0.5–3 GHz is incident on the surface using EMW module for experimentation. With various geometrical parameters from Table 1 the model is analyzed and simulated to capture scattering loss. Figure 6 shows the 2D prototype model for IED detection.

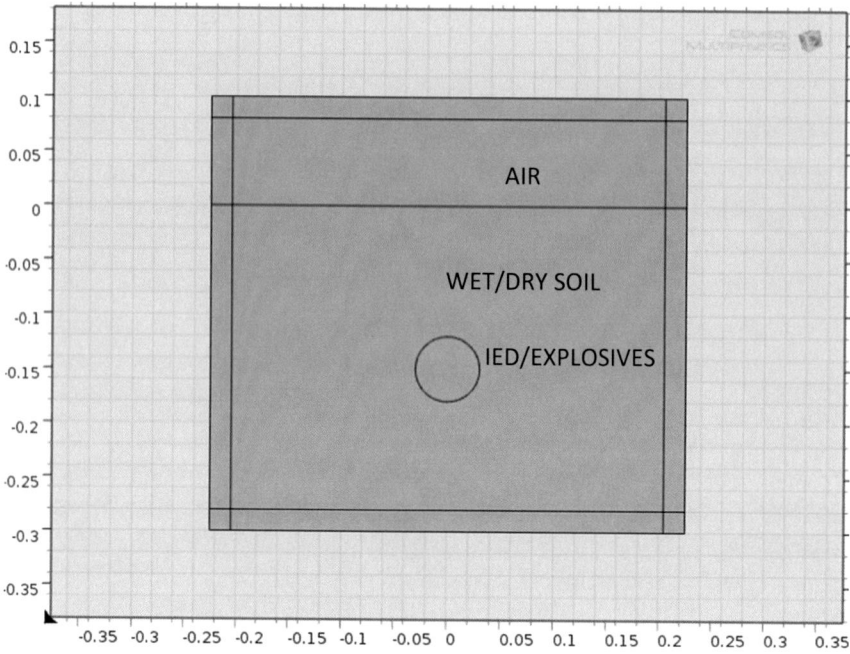

Fig. 6 A 2D model representation for IED detection

3.3 Performance Requirements

The performance factor depends on randomness or probability of detecting IED. Probability Detection (PD) and its false alarm probability (PFA) are the two main factors to measure performance. For example, the maximum performance achieved for detecting IED using handheld metal detector is measured as PD = 1 and PFA = 0. The Receiver Operating Curve (ROC) in Fig. 7 is plotted between probability of false IED to sensitivity of RADAR signal. PD is inversely proportional to PFA [6]. The ROC is used to measure both human and machine performance. True positive (PD) and True negative (PFA) depends on step frequency response. Figure 6 represents the ROC based on TP and TD based on step frequency approach.

The dielectric constant of the soil and explosives is also an important parameter to measure the performance of GPR. For example, the relative dielectric constant of TNT explosives is 2.70 and for RDX it is 2.90. Based on the dielectric constant, the radiated power is measured in the order of milliwatts.

The propagation loss of soil is a performance factor that depends on conductivity of soil and the frequency of the detected signal. The moisture content ensures the relative dielectric constant of soil. From study, the relative dielectric constant for dry to wet soil is 3–16, respectively. Figure 8 represents the spectrum of transmitted and

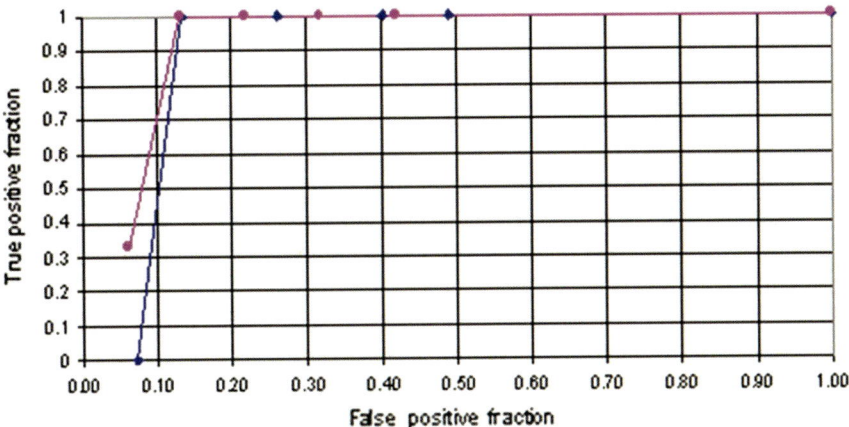

Fig. 7 ROC characteristics (TP vs. TN)

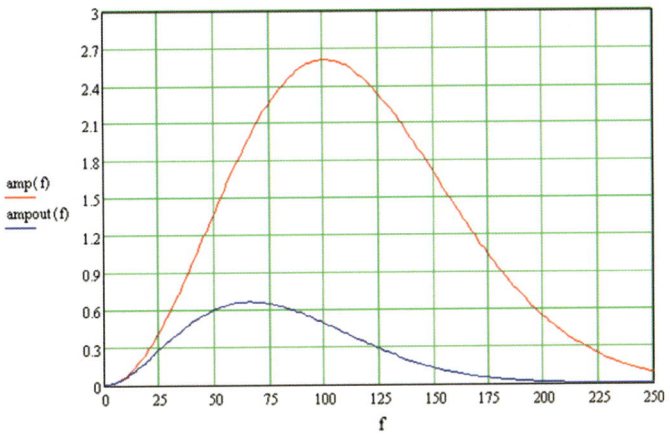

Fig. 8 Transmitted and received signal spectrum after penetrating the lossy soil. (red-transmitted signal, blue-received signal)

received signal after penetrating lossy soil. From Fig. 7, the nature of received signal from lossy dielectric soil is attenuated due to penetration loss.

The attenuation loss is measured using the following formula:

$$La = 8.686 * 2R * 2\pi \ f\left(((\mu_0\mu_r\varepsilon_0\varepsilon_r)/2) * (1 + \tan^{2\delta})^{1/2} - 1\right)^{1/2}$$

where

f Frequency (Hz)
tan d   Loss tangent of material
ε_r Relative permittivity of material

ε_0 Permittivity of free space
μ_r Relative magnetic susceptibility of material
μ_0 Magnetic susceptibility of free space
R Range (meters).

From the equation, the performance is featured as follows: the propagation losses are proportional to frequency rates and the near-field regions [7]. The attenuation losses increase with frequency, and it falls in the range 300 MHz–1.5 GHz at receiving end. The attenuation losses will reduce the directivity and gain of multiple array antennas connected to the RADAR.

4   Experimental Results

From the experimental research through simulation, the scattering effects are directly proportional to the IED depth in perfectly matched layer. The step frequency of 0.5–3 GHz is incident on surface using EMW module. From Table 1, the TNT explosive was taken for reference. With the dielectric constant of 2.70, the scattering effects are simulated with the prototype model. From simulation, the dry soil encounters more interference than the wet soil as shown in Fig. 9. Therefore, conductivity is more in the wet soil which can retain waves more than dry soil. The depth of the landmine is proportional to the amplitude of scattered waves. The step of frequencies shows that the optimal frequency to detect IDE in the subsurface scan was 2 GHz.

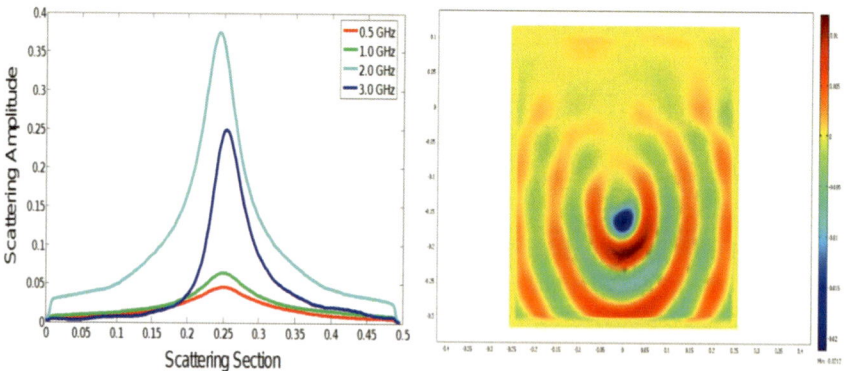

Fig. 9 Scattering amplitude cross at different frequencies for the air/wet soil/dry soil model

4.1   Other Applications of GPR

GPR systems are widely used in demining and detecting IED. It is also used in various capacities with high-resolution electromagnetic imagery such as

- Inspection of Bridge deck.
- Road surveying.
- Delimitation identification.
- Ballast inspection.
- Utility mapping and unmanned missions.
- Archeology.

5   Conclusion

In this paper, an experimental homogenous prototype model for the subsurface detection of IEDs using GPR technology was designed. By developing this prototype model, the nature of radio waves at various dielectric constants is identified. The relation between various parameters which affects the performance of incident radio waves is analyzed. The scattering and attenuation losses of incident waves with step frequency input signal is simulated. This technique provides the limelight to characterize, locate, and identify the IED and explosives all across the globe to counter terrorism.

References

1. Ghazali KH, Jadin MS (2014) Detection improvised explosive devices (IED) emplacement using infrared image: ieee computer society—16th international conference on computer modeling and simulation https://doi.org/10.1109/UKSim.2014.111
2. Muthukumar K, DeviMala E (2011) Detection of IED using nanotechnology: Int J Sci Eng Res 2(12) ISSN 2229-5518
3. Liu R, Wang H: Detection and localization of improvised explosive devices based on 3-axis magnetic sensor array system: 2010 symposium on security detection and information processing: Elsevier. Proc Eng 1877–7058 doi:10.1016
4. Lopera O, Milisavljevie N, Daniels D, Macq B (2007) Time-frequency domain signature analysis of GPR data for landmine identification, advanced ground penetrating RADAR. In: 2007 4th international workshop 27–29 2007, pp 159–162
5. Daniels DJ (2004) Ground penetrating RADAR: ISBN 0863413609: IEEE (RADAR, Sonar and Navigation):year-2004
6. Tripe Wires: Introduction for explosives. Homeland Security Department, USA
7. The White House Countering improvised explosives devices. Washington DC, USA

A Study on Various Deep Learning Algorithms to Diagnose Alzheimer's Disease

M. Deepika Nair, M. S. Sinta and M. Vidya

Abstract Alzheimer's disease (AD) is one of the most frequent types of dementia, which is deterioration in mental ability severe enough to interfere with daily life and gradually affect the human's brain, its capability to learn, think and communicate. The symptoms of AD develop over time and become a major brain disease over the course of several years. To bring out patterns from the brain neuroimaging data, different statistical and machine learning approaches have been used to find the Alzheimer's disease present in older adults at clinical as well as research applications; however, differentiating the phases of the Alzheimer's and healthy brain data has been difficult due to the similarity in brain atrophy patterns and image intensities. Recently, number of deep learning methods has been expeditiously developing into numerous areas, which consist of medical image analysis. This survey gives out the idea of deep learning-based methods which used to differentiate between Alzheimer's Magnetic Resonance Imaging (MRI) and functional MRI from the normal healthy control data.

1 Introduction

The research in Deep learning has been growing fast in the medical imaging field, including Computer-Aided Diagnosis (CAD), medical image analysis, and radiomics. CAD is an expeditiously developing area of research in the medical industry. The recent researchers in machine learning guarantee the enhanced efficiency in detection of disease. Here, the computers are enabled to think by

M. Deepika Nair (✉) · M. S. Sinta · M. Vidya
Department of Computer Science Engineering, Vidya Academy of Science and Technology,
Thrissur, Kerala 680501, India
e-mail: deepikamnair1995@gmail.com

M. S. Sinta
e-mail: sintamsuresh@gmail.com

M. Vidya
e-mail: vidya.m@vidyaacademy.ac.in

© Springer Nature Switzerland AG 2019
D. Pandian et al. (eds.), *Proceedings of the International Conference on ISMAC in Computational Vision and Bio-Engineering 2018 (ISMAC-CVB)*, Lecture Notes in Computational Vision and Biomechanics 30,
https://doi.org/10.1007/978-3-030-00665-5_157

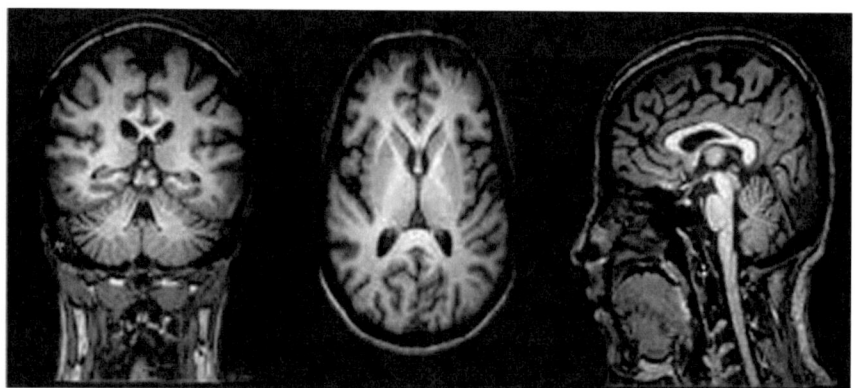

Fig. 1 T1-weighted MRI scans acquired in coronal (left), axial (center), and sagittal (right) planes with 3 T

developing intelligence by learning [1]. There are distinct types of deep learning techniques and which are used to analyze the data sets (Fig. 1).

One of the symptoms of Alzheimer's disease is the enlargement of amyloid plaques among brain nerve cells [2]. To diagnosis the Alzheimer's disease, it is required to know the internal structures hidden in the brain which can be obtained through various types of scanning techniques existed.

These scanning methods in the medical field include Angiography and Computed Tomography (CT), MRI Angiography (MRA), Dynamic CT or Dynamic MRI, Magneto Encephalography (ME), MRI, Flow-sensitive MRI (FSMRI), functional MRI (fMRI), etc. Structural MRI is a chief imaging biomarker in AD as the cerebral atrophy is shown to closely link with cognitive symptoms.

Deep Learning (DL) is a member of machine learning research inspired by the function and structure of human brain, and which aim at discovering multiple levels of distributed representations. Recently, many deep learning procedures have been used to fix conventional artificial intelligence problems. Mainly, there are two groups of deep learning models which are dissimilar with respect to how the data flow over the network. The information in the feed-forward networks flows over the network in just one direction, from the input layer to the output layer. Compared to feed-forward network, recurrent networks have feedback connections that allow the information from past inputs to affect the current output. In the framework of supervised classification problems, the implementation of Deep Learning requires two main steps. The first stage is called training and this phase uses the training set which is a portion of the dataset available to correct the networks parameters to perform the classification. The next step is testing phase, which is used the rest of the subset called as the test set to determine whether model that is trained can correctly predict the new observations class. When the number of available data is less, it is also possible to run the training and testing phases several times on different training and test splits of the original data and then estimate the average performance of the model. This approach is known as cross-validation. The training phase and testing phase is not a unique feature of Deep Learning but are used in conventional ML methods [3].

The paper is organized is as follows. Section 2 describes various deep learning techniques to detect Alzheimer's disease. Section 3 concludes the paper.

2 Related Works

This section reviews existing Deep Learning approaches used to detect Alzheimer's disease from structural and functional MRI scans.

2.1 Autoencoder

An autoencoder is an artificial neural network which can grasp the features in an unsupervised way by minimizing reconstruction errors [4]. The intention of autoencoder is to train a model for the collection of data, with the typical purpose of data compression. The process of training in autoencoder is based on the development of a cost function. The cost function estimates the error through backpropagation and it is rebuilding at the output. An autoencoder constitutes of an encoder followed by decoder. The encoder and decoder can possess multiple layers; nevertheless for simplicity, we consider that all of them have only one layer. The basic structure of autoencoder is shown in Fig. 2.

Debesh Jha and Goo-Rak Kwon Debesh Jha proposed a method based on autoencoder and this framework use structural MRI data provided from Open Access Series of Imaging Studies (OASIS) database [5]. Here use deep learning architecture, which encompass of sparse autoencoders, Scale Conjugate Gradient (SCG), stacked autoencoder, and a softmax output layer to overcome the bottleneck and support the analysis

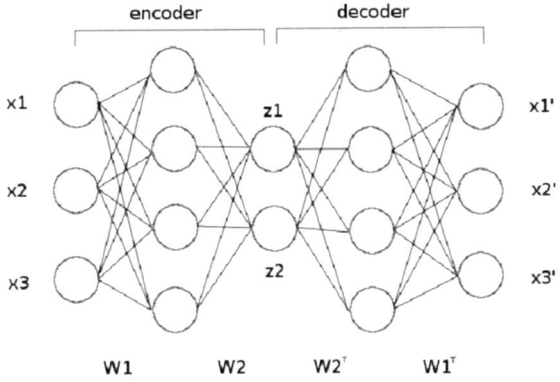

Fig. 2 Basic structure of autoencoder

of AD and normal healthy controls. Compared to the former workflows, this technique requires fewer labeled training examples and minimum prior knowledge and also performs dimensionality reduction and data fusion at the same moment. A performance gain is achieved with the binary classification and gets 91.6% accuracy.

Bhatkoti and Paul [6] studied the effectiveness of the k-sparse Autoencoder (KSA) algorithm in deep learning structure for the diagnosis of Alzheimer's disease. It compared the modified approach to non-modified k-sparse approach in this application. The research used MRI scan data, CSF, and PET images each of 150 patients. MRI images for comparison with research images together with CSF and PET data were obtained from Alzheimer's disease Neuroimaging Initiative (ADNI). The MRI scans were reprocessed by correcting orientation errors and by skull strip to obtain underlying tissues. The images were normalized and smoothened. Patch extraction was then done and masks for different brain subregions were obtained and transformed during registration in the Automatic Anatomical Labeling (AAL) template. A feed-forward convolutional pair predictor neural network was developed. Flattening, concatenation, and sorting were done on feature vectors which were in turn input into feed-forward multi-layer perceptron. Prediction of output was carried out using probability function. Three-dimensional convolutional neural network with 384 input neurons and 200 hidden neurons were used in multi-layer perceptron. Cross-validation algorithm with 20-fold cross-validation was used in training. A practical approach with actual MRI images from patient screening was used in this research and compared with data from ADNI as well as those employed in previous studies, and this method contributes to efficiency of 63.24% early diagnosis of Alzheimer's disease and confirms that KSA enhances the efficiency as 74.05%.

2.2 Convolutional Neural Network

A Convolutional Neural Network (CNN) is made up of convolutional layers, pooling layers, normalization layers and then the fully connected layers. The construction of a CNN is described to yield the advantage of the 2D format of an input image. CNN use little preprocessing operation rather than other image processing operations. The advantage of CNNs is that it cut down the number of parameters with the same number of hidden units and the training of CNN is straightforward. The convolutional layer takes an MxMxR input image, which is corresponding to the height, width, and the number of channels of an image, respectively.

The convolutional layer plays a crucial role in CNN architecture and is the basic building block in this network. The CONV layers parameters contains a group of learnable filters. Spatially, the size of each filter is small but enhance through the complete input volume depth. For each forward pass, it is multiplying the original image pixel values with the values in the filter which is used by CONV layer. These multiplications are all summed up and producing a two-dimensional activation map of that filter. Next, all filters activation maps are stacked and produce output volume. A pooling layer is mostly added in-between subsequent Conv layers. Its function is to

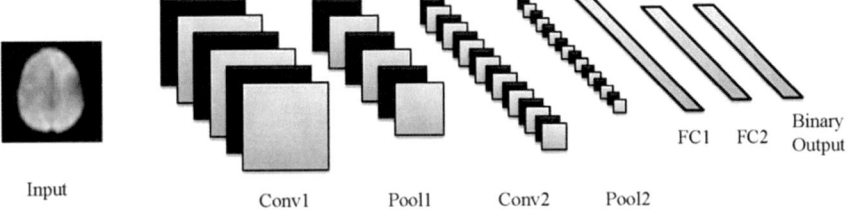

FC1   FC2   Binary Output

Input   Conv1   Pool1   Conv2   Pool2

Fig. 3 LeNet-5 architecture

reduce the dimensionality of each feature map but preserve the important information. The pooling Layer operates on each feature map independently and resizes the input spatially. Spatial pooling can be of different types Max, Average, Sum, etc. Recent researchers have developed more successful CNN such as AlexNet, GoogleNet, ResNet, ZF Net, VGGNet, and LeNet. The major problem of constructing ConvNet architectures is the memory restrictions of GPU [7–9].

Sarraf and Tofighi [10] in their paper used CNN deep learning architecture (LeNet) that was trained and tested with huge number of images and classified the AD data from normal control with 96.86% accuracy. The architecture of LeNet-5 is shown in Fig. 3.

Ciprian et al. [11] use DemNet architecture which is a modified version of the 16-layer CNN made by the Oxford University Visual Geometry Group (VGG) for the ImageNet Large-Scale Visual Recognition Challenge (ILSVRC). It is a collection of 13 convolutional layers followed by 3 fully connected layers. This architecture successfully classifies AD and MCI from HC on the Alzheimer's disease Neuroimaging Initiative (ADNI) dataset with an accuracy of 91.85%.

Glozman and Liba [12] propose a method for AD classification using AlexNet architecture includes five convolutional layers and three fully connected layers. Here, the network pretrained on natural images can be fine-tuned to classify neuroimaging data in which the difference between the different classes are very subtle, even for the human eye. This method results suggest that with the available data, the network can learn to classify the two extreme classes (NC vs. AD), but when faced with a three-way classification task, it will not achieve good accuracy.

3   Conclusion

In this paper, a review on different deep learning methods for diagnosis of Alzheimer's disease is discussed. Generally, the disease developing through three stages: Normal control, MCI, and AD. The Alzheimer's disease causes some changes in the brain, and these changes mainly developed on both the structures which are larger and small cells in the brain. Alzheimer's disease affects some parts of limbic system mainly the hippocampus, then the cerebral cortex, finally the brain stem. Most of the methods use MRI images because it is considered as the favored neuroimaging examination for Alzheimer's disease.

Early diagnosis of AD and MCI based on deep learning methods needs only minimal prior knowledge dependency in the model optimization. The advantage of autoencoder technique is it requires fewer labeled training examples and minimal prior knowledge. Compared to autoencoder technique, CNN deep learning architecture which was trained and tested with a huge number of images classified the AD more accurately.

References

1. Bishop C (2006) Pattern recognition and machine learning. Springer-Verlag, New York
2. American Society of Health-System Pharmacists (2001) Alzheimer's disease education and referral center. Am J Health-Syst Pharm 58(9):826
3. Vieira S, Pinaya WHL, Mechelli A, Serif L (2017) Using deep learning to investigate the neuroimaging correlates of psychiatric and neurological disorders: methods and applications. Neurosci Biobehav Rev 74(Part A)
4. Shin HC, Orton MR, Collins DJ, Doran SJ, Leach MO (2013) Stacked autoencoders for unsupervised feature learning and multiple organ detection in a pilot study using 4d patient data. IEEE Trans Pattern Anal Mach Intell 35:19301943
5. JhaD, Kwon G-R (2017) Alzheimer's disease detection using sparse autoencoder, scale conjugate gradient and softmax output layer with fine tuning. Int J Mach Learn Comput 7(1)
6. Bhatkoti P, Paul M (2016) Early diagnosis of alzheimer's disease: a multi-class deep learning framework with modified k-sparse autoencoder classification. IEEE 2016
7. Arel I, Rose DC, Karnowski TP (2010) Deep machine learning a new frontier in artificial intelligence research [research frontier]. Comput Intell Mag IEEE 5(4):1318
8. Jia SE, Donahue J, Karayev S, Long J, Girshick R, Guadarrama S, Darrell CT (2014) Convolutional architecture for fast feature embedding. In: Proceedings of the ACM international conference on multimedia, pp 675678, ACM
9. LeCun Y, Bottou L, Bengio Y, Haffner P (1998) Gradient based learning applied to document recognition. Proc IEEE 86(11):22782324
10. Sarraf S, Tofighi G (2016) Classification of alzheimer's disease using fMRI Data and deep learning convolutional neural networks
11. Ciprian D, Billones Jr, Louville OJ, Demetria D, Earl D, Hostallero D, Prospero, Naval Jr. C (2016) A convolutional neural network for the detection of alzheimer's disease and mild cognitive impairment. IEEE 2016
12. Glozman T, Liba O (2016) Cues: deep learning for alzheimer's disease classification. CS331B project final report, 2016

Performance Analysis of Image Enhancement Techniques for Mammogram Images

A. R. Mrunalini and J. Premaladha

Abstract Mammography is a technique which uses X-rays to take mammographic images of the breast, but identifying abnormalities from a mammogram is a challenging task. Many Computer-Aided Diagnosis (CAD) systems are developed to aid the classification of mammograms, as they search in digitized mammographic images for any abnormalities like masses, microcalcification which is difficult to identify especially in dense breasts. The first step in designing a CAD system is preprocessing. It is the process of improving the quality of the image. This paper focuses on the techniques involved in preprocessing the mammogram images to improve its quality for early diagnosis. Preprocessing involves filtering the image, applying image enhancement techniques like Histogram Equalization (HE), Adaptive Histogram Equalization (AHE), Contrast-Limited Adaptive Histogram Equalization (CLAHE), Contrast Stretching, and Bit-plane slicing; filtering techniques like mean, median, Gaussian and Wiener filters are also applied to the mammogram images. The performance of these image enhancement techniques are evaluated using quality metrics, namely Mean Square Error (MSE), Peak Signal-to-Noise Ratio (PSNR), and Contrast-to-Noise Ratio.

1 Introduction

1.1 Mammogram

Cancer is caused when abnormal tissues multiply uncontrollably. Breast cancer develops in the breast tissue and can be detected using mammography, which is an effective breast imaging technique to detect cancer at an early stage through low-dose X-rays, even before the symptoms are experienced. Mammography requires higher exposure of radiation for detecting any signs of tumor as the breast contains soft tissues. The

A. R. Mrunalini · J. Premaladha (✉)
School of Computing, SASTRA Deemed-to-be-University, Thanjavur, India
e-mail: premi.ph.d@gmail.com

© Springer Nature Switzerland AG 2019
D. Pandian et al. (eds.), *Proceedings of the International Conference on ISMAC in Computational Vision and Bio-Engineering 2018 (ISMAC-CVB)*, Lecture Notes in Computational Vision and Biomechanics 30,
https://doi.org/10.1007/978-3-030-00665-5_158

mammographic lesions can be diagnosed using a Computer-Aided Diagnosis (CAD) system and can be classified according to its severity.

1.2 Image Enhancement

Mammograms are gray scale images with low contrast and hence, it is difficult to visually distinguish between normal and lesion tissues. Mammogram enhancement is an important step to detect abnormalities and increase the reliability of CAD system. Enhancement of mammograms without blurring fine details and edges is a challenging task as it involves highlighting the suspicious regions without affecting the normal breast tissues. Image enhancement enhances the interested regions to detect the defects in a simple and efficient manner.

Preprocessing significantly enhances the image quality by suppressing uninterested areas. It is the basic step and its accuracy determines the probability of success of segmentation and classification. Filters like mean, median, Gaussian, and Wiener are used to remove salt-and-pepper noises from a mammogram image. The results are compared using parameters such as PSNR, MSE, and SNR to determine the better filter.

1.3 Literature Survey

Mean, median, adaptive median, Gaussian, and Wiener de-noising filters are used to remove salt-and-pepper noises, Gaussian noise, and speckle noise from mammogram images and comparison of the performance is done based on PSNR, MSE, and CNR parameters and the better filter is finalized [1]. Histogram Equalization (HE) is an efficient image enhancement technique. Adaptive Histogram Equalization (AHE) is an efficient contrast enhancement technique that is suitable for medical images and other initially nonvisual images. Contrast-Limited Adaptive Histogram Equalization (CLAHE) prevents local overenhancement using an amplitude limiting method [2]. The objective of image processing field is image de-noising. Preprocessing or image de-noising is necessary for noise suppression and image quality improvement [3]. Histogram equalization is an effective technique to achieve contrast enhancement. Digital mammography deals with gray scale images [4]. Enhancing the mammograms helps the segmentation of breast region for automated analysis of digital mammogram images to be done easier. The Region of Interests (ROI) in the image can be determined by eliminating background noise and improving the image quality [5]. The number of false positives can be reduced and the accuracy of determining the abnormalities can be increased through preprocessing hence making segmentation easier [6]. Image preprocessing is essential for training the CNNs. The performance of CNN can be improved particularly through including brightness and contrast variations which are among the preprocessing techniques [7]. Contrast-

Limited Adaptive Histogram Equalization (CLAHE) and morphology methods are used in particular for mammograms [2].

2   Image Enhancement Techniques

Image enhancement techniques accentuate or sharpen image features such as edges or boundaries to make it helpful for analysis. Image enhancement techniques include:

1. Intensity transformation functions
2. Histogram processing
3. Spatial filtering

   - Linear
   - Nonlinear

2.1   Intensity Transformation Functions

Intensity transformation technique is done to increase the contrast between certain intensity values to analyze particular features in an image. Some of the intensity transformation techniques are:

- Contrast stretching
- Bit-plane slicing.

2.1.1   Contrast Stretching

Contrast stretching is a technique where the intensity range of values in an image is stretched to improve the contrast in an image [8]. It is a measure of the complete range

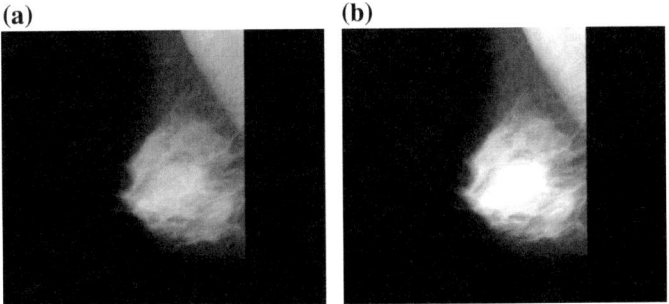

Fig. 1 Contrast stretching http://peipa.essex.ac.uk/pix/mias/all-mias.tar.gz

of intensity values contained within an image and can be calculated by subtracting minimum pixel value from the maximum pixel value. Contrast stretching was applied using MATLAB to the mammogram image obtained from the mini-MIAS database. Figure 1 shows the original image (a) and the enhanced image (b).

2.1.2   Bit-Plane Slicing

Bit-plane slicing is done to convert a gray level image to a binary image. It represents an image with fewer bits and enhances the image by focusing. The contribution made to the total image appearance is highlighted by specific bits. It is based on the assumption that each pixel in the image is represented by 8 bits and the image is composed of 8, 1-bit planes. The least significant bit is composed by plane 1 and the most significant bit by plane 8. The relative importance of each bit can be analyzed through this technique. Bit-plane slicing was applied to the image taken from mini-MIAS database named as "mdb001.pgm". Figure 2a is the original image, (b) is the second bit, (c) is the sixth bit, and (d) is the eighth bit.

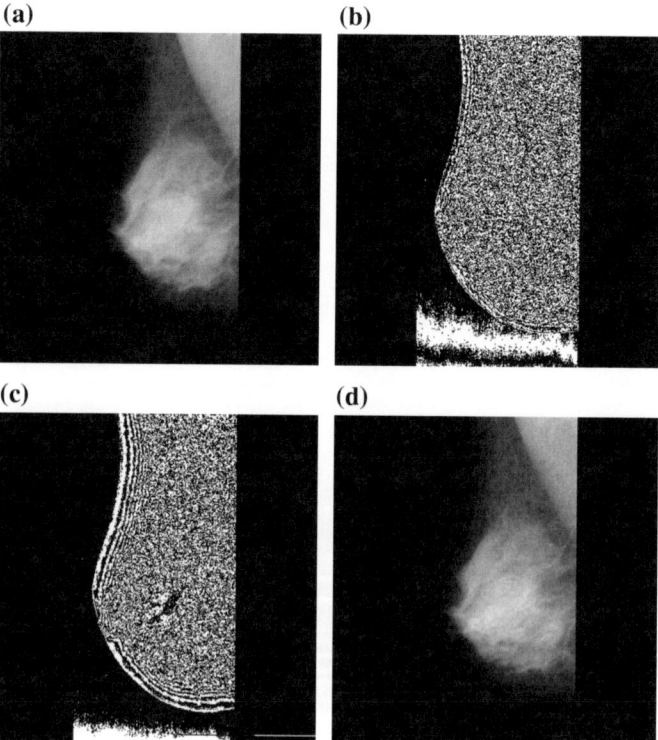

Fig. 2 Bit-plane slicing http://peipa.essex.ac.uk/pix/mias/all-mias.tar.gz

2.2 *Histogram Processing*

Histograms can be used for image enhancement. It is a plot of the frequency of occurrence of an event. It is a spatial domain technique. Image intensities can be adjusted through Histogram Equalization technique to enhance the contrast. There are three Histogram Equalization techniques:

- Histogram Equalization (HE)
- Adjusted Histogram Equalization (AHE)
- Contrast Adjusted Histogram Equalization (CLAHE).

2.2.1   Histogram Equalization (HE)

Histogram Equalization produces an output whose histogram is uniform by altering the input histogram. The various pixel intensities in the output histogram are equally distributed over the entire dynamic range. "histeq" is the command used in MATLAB for Histogram Equalization.

Histogram Equalization was applied to the original image taken from the mini-MIAS database. Figure 3a is the original image, (b) is the histogram of the original

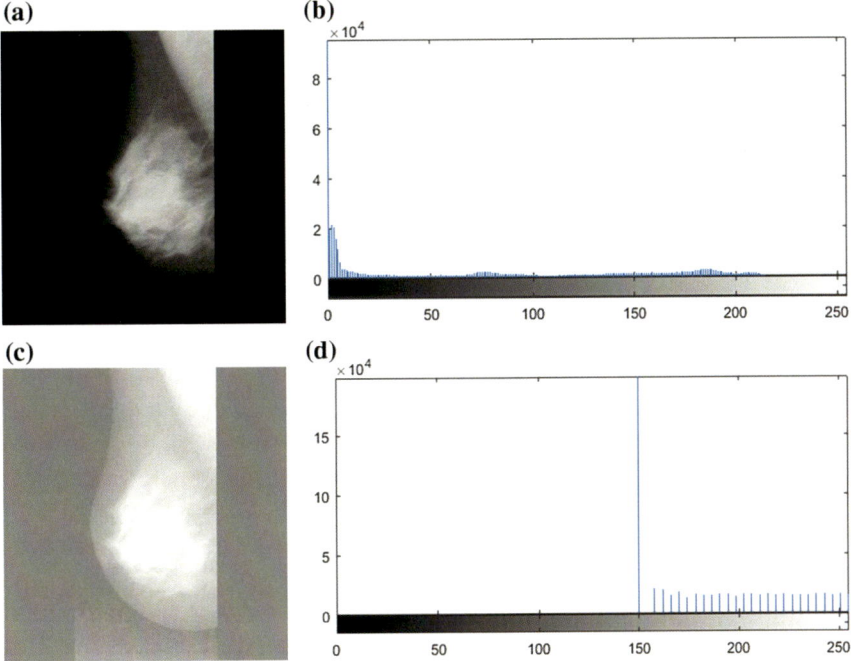

Fig. 3 Histogram equalization (HE) http://peipa.essex.ac.uk/pix/mias/all-mias.tar.gz

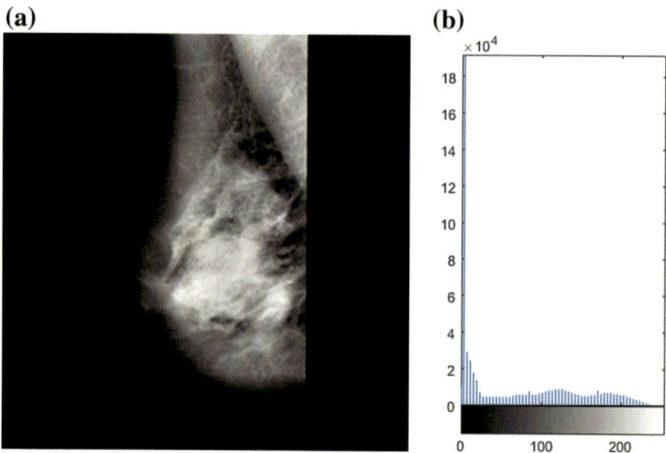

Fig. 4 Adjusted histogram equalization (AHE) http://peipa.essex.ac.uk/pix/mias/all-mias.tar.gz

image, (c) is the contrast adjusted image using Histogram Equalization, and (d) is the histogram of the contrast adjusted image.

2.2.2 Adjusted Histogram Equalization (AHE)

Figure 4a shows the contrast adjusted image and Fig. 4b shows the histogram obtained using adaptive histogram equalization "adapthisteq" applied to the image "mdb001.pgm" from the mini-MIAS database.

2.2.3 Contrast-Limited Adaptive Histogram Equalization

Contrast-Limited Adaptive Histogram Equalization (CLAHE) is applied to the image "mdb001.pgm" obtained from mini-MIAS database and the results are displayed in Fig. 5.

2.3 Spatial Filtering

Filtering is done to smoothen the image or upgrade or distinguish the edges in it. There are two filtering techniques:

- Linear
- Nonlinear

Fig. 5 Contrast-limited adaptive histogram equalization http://peipa. essex.ac.uk/pix/mias/all-mias.tar.gz

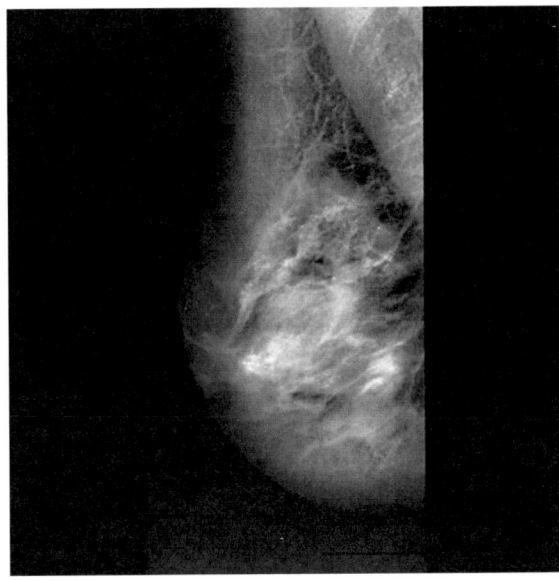

2.3.1 Linear

In a linear filter, the output will change linearly with a change in the input.
For Example,

(i) Gaussian.

Gaussian

Gaussian filtering is used to blur images and remove noise and detail. Figure 6a is the original image, (b) is the image with Gaussian noise, and (c) is the filtered image.

2.3.2 Nonlinear

There are various kinds of nonlinear filtering techniques such as

1. Mean filter
2. Median filter
3. Wiener filter.

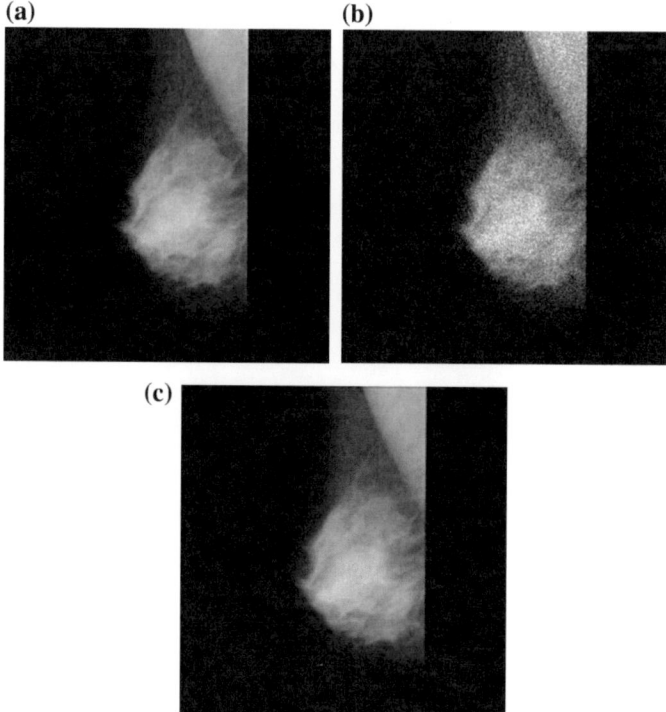

Fig. 6 Gaussian filter http://peipa.essex.ac.uk/pix/mias/all-mias.tar.gz

Mean Filter

Data set from mini-MIAS database has been obtained through PEIPA. In mean filter, at each position, the average value is assigned to the place of the center pixel. Mean filter was applied using MATLAB to the image ("mdb001.pgm") obtained from mini-MIAS database and the result is displayed in Fig. 7. Figure 7a is the original image, (b) is the image with salt-and-pepper noise introduced, and (c) is the filtered image.

Median Filter

Median filter is used to remove salt-and-pepper noise from the image using a nonlinear filtering technique. But Median filter cannot differentiate fine detail from noise. Figure 8a is the original image obtained from the mini-MIAS database, (b) is an image with salt-and-pepper noise. After applying median filter, the resulting image is displayed in (c).

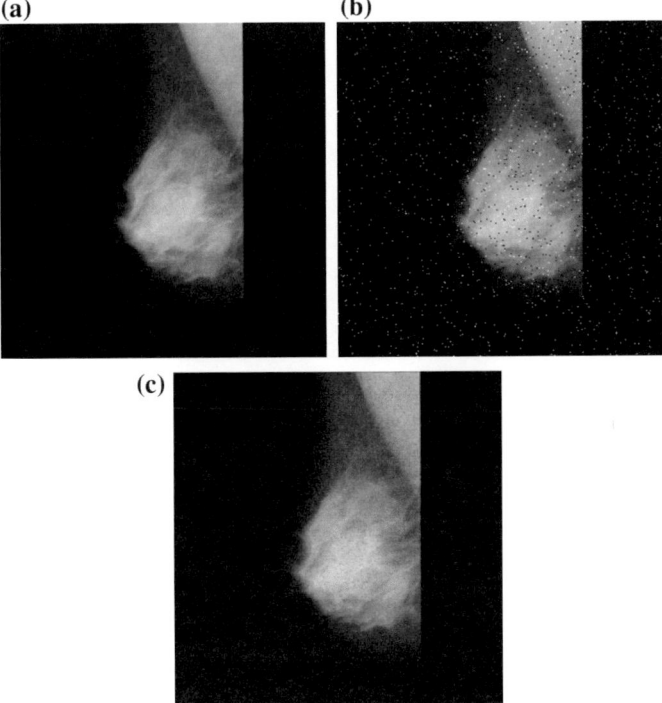

Fig. 7 Mean filter http://peipa.essex.ac.uk/pix/mias/all-mias.tar.gz

Wiener Filter

Wiener filter is used to remove the additive noise while inverting the blurring simultaneously. Figure 9a shows the original image taken from the mini-MIAS database, (b) is the image after introducing salt-and-pepper noise, (c) shows the filtered image.

3 Performance Analysis

3.1 Mean Square Error (MSE)

MSE of a filter is one of the ways to evaluate the difference between true value and the value implied by a filter. The difference exists as a consequence of randomness or because the filters may not consider information that could bring about a more veracious estimate. It corresponds to the expected value of square error loss. The average of squares of the error is gaged by MSE.

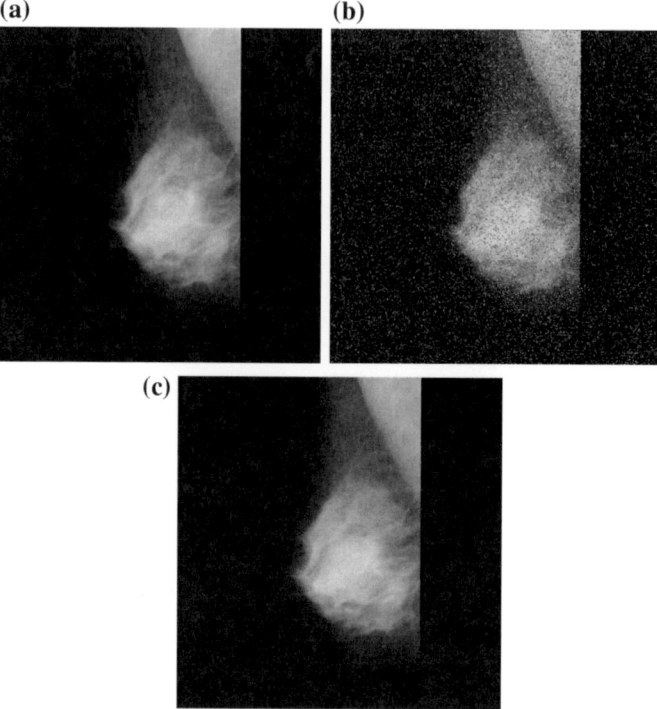

Fig. 8 Mean filter http://peipa.essex.ac.uk/pix/mias/all-mias.tar.gz

3.2   Peak Signal-to-Noise Ratio (PSNR)

PSNR is expressed in decibels and pixels in the image are assumed to be independent of their neighbors. Noise can get introduced to an original image due to compression, and PSNR is used to determine the image quality. For identical images, MSE is zero and PSNR becomes infinite. Higher value of PSNR implies that the image is of higher quality.

3.3   Contrast-to-Noise Ratio (CNR)

The quality of an image can be assessed using the CNR parameter. By improving CNR, the difference between two regions of interest is increased. Thus, the CNR is a variable index for evaluating detectability in digital radiography. Low-contrast lesion detection can be done using CNR between lesion and background. If the CNR is high, the lesion can be detected with higher probability.

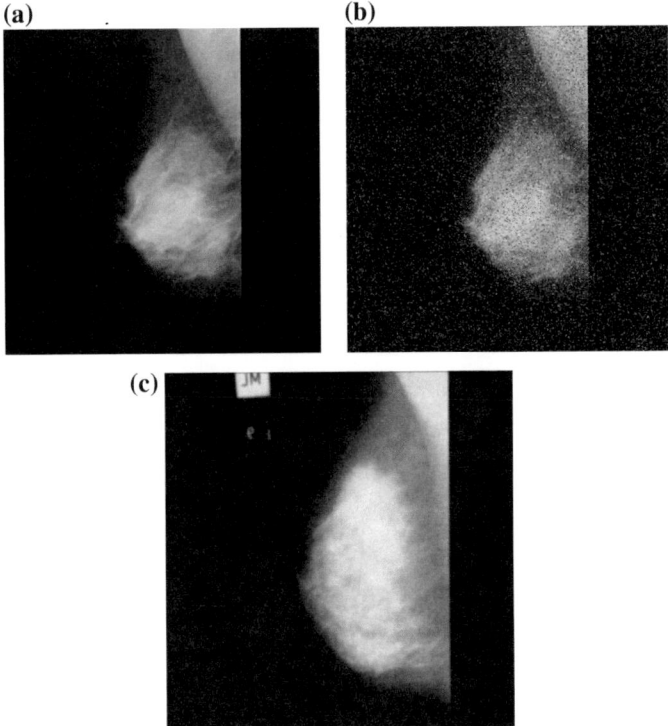

Fig. 9   Wiener filter http://peipa.essex.ac.uk/pix/mias/all-mias.tar.gz

4   Results and Discussion

Table 1 shows the values of MSE, PSNR, and CNR values of filters and image enhancement techniques applied. By comparing the performance of the filters, Gaussian filter has low MSE of 1.3935 and high PSNR of 36.6896. Among the image enhancement techniques used, contrast stretching has an MSE value of 2.8426 and PSNR value as 23.5935.

5   Conclusion

Hence, the preprocessing methods are discussed in this paper and Gaussian filter and contrast stretching are found to perform better when compared using MSE, PSNR, and CNR parameters with values 1.3935, 36.6896, and 6.8582 for Gaussian and 2.8426, 23.5935, and 1.6703 for contrast stretching, respectively.

Table 1 The values of MSE, PSNR, and CNR of filters and image enhancement techniques are applied

S.No	Technique	MSE	PSNR	CNR
1	Wiener filter	1.4194	26.9395	6.9742
2	Mean filter	5.8467	42.3309	6.7085
3	Average filter	4.2955	21.8138	1.4872
4	Median filter	4.6552	42.6932	1.2423
5	Rank filter	3.7373	2.416	1.7852
6	Gaussian filter	1.3935	36.6896	6.8582
7	Bit-plane slicing	2.1462	6.6833	6.9352
8	Histogram equalization (HE)	1.9396	5.253	1.3077
10	Adaptive HE (AHE)	3.8396	22.2879	2.7456
11	Contrast limited (AHE)	9.0199	18.5787	2.9872
12	Contrast stretching	2.8426	23.5935	1.6703

Compliance to Ethical Standards Conflict of Interest

Author A. R. Mrunalini, Author J. Premaladha declares that they have no conflict of interest.

Funding

We the authors would like to thank the Department of Science and Technology, India for their financial support through Fund for Improvement of S&T Infrastructure (FIST) programme (SR/FST/ETI-349/2013).

Ethical approval

This article does not contain any studies with human participants or animals performed by any of the authors.

Acknowledgement

We the authors sincerely thank the SASTRA Deemed to be University for providing an excellent infrastructure to carry out the research work.

References

1. Patel BK, Ranjbar S, Wu T, Pockaj BA, Li J, Zhang N, Lobbes M, Zhang B, Mitchell JR (2018) Computer-aided diagnosis of contrast-enhanced spectral mammography: a feasibility study. Eur J Radiol 31(98):207–213
2. Singh B, Kaur M (2018) An approach for classification of malignant and benign microcalcification clusters. Sādhanā 43(3):39
3. Khan KB, Khaliq AA, Jalil A, Shahid M (2018) A robust technique based on VLM and Frangi filter for retinal vessel extraction and denoising. PLoS ONE 13(2):e0192203
4. Shastri AA, Tamrakar D, Ahuja K (2018) Density-wise two stage mammogram classification using texture exploiting descriptors. Expert Syst Appl 1(99):71–82

5. Salem MA, Atef A, Salah A, Shams M (2018) Recent survey on medical image segmentation. In: Computer vision: concepts, methodologies, tools, and applications: concepts, methodologies, tools, and applications 2:129
6. de Moor T, Rodriguez-Ruiz A, Mann R, Teuwen J (2018) Automated soft tissue lesion detection and segmentation in digital mammography using a u-net deep learning network. ArXiv preprint arXiv:1802.06865
7. Diniz JO, Diniz PH, Valente TL, Silva AC, de Paiva AC, Gattass M (2018) Detection of mass regions in mammograms by bilateral analysis adapted to breast density using similarity indexes and convolutional neural networks. Comput Methods Programs Biomed
8. George MJ, Sankar SP. Efficient preprocessing filters and mass segmentation techniques for mammogram images. In: 2017 IEEE international conference on circuits and systems (ICCS). IEEE pp 408–413

A Study on Preprocessing Techniques for Ultrasound Images of Carotid Artery

S. Mounica, S. Ramakrishnan and B. Thamotharan

Abstract Ultrasound imaging has been widely used in the diagnosis of atherosclerosis. To precisely diagnose the carotid plaque, the affected region should be segmented from the ultrasonic image of carotid artery. Many techniques have been used to identify the plaque in ultrasound images. Image enhancement and restoration are the important processes to acquire high-quality images from the noisy images. When the artery images are captured, noise occurs due to high-frequency rate. To acquire a high-quality image, preprocessing is the first step to be done. The quality of the image is improved in this process. The techniques involved in preprocessing are dealt in this paper. Preprocessing involves filtering the image and removing the noise by various filtering techniques. Salt-and-pepper and Gaussian noise in ultrasound images can be filtered using techniques like mean, median and Wiener filters. Salt-and-pepper noise is multiplicative in nature and it is introduced by the image acquisition mechanism. The quality of the input sensor is reflected by the Gaussian noise. In this paper, the performance of image enhancement techniques on ultrasound images are evaluated using quality metrics, namely Mean Square Error (MSE) and Peak Signal–to-Noise Ratio (PSNR).

S. Mounica · S. Ramakrishnan · B. Thamotharan (✉)
Computer Vision & Machine Learning Laboratory, School of Computing, SASTRA Deemed University, Thanjavur, India
e-mail: balakrishthamo@gmail.com

S. Mounica
e-mail: sgammumounika@gmail.com

S. Ramakrishnan
e-mail: srk@ict.sastra.edu

© Springer Nature Switzerland AG 2019
D. Pandian et al. (eds.), *Proceedings of the International Conference on ISMAC in Computational Vision and Bio-Engineering 2018 (ISMAC-CVB)*, Lecture Notes in Computational Vision and Biomechanics 30, https://doi.org/10.1007/978-3-030-00665-5_159

1 Introduction

1.1 Ultrasound Carotid

Ultrasound imaging is a non-invasive procedure which uses sound waves to capture images that helps physicians to identify and treat illnesses conditions. Carotid artery is located on both sides of the neck which carries blood from the heart to the brain. Ultrasound image of the carotid arteries provides a clear understanding of these blood vessels and information about the blood flowing through them. The ultrasound imaging of carotid arteries in the neck is safe and painless. A Doppler ultrasound is one of the techniques that evaluates the blood flow through a blood vessel. A Doppler ultrasound study is a presurgical study used for examination of carotid artries. It is used to screen patients for Stenosis, a condition which increases the risk of stroke. It is also used for screening atherosclerosis which is caused by accumulation of lipids in carotid artries. Sample ultrasound image of the carotid artery is given in Fig. 1.

1.2 Preprocessing

The preprocessing of ultrasound images is done for removing speckle noise and noise due to wave interferences. Better segmentaiton can be achieved only after removing the above noises from the ultrasound image. Average and Bilateral filters not only

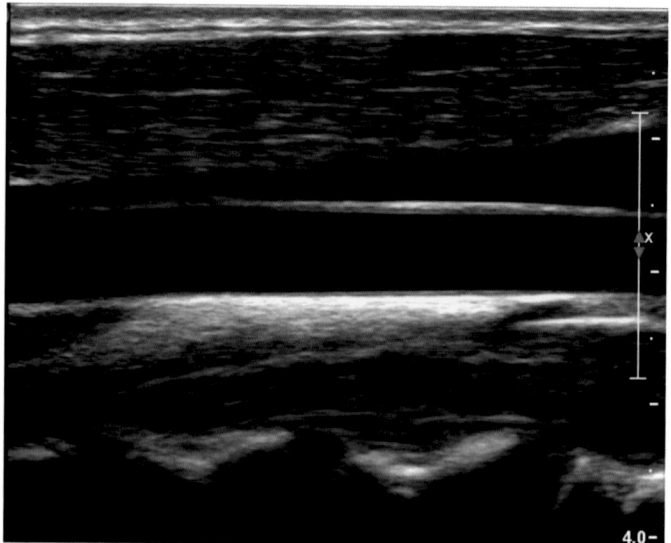

Fig. 1 Ultrasound image of carotid artery http://splab.cz/en/download/databaze/ultrasound

removes noise but also preserves the details of the ultrasound images of carotid artery [1].

1.3 *Image Enhancement*

Image enhacement is required to improve the quality of the input image. The image quality has some essential factors such as contrast, brightness, spatial resolution and noise. Image enhancement techniques are used for improving the above said quality factors of a image. A histogram gives the summary of the distribution of grey levels in an image by plotting the frequency of occurrence of grey levels. This plot provides a global distribution of appearance of the image that is used for image quality assessment. The image quality metrics are divided into objective and subjective fidelity criteria. The objective fidelity criteria is used to quantify the error that characterizes image quality. Subjective fidelity criteria depends on perception of the human observers and their visual systems.

1.4 *Related Work*

Datasets
The signal processing lab research group is a part of the Brno University of Technology and is involved in fields of voice, image, video and text processing which is used in fundamental and applied research. They have long-term experience in cooperation with many small, medium and supranational companies, and can offer advanced technologies and customized solutions in a variety of industrial fields. The biomedical signal processing focuses on signal analysis, which leads to the facilitation or improvement of diagnostic procedures and a better description or modelling of the human body as a complex dynamic spatial object. The key signals to be processed are image data produced by different instruments which are often in the form of temporal or spatial sequences (cuts). One-dimensional signals produced by different sensors are also used.

References	Techniques	Result
Chen et al. [2]	1. Butterworth filtering low pass and high pass 2. Adaptive weighted median filtering	Median filter removes the noise in the ultrasound image and helps to retain more information
Loizou et al. [3]	Median filter, Geometric filter, homomorphic filter, nonlinear coherence diffusion, Wavelet filter	The best performance is achieved by first-order statistics filter lmsv, geometric filter gf4d, homogeneous mask area filter
Shruthi et al. [4]	Median filter, Wiener filter, Gaussian low-pass filter, wavelet filter	Wavelet-based thresholding method gives better results in speckle noise reduction
Loizou et al. [5]	Normalization and speckle reduction filtering, a mean and variance of a pixel is utilized by Linear Scaling Mean Variance (LSMV)	Normalization and speckle reduction filtering give high quality image that can be used for accurate segmentation.
Abd-Elmoniem et al. [6]	NCD model, AWMF model	The article presents models that assists the segmentation technique and also area/volume calculation methods
Latha et al. [7]	Kaun filter Gabor filter Wavelet transform Adaptive filter	Kaun filter gives better results than average and bilateral filtering techniques
Noble et al. [8]	Speckle, K-distribution, Rician inverse of Gaussian distribution	The paper reviews ultrasound segmentation methods and highlighted the segmentation methods developed for medical B-mode ultrasound images.
Jeyalakshmi et al. [9]	Morphological Image Cleaning algorithm (MIC)	MMIC algorithm performs better in removing speckle noise and also preserves the features of the image.
Sudha et al. [10]	Wavelet domain noise filtering—Discrete Wavelet Transform (DWT)	When compared to other techniques, Wavelet based noise filtering provides image with improved visual quality and high Signal-to-Noise Ratio
Tang et al. [11]	ORIGIN, ENN, TOMEK	Filtering algorithm proposed in this article performs better than the other state of the art filtering techniques

2 Image Enhancement Techniques

2.1 Spatial Domain

In this domain, the point transforms or grey level scaling transformations to create the corresponding pixels in the out-take image depends only on the pixels of the intake image.

2.1.1 Linear Point Transformations

Inversion is one of the important transformations which performs a digital negative operation. In a binary image, inverse transformation reverses the image by changing a black pixel to a white one and vice versa.

2.1.2 Nonlinear Transformations

The relationship between both the input and output variables is not linear. So, to enhance the contrast of the given image, squared function is used and to compress the dynamic range of images, logarithmic function and exponential function are used to enhance the details in high-value regions of image and decrease the dynamic range in low-value regions, and power transformation is used to calculate the c which is positive constants.

2.2 Histogram Processing

Histogram techniques are effective and useful in image processing applications. It is used in manipulating the contrast and brightness of an image. It can be visualized as an intensity distribution or probability density function.

2.3 Spatial Filtering

Spatial filtering is the term which defines the filtering operations that are performed directly on the pixels of an image. It is used in image smoothing, sharpening, and noise removal. A filtered image is generated as the centre of the mask moves over every pixel in the input image.

2.3.1 Linear Filter

It is used to remove certain types of noise. If the out-put is a weighted sum of the input pixels, then the filtering method is linear. It works best for Gaussian noise. It is simple to execute and blurs the sharp edges. It is used to remove the lines and other details of the image. Gaussian and mean filters are commonly used linear filters.

2.3.2 Non-linear Filters

It is used to preserve edges and is an effective way to discard impulse noise. When compared to linear filters, non-linear filters provide better results. Non-linear filters remove noises without blurring the edges of the image. Median filter is the commonly used non-linear filter.

2.3.3 Gaussian Filter

Gaussian noise occurs when incidental values are added to an image due to incidental variations in the signal. This noise has normal distribution of the Probability Density Function (PDF). In digital images, it arises during acquisition and transmission. The techniques used to remove Gaussian noise are mean (convolutional), median and Gaussian filters (Fig. 2).

2.3.4 Salt-and-Pepper Noise

It is also called as impulse noise. The impulse noise is caused due to a pointed and abrupt disruption. Image with impulse noise will possess sparse black and white pixels as shown in Fig. 3.

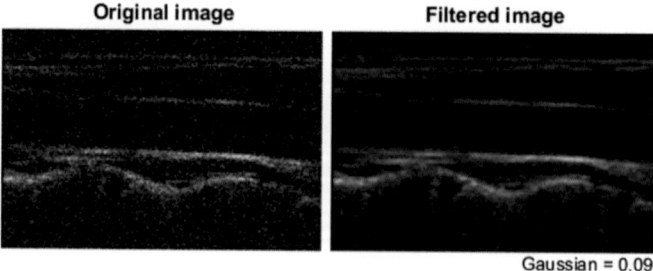

Original image   Filtered image

Gaussian = 0.09

Fig. 2 Gaussian filter http://splab.cz/en/download/databaze/ultrasound

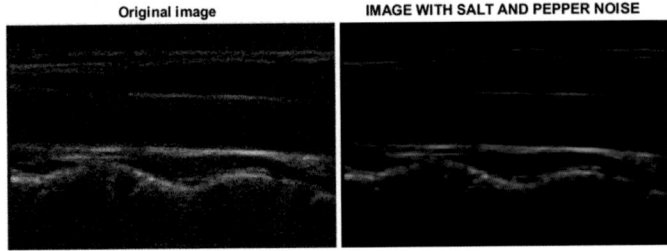

Fig. 3 Salt-and-pepper noise http://splab.cz/en/download/databaze/ultrasound

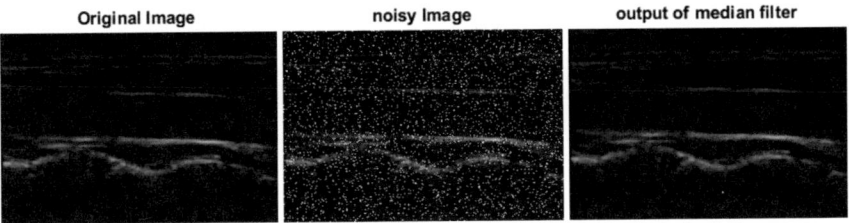

Fig. 4 Median filter http://splab.cz/en/download/databaze/ultrasound

2.3.5 Speckle Noise

The incidental values can be modelled by multiplying the pixel values of an image. Incidental fluctuation is a result that returns signal from an object. It increases the local area level of a mean grey. The filtering techniques used to remove speckle noise are mean and median filtering.

2.3.6 Median Filter

To preserve a useful detail of an image, median filter is often used than the mean filter. It is a nonlinear filtering technique that is used to remove noise from an image or signal by reducing the amount of intensity variation between one pixel and the other pixel. The median filter allows a high spatial frequency to pass through whereas remaining is used in removing noise on images. The affected one is less than half of the pixels in a smoothing neighbourhood [12]. When an image is corrupted by Gaussian noise, median filter is less effective at removing noise. Also, median filter is is relatively complex to compute and expensive. (Fig. 4).

2.3.7 Mean Filter

Mean filtering often uses averaging kernel. It is a simple linear filter that replaces each pixel value in an image with the mean value of its neighbourhood to remove the impulse noise. Mean filter is often used for smoothing. Nath et al. says that the convolution filter is used in mean filtering. Like other convolutions, the shape and size of the neighbourhood to be sampled is based on the kernel while calculating the mean [13] (Fig. 5).

2.3.8 Wiener Filter

Weiner filter remove the noise that corrupts the image. It is based on statistical approach. Reducing the mean square error is the main goal [12].

The spectral property of noise and the original signal is based on knowledge of

$$G(u, v) = H * (u, v) \frac{H * (u, v)}{|H(u, v)|^2 Ps(u, v) + Pn(u, v)}$$

The Wiener filter Fourier domains are

$H*(u, v)$ complex conjugate of degradation function,
$Pn(u, v)$ power spectral density of noise,
$Ps(u, v)$ power spectral density of non-degraded image (Fig. 6).

2.3.9 Fourier Transform

See Figs 7, 8.

Fig. 5 Mean filter

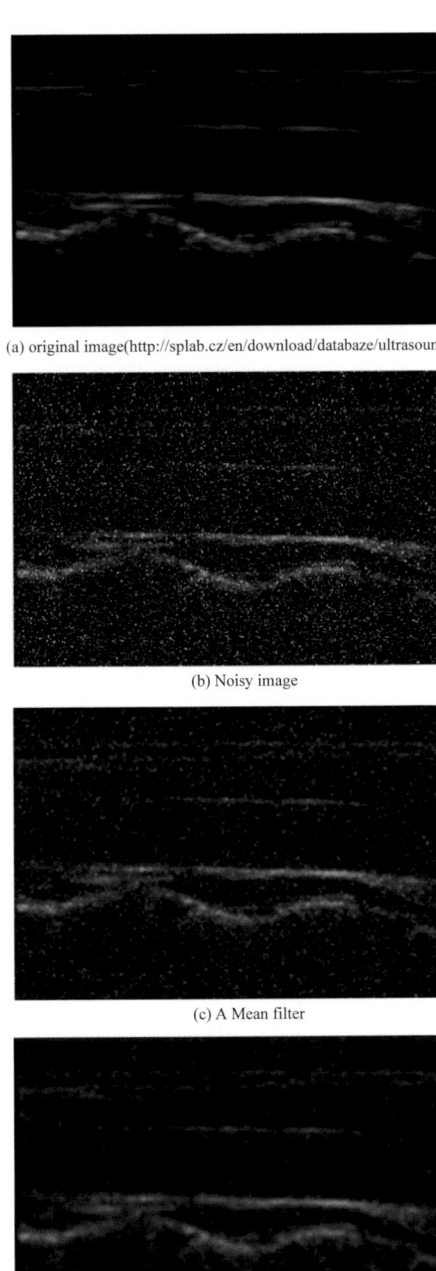

(a) original image(http://splab.cz/en/download/databaze/ultrasound)

(b) Noisy image

(c) A Mean filter

(d) B Mean filter

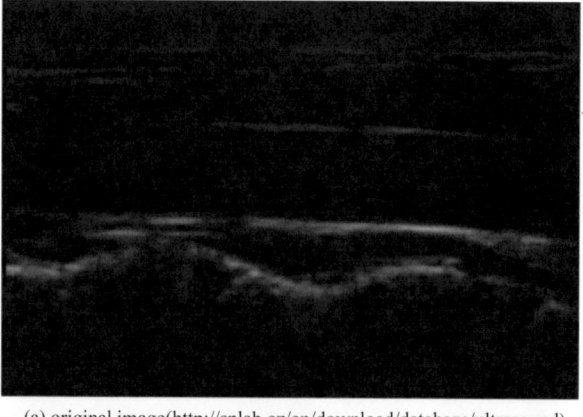

(a) original image(http://splab.cz/en/download/databaze/ultrasound)

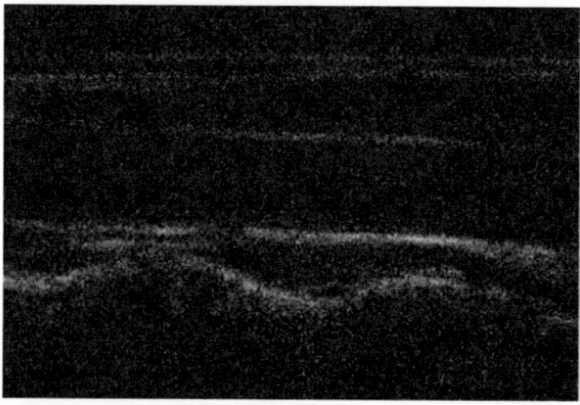

(b) Noisy Image

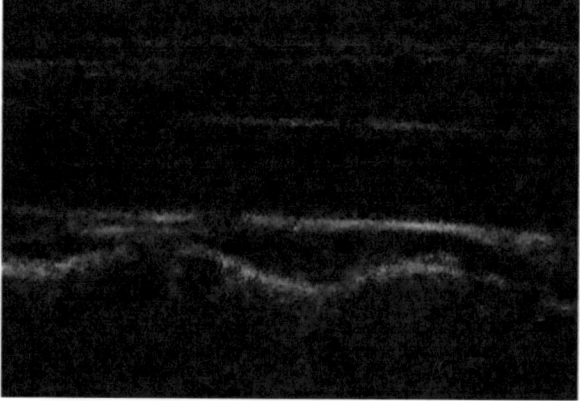

(c) Weiner filter

Fig. 6 Weiner filter

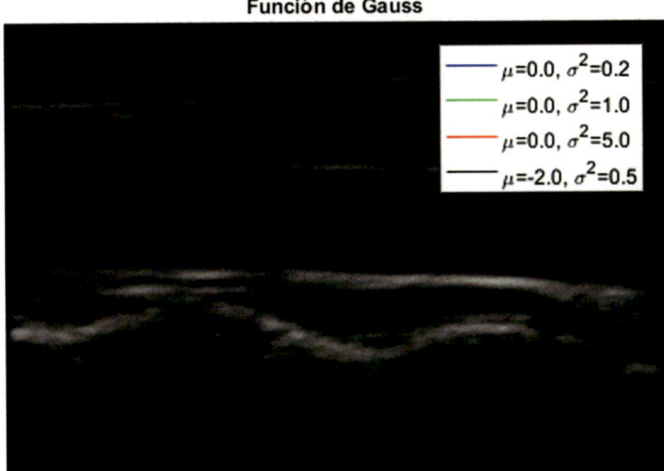

Fig. 7　Gaussian http://splab.cz/en/download/databaze/ultrasound

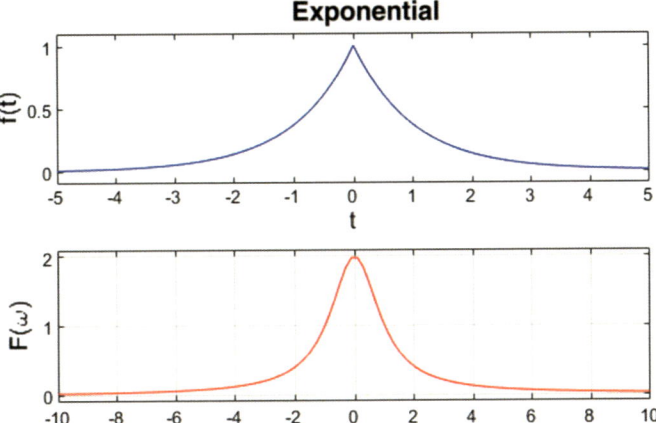

Fig. 8　Exponential

3 Performance Analysis

3.1 Mean Square Error (MSE)

Mean Square Error (MSE) of an image measures the average of squares of errors or deviations i.e., difference between original image and filtered image.

The MSE is measured between the encoded and the original image which is defined by a cumulative square error

$$MSE = \frac{1}{mn} \sum_{0}^{m-1} \sum_{0}^{n-1} \| f(i, j) - g(i, j) \|^2$$

3.2 Peak Signal-to-Noise Ratio (PSNR)

The ratio between the maximal available value (power) of a signal and the power of deceiving noise is termed as peak signal-to-noise ratio which affects the condition of its representation. It is usually expressed in logarithmic decibel scale. It is an approximation of reconstruction quality in human perception.

$$PSNR = 20 \log_{10} \left(\frac{MAX_f}{\sqrt{MSE}} \right)$$

4 Results and Discussion

S.No	Technique	MSE	PSNR
1	Guassian filter	23.0816	17.6008
2	Mean filter	37.79	32.38
3	Median filter	33.17	32.96
4	Weiner filter	38.87	32.25
5	Salt-and-pepper noise	35.11	32.72
6	Fourier transform	36.20	32.60

5 Conclusion

The most necessary task in image processing is the enhancement of an input image. For removing the noise from the images, filters are required. In this paper, different filters are implemented for different types of noisy ultrasound images and it is concluded that median filter performs better. Higher the PSNR value, lower the MSE value, higher the quality of the image. Ultrasound image enhanced with median filter results in higher PSNR value and lower MSE value. Enhanced image can be used for further image processing steps like segmentation, feature extraction, etc.

Compliance to Ethical Standards Conflict of Interest
Author S. Mounica, Author S. Ramakrishnan and Author B. Thamotharan declares that they have no conflict of interest.

Funding
We the authors would like to thank the Department of Science and Technology, India for their financial support through Fund for Improvement of S&T Infrastructure (FIST) programme (SR/FST/ETI-349/2013).

Ethical approval
This article does not contain any studies with human participants or animals performed by any of the authors.

Acknowledgement
We the authors sincerely thank the SASTRA Deemed to be University for providing an excellent infrastructure to carry out the research work.

References

1. Kazubek M (2003) Wavelet domain image denoising by thresholding and wiener filtering. IEEE Signal Process Lett 10(11):324–326
2. Chen J, Li F, Fu Y, Liu Q, Huang J, Li K (2017) A study of image segmentation algorithms combined with different image preprocessing methods for thyroid ultrasound images. IEEE ICIST:1–5
3. Loizou CP, C.S. Pattichis, C.I. Christodoulou, R.S.H. Istepanian, M. Pantziaris, A. Nicolaides(2005) Comparative evaluation of despeckle filtering in ultrasound imaging of the carotid artery. IEEE Trans Ultrason Ferroelectr Freq Control 52 (10):1653–1669
4. Shruthi B, Renukalatha S, Siddappa DM (2015) Speckle noise reduction in Ultrasound images—A review. IJERT 4(02):1402–1406
5. Loizou CP, Pattichis CS, Pantziaris M, Tyllis T, Nicolaides A (2006) Quality evaluation of ultrasound imaging in the carotid artery based on normalization and speckle reduction filtering. Med Biol Eng Compu 44(5):414
6. Abd-Elmoniem KZ, A.-B.M. Youssef, Y.M. Kadah, (2002) Real-time speckle reduction and coherence enhancement in ultrasound imaging via nonlinear anisotropic diffusion. IEEE Trans Biomed Eng 49(9):997–1014
7. Latha S, Dhanalakshmi S, Muthu P (2016) A Review and Comparative Study of Methods used in Finding Carotid Artery Abnormalities using Ultrasound Images. IJCTA 9(10):4891–4898
8. Noble JA, Boukerroui D (2006) Ultrasound image segmentation: a survey. IEEE Trans Med Imaging 25(8):987–1010

9. Jeyalakshmi TR, Ramar K (2010) A modified method for speckle noise removal in ultrasound medical images. Int J Comput Electr Eng 2(1):54
10. Sudha S, Suresh GR, Sukanesh R (2009) Speckle Noise Reduction in Ultrasound Images by Wavelet Thresholding based on Weighted Variance. Int J of Comput Theory Eng 7–12
11. Tang S, Chen SP (2008) An effective data preprocessing mechanism of ultrasound image recognition. IEEE ICBBE 2708–2711
12. Kumar S, Kumar P, Gupta M, Nagawat AK (2010) Performance comparison of median and wiener filter in image de-noising. Int J Comput Appl 12:0975–8887
13. Nath A (2013) Image denoising algorithms: a comparative study of different filtration approaches used in image restoration. In: 2013 International conference on communication systems and network technologies (CSNT). IEEE, pp 157–163
14. Paini A, Boutouyrie P, Calvet D, Zidi M, Agabiti-Rosei E, Laurent S Multiaxial mechanical characteristics of carotid plaque: analysis by Multiar

Fractional Reaction Diffusion Model for Parkinson's Disease

Hardik Joshi and Brajesh Kumar Jha

Abstract Calcium (Ca^{2+}) ion known as a second messenger, involve in variety of signalling process, and directly link with the intracellular calcium concentration ($[Ca^{2+}]$) that are continuously remodelled for the survival of the nerve cell. Buffer, also refer as a protein, react with Ca^{2+} and significantly lower down the intracellular $[Ca^{2+}]$ in nerve cell. There are numerous signalling processes in mammalian brain which can initiate at the high level of intracellular $[Ca^{2+}]$. Voltage gated calcium channel (VGCC), and ryanodine receptor (RyR) are work as an outward source of Ca^{2+} which initiate, and sustain the signalling process for smooth functioning of the cells. Parkinson's disease (PD) is a brain disorder of the central nervous system accompanied with the alteration of the signalling process. In present paper, a one dimensional fractional reaction diffusion model is consider to understand the physiological role of buffer, VGCC, and RyR in view of the PD.

1 Introduction

Parkinson's disease (PD) is second most common neurodegenerative disorder spread in the world after the Alzheimer's disease [1]. The movement symtoms of PD is tremor, rigidity, bradykinesia, and postural instability caused by the loss or dysfunction of the neurotransmitter dopamine in the midbrain [1]. The alteration in the signalling process resultant transient changes in spatial as well as the temporal intracellular concentration. The high level of intracellular concentration is responsible for the apoptosis or the dysfunction of nerve cells [2] which produce the neurotransmitter dopamine [3]. Ca^{2+} being a second messenger play a vital role in variety of physiological function which is essential for the signalling process, like movement

H. Joshi (✉) · B. K. Jha
Department of Mathematics, School of Technology,
Pandit Deendayal Petroleum University, Gujarat, India
e-mail: hardik.joshi8185@gmail.com

B. K. Jha
e-mail: brajeshjha2881@gmail.com

© Springer Nature Switzerland AG 2019
D. Pandian et al. (eds.), *Proceedings of the International Conference on ISMAC in Computational Vision and Bio-Engineering 2018 (ISMAC-CVB)*, Lecture Notes in Computational Vision and Biomechanics 30,
https://doi.org/10.1007/978-3-030-00665-5_160

of the muscle, fertilization, proliferation, cellular motility, memory of the cell, etc. [4]. Ca^{2+} make a place making role in the presence of intracellular and extracellular source of Ca^{2+}. Buffer is one of the intracellular source of Ca^{2+}. Calbindin-D_{28k} is a Ca^{2+} binding buffer, they react with the free Ca^{2+} in the cytosol, and make a Ca^{2+} bound buffer [5]. The overexpression of Ca^{2+} bound buffer helps to decrease the peak of intracellular $[Ca^{2+}]$ in the nerve cell [6–8]. But there are various process inside the cell which can be initiate at the threshold level or at the presence of high intracellular $[Ca^{2+}]$. Voltage gated calcium channel (VGCC) and ryanodine receptor (RyR) are the extracellular source of Ca^{2+} which are normally closed at resting position of the cells. The opening of VGCC and RyR initiate as well as sustain many of physiological function of the nerve cells [8]. The alteration or dysfunction of these parameters may lead to resultant transient changes in spatially as well as temporally, which is the pathological symptoms of PD [2].

Literature survey evidently show that, the effect of buffer on Ca^{2+} distribution have been reported on the different cells like neuron, astrocytes, oocytes, fibroblast, myocytes, etc. [4, 7–12]. Tewari and Pardasani have studied the cytosolic Ca^{2+} diffusion in presence of excess buffer [9]. Jha et al. have studied Ca^{2+} distribution in presence of excess buffer on astrocytes cell [4]. Panday and Pardasani have studied Ca^{2+} distribution in oocytes cell [10]. In the same year, Kotwani et al. have studied Ca^{2+} distribution in fibroblast cell [11]. Pathak and Adlakha have studied Ca^{2+} distribution in cardiac myocytes [12]. All the authors [4, 8–12] have employed the finite element method to obtain the numerical results. In recent year, Dave and Jha have studied the effect of buffer on Ca^{2+} diffusion in view of neurodegenerative disease. They have adopted two dimensional advection reaction equation and applied Laplace and similarity transformation to obtain the analytical solutions. Finally, they interpreted the obtained results with the physiology of Alzheimer's disease [13].

Here an attempt has been made to study the one dimensional fractional reaction diffusion model in presence of calbindin-D_{28k}, VGCC, and RyR. The obtained analytical results are stimulate in the MATLAB, and incorporate with the physiology of the Parkinson's brain. The main motive to study reaction diffusion model by applying fractional approach is, due to its non-local property. The integer order ordinary or partial differential equation has a local property. It means that to obtain the results at any stage of the entire process, is considered the input at that particular stage. On the other hand in fractional differential equation it compute the results at any stage of the process by taking all the input from the resting stage of the process [14, 15]. Ca^{2+} diffusion in the nerve cell is a complex dynamics process occurs in mammalian brain, and calculated in μs, so in view of this reaction diffusion model is studied fractionally.

2 Mathematical Model

Ca^{2+} kinetics in the human nerve cells is governed by a set of reaction diffusion equations which can be given by the bimolecular reaction between Ca^{2+} and buffer species [4, 7, 16]

$$[Ca^{2+}] + [B] \underset{k^-}{\overset{k^+}{\rightleftharpoons}} [CaB] \tag{1}$$

where $[Ca^{2+}]$, $[B]$, and $[CaB]$ are intracellular $[Ca^{2+}]$, free and bound buffers respectively.

$$\frac{\partial[Ca^{2+}]}{\partial t} = D_{Ca}\nabla^2[Ca^{2+}] + \sum_j R_j \tag{2}$$

$$\frac{\partial[B]}{\partial t} = D_B\nabla^2[B] + R_j \tag{3}$$

$$\frac{\partial[CaB]}{\partial t} = D_{CaB}\nabla^2[CaB] - R_j \tag{4}$$

where

$$R_j = -k^+[B][Ca^{2+}] + k^-[CaB] \tag{5}$$

D_{Ca}, D_B, D_{CaB} are diffusion coefficients of free Ca^{2+}, free buffer and Ca^{2+} bound buffers respectively. k^+ and k^- are association and dissociation rate constants for buffer 'j' respectively. Then by combining Eqs. (1–5) the mathematical model can be frame as

$$\frac{\partial^\alpha[Ca^{2+}]}{\partial t^\alpha} = D_{ca}\frac{\partial^\beta[Ca^{2+}]}{\partial x^\beta} - k_j^+[B]_\infty([Ca^{2+}] - [Ca^{2+}]_\infty) + J_{VGCC} + J_{RyR} \tag{6}$$

where $\frac{\partial^\alpha[Ca^{2+}]}{\partial t^\alpha}$ and $\frac{\partial^\beta[Ca^{2+}]}{\partial x^\beta}$ is the Caputo time fractional derivative of order $\alpha(0 < \alpha \leq 1)$ and Caputo space fractional derivative of order $\beta(1 < \beta \leq 2)$ respectively [17].

J_{VGCC} is the Ca^{2+} influx through VGCC and modelled using the Goldman-Hodgkin-Katz (GHK) current equation given by Jha et al. [6]

$$I_{Ca} = P_{Ca}z_{Ca}^2\frac{F^2 V_m}{RT}\frac{[Ca^{2+}]_i - [Ca^{2+}]_o \exp(-z_{Ca}\frac{FV_m}{RT})}{1 - \exp(-z_{Ca}\frac{FV_m}{RT})} \tag{7}$$

where $[Ca^{2+}]_i$ and $[Ca^{2+}]_o$ are the intracellular and extracellular $[Ca^{2+}]$ respectively; P_{Ca} is the permeability of Ca^{2+} ion; z_{Ca} is the valance of Ca^{2+} ion; F is Faraday's constant; V_m is the membrane potential; R is the real gas constant; T is the absolute temperature. The current equation is converted into molar/second by using [6]

$$\sigma_{Ca} = -\frac{I_{Ca}}{z_{Ca} F V_{Nervecells}} \tag{8}$$

where negative sign indicated that the inward current is negative.

J_{RyR} is the Ca^{2+} influx through RyR given as [8]

$$J_{RyR} = V_{RyR} P_o([Ca^{2+}]_{ER} - [Ca^{2+}]) \tag{9}$$

Combining Eqs. (6–9), the proposed mathematical model converted as,

$$\begin{aligned}
\frac{\partial^\alpha [Ca^{2+}]}{\partial t^\alpha} &= D_{ca} \frac{\partial^\beta [Ca^{2+}]}{\partial x^\beta} - k_j^+[B]_\infty([Ca^{2+}] - [Ca^{2+}]_\infty) \\
&+ P_{Ca} z_{Ca}^2 \frac{F^2 V_m}{RT} \frac{[Ca^{2+}]_i - [Ca^{2+}]_o \exp\left(-z_{Ca}\frac{F V_m}{RT}\right)}{1 - \exp\left(-z_{Ca}\frac{F V_m}{RT}\right)} \\
&+ V_{RyR} P_o([Ca^{2+}]_{ER} - [Ca^{2+}])
\end{aligned} \tag{10}$$

Along with initial and boundary conditions as,

$$[Ca^{2+}](x, 0) = g(x), \quad [Ca^{2+}](\pm\infty, t) = 0 \tag{11}$$

The Eq. (10) can be rewritten as

$$\frac{\partial^\alpha C}{\partial t^\alpha} = D_{ca} \frac{\partial^\beta C}{\partial x^\beta} - aC + b \tag{12}$$

where $C = [Ca^{2+}]$,

$$a = k_j^+[B]_\infty + \frac{P_{Ca} z_{Ca} F V_m}{RT V_{Nervecells}} \frac{\exp\left(z_{Ca}\frac{F V_m}{RT}\right)}{1 - \exp\left(z_{Ca}\frac{F V_m}{RT}\right)} + V_{RyR} P_o \tag{13}$$

and

$$b = k_j^+[B]_\infty C_\infty + \frac{P_{Ca} z_{Ca} F V_m}{RT V_{Nervecells}} \frac{[Ca^{2+}]_o}{1 - \exp(z_{Ca}\frac{F V_m}{RT})} + V_{RyR} P_o C_{ER} \tag{14}$$

Applying temporal fractional Laplace transform and spatial fractional Fourier transform technique on Eq. (12) along with Eq. (11), the solution of Eq. (12) is obtained in terms of green function

$$G_{\alpha,\beta}(x,t) = \frac{1}{2\pi} \int_{-\infty}^{\infty} e^{-ikx} E_{\alpha}\left[\{D_{Ca}(-ik)^{\beta} - a\} \cdot t^{\alpha}\right]dk$$

$$+ \frac{bt^{\alpha}}{2\pi} \int_{-\infty}^{\infty} e^{-ikx} E_{\alpha,\alpha+1}\left[\{D_{Ca}(-ik)^{\beta} - a\} \cdot t^{\alpha}\right]dk \qquad (15)$$

where

$$E_{\alpha}(z) = \sum_{k=0}^{\infty} \frac{z^k}{\Gamma(\alpha k + 1)}, \quad R(\alpha) > 0, \quad \alpha, z \in \varsigma \text{ and}$$

$$E_{\alpha,\beta}(z) = \sum_{k=0}^{\infty} \frac{z^k}{\Gamma(\alpha k + \beta)}, \quad R(\alpha), R(\beta) > 0, \quad \alpha, \beta, z \in \varsigma$$

are the Mittag-Leffler function for one and two parameter respectively [17].

The solution of time fractional reaction diffusion equation can be obtained by varying fractional order α in Eq. (15), keeping fix space derivative $\beta = 2$. Again by applying fractional Laplace and Fourier transform technique, the solution corresponding to time fractional reaction diffusion equation is

$$G_{\alpha,2}(x,t) = \frac{1}{2\sqrt{\pi D_{Ca}t^{\alpha}}} \int_{0}^{\infty} e^{-\frac{x^2}{4D_{Ca}t^{\alpha}k} - at^{\alpha}k} k^{-\frac{1}{2}} M_{\alpha}(k)dk$$

$$+ \frac{bt^{\alpha}}{2\sqrt{\pi D_{Ca}t^{\alpha}}} \int_{0}^{\infty} e^{-\frac{x^2}{4D_{Ca}t^{\alpha}k} - at^{\alpha}k} k^{-\frac{1}{2}} \phi(-\alpha, 1; -k)dk \qquad (16)$$

where

$$M_{\alpha}(z) = \sum_{k=0}^{\infty} \frac{(-1)^k z^k}{\Gamma(-\alpha k + (1-\alpha)) \cdot k!}, \quad 0 < \alpha < 1 \text{ and}$$

$$\phi(\alpha, \beta; z) = \sum_{k=0}^{\infty} \frac{z^k}{\Gamma(\alpha k + \beta) \cdot k!}, \quad \alpha > -1, \quad \beta \in \varsigma$$

are the Mainardi function and Wright function respectively [17].

The solution of space fractional reaction diffusion equation can be obtained by varying fractional order β in Eq. (15), keeping fix time derivative $\alpha = 1$. By applying same fractional Fourier transform technique, we get the solution corresponding to space fractional reaction diffusion equation in terms of Levy stable distribution [18]

$$G_{1,\beta}(x,t) = \frac{e^{-at}}{\left(D_{Ca_\alpha}t\right)^{1/\beta}} S\left(\frac{x}{\left(D_{Ca_\alpha}t\right)^{1/\beta}}/\beta, 1, 1, 0; 1\right)$$

$$+ \frac{bt}{\left(D_{Ca_\alpha}t\right)^{1/\beta}\left\{\ln S\left(\frac{x}{\left(D_{Ca_\alpha}t\right)^{1/\beta}}/\beta, 1, 1, 0; 1\right) - at\right\}}$$

$$\times \left[\left\{\frac{e^{-at}}{\left(D_{Ca_\alpha}t\right)^{1/\beta}} S\left(\frac{x}{\left(D_{Ca_\alpha}t\right)^{1/\beta}}/\beta, 1, 1, 0; 1\right)\right\} - 1\right] \qquad (17)$$

3 Results and Discussion

The values of physiological parameters used to simulate the results are given in Table 1 or stated along with the figure. Figure 1 show the temporal distribution of $[Ca^{2+}]$ in presence of buffer, VGCC, and RyR. The value of the temporal and spatial order derivative in Fig. 1a–e is, in figure (a) $\alpha = 1$ and $\beta = 2$, (b) $\alpha = 1$ and $\beta = 1.9$, (c) $\alpha = 1$ and $\beta = 1.8$, (d) $\alpha = 0.9$ and $\beta = 2$, and (e) $\alpha = 0.8$ and $\beta = 2$ respectively. In Fig. 1, the value of $x = 0.001$ μm i.e. near the source of the channel. In all Fig. 1a–e, it can be observed that increase the amount of buffer concentration, lower down the profile of $[Ca^{2+}]$. Physiologically, it happens due to more amount of buffer react with free Ca^{2+} ion in the cell and make Ca^{2+} bound buffer to control the intracellular $[Ca^{2+}]$. The Fig. 1a is corresponding to $\alpha = 1$ and $\beta = 2$, it means

Table 1 Values of physiological parameters [5, 6, 8]

Symbol	Parameter	Value
D_{Ca}	Diffusion coefficient	250 μm²/s
$[B]$	Buffer concentration	100–350 μM
$[Ca^{2+}]_\infty$	Background $[Ca^{2+}]$	0.1 μM
k^+	Buffer association rate	75 μM⁻¹s⁻¹
$V_{Nervecells}$	Volume of nerve cells	$5.233 * 10^{13}$ l
R	Gas constant	8.31 J/(mole · K)
T	Temperature	300 K
Z_{Ca}	Valence of Ca^{2+} ion	2
P_{Ca}	Permeability of Ca^{2+}	$4.3 * 10^{-8}$ m/s
V_m	Membrane potential	−0.05 V
V_{RyR}	RyR rate	0.5 μM/s
P_o	Rate of Ca^{2+} efflux	0.5 M/s
$[Ca^{2+}]_{ER}$	ER $[Ca^{2+}]$	500 μM

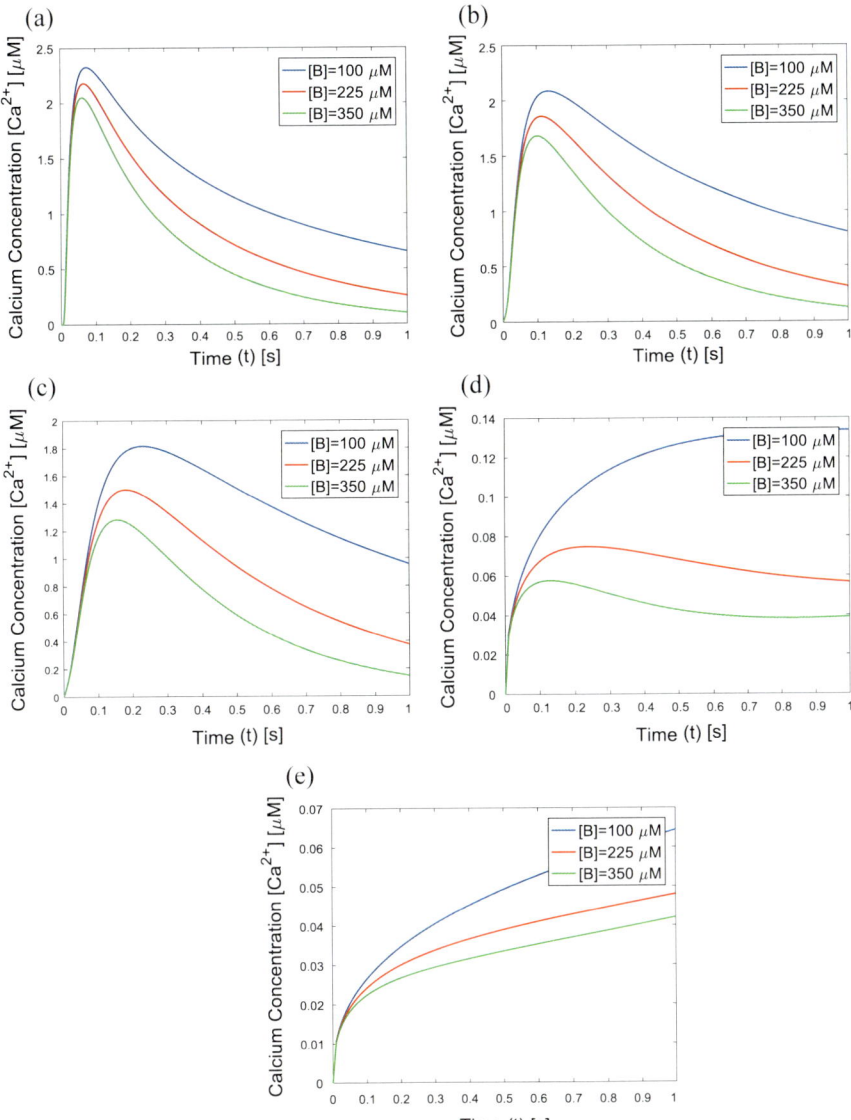

Fig. 1 Temporal distribution of $[Ca^{2+}]$ in presence of Buffer, VGCC, and RyR with **a** $\alpha = 1$ and $\beta = 2$, **b** $\alpha = 1$ and $\beta = 1.9$, **c** $\alpha = 1$ and $\beta = 1.8$, **d** $\alpha = 0.9$ and $\beta = 2$, and **e** $\alpha = 0.8$ and $\beta = 2$

the standard diffusion model for Ca^{2+} dynamics, which can be easily validate for the previous obtained results [8]. The main advantage of this study is obtained the results of Ca^{2+} dynamics in nerve cells at any fraction of time, which can be easily visualized in Fig. 1b–e.

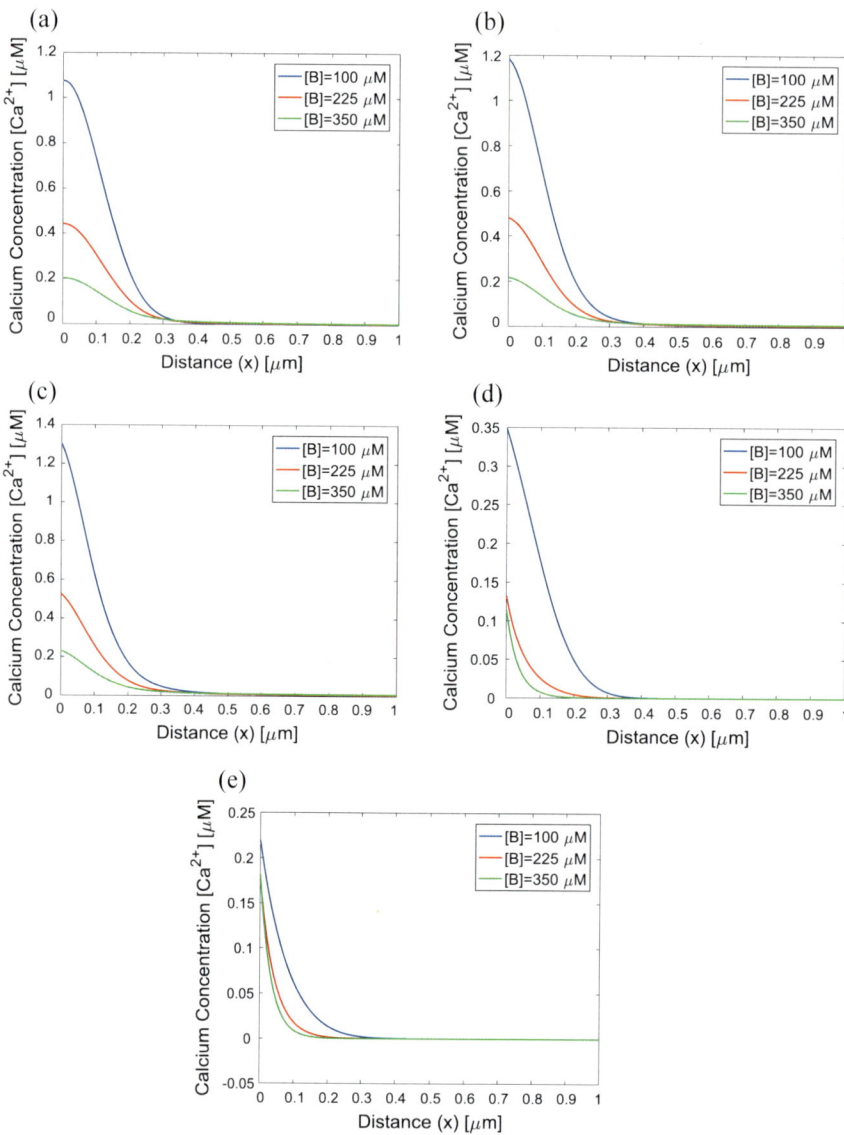

Fig. 2 Spatial distribution of [Ca^{2+}] in presence of Buffer, VGCC, and RyR with **a** $\alpha = 1$ and $\beta = 2$, **b** $\alpha = 1$ and $\beta = 1.9$, **c** $\alpha = 1$ and $\beta = 1.8$, **d** $\alpha = 0.9$ and $\beta = 2$, and **e** $\alpha = 0.8$ and $\beta = 2$

Figure 2 show the spatial distribution of [Ca^{2+}] in presence of buffer, VGCC, and RyR. The value of the temporal and spatial order derivative in Fig. 2a–e is same as in Fig. 1a–e. In Fig. 2, the value of $t = 0.01$ s. In all Fig. 2a–e, it can be observed that increase the amount of buffer concentration, reduce the profile of [Ca^{2+}] due to the same physiologically as mention above. The Fig. 2a is corresponding to $\alpha = 1$

and $\beta = 2$, it means the standard spatial diffusion for Ca^{2+} dynamics, which can be easily validate the previously obtained results [8]. It is observed from the figure that for smaller value of fractional order the solution become to much closer and achieve the steady state as soon as possible.

4 Conclusion

The fractional model studied in this paper provide interesting results in presence of buffers, VGCC, and RyR in view of the progression of the disease. The buffer play a vital role in gradually reduced the amount of the intracellular $[Ca^{2+}]$. Whereas VGCC, and RyR have significant effect in the presence of low amount of calbindin-D_{28k}. There are complex interplay among the buffer, VGCC, and RyR for maintaining the adequate level of intracellular concentration for the nerve cells. The dysfunctioning or alteration in these complex interplay associated with the brain disorder, PD. The obtained results can be helps to mediate or slower down the progression of the Parkinson's.

Acknowledgements We are very grateful to the reviewers for their fruitful comments and suggestions for making updation in the research article.

References

1. Surmeier DJ, Schumacker PT, Guzman JD, Ilijic E, Yang B, Zampese E (2017) Calcium and parkinson's disease. Biochem Biophys Res Commun. 483:1013–1019
2. Jha BK, Joshi H, Dave DD (2016) Portraying the effect of calcium-binding proteins on cytosolic calcium concentration distribution fractionally in nerve cells. Interdiscip Sci
3. Zaichick SV, McGrath KM, Caraveo G (2017) The role of Ca^{2+} signaling in parkinson's disease. Dis Model Mech 10:519–535
4. Jha BK, Adlakha N, Mehta MN (2014) Two-dimensional finite element model to study calcium distribution in astrocytes in presence of excess buffer. Int J Biomath 7:1–11
5. Schmidt H (2012) Three functional facets of calbindin D-28k. Front Mol Neurosci 5:25
6. Jha BK, Adlakha N, Mehta MN (2013) Two-dimensional finite element model to study calcium distribution in astrocytes in presence of VGCC and excess buffer. Int J Model Simulation, Sci Comput 4
7. Jha A, Adlakha N (2014) Analytical solution of two dimensional unsteady state problem of calcium diffusion in a neuron cell. J Med Imaging Heal Informatics 4:547–553
8. Naik PA, Pardasani KR (2015) One dimensional finite element model to study calcium distribution in oocytes in presence of VGCC, RyR and buffers. J Med Imaging Heal Informatics 5:471–476
9. Tewari SG, Pardasani KR (2010) Finite element model to study two dimensional unsteady state cytosolic calcium diffusion in presence of excess buffers. IAENG Int J Appl Mathe 40:40-3-01
10. Panday S, Pardasani KR (2014) Finite element model to study the mechanics of calcium regulation in Oocyte. J Mech Med Biol 14:1450022-1-16 (2014)
11. Kotwani M, Adlakha N, Mehta MN (2014) Finite element model to study the effect of buffers, source amplitude and source geometry on spatio-temporal calcium distribution in fibroblast cell. J Med Imaging Health Inform. 4:840–847

12. Pathak K, Adlakha N (2016) Finite element model to study two dimensional unsteady state calcium distribution in cardiac myocytes. Alexandria J Med 52:261–268

13. Dave DD, Jha BK (2018) Delineation of calcium diffusion in Alzheimeric brain. J Mech Med Biol 18:1850028-1-15

14. Magin RL (2010) Fractional calculus models of complex dynamics in biological tissues. Comput Math Appl 59:1586–1593

15. Agarwal R, Jain S, Agarwal RP (2018) Mathematical modeling and analysis of dynamics of cytosolic calcium ion in astrocytes using fractional calculus. J Fract Calc Appl 9:1–12

16. Smith GD (1996) Analytical steady-state solution to the rapid buffering approximation near an open Ca^{2+} channel. Biophys J 71:3064–3072

17. Podlubny I (1999) Fractional differential equations. Academic Press, New York, vol 198, pp 1–366

18. Borak S, Härdle W, Weron R (2005) Stable distributions. In: Statistical tools for finance and insurance. Diss Paper Springer pp. 21–44

Prediction-Based Lossless Image Compression

Mohamed Uvaze Ahamed Ayoobkhan, Eswaran Chikkannan,
Kannan Ramakrishnan and Saravana Balaji Balasubramanian

Abstract In this paper, a lossless image compression technique using prediction errors is proposed. To achieve better compression performance, a novel classifier which makes use of wavelet and Fourier descriptor features is employed. Artificial neural network (ANN) is used as a predictor. An optimum ANN configuration is determined for each class of the images. In the second stage, an entropy encoding is performed on the prediction errors which improve the compression performance further. The prediction process is made lossless by making the predicted values as integers both at the compression and decompression stages. The proposed method is tested using three types of datasets, namely CLEF med 2009, COREL1 k and standard benchmarking images. It is found that the proposed method yields good compression ratio values in all these cases and for standard images, the compression ratio values achieved are higher compared to those obtained by the known algorithms.

1 Introduction

In recent years, the availability of huge volume of image data has made the storage and transmission problems very challenging [1]. Many image compression techniques have been reported to solve these problems. Compression can be lossy or

M. U. A. Ayoobkhan (✉) · S. B. Balasubramanian
Computer Science Department, Cihan University-Erbil, Erbil, Iraq
e-mail: mohamed.sha33@gmail.com

S. B. Balasubramanian
e-mail: saravanabalaji.b@gmail.com

E. Chikkannan · K. Ramakrishnan
Centre for Visual Computing, Faculty of Computing and Informatics,
Multimedia University, Cyberjaya, Malaysia
e-mail: eswaran@mmu.edu.my

K. Ramakrishnan
e-mail: kannan.ramakrishnan@mmu.edu.my

S. B. Balasubramanian
Department of Information Technology, Lebanese French University, Erbil, Iraq

© Springer Nature Switzerland AG 2019 1749
D. Pandian et al. (eds.), *Proceedings of the International Conference on ISMAC in Computational Vision and Bio-Engineering 2018 (ISMAC-CVB)*, Lecture Notes in Computational Vision and Biomechanics 30,
https://doi.org/10.1007/978-3-030-00665-5_161

lossless. Lossy compression is mostly used for applications where loss of data may be acceptable [2, 3]. However, lossless compression is needed for many applications such as telemedicine and satellite communication where any degradation of the image quality is not acceptable.

The traditional lossless compression methods normally make use of three steps, namely transformation, data-to-symbol mapping and encoding [4–6]. In this paper, a new lossless compression scheme involving classification, prediction and entropy encoding is presented. The main goal of the proposed lossless compression scheme is to achieve high compression ratio in order to reduce the storage space and bandwidth requirement for transmission and at the same time maintaining the original quality of the images.

As a preprocessing step, the image database is first grouped into sets of similar images. This step leads to a more accurate prediction process which results in achieving a high compression ratio. For classification of images, wavelet-based contourlet transform (WBCT) [7, 8] and Fourier descriptor (FD) [9, 10] is used to extract image features which are then applied to a fuzzy c-means (FCM) classifier [11–13]. After classification, each class or group of images is compressed separately using an artificial neural network (ANN) predictor [14, 15]. The predicted pixel values are subtracted from the original values to obtain the residue or prediction error sequence (E). By using an integer format for the predicted and error values, we ensure that no loss is involved in the prediction process. Since these error values only are stored or transmitted, less number of bits are required compared to the original pixel values, thereby achieving compression. In order to improve the compression ratio further, we apply a lossless entropy encoding scheme, namely Huffman coding to the error sequence. For decompression, the error sequence is combined with the predicted values obtained by using an ANN which is identical to the one employed in the compression process.

The proposed compression scheme is evaluated using three datasets, namely COREL-1 k database [16], CLEF med 2009 medical database [17] and standard images such as LENA, BARBARA, etc. An additional feature of the proposed method is that it ensures data security as only prediction errors are involved in the storage and transmission processes instead of the actual image pixels.

2 Proposed Method

The proposed compression method makes use of classification and prediction techniques. For classification, features obtained using wavelet-based countrolet transformation (WBCT) [7, 8] and Fourier descriptor (FD) [9, 10] is employed. Based on these features, the images are classified first into different groups using a fuzzy c-means (FCM) classifier [11–13]. The prediction error sequence (E) which is the difference between the original and the predicted pixel values is obtained for each group of images separately [14, 15]. The preprocessing step, namely classification improves the compression ratio as it leads to better prediction thereby reducing the

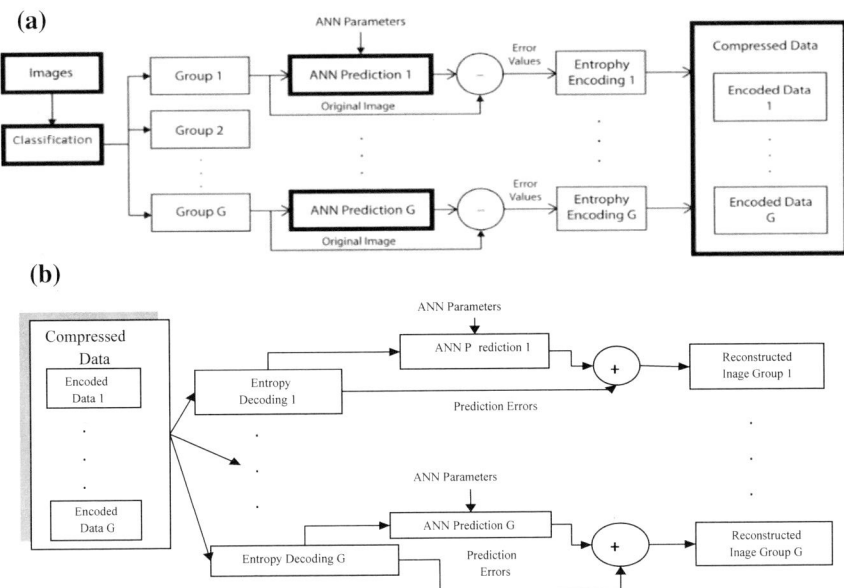

Fig. 1 Block diagram **a** compression stage **b** decompression stage

magnitude of prediction error values. To increase the compression ratio further, the error values are subsequently entropy encoded using Huffman coding technique. For decompression, the reverse steps are adopted. First entropy decoding is done to get back the error values. These error values are then added to the predicted values obtained using an identical predictor (as in the compression stage) to reconstruct the original images.

In the proposed method, the prediction is implemented using artificial neural networks (ANNs). It is assumed that identical ANNs are used for compression and decompression processes. A detailed explanation about the classification phase and the selection of optimum neural network configuration has already been reported [18]. The block diagram of the proposed system is shown in Fig. 1. In the compression scheme (Fig. 1a), the images are first classified into different groups say 1, 2..., G. An optimum ANN network is employed for each group. The error values are then entropy encoded using Huffman coding technique before storing in the memory. The reverse steps are used for decompression of images as shown in Fig. 1b. The error values are combined with the ANN predicted values to obtain the original or reconstructed images. It is assumed that the ANN parameters and the codebooks are available at both the compression and decompression stages.

The ANN parameters shown in Fig. 1 represents the training parameters of each ANN such as the number of inputs, weight values, number of hidden layer neurons, activation functions, etc.

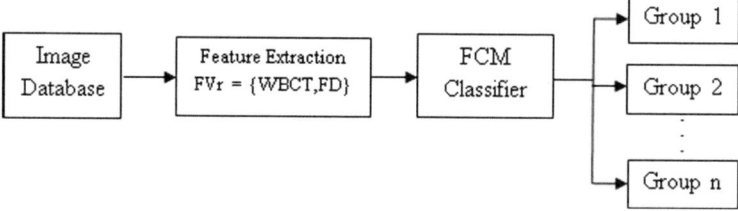

Fig. 2 FCM classification

2.1 Fuzzy c-Means (FCM) Classifier

As a preprocessing step, classification of images is carried out in order to increase the compression ratio since classification improves prediction accuracy. A fuzzy c-means (FCM) [11–13, 18] classifier is employed for this purpose. The feature vector FV_r used for FCM comprises the features obtained using the wavelet-based contourlet transform (WBCT) [7, 8] and the Fourier descriptor (FD) [9, 10]. Figure 2 shows the process of classification in the proposed method.

FCM classification is obtained by minimizing the objective function shown in Eq. (1) [12].

$$A = \sum_{j=1}^{J} \sum_{k=1}^{K} \mu_{ik}^{m} |p_i - v_k|^2 \tag{1}$$

where A is the objective function, J is the number of elements in the feature vector FV_r extracted from the image I, K denotes the number of clusters, μ represents the degree of membership, m is any real value greater than 1, p_i is the ith element in FV_r and $|p_i - v_k|$ is the Euclidean distance between p_i and v_k (centroid of the kth cluster).

2.2 Feature Extraction

2.2.1 Wavelet-Based Contourlet Transform (WBCT)

Wavelet-based contourlet transform (WBCT) [7, 8 and 18] is similar to the contourlet transform [19]. WBCT comprises two stages, namely sub-band and angular decompositions. In the sub-band decomposition, separable wavelet filters are used to separate the frequency bands. These filters extract three high-pass bands corresponding to the HH, LH and HL bands. Since wavelet filters are not perfect in splitting high-frequency from the low-frequency components, a fully decomposed directional filter bank (DFB) is used in each band at the second stage. DFB provides angular decomposition using non-separable filter banks [8]. DFB is applied in each high-pass

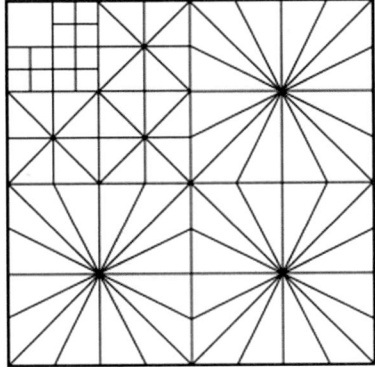

Fig. 3 WBCT with three wavelet levels and eight directions

band at a given level. Starting from the maximum number of directions $D = 2^i$ on the finest level i, where i represents the maximum level of wavelet decomposition. The number of directions depends on the dyadic scale as shown in Fig. 3.

The wavelet coefficients corresponding to the three high-pass bands HH, LH and HL are obtained and used as texture discrimination features. In the proposed method, the coefficients corresponding to LH and HH bands only are employed as texture features. For an image of size $M \times N$, the length of WBCT feature vector WT_n is $N/2$ [8]. In order to improve the classification accuracy, the texture features WT_n obtained using WBCT are combined with the Fourier descriptors (FD_n) which are determined using Fourier transform [5].

2.2.2 Fourier Descriptors

Fourier descriptors (FD_n) are derived by applying Fourier transform on shape signatures [9]. Following are the steps involved in computing the FD_n for an image [10]:

- Boundary coordinates are determined by detecting the boundary points (L).
- Centroid distance function $r(t)$ is obtained from the boundary points.
- Fourier transformation is applied on $r(t)$ to obtain the Fourier coefficients.
- FD_n is determined by using the transformed and original Fourier coefficients.
- The FD_n feature vector of an image with L boundary points comprises only $L/2$ elements because of the symmetry property.

The feature vector FV_r which is applied to FCM classifier is obtained by concatenating WT_n and FD_n. The length of the feature vectors WT_n and FD_n will depend on the size of the image and on the number of boundary points of the image, respectively [5]. For an image of size $M \times N$ with L boundary points, the FV_r is formulated as $FV_r = \{WT_1, WT_2, \ldots, WT_{N/2}, FD_1, FD_2, \ldots, FD_{L/2}\}$.

2.3 Prediction Process

Figure 4 illustrates the steps followed in the prediction process. Figure 4a shows an $M \times N$ image matrix denoted as I where I_{11} to I_{1N} represent the first row and I_{11} to I_{M1} represent the first column. The pixel I_{22} is predicted using the neighbouring pixels I_{11}, I_{12} and I_{21}. Let us denote the predicted pixel as I'_{22}. In the next step, I_{23} is predicted using I_{12}, I_{13} and I'_{22}. This process is continued for the remaining pixels. In effect, the entire image is predicted using the first row and first column pixels as the initial values. Figure 4b shows the predicted image matrix I' corresponding to I. The error matrix of the image denoted as E is obtained as shown in Eq. (2) [18, 20].

$$E = I - I' \tag{2}$$

In I', we make the first row and first column values as zeros so that the first row and first column values of E contain the original pixel values. Figure 4c shows the error matrix E corresponding to I. In order to implement a lossless compression scheme, the predicted values are rounded off to the nearest integers so that the errors (difference between the original and predicted pixel values) will also be integers. This rounding off to the nearest integer operation is carryout both at the compression and decompression stages. This process will ensure that the reconstructed and original images are identical.

2.4 Artificial Neural Network as Predictor

An artificial neural network (ANN) comprising an input layer, hidden layer and an output layer [14, 18] is used as the predictor in the proposed method. The optimum ANN configuration is identified to achieve good prediction accuracy or to reduce the difference between the original and the predicted pixel values. Figure 5 shows the ANN structure with three layers.

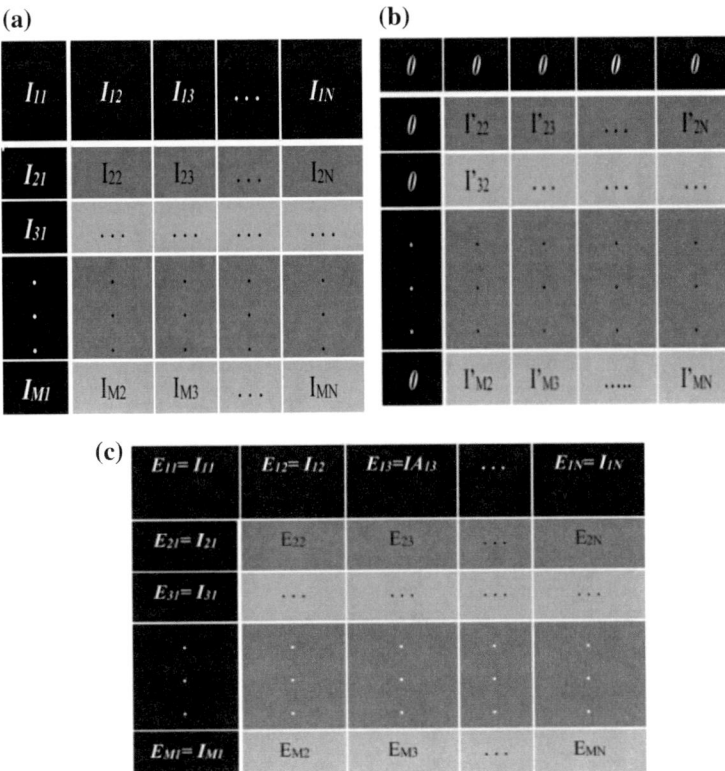

Fig. 4 Prediction steps: **a** original image matrix (I) **b** predicted image matrix (I') **c** error matrix (E)

Selecting an optimum ANN configuration for each group of images plays an important role in improving the prediction accuracy. This process involves finding the optimum number of input and hidden layer neurons, hidden and output layer activation functions and training algorithms. Based on the experimental results, it is found that the optimum number of input and hidden layer neurons is 3 and 10, respectively. The optimum training algorithms, as well as the activation functions, are identified for each group on trial and error basis. The ANN is trained for each group of images separately and for this purpose, 60% of the images in each group are used for training and the remaining images are used for testing [18].

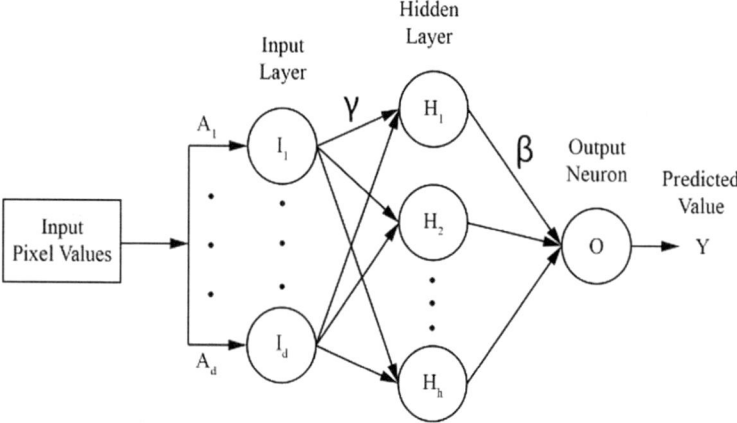

Fig. 5 ANN structure

2.5 *Entropy Encoding*

In order to improve the compression ratio, entropy encoding is used to compress the prediction errors further. This process, being lossless, the quality of the reconstructed image is not affected. The codebooks used in the encoding and decoding scheme are assumed to be available both at the compression and decompression stages. From our experimental results, it is found that Huffman coding yields better results compared to arithmetic coding. Hence in the proposed method, Huffman encoding is adopted [21].

3 Image Datasets Used for Testing

The proposed system is tested using the following three different sets of image databases.

3.1 *CLEF Med 2009 Database*

The medical dataset, namely CLEF med 2009 [17] contains more than 15 K images, in which manually 1000 images are selected with 12 different group of images. Figure 6 shows a random sample images from the chest and pelvic girdle group of images from the medical database.

Fig. 6 Sample chest and pelvic girdle images of the CLEF med 2009 database

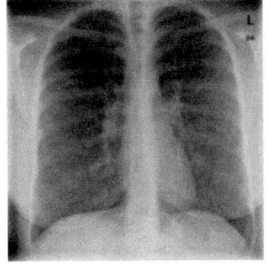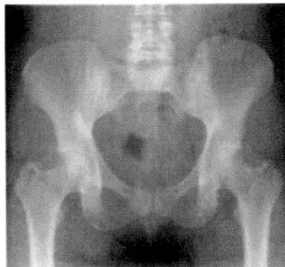

Fig. 7 Sample images of the COREL database

Fig. 8 Standard benchmarking images

3.2 Corel Database

The general purpose dataset, namely COREL dataset [16] contains 1000 images of 10 groups with 100 images in each group. Figure 7 shows some sample images in the database.

3.3 Standard Images

Figure 8 shows some standard benchmarking images such as Lena, Barbara, Cameraman, Peppers, Man, Goldhill and Boats that are used for testing the proposed system.

Table 1 Compression result of CLEF med 2009 database

Image class	Compression ratio
Chest	5.06
Leg	4.85
Radius and ulna bones	4.95
Pelvic girdle	5.13
Skull front view	5.27
Skull right view	5.76
Skull left view	5.29
Mammogram left	5.95
Mammogram right	6.02
Hand X-ray images	4.85
Pelvic + Back bone	4.87
Neck	4.92
Average	5.24

Table 2 Compression result of Corel 1 K database

Image class	Compression ratio
African people	4.83
Beach	4.95
Building	4.75
Bus	4.65
Dinosaur	5.22
Elephants	4.85
Flower	4.90
Horse	4.73
Mountain	4.69
Food	4.77
Average	4.83

4 Experimental Results

The compression efficiency is calculated using the parameter compression ratio (CR) which is defined as in Eq. (3). The results obtained for the medical and Corel database are shown in Tables 1 and 2, respectively.

$$CR = \frac{\text{Size of original image (Bytes)}}{\text{Size of compressed image (Bytes)}} \tag{3}$$

The CR value given for each group in Tables 1 and 2 represent the average CR value obtained for the images of that group. Table 3 shows the CR values obtained for standard images using the proposed method. The CR value given for each group

Table 3 Compression result of benchmarking images

STD image name	Image dimension	Compression ratio
Lena	128*128	2.89
	256*256	3.95
	512*512	5.10
Barbara	128*128	3.90
	256*256	5.02
	512*512	6.78
Cameraman	128*128	2.98
	256*256	3.96
	512*512	4.76
Peppers	128*128	2.95
	256*256	4.02
	512*512	5.13
Man	128*128	2.93
	256*256	3.65
	512*512	4.80

Table 4 Comparison of existing methods with the proposed method

STD image name	CALIC	RPC	ETC	Proposed method
Lena	4.48	5.05	4.09	5.10
Barbara	–	6.12	4.58	6.78
Goldhill	4.39	5.10	4.60	5.14
Peppers	4.42	5.05	4.39	5.13
Man	–	–	4.34	4.80
Boats	3.83	4.91	4.12	4.98

in Tables 1 and 2 represent the average CR value obtained for the images of that group. Table 3 shows the CR values obtained for standard images using the proposed method. Table 4 shows a comparison of the proposed method with four existing methods, namely, CALIC [22], ETC [23] and RPC [24]. From Table 4, it is clear that the proposed method yields better CR values compared to other known methods.

5 Conclusions

In this paper, a lossless image compression scheme based on prediction error concept has been presented. Artificial neural network has been employed as the predictor. The prediction process is made lossless by making the predicted values integers. As a preprocessing step, a fuzzy c-means classifier using wavelet and Fourier descriptor features has been used. To improve the compression efficiency further, Huffman encoding is performed on the prediction errors. The experimental results show that the compression ratio values obtained by the proposed method are higher compared to those obtained by the known algorithms. Though the method presented in this paper focuses only on grey scale images, the proposed concept can be easily extended to RGB images also.

Ethical Approval For this retrospective type of study formal consent is not required.

References

1. Celik M, Tekalp AM, Sharma G (2003) Level-embedded lossless image compression. Acoustics, speech, and signal processing, 2003. In: 2003 IEEE international conference on Proceedings (ICASSP'03), vol 3, pp III–245, IEEE
2. Ahamed A, Eswaran C, Kannan R (2018) Lossy image compression based on vector quantization using artificial bee colony and genetic algorithms. Adv Sci Lett 24(2):1134–1137
3. Ayoobkhan MUA, Chikkannan E, Ramakrishnan K (2017) Lossy image compression based on prediction error and vector quantisation. EURASIP J Image Video Proc 2017 1:35
4. Bovik AC (2009) The essential guide to image processing. Academic Press
5. Nasir DM, Sayood K (1995) Lossless image compression: a comparative study. In: IS&T/SPIE's symposium on electronic imaging: science and technology. In: International society for optics and photonics, pp 8–20
6. Ramakrishnan K, Eswaran C (2007) Lossless compression schemes for ECG signals using neural network predictors. EURASIP J Appl Sig Proc 2007, 1:102–102
7. Hung CH, Hang HM (2012) A reduced-complexity image coding scheme using decision directed wavelet-based contourlet transform. J Vis Commun Image Represent 23(7):1128–1143
8. Venkateswaran K, Kasthuri N, Alaguraja R (2015) Performance comparison of wavelet and contourlet frame based features for improving classification accuracy in remote sensing images. J Indian Soc Remote Sens 43(4):729–737
9. Zhang D, Lu G (2003) A comparative study of curvature scale space and fourier descriptors for shape-based image retrieval. J Vis Commun Image Represent 14(1):39–57
10. Sokic E, Konjicija S (2016) Phase preserving fourier descriptor for shape-based image retrieval. Sig Process Image Commun 40:82–96
11. Hung CC, Kulkarni S, Kuo BC (2011) A new weighted fuzzy c-means clustering algorithm for remotely sensed image classification. IEEE J Sel Top Sign Proces 5(3):543–553
12. Pandey D, Kumar R (2012) Hybrid algorithm using fuzzy c-means and local binary patterns for image indexing and retrieval. In: Soft computing techniques in vision science. Springer, pp 115–125
13. Fabija´nska A (2009) A fuzzy segmentation method for images of heat-emitting objects. In: Iberoamerican congress on pattern recognition. Springer, pp 217–224
14. Oliveira FD, Haas HL, Gomes JGR, Petraglia A (2013) Cmos imager with focal-plane analog image compression combining dpcm and vq. IEEE Trans Circuits Syst I Regul Pap 60(5):1331–1344
15. Mielikainen J, Huang B (2012) Lossless compression of hyperspectral images using clustered linear prediction with adaptive prediction length. IEEE Geosci Remote Sens Lett 9(6):1118–1121
16. Li J, Wang JZ (2003) Automatic linguistic indexing of pictures by a statistical modelling approach. IEEE Trans Pattern Anal Mach Intell 25(9):1075–1088
17. Tommasi T, Caputo B, Welter P, Guld MO, Deserno TM (2009) Overview of the CLEF 2009 medical image annotation track. In: Workshop of the cross-language evaluation forum for european languages. Springer, pp 85–93

18. Mohamed Ahamed A, Uvaze C, Eswaran R Kannan (2017) CBIR system based on prediction errors. J Inf Sci Eng 33(2):347–365

19. Li H, Chai Y, Li Z (2013) Multi-focus image fusion based on nonsubsampled contourlet transform and focused regions detection. Optik-Int J Light Electron Opt 124(1):40–51

20. Ayoobkhan MUA, Chikkannan E, Ramakrishnan K (2018) Feed-forward neural network-based predictive image coding for medical image compression. Arabian J Sci Eng 43(8):4239–4247

21. Abo-Zahhad M, Gharieb RR, Ahmed SM, Abd-Ellah MK (2015) Huffman image compression incorporating dpcm and dwt. J Sig Inf Proc 6(02):123

22. Wu X, Memon N (1997) Context-based, adaptive, lossless image coding. IEEE Trans Commun 45(4):437–444

23. Zhou J, Liu X, Au OC, Tang YY (2014) Designing an efficient image encryption-then compression system via prediction error clustering and random permutation. IEEE Trans Inf Forensics Secur 9(1):39–50

24. Liu W, Zeng W, Dong L, Yao Q (2010) Efficient compression of encrypted grayscale images. IEEE Trans Image Process 19(4):1097–1102

Audio and Video Streaming in Telepresence Application Using WebRTC for Healthcare System

Dhvani Kagathara and Nikita Bhatt

Abstract Currently, health care in Lower- and Middle-Income Countries (LMICs) is suffering from shortage of trained physicians in rural areas. The development of certain technologies, particularly in sensors has yielded the birth of mHealth. The main objective of this paper is to develop a telepresence application for a mobile platform. The functionality of the system will include the ability for the doctor to connect with a doctor/patient from a remote location using mobile or web application, and the ability to diagnose patient disease remotely and recommend follow-up care. There is a requirement for low price, well suited, and easy-to-use video communication system. The web real-time communication (WebRTC) permits browsers to set up a peer connection to deliver information and media with real-time conferencing capabilities through simple JavaScript APIs.

1 Introduction

The improvement of financial and social elements regarding a country is said to be complementary to each other. The fields including education or health care are suffering from the social need which sooner or later depicts the financial growth and in the long run, the excellent of lifestyles. India has expected digital adoption, the Indian Healthcare marketplace, that is certainly worth around US$ a 100 billion, will probably develop at a CAGR of 23% to US$ 280 billion through means of 2020 [1]. Including infrastructure or medical specialists via myself will now not be capable about unravel India's considerable unmet desires among health care. It desires to be supported through the era. Scientific technology innovation may remain the tool to redact current care accessible, available then less costly to all by way on lowering the cost of the product or delivery.

D. Kagathara (✉) · N. Bhatt
Chandubhai. S. Patel Institute of Technology, Charotar University of Science and Technology, Nadiad, India
e-mail: kagatharadhvani@gmail.com

© Springer Nature Switzerland AG 2019
D. Pandian et al. (eds.), *Proceedings of the International Conference on ISMAC in Computational Vision and Bio-Engineering 2018 (ISMAC-CVB)*, Lecture Notes in Computational Vision and Biomechanics 30,
https://doi.org/10.1007/978-3-030-00665-5_162

The characteristic of playing back audio and video also as the record is Wight received is recognized as streaming that is a point in accordance with point and multipoint. Unlike everyday data file, a streaming media file for consideration is big, as a consequence calls because of excessive channel bandwidth. Furthermore, streaming media moreover contains severe demand inside the timing of bundle delivery. It has some constraints consisting of bandwidth, latency, or congestion. Video streaming may stay categorized into two categories such as live and on-demand. In this paper, we observe nearly used streaming protocols for mobile utility or web [2]. They may additionally be lower layer as TCP or UDP and upper layer so RTP, RTCP or RTSP [3]. In video streaming, lower layer protocol such as Transmission Control Protocol (TCP) or User Datagram Protocol (UDP) is not feasible to use [3]. So far, we will use the higher layer protocols such as Real-time Transport Protocol, (RTP) Real-time Control Protocol (RTCP), and Real-time Streaming Protocol (RTSP) [3]. Real-Time Communication (WebRTC), a project maintained by way of means over Google among 2011 primarily based on RTP is analyzed or related according to such through IETF and its browses API by the usage of W3C [4]. This gadget allows ongoing correspondences by means of a couple which use Programming Interfaces (API) alongside a top-notch, low idleness, and low transmission capacity utilization [2].

2   WebRTC

WebRTC is the collection of protocols, standards, and APIs [2]. Audio, video and information shares between peers using the connection [2]. There is no software or any extra plug-ins required [5]. The WebRTC implementation is created through Google and open source launched it [2]. World Wide Web Consortium (W3C) organization standardized the WebRTC at the application level and Internet Engineering Task Force (IETF) organization standardized the WebRTC at the protocol level [4].
    The major components of WebRTC are:

2.1   Media Stream

Get user media allows to acquiring audio or video from the hardware such as camera, microphone, screen capturing, or display [6]. Using get User media API, we can get the screenshots and share the screen with different parties also [6] (Table 1).

Table 1 Media engine performance [10]

Module	Measured statistics
Video capture	Captured video resolution, and a number of frames captured per second [10]
Video encoding	Encoded stream bit rate, frame rate and resolution also encoder ramp-up speed [10]
Video decoding/rendering	Frame numbers decoded/rendered per second; resolution of the rendered stream [10]
Networking	Bytes/packets sent, received and dropped, RTT [10]
Jitter buffer	Jitter, A/V Sync [10]

2.2 RTC Peer Connection

RTC peer connection is used for setting the audio/video streams [2]. It consists a lot of special tasks like codec execution, signal processing, bandwidth administration, safety streaming, and many more which represents the contain into MediaStream and DataChannel by presenting a handshake mechanism because of pair machines according to exchange fundamental data so a Peer-to-Peer connection may stay set up [6] (Table 2).

2.3 RTC DataChannel

Connected users can share audio, video, and data using RTC DataChannel [2]. Which is used for Bidirectional communication among peers and exchange the data which is in any format [2]. Which is share arbitrary data between peers [6] (Table 3).

2.4 GeoStats

API call that hat permits getting diverse insights about a WebRTC session [2].

3 WebRTC Architecture

Figure 1 illustrates the WebRTC overall structure which API used and different audio and video codecs used in the streaming application. Also, it shows the features supported by browsers and application for real-time communication [7]. WebRTC offers net software builders the potential to put in writing wealthy, real time multimedia

Table 2 Connection flow measurements [10]

	Parameters	Measurement statistics
Initialization	New peer connection	Time for instantiating a new RTC peer connection object [10]
	Get user media	Time spent capturing user media excluding time waiting for user permission [10]
	Can Play Local Media	Time spent after media request until local media becomes available [10]
	Setup signaling channel	Time to setup signaling channel using web sockets [10]
Communication	Make call	Time for setting session descriptions and creating an offer [10]
	ICE hole punching	Time spent on ICE hole punching process [10].
	Make answer	Time for setting session descriptions and creating an answer [10]
	Signaling	Time spent propagation of answer/offer messages and signals across the network [10]
	Open dataChannel	Time for the opening dataChannel [10]
	Play remote rtream	Time for play remote media stream to be available [10]

Table 3 DataChannel measurements [10]

Category	Measured statistics
Network	Bytes sent/received/RTT [10]
Application	Messages sent/received/RTT [10]

applications (suppose video chat) at the web without requiring plug-ins, downloads or installs. Its cause is to help construct a robust RTC platform that works throughout more than one web browsers throughout a couple of platforms [8].

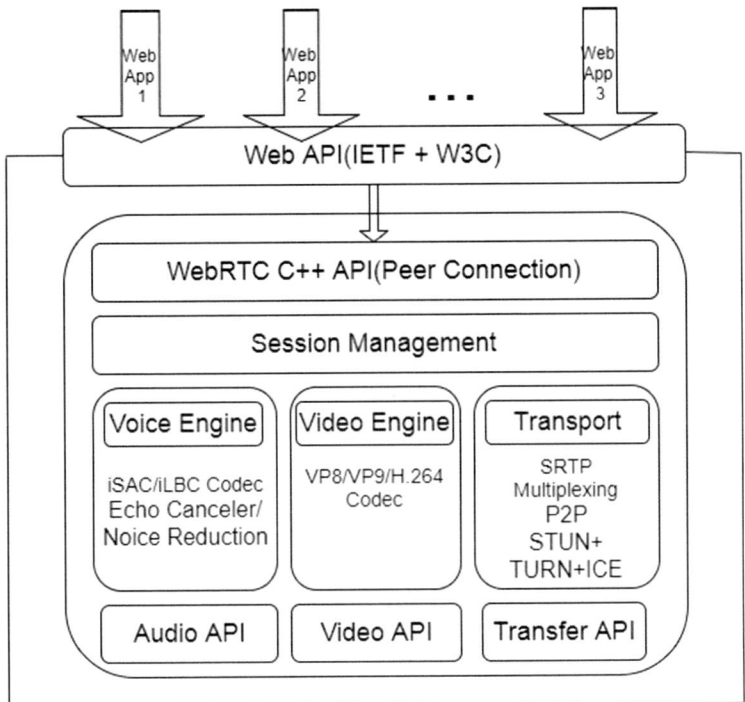

Fig. 1 WebRTC structure

3.1 Voice Engine

The voice engine is usually called sound engine [7]. They are processed for echo cancelation and noise reduction [9]. The sound machines facilitate the audio codecs which is optimized by narrowband and wideband [9]. The concurred sound codecs up to desire require in congruity with be upheld by method for WebRTC all around coordinated programs comprise of iLBC (i.e., a narrowband discourse codec for VoIP or gushing sound), iSAC (i.e., a wideband voice codec in light of the fact that VoIP at that point spilling), or Opus (i.e., a codec so much aides bit rate encoding) [7]. In sender side, they process the raw streams to enhance the quality and match the bandwidth and latency which is continuously fluctuating [9]. From receiver side, received streams decode and adjust the latency delay and network jitter [9].

3.2 Video Engine

The video engine is a framework as offers together with video streams that come beside the digital camera in conformity with the community and from the community in accordance with the show display [7]. Just as the frame engine, the video engine helps the video codecs. VP8 is writing in time agreed with the video codec [7]. It helps potential jitter buffer because of the video who allows in accordance with hide the outcomes concerning get rid of jitter then packet loss in imitation of enhance the overall video top-notch [7]. It additionally allows image enhancement, as much an instance, eliminating video noise from the image capture by using the usage of behavior regarding the webcam [7]. It is designed for low latency so that it is well suited for RTC [9].

3.3 Transport Support

At the transport layer pass, the audio and video streams to the peers using this activity [7]. TCP buffers all packets after th intermediate packet is lost and wait for retransmission after delivering the stream [9]. A secure Real-Time Protocol is used with WebRTC to secure the streams of audio or video using encrypts of all the records [7]. UDP delivers each packet at the moment it arrives which is no promises on ordered data [9].

3.4 Session Management

The session management defines as beneficial to WebRTC as abstract signaling approach and the protocol according to session data as explained or leaves the statistics about the signaling implementation on the hand concerning application builders [7]. Session description protocols are most affected by WebRTC [7]. C++ native API of WebRTC at this API layer allows browser markers in accordance with except issues put into effect the web API because of his or her browses [7]. Firewalls deals the parameters for every stream and implement congestion control, flow control and encryption of data, and more [9].

4 Implementation

4.1 Environment Setup [7]

We are using the WebRTC functionalities for establishing the environment which is established previously by EasyRTC APIs and JavaScript APIs for Web Browsers. EasyRTC is the open-source project [7]. We used the node.js server and JavaScript APIs to run the application in web browser and for the mobile application, we used the WebView [7].

4.2 Audio and Video Streams Gathering [7]

For audio and video streams gathering found complexity [7]. Call EasyRTC. getLo-calStreame and EasyRTC [7]. set video ObjectSrc [7]. We have to invoke the method for ask the permission for Local camera and microphone [7]. After giving the camera and microphone permission, the user can see their local screen. For incoming streams, we have to call the callback method [7]. For setting the bandwidth for each video track we can use EasyRTC setVideoBandwidth() [7].

4.3 Signaling and Peer Connection [7]

To establish the communication, first exchange the data between peers [7]. As an example, for making the connection in the same network, only public IP address is necessary [7]. But the process is complicated when peers were far to each other and peer's public IP addresses and port address are hidden to each other [7]. If one of the peers is not available, we have to first acquire the viable IP addresses and port candidates for each peer, traverse the NATs, and then run the connectivity [7]. The single interface encapsulating all the connection setup, control, and state PeerConnection API within the WebRTC is dealing with the life cycle for every peer connection [7].

Figure 2 explains the signaling interactions and the requirements between two peers.

(a) Whenever the user from one side which uses browser 1 clicks the "join" button, the browser of user obtains and presentation the local media [7].
(b) After a user of one side wants to connect to the user of another side which uses browser 2 for video conference of the secure channel using the signal server [7].
(c) After the session is mounted between both users [7].
(d) For exchange, the media, codecs and setting their bandwidth information, a user from one side uses the Session Description Protocol (SDP) [7]. Real media itself is not connected via offer [7].
(e) As soon as the offer is gotten, a client from opposite side makes the offer to associate with the client in one side and furthermore sends the comparing session profile utilizing the SDP [7]. After that, the client from one side realizes that client from the opposite side is prepared to associate utilizing shared connection [7].
(f) User from the opposite side makes an intuitive network foundation (ICE) operator, (ICE A) to the client from program 1 side [7]. The ICE specialist finds the closer IP with the port tuple of companion and uses outer STUN server for get general society IP and port tuple [7].
(g) The client from program 2 side accepting the ICE an indistinguishable activity from a client from program 1 side to make and send ICE specialist (ICE B) [7].

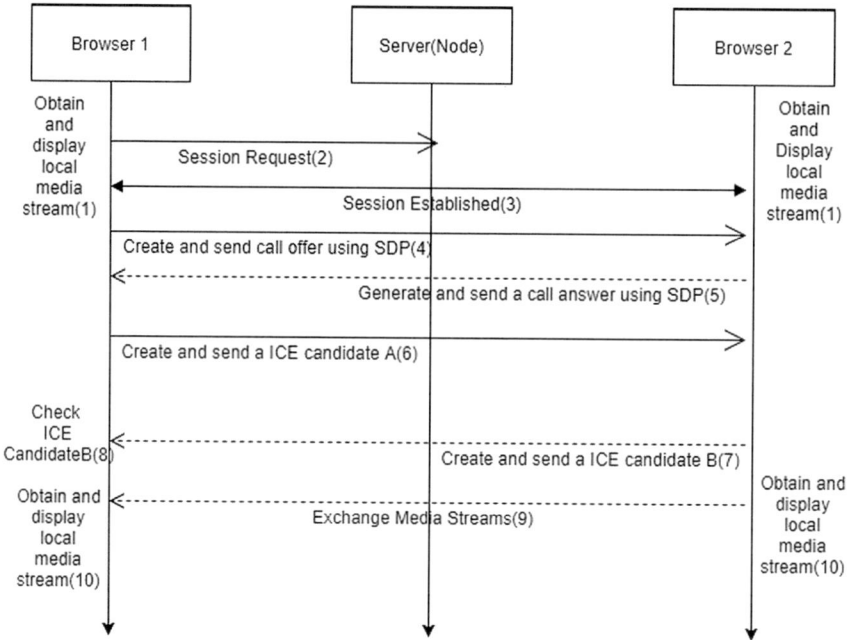

Fig. 2 Peer-to-peer interaction

(h) TURN server utilized as an intermediary to hand off movement if clients cannot make peer association utilizing STUN server [7]. It will get data ahead of time which is suits in ICE B [7].

(i) User from one side sends the video stream to another side user [7]. Both side data and nearby media streams established [7]. In the wake of getting general society IP locations and port assortment tuples ahead of time from ICE A [7].

(j) In the end, media content is shown by both the browsers and the video conferencing is operating [7].

STUN/TURN servers can be used to handle a small number of participants. Javascript code is added to make the peer connection [7].

4.4 Chatting Room Establishment [7]

This is used for the patients and doctors to conduct the meetings [7]. Codecs for client and server needs to be implemented. Using Socket.io, the first client is getting the chat channel [7]. As the connection is established, the client can send the text. After every user joins the room. Display the incoming messages and outgoing messages [7].

5 Limitations

Limitations of the existing telepresence applications are: Present video streaming application used for telepresence criticized for plenty of reasons: (a) they are often too high priced to buy and preserve, (b) using technologies which is proprietary cannot be compatible with each other, and (c) to keep the device they require fairly skilled IT personnel [7].

6 Conclusion and Future Extension

In this paper, we have discussed the WebRTC standard-based application for tele-health which is simple, interoperable, and inexpensive for video conferencing. The proposed healthcare system will be helpful for communication of doctors and patients. The system would be developed using web real-time communication with better quality audio and video streams using different protocols and standards to evaluate the need for Telehealth. Real-time communication use for all the platforms of a web on cloud computing and Android/iOS devices. We have discussed related implementation of a WebRTC based system which is emerging different protocols and architecture. Which is provide the good user experience in web and mobile both. We have used different mechanisms like RTP based media exchange and peer-to-peer data exchange. WebRTC provides more developer friendly environment that does not require specific software and plug-ins, and it will be available anytime and anywhere. WebRTC requires permission to use camera and microphone to access from the user. WebRTC reduces energy intake and bandwidth utilization and accordingly enhances the effectiveness of the system. We have used different approaches for real-time communication but the methods have some limitations which may be overcome by improving an existing system. In the future, healthcare device can handle multiple connections with low bandwidth and high latency. Audio and video must of good quality.

References

1. https://www.ibef.org/industry/healthcare-india.aspx
2. Ivan santos-gonzalez A (2016) R.-G.-B.-G.-G: real-time streaming: a comparative study between RTSP and WebRTC. Springer, pp 313–325
3. Arsan T (2013) Review of bandwidth estimation tools and application to bandwidth adaptive video streaming
4. https://en.wikipedia.org/wiki/WebRTC
5. Jansen BA (2016) Performance analysis of WebRTC-based video conferencing. Springer
6. Laurent Lucas HD (2017) USE together, a WebRTC-based solution for multi-user presence desktop. Springer
7. Julian Jang-Jaccard SN (2014) WebRTC-based video conferencing service for telehealth. Springer, CSIRO Computational Informatics (CCI), Marsfield, Australia

8. https://webrtc.org/architecture/
9. https://princiya777.wordpress.com/2017/08/19/webrtc-architecture-protocols/
10. Sajjad Taheri LA (2015) WebRTCBench: a benchmark for performance assessment of WebRTC implementations. IEEE

Review on Image Segmentation Techniques Incorporated with Machine Learning in the Scrutinization of Leukemic Microscopic Stained Blood Smear Images

Duraiswamy Umamaheswari and Shanmugam Geetha

Abstract This paper is a contemplated work of N-different methods that have been employed in the area of revealing and classifying leukocytes and leukoblast cells. Blood cell images obtained through digital microscopes are taken as input to the algorithms reviewed. In bringing out the nucleus and cytoplasm of White Blood Cells (WBCs), the images have been undergone by a variety of image segmentation techniques along with filtering, enhancement, edge detection, feature extraction, classification, and image recognition steps. Apart from image processing, the analysis and categorization of the leukemic images are handled using some other machine learning techniques of computer science discipline. Assessment of accuracy and correctness of the proposals were done by applying texture, color, contour, morphological, geometrical, and statistical features.

1 Introduction

Biomedical image processing is a prominent field of computer science for processing medical images which are digital in character. Healthcare industries are facing hard issues in detecting the cancerous malignant cells in early stages. In that way, leukemia is a grievous disease that threatens human by reducing their lives.

Leukemia damages the leukocytes of blood without exhibiting the causes in prior. White blood cells otherwise called as leukocytes, the antibodies of the human immune system, get spoiled by leukemia, which in turn causes continuous illness. Numerous research processes have been performed to detect and classify malignant leukocytes

D. Umamaheswari (✉)
Department of Computer Science, Vidyasagar College
of Arts and Science, Udumalpet, Tamil Nadu, India
e-mail: ums1082@gmail.com

S. Geetha
Department of Computer Science, Government Arts College,
Udumalpet, Tamil Nadu, India
e-mail: geet_shan@yahoo.com

© Springer Nature Switzerland AG 2019
D. Pandian et al. (eds.), *Proceedings of the International Conference on ISMAC in Computational Vision and Bio-Engineering 2018 (ISMAC-CVB)*, Lecture Notes in Computational Vision and Biomechanics 30,
https://doi.org/10.1007/978-3-030-00665-5_163

1773

and regular WBC. Cytoplasm and nuclei of the WBC are needed to be separated, for processing in the recognition of leukemia cells. For this cause, different sorts of computer-aided techniques are proposed. Eventually, it is obvious for the researchers getting assistance from pathologists and oncologists for the error-free classification. Microscopic images of blood smear are being the major source of input and because there are no standard datasets available, it is must perform sensitive evaluations and comparisons of one's method with others in terms of accuracy.

In this paper, a review work is made about various techniques that have been incorporated in conjunction with Digital Image Processing (DIP) methodologies in the arena of leukemia detection. The paper describes leukemia and its types in Sect. 2, various stages of DIP are discussed in Sect. 3, review of the research activities made in the discipline of leukemia diagnosis is done in Sect. 4, Sect. 5 explains the issues faced by the researchers, and the manifestation of the review work is concluded in Sect. 6.

2 Leukemia and Its Subtypes

Leukemia is also acknowledged as the blood cancer that affects the spongy substance surrounded with the bone marrow, which is in charge of producing blood cells. Leukemic bone marrow secretes surplus amount of blood cells rather what is needed. Leukemia is recognized with diverse types. With the conduct of the proper examination of leukocytes, the nature of leukemia can be discriminated. Research activities of this line will be useful to the hematologists in reporting the results of blood tests. Major categorization of leukemia is of four types depending on whether it is acute otherwise chronic, leukoblastic, or else myeloid: Acute Lymphoblastic (Lymphocytic) Leukemia (ALL), Acute Myeloblastic (Myeloid) Leukemia (AML), Chronic Lymphoblastic Leukemia (CLL), and Chronic Myeloblastic (Myeloid) Leukemia (CML).

Most of the diseases are diagnosed by examining the blood samples through the microscopes. Hematologists sense the illness with the application of multifarious features of blood; one among them is counting of blood cells. In the medicinal field, numerous research activities are being in progression to bring out the exact count of blood cells. The differential count is the common label used to represent the count of WBC. Blood is the combination of various components: plasma, erythrocytes, leukocytes, and platelets. The last three are contained in plasma, a cell-less watery substance. Erythrocytes are widely referred to the character name red blood cells. The immune system of human being provides resistance from diseases by means of the antibodies referred to as leukocytes or WBC. WBCs are categorized about more than 20 types. Five significant subtypes are, namely, neutrophil, basophil, eosinophil, monocyte, and lymphocyte as shown in Fig. 1.

Technology developments in the counting of WBC and their subtypes also help in the healthcare industry in diagnosing the health disorders. Researchers support the biomedical technicians to perceive the leukemia in the way of proposing accurate

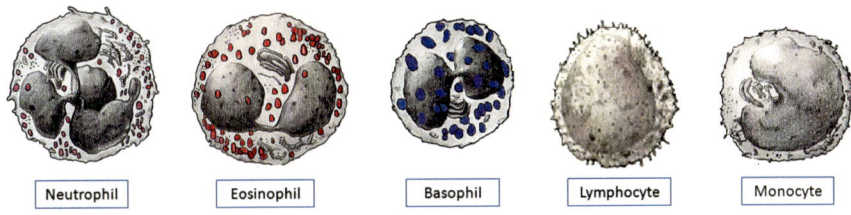

| Neutrophil | Eosinophil | Basophil | Lymphocyte | Monocyte |

Fig. 1 Subtypes of WBC

algorithms intended for partitioning the nuclei or cytoplasm of the blood cells as well as classifying them as normal or malignant cells.

3 Stages of Image Processing

Image processing is about processing digital images in a series of stages. According to our current study of leukemia detection and classification, the stages of DIP shown in Fig. 2 are carried out in the subsequent manner:

1. Image Acquisition: It is the task of getting microscopic blood images or biopsy microscopic images.
2. Image Transformation: The images are converted to a standard format and color space.
3. Image Restoration: Microscopic images with varying noise types are made better, without taking into account the intended features of the image.
4. Image Filtering: The biological or acquisitional artifacts are cleaned up with the aid of filters.
5. Image Enhancement: Images can be viewed better by enhancing their color, contrast, and brightness like features.
6. Edge Detection: Edges are the boundaries of the components which are found with edge detectors.
7. Morphological Operations: To extract the shape and size features of cells of blood, opening, closing, dilation, and erosion like operations are carried out.
8. Segmentation: This is the process of partitioning nucleus and cytoplasm of WBCs.
9. Feature Extraction: Features are the characters of an image. After segmentation of the sub-images, the texture, color, contours, geometrical, and statistical features are extracted from them.
10. Object Recognition: This process results in naming the parts of microscopic images.

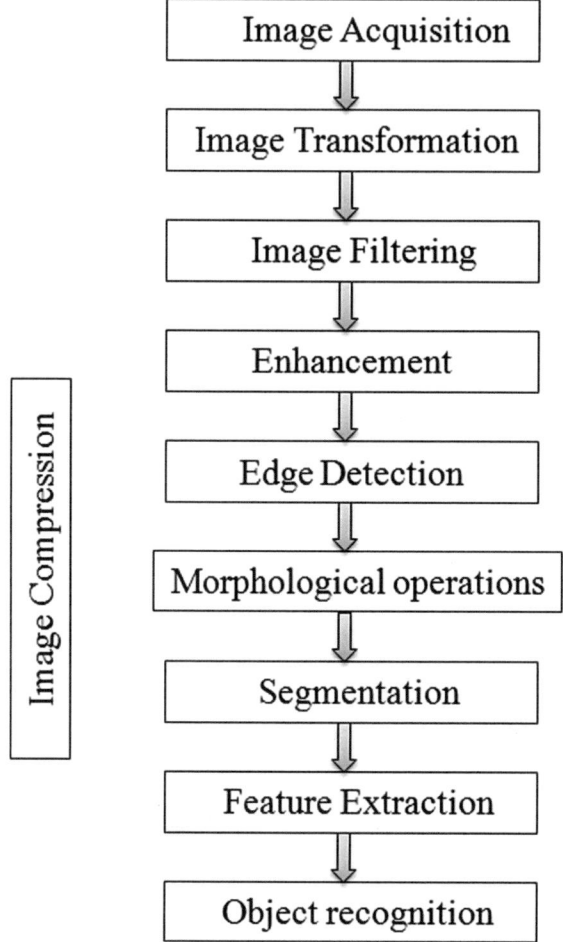

Fig. 2 Stages of digital image processing

4 Review of Literature

The baseline of the research activities on identifying leukemia is partitioning the nucleus and cytoplasm of the WBC. This work is implemented with the support of number of image segmentation algorithms. The segmented regions are taken into contemplation for categorizing different leukemia types with the help of machine learning techniques like classification.

4.1 Image Segmentation

Segmentation does the vital task of separating the nucleus and cytoplasm for analysis. The algorithms come up with several flavors for effective segmentation.

4.1.1 Segmentation by Thresholding

Cseke [1] projected a method of segmentation for the automatic differential count of WBC with the application of Otsu's threshold and automatic thresholding with the segmentation probability of 0.92. Liao and Deng [2] have presented a segmentation scheme seeking the count of WBC by applying thresholding with shape detection method in a step-by-step manner. Prasad et al. [3] presented an algorithm which performs recursive segmentation by comparing an arbitrary threshold value with the histogram of pixel intensities, which stripes off background pixels by segmenting only relevant pixels. Then, single cells are classified from cell clusters using signature plots of cluster classification; lastly, cell size, count, and location are determined and showed an accuracy of 97%. The authors Prasad with Badawy [4] proposed an algorithm for identifying, classifying, and quantifying the count of leukemia cells. Here, the pixels under ROI are recursively segmented by comparing the intensity values with a threshold and gives 95% accuracy for poor quality images yet.

Biswas and Ghoshal [5] made an increase in the performance of Sobel edge detector by a succeeding thresholding assessment-based watershed algorithm. The evaluation of this algorithm is done on 30 microscopic images and returns the accuracy of 93%. Kit et al. [6] proposed a framework for the Complete Blood Count (CBC). In this work, red and WBCs are extracted by the way of color space analysis, thresholding, and Otsu's method. Afterward, the appropriate cells are counted using topological structure analysis; then, cell mass regions are predicted through Hough Circle Transform (HCT), and the accuracy level is 92.93% in RBC count and 100% in WBC count.

Scotti [7] reported methods to measure the attributes of WBC. In their proposal, segmentation of WBC is done by gray-level thresholding. The combination of Gram-Schmidt technique and parametric deformable representation was introduced by Rezatofighi et al. [8] for segmenting the WBC's nuclei and cytoplasm, and also a proper thresholding technique was applied to improve the outcome of segmenting cytoplasm. In [9], Bergen et al. reported a scheme to segment the erythrocytes (RBCs) and leukocytes (WBCs). Otsu's thresholding is applied to separate nucleus. Rezatofighi et al. [10] proposal is based on orthogonality principle added with the Gram-Schmidt method which amplifies the desired color vector and suppresses the undesired color vectors for segmenting the nucleus of WBC with suitable threshold. Sadeghian et al. have designed a framework [11] with the intention of segmenting nucleus and cytoplasm. Nucleus segmentation was carried out through GVF snake, and cytoplasm was found through Zack thresholding with the segmentation accuracy of 92 and 78%, respectively.

In the paper [12], Nor Hazlyna et al. mentioned a segmentation methodology to segment blood image of leukemia into blast region and background region by means of the color spaces HSI and RGB. This is implemented with a threshold value. Mohamed and Far [13] proposed a new automatic segmentation algorithm that used Otsu's threshold to convert the contrast-enhanced gray image into twofold binary image. Its segmentation accuracy is 80.6%. Di Ruberto and Putzu [14] have presented an efficient segmentation method based on fuzzy set thresholding. Ahasan et al. proposed an algorithm [15] for segmenting the nuclei of blood white cells; it has a number of stages such as color space conversion, color thresholding, binary imaging by Otsu, average filter, marker-controlled watershed, different morphological operations, and eliminating borders with an accuracy of 88.57%. Lina et al. [16] presented a scheme to pin out WBCs. In this work, color filtering is prepared by comparing with a threshold value. The method works with the leukocyte detection accuracy rate of 82.12%. In [17], the authors Ananthi and Balasubramaniam implemented the segmentation by choosing a threshold from a fuzzy set with maximum similarity.

Sandhu [18] made a study on leukemia detection and concluded that watershed segmentation with optimal thresholding is best with the segmentation accuracy of 99.85, 99.2, and 99.63% for segmenting cells, nucleus, and cytoplasm, correspondingly. Putzu together with, Di Ruberto [19] introduced a methodology to find out WBCs and their groups by using triangle-based thresholding method of segmentation. Further, the cytoplasms along with nuclei of WBC are found. Gosh et al. demonstrated [20] a scheme of revealing leukemia with fuzzy deviation and adapted thresholding methods. Khobragade et al. [21] used Otsu's thresholding to convert the gray image into binary. From this, it could be possible to classify leukemia types with 91% accuracy. Abbas and Mohamad [22] proposed a new method of segmentation. In this method, the necessary areas are found by binarizing the image using Otsu's threshold with the segmentation accuracy of 96.56% with a reduction in computation time about 50%. Khashman and Al-Zgoul [23] proposed a scheme of segmenting the cytoplasm and nucleus regions with the assistance of bimodal thresholding. It assures the segmentation accuracy ratio of 98.33%. Abbas et al. [24] introduced a new scheme of segmentation assisted with a threshold value. The method improved the accuracy to 0.8955%. Li et al. [25] presented a technique of WBCs segmentation in ALL images with the application of the dual-threshold method. This proposal attained 97.85% of segmentation accurateness.

4.1.2   Segmentation by Watershed Algorithm

Nilsson and Heyden [26] introduced a model of segmentation method using watershed algorithm and region growing for segmenting the nucleus and accordingly its cytoplasm. This work gives the differential counts of WBC. Jiang et al. [27] offered a scheme to segment WBCs. In this scheme, watershed clustering is applied in separating cytoplasm. Shankar et al. [28] explained a method to identify ALL by a set of image processing steps including segmentation with watershed algorithm; afterward, abnormal cells are separated. This method attains more than 90% of accuracy. Ghane et al. [29] found a technique combining thresholding, k-means clustering, and watershed algorithms to segment both the blood cells and nucleus of lymphocytes. The work was compared with manual segmentation by three factors, namely, similarity measures, sensitivity, and precision.

Belekar and Chougule in [30] explored a scheme to segment nuclei and cytoplasm of leukocytes by utilizing watershed transform along with level set. Liu et al. [31] demonstrated a method to segment WBC using average shift clustering along with watershed algorithm.

4.1.3 Segmentation by k-means Clustering

Sinha and Ramakrishnan [32] worked on color images of discolored peripheral blood smears to reveal various classes of WBCs. The system does the segmentation in two stages: Classification by k-means clustering and feature extraction by EM algorithm. Basima and Panicker [33] presented an automated method of detecting, counting, and classifying blood white cells. In this system, WBCs are detected through k-means segmentation; features are acquired by using gray-level co-occurrence matrices, and then classification of healthy WBCs and blast WBCs by SVM with the accuracy range of 94.56%.

Foran et al. [34] introduced a computer-aided medical decision support system that differentiates the hematological malignancies using. In this system, segmentation is done by clustering. Su et al. [35] proposed a technique to count blast, erythroid, lymphocyte, monocyte, myeloid, and neutrophil cells in separate in addition to the overall cell count. Using these details, AML is detected by segmenting the cells using k-means clustering. Laosai and Chamnongthai [36] developed a system to segment and classify lymphoid and myeloid blood cell. Segmentation of nucleus is done by k-means clustering and SVM is used for classification. The accuracy offered by this system is 92%. In [37], a framework was proposed by Kumar et al. with the purpose of identifying and classifying cancer using biopsy microscopic images. It involves a number of steps including k-means segmentation.

Neoh et al. [38] introduced a decision support system to identify ALL by using a clustering algorithm in conjunction with Simulating Discriminant Measures (SDMs) in segmenting the nucleus and cytoplasm of both lymphocytes and lymphoblasts. The system has used classifiers such as Multi-Level Perceptron (MLP), SVM, and Dempster–Shafer. Among these, Dempster–Shafer brought out the accuracy rate of 96.72%. Agaian et al. [39] presented a technique to make known the ALL by segmenting the nucleus with the application of k-means clustering as a segmentation methodology. This application showed the accuracy rate of 98%. Moradi Amin et al. [40] give away a scheme of revealing ALL with the help of k-means clustering for segmentation and SVM for classification. Hazra et al. [41] achieved 90% of accuracy in segmentation with the application of k-means clustering. Indira et al. [42] exercised k-means clustering algorithm for the segmentation of WBCs' nucleus to ascertain AML. SVM classifies the cell with the assistance of features gathered by LBP and

LTP. Madhukar et al. [43] proposed a method to find AML; in that, segmentation is done with k-means clustering and classification by SVM with the accuracy rate of 93.5%. Soltanzadeh et al. [44] used k-means segmentation for separating the nuclei of leukocytes and curvelet transforms to enhance the extracted region. This method provides 80.2 and 84.3% of specificity and sensitivity.

Sajjad et al. [45] and Selvaraj and Kanakaraj [46] partitioned nucleus of WBC using k-means clustering algorithm. Sivakumar and Ramesh [47] used the k-means clustering along with morphological actions to segment the leukocytes. Kazemi et al. [48] proposed a method to reveal AML by applying k-means technique of segmentation of WBCs' nucleus and SVM for classification of subcategories of AML. Sharma and Kinra [49] examined k-means method and histogram equalization to extract and count the healthy leukocyte and malignant leukoblast cells for the early detection of leukemia. Harun et al. [50] combinedly used k-means and fuzzy c-means along with moving k-means clustering algorithms in segmenting the injured blast cells. Among the three methods, MKM produces the best segmentation sensitivity rate of 86.18%.

4.1.4 Segmentation by Morphological Operations

Piuri with Scotti [51] both worked on microscopic color images for automatic detection and categorization of leukocytes. In their work, leukocytes separated from all the other cells of blood using morphological operations, and the subtypes of WBC such as basophile, eosinophil, neutrophil, lymphocyte, and monocyte are classified using neural classifier. Theera-Umpon along with Dhompongsa [52] examined that the nucleus of WBC is with mathematical morphology. Bayes' classifiers and neural networks are functioned as classifiers in fivefold, which achieved 77% of classification rate.

Vogado et al. [53] proposed a leukemia segmentation technique using morphological processes. Results of experiments have figured out the kappa index of 0.9306, 0.8603, and 0.9119 with the databases ALL-IDB 2, BloodSeg, and leukocyte, correspondingly. In [54], Bouzid-Daho et al. introduced a system; in that, the abnormal cells are detected by applying mathematical morphological operators for segmentation. Bhattacharjee [55] with Saini developed a GUI using the morphological operators for partitioning nucleus through segmentation. The classification was implemented through K-nearest neighbor, k-means clustering, Artificial Neural Networks (ANNs), and SVM. In [56], Scotti proposed a system that works on peripheral blood microscopic images. Here, the leucocytes are separated from other blood cells, lymphocyte cells are extracted, and morphological indexes are evaluated from them; finally, the existence of leukemia is classified. Vishwanathan [57] presented a new approach for detecting leukemia by fuzzy C-means cluster optimization based on morphological contour segmentation.

Gumble and Rode [58] have investigated DIP techniques such as median filtering, Sobel edge detector, morphological erosion and dilation for segmentation of nuclei and cytoplasm of WBC, morphological features extraction, and KNN classifier for classifying normal and malignant blood cells. Polyakov and Nikitaev proposed [59] a morphological feature extraction method to differentiate blast and normal lymphocytes. The features are measured using statistical measures like standard deviation and mean. Grimwade et al. reported [60] the application of image flow cytometry with microscopic blood smear images to detect acute leukemia. In this streamline, the components of cell are identified by morphology. Singhal along with Singh [61] distinguishes normal leukocyte cells and leukoblast cells by applying morphological dilation. Then Local Binary Pattern (LBP) features of texture are acquired. It brought out fine accuracy while classification.

Warude and Singh [62] developed a hybrid approach of segmenting nuclei of WBC including the fuzzy c-means clustering, morphological operations, and gradient-based watershed algorithm. Bhukya et al. [63] used Otsu's thresholding for binarization and the morphological operations to segment the nucleus of leukocyte cells in detecting ALL. Then, SVM is utilized for discriminating fit and injurious cells with the accuracy of 92.7% by using the shape features. Vaghela et al. [64] applied different segmentation methods like watershed transformation, k-means clustering, histogram equalization, and morphological operations to reveal RBCs and immature WBCs. Amidst of these, morphological operations outperform with 97.8% of accuracy.

Nee et al. [65] combinedly used thresholding, morphological operations, and watershed algorithm for segmenting the leukocyte and leukoblast cells. This is helpful to categorize various types of acute leukemia, like M2, M5, and M6 with an accurateness level of 94.5%. Dorini et al. [66] adopted morphological operations powered with scale-space features for separating leukocytes with assured results.

4.1.5 Wavelet Transform-Based Segmentation

Sarode et al. [67] introduced an algorithm to detect leukemia using digital cosine transform that outperforms the previous image segmentation algorithm in terms of accuracy by 8.5%. In Sarode et al. [68] proposal, Haar transform is used to identify leukemia. It achieves 13.24% of segmentation accuracy and 23.32% of precision than the former methods. Lakshmikanth and Abdul Khayum [69] developed an AML detection system by taken up Haar wavelet transform and Daubechies wavelet transform techniques to partition the cells. Then, ANN is used as classifier. Mazalan et al. [70] defined an RBC counting system using Circular Hough Transform (CHT) with the accuracy rate of 91.87%.

4.1.6 Active Contour-Based Segmentation

Gim et al. [71] offered a method of segmenting the nucleus and cytoplasm of WBC based on the shapes by the way of mean shift clustering and active contour-based gradient-vector flow snake method, respectively. Mathur et al. [72] used active contour for segmenting nucleus and cytoplasm in reporting the WBC subtypes. The images are processed in the color space called Hue saturation value to fix the seed points.

4.1.7   Region Growing Based Segmentation

Madhloom et al. [73] introduced a scheme of segmentation with the intent to detect ALL, which comprised of mathematical morphology, watershed method, and seeded region growing influenced by histogram equalization for segmenting the leukemic cells, separating the overlying cells and partitioning the cytoplasm as well as nucleus portions correspondingly. This scheme is 96% accurate in parting the leukemic cells and 94% accurate in cytoplasm and nuclei segmentation. Gomez et al. [74] proposed a novel instance-based seeded region growing algorithm. It has been compared with the seeded region growing, and auto-threshold algorithms. Among these, the projected method has shown better results. Abd Halim et al. [75] implemented a method to discover and to discriminate AML and ALL cells. For that, the acquired RGB figure is transformed into HSI color space, s-component is selected to attain the threshold value, and then the region growing is applied to segment the regions.

4.1.8   Segmentation Based on Other Mathematical and Statistical Methods

Biji and Hariharan [76] worked on the research identification of WBC using heuristic-based Electromagnetism like Optimization (EMO) method acting as a circle detecting operator. In [77], Gibson et al. derived a sample size formula to estimate the dissimilarities in segmentation accuracy of two algorithms. Raje and Rangole [78] proposed a segmentation scheme in which statistical values like mean and standard deviation are used to partition WBC from erythrocytes and platelets. Geometrical features of WBC are used in classifying regular and blast cells. Nikitaev et al. [79] examined the color components based on texture features of the nucleus of WBCs present in bone marrow images for separating the blast cells and lymphocytes. A mathematical procedure is used to show that component (V) color difference signal is most important in YUV color model.

Reta et al. [80] introduced a contemporary application in classifying the five different types of ALL cells. In this, the Markov random field algorithm is applied to get the color and texture information for acquiring the cytoplasm and nucleus regions of cells. It gains the accuracy of 90%. Mohammed et al. [81] introduced a segmentation technique for separating the nucleus by converting RGB color space into the color space C-Y and using the brightness element with 98.38% accuracy in segmentation.

4.2   Machine Learning

Leukemic malignancies are classified from healthy leukocytes with the aid of machine learning approaches. Decision trees, neural networks, support vector machines, and K-nearest neighboring methods are playing the significant role of classification in machine learning realm.

4.2.1 Classification Based on Decision Trees

In [82], Serbouti et al. proposed a work to analyze blood malignancies and used Classification And Regression Trees (CART) tool for classification. It gives a fine differentiation between ALL with Acute Undifferentiated Leukemia (AUL). Ko et al. [83] proved that nucleus is enough to classify WBC and they also proved that random forest is a sensible classifier to classify the types of WBC. This is another variation of decision tree kind of machine learning.

4.2.2 Classification with Neural Networks and Fuzzy Logic

In [84], a blood cell diagnosing and classification system was proposed by Kim et al. for investigating and diagnosing blood cells. In their proposal, a neural network-based model is utilized for proficient classification of RBCs and WBCs. Jati et al. [85] proposed an involuntary partitioning method to separate the nucleus of leukocytes on regular and noisiest environmental images. This task is executed with the application of exponential intuitionistic fuzzy deviation method. This method gives 98.52% of segmentation accuracy in the regular noiseless environment and 93.90 and 94.93% accuracy rates for Speckle and Gaussian noises, respectively. The region that comes under Operator Characteristic Curve (ROC) is about 0.9514. Asl and Zarandi [86] introduced a type-2 fuzzy expert system to diagnose leukemia using the Mamdani-style inference that provides 94% of classification accuracy.

Fatma and Sharma [87] designed for the cause of search out and classifying leukemia through extracting the features from blood smear images. Taken features are given as input to the neural (semantic networks) network in sorting out the leukemia categories. Colantonio et al. proposal is [88] segmentation of nuclei of leukocytes in finding the malignancy through the application of the fuzzy c-means method in the first stage and ANN in the second stage of uncovering subcomponents of blood cells. Singh et al. [89] introduced a procedure to detect leukemia that uses SIFT method to take out the features; extracted features are modified to get maximum efficiency with Bacterial Foraging Optimization (BFO); and the neural network is employed in classifying the feature with the accuracy of 97.22%. Garg and Kaur [90] exercised k-means technique for segmentation and backpropagation neural networks for classifying leukemic and non-leukemic cells.

4.2.3 Classification with Support Vector Machines

Ramose et al. [91] presented a method to segment and classify leukocytes influenced by cell appearance and image quality. In this proposal, pair-wise Support Vector

Machines (SVMs) classification is used to distinguish the cells of different types. Tai et al. [92] introduced a segmentation and classification scheme to distinguish various categories of blood cells. Classification is accomplished hierarchically started from the geometric features of nucleus and cytoplasm through multi-class SVM.

The procedure proposed by Rawat et al. [93] distinguishes ALL cells and healthy WBCs using shape features and GLCM. Auto SVM classifies the extracted features. It shows the accuracy in classification as 86.7% for the nucleus, 72.4% for cytoplasm, and together it shows the accuracy of 89.8%. Their research reports that in differentiating normally matured and immature lymphocytes, the nucleus shape is more crucial than cytoplasms. Mohapatra and Patra [94] have used the features Hausdorff dimension and contour signature to make out the shape irregularities in the boundary of leukocytes nuclei. SVM is used as the classifier for classifying the leukemic cells. Bigorra et al. [95] presented a technique of distinguishing Reactive Lymphoblastic Cells (RFCs), lymphoblastic, and myeloid blast cells from blood cell images with SVM. This method provides the classification accuracy of 90.1% with 220 test images.

Patil and Raskar [96] suggested a scheme to detect blast cells by means of segmenting the nuclei using Otsu's threshold, and SVM is applied for classification. Moradi Amin et al. [97] proposed a system to categorize the four types of leukemia by SVM. Meera and Mathew [98] have segmented the nuclei region of WBC using fuzzy local information-based c-means clustering, features extracted using GLCM combined with fractal dimensions, and then classified with SVM. Pan et al. [99] applied mean shift scheme and learning by training kind of trained SVM to partition the nucleus of WBCs. Mohapatra et al. [100] proposed a color-based clustering scheme of segmentation that includes k-medoid, k-means, Fuzzy Possibilistic C-Means (FPCM) along with Gustafson Kessel to separate nuclei of WBCs in perceiving ALL cells and after that SVM is employed as a classifier. Ravikumar and Shanmugam [101] proposed an algorithm named relevance vector machine a change of SVM for classifying the leukocytes and it is efficient of 91%. James and Nair [102] have done segmentation of WBC in revealing AML using k-means algorithm. In this work, healthy and cancerous cells are classified with SVM.

4.2.4 *K*-Nearest Neighbor Classification

Asadi et al. [103] offered a holographic system to identify and classify an unknown leukemia cell. Feature extraction has been done with Zernike moments. To classify leukemia, KNN and Minimum Mean Distance (MMD) methods are applied. Joshi et al. [104] proposed an automatic segmentation combined with KNN classifier to classify normal lymphocytes and lymphoblast cells with the classification accuracy rate of 93%. Abdeldaim et al. [105] investigated a method to identify ALL by converting the RGB image to CMYK color space, threshold by Zack technique, classification is done using different classifiers, and finally KNN classifier achieved good classification accuracy. Di Ruberto et al. [106] presented a new technique combining nearest neighbor along with SVM classifiers in segmenting various components of blood cytoplasm and nucleus of WBC, RBCs, and background. This scheme ensures accuracy in the segmentation about 99% with the dataset ALL-IDB.

5 Research Issues

To substantiate the researchers of medical image processing, several standard datasets of images of different types of diseases are made freely available. In these years, lots of efforts have been put on analyzing and finding new methodologies in recognition and classification of leukemia by using the images in benchmark leukemia datasets such as ALL-IDB 1, ALL-IDB 2, HistologyDS2828, LYMPH, BloodSeg, leukocytes, LISC, and MLL_leukemia. Henceforth, researchers have to explore new datasets of images to show accuracies. All the investigations have not been made on the same dataset. Resultantly, comparisons about accuracy and efficiency of algorithms cannot be made easily. Though the developed methods work well, eventually it is necessary to get substantiation from a hematologist or pathologist. For the cause of saving lives of human, the algorithms in medical image processing must be designed with care and tested on datasets of a large number of images. In view of the fact that medical images are not in easy access, complications are faced in experimenting and showing accuracy with a huge real-time dataset.

6 Conclusion

Research activities accomplished in medical image processing make an effort to reduce the percentage of mortality among living beings. With this insight, an analysis and study over the area of diagnosis of blood cells and leukemia detection are made by referring the research works from the year 1991 to till date. It is observed that by applying multifarious combination of image processing and machine learning techniques on microscopic blood smear images, the leukemia detection and classification can be done through segmenting the cytoplasm and the nuclei of WBC. Some of the research activities have been done on classifying the four major subtypes of leukemia. Derived methods are evaluated and verified by means of extracting the texture, contour, and color features. Though sufficient benchmark datasets of microscopic images of blood cells exist, all the studies have not been undergone with the standard identical datasets of same size and resolution. As a result, the proposed methods cannot be easily compared with each other by means of accuracy and efficiency. Therefore, it is obvious that improvements could be shown by the way of finding out innovative segmentation and classification methodologies with new datasets, and it is ought to focus on categorizing the types of leukemia. In our future work, better segmentation algorithm is going to be proposed with improved accuracy in classification. The majority of the research applications have been implemented with the tools MATLAB, Python, and LabVIEW.

References

1. Cseke I (1992) A fast segmentation scheme for white blood cell images. IEEE conference
2. Liao Q, Deng Y (2002) An accurate segmentation method for white blood cell images. IEEE
3. Prasad B, Choi J-SI, Badawy W (2006) A high throughput screening algorithm for leukemia cells. In: IEEE Canadian conference on electrical and computer engineering, Ottawa, pp 2094–2097
4. Prasad B, Badawy W (2007) High throughput algorithm for leukemia cell population statistics on a hemocytometer. In: IEEE biomedical circuits and systems conference, pp 27–30
5. Biswas S, Ghoshal D (2016) Blood cell detection using thresholding estimation based watershed transformation with Sobel filter in frequency domain. In: Twelfth international multi-conference on information processing (IMCIP-2016) (Elsevier)
6. Kit CY, Tomari R, Nurshazwani W, Zakaria W, Othman N, Safuan SNM, Yi JAJ, Sheng NTC (2017) Mobile based automated complete blood count (Auto-CBC) analysis system from blood smeared image. Int J Electr Comput Eng (IJECE) 7(6):3020–3029. ISSN 2088-8708
7. Scotti F (2006) Robust segmentation and measurements techniques of white cells in blood microscope images. In: Proceedings of the IEEE instrumentation and measurement technology conference, pp 43–48
8. Rezatofighi SH, Zoroofi RA, Sharifian R, Soltanian-Zadeh H (2008) Segmentation of nucleus and cytoplasm of white blood cells using gram-schmid orthogonalization of deformable models. In: 9th international conference on signal processing, IEEE Explore
9. Bergen T, Steckhan D, Wittenberg T, Zerfass T (2008) Segmentation of leukocytes and erythrocytes in blood smear images. In: 30th annual international IEEE EMBS conference Vancouver, British Columbia, Canada
10. Rezatofighi SH, Soltanian-Zadeh H, Sharifian R, Zoroofi RA (2009) A new approach to white blood cell nucleus segmentation based on gram-schmidt orthogonalization. In: International conference on digital image processing, IEEE
11. Sadeghian F, Seman Z, Ramli AR, Kahar BHA, Saripan M-I (2009) A framework for white blood cell segmentation in microscopic blood images using digital image processing. Biological Procedures Online, Springer, New York, pp 196–206
12. Nor Hazlyna H, Mashor MY, Mokhtar NR, Aimi Salihah AN, Hassan R, Raof RAA, Osman MK (2010) Comparison of acute leukemia image segmentation using HSI and RGB colorspace. In: 10th international conference on information sciences signal processing and their applications (ISSPA), pp 749–752
13. Mohamed M, Far M (2012) An enhanced threshold based technique for white blood cells nuclei automatic segmentation. In: 14th international conference on e-health networking, applications and service
14. Di Ruberto C, Putzu L (2014) Accurate blood cells segmentation. In: 2014 tenth international conference on signal-image technology and internet-based systems
15. Ahasan R, Ratul AU, Bakibillah ASM (2016) White blood cells nucleus segmentation from microscopic images of strained peripheral blood film during leukemia and normal condition. In: 5th international conference on informatics, electronics and vision (ICIEV)
16. Lina, Chris A, Mulyavan B, Dharmawan AB (2016) Leukocyte detection using image stitching and color overlapping windows. Int J Comput Electr Autom Control Inf Eng 10(5) (World Academy of Science, Engineering and Technology)
17. Ananthi VP, Balasubramaniam P (2016) A new thresholding technique based on fuzzy set as an application to leukocyte nucleus segmentation. Comput Methods Programs Biomed
18. Sandhu RK (2017) Comparative study of various techniques for leukemia detection. Int J Comput Sci Eng (IJCSE) 9(5). ISSN 0975-3397
19. Putzu L, Di Ruberto C (2013) White blood cells identification and counting from microscopic blood image. Int J Med Health Biomed Bioeng Pharm Eng 7(1) (World Academy of Science, Engineering and Technology)
20. Gosh M, Das D, Chakraborty C, Ray AK (2010) Automated leukocyte recognition using fuzzy divergence. Micron 41:840–846 (Elsevier Ltd.)

21. Khobragade S, Mor DD, Patil CY (2015) Detection of leukemia in microscopic white blood cell images. Int Conf Image Process (ICIP)
22. Abbas N, Mohamad D (2014) Automatic color nuclei segmentation of leukocytes for acute leukemia. Res J Appl Sci Eng Technol 7(14):2987–2993
23. Khashman A, Al-Zgoul E (2010) Image segmentation of blood cells in leukemia patients. Recent Adv Comput Eng Appl 104–109 (ACM Digital Library)
24. Abbas N, Mohamad D, Abdullah AH, Saba T, Al-Rodhaan M, Al-Dhelaan A (2015) Nuclei segmentation of leukocytes in blood smear digital images. Pakistan J Pharm Sci 28(5):1801–1806
25. Li Y, Zhu R, Mi L, Cao Y, Yao D (2016) Segmentation of white blood cell from acute lymphoblastic leukemia images using dual-threshold method. Comput Math Methods Med 2016 (Hindawi)
26. Nilsson B, Heyden A (2001) Segmentation of dense leukocyte clusters. In: IEEE workshop on mathematical methods in biomedical image analysis (MMBIA01), pp 221–227
27. Jiang K, Liao Q-M, Dai Mach S-Y (2003) A novel white blood cell segmentation using scale-space filtering and watershed clustering (vol 5, pp 2820–2825). In: Proceedings of the second international conference on machine learning and cybernetics, Xi'an, IEEE, pp 2–5
28. Shankar V, Deshpande MM, Chaitra N, Aditi S (2016) Automatic detection of acute lymphoblastic leukemia using image processing. In: IEEE international conference on advances in computer applications
29. Ghane N, Vard A, Talebi A, Nematollahy P (2017) Segmentation of white blood cells from microscopic images using a novel combination of K-means clustering and modified watershed algorithm. J Med Signals Sensors 7(2):92–101
30. Belekar SJ, Chougule SR (2015) WBC segmentation using morphological operation and SMMT operator—a review. Int J Innov Res Comput Commun Eng 3(1). ISSN 2320-9801
31. Liu Z, Liu J, Xiao X, Yuan H, Li X, Chang J, Zheng C (2015) Segmentation of white blood cells through nucleus mark watershed operations and mean shift clustering. Sensors 15:22561–22586
32. Sinha N, Ramakrishnan AG (2003) Automation of differential blood count, IEEE
33. Basima CT, Panicker JR (2016) Enhance leukocyte classification for leukemia detection. IEEE
34. Foran DJ, Comaniciu D, Meer P, Goodell LA (2000) Computer-assisted discrimination among malignant lymphomas and leukemia using immunophenotyping. In: Intelligent image repositories and telemicroscopy. IEEE Transactions on Information Technology in Biomedicine
35. Su J, Liu S, Song J (2017) A segmentation method based on HMRF for the aided diagnosis of acute myeloid leukemia. Comput Methods Programs Biomed. https://doi.org/10.1016/j.cmpb.2017.09.011
36. Laosai J, Chamnongthai K (2014) Acute leukemia classification by using SVM and K-means clustering. In: Proceedings of the international electrical engineering congress, IEEE
37. Kumar R, Srivastava R, Srivastava S (2015) Detection and classification of cancer from microscopic biopsy images using clinically significant and biologically interpretable features. J Med Eng 2015 (Hindawi Publishing Corporation)
38. Neoh SC, Srisukkham W, Zhang L, Todryk S, Greystoke B, Lim CP, Hossain MA, Aslam N (2015) An intelligent decision support system for leukemia diagnosis using microscopic blood images. J Sci Reports
39. Agaian S, Madhukumar M, Chronopoulos AT (2014) Automated screening system for acute myelogenous leukemia detection in blood microscopic images. IEEE Syst J 8(3)
40. Moradi Amin M, Kermani S, Talebi A, Oghli MG (2015) Recognition of acute lymphoblastic leukemia cells in microscopic images using K-means clustering and support vector machines. J Med Signals Sensors 5(1):49–58
41. Hazra T, Kumar M, Tripathy SS (2017) Automatic leukemia detection using image processing technique. Int J Latest Technol Eng Manag Appl Sci (IJLTEMAS) 6(4). ISSN 2278-2540
42. Indira P, Ganesh Babu TR, Vidhya K (2016) Detection of leukemia in blood microscope images. Int J Control Theory Appl 9(5):2147–2151

43. Madhukar M, Agaian S, Chronopoulos AT (2012) Deterministic model for acute myeloge-
    nous leukemia classification. In: 2012 IEEE international conference on systems, man, and
    cybernetics, COEX, Seoul, Korea, 14–17 Oct 2012
44. Soltanzadeh R, Rabbani H, Talebi A (2012) Extraction of nucleolus candidate zone in white
    blood cells of peripheral blood smear images using curvelet transform. Comput Math Methods
    Med 2012
45. Sajjad M, Khan S, Jan Z, Muhammad K, Hyeonjoon Moon, Kwak JT, Rho S, Baik SW,
    Mehmood I (2017) Leukocytes classification and segmentation in microscopic blood smear:
    a resource-aware healthcare service in smart cities. In: Special section on advances of multi-
    sensory services and technologies for healthcare in smart cities, vol 5, IEEE
46. Selvaraj S, Kanakaraj B (2015) Naïve Bayesian classifier for acute lymphocytic leukemia
    detection. ARPN J Eng Appl Sci 10(16). ISSN 1819-6608
47. Sivakumar S, Ramesh S (0215) Automatic white blood cell segmentation using K-means
    clustering. Int J Sci Eng Res 3(4)
48. Kazemi F, Najafabadi TA, Araabi BN (2016) Automatic recognition of acute myelogenous
    leukemia in blood microscopic images using K-means clustering and support vector machines.
    J Med Signals Sens
49. Sharma N, Kinra N (2014) Detecting and counting the no of white blood cells in blood sample
    images by color based K-means clustering. Int J Electr Electron Eng 1(3). ISSN 1694–2310
50. Harun NH, Absdul Nasir AS, Mashor MY, Hassan R (2015) Unsupervised segmentation
    technique for acute leukemia cells using clustering algorithms. Int J Comput Inf Eng 9(1)
    (World Academy of Science, Engineering and Technology)
51. Piuri V, Scotti F (2004) Morphological classification of blood leukocytes by microscopic
    images. In: IEEE international conference on computational intelligence for measurement
    systems and applications
52. Theera-Umpon N, Dhompongsa S (2007) Morphological granulometric features of nucleus
    in automatic bone marrow white blood cell classification. IEEE Trans Inf Technol Biomed
    11(3):353–359
53. Vogado LHS, Rodrigo de M. S. Veras, Andrade AR, Romuere R. V. e Silva, Flavio H. D.
    de Araujo, de Medeiros FNS (2016) Unsupervised leukemia cells segmentation based on
    multi-space color channels. In: IEEE international symposium on multimedia
54. Bouzid-Daho A, Boughazi M, Tanouast C (2017) Algorithmic processing to aid in leukemia
    detection. Med Technol J 1(1):10–11
55. Bhattacharjee R, Saini LM (2015) Robust technique for the detection of acute lymphoblastic
    leukemia. In: Communication and information technology conference (PCITC) Siksha 'O'
    Anusandhan University, Bhubaneswar, India, IEEE Power
56. Scotti F (2005) Automatic morphological analysis for acute leukemia identification in periph-
    eral blood microscopic images. In: IEEE international conference on computational Intelli-
    gence for measurement systems and Applications
57. Vishwanathan P (2015) Fuzzy C means detection of leukemia based on morphological contour
    segmentation. Procedia Comput Sci 58:84–90 (Elsevier)
58. Gumble P, Rode SV (2017) Study and analysis of acute lymphoblastic leukemia blood
    cells using image processing. Int J Innov Res Comput Commun Eng 5(1). ISSN (online):
    2320–9801
59. Polyakov EV, Nikitaev VG (2017) A method for estimating the accuracy of measurements of
    optical characteristics of the nuclei of blood cells in the diagnosis of acute leukemia. J Phys
    Conf Ser 784:012042
60. Grimwade LF, Fuller KA, Erber WN (2016) Applications of imaging flow cytometry in the
    diagnostic assessment of acute leukemia. Elsevier
61. Singhal V, Singh P (2014) Local binary pattern for automatic detection of acute lymphoblastic
    leukemia. IEEE
62. Warude D, Singh R (2016) Automatic detection method of leukemia by using segmentation
    method. Int J Adv Res Comput Commun Eng 5(3)

63. Bhukya R, Prasanth B, Sasank Vihari V, Ajay Y (2017) Detection of acute lymphoblastic using microscopic images of blood. Int J Adv Appl Sci 4(8):74–78
64. Vaghela HP, Modi H, Pandya M, Potdar MB (2015) Leukemia detection using digital image processing techniques. Int J Appl Inf Syst (IJAIS) 10(1). ISSN: 2249-0868
65. Nee LH, Mashor MY, Hassan R (2012) White blood cell segmentation for acute leukemia bone marrow images, In: 2012 international conference on biomedical engineering (ICoBE), IEEE
66. Dorini LB, Minetto R, Leite NJ (2007) White blood cell segmentation using morphological operators and scale-space analysis. Comput Graph Image Process. ISSN 1530-1834 (IEEEXplore)
67. Sarode TK, Thakkar BK, Purandare SJ, Gupta VM (2016) Cancerous cell detection in bone marrow smear. Int J Comput Appl
68. Sarode TK, Thakkar B, Purandare SJ, Gupta VM (2016) Cancerous cell detection in bone marrow smear using Haar transform. In: International conference and workshop on electronics & telecommunication engineering (ICWET 2016), IEEE
69. Lakshmikanth BK, Abdul khayum P (2017) Acute myelogenous leukemia detection in blood microscopic images using different wavelet family techniques. Int J Eng Manag Res 7(4):174–182
70. Mazalan SM, Mahmood NH, Razak MAA (2013) Automated red blood cells counting in peripheral blood smear image using circular hough transform. In: 2013 first international conference on artificial intelligence, modeling & simulation, IEEEXplore
71. Gim J-W, Park J, Lee J-H, Ko BC, Nam J-Y (2011) A novel framework for white blood cell segmentation based on stepwise rules and morphological features. In: Proceedings of SPIE-IS&T 7877, image processing machine vision applications, San Francisco, pp 1–6
72. Mathur A, Tripathi AS, Kuse M (2012) Scalable system for classification of white blood cells from leishman stained blood stain images. J Pathol Inform (HIMA workshop at MICCAI, Nice, France)
73. Madhloom HT, Kareem SA, Ariffin H (2015) Computer-aided acute leukemia blast cells segmentation in peripheral blood images. J VibroEng 17(8):4517–4532
74. Gomez O, Gonzalez JA, Morales EF (2007) Image segmentation using automatic seeded region growing and instance-based learning. In: Progress in pattern recognition, image analysis and applications, CIARP 2007, Springer, Berlin, Heidelberg
75. Abd Halim NH, Mashor MY, Abdul Nasir AS, Mokhtar NR, Rosline H (2011) Nucleus segmentation technique for acute leukemia. In: 2011 IEEE 7th international colloquium on signal processing and its applications
76. Biji G, Hariharan S (2017) An efficient peripheral blood smear image analysis technique for leukemia detection. In: International conference on I-SMAC (IOT in Social, Mobile, Analytics, and Cloud), I-SMAC, IEEE
77. Gibson E, Hu Y, Huisman HJ, Barratt DC (2017) Designing image segmentation studies: statistical power, sample size, and reference standard quality. Med Image Anal 42:44–59 (Elsevier)
78. Raje C, Rangole J (2014) Detection of leukemia in microscopic images using image processing. In: International conference on communication and signal processing
79. Nikitaev VG, Pronichev AN, Polyakov EV, Dmitrieva VV, Tupitsyn NN, Frenkel MA, Mozhenkoa AV (2017) The influence of physical factors on recognizing blood cells in the computer microscopy systems of acute leukemia diagnosis. J Phys Conf Series 798:012128
80. Reta C, Altamirano L, Gonzalez JA, Diaz-Hernandez R, Peregrina H, Olmos I, Alonso JE, Lobato R (2015) Segmentation and classification of bone marrow cells images using contextual information for medical diagnosis of acute leukemias. PLOS One 10(6)
81. Mohammed R, Nomir O, Khalifa I (2014) Segmentation of acute lymphoblastic leukemia using C-Y color space. Int J Adv Comput Sci Appl (IJACSA) 5(11)
82. Serbouti S, Duhamel A, Harms H, Gunzer U, Aus HM, Mary JY, Beuscart R (1991) Image segmentation and classification methods to detect leukemias. In: Annual international conference of IEEE engineering in medicine and biology society

83. Ko BC, Gim JW, Nam JY (2011) Cell image classification based on ensemble features and random forest. Electron Lett 47(11):638–639
84. Kim KS, Kim PK, Song JJ, Park YC (2002) Analyzing blood cell images do distinguish its abnormalities. In: ACM international conference on multimedia, Los Angeles, CA, USA, pp 395–397
85. Jati A, Singh G, Mukherjee R, Gosh M, Konar A, Chakraborty C, Nagar AK (2014) Automatic leukocyte nucleus segmentation by intuitionistic fuzzy divergence based thresholding. Micron 58:55–65 (Elsevier Ltd.)
86. Asl AAS, Zarandi MHF (2018) A type-2 fuzzy expert system for diagnosis of leukemia. In: Fuzzy logic in intelligent system design, advances in intelligent systems and computing, vol 648. Springer International Journal AG (in press)
87. Fatma M, Sharma J (2014) Identification and classification of acute leukemia using neural networks. In: 2014 international conference on medical imaging, m-health and emerging communication systems (MedCom), IEEE
88. Colantonio S, Gurevich IB, Salvetti O (2008) Automatic fuzzy-neural based segmentation of microscopic cell images. Int J Signal Image Syst Eng 1(1)
89. Singh G, Bathla G, Kaur S: (2016) Design of new architecture to detect leukemia cancer from medical images. Int J Appl Eng Res 11(10):7087–7094. ISSN 0973-4562 (Research India Publications)
90. Garg J, Kaur D (2016) Automated blood cancer detection (leukemia) using artificial intelligence by ACO algorithm with BPNN classifier under soft computing 4(3). ISSN 2321-2632
91. Ramoser H, Lauria V, Bischof H, Ecker R (2008) Leukocyte segmentation and SVM classification in blood smear images. Graph Vis 17(1):187–200
92. Tai W-L, Hu R-M, Hsiao HCW, Chen R-M, Tsai JJP (2011) Blood cell image classification based on hierarchical SVM. Department of Biomedical Informatics, Sia University, Taiwan, IEEE
93. Rawat J, Singh A, Bhadauria HS, Virmani J (2015) Computer aided diagnostic system for detection of leukemia using microscopic images. Procedia Comput Sci 70:748–756 (Elsevier)
94. Mohapatra S, Patra D (2010) Automatic cell nucleus segmentation and acute leukemia detection in blood microscopic images. In: International conference on systems in medicine and biology, IIT Kharagpur India
95. Bigorra L, Merino A, Alferez S, Rodellar J (2016) Feature analysis and automatic identification of leukemic lineage blast cells and reactive lymphoid cells from peripheral blood cell images. J Clin Lab Anal 00:1–9
96. Patil TG, Raskar VB (2015) Automated leukemia detection by using contour signature method. Int J Adv Found Res Comput (IJAFRC) 2(6). ISSN 2348-4853
97. MoradiAmin M, Memari A, Samadzadehaghdam N, Kermani S, Talebi A (2016) Computer aided detection and classification of acute lymphoblastic leukemia cell subtypes based on microscopic image analysis. Microsc Res Techn 908–916 (Wiley Periodicals, Inc.)
98. Meera V, Mathew SA (2014) Fuzzy local information C means clustering for acute myelogenous leukemia image segmentation. Int J Innov Res Sci, Eng Technol 3(5). ISSN 2319-8753
99. Pan C, Lu H, Cao F (2009) Segmentation of blood and bone marrow cell images via learning by sampling. In: Emerging intelligent computing technology and applications, ICIC 2009, vol 5754. Springer, pp 336–345
100. Mohapatra S, Patra D, Satpathy S (2012) Unsupervised blood microscopic image segmentation and leukemia detection using color based clustering. Int J Comput Inf Syst Ind Manag Appl 4:477-485. ISSN 2150-7988
101. Ravikumar S, Shanmugam A (2014) WBC image segmentation and classification using RVM. Appli Math Sci 8(45):2227–2237 (HIKARI Ltd.)
102. James J, Nair KN (2013) Automated acute myelogenous leukemia detection in blood microscopic image. Int J Sci Res (IJSR). ISSN 2319-7064
103. Asadi MR, Vahedi A, Amindavar H (2006) Leukemia cell recognition with zernike moments of holographic images. In: IEEE proceedings of 7th nordic signal processing symposium, Iceland, pp 214–217

104. Joshi MD, Karode AH, Suralkar SR (2013) White blood cells segmentation and classification to detect acute leukemia. Int J Emerg Trends Technol Comput Sci
105. Abdeldaim AM, Sahlol AT, Elhoseny M, Hassanien AE (2018) Computer-aided acute lymphoblastic leukemia diagnosis system based on image analysis. In: Advances in soft computing and machine learning in image processing. Springer International Publishing AG (in press)
106. Di Ruberto C, Loddo A, Putzu L (2015) A multiple classifier learning by sampling system for white blood cells segmentation. In: International conference on computer analysis of images and patterns, CAIP 2015, vol 9257. Springer, pp 415–425

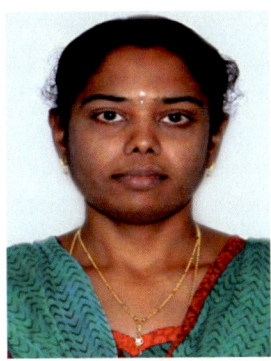

D. Umamaheswari has completed her M.Sc. and M.Phil. degrees in Computer Science and published five papers in international journals. She has 12 years of teaching experience and qualified in State Level Eligibility Test (SLET). At present, she is working in Vidyasagar College of Arts and Science, Udumalepet, Tamilnadu. Currently, she is pursuing Ph.D. (Part-time) in Computer Science, Government Arts College, Udumalpet, Tamilnadu, India. Her area of interest is image processing.

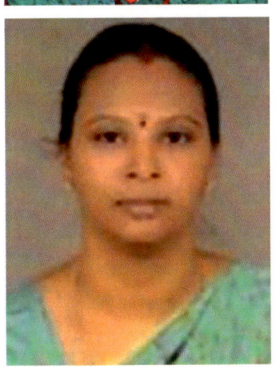

Dr. S. Geetha has completed her M.Sc., M.Phil., and Ph.D. degrees in Computer Science. She has published various papers in international journals of repute. Her area of specialization is soft computing. She has 20 years of experience in PSG College of Technology, Coimbatore, Tamilnadu. Presently, she is working at LRG Government Arts College for Women, Tiruppur, Tamilnadu, India.

Detection of Gaze Direction for Human–Computer Interaction

G. Merlin Sheeba and Abitha Memala

Abstract Eye guide is an assistive specialized apparatus intended for the incapacitated or physically disabled individuals who were not able to move parts of their body, especially people whose communications are limited only to eye movements. The prototype consists of a camera and a computer. The system recognizes gazes in four directions and performs required user actions in related directions. The detected eye direction can then be used to control the applications. The facial regions which form the images are extracted using the skin color model and connected-component analysis. When the eye regions are detected, the tracking is performed. The system models consist of image processing, face detector, face tracker, and eyeblink detection. The eye guide system potentially helps as a computer input control device for the disabled people with severe paralysis.

1 Introduction

Human–Computer Interaction (HCI) is the study of how people interact with computers and to what extent computer is not developed for successful interaction with human beings. The goals of HCI are to produce usable and safe systems, as well as functional systems. The human–computer interface can be portrayed as the purpose of correspondence between the human client and the Personal Computer (PC) [1–3]. The surge of information between the human and PC is described as the hover of correspondence. The circle of cooperation has a few angles to it, including the following:

- Visual Based: The visual-based human PC between activities likely the most across the board region in HCI look into.
- Audio Based: The sound-based collaboration between a PC and a human is another critical territory in HCI frameworks. This territory manages data gained by various sound signs.

G. M. Sheeba (✉) · A. Memala
Sathyabama Institute of Science and Technology, Chennai, TamilNadu, India
e-mail: merlinsheebu@gmail.com

© Springer Nature Switzerland AG 2019
D. Pandian et al. (eds.), *Proceedings of the International Conference on ISMAC in Computational Vision and Bio-Engineering 2018 (ISMAC-CVB)*, Lecture Notes in Computational Vision and Biomechanics 30,
https://doi.org/10.1007/978-3-030-00665-5_164

- Task Environment: The conditions and objectives set upon the client.
- Machine Environment: The environment that the PC is associated with.
- Output: The stream of data that begins in the machine environment.
- Feedback: Loops through the interfaces that assess, direct, and affirm forms as they go from the human through the interface to the PC and back.

2 System Description

2.1 System Model

The components of the system shown in Fig. 1 are explained as follows: The video input is converted into frames and the image is processed further.

2.1.1 Conversion of Grayscale

Each frame is processed into a grayscale image. The grayscale is a picture in which the estimation of pixel is a solitary example, i.e., it conveys just force data. The dark scale pictures are likewise called as highly contrasting pictures. Here, the weakest part will be in dark and most grounded part will be white as shown in Fig. 2.

2.1.2 Scanning

It is the process of translating the images or photographs into a digital form that can be recognized by a computer.

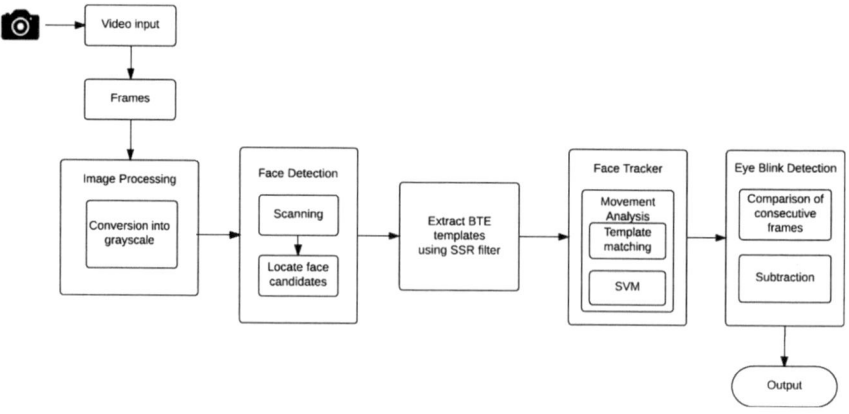

Fig. 1 The human–computer interaction system model

2.1.3 Six-Segmented Rectangular Filter (SSR)

A 2×3 rectangle is considered to examine on the selected frame. This rectangle is fragmented into six sections as demonstrated as follows.

The SSR channel is utilized to distinguish between the eyes in light of two qualities of face geometry.

- The nose zone (Sn) is brighter than the privilege and left eye region (Ser and Sel, separately) as appeared in Fig. 2b, where

$$Sn = Sb2 + Sb5 \tag{1}$$
$$Ser = Sb1 + Sb4 \tag{2}$$
$$Sel = Sb3 + Sb6 \tag{3}$$

At that point,

$$Sn > Ser \tag{4}$$
$$Sn > Sel \tag{5}$$

- The eye range (both eyes and eyebrows) (Se) is generally darker than the cheekbone region (counting nose) (Sc) as appeared in Fig. 2c, where

$$Se = Sb1 + Sb2 + Sb3 \tag{6}$$
$$Sc = Sb4 + Sb5 + Sb6 \tag{7}$$

At that point,

$$Se < Sc \tag{8}$$

Whenever expressions (4), (5), and (8) are altogether fulfilled, the focal point of the rectangle can be a contender for between the eyes.

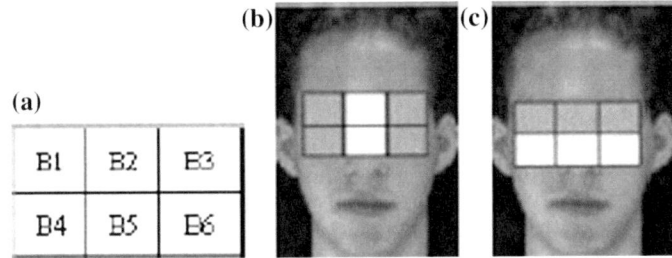

Fig. 2 SSR filter

Fig. 3 Determination of
BTE

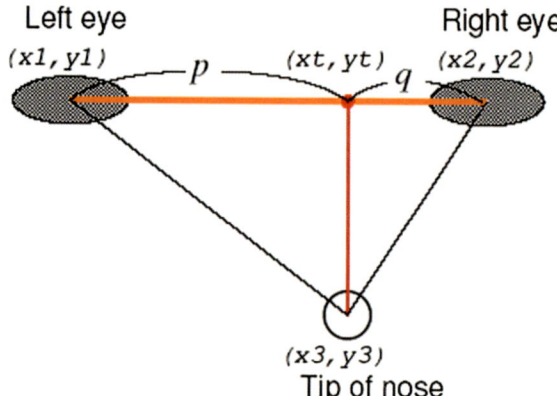

2.1.4   Tracking

The raised shaped elements in the facial features do not have unique property. Hence, using skin shading, outline subtraction, and learning of the component positions, the initial step is to ascertain the region for nearby pursuit of the nose highlight. The second step comprises checking the nearby hunt territory and discovering pixel "u". Before continuing to this progression, the picture is reprocessed with the Gaussian channel to smooth the deformities of pictures brought about by low nature of the cameras [4]. The last stride is to refine the position of the best match, utilizing the confirmation-based convolution channel. The acquired match of the raised shape highlight is both strong (to turn and scale) and exact (processed with sub-pixel exactness).

2.1.5   Between the Eye (BTE)

The identification of BTE depends on the property of the picture quality specifically the region on face. The BTE is the seat point on the hyperbolic surface [5]. A rotationally invariant channel could be formulated for identifying the BTE region as shown in Fig. 3.

The basic BTE region on human face looks like a hyperbolic surface. For example, the hyperbolic surface focal and seat points are indicated in the facial image in Fig. 4.

2.1.6   Template Matching

It is a method in computerized picture preparing for discovering little parts of a picture which coordinate a format picture. It can be utilized as a part of assembling as a piece of value control, an approach to explore and to distinguish the edge of image [6].

Fig. 4 Indication of the
focal and seat points in a
facial image

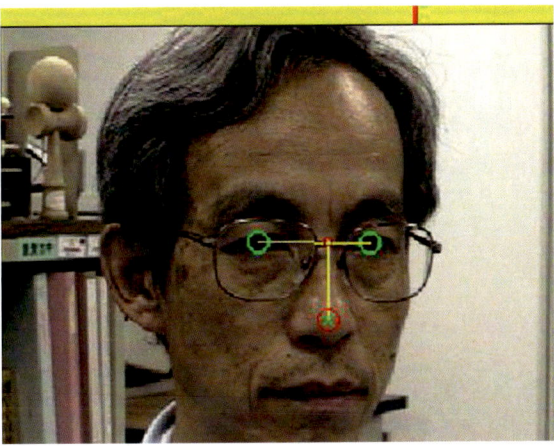

2.1.7   Support Vector Machine

Support vector machines (SVMs) are an arrangement of related regulated learning techniques utilized for order and relapse. SVM takes as information preparing
information tests, where each example comprises characteristics and a class mark
(positive or negative). The information tests that are nearest to the hyperplane are
called support vectors. The hyperplane is characterized by adjusting its separation
among positive and negative support vectors with a specific end goal to get the
maximal edge of the preparation informational index.

A support vector machine develops a hyperplane or set of hyperplanes in a high
or vast dimensional space, which can be utilized for grouping, relapse, or different
undertakings [7, 8]. To discover the biggest hyperplane it is required to isolate the
positive examples from the negative specimens, which are the two unique classes
of information. A decision rule is formulated as below to determine the classes of
positive and negative examples:

$$(w \cdot x_i + b) \geq 1, \quad \text{if } y_i = 1 \tag{9}$$

$$(w \cdot x_i + b) \leq -1, \quad \text{if } y_i = -1. \tag{10}$$

2.1.8   Detection of Eyeblink

The idea of the second request change discovery, which permits one to segregate
the neighborhood (latest) change in picture, for example, squint of the eyes, from
the worldwide (durable) change, for example, the movement of head, is presented.
This idea sets the base for planning complete face-worked control frameworks, in
which, utilizing the similarity with mouse, "indicating" is performed by nose and
"clicking" is performed by twofold squinting of the eyes. Utilizing similarity with

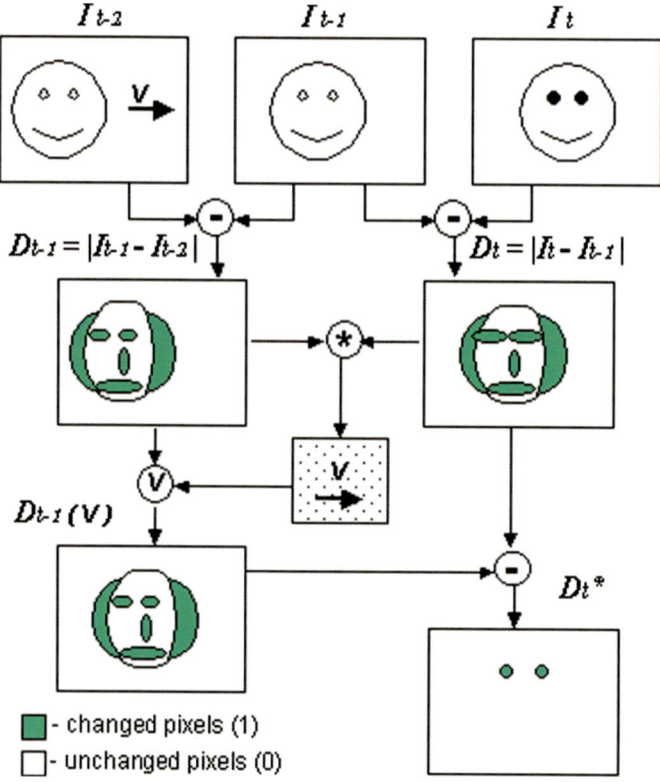

Fig. 5 A flowchart for detecting eyeblinks using second-order change detection

second request subordinates of capacities, the second request changes as a change of a change. The strategy utilized a part of numerical investigation for figuring the second request subordinate of a capacity in light of considering three neighboring focuses, and the second request change in video can be registered by considering three back-to-back edges of the video as shown in Fig. 5.

After comparing the pixels, the head movement has been shifted through the second request change called location methodology; the rest of the pixels are utilized to choose whether a squint has occurred. To begin with, the quantity of the staying changed pixels is analyzed and on the off chance that it is too little, implying that there is no movement, or in the event that it is too huge, implying that the foundation has changed, the eye location procedure is skipped. Something else, the vector quantization system is utilized keeping in mind the end goal to characterize splendid (changed) pixels into two classes, relating to the places of two potential "eyes" as shown in Fig. 6.

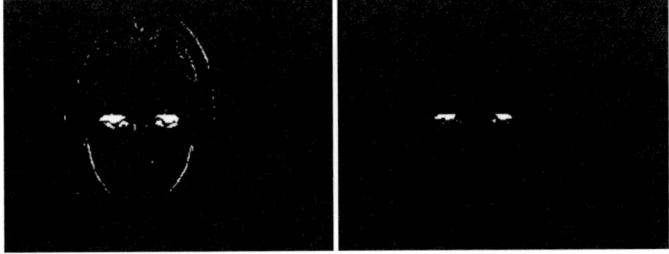

Fig. 6 Threshold difference image prior to erosion (left), and same image after erosion (right). Erosion removes noise caused by insignificant motion in the scene

2.1.9 Human Skin Model

A human skin shading model is utilized to choose if shading is a skin or non-skin shading. It is trusted that individuals from various races have same skin shading; yet, they contrast in lighting force of that shading keeping in mind the end goal to locate an appropriate skin shading model; as indicated by the already specified thought, we will utilize the immaculate r and g values which are the R and G estimations of the RGB shading model without shine and they are computed with the accompanying conditions:

$$r = R/(R + G + B) \tag{11}$$

$$g = G/(R + G + B) \tag{12}$$

We move the r and g qualities to the (a, b) shading space with the accompanying conditions:

$$a = r + g/2 \tag{13}$$

$$b = \sqrt{3}/2\,g \tag{14}$$

The scope of "an" is from 0 to 1, while the scope of "b" is from 0 to $\sqrt{3}/2$.

To digitize the (a, b) space, we round qualities to multiplicands of 0.01.

At last, to discover the skin shading model, we removed 735 skin pixel tests from each of 771 face pictures which were taken from, so the aggregate number of skin pixel tests was 566,685 examples. For each of the examples, we ascertained the "an" and "b" values, and expanded the counter of that esteem ("a" has 101 counters, "b" has 88 counters). The qualities that were most successive are considered as skin pixel values, so subsequent to plotting the "an" and "b" comes about we reasoned the accompanying limits for "an", and "b" to consider a pixel as a skin pixel:

$$0.49 < \text{"a"} < 0.59 \tag{15}$$

$$0.24 < \text{"a"} < 0.29 \tag{16}$$

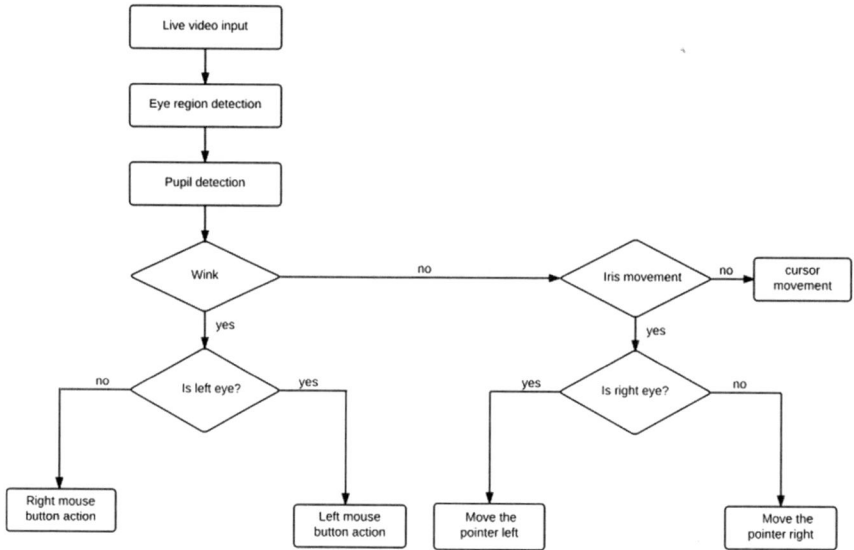

Fig. 7 Flowchart

2.1.10   Model Flow Diagram

The proposed framework utilizes a camcorder to track client's face position in 3D keeping in mind the end goal to change over it to a place of a cursor or another virtual protest in 2D screen [9]. The eyes and nose are found and after that, we begin to track them by breaking down the development of these elements in resulting outlines, utilizing some format coordinating, and heuristics. The framework utilizes the light appeared in the clients' eyes keeping in mind the end goal to find the understudies. The framework then checks if there is any wink in clients' eye. On the off chance that there is any wink recognized, then it checks if the left eye has winked and plays out the left mouse catch activity, else the correct mouse catch activity is performed correspondingly. The framework then checks if there is any development in clients' understudy. On the off chance that any development is identified, then it checks if the left eye has moved and moves the pointer left, else moves the pointer right correspondingly as shown in Fig. 7.

3   Result and Discussions

The eyes and nose are located, and then the systems start to track them by analyzing the movement of these features in subsequent frames, using some template matching, and heuristics. The proposed system uses the light reflection in the user's eyes in order to locate the pupils. The system then checks if there is any wink in user's eye. If there

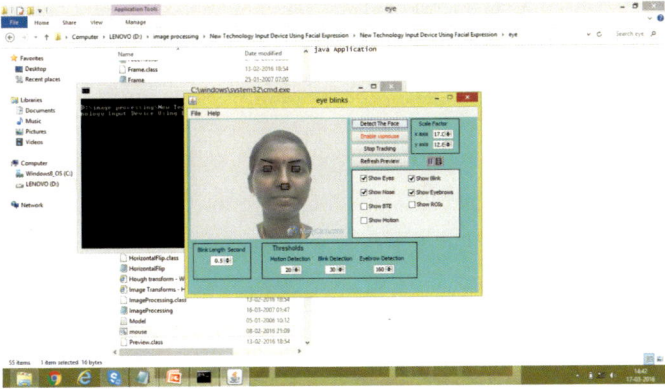

Fig. 8 GUI output

is any wink detected, then it checks if the left eye has winked and performs the left mouse button action, else the right mouse button action is performed correspondingly. The system then checks if there is any movement in user's pupil. If any movement is detected, then it checks if the left eye has moved and moves the pointer left, else moves the pointer right correspondingly. The GUI MATLAB output is shown in Fig. 8.

Reproducing the same movements of eyeblink videos is difficult in eye tracking. In this proposed system, extensive experimental measurements are carried out for more samples to evaluate precisely the nose regions and eyebrow movements also. The web camera is assisted with the node-based tracking also [10]. The speed of tracking is high as universal serial bus cameras are fixed to track the blinks.

3.1 Performance Metrics

- Accuracy: It refers to closeness of a measured value to standard value of eye recognition. The proposed algorithm is 7% greater than the Fisher vector, which is given in percentage as shown in Fig. 9a.
- False Positive: It refers to the probability of falsely rejecting the null hypothesis for a particular test [11]. It is also calculated as the ratio between the number of negative events wrongly categorized as positive and total number of actual negative events. The proposed algorithm is 22% lesser than the Fisher vector, which is used for eye detection as shown in Fig. 9b.
- Response Rate: It is also known as completion rate to return rate. In this, system calculates the time for eye recognition. The proposed algorithm is 17% high in their speed (ms) than the Fisher vector as shown in Fig. 9c.

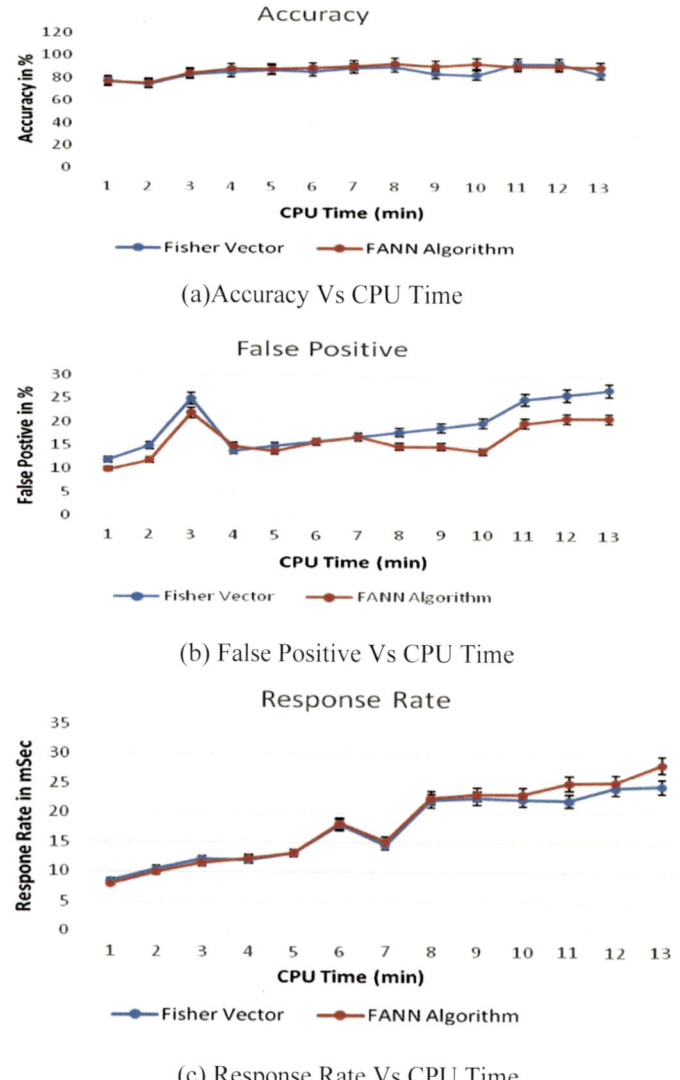

(a)Accuracy Vs CPU Time

(b) False Positive Vs CPU Time

(c) Response Rate Vs CPU Time

Fig. 9 Performance evaluation

4   Conclusion

This proposed system based on the face tracking with convex-shape nose feature and eyeblink detection using subtraction methods allows a user the flexibility and convenience of motions. The system serves as an alternative communication method that is suitable for people with severe disabilities. In previous tracking schemes, the disables have to wear special transmitters, sensors, or markers which resulted in discomfort, whereas in this method the eyeblinks use only a web camera and it is completely nonintrusive. The absence of any accessories on the user makes the system easier to configure and therefore more user-friendly in a clinical or academic environment.

Acknowledgements Ethical Compliance Comments
Figure 4 is a facial image taken from UCI repository dataset as an example to indicate the focal and seat points in the face.

References

1. Ebisawa Y (1998) Improved video-based eye-gaze detection method. IEEE Trans Instrum Meas 47(4):948–955
2. Kim KN, Ramakrishna RS (1999) Vision-based eye-gaze tracking for human computer interface. In: 1999 IEEE international conference on systems, man, and cybernetics, 1999. IEEE SMC'99 conference proceedings, vol 2. IEEE, pp 324–329
3. Galante A, Menezes P (2012) A gaze-based interaction system for people with cerebral palsy. Procedia Technol 5:895–902
4. Mahalakshmi E, Sheeba GM (2015) Enhancement of CFA in single sensor camera using laplacian projection technique. Res J Pharm Biol Chem Sci 6(3):1529–1536
5. Magee JJ, Betke M, Gips J, Scott MR, Waber BN (2008) A human–computer interface using symmetry between eyes to detect gaze direction. IEEE Trans Syst, Man, Cybern Part A Syst Humans 38(6):1248–1261
6. Al-Rahayfeh AMER, Faezipour MIAD (2013) Eye tracking and head movement detection: a state-of-art survey. IEEE J Transl Eng Health Med 1:2100212
7. Rantanen V, Vanhala T, Tuisku O, Niemenlehto PH, Verho J, SurakkaV, Juhola M, Lekkala JO (2011) A wearable, wireless gaze tracker with integrated selection command source for human–computer interaction. IEEE Trans Inf Technol BioMedicine 15(5):795–801
8. Carbone A, Martínez F, Pissaloux E, Mazeika D, Velázquez R (2012) On the design of a low cost gaze tracker for interaction. Procedia Technol 3:89–96
9. Zhang L, Vaughan R (2016, October) Optimal robot selection by gaze direction in multi-human multi-robot interaction. In: 2016 IEEE/RSJ international conference on intelligent robots and systems (IROS).IEEE, pp 5077–5083)
10. Miyake T, Asakawa T, Yoshida T, Imamura T, Zhang Z (2009, November). Detection of view direction with a single camera and its application using eye gaze. In: 35th annual conference of IEEE industrial electronics, 2009. IECON'09, pp 2037–2043
11. Merlin Sheeba G (2016) Enhanced wavelet OTSU tracking method for carcinoma cells. Int J Pharm Technol 8(2):11675–11684
12. Santos R, Santos N, Jorge PM, Abrantes A (2014) Eye gaze as a human-computer interface. Procedia Technol 17:376–383
13. Sun L, Liu Z, Sun MT (2015) Real time gaze estimation with a consumer depth camera. Inf Sci 320:346–360

Road Detection by Boundary Extraction Technique and Hough Transform

Namboodiri Sandhya Parameswaran, E. Revathi Achan, V. Subhashree
and R. Manjusha

Abstract Visual perception of road images captured by cameras mounted within a vehicle is the main element of an autonomous vehicle system. Road detection plays a vital role in a visual routing system for a self-governing vehicle. Effective detection of roads under varying illumination conditions plays a vital role to prevent majority of the road accidents that occur currently. In the current study, a new method using "boundary extraction" technique along with "Hough transform" is proposed for effective road detection. Here, two different algorithms, one using "Canny edge detection" and "Hough transform" and another using "boundary extraction" technique and "Hough transform" were implemented and tested on the same dataset. The comparison of the results of both the techniques showed that the algorithm using "boundary extraction" technique worked better than that which used "Canny edge" detection technique.

1 Introduction

Massive expansion in road network over the recent years has lead India to achieve the highest growth rate in the world. Our country is facing serious influences on road safety levels. A survey conducted by the Ministry of Road Transport and Highways Transport Research wing, Gov. of India (2015) says that the accident rate has increased by 2–5% from 4,89,400 in 2014 to 5,01,423 in 2015. An automated road

N. S. Parameswaran · E. Revathi Achan · V. Subhashree · R. Manjusha (✉)
Department of Computer Science and Engineering,
Amrita School of Engineering, Amrita Vishwa Vidyapeetham, Coimbatore, India
e-mail: r_manjusha@cb.amrita.edu

N. S. Parameswaran
e-mail: cb.en.p2cvi16004@cb.students.amrita.edu

E. Revathi Achan
e-mail: cb.en.p2cvi16006@cb.students.amrita.edu

V. Subhashree
e-mail: cb.en.p2cvi16008@cb.students.amrita.edu

© Springer Nature Switzerland AG 2019
D. Pandian et al. (eds.), *Proceedings of the International Conference on ISMAC in Computational Vision and Bio-Engineering 2018 (ISMAC-CVB)*, Lecture Notes in Computational Vision and Biomechanics 30,
https://doi.org/10.1007/978-3-030-00665-5_165

detection system helps to keep driver alert, by showing the forthcoming road. Such a system would help to support drivers in recognizing any unsafe situations prior, to avoid road accidents, by sensing and understanding of the environment around and also by reducing the risk of increasing the speed of the vehicle.

Different techniques have been proposed to detect roads using image processing techniques. Majority of the methods use the edge detection done by using Canny filter, and finally the road boundary detection by Hough transform. This paper analyzes and compares two different methods for road detection. First method uses the widely used "Canny" edge detection technique with "Hough transform" for road detection. The second method is a slight variation of the first method. Instead of "Canny" edge detection, this algorithm uses "boundary extraction" technique for detection of road boundaries along with "Hough transform". This paper contains five sections; Sect. 1 is the introduction, and Sect. 2 talks about the related work done in this area. Section 3 explains the two proposed algorithms and their outputs in detail. Section 4 gives the experimental results of the system. The conclusions and future works are in Sect. 5.

2   Related Work

Kumar and Kaur [1] deal with a comprehensive review of different works in lane detection techniques. First, an image of road is taken with the help of a camera and converted to a grayscale image; then, the filters are applied to remove noise present in the image and the edge detection is done by using Canny filter. Finally, the lane boundary is detected by using Hough transform. Rasmussen [2] explains the unsupervised algorithm for ill-structured roads. Here, texture orientation is computed using "Gabor wavelet" filters. This technique is used to estimate straight roads and also for curved and undulating roads by vanishing point detection. Kong et al. [3] implement two main steps. First is the estimation of vanishing point of straight roads and curved roads and second is the constrained road segmentation, done using the detected vanishing point. The proposed approach adapts a Gabor filter-based adaptive soft voting scheme. Then, edge detection technique is used to detect road boundaries. The technique used here is AdaBoost-based region segmentation and boundary detection. Aly [4] proposed the lane detection process using Hough transform that detects edges to identify different lines separated over an image. Saha et al. [5] describe an automated approach of the road lane detection process. "Labeling" and "flood fill" algorithms are used for lane detection. Hu et al. [6] propose a system to check for cracks located at the surface of roads. Danti et al. [7] made an attempt to invent an automated driver guidance mechanism to make driving safe and easier on Indian roads. The proposed method is based on detecting the issues of Indian roads like potholes and then recognizes road signs in Indian roads. Xue et al. [8] proposed a real-time positioning method for robotic cars in urban areas using an efficient lane marking algorithm. An efficient shape registration algorithm is proposed to detect the distance between the detected lane markings which improves the accuracy of global localization of robotic car. Moeves and Kruse [9] proposed the usefulness of two recent trends in fuzzy methods of machine learning. First, fuzzy support vector machine is identical to a special type of SVM. Then, they categorize the existing approaches to develop fuzzy rules from SVM. Guan et al. [10] proposed a system for detecting road lanes in automotive applications. Here, they have explained an FPGA hardware implementation of novel Hough transform (HT) architecture. Aminuddin et al. [11] proposed videos footages that are considered for road detection dynamically. From the point of intersection of the road lane, the vanishing point is found out, and then, road triangles are formed. Then, using image multiplication technique between the original image and triangle-shaped mask image, road is detected by eliminating the area outside of the road boundary. Fernandes and Oliveira [12] claim Hough transform as a popular tool for detection of missing data. The proposed method presents an improved voting scheme for Hough transform in large images to achieve real-time performance on large images. This method improves the performance of the voting scheme significantly and creates a voting map which makes the transform more robust for detecting the lines. Hari et al. [13] proposed a new method called "midpoint Hough transform" which calculates the midpoints of the two selected random points of the input image which is in binary format. If the midpoint is a bright point of the same image, then the point is transformed to the parameter space by using standard Hough transform equation. This process is repeated until no bright points are left in the image.

3 Proposed Algorithm

In this paper, mainly comparison and analysis of two different algorithms for road detection are being done. The first algorithm deals with the detection of roads using "Canny" edge detection and then applying "Hough" transform for detection of the boundary lines of the road. The second algorithm deals with the detection of straight roads by applying "boundary extraction" first to the input image and then applying the "Hough" transform for detecting the straight lines of the boundary of the roads. Here, the input dataset consists of 200 jpeg images of "straight urban" roads with minimum vehicles or obstacles. The dataset images are taken from a camera mounted inside front of a car. Both the algorithms are implemented using the "Matlab 2016b" tool.

3.1 Algorithm I: Straight Road Detection Using "Canny" Edge Detection and "Hough" Transform

The algorithm works in the following five steps:

Step 1: 3D Gaussian Filtering: The input image is first smoothed using 3D Gaussian filtering for the removal of existing noise present in the image. The presence of noise will lead to inappropriate detection of edges. The 3D Gaussian filter filters 3-D input

(a) **(b)**

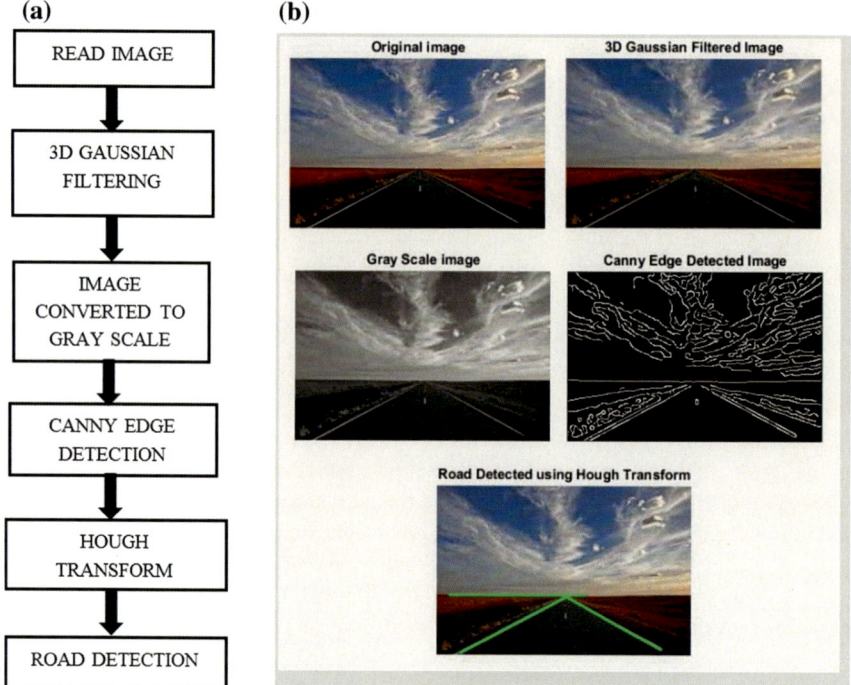

Fig. 1 **a** Architecture diagram for algorithm I; **b** output at each step

image with a 3-D Gaussian smoothing kernel. The filter is applied with a standard deviation of 0.5.

Step 2: Conversion to Grayscale: The image is then converted to grayscale image. This is done so that the time taken to process the image decreases.

Step 3: Canny Edge Detection: Then "Canny edge detection" technique is applied to the image to detect the edges present in the image. The Canny edge detector algorithm is a highly efficient algorithm that detects a large range of edges in a given image. It suppresses all the edges that are weak and outputs only the strong edges.

Step 4: Applying Hough Transform: To the Canny edge detected image, Hough transform is applied to detect the straight lines present in the image. Hough transform is a technique to extract features of objects that fall in a certain shape class by a procedure of voting.

Step 5: Road Detection: The Hough transform detects the lines that are present in the image. Here, in this algorithm, it detects the lines along the road boundaries and outputs an image with the detection of road in the input image.

The working of this algorithm can be depicted using the architectural diagram in Fig. 1a. The output at each step of the algorithm is given in Fig. 1b.

3.2   Algorithm II: Straight Road Detection Using "Boundary Extraction" and "Hough" Transform

The algorithm works in the following six steps:

Step 1: 3D Gaussian Filtering: As done in algorithm I, first the image is filtered or the noise is removed using 3D Gaussian filtering.

Step 2: Extracting the "Blue" Channel of the RGB Image: The filtered image is then split into its three channels: red channel, blue channel, and green channel. Our aim here is to detect roads, whose majority of the color information lies in the blue channel of the image. Hence, here we discard the red and the green channels. Further processing is done only on the "blue" channel of the image.

Step 3: Threshold the Blue Channel to Create a Blue Mask Image: The extracted blue band of the image is then further subjected to threshold to form the threshold image which we will call the "blue mask".

Step 4: Boundary Extraction: The "boundary extraction" algorithm is applied to this "blue mask" image to extract the boundaries of different regions present in the image.

Step 5: Applying Hough Transform: Finally, "Hough transform" for straight line detection is applied to the boundary extracted image to detect the different lines present in the image.

Step 6: Detection of Road Boundaries: The "Hough transform" algorithm detects the lines along the road boundaries and outputs an image with the detection of road in the input image.

The architecture diagram is shown in Fig. 2a. The intermediate results of the algorithm are shown in Fig. 2b.

4   Experimental Results and Analysis

Both the algorithms explained here were implemented and compared parallelly. A dataset of 200 different road images was given as input to each of the algorithms. The input image dataset did not contain any images of curved roads or roads with shadows. Road detection for each image was done using both the algorithms and the Hough transform lines formed for the road boundary detection, which were compared for each image of the dataset. The resultant "road-detected" images were manually analyzed and compared. Only those images where both the boundaries of the road were detected were considered as correct detection of the road. All the other images, where even if one of the road boundaries was detected, were not considered as correct road detection. Out of the total number of lines detected by Hough transform, only three lines were displayed on the image for road detection. Only those images which had all the three line aligned properly over the road boundaries were considered as correct detection of the road. All the other images were considered as wrong

(a)   **(b)**

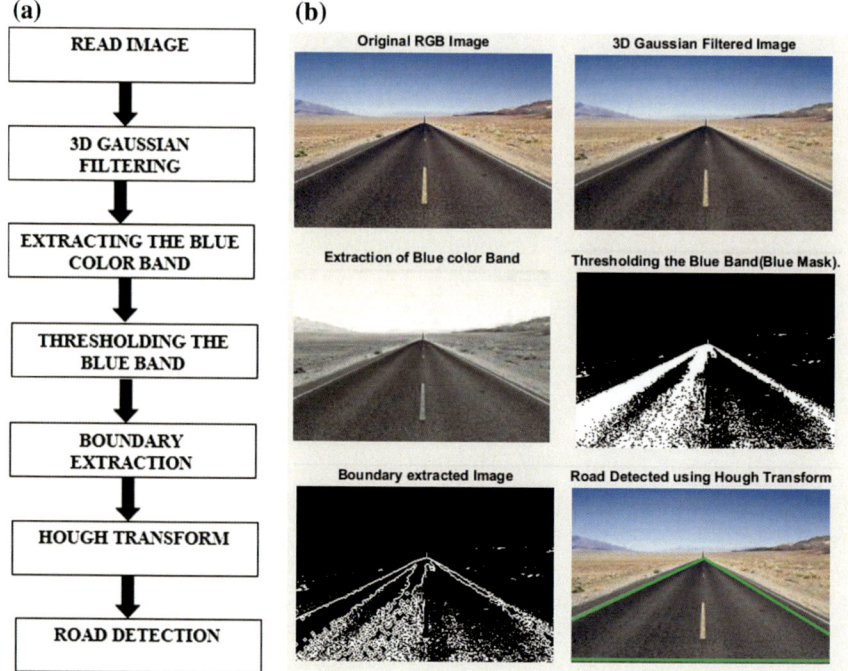

Fig. 2 **a** Architecture diagram of algorithm II; **b** output at each step

detection of road boundary. The algorithms were implemented using Matlab 2016b. The output of each algorithm for five different images is shown in Fig. 3.

The experiment results are shown in Table 1:

Table 1 shows the working of the two algorithms for 200 different road images. The dataset consists of images that are of straight roads taken from a camera mounted inside front of a car. The images are first subjected to Gaussian filtering for noise removal as preprocessing technique. The image dataset did not contain any images of any images of curved roads or roads with shadows.

Among the input images, those images where both the boundaries of the road were detected were considered as correct detection of the road. Out of the total number of lines detected by Hough transform, only three lines were displayed on the image for road detection. Only those images which had all the three lines aligned properly over the road boundaries were considered as correct detection of the road. All the other images, where even if one of the road boundaries was detected, were considered as wrong detection of road boundary.

As we can see in Table 1, the algorithm I which uses Canny edge detection along with Hough transform correctly detects roads for 151 images out of 200. In the remaining 49 images, the road was not detected properly or only a single boundary of the road was detected. In the second algorithm using Boundary extraction instead of Canny edge detection and Hough transform, road was correctly detected in 173

Table 1 Working of the algorithms for total of 200 images

Road detected correctly—Alg-I	Road detected wrongly—Alg-I	Accuracy = (correct detection/total images) * 100—for Alg-I (%)	Road detected correctly—Alg-II	Road detected wrongly—Alg-II	Accuracy = (correct detection/total images) * 100—for Alg-II (%)
151	49	75.5	173	27	86.5

Original Image	Algorithm I output	Algorithm II output

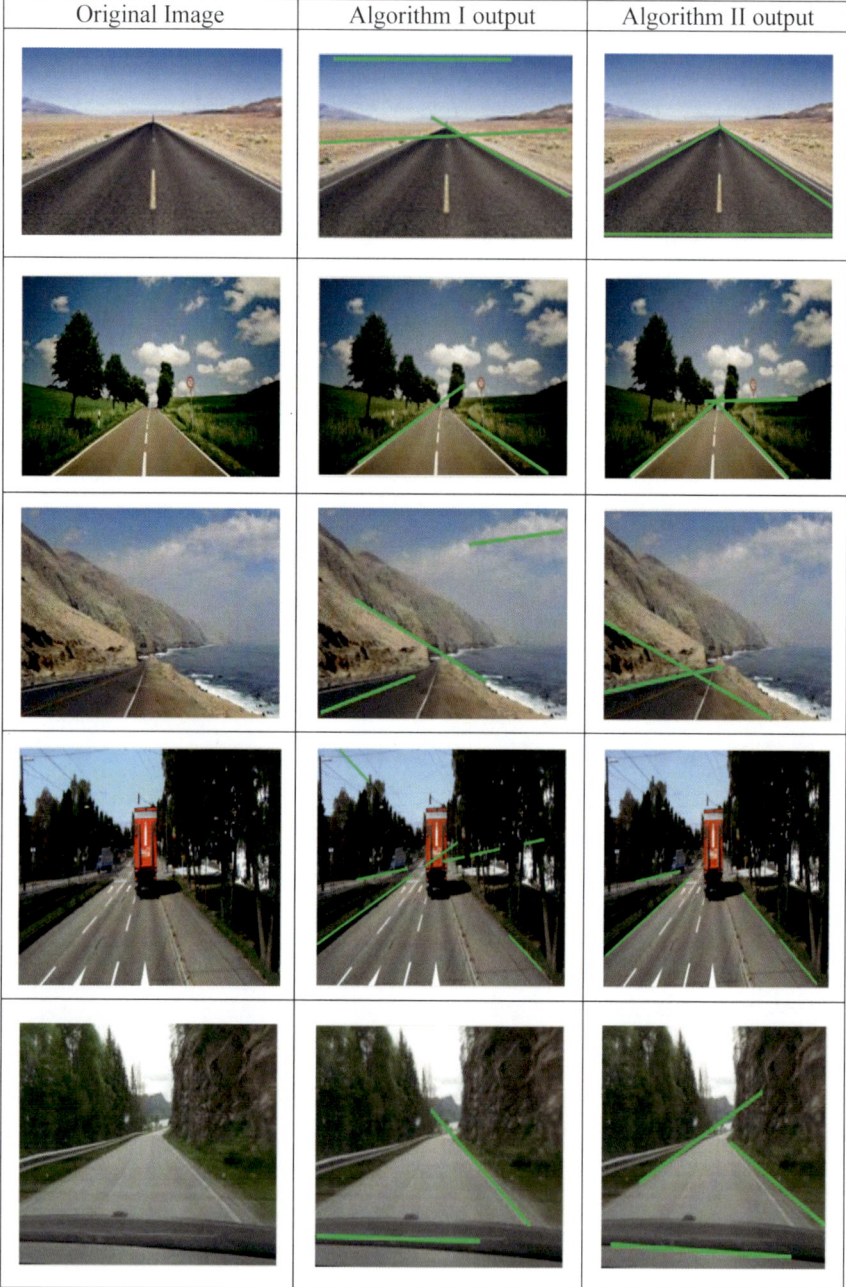

Fig. 3 Output of both algorithms for five different images

images out of 200 images. In 27 images, the road was not detected properly by algorithm II.

The accuracy of both the algorithms was calculated by the following formula:

$$Accuracy = (correct\ detection/total\ images) * 100 \qquad (1)$$

Accuracy got for algorithm I is 75.5% which is less as compared to the accuracy of 86.5% for algorithm II. This shows that the algorithm II works better for road detection, as compared to the algorithm I.

5 Conclusion

The extraction of straight roads from an image is very important requirement for preprocessing of images captured by surveillance cameras mounted within a vehicle. Existing road extraction methods widely make use of "Canny" edge detection technique for extraction of the edges in the roads. A new algorithm using "boundary extraction" with "Hough transform" for detection of boundaries of the roads is implemented in this paper. Road extraction using "Canny edge detector" yielded less accuracy than the "boundary extraction" based road detection. Thus, for straight urban roads, it is found that boundary extraction-based Hough transform produced better results.

Future work would include the extension of this algorithm for road images having shadows and for curved and muddy roads.

References

1. Kumar D, Kaur G (2008) Lane detection techniques: a review. IEEE Trans Intell Transp Syst 9(1):16–26
2. Rasmussen C (2004) Texture-based vanishing point voting for road shape estimation. In: BMVC, pp 1–10
3. Kong H, Audibert JY, PonceJ (2009) Vanishing point detection for road detection. In: IEEE conference on computer vision and pattern recognition, 2009. CVPR 2009. IEEE, pp 96–103
4. Aly M (2008) Real time detection of lane markers in urban streets. In: Intelligent vehicles symposium, 2008 IEEE, pp 7–12
5. Saha A, Roy DD, Alam T, Deb K (2012) Automated road lane detection for intelligent vehicles. Glob J Computer Sci Technol
6. Hu H, Gu Q, Zhou J (2010) HTF: a novel feature for general crack detection. In: 2010 17th IEEE international conference on image processing (ICIP). IEEE, pp 1633–1636
7. Danti A, Kulkarni JY, Hiremath PS (2012) An image processing approach to detect lanes, pot holes and recognize road signs in Indian roads. Int J Model Optim 2(6):658
8. Cui D, Xue J, Zheng N (2016) Real-time global localization of robotic cars in lane level via lane marking detection and shape registration. IEEE Trans Intell Transp Syst 17(4):1039–1050
9. Moewes C, Kruse R (2011) On the usefulness of fuzzy SVMs and the extraction of fuzzy rules from SVMs. In: EUSFLAT Conf., pp 943–948

10. Guan J, An F, Zhang X, Chen L, Mattausch HJ (2017) Parallelization of Hough transform for high-speed straight-line detection in XGA-size videos. In: 2017 IEEE international conference on consumer electronics-Taiwan (ICCE-TW). IEEE, pp 313–314

11. Aminuddin NS, Masrullizam MI, Ali NM, Radzi SA, Saad WHM, Darsono AM (2017) A new approach to highway lane detection by using hough transform technique. J Inf Commun Technol 16(2):244

12. Fernandes LAF, Oliveira MM (2008) Real-time line detection through an improved Hough transform voting scheme. Pattern Recogn 41(1):299–314

13. Hari CV et al (2009) Mid-point hough transform: a fast line detection method. In: India conference (INDICON), 2009 Annual IEEE

Investigating the Impact of Various Feature Selection Techniques on the Attributes Used in the Diagnosis of Alzheimer's Disease

S. R. Bhagyashree and Muralikrishna

Abstract According to the Dementia India report 2010, it is estimated that over 3.7 million people are affected by dementia and is expected to be double by 2030. Around 60–80% of the demented are suffering from Alzheimer's disease. Neuropsychological tests are useful tools for diagnosis of dementia. Diagnosis of dementia using machine learning for low- and middle-income setting is a rare study. Various attributes are used for diagnosing dementia. Finding the prominent attributes among them is a tedious job. Chi-squared, gain ratio, info gain and ReliefF filtering techniques are used for finding the prominent attributes. Cognitive score is identified as the most prominent attribute.

1 Introduction

Dementia is a neurodegenerative disease. It causes loss of cognitive functions such as reasoning, memory and other mental abilities which may be due to trauma or normal ageing [1]. It is estimated that 10–20% of people aged above 65 are having mild cognitive impairment [2]. It has been estimated that worldwide, around 4.6 million new dementia cases are identified every year [3]. Approximately, 473,000 people with age 65 and above will develop Alzheimer's disease [4]. Neuropsychological assessment is used for diagnosis of the disease. Alzheimer Disease International has designed a battery named Community Screening Instrument for Dementia (CSID) for diagnosis of dementia [5]. For the same dataset, classification using Naïve Bayes is performed and it has fetched a classification accuracy of 96.69%. Wrapper method is used as feature selection technique. The Cog-Score is identified as the prominent

S. R. Bhagyashree (✉)
E & C Department, ATME College of Engineering, Mysore, Karnataka, India
e-mail: srbhagyashree@yahoo.co.in

Muralikrishna
Wellcome DBT Allianz,
CSI Holdsworth Memorial Hospital, Mysore, India
e-mail: muralidoc@gmail.com

© Springer Nature Switzerland AG 2019
D. Pandian et al. (eds.), *Proceedings of the International Conference on ISMAC in Computational Vision and Bio-Engineering 2018 (ISMAC-CVB)*, Lecture Notes in Computational Vision and Biomechanics 30,
https://doi.org/10.1007/978-3-030-00665-5_166

attribute [6]. In this work, the feature selection is performed using filter method. In this work, the impact of various feature selection techniques on the accuracy of classification before and after feature selection techniques is studied. Along with that, the impact of number of attributes on classification accuracy is also studied. This work also focuses on the most prominent attribute that is needed to diagnose dementia. The paper is organized as follows. Section 2 explains the architecture of the proposed work which includes details of dataset, preprocessing with related literature survey and classification. Section 3 has results and discussion. Section 4 focuses on conclusion.

2 Related Work

Dementia is classified into Alzheimer's disease, dementia with Lewy bodies, Parkinson's disease, normal pressure hydrocephalus, cascular dementia and front temporal labour degeneration dementia. Alzheimer's disease is the major stakeholder [7]. The disease can be diagnosed by consulting general physician, by conducting neuropsychological assessment and by doing magnetic resonance imaging. Common brief assessments include the Mini-Mental State Examination (MMSE), the Brief Cognitive Rating Scale and Alzheimer's Disease Assessment Scale-Cognitive (ADAS-Cog), mini-cog, Montreal Cognitive Assessment (MOCA) etc. These neuropsychological assessments have their own disadvantages [8]. Alzheimer's Disease International (ADI) has founded 10/66 research group with an objective to design a battery which will overcome all the above problems. The group has studied the subjects of various age groups in different developing countries and designed a battery called Community Screening Instrument for Dementia (CSID) [5]. In this work, CSID battery is used for diagnosis.

Availability of specialist human resource including psychologists is less than one for one lakh population, whereas the demented is around 4 million and it is increasing with time [9]. This infers that at no point of time all the demented get diagnosed on time. Hence, there is a need of method which covers maximum population and diagnoses them on time.

Progress in data mining applications and its implications are manifested in the areas of information management in healthcare organizations, health informatics, epidemiology, patient care and monitoring systems, assistive technology, large-scale image analysis to information extraction and automatic identification of unknown classes. According to the most recent and sound systematic literature overview, performed by Esfandiari et al., four main application areas of DM application in medicine can be defined [10].

Increasing the efficiency and elimination of the human factor which deals with tasks for diagnosis of certain diseases where accuracy is essential is one such application. Researchers have used machine learning algorithms for diagnosis of various diseases like liver disorder, heart diseases etc. [11, 12].

3 Proposed Work

The architecture of the work is shown in Fig. 1. The dataset comprises details of 466 subjects having 50 attributes. In preprocessing, the feature selection is done using various filtering techniques. The classification is done using Naïve Bayes. The model evaluation is performed using 10-fold cross-validation.

3.1 Collection of Dataset

Global cognitive function is measured by administering the Community Screening Instrument for Dementia (CSID) which includes a 32-item cognitive test comprises 50 attributes. This includes assessing, orientation, comprehension, memory, naming, language expression, etc. Dataset consists of details of 282 men and 184 women, aged between 60 and 90 with different levels of education starting from illiterates to postgraduates, belonging to different religions.

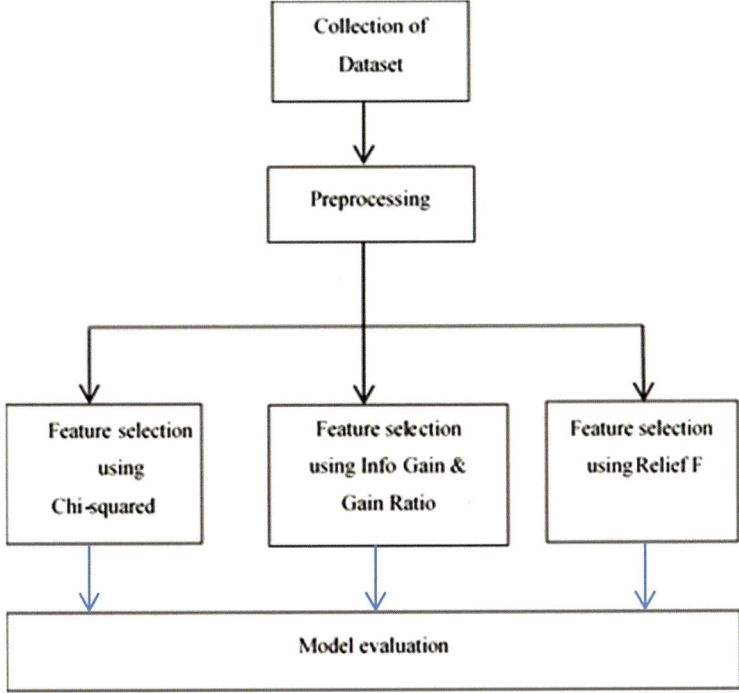

Fig. 1 Architecture

3.2 Preprocessing

Preprocessing is very important step in discovering knowledge from data. The initial steps like data cleaning and data integration are not performed as the data is a manually collected and is free from missing of any information. The data reduction is done using feature selection. The filter approach is used for implementing the same.

1. Attribute evaluation algorithms
2. Subset evaluation algorithms.

   The algorithms are categorized based on whether they rate the relevance of individual features or feature subsets. Attribute evaluation algorithms rank the features individually and assign a weight to each feature according to each feature's degree of relevance to the target feature. The attribute evaluation methods are likely to yield subsets with redundant features since these methods do not measure the correlation between features. Subset evaluation methods, in contrast, select feature subsets and rank them based on certain evaluation criteria and hence are more efficient in removing redundant features [13, 14].

3.2.1 Feature Selection Using Filter Method

The filter attribute selection method is independent of the classification algorithm. Filter method is further categorized into the following techniques.

3.2.2 Information Gain Attribute Evaluation

The entropy and the information gain are attribute measures which indicate how much percentage the given attribute separates the training dataset according to their final classification. The entropy and gain for a set S are calculated as

$$\text{Entropy}(S) = \sum_{i=1}^{n} -P_i \log_2 P_i \tag{1}$$

$$\text{Gain}(A) = \text{Entropy}(S) - \sum_{k=1}^{m} \frac{|S_k|}{|S|} \times \text{Entropy}(S_k) \tag{2}$$

where

"n" is the number of class.
P_i is the probability of S belongs to class i.
S_k is the subset of S [8].

3.2.3   Gain Ratio (GR) Attributes Evaluation

The gain ratio is the non-symmetrical measure that is introduced to compensate for the bias of the IG [15]. GR is given by

$$GR = \frac{IG}{H(X)} \tag{3}$$

As Eq. (3) presents, when the variable Y has to be predicted, we normalize the IG by dividing the entropy of X, and vice versa. Due to this normalization, the GR values always fall in the range [0, 1]. A value of $GR = 1$ indicates that the knowledge of X completely predicts Y and $GR = 0$ means that there is no relation between Y and X. In opposition to IR, the GR favours variables with fewer values [16].

3.2.4   Chi-Squared Statistic

This method measures the lack of independence between a term and the category. Chi-squared is the common statistical test that measures divergence from the distribution expected, if one assumes that the feature occurrence is actually independent of the class value. As a statistical test, it is known to behave erratically for very small expected counts, which are common in text classification. It is because of having rarely occurring word features and having few positive training examples for a concept. In statistics, the X^2 test is applied to test the independence of two events, where two events A and B are defined to be independent if $P(AB) = P(A)P(B)$ or, equivalently, $P(A|B) = P(A)$ and $P(B|A) = P(B)$. In feature selection, the two events are occurrence of the term and occurrence of the class. Feature selection using the X^2 statistic is analogous to performing a hypothesis test on the distribution of the class as it relates to the values of the feature in question [17].

3.2.5   ReliefF

ReliefF attribute evaluation evaluates the worth of a feature by repeatedly sampling an instance and considering the value of the given feature for the nearest instance of the same and different classes. This attribute evaluation assigns a weight to each feature, based on the ability of the feature to distinguish among the classes and then selects those features whose weights exceed a user-defined threshold as relevant features. Chi-squared, entropy and CFS filters are used for feature selection for two different datasets. The first group comprises 327 samples of acute lymphoblastic leukaemia, and the second group consists of 162 samples of ovarian cancer. The dataset is classified using classification algorithms, namely, k-NN, C4.5 and Naïve Bayes SVM [18].

3.3 Classification

Features are selected using different filtering techniques. Naïve Bayes and C4.5 are part of top ten algorithms used in data mining [19]. In the current study, the data is classified using Naïve Bayes algorithm. Naive Bayesian classifier is a selective classifier which calculates the set of probabilities by counting the frequency and combination of values in a given dataset. It assumes that all variables which contribute towards classification are mutually independent [20].

4 Results and Discussion

The details of 466 subjects are collected by conducting neuropsychological test. The data so collected is realistic, manually collected and hence the data is free from noise. The data is converted to ARFF format. The feature selection is done by using filter approach. Chi-squared, gain ratio, info gain and ReliefF are used as attribute evaluators. The dataset is classified using Naïve Bayes. The model evaluation is done by applying 10-fold cross-validation.

Table 1 shows the classification accuracy after applying Naïve Bayes algorithm for datasets using different feature selection filters. The number of attributes considered is 50, 35, 25, 15 and 8, respectively.

Table 2 shows the results before and after feature selection.

Table 1 Results after applying Naïve Bayes classification

Classification accuracy (%)

Filters	Number of attributes				
	50	35	25	15	8
Chi-squared	96.78	96.35	96.56	96.56	98.06
Gain ratio	96.78	96.35	97.21	96.78	98.08
Info gain	96.78	96.35	96.99	97.42	96.99
ReliefF	96.78	96.13	97.85	96.99	94.63

Table 2 Classification Accuracy before and after feature selection

Filters	Classification accuracy (%) Naïve Bayes	
	Before feature selection	After feature selection
Chi-squared	96.78	98.06
Gain ratio	96.78	98.06
Info gain	96.78	96.99
ReliefF	96.78	94.63

Table 3 Results before and after feature selection

Rank	Chi-squared	Gain ratio	Info gain	ReliefF
1	Cog-Score	Cog-Score	Cog-Score	Sentence repeat
2	Orientation with time	City	Orientation with time	Year
3	Story	Pray	Story	Long mem
4	Month	Watch	Semantic memory	Month
5	Semantic memory	Chemis	Month	Elbow
6	WLR	Learn2	Year	Nrecal
7	Learn2	Orientation with time	WLR	Street

Table 3 shows the list of high ranking attributes that are selected with the application of different feature selection techniques. Top seven attributes are listed here with Class as eighth attribute.

5 Conclusion

In this work, the details of 466 subjects comprising orientation, comprehension, memory, naming and language expression are considered. In preprocessing, the feature selection is performed using filter method. In this work, chi-squared, gain ratio, info gain and ReliefF are used as attribute evaluators. The features were selected using ranker approach. The attributes with lower ranking were removed and the classification accuracy is calculated at every stage.

The Naïve Bayes classifier shows a variation in the classification accuracy with variation in number of attributes. In chi-squared, the attributes 25 and 15 act as threshold values as there is change in a classification accuracy below and above these values. In chi-squared and gain ratio, maximum classification accuracy is achieved with minimum number of attributes. When the number of attributes is decreased from 50 to 35, the classification accuracy has decreased in all the evaluator schemes. Further, with an additional decrease of 10 attributes, there is an increase in the accuracy of the classification. The classification accuracy is more for less number of attributes and vice versa except for ReliefF. This infers that to achieve higher efficiency, the number of attributes needs to be minimal.

Table 3 clearly shows the significance of Cog-Score and orientation with time as prominent attributes in most of the feature selection techniques. The Cog-Score represents the entire set of attributes; hence, it is considered as the vital attribute for the accurate diagnosis of dementia. By considering the previous work [6] and the present work, it is very much clear that, irrespective of the feature selection techniques used, Cog-Score acts as the prominent attribute in the diagnosis of Alzheimer's disease.

Acknowledgements We are grateful to the participants and their family members for taking part in this study. Our sincere thanks to Principal and Management of ATME College of Engineering.

Ethics The ethics approval for this study was obtained from the Ethics Committee of CSI Holdsworth Memorial Hospital, Mysore.

Only those participants who were able to provide fully informed consent participated in this study and the informed consent was obtained from the participants.

Conflict of Interest None of the authors have any conflict of interest to declare.

References

1. Bhagyashree SR, Sheshadri HS (2014) An initial investigation in the diagnosis of Alzheimer's disease using various classification techniques. IEEE international conference on computational intelligence and computing research (ICCIC), pp 1–5
2. Gaugler J, James B, Johnson T, Scholz K, Weuve J, (2015) Sc.D Alzheimer's Association 2015 Alzheimer's disease facts and figures. Alzheimer's & Dementia 2015, pp 1–83
3. Prince M, Bryce R, Albanese E, Wimo A, Ribeiro W, Ferri CP (2013) The global prevalence of dementia: a systematic review and metaanalysis. Alzheimer's dement 9(1):63–75.e2. https://doi.org/10.1016/j.jalz.012.11.007
4. Bhagya Shree SR, Sheshadri HS (2014) An approach to preprocess data in the diagnosis of Alzheimer's disease. IEEE international conference on cloud computing and internet of things (CCIOT 2014), pp 135–139
5. Luisa AL et al (2009) Population normative data for the 10/66 Dementia Research Group cognitive test battery from Latin America, India and China: across sectional survey. Access NIH public, PubMed central, BMC Neurol 9:1–11
6. Bhagya Shree SR, Sheshadri HS (2016) Diagnosis of Alzheimer's disease using Naive Bayesian Classifier. NCAA, Neural Comput Appl 29:1–10
7. Bhagya Shree SR, Sheshadri HS, Joshi S (2014) A review on the method of diagnosing Alzheimer's disease using data mining. Int J Eng Res Technol 3(3):2417–2420. ISSN: 2278–0181
8. Shree SB, Sheshadri H (2014) An initial investigation in the diagnosis of Alzheimer's disease using various classification techniques. In: 2014 IEEE international conference on computational intelligence and computing research (ICCIC). IEEE, pp 1–5
9. Shaji K, Jotheeswaran A, Girish N, Bharath S, Dias A, Pattabiraman M, Varghese M (2010) Alzheimer's and related disorders society of India. The Dementia India Report: prevalence, impact, costs and services for Dementia
10. Esfandiari N, Babavalian MR, Moghadam A-ME, Tabar VK (2014) Knowledge discovery in medicine: current issue and future trend. Expert Syst Appl 41(9):4434–4463
11. Taneja A et al Heart disease prediction system using data mining techniques. Orient J Comput Sci Technol 6(4):457–466
12. Rajeswari P, Reena GS (2010) Analysis of liver disorder using data mining algorithm. Global J Comput Sci Technol 10(14):48
13. Henderson AS, Jorm AF (2000) Definition of epidemiology of dementia: a review. Dementia 2:1–33
14. Vanaja S, Kumar KR (2014) Analysis of feature selection algorithms on classification: a survey. Int J Comput Appl 96(17):975

15. Hussain L, Aziz W, Nadeem S, Abbasi A (2014) Classification of normal and pathological heart signal variability using machine learning techniques
16. Novaković J, Štrbac P, Bulatović D (2011) Toward optimal feature selection using ranking methods and classification algorithms. Yugoslav J Oper Res 21(1). ISSN: 0354-0243 EISSN: 2334-6043
17. Ladha L, Deepa T (2011) Feature selection methods and algorithms. Int J Comput Sci Eng 1(3):1787–1797
18. Yu J, Hu S, Wang J, Wong GKS, Li S, Liu B, Deng Y, Dai L, Zhou Y (2002) A draft sequence of the rice genome (Oryza sativa L. ssp. indica). American Association for the Advancement of Science
19. Wu X, Kumar JV, Quinlan R, Ghosh J, Yang Q, Motoda H, Geoffrey J, Ng MN, Liu B, Yu PS, Zhou Z-H, Steinbach M, Hand DJ, Steinberg D (2008) Top 10 algorithms in data mining. Springer, pp 1–37
20. Ferreira D et al (2012) Applying data mining techniques to improve diagnosis in neonatal jaundice. BMC Med Inform Decis Making 1–5

Detection of Sleep Apnea Based on HRV Analysis of ECG Signal

A. J. Heima, S. Arun Karthick and L. Suganthi

Abstract Sleep apnea is a breathing disorder which occurs during sleep. Sleep apnea causes more health-threatening problems such as daytime sleepiness, fatigue and cognitive problems, coronary arterial disease, arrhythmias, and stroke. However, there is an extremely low public consciousness about this disease. The most common type of sleep apnea is *obstructive sleep apnea (OSA)*. Polysomnography (PSG) is the widely used technique to detect OSA. *Obstructive sleep apnea* is extremely undiagnosed due to the inconvenient and costly polysomnography (PSG) testing procedure at hospitals. Moreover, a human expert has to monitor the patient overnight. Hence, there is a requirement of new method to diagnose sleep apnea with efficient algorithms using noninvasive peripheral signal. This work is basically aimed at detection of sleep apnea using a physiological signal electrocardiogram (ECG) alone which is taken from free online apnea ECG database provided by PhysioNet/PhysioBank. This database consists of 70 ECG recordings. A detailed time- and frequency-domain features and nonlinear features extracted from the RR interval of the ECG signals for observing minutes of sleep apnea are occurred in this work. Time-domain features mean HR ($P = 0.0093$, $r = 0.3593$) and RR interval mean (ms) ($p = 0.0003$, $r = 0.376$), frequency-domain features VLF power (%) ($P = 0.00659$, $r = 0.1081$) and HF power (%) ($P = 0.00135$, $r = 0.41138$), and nonlinear analysis feature SD1 ($P = 0.00039$, $r = 0.18998$), significantly different for normal and apnea ECG. Further supervised learning algorithms have been used to classify ECG signal to differentiate normal and apnea data. The overall efficiency is 90.5%. Algorithms which deal ECG signal along with respiratory signal will give more incite about apnea disease and also the classification accuracy may be improved.

A. J. Heima · S. Arun Karthick (✉) · L. Suganthi
Department of Biomedical Engineering, SSN College of Engineering, Chennai, India
e-mail: arunkarthicks@ssn.edu.in

A. J. Heima
e-mail: heima16455004@bme.ssn.edu.in

L. Suganthi
e-mail: suganthil@ssn.edu.in

© Springer Nature Switzerland AG 2019
D. Pandian et al. (eds.), *Proceedings of the International Conference on ISMAC in Computational Vision and Bio-Engineering 2018 (ISMAC-CVB)*, Lecture Notes in Computational Vision and Biomechanics 30,
https://doi.org/10.1007/978-3-030-00665-5_167

1   Introduction

Sleep constitutes a naturally send back to the state of rest for the body and mind. The central nervous system is restored during sleeping period. Therefore, sleep plays an important role in good physical condition by defending the mental and physical health. It also increases the quality of life. Human sleeps for about one-third of their whole lifetime.

People with sleeping disorder could not sleep peacefully. Deficiency of sleep can affect the body physically, emotionally, and psychologically. Eighty-four kinds of sleep disorders have been found till date. The most common sleep disorders are sleeplessness, sleep apnea, narcolepsy, and restless leg syndrome. A common type of sleep disorder is constituted by sleep apnea. Sleep apnea is defined as a pause in breathing for 10 s or more during sleep. The three types of sleep apnea are obstructive sleep apnea (OSA), central sleep apnea (CSA), and mixed apnea. The restriction on airflow, which conducts snoring during sleep, causes OSA. The brain failed to send signal breathing demand to the muscle during sleep which causes CSA. Mixed apnea is a combination of OSA and CSA. Obstructive sleep apnea is the most common type of sleep apnea. The effects of OSA are tiredness during the day, reduced reaction times, impaired vision or cognitive problems, and memory disorders [1]. Moreover, some studies show an enlarged number of apnea episodes that can enlarge the risk of diabetes [1]. OSA is also linked to cardiovascular diseases. Apnea episodes may cause cardiac ischemia [1].

In general population, 14% men and 5% women have obstructive sleep apnea [2]. Most OSA cases are undiagnosed and cause 980 deaths each year, because of the inconvenience, expenses, and unavailability of testing. Polysomnography (PSG) is the usual testing process, which is a standard procedure for all sleep disorder diagnosis. It records the breath airflow, respiratory movement, oxygen saturation, body position, electroencephalogram (EEG), electrooculogram (EOG), electromyogram (EMG), and electrocardiogram (ECG). The dependency on PSG needs to be taken away from the home for effective detection and treatment of sleep apnea. Therefore, it is essential to develop simple and efficient methods which diagnose OSA [3].

ECG signals have been used in earlier research works for the detection of sleep apnea [4–7]. Heart rate variability in a huge sample of healthy subjects was analyzed using time- and frequency-domain methods and a Poincaré plot. In the Fourier transform of ECG-RR interval, power of ultralow frequency (ULF), power of very low frequency (VLF), and power of low frequency (LF) are good indicators of health in cardiac patients [8]. Mean, standard deviation, median interquartile, and mean absolute deviation are other features which are extracted from RR interval of ECG signal [9].

This work is mainly aimed at diagnosis of sleep apnea from physiological signal electrocardiogram (ECG) by using time- and frequency-domain parameters of ECG–RR interval and nonlinear parameter derived from the Poincaré plot of RR interval.

2 Materials and Methods

The vital features of ECG and OSA show a high interaction among all biosignals. In this research, we use the free online apnea–ECG database provided by PhysioNet/PhysioBank. It contains 70 overnight ECG recordings provided by Philipps University, Marburg, Germany. The ECG signal is acquired at 100 samples per second with an amplitude resolution of 5 μV per least significant bit. Each record contains 8 h of ECG signal, and a record of diagnoses of sleep apnea per minute by clinical experts. In the ECG recordings, each apnea group recording has 100 or more minutes with tangled breathing. Twenty-five recordings were selected in this category (24 male, 1 female) with mean and standard deviation of age 51.84 ± 6.19. In control group, recordings have fewer than 5 min of disordered breathing. Seventeen recordings were selected in this category (10 male, 7 female) with mean and standard deviation of age 33.12 ± 5.74. Figure 1 shows the block diagram of methodology.

The ECG signal was affected by different noises during its acquisition and transmission. So, ECG signal has to be preprocessed. Filtering techniques were used to remove these noises. The AC component of the ECG signal is taken out using an analog low-pass filter with a cutoff frequency of 20 Hz. Then, the DC component of the signal is removed using a high-pass filter with a cutoff frequency of 0.15 Hz. We have considered 1 min data (at least 60 successive cardiac cycles) in order to reduce the influence of the respiratory system and other movement artifacts, and then the parameters are extracted and averaged. An averaging period covering at least 60 heartbeats (1 min data) has been used in earlier studies to improve conviction in the RR interval parameter extracted from the ECG signal [9]. After the preprocessing of ECG signal, QRS detection algorithm has been implemented for the detection of RR interval. The following operations are carried out in QRS detection algorithm such as cancelation DC drift and normalization, low-pass filtering (cutoff frequency 35 Hz),

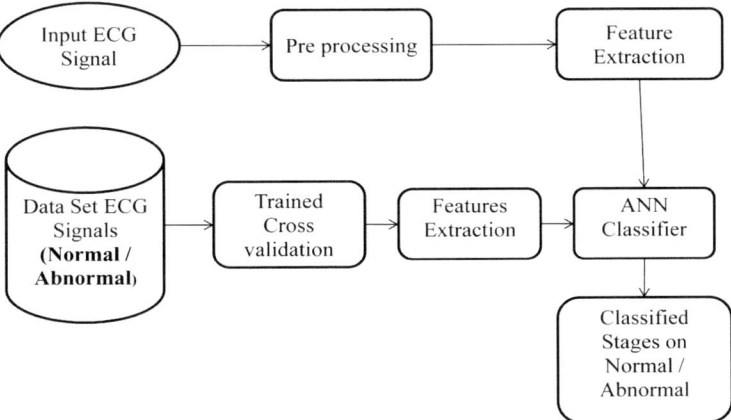

Fig. 1 Block diagram of methodology

high-pass filtering (cutoff frequency 5 Hz), derivative filter, squaring, and moving window integration [10]. In the first step of the algorithm, the signal is filtered by a low-pass and a high-pass filter in order to remove the influence of the muscle noise, the power line interference, the baseline wander, and the T-wave interference. Differentiation is performed to find the slope of the R-wave in the QRS complex. After finding the derivative, squaring operation is performed to find the characteristics of the signal. After the squaring of derivative of the signal, the start of the QRS complex is found with an adjusting threshold which is shown in Fig. 2. The peak of the squared signal is identified as the R-peak of the ECG data.

Kubios HRV has been used for performing time-domain, frequency-domain, and nonlinear analysis of RR intervals. The RR interval series is converted into equidistantly sampled series by interpolation (cubic spline interpolation) methods. There are many features which are extracted from ECG signal in time domain such as mean HR, SDNN, RR interval mean, and NN50.

In the frequency-domain analysis of RR interval, the power spectral density estimation is performed using fast Fourier transform (FFT) of RR interval. FFT spectrum is found based on Welch's periodogram method with 256 window size and 50% overlap. The frequency-domain features are very low-frequency (VLF) power, high-frequency (HF) power, and low-frequency (LF) power.

Poincaré plot is a nonlinear method to find correlation between successive RR intervals, (plot of RR_{j+1} as a function of RR_j). The parameters standard deviation 1 (SD1) and standard deviation 2 (SD2) are derived from Poincaré plots.

MATLAB pattern recognition tool is used for the classification of normal and apnea ECG. The pattern recognition tool uses supervised learning algorithms. The proposed algorithm classifies the signal on the basis of supervised learning algorithms. Extracted features from ECG signal are used for apnea classification.

Due to the non-uniformity of data, for nonparametric data analysis, Mann–Whitney U test is selected for the finding significant difference between the mean parameters of control and apnea group. We have calculated Spearman's rank correlation

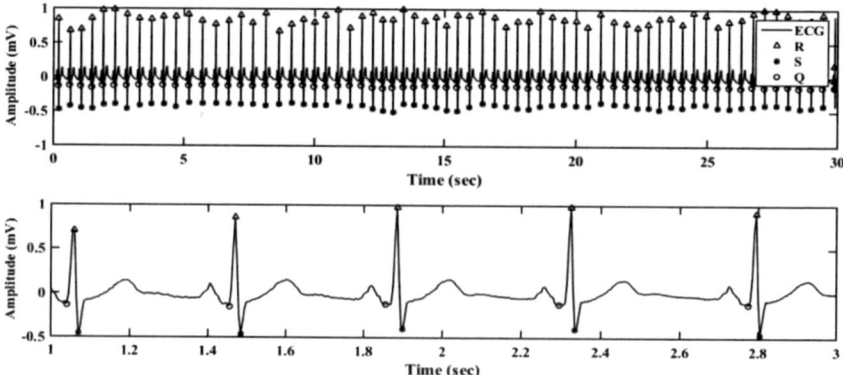

Fig. 2 ECG QRS detection

coefficient and probability using MATLAB 2007. The significant result was considered as $P > 0.05$.

3 Results

Using box plots, we have found the outliers for each parameter of normal and apnea ECG signals which are shown in Figs. 3, 4, and 5.

Without considering these outliers, the mean and standard error of the time-domain, frequency-domain, and nonlinear parameters are calculated which are shown in Figs. 6, 7, and 8, respectively. Among those features, mean HR ($p = 0.0093$, $r =$

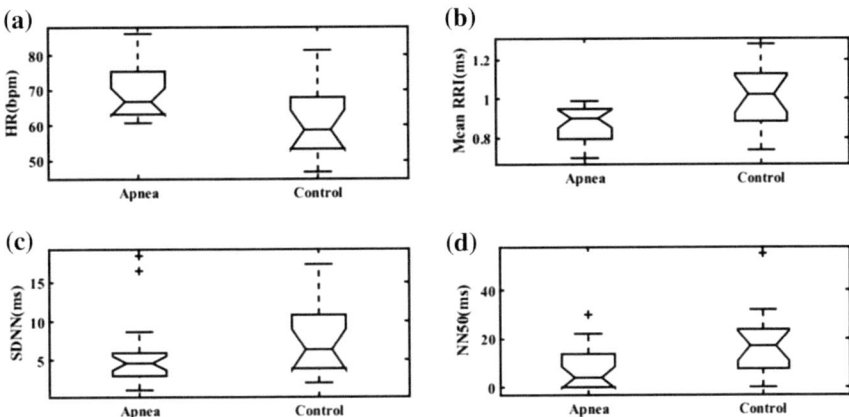

Fig. 3 Box plot for time-domain features, **a** heart rate (bpm), **b** mean RR interval (ms), **c** SDNN, **d** NN50

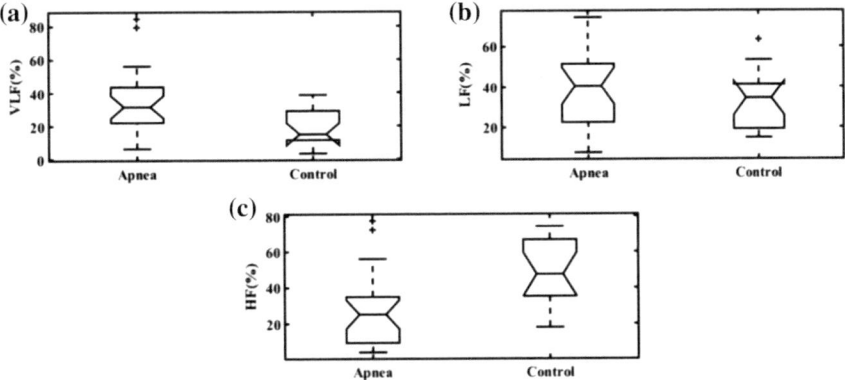

Fig. 4 Box plot for frequency-domain features, **a** very low-frequency (VLF) power, **b** high-frequency (HF) power, and **c** low-frequency (LF) power

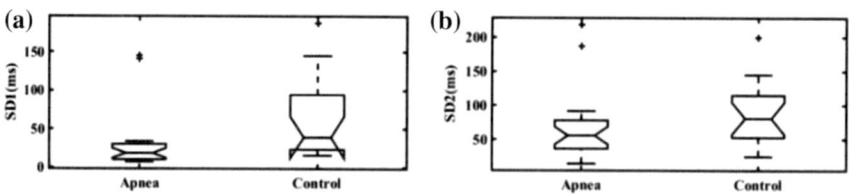

Fig. 5 Box plot for nonlinear features, **a** standard deviation 1 (SD1), **b** standard deviation 2 (SD2)

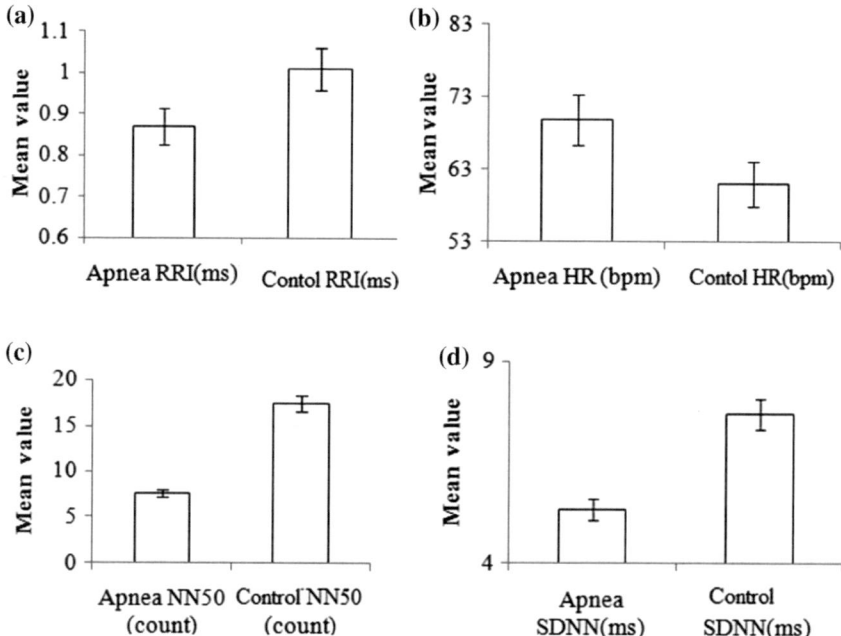

Fig. 6 Mean and standard error plot for time-domain feature. **a** RR interval, **b** Heart rate, **c** NN50, **d** SDNN

0.3593), RR interval mean (ms) ($p = 0.0003$, $r = 0.376$), VLF power (%) ($p = 0.00659$, $r = 0.1081$), HF power (%) ($p = 0.00135$, $r = 0.41138$), and SD1 ($P = 0.00039$, $r = 0.18998$) show significant difference between normal and apnea ECG.

Mean HR of normal person is lesser than apnea patient. HF of apnea person is lesser than normal patient. SD1 of apnea patient is lesser than normal person. The statistical analysis has been performed by using Mann–Whitney U test. The Spearman's rank correlation coefficient (r_c) and probability (P) which we obtained are given in Table 1. Artificial neural network algorithm which is written in MATLAB software classifies the significant features of test input. Classification is performed on the basis of supervised learning algorithms. Seventy percentage of input data is considered for the training input of ANN algorithm, and 30% of data is considered

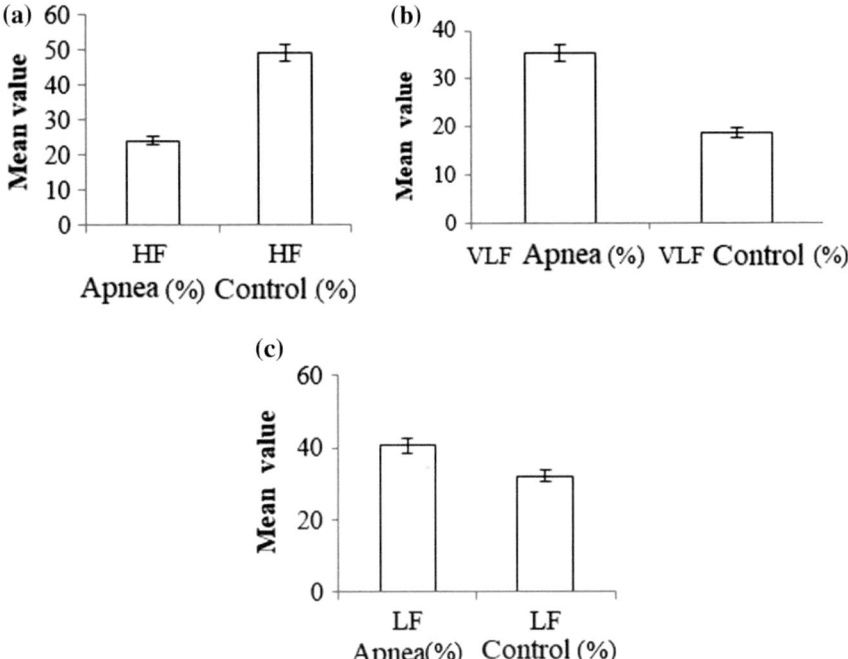

Fig. 7 Mean and standard error plots for frequency-domain features, **a** high-frequency (HF) power, **b** very low-frequency (VLF) power, and **c** low-frequency (LF) power

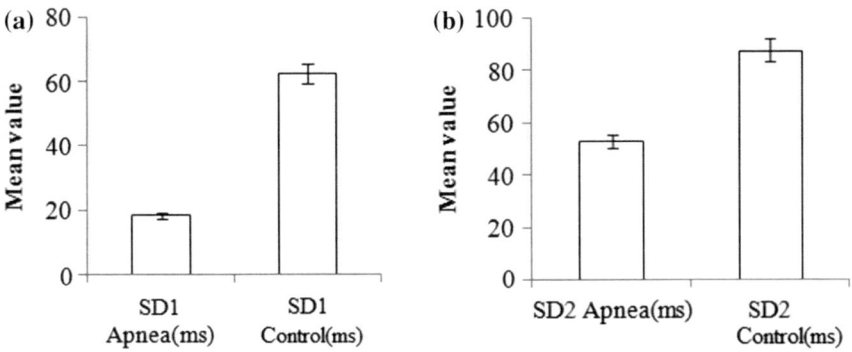

Fig. 8 Mean and standard error plot for nonlinear feature such as **a** standard deviation 1 (SD1), **b** standard deviation 2 (SD2)

for testing. The results are derived from the confusion matrix of normal and apnea patients as shown in Table 2. Performance measure of classification of normal and apnea signal is determined using accuracy, precision, sensitivity, and specificity. Percentage of each parameter describes the accuracy and efficiency for normal and apnea signal. The overall classification efficiency is 90.5% as shown in Table 2.

Table 1 Mean parameter values, Pearson correlation coefficient (r) and probability (P-value) using Mann–Whitney U test for control and apnea group

Feature	Normal	Apnea	Pearson correlation coefficient	P-value
Mean HR (bpm)	67.30116	85.84568	0.35931	0.0093*
SDNN (ms)	4.983797	4.615369	0.354	0.0646
RR interval mean(ms)	0.891515	0.698929	0.376	0.000*
NN50 (count)	13	6	0.2251	0.0646
Very low-frequency power (%)	23.7	10.1	0.1081	0.00659*
Low-frequency power (%)	69.4	45.1	−0.264	0.23846
High-frequency power (%)	7	44.9	0.41138	0.00135*
Standard deviation 1 (ms)	10.3	32.5	0.18998	0.00039*
Standard deviation 2 (ms)	34.1	62.9	0.08017	0.08606

*P-value < 0.05

Table 2 Classification accuracy of apnea and normal patient

	Prediction			Measurement
		Apnea	Control	
Actual	Apnea	True positive (TP) 24(57.1%)	False negative (FN) 3(7.1%)	Sensitivity 88.9%
	Control	False positive (FP) 1(2.4%)	True negative (TN) 14(33.3%)	Specificity 93.3%
Measurement		Positive predictive value 96%	Negative predictive value 82.4%	Accuracy 90.5%

4 Discussions

In apnea, respiratory rate varies with respect to the effect of disorder. The movement of diaphragm decides the position of the heart. Because the central tendon of the diaphragm is attached to the outer layer of heart (pericardium), the systemic or pulmonary arterial pressure or both increase during apnea. OSA patients have higher degree of pulmonary hypertension. Hemodynamic changes affect right and left ventricle during OSA episodes [11]. Because of these conditions, features derived from ECG signal vary during apnea [12].

The detailed study of HRV signal leads to better understanding of apnea. Recent studies show that an efficient algorithm that combines magnitude and frequency information from the power spectrum is more suitable for OSA detection [13]. The time-domain features related to OSA are proposed by de Chazal et al. [14]. Heart rate variability contains many features in both time and frequency domains and needs a different method in calculation of these features [15].

Changes in respiratory patterns can also vary HR and HRV independent of cardiac autonomic activity. In general, a decrease in respiratory frequency is associated with an increase in the heart period. In time-domain features, mean HR and RR interval shows more variability between normal and apnea patient which is shown in Fig. 3. The mean RR interval of normal case is high compared to apnea patient, because during apnea depolarization of ventricle is low compared to normal case.

The heart rate variability analysis approach uses Fourier transforms. The HRV spectrum contains two major components: First is high-frequency (0.18–0.4 Hz) HRV that has been used as an index of parasympathetic activity and is consecutive with respiration which is same as the respiratory sinus arrhythmia (RSA). The second is a low-frequency (0.04–0.15 Hz) component that appears to be mediated by both the vagus and cardiac sympathetic nerves. The area below the relevant frequencies presented in absolute units (square milliseconds) is the power of spectral components. The total power of a signal is equal to the variance of the entire signal, integrated over all frequencies. Parasympathetic–sympathetic balance is used as the ratio of the low-to-high frequency spectra [11]. In the control group, high-frequency (HF) band power is higher than apnea group which is shown in Fig. 4.

The time-domain feature SD1 shows higher variation between normal and apnea patient which is shown in Fig. 5. SD1 of normal case is high as compared to apnea patient. In case of heart rate variability (HRV), SD1 reveals both short- and long-term variations of the signal which is a visual pattern of the RR interval data [16].

The overall classification accuracy using this algorithm is 90.5%. Data sample size is small. In future, including larger dataset and use of different algorithms may increase the efficiency. In order to increase the accuracy of different methods like principal component analysis [5], Rusboost along with TQWT can be used for the efficient classification of OSA [4].

5   Conclusions

In this work, sleep apnea is detected from more commonly available physiological signal such as electrocardiogram (ECG). Some effective features are extracted from ECG signal. In time domain, the features such as RR interval mean, mean HR, NN50, and SDNN are extracted from ECG signal. In frequency domain, the features such as VLF power, HF power, LF power, standard deviation 1, and standard deviation 2 are extracted from ECG signal. Among those features, HF power (%), mean HR (bpm), standard deviation 1 (ms) show higher variation between normal and apnea ECG signals. The neural network classifier is used to classify an ECG signal normal data and apnea data on the basis of above features. The overall classification efficiency is 90.5%. In the future work, inclusion of large dataset and the use of different algorithms may increase the classification efficiency higher.

References

1. Rotariu C, Cristea C, Arotaritei D, Bozomitu RG, Pasarica A (2016) Continuous respiratory monitoring device for detection of sleep apnea episodes. In: 2016 IEEE 22nd international symposium for design and technology in electronic packaging (SIITME), pp 106–109
2. Martín-González S, Navarro-Mesa JL, Juliá-Serdá G, Kraemer JF, Wessel N, Ravelo-García AG (2017) Heart rate variability feature selection in the presence of sleep apnea: an expert system for the characterization and detection of the disorder. Comput Biol Med 91:47–58
3. Fan SH, Chou CC, Chen WC, Fang WC (2015) Real-time obstructive sleep apnea detection from frequency analysis of EDR and HRV using Lomb Periodogram. In: 2015 37th annual international conference of the IEEE engineering in medicine and biology society (EMBC), pp 5989–5992
4. Hassan AR, Haque MA (2017) An expert system for automated identification of obstructive sleep apnea from single-lead ECG using random under sampling boosting. Neurocomputing 235:122–130
5. Saxena N, Shinghal K (2015) Extraction of various features of ECG signal. Int J Eng Sci Emerg Technol 7(4):707–714
6. Rodríguez R, Mexicano A, Bila J, Cervantes S, Ponce R (2015) Feature extraction of electrocardiogram signals by applying adaptive threshold and principal component analysis. J Appl Res Technol 13(2):261–269
7. Mannurmath JC, Raveendra M (2014) MATLAB based ECG signal classification. Int J Sci Eng Technol Res (IJSETR) 3(7):1946–1950
8. Corrales MM, de la Cruz Torres B, Esquivel AG, Salazar MAG, Orellana JN (2012) Normal values of heart rate variability at rest in a young, healthy and active Mexican population. Health 4(07):377
9. Ma HT, Liu J, Zhang P, Zhang X, Yang M (2015) Real-time automatic monitoring system for sleep apnea using single-lead electrocardiogram. In: TENCON 2015-2015 IEEE region 10 conference, pp 1–4
10. QRS Detection Using Pan-Tompkins algorithm. https://archive.cnx.org/contents/611d4152-cf7f-4344-b56d-15775f92fbac@1/qrs-detection-using-pan-tompkins-algorithm
11. Khalil MM, Rifaie OA (1998) Electrocardiographic changes in obstructive sleep apnoea syndrome. Respir Med 92(1):25–27
12. BIOPAC Systems, Inc., https://blog.biopac.com/respiration-and-ecg/
13. Quiceno-Manrique AF, Alonso-Hernandez JB, Travieso-Gonzalez CM, Ferrer-Ballester MA, Castellanos-Dominguez G (2009) Detection of obstructive sleep apnea in ECG recordings using time-frequency distributions and dynamic features. In: Annual international conference of the IEEE engineering in medicine and biology society, 2009, EMBC 2009, pp 5559–5562
14. de Chazal P, Penzel T, Heneghan C (2004) Automated detection of obstructive sleep apnoea at different time scales using the electrocardiogram. Physiol Meas 25(4):967
15. Kocak O, Bayrak T, Erdamar A, Ozparlak L, Telatar Z, Erogul O (2012) Automated detection and classification of sleep apnea types using electrocardiogram (ECG) and electroencephalogram (EEG) features. In: Advances in electrocardiograms-clinical applications. Turkey
16. Behbahani S, Moridani MK (2015) Non-linear Poincaré analysis of respiratory efforts in sleep apnea. Bratisl Lek Listy 116(7):426–432

Performance Comparison of SVM Classifier Based on Kernel Functions in Colposcopic Image Segmentation for Cervical Cancer

N. Thendral and D. Lakshmi

Abstract Cervical cancer is the second most common cancer affecting women worldwide. It can be cured in almost all patients if detected and treated in time. Pap smear test has been broadly used for detection of cervical cancer. The conventional Pap smear test has several shortcomings including subjective nature, low sensitivity, and frequent retesting. In order to overcome this issue, colposcopy method is used for visual inspection of cervix with the aid of acetic acid and with proper magnification, abnormal cells to be identified. Thus, we propose a method for automatic cervical cancer detection using segmentation and classification. In this work, several methods used for detecting cervical cancer is discussed which uses different classification techniques like K-means clustering, texture classification and Support Vector Machine (SVM) to detect cervical cancer. The proposed work compares and determines accuracy for five types of kernel functions, namely Polynomial kernel, Quadratic kernel, RBF kernel, linear kernel, and Multi-Layer Perceptron kernel. Analysis shows that Multi-layer Perceptron kernel in SVM classifier provides the best performance with an accuracy of 98%.

1 Introduction

Cervical cancer is a cancer arising from the cervix. It is due to the abnormal growth of cells that have the ability to invade or spread to other parts of the body. Cervix region is made up of three types of tissues, namely Columnar Epithelium (CE), Squamous Epithelium (SE), and Aceto White (AW) region. Cervical cancer typically develops from precancerous changes from 10 to 20 years. About 90% of cervical cancer cases

N. Thendral (✉)
Faculty of Information and Communication, Anna University, Chennai, India
e-mail: thendral2082@gmail.com

D. Lakshmi
Department of Electronics and Communication, St. Joseph's College of Engineering, Chennai, India
e-mail: lakhramdevan@gmail.com

© Springer Nature Switzerland AG 2019
D. Pandian et al. (eds.), *Proceedings of the International Conference on ISMAC in Computational Vision and Bio-Engineering 2018 (ISMAC-CVB)*, Lecture Notes in Computational Vision and Biomechanics 30,
https://doi.org/10.1007/978-3-030-00665-5_168

are squamous cell carcinomas, 10% are adenocarcinoma, and a small number are other types. Early on, typically no symptoms are seen. Later, symptoms may include abnormal vaginal bleeding, pelvic pain, or pain during sexual intercourse. Diagnosis is typically by cervical screening followed by a biopsy [1]. Medical imaging is then done to determine whether or not the cancer has spread. Several methods are available for diagnosis of cervical cancer. Some of them are Pap test, LCB test, and colposcopy test.

The Papanicolaou test (abbreviated as Pap test, also known as Pap smear, cervical smear, or smear test) is a method of cervical screening which is used to detect potentially precancerous and cancerous processes in the cervix (opening of the uterus or womb) [2]. A Pap smear is performed by opening the vaginal canal with a speculum, and then collecting cells at the outer opening of the cervix at the transformation zone. The collected cells are examined under a microscope to look for abnormalities. Pap test results report a normal (negative) when all the cells are of healthy size and shape. It reports an abnormal (positive) test if there are any changes in size and texture. An abnormal Pap test does not mean that women patients have cancer as the observations might tend to go wrong due to improper visualization. Pap test method is simple in operation but it takes more time to give the observation for a patient as it requires investigation in clinical laboratory. The next manual screening method is the Liquid-Based Cytology (LCB) test used for detecting cervical cancer uses 5% acetic acid in the biopsy of the cervical tissues which changes the AW region into white color is the way of diagnosing cervical cancer.

On the other hand, colposcopy is a method which uses a device called as colposcopy for visual inspection of cervix with the aid of acetic acid. Colposcopy is a microscope-like device used by doctors to examine women external genital area (vulna), vagina, and cervix with proper magnification so that abnormal cells are easily identified. Images obtained with colposcopy referred to as colposcopy images or cervical images are widely used as diagnostic and screening tool for cervical cancer. The accuracy of colposcopy is highly dependent on the physician's individual skills. In expert hands, colposcopy has been reported to have a high sensitivity (96%) and low specificity (48%) when differentiating abnormal tissues. These manual screening methods suffers from accuracy and is also time-consuming. So, the computerized methods of classification are used for detecting normal and abnormal cells [3]. The most common classification methods are Support Vector Machine (SVM), k-means clustering and texture classification. The input image is fed where the preprocessing stage is performed by using filters to eliminate unwanted noise from the image and to enhance its quality [4]. This image is used for feature extraction so as to extract the feature of nucleus, cytoplasm and background. After the features are extracted, the nucleus and cytoplasm are segmented from the image. Finally, the image is classified as normal and abnormal cells.

Fig. 1 Block diagram of the proposed system

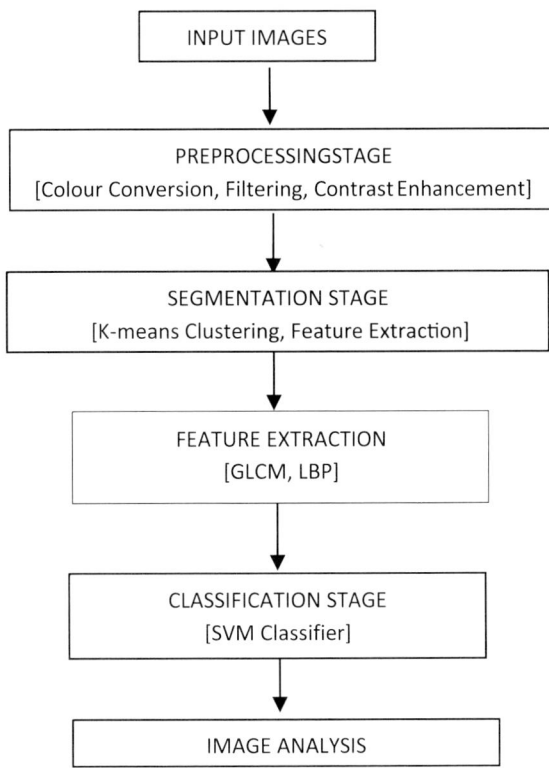

2 Methodologies

The methodologies adopted in this paper are depicted in Fig. 1.

2.1 Preprocessing

Preprocessing of an image is the preparation of the sample to introduce it to an algorithm for the specified task. The aim of preprocessing is an improvement of image data that suppresses unwanted distortions or enhances some image features for further processing.

2.1.1 Color Conversion

Here, the RGB image is converted into gray scale image. Conversion of color image to gray scale image is one of the image processing applications used in different fields effectively. To convert the color images, the new algorithm performs RGB approximation, reduction, and addition chrominance and luminance. The gray scale conversion images generated using the algorithm preserved the salient features of the color image such as contrasts, sharpness, shadow, and image structure.

By emitting a restricted combination of three colors (red, green, and blue), we are able to generate almost any detectable color. This is the reason color images are often stored as three separate image matrices; one storing the amount of red (R) in each pixel, one the amount of green (G), and one the amount of blue (B).

Gray scale images do not differentiate gives dark pixels and much light is perceived as bright pixels. When converting an RGB image to gray scale, RGB values are considered for each pixel and make output as a single value reflecting the brightness of that pixel.

2.1.2 Image Filteration

Image filteration is carried out using median filter. Median filtering is a nonlinear method used to remove noise from images. It is widely used as it is very effective at removing noise while preserving edges. The median filter works by moving through the image pixel by pixel, replacing each value with the median value of neighboring pixels. It suppresses the noise or other small fluctuations in the image; equivalent to the suppressions of high frequencies in the frequency domain. For smoothing the image, Median Filter is used which is a nonlinear operation often used in image processing to reduce "salt and pepper" noise. A median filter is more effective than convolution when the goal is to simultaneously reduce noise and preserve edges.

2.1.3 Contrast Enhancement

Image enhancement technique makes an image easier to analyze and interpret. The range of brightness values present on an image is referred to as contrast. Contrast enhancement is a process that makes the image features stand out more clearly by making optimal use of the colors available on the display or output device.

2.2 Segmentation

2.2.1 K-Means Clustering

K-means clustering is a type of hard clustering algorithm. It belongs to unsupervised cluster analysis algorithm and achieves partition clustering method. It is a key technique in pixel-based methods, where pixel-based methods based on K-means clustering are simple and the computational complexity is relatively low compared with other region-based or edge-based methods, the application is more practical. In the K-means algorithm, it initially defines the number of clusters k. Then, k-cluster center are chosen randomly. The distance between the each pixel to each cluster centers is calculated. The distance is the simplified Euclidean function. A single pixel is compared to all cluster centers using the distance formula. The pixel is moved to that particular cluster, which has the shortest distance amongst all. Then the center is recalculated. Again, each pixel is compared to all centers. The process continues until the center converges.

2.3 Feature Extraction

Feature extraction involves reducing the amount of resources required to describe a large set of data. Features are obtained using Local Binary Pattern (LBP), Gray Level Co-occurrence Matrix (GLCM), and LBP counts several binary patterns that occur in the image. GLCM is a kind of statistical approach that uses homogeneity, contrast, energy, and correlation information from pixels.

2.3.1 Gray Level Co-Occurrence Matrix (GLCM) Method

Gray Level Co-Occurrence Matrix has proved to be a popular statistical method of extracting textural feature from images. According to co-occurrence matrix, Haralick defines 14 textural features measured from the probability matrix to extract the characteristics of texture statistics of remote sensing images. The features like energy, correlation, homogenity, and contrast are obtained using GLCM method.

2.3.2 Local Binary Pattern (LBP) Method

LBP is the particular case of the texture spectrum model. It has further been determined that when LBP is combined with the Histogram of Oriented Gradients (HOG) descriptor, and it improves the detection performance considerably on some datasets.

LBP method divides the examined window into cells (e.g., 16×16 pixels for each cell). For each pixel in a cell, compare the pixel to each of its eight neighbors (on its left-top, left-middle, left-bottom, right-top, etc.). Follow the pixels along a circle, i.e., clockwise or counterclockwise. Where the center pixel's value is greater than the neighbor's value, write "0". Otherwise, write "1". This gives an 8-digit binary number (which is usually converted to decimal for convenience). Compute the histogram over the cell of the frequency of each "number" occurring (i.e., each combination of which pixels are smaller and which are greater than the center). This histogram can be seen as a 256-dimensional feature vector. In the computation of the LBP histogram, the histogram has a separate bin for every uniform pattern, and all nonuniform patterns are assigned to a single bin. Using uniform patterns, the length of the feature vector for a single cell reduces from 256 to 59.

2.4   Classification

2.4.1   Support Vector Machines (SVMs)

Support Vector Machine (SVM) was first heard in 1992, introduced by Boser, Guyon, and Vapnik in COLT-92. Support vector machines (SVMs) are extreme learning machines. It has a set of related supervised learning methods used for classification and regression. They belong to a family of generalized linear classifiers. In another term, Support Vector Machine (SVM) is a classification and regression prediction tool that uses machine learning theory to maximize predictive accuracy. Support Vector Machines can be defined as systems which use hypothesis space of a linear functions in a high dimensional feature space, trained with a learning algorithm from optimization theory that implements a learning bias derived from statistical learning theory [5].

2.4.2   Kernal Functions in SVM

Kernel function implicitly maps data from its original space to a higher dimensional feature space. Kernel function parameter selection is one of the important parts of support vector machine modeling. We propose emphasizes the classification task with support vector machine with different kernel function. It has several kernel functions including linear, polynomial, radial basis function, quadratic kernel, and multi-layer perception. The following are the kernel functions which are used to find out the accuracy [5–7].

RBF Kernel

The radial basis function network is one approach which has shown a great promise in this sort of problems because of its faster learning capacity. It has better capability of approximation to underlying functions, faster learning speed, better size of network, and high robustness to outliers. A radial basis function produces a piecewise linear solution which can be attractive. The equation for RBF Kernel is given by

$$K(x, y) = \exp\left(-\frac{\|x - y\|^2}{2\sigma^2}\right) \tag{1}$$

where > 0 is constant which denotes kernel width.

Linear Kernel

A linear kernel is often recommended for text classification with SVM because text data has lot of features and is often linearly separable. In text classification, both the number of instances and features are large. The equation for linear kernel is

$$k(x, y) = X_i.Y_j \tag{2}$$

Polynomial Kernel

Polynomial is one of the most powerful functions that have been used in many fields of mathematics such as fitting and regression. Low-order polynomial is desired for their smoothness, good local approximation and interpolation. A polynomial mapping is a popular method for nonlinear modeling. The equation of polynomial kernel is

$$K(x, y) = (x^T y + 1)^p \tag{3}$$

where $p > 0$ is a constant that defines the kernel order.

Quadratic Kernel

A new quadratic kernel the support vector machine is introduced to separate the data nonlinearly. A quadratic decision of separating nonlinearly the date is used. The equation for quadratic kernel is

$$K(x, y) = (x^T y + 1)^2 \tag{4}$$

Multi-layer Perceptron

MLP network is trained with cervical colposcopic images database with original features and then only extracted features. Classification performance of MLP in two cases is calculated and analyzed with help of network measures such as classification accuracy, recall, precision, mean squared error, and time [8, 9].

$$K(x) = \frac{(\tanh(x) + 1)}{2} - \frac{1}{1 + \exp(-2x)} \tag{5}$$

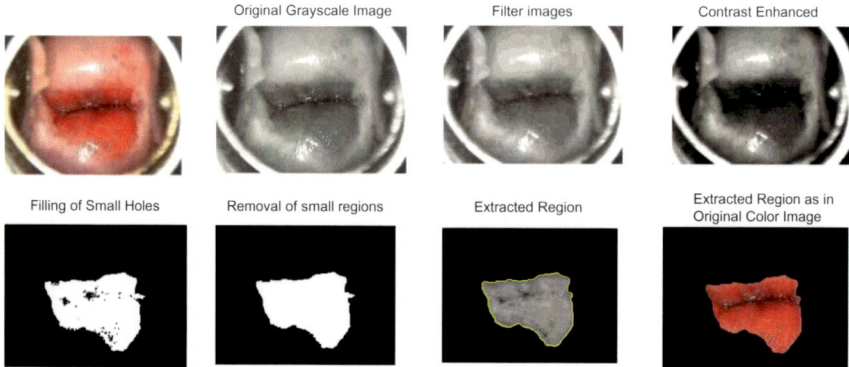

Fig. 2 Test image and its corresponding outcome

Fig. 3 Bar chart representation of accuracy by comparing five kernel functions

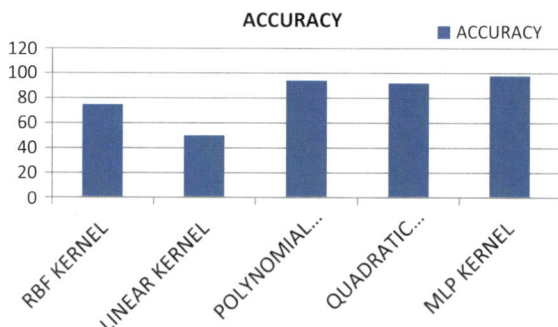

3 Results and Discussions

The extracted region of the affected area is identified using the proposed methodology for one sample image and is shown in Fig. 2.

The accuracy values of all the kernel functions is compared and represented in bar chart (Fig. 3).

The texture features, like contrast, correlation, energy, and homogenity are extracted from the segmented sample images by using GLCM are listed out in Table 1.

Finally, we compare the five types of kernels and the accuracy values of the kernel functions for the sample images are list out in Table 2.

4 Conclusion

Cervical cancer is one of the most death causing diseases around the world. Early detection is one of the most promising approaches in reducing the burden of growing cancer. In order to detect the cancer in early stage, we propose a method for auto-

Table 1 GLCM

Samples	Contrast	Correlation	Energy	Homogeneity
Sample 1	0.212779	0.97516	0.096202	0.916656
Sample 2	0.206783	0.973301	0.121533	0.924258
Sample 3	0.070206	0.982737	0.196052	0.966464
Sample 4	0.097927	0.984116	0.185056	0.954634
Sample 5	0.59759	0.992685	0.214383	0.974872
Sample 6	0.78818	0.989015	0.126878	0.964230
Sample 7	0.059680	0.991099	0.153083	0.975513
Sample 8	0.212045	0.956254	0.167082	0.951011
Sample 9	0.154843	0.974881	0.126620	0.933395
Sample 10	0.301584	0.947365	0.152555	0.942753

Table 2 Support Vector Machine techniques (values in %)

Samples	RBF kernel	Linear kernel	Polynomial kernel	Quadradic kernel	MLP
Sample 1	75	75	94	93	98
Sample 2	75	75	94	93	98
Sample 3	75	75	94	93	98
Sample 4	75	25	94	93	98
Sample 5	75	50	94	93	98
Sample 6	75	50	94	93	98
Sample 7	75	75	94	93	98
Sample 8	75	75	94	93	98
Sample 9	75	75	94	93	98
Sample 10	75	75	94	93	98

matic cervical cancer detection using segmentation and classification. The algorithms which are used in the four stages of preprocessing, segmentation, feature extraction, and classification are summarized [10, 11]. Another significant contribution of the proposed work is that it compares the five types of kernel function in SVM classifier. Analysis shows that MLP kernel in SVM classifier provides the best performance with an accuracy of 98%. Thus with the help of maximum accuracy, the cancer can be detected in early stage.

References

1. Kashyap D et al (2016) Cervical cancer detection and classification using Independent Level sets and multi SVMs. In: 2016 39th international conference on telecommunications and signal

processing (TSP), pp 523–528

2. Kaaviya S, Saranyadevi V, Nirmala M (2015) PAP smear image analysis for cervical cancer detection. In: 2015 IEEE International conference on engineering and technology (ICETECH), pp 1–4

3. Obukhova NA, Motyko AA, Kang U, Bae S-J, Lee D-S (2017) Automated image analysis in multispectral system for cervical cancer diagnostic. In: 2017 20th conference of open innovations association (FRUCT), pp 345–351

4. Njoroge E, Alty SR, Gani MR, Alkatib M (2006) Classification of cervical cancer cells using FTIR data. In: 2006 international conference of the ieee engineering in medicine and biology society, pp 5338–5341

5. Teeyapan K, Theera-Umpon N, Auephanwiriyakul S (2015) Application of support vector based methods for cervical cancer cell classification. In: 2015 IEEE international conference on control system, computing and engineering (ICCSCE), pp 514–519

6. Rajinikanth V, Dey N, Satapathy SC, Ashour AS (2018) An approach to examine magnetic resonance angiography based on Tsallis entropy and deformable snake model. Futur Gener Comput Syst 18:160–172

7. Rajinikanth V, Raja NSM, Kamalanand K (2017) Firefly algorithm assisted segmentation of tumor from brain MRI using Tsallis function and markov random field. Control Eng Appl Informatics 19(3):97–106

8. Karthick J, Lakshmi D (2016) A Sub-Threshold SRAM design for stability improvement in low power. Int J Emerg Trends Sci Technol 3(5):383–387

9. Muthumaheswaran N, Kabilamani P, Lakshmi D (2016) A high gain e-band power amplifier using 45 nm CMOS technology. Int J Emerg Trends Sci Technol 3(5):353–354

10. Lakshmi D, Roy S, Ranganathan H (2014) Automated texture based characterization of fibrosis and carcinoma using low-dose lung CT images. Int J Imaging Syst Technol 24(1):39–44

11. Lakshmi D, Roy S, Ranganathan H (2014) ANOVA of texture based feature set for lung tissue characterization using CT Images. J Comput Appl 7(1):1–5

GPU Based Denoising Filter for Knee MRI

Shraddha Oza and Kalyani R. Joshi

Abstract MRI is a popularly used technique for diagnosing muscle and skeletal disorders, especially of the knee. For accuracy in diagnosis, the rician noisy knee image needs to be filtered using efficient denoising algorithm. In recent years, the spatial neighborhood bilateral filter is being explored by researchers for its capacity to retain edges and tissue structures. It is noted that increase in image resolution slows down performance of the bilateral filter effectively discouraging its use. The research work proposes a cost-effective accelerated solution to the problem by implementing CUDA-based bilateral filter as applied to T2-weighted sagittal knee MRI slice. The work suggests use of GPU shared memory for optimized implementation and better speedup. The speedup achieved for 3.96 Mpixel knee MR image is 114.27 times more than that of its CPU counterpart. The results indicate average occupancy of 90.15% for image size of 630^2 pixels, indicating effective parallelization. Also, over varying rician noise levels, the average PSNR achieved is 21.83455 dB indicating good filter performance.

1 Introduction

Magnetic Resonance Imaging (MRI) is one of the leading imaging modalities that employs strong magnetic field along with radio frequency pulses to generate images of any internal organ of the body. The noninvasive diagnostic technique is frequently used to investigate cartilage or muscle tear in the knee leading to decisions of appropriate treatment. The MR image of knee provides high-resolution images from all possible angles of bones, ligaments, cartilages, and muscles in the knee joint. The

S. Oza (✉)
E&Tc Department, Army Institute of Technology, Pune, India
e-mail: sdoza@aitpune.edu.in

K. R. Joshi
E&Tc Department, PES Modern College of Engineering, Pune, India
e-mail: krjpune@gmail.com

© Springer Nature Switzerland AG 2019   1845
D. Pandian et al. (eds.), *Proceedings of the International Conference on ISMAC in Computational Vision and Bio-Engineering 2018 (ISMAC-CVB)*, Lecture Notes in Computational Vision and Biomechanics 30,
https://doi.org/10.1007/978-3-030-00665-5_169

MRI technique may also help to effectively articulate cartilage deterioration and thus avoid unnecessary knee arthroscopy which is generally painful to patient [1, 2].

The process of MR imaging introduces a signal-dependent rician noise which is multiplicative in nature. The noise hinders visibility of the tissue structures in the image adversely affecting accuracy of diagnosis. In recent years, bilateral denoising filter has become popular with the researchers, due to its edge preserving ability of tissues and simplicity of implementation. The filter was proposed by Tomasi and Manduchi in the year 2005 and recently has found significant place in the domain of medical image processing [3–7]. The only disadvantage of the filter is its computational complexity which increases with increase in the image size and resolution. The MR imaging technique being digitized generates scans of knee joint with high resolution of the order of megapixels [8, 9]. The performance of bilateral filter slows down for such images which discourages its use for medical images. The research work proposes a solution in the form of accelerated bilateral filter using NVIDIA's CUDA GPU. The GPU devices are basically single-instruction multiple data architectures, thereby making them suitable for implementing image processing applications [10–12].

The GPU devices are extremely powerful parallel architectures which make it possible to achieve large concurrency and speedup. In the present research work, the bilateral filter is implemented on CUDA GPU GTX 830 M which is a blend of Maxell as well as Kepler architecture. It has a compute capability of 5.0 and in the present work, it is hosted by Intel i5 Quad-Core Processor @ 2.2 GHz with 12 GB RAM. Using this system, speedup achieved is 114.27 for a knee T2-weighted image slice with a resolution of 630×630. In the proposed work, the filter performance for rician noisy knee MRI is quantified using PSNR metric. For a maximum noisy image slice, the PSNR value achieved is 16.9 dB while is observed to be 27.88 dB for a minimum noisy image with $s = 0.1$ indicating good filter performance.

The sections are distributed as follows. Section 2 briefly gives details of knee MRI. Section 3 elaborates bilateral filter theory. Section 4 discusses CUDA programming model. Section 5 gives details of experimentation and results specific to implementation on GTX 830 M GPU device, while Sect. 6 states the conclusions of the work done.

2 Knee MRI

Magnetic resonance imaging (MRI) of the knee is performed to identify and analyze problems in the knee joint, such as bone tumors, damaged cartilage or ligaments, and arthritis. There exist different types of MRI sequences out of which T1-weighted and T2-weighted are most commonly used. The T1-weighted image is characterized by short echo time (TE) and pulse repetition time (TR) and is used to evaluate anatomy of the knee. In this image sequence, fatty bone marrow appears bright while the knee joint fluid appears dark. The reverse is true for T2-weighted and is more sensitive to pathological information of the knee joint internals. The choice of the image sequence depends on the type of disease or problem of the patient's knee [2, 3].

The MR image is corrupted with noise that is introduced from the molecular movement within body parts and the electrical resistance of the RF receiver coils. The MR image generated needs to have good quality with high signal-to-noise ratio for accurate diagnosis. MRI machine with higher magnetic field strength (>3 T) is able to generate images with higher SNR as well as of higher resolution in minimum time. In addition to magnetic field strength, relaxation time, TE/TR, and field of view are some of the key parameters affecting SNR of the MR Image. The noise in the MR image is modeled to be rician which is signal-dependent noise and is multiplicative in nature. The probability distribution of this noise appears to be rayleigh for SNR < 2 and is close to gaussian noise for SNR > 2 [3, 13, 14]. The present work uses T2-weighted image of the knee joint to build Rician noisy database for experimentation. The details of the same are discussed in Sect. 5. The next section briefly discusses the bilateral filter.

3   Bilateral Filter

Bilateral filtering is appropriate denoising algorithm for a Rician noisy image as it has the ability to retain contours. It works on the assumption that a pixel may influence another pixel in the neighborhood as well as the one which may not be in the neighborhood but anywhere in the image which has similar intensity value. The filter uses a spatial Gaussian which accounts for spatial proximity of pixels and a range Gaussian to identify photometric proximity between the pixels. The filter is defined as

$$BF[Ix] = \frac{1}{Wp} \sum_{y \in R} G_{\sigma_s}(\|x - y\|) G_{\sigma_r}(|Ix - Iy|) Iy \tag{1}$$

Here, R is a spatial neighborhood of pixel x. Spatial Gaussian weight G_{σ_s} is minimum for distant pixels, while G_{σ_r} is the range gaussian which is minimum for two dissimilar pixels. Filter performance is controlled by the values of standard deviation thresholds of space σ_s and range σ_r. In executing the filter equation, computations are done on each pixel with reference to all the pixels in the search area within the image. Thus, the computational complexity increases with the image and size of the neighborhood R. This also increases execution time of the filter which is unwanted. Therefore, bilateral filter may be discouraged to use for MR images with high resolutions as its performance would slow down [15–17]. The problem can be solved by porting the filter implementation to parallel domain. CUDA GPU avails an extensively parallel SIMD architecture on desktops and laptops making itself a cost-effective solution [18, 19]. The next section briefly discusses programming model of the CUDA GPU and implementation of bilateral filter using the same.

4 CUDA Programming Model

CUDA GPU was launched by NVIDIA Corporation in 2007 for general-purpose computing, and since then it has revolutionized parallel programming. Though parallel architectures like Cray, IBM cell were available, they required specialized programmers to use them. Also, GPUs were used especially for graphics applications and required OpenGL, OpenCL programming expertise. With CUDA compliant GPU, a professional C programmer could easily develop an application using CUDA C library which merely is an extension of C which brought the breakthrough.

CUDA compliant GPU architecture consists of multiple streaming multiprocessors (SMP), wherein each SMP has hundreds of cores (Fig. 1). A unique programming entity thread is mapped to a core and 32 such concurrently running threads are identified as a warp. A group of 16 or 32 warps are identified as a block, and a matrix of such blocks is defined as grid. In CUDA C programming environment, the kernel function is mapped to a grid of blocks which executes the function by launching many concurrent threads being executed by the cores. The CUDA programmer needs to predefine the concurrency in terms of number of blocks and threads with reference to the image matrix size. This concurrency factor defines the speedup achieved [20–22].

The data bandwidth between CPU memory and GPU memory is limited by that of PCI bus specification and often is a bottleneck in overall performance of GPU. Thus, these data transfers should be kept to a minimum and the GPU on-chip memory model should be made use of to achieve maximum performance. The GPU registers allocated per thread make up for the fastest storage while access to device memory slows down the performance speedup. The device allocates fast shared memory per block which when used optimizes the memory access to a great extent. The architecture also includes cached texture and constant memory which is read only and is of great use to accelerate the speed up.

The research work presented here uses shared memory to achieve the required speedup for bilateral filter implementation. It is one of the effective memory optimization techniques used to achieve efficient GPU performance [22]. The details are briefly discussed in Sect. 5.

Performance of GPU is generally quantified using occupancy index, number of active warps, branch diversions, and memory access. The occupancy index indicates the on-chip resources utilized by the kernel in terms of number of blocks instantiated, registers, and shared memory used [23]. For any CUDA application, these details are made available by the profiling tools such as NSight Visual Profiler. The details of experimentation and the profiler tool are discussed in the next section.

5 Experimentation

In the present work, the bilateral filter (BF) is implemented on CUDA GTX 830 M device (Fig. 2) hosted by Intel i5 Quad-Core processor @ 2.2 GHz with 12 GB RAM.

Fig. 1 CUDA GPU
programming model

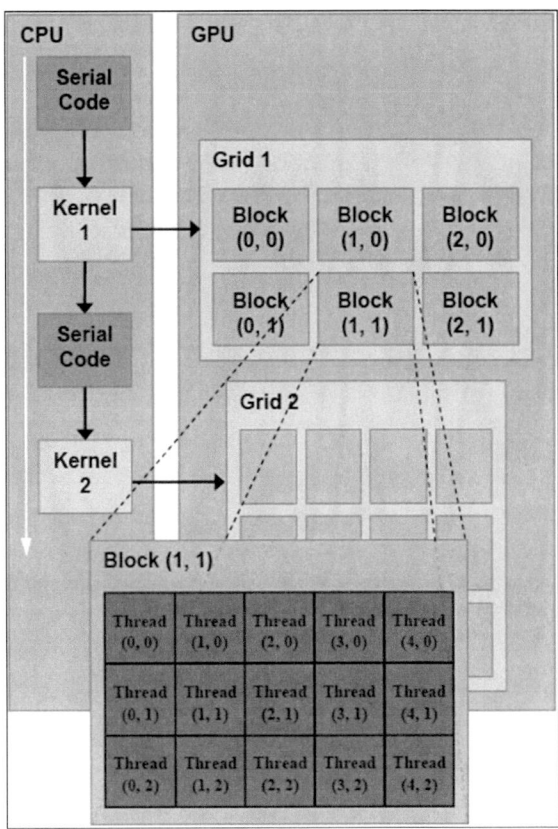

For building the test image database, T2-weighted sagittal knee MR image of 630 × 630 resolution is corrupted with varying levels of Rician noise from $s = 0.02$–0.8 in steps. The BF is applied to each test image, and the filter performance parameters such as MSE and PSNR are noted (Table 1). For quantifying the GPU performance, the execution time and memory access time for each test image are noted down along with number of blocks and occupancy index.

The proposed BF implementation uses shared memory which is very fast next to register access and is shared by all the threads in a block. In the present work, a tile of input MR image is loaded in shared memory for BF convolution. The scheduled threads are synchronized, and each thread operates on pixels in shared memory concurrently. On completion of computations, the tile is written back to global memory. This method accelerates the speedup by optimizing memory access. The observations of execution time and memory access are given in Table 1.

The CUDA C filter implementation uses OpenCV library for image read and display/write functions. OpenCV being an open-source image toolkit is extremely optimized and thus contributes to the overall speedup [24]. The filter radius is defined to be 5 × 5 and the filter parameters, namely, sigma domain and sigma range for BF

Fig. 2 Device properties of
GTX 830M GPU

Cuda Device ID [Address]	[0] [0x0]
Device	GPU 0 - GeForce 830M
Process	T13.exe [12344]
MAX_THREADS_PER_BLOCK	1024
MAX_BLOCK_DIM_X	1024
MAX_BLOCK_DIM_Y	1024
MAX_BLOCK_DIM_Z	64
MAX_GRID_DIM_X	2147483647
MAX_GRID_DIM_Y	65535
MAX_GRID_DIM_Z	65535
MAX_SHARED_MEMORY_PER_BLOCK	49152
TOTAL_CONSTANT_MEMORY	65536
WARP_SIZE	32
MAX_PITCH	2147483647
MAX_REGISTERS_PER_BLOCK	65536
CLOCK_RATE	1150000
TEXTURE_ALIGNMENT	512
GPU_OVERLAP	1
MULTIPROCESSOR_COUNT	2
KERNEL_EXEC_TIMEOUT	1
INTEGRATED	0
CAN_MAP_HOST_MEMORY	1
COMPUTE_MODE	0
MAXIMUM_TEXTURE1D_WIDTH	65536
MAXIMUM_TEXTURE2D_WIDTH	65536
MAXIMUM_TEXTURE2D_HEIGHT	65536
MAXIMUM_TEXTURE3D_WIDTH	4096

kernel (GPU), are varied from 20 to 200 with increase in noise level so as to achieve optimum filter performance. The readings for PSNR and MSE are given in Table 1. The GPU implementation is done with block size of 1024 threads and grid size of maximum {32, 32, 1} blocks. The observations for actual GPU utilization summary are given in Fig. 5 Also, the work here implements bilateral filter in sequential domain (BF_CPU) in C language using the same host i5 machine for analysis of speedup. The comparative execution time is shown in Fig. 3 graphically.

Table 1 Observations for BF_GPU filter performance in terms of MSE, PSNR, execution time, and memory access time, for 630 × 630 sized knee MRI with varying rician noise level from $s =$ 0.02 to 0.8

Noise Level (s)	MSE	PSNR (dB)	Execution time (msec)	Memory access time (msec) Host→Device	Memory access time (msec) Device→Host
0.02	115.5874	27.5017	1.412	0.28774	0.27136
0.05	123.4603	27.2155	1.414	0.28774	0.27546
0.1	148.8829	26.4024	1.406	0.29552	0.24269
0.2	246.4508	24.2135	1.384	0.28803	0.27136
0.3	398.2268	22.1295	1.371	0.30634	0.27443
0.4	588.8214	20.431	1.356	0.13414	0.27034
0.5	804.6602	19.0747	1.353	0.13517	0.27341
0.6	1042.385	17.9505	1.345	0.25088	0.23654
0.7	1273.116	17.0821	1.347	0.25091	0.23552
0.8	1508.768	16.3446	1.351	0.10854	0.26112

Fig. 3 Execution time of BF_GPU (1.3739 ms) versus BF_CPU (157 ms) showing 114.27 times speedup achieved

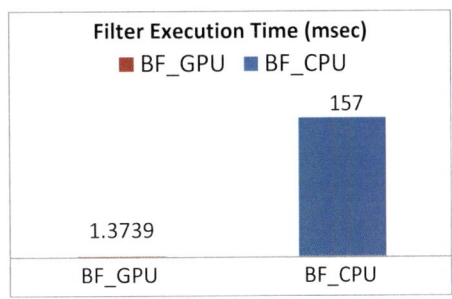

5.1 Observations and Discussion

It can be seen from Table 1 that, for minimum noise level of 0.02, the PSNR is 27.5 dB with MSE value as 115.5874. With increase in noise level from 0.02 to 0.8, the MSE increases to 1508.768 and PSNR drops to 16.3446 dB. The same is also shown graphically in Fig. 4a and b. The noisy images for $s = 0.02$, 0.5 and the respective filtered output images are given in Fig. 5 clearly indicating good filter performance.

The execution time with an average value of 1.3739 ms is achieved by GPU based bilateral filter while that achieved by its CPU counterpart is 157 ms indicating the GPU speedup of almost 114.27 (Fig. 3). The memory access time to copy the image from and to the CPU memory is tabulated in Table 1. The readings indicate total memory access time with an average value of 0.4957 ms which is almost one-third of the kernel execution time. This ratio is optimum because of shared memory being used.

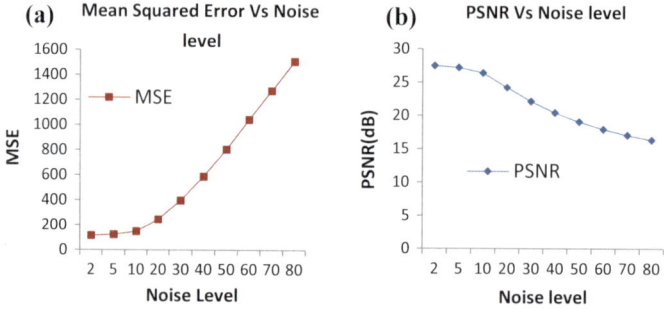

Fig. 4 Graphical representation of **a** MSE versus noise level, **b** PSNR (dB) versus noise level

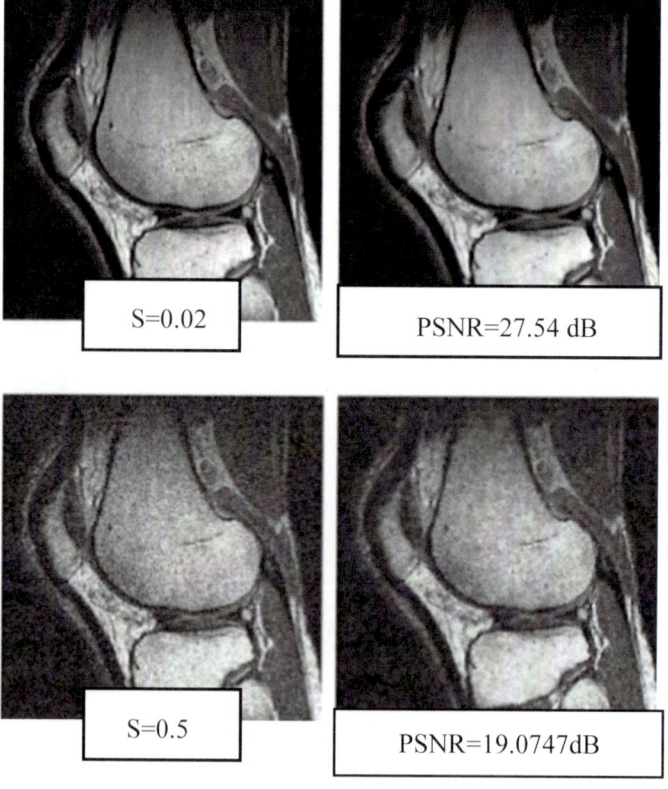

Fig. 5 Noisy image with s = 0.02, 0.5 and corresponding filtered output image with PSNR = 27.54 dB and 19.0747 dB, respectively

The observations from the report generated by nSight profiler are given in Fig. 6 and Table 2. It can be seen that the grid size, i.e., total number of blocks used, is 221(17, 13, 1) with 1024 threads being launched per block. The shared memory

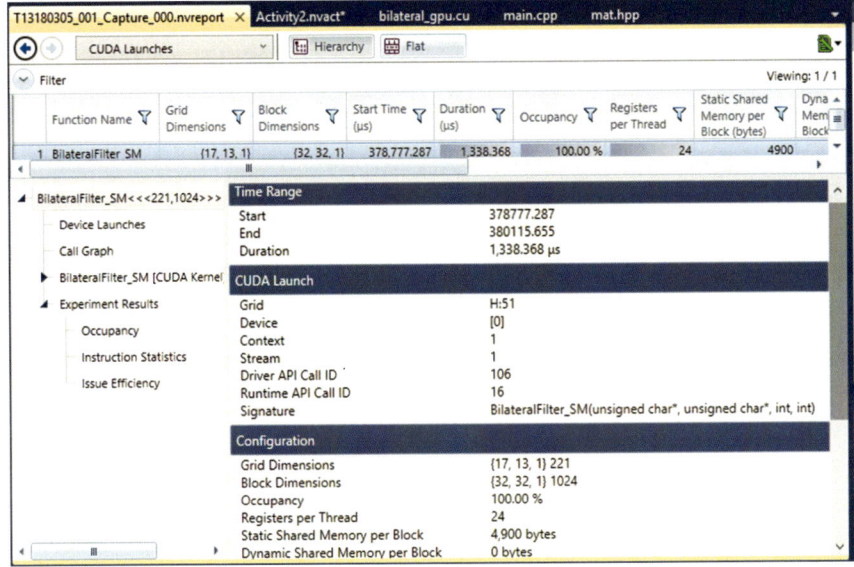

Fig. 6 Report of CUDA application BF profile generated by NSight visual profiler indicating execution time, memory access time, and device resource utilization in terms of blocks, threads, shared memory, and registers

Table 2 Observations of GPU device utilization for input image 630×630 pixels

Image Resolution (pixel s)	No. of Blocks/Grid	No. of Threads/Block	No. of Registers/Thread	Shared Memory used/Block (bytes)	Active warps (64 max)	Occupancy index (%)
630×630	221 {17,13,1}	1024	24	4900	64	90.15

used is 4900 bytes, and registers allocated per thread are 24. The usage of number of blocks and shared memory is unique to the test image size used here (630×630) and will change with change in image size. The number of active warps used for the filter implementation is 64 out of maximum 64 available (Table 2). In other words, the implementation is able to launch maximum possible number of threads indicating extremely efficient parallelization achieved. The occupancy index which is an index of utilization of resources, such as active warps, registers, and shared memory used, achieved here is 90.15%, again indicating the efficient parallelization.

6 Conclusion

The MRI of knee is significantly used by orthopedics for analyzing and diagnosing different problems such as knee pain, cartilage/muscle tear, and arthritis. The quality of the image needs to be with high SNR as it directly affects visibility of structures and tissues internal to knee. With digitization, the MR image size has increased to megapixels, thereby discouraging the use of bilateral filter, an efficient denoising algorithm.

The work presented here provides a cost-effective accelerated solution in the form of GTX 830 M CUDA GPU-based bilateral filter for T2-weighted sagittal knee MRI. The GPU implementation here uses shared memory technique giving speedup of 114.27 times over its CPU implementation. The GPU performance is characterized by occupancy index of 90.15%, clearly indicating efficient parallelization. The denoising performance of the BF measured in terms of PSNR is average 21.8345 dB indicating efficient noise removal for Rician noise level varying from 0.02 to 0.8. The results obtained here clearly underline the contribution of GPU to fast denoising of knee MR image. Further, it can be said that the CUDA GPU certainly would lead the research in coming years in the medical imaging domain.

Acknowledgements The research work acknowledges the extensive support provided by the management of Army Institute of Technology, Pune, India and of the research center, College of Engineering, Pune, India.

References

1. https://www.slideshare.net/benthungoe/mri-knee-of-orthopedic-importance-67500963
2. Rodrigues MB, Camanho GL (2010) Mri evaluation of knee cartilage. Rev Bras Orthop 45(4):340–346
3. Joshi KR, Oza SD (2018) MRI denoising for healthcare In: Kolekar MH, and Vinod Kumar (eds) Biomedical signal and image processing in patient care. IGI Global, pp 65–85
4. Oza SD, Joshi KR (2016) Performance analysis of denoising filters for MR images. In: Chakrabarti A, Sharma N, Balas VE (eds) Advances in computing applications. Springer, Singapore, pp 86–96
5. Tomasi C, Manduchi R (1998) Bilateral filtering for gray and colour images. In: Proceedings of the international conference on computer vision. IEEE, pp 839–46
6. Dougherty G (ed) (2011) Medical image processing: techniques and applications. Springer-Verlag, New York
7. Mohana J, Krishnavenib V, Guoc Y (2014) A survey on the magnetic resonance image denoising methods. Biomedical Signal Processing and Control, Elsevier, vol 9, pp 56–69
8. Rakhshan V (2014) Image resolution in the digital era: notion and clinical implications. J Dent Shiraz Univ Med Sci 15(4):153–155
9. Scholl I, Aach T, Deserno TM et al (2011) Challenges of medical image processing. Comput Sci Res Dev 26(5), Springer-Verlag
10. Shi L, Liu W, Zhang H, Xie Y, Wang D (2012) A survey of GPU-based medical image computing techniques. Quant Imaging Med Surg 2(3):188–206. AME Publishing Company
11. Moulika S, Boonna W (2011) The role of GPU computing in medical image analysis and visualization. In: Proceeding of Medical Imaging, vol 7967, SPIE
12. Kalaiselvi T, Sriramakrishnan P, Somasundaram K (2017) Survey of using GPU CUDA programming model in medical image analysis. Inf Med Unlocked 9:133–144. Elsevier
13. Gravel P, Beaudoin G, De Guise JA (2004) A method for modeling noise in medical images. IEEE transactions on medical imaging 23(10)

14. Gudbjartsson H, Partz S (2008) The rician distribution of noisy MRI data. In: HHS public access manuscript, PMC
15. Paris S, Kornprobst P, Tumblin J, Durand F (2008) Bilateral filtering: theory and applications. Comput Graph Vision 4(1):1–73
16. Zhang M (2006) Bilateral filtering in image processing, a thesis. Beijing University of Posts and Communications
17. Lekan M (2009) Impact of bilateral filter parameters on medical image noise reduction and edge preservation, a thesis. University of Toledo, Health Sciences Campus
18. Agarwal D, Wilf S, Dhungel A, Prasad SK (2012) Acceleration of bilateral filtering algorithm for many core and multicore architectures. In: Proceedings of IEEE 41st international conference on Parallel Processing (ICPP), pp 352–359
19. Larsson J (2015) A case study of parallel bilateral filtering on the GPU. Master Thesis, Malardalen University
20. Nvidia CUDA C (2012) Programming guide version 4.2
21. CUDA C (2012) Best practices guide DG-05603-001_v5.0. Design guide
22. http://www.training.praceri.eu/uploads/tx_pracetmo/GPU_Optimisation.pdf
23. CUDA Kernel occupancy : a detailed outlook. Application note, einfochips
24. www.opencv.org

Performance Evaluation of Audio Watermarking Algorithms Using DWT and LFSR

Ramesh Shelke and Milind Nemade

Abstract Advancement in Internet technology has compelled encryption, copyright protection and authentication of audios, videos, images, and documents. Watermarking techniques have been widely employed in copyright protection and authentications of digital media. Digital image watermarking techniques are extensively explored and found suitable for copyright protection of images, whereas audio watermarking techniques are less studied and need extensive research due to superior hearing ability of the human beings. In this paper, we present audio watermarking technique using discrete wavelet transform (DWT) and linear feedback shift registers (LFSR). Watermark signal is obtained using LFSR technique before embedding into audio signal. Dispersion of maximum power spectral density property of LFSR is explored that makes it suitable as scramblers. LFSR does not require secret key for scrambling and descrambling of watermark signal. Sequences were embedded into the DWT coefficients of the audio signal. Experimental simulations were performed to evaluate the performance of the audio watermarking technique. Audio watermarking finds applications in the area of ownership protection, tamper detection and authentication of music, military communication, voice-activated machines, and robots.

R. Shelke (✉)
Faculty of Engineering, Pacific Academy of Higher Education and Research University, Udaipur, India
e-mail: mail2shelke@gmail.com

M. Nemade
Department of Electronics Engineering, K. J. Somaiya Institute of Engineering and Information Technology, Mumbai, India
e-mail: mnemade@somaiya.edu

1 Introduction

Tremendous development in Internet, networking, and digital communication technology has fueled rapid sharing of digital media such as images, audios, videos, and documents. It has resulted in extensive demand for copyright protection and authentication of digital media. Due to sophisticated signal and image processing techniques, manipulation and duplication of digital data are much easier. Also, transmitting information over the communication networks is vulnerable for malicious attacks, and hence networks are insecure. Therefore, it is very much necessary to protect copyright information using watermarking techniques. The objective of the watermarking technique is to hide some information such as name, logo, signature, ID number, etc. into the actual media file without significant change in the original file [1, 2]. Whenever necessary, the hidden information can be extracted from the media to prove ownership or copyright. Audio watermarking is more difficult than image watermarking due to superior audio perception of human beings. Imperceptibility, robustness, and security are some of the most important characteristics of audio watermarking techniques. Security means that the embedded watermark cannot be extracted, manipulated, and changed by the unauthorized person, whereas robustness means least damage to the embedded watermark under various signal processing attacks and imperceptibility means embedded watermark must be inaudible. Furthermore, the audio watermarking technique must be computationally less complex and minimum time for embedding and extraction.

Many algorithms have been proposed for audio watermarking, and several of them are found to be too complex or require significant time for embedding/extraction or vulnerable to several forms of attacks [3]. However, complete security to the watermark signal is desirable without measurable distortion in the original audio signal. Patchwork techniques were proposed for image watermarking and extended further for audio watermarking. Patchwork techniques employ discrete cosine transform (DCT) and Fourier transform (FT) coefficients of original audio signals to obtain patches [4, 5]. Mostly, these patches do not have similar statistical properties. Similar technique was explored using wavelet transform coefficients to obtain two patches per audio segment. But it was unsuitable since many such audio patches do not have similar statistical properties. Hence, few patches were selected for watermarking. It also became difficult to identify that which patches were watermarked. The watermark embedding and extraction were explored using spread spectrum (SS) technique. SS sequence can be added to the host audio samples in time domain [5], FFT coefficients, in sub-band domain [6, 7], cepstral coefficients [8], and in a compressed domain [9]. Watermark signal is scrambled using LFSR technique before embedding into audio signal. Dispersion of maximum power spectral density property of LFSR is explored that makes it suitable as scramblers [10].

In this paper, performance of audio watermarking algorithms using DWT and linear feedback shift registers (LFSR) is evaluated to determine its suitability for audio watermarking techniques. Watermark signal is obtained using LFSR technique before embedding into audio signal. Dispersion of maximum power spectral density property of LFSR is explored that makes it suitable as scramblers. LFSR does not require secret key for scrambling and descrambling of watermark signal. Sequences were embedded into the DWT coefficients of the audio signal. The paper is organized as follows: Sect. 1 introduces to watermarking; Sect. 2 describes watermarking embedding and extraction algorithm in details. Results are discussed in Sect. 3 and finally concluded in Sect. 4.

2 Watermark Embedding and Extraction Algorithm

In this section, watermark embedding algorithm is discussed along with extraction algorithm. Extraction is exactly the reverse process of embedding.

2.1 Embedding Algorithm

Let $x[n]$ and $w[n]$ represent the original input audio signal and watermark signal, respectively, with m number of samples. The steps involved in the embedding of the audio signals are as follows:

1. Read $x[n]$ of the length m from the input audio signal.
2. Characteristic polynomial is

$$p(x) = x^4 + x^3 + x + 1 \tag{1}$$

   a. Compute the sequences by shifting the elements in circular manner.
   b. The sequences are obtained as

$$p = \{p_1, \ p_2, \ p_3, \ - - -, \ p_m\} \tag{2}$$

3. We apply DWT to input audio signal $x[n]$

$$\varphi(t) = \frac{1}{\sqrt{a}} \varphi\left(\frac{t - b}{a}\right) \tag{3}$$

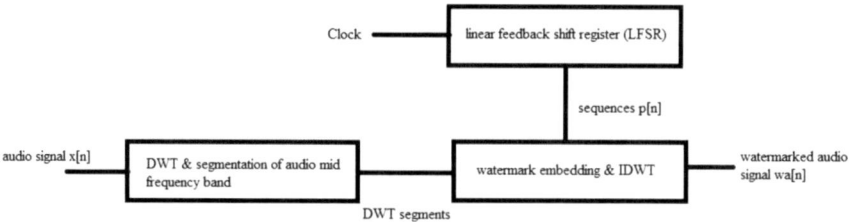

Fig. 1 Watermark embedding procedure

It is well known that very low-frequency and high-frequency components are vulnerable to signal processing attacks such as filtering and compression. Therefore, only mid-frequency range band can be used for watermark embedding.

4. We select the mid-frequency DWT coefficients for watermark embedding. These coefficients were divided into m segments of length N where m is the number of samples and N is the number of bits per watermark sequence sample.

$$a_i(k) = \{a_i(0), a_i(2), - - -, a_i(N - 1)\} \tag{4}$$

where $i = 0, 1, 2, 3, —, m-1$

5. We have m number of sequences each with N bits for watermark embedding. We now embed watermark bit into the segments of DWT coefficients using (5).

$$wa_{i,k}(0) = a_{i,k}(0) \, x \text{ or } p_i(k) \tag{5}$$

where $i = 0, 1, 2, 3, —, m-1$ and $k = 0,1, 2, —N-1$

6. After obtaining all watermark segments, inverse discrete wavelet transform (IDWT) is applied. Figure 1 shows the watermark embedding procedure.

2.2   Extraction Algorithm

The algorithm extracts the watermark from the received audio signal, with the use of original sequence p_1. Let $y[n]$ is the received watermarked audio signal with or without signal processing attacks.

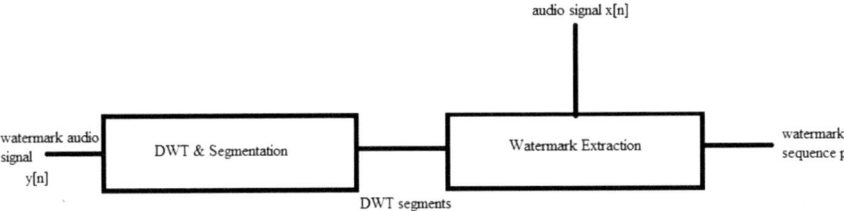

Fig. 2 Watermark extraction procedure

1. Read $y[n]$ of the length m from the input watermarked audio signal.
2. Characteristics polynomial is

$$p(x) = x^4 + x^3 + x + 1 \qquad (6)$$

   a. Compute the sequences by shifting the elements in circular manner.
   b. The sequences starting from p_1 are obtained as

$$p = \{p_1,\ p_2,\ p_3,\ ---,\ p_m\} \qquad (7)$$

3. We apply DWT to input audio signal $y[n]$

$$\varphi(t) = \frac{1}{\sqrt{a}} \varphi\left(\frac{t-b}{a}\right) \qquad (8)$$

Only mid-frequency range band was embedded with watermark; therefore, those coefficients are obtained.

4. We select the mid-frequency DWT coefficients for watermark embedding. These coefficients were divided into m segments of length N where m is the number of samples and N is the number of bits per watermark sequence sample.

$$ya_i(k) = \{ya_i(0),\ ya_i(2),\ ---,\ ya_i(N-1)\} \qquad (9)$$

5. We have m number of sequence with N bits each. The watermark is now extracted from the segments of the DWT coefficients using (10)

$$p_i(k) = ya_{i,k}(0)\ xor\ a_{i,k}(0) \qquad (10)$$

Figure 2 shows the watermark extraction procedure.

3   Experimental Results

Simulation results are depicted in this section using computer programming language MATLAB. Nine different audio signals were recorded using mobile phone with the sampling frequency of 44 kHz. Sound section was manually extracted from the audio, and inaudible section was removed. The audio signal was downsampled to obtain 4000 samples and quantized with 16 bits. Following types of attacks were used in the evaluation of the performance of the watermarking algorithm.

1. Low-pass filtering (LPF): low-pass filter with 12 kHz cut off frequency is applied to watermarked signal.
2. High-pass filtering (HPF): high-pass filter with 100 Hz cutoff frequency is applied to watermarked signal
3. MP3 attack (Compression): layer III compression was performed on watermarked signal.
4. Noise attack (NA): random noise of signal-to-noise ratio of 20 dB was added to watermarked signal.
5. Amplitude attack (AA): amplitude of the watermarked signal was increased by 1.25 times.

The robustness of the watermarking algorithm was measured using extraction rate (ET) given as (11)

$$ET = \frac{\text{correctly extracted watermark}}{\text{total embedded watermark}} \times 100\ (\%) \qquad (11)$$

Table 1 depicts the measured ET for all the nine audio signals under various signal processing attacks. Figure 3 shows the three audio signals before watermark embedding.

Table 1 Extraction rate

Audio signal/attacks	Extraction rate (ET) %								
	A1	A2	A3	A4	A5	A6	A7	A8	A9
LPF	91	92	94	93	94	92	90	91	93
HPF	96	94	95	97	96	94	93	95	94
MP3	82	81	83	81	83	83	80	81	83
NA	89	88	92	91	93	91	89	90	92
AA	98	97	99	98	99	98	96	97	98

Fig. 3 Audio signals **a** A1, **b** A7, **c** A9

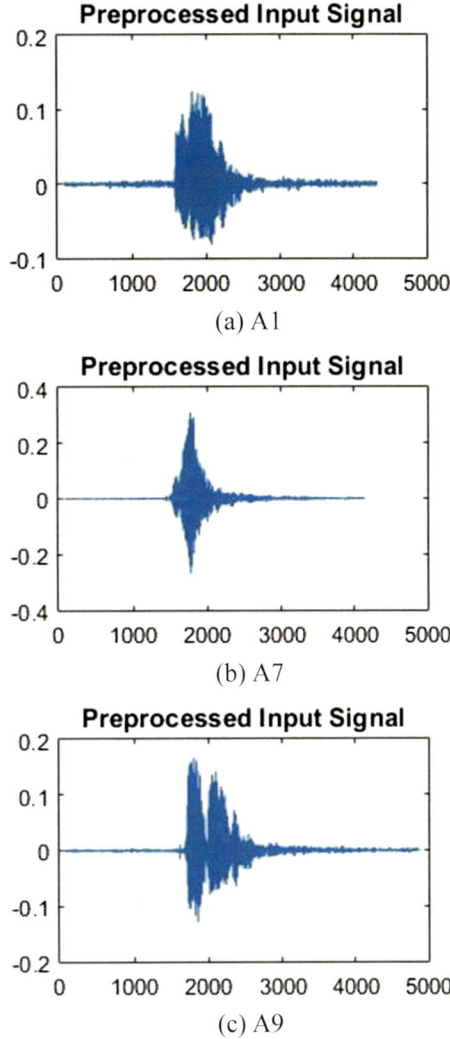

(a) A1

(b) A7

(c) A9

4 Conclusion

In this paper, performance of audio watermarking algorithms using DWT and linear feedback shift registers (LFSR) is evaluated to determine its suitability for audio watermarking techniques. Watermark signal is obtained using LFSR technique before embedding into audio signal. Experimental results clearly indicates suitable extraction rate for various types of attacks. Furthermore, extraction rate needs to be improved for MP3 or compression attack, LPF, and HPF. LFSR technique provides security to watermark signal by introducing randomness that makes it unpredictable.

References

1. Shanshan Li, Tingting Liu, Kun Bao, Ruijuan Qiao (2014) The study on the validity and reliability of digital audio watermarking. Comput Secur 2:34–36
2. Kob S, Nishimura R, Suzuki Y (2005) Time-spread echo method for digital audio watermarking. IEEE Trans Multimedia 7(2):212–221
3. Radha N, Venkatesulu M (2012) A chaotic block cipher for real-time multimedia. J Comput Sci 8(6):994–1000
4. Rahman MdM, Saha TK, Bhuiyan MdAA (2012) Implementation of RSA algorithm for speech data encryption and decryption. Int J Comput Sci Netw Sec 12(3):74–82
5. Kumar Singh, B, Tsegaye A, Singh J (2017) Probabilistic data encryption using elliptic curve cryptography and arnold transformation. In: IEEE international conference on I-SMAC (IoT in Social, Mobile, Analytics and Cloud), pp 644–651
6. Elshazly AR et al (2016) Synchronized double watermark audio watermarking scheme based on a transform domain for stereo signals. In: IEEE fourth international ja-pan-egypt conference on electronics, communications and computers
7. Balamurugan R, Kamalakannan V, Rahul Ganth D, Tamilselvan S (2014) Enhancing security in text messages using matrix based mapping and elgamal method in El-liptic curve cryptography. In: IEEE international conference on contemporary computing and informatics, pp 103–106
8. Xiang Y, Natgunanathan I, Peng D, Zhou W, Yu S (2012) A dual-channel time-spread echo method for audio watermarking. IEEE Trans Inf Forensics Sec 7(2):383–392
9. Kang H, Yamaguchi K, Kurkoski B, Yamaguchi K, Kobayashi K (2008) Full-index-embedding patchwork algorithm for audio watermarking. IEICE Trans Inf Syst, E91–D(11):2731–2734
10. Menezes AJ, van Oorschot PC, Vanstone SA (2001) Handbook of applied cryptography, 5th edn, CRC Press

Biometric Image Encryption Algorithm Based on Modified Rubik's Cube Principle

Mahendra V. Patil, Avinash D. Gawande and Dilendra

Abstract Multi-biometric systems are being mostly installed in many large-scale biometric applications most importantly UIDAI system in India. However, multi-biometric systems require storage of multiple biometric templates for each user that results in increased risk to user privacy and cybercrimes. Among all the biometrics, fingerprints are widely used in personal verification systems especially from small-scale to large-scale organization. Encryption of the biometric images is very important due to increasingly use of insecure systems and networks for storage, transmission, and verification. We proposed encryption algorithm based on modified Rubik's cube principle for fingerprint biometric images in this paper. Secret key for the encryption is obtained from the same fingerprint biometric image that assists in key management and storage for further decryption and verification process. The principle of Rubik's cube involves row and column rotation or changing of the pixels position. Further, bitwise XOR operation through modifications in secret key adds to security of the biometric images. Experimental simulations were performed to obtain the encryption parameters and validate the algorithm through common attacks such as noise and compression. Thus, the proposed encryption algorithm may be applied to fingerprint images used in attendance monitoring systems since it provides key management and does not involve complex calculations.

M. V. Patil (✉) · Dilendra
Faculty of Engineering, Pacific Academy of Higher
Education and Research University, Udaipur, India
e-mail: apmahendra@yahoo.com

A. D. Gawande
Department of Computer Science and Engineering, Sipna College
of Engineering and Technology, Amravati, India

© Springer Nature Switzerland AG 2019
D. Pandian et al. (eds.), *Proceedings of the International Conference on ISMAC in Computational Vision and Bio-Engineering 2018 (ISMAC-CVB)*, Lecture Notes in Computational Vision and Biomechanics 30,
https://doi.org/10.1007/978-3-030-00665-5_171

1 Introduction

With the tremendous development in Internet and communication technology, sharing of audio, images, video, and biometric images has also increased. Users of the multimedia data are able to carry out real-time streaming video clips, video conferencing, listening to music, etc. However, most of the communication networks and systems are insecure and vulnerable to virus attacks. Therefore, these are unsuitable for communicating sensitive and important multimedia content such as defensed, commercial, or private information, and biometric images and videos. Thus, it necessitates the usage of encryption algorithm in the protection of valuable information.

Encryption algorithms have been extensively studied by the researchers for the data previously. This standard has resulted in algorithms like Data Encryption Standard (DES), Advanced Encryption Standard (AES), RC4, and RC6. However, application of these algorithms for images may not result in practically beneficial image encryption schemes. Image encryption schemes are different from data encryption schemes because image data are usually large in size, require more time for encryption, and also require a comparable amount of time for decryption. Computational overhead makes the traditional encryption algorithm inappropriate for real-time image encryption and decryption [1–3].

Multi-biometric systems are being mostly installed in many large-scale biometric applications most importantly UIDAI system in India. However, multi-biometric systems require storage of multiple biometric templates for each user that results in increased risk to user privacy and cybercrimes. Among all the biometrics, fingerprints are widely used in personal verification systems especially from small-scale to large-scale organization. Encryption of the biometric images is very important due to increasingly use of insecure systems and networks for storage, transmission, and verification. This provides a strong motivation for developing techniques that can protect the biometrics fingerprint data particularly during communication and transmission over insecure network channels. Many solutions based on spatial-domain and frequency- or transform-domain approach are proposed [4–7]. Spatial-domain technique uses traditional cryptography algorithms, whereas frequency-domain technique uses transforms such as Wavelet and Fourier. In both the cases, the basic idea is to shuffle pixel positions in the biometric images and chaotically generate transform orders [8–10].

Two important attributes of a powerful encryption technique are diffusion and confusion. Hence, the motivation is to generate a scheme which has more diffusion and confusion. Due to some intrinsic features of biometric images, some of the traditional encryption schemes are not very suitable for image encryption. Hence, the motivation is to generate schemes to encrypt images efficiently. To enhance the security of images, a biometric cryptosystem approach combines cryptography and biometrics. Under this approach, the biometric image is encrypted with the help of key. A key generated with the combination of fingerprint and password is used for image encryption. This mechanism is seen to enhance the security of biometrics images during storage, transmission, and verification.

We proposed encryption algorithm based on modified Rubik's cube principle for fingerprint biometric images in this paper. Secret key for the encryption is obtained from the same fingerprint biometric image that assists in key management and storage for further decryption and verification process. The principle of Rubik's cube involves row and column rotation or changing of the pixels position. Further, bitwise XOR operation through modifications in secret key adds to security of the biometric images. The paper is organized as follows: Sect. 1 introduces encryption; Sect. 2 describes encryption and decryption algorithm in details. Results are discussed in Sect. 3 and finally concluded in Sect. 4

2 Modified Rubik's Cube Algorithm

In this section, modified Rubik's cube encryption algorithm is discussed along with decryption algorithm. Decryption is exactly the reverse process of encryption.

2.1 Encryption Algorithm

Let I represent the 8-bit grayscale input biometric image of the size m x n. The steps involved in the encryption of the biometric image are as follows:

1. Generate two vectors k_r and k_c of the length m and n, respectively, from the input biometric images randomly from any one row and column.
2. Decide the number of iterations *i*max and initialized counter $i = 0$.
3. For each row x in I,

    a. Calculate the sum of all elements in the row x represented by ar(x)
    b. Compute modulo 2 of ar(x) represented by mar(x)
    c. Left or right circular shift each pixel elements $k_r(x)$ times according to the following if mar(x) $= 0$ right shift else left shift.

4. For each column y in I,

    a. Calculate the sum of all elements in the column y represented by ac(y)
    b. Compute modulo 2 of ac(y) represented by mac(y)
    c. Up or down circular shift each pixel elements $k_c(y)$ times according to the following if mac(y) $= 0$ upshift else downshift.
    Steps 3 and 4 will create a scrambled image represented by I_s.

5. Use vector k_r to apply bitwise XOR operator on each row of the partial resultant image I_s using the following equations:

$$I_1(x, y) = I_s(m - x, n - y)\text{XOR } k_r(x) \tag{1}$$

where XOR is the XOR operator applied on flipped scrambled image upside down with vector $k_r(x)$.

6. Use vector k_c to apply bitwise XOR operator on each column of the partial resultant image I_1 using the following equations:

$$I_2(x, y) = I_1(m - x, n - y)\text{XOR } k_c(y) \tag{2}$$

where XOR is the XOR operator applied on flipped partially encrypted image upside down with vector $k_c(y)$.

7. Increment the counter $i = i + 1$.
8. Repeat steps 3–7 if $i <$ imax else stop.

The encrypted image Ie is obtained, and thus the process of encryption is completed. Vectors k_r and k_c are designated as secret key and obtained randomly from the input biometric image. Also, vectors k_r and k_c are randomly hidden into the final encrypted image and thus reduce key management. Vectors k_r and k_c can be again obtained from the encrypted image during decryption for further processing of biometric image.

2.2 Decryption Algorithm

Let Ie represent the 8-bit grayscale input encrypted biometric image of the size $m + 1 \times n + 1$ due to hidden vectors k_r and k_c. The steps involved in the decryption of the encrypted biometric image are as follows:

1. Obtain two vectors kr and kc of the length m and n, respectively, from the encrypted biometric image and restore the size of the encrypted image to $m \times n$.
2. Determine the number of iterations imax and initialized counter $i = 0$.
3. Use vector k_c to apply bitwise XOR operator on each column of the encrypted image I_e using the following equations:

$$I_{11}(x, y) = I_e(m - x, n - y)\text{XOR } k_c(y) \tag{3}$$

where XOR is the XOR operator applied on flipped encrypted image upside down with vector $k_c(y)$.

4. Use vector k_r to apply bitwise XOR operator on each row of the partial resultant image I_{11} using the following equations:

$$I_{21}(x, y) = I_{11}(m - x, n - y)\text{XOR } k_r(x) \tag{4}$$

where XOR is the XOR operator applied on flipped scrambled image upside down with vector $k_r(x)$.

5. For each column y in I_{21},

   a. Calculate the sum of all elements in the column y represented by ac(y)
   b. Compute modulo 2 of ac(y) represented by mac(y)
   c. Up or down circular shift each pixel elements $k_c(y)$ times according to the following if mac(y) = 1 upshift else downshift.

6. For each row x in I_{21},

   a. Calculate the sum of all elements in the row x represented by ar(x)
   b. Compute modulo 2 of ar(x) represented by mar(x)
   c. Left or right circular shift each pixel elements $k_r(x)$ times according to the following if mar(x) = 1 right shift else left shift.
      Steps 5 and 6 will create a final decrypted image represented by Id.

7. Increment the counter $i = i + 1$.
8. Repeat steps 3–7 if $i < i$max else stop.

The complete process of encryption and decryption with secret key management policy facilitates security of the biometric images during the process of transmission, storage, and verification. Also, the modified algorithm overcomes the disadvantage of not having constant values in vectors k_r and k_c. The complete process is based on changing pixel position in more confused manner with no secret key generation and management policy.

3 Experimental Results

Visual testing, entropy analysis, noise attack, and compression attack was performed on the obtained experimental results through modified Rubik's cube principle and algorithms.

3.1 Visual Testing

In this experiment, biometric fingerprint image of size 125×125 pixels was used for visual testing. Figure 1 illustrates the input biometric fingerprint image and its encrypted output image through the proposed algorithm. It is clearly understood that there is absolutely no pictographic similarity between these images. It is desired that the encrypted image should mostly be different from its original input image. Mostly two parameters are obtained to compute the differences between input and encrypted image. First, the measure is the number of pixels change rate (NPCR) that specifies the percentage of dissimilar pixels between two images. The second one is the unified average changing intensity (UACI), which measures the average intensity of dissimilarities in pixels between two images [11]. To achieve the performances of an ideal image encryption algorithm, NPCR values must be as large as possible, and

Fig. 1 Result of visual
testing

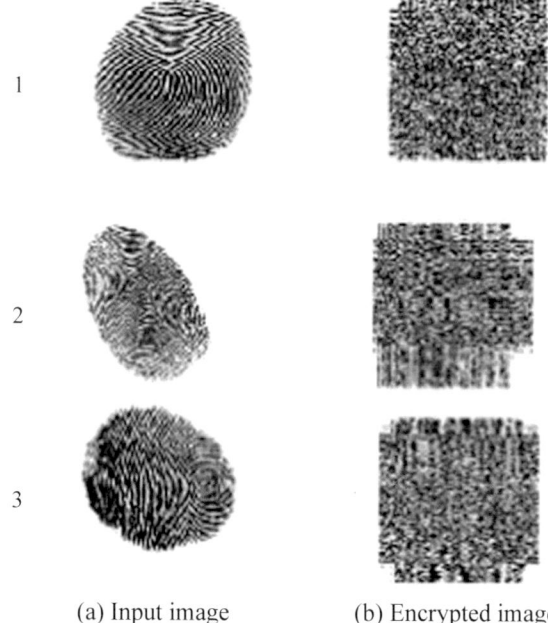

(a) Input image   (b) Encrypted image

Table 1 NPCR and UACI

Biometric images	NPCR	UACI
Fingerprint 1	99.18	25.02
Fingerprint 2	98.17	28.87
Fingerprint 3	98.98	25.32

UACI values must be around 30%. Table 1 indicates the obtained values of NPCR and UACI for the original input image and their encrypted counterpart. Thus, the values are very close to 100% for the NPCR measure and also the UACI values are also acceptable. The high percentage values of the NPCR measure clearly indicates that the pixels locations have been arbitrarily changed. Furthermore, the UACI values show that practically all pixel grayscale values of encrypted image have been altered from their values in the original input image.

3.2 *Entropy Analysis*

For 8-bit grayscale images having 256 levels, if each level of gray is assumed to be equiprobable, then the entropy of this image will be theoretically equal to 8 (no. of bits). Thus, it is understood that ideally an encryption algorithm for biometric images should give an encrypted image having entropy nearly equal to 8. Table 2 shows the entropy values of the three input fingerprint images and their respective encrypted images.

Table 2 Entropy values

Number	Input image	Encrypted image
Fingerprint 1	6.5	7.96
Fingerprint 2	5.5	7.89
Fingerprint 3	6.8	7.97

3.3 Noise Attack

To measure the robustness of the proposed modified Rubik's cube algorithm, random noise was added into the encrypted image. Thereafter, the noisy encrypted image was decrypted using algorithm. In this experiment, salt-and-pepper noise of density 0.01 was added into the encrypted image. Figure 2 shows the noisy input image and its encrypted and decrypted counterpart images. Results clearly indicate that noise attack such as salt-and-pepper noise has significant effect on the decrypted image.

Fig. 2 Result of noise attack

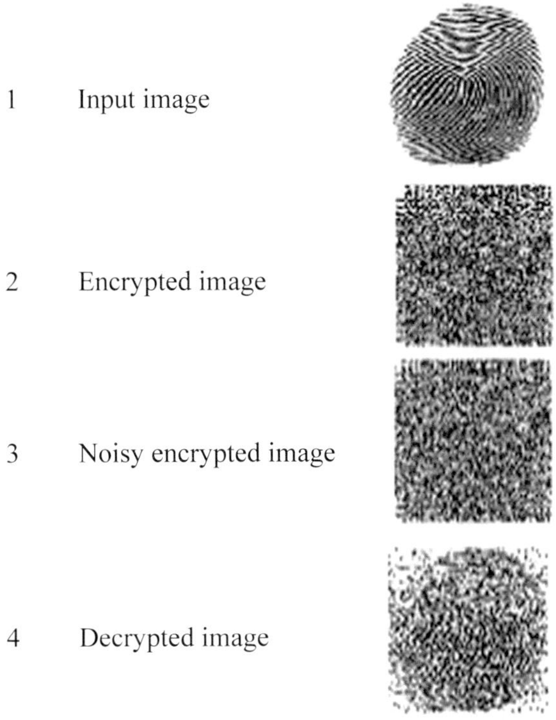

1   Input image

2   Encrypted image

3   Noisy encrypted image

4   Decrypted image

Fig. 3 Result of
compression attack

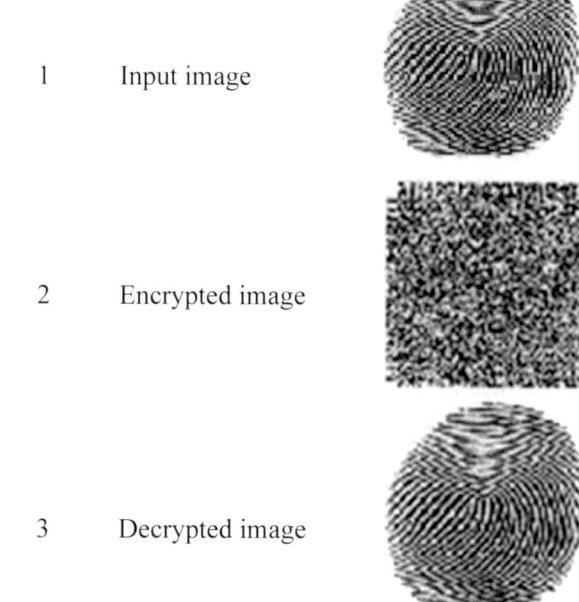

1   Input image

2   Encrypted image

3   Decrypted image

3.4 Compression Attack

Mostly, in the biometric image systems, the original input fingerprint image will be scanned, encrypted, and stored. Therefore, it is mandatory to evaluate the effect of compression on the decryption process. In this experiment, the original encrypted image was compressed using JPEG format and then decrypted. Figure 3 shows the input image, encrypted image, and decrypted image obtained after encoding and decoding using JPEG algorithm. Thus, it is clearly illustrated that compression does not significantly affect the decryption process and is robust against it.

4 Conclusion

We proposed modified encryption algorithm based on Rubik's cube principle for fingerprint biometric images in this paper. Secret key for the encryption is obtained from the same fingerprint biometric image that assists in key management and storage for further decryption and verification process. Experimental simulations were performed to obtain the encryption parameters and validate the algorithm through common attacks such as noise and compression. The simulation was done using MATLAB software on Intel i5 processor with 3 GHz speed and 4 GB RAM. Visual testing, entropy analysis, noise attack, and compression attack were performed on the obtained experimental results. The proposed algorithm is best suitable for biometric image encryption especially fingerprint. It is robust to compression attack and satisfies all encryption parameters such as NPCR, UACI, and entropy. Further, the robustness of the proposed algorithm needs to be improved for noise attack.

Acknowledgements Fingerprint images are used the author and co-authors. The Consent of all participants was taken.

References

1. Murugan B, Gounder AGN (2016) Image encryption scheme based on blockbased confusion and multiple levels of diffusion. IET Comput Vision 10(6):593–602
2. Bringer J, Chabanne H, Patey A (2013) Privacy-preserving biometric identification using secure multiparty computation: An overview and recent trends. Sig Process Mag 30(2):42–52
3. Bhatnagar G, Wu QJ, Raman B (2012) A new fractional random wavelet transform for fingerprint security. IEEE Trans Syst Man Cybern Part A Syst Hum 42(1):262–275
4. Bhatnagar G, Wu QJ (2012) Chaos-based security solution for fingerprint data during communication and transmission. IEEE Trans Instrum Meas 61(4):876–887
5. Norouzi B, Seyedzadeh SM, Mirzakuchaki S, Mosavi MR (2013) A novel image encryption based on row-column, masking and main diffusion processes with hyper chaos," Springer Multimedia Tools and Applications, pp 1–31
6. Nagar A, Nandakumar K, Jain AK (2012) Multibiometric cryptosystems based on feature-level fusion. IEEE Trans Inf Forensics Secur 7(1):255–268
7. Hongjuna L, Xingyuana W (2010) Color image encryption based on one-time keys and robust chaotic maps. Elsevier J Comput Math Appl 5:3320–3327
8. Philip Chen CL, Zhang T, Zhou Y (2012) Image encryption algorithm based on a new combined chaotic system. In: IEEE International conference on systems, man, and cybernetics
9. Wang Y et al (2011) A new chaos-based fast image encryption algorithm. Elsevier J Appl Soft Comput 11:514–522
10. Wu X, Wang Z (2015) A new DWT-based lossless chaotic encryption scheme for color images. In: IEEE international conference on computer and computational sciences
11. Chen G, Mao Y, Chui CK (2004) A symmetric image encryption scheme based on 3D chaotic cat maps. Chaos, Solitons Fractals 21(3):749–761

Leaf Disease Detection Based on Machine Learning

Anish Polke, Kavita Joshi and Pramod Gouda

Abstract Farming is a standout among the most imperative elements in light of which a nation's economy is chosen. Alignments in crops are very regular, which is one of the prominent factors that leads to the disease location and detection in plant's parts which is of high importance in agroindustry. In this way, it is imperative to effectively distinguish the maladies from the harvest to specifically shower herbicides and treat to diminish wastage utilization of concoction. In this work, we display an approach that coordinates picture handling and machine figuring out how to permit diagnosing infections from leaf pictures and to contemplate quantitative plant physiology imaging and PC vision which is utilized. Wavelet is extremely mainstream apparatus in picture preparing calculation. Surface highlights are utilized for recognition of yield. These features were mean, standard deviation, skewness, and kurtosis which we have used in this paper. We are proposing an approach, which used to identify the plant infection, i.e., plant disease. Here, we are using minimum distance classifier for the classification of the disease. The proposed approach displays a way toward robotized plant sicknesses finding on a gigantic scale.

1 Introduction

Indian economy is profoundly reliant of agrarian efficiency. One of the major factors responsible for the crop destruction is plant disease. Display strategy for infection discovery of plant is perception by specialists through their stripped eye which required

A. Polke (✉) · K. Joshi · P. Gouda
Department of Electronics and Telecommunication, G.H. Raisoni
College of Engineering and Management, Pune, India
e-mail: polke_anish.ghrcemmevlsi@raisoni.net; anishpolke@gmail.com

K. Joshi
e-mail: kavita.joshi@raisoni.net

P. Gouda
e-mail: pramod.gouda@raisoni.net

© Springer Nature Switzerland AG 2019
D. Pandian et al. (eds.), *Proceedings of the International Conference on ISMAC in Computational Vision and Bio-Engineering 2018 (ISMAC-CVB)*, Lecture Notes in Computational Vision and Biomechanics 30,
https://doi.org/10.1007/978-3-030-00665-5_172

a substantial group of specialists and also consistent observing of plant is required, which is expensive undertaking. In created nations, they are utilizing the bleeding edge advancements like data and space innovation, and biotechnology. It is important to change and use the exploration for eco-accommodating yield administration and expanded part of research-based innovations and supportable harvest generation. In plants, some broad sicknesses seen are dark colored and yellow spots, early and late singe, and others are parasitic, viral, and bacterial illnesses. Picture preparing [1] can be utilized for estimating influenced territory of ailment and to decide the distinction in the shade of the influenced zone. Programmed location of the illnesses by simply observing the indications on the plant leaves makes it simpler and additionally less expensive. Accordingly, in the field of farming, utilization of programmed illness recognition strategy in plants assumes a critical part. Recognition of a plant malady in extremely introductory stage would be exceptionally gainful. This additionally underpins machine vision to give picture-based programmed process control, examination, and robot direction. The straightforward standard of the activity of proposed framework is to take the reactions from different parameters that choose the profitability, handling them as per the calculation, and anticipate the appropriate yield for the land alongside the present status of product if tainted or not. Additionally, recommendation is given of a few manures that could be utilized to enhance the richness. For making framework accessible to average citizens, we propose that this framework is made as an Android application, where the agriculturist could bolster the information sources and get the essential outcome.

2   Literature Survey

After auditing the various reports, we can infer that the machine vision examined differs broadly in relying upon the intricacy of the visual scene. With the progression of the PC, the greater part of the on-field machine vision recognition work has been performed and come about, that is, beginning from singular leaves kept on plain foundation to limit leaves scenes of field plants developing in on-field characteristic conditions. Surfaces and size component are utilized for weed discovery and mechanization of manure splashing [2]. Wavelet change is utilized to extricate the surface highlights of the yield and weed pictures for order [3].

Singh and Misra [4] present the review on various diseases sorting methods utilized in plant leaf illness location and a calculation for picture division strategy that can be utilized for programmed identification and additionally characterization of plant leaf ailments. With less computational endeavors, the ideal outcomes were acquired, which likewise demonstrates the proficiency of proposed calculation in acknowledgment and characterization of the leaf illnesses. Another favorable position of utilizing this technique is that the plant's illnesses can be distinguished at the beginning period or the underlying organize. To enhance acknowledgment rate in grouping process, simulated neural network, Bayes classifier, fuzzy logic, and crossbreed calculations can likewise be utilized.

Raj Kumar and Sowrirajan [5] have proposed image-processing-based way to deal with consequent order of the ordinary or infected leaves (Early leaf spot, late leaf spot, alternaria leaf spot), and furthermore give the cure to a similar which would be gainful to novices in cultivating or planting (as these maladies are basic in blossoming plants like rose too). In this approach, they have joined all the mixture highlights of a leaf to prepare the ANN (BPN-FF) and have made utilization of Lloyd's bunching which is more productive than the customary K-implies grouping to section the test pictures.

Dheeb Al Bashish et al. [6] proposed that the RGB images are converted into HSI plane, and then the color features are extracted (by SGDM generation). The texture features are extracted by obtaining Gray-Level Co-occurrence Matrix (GLCM). The information pictures are divided utilizing K-means clustering system, and afterward the sectioned pictures are broke down by a preprepared BPN arranged for discovery and order of plant leaf sicknesses. The author also compares between various models incorporating various components such as HS, H, S, I, and HSI and found that model HS provides the best efficient output among all other models with the efficiency of 92.7%.

Niket et al. [7] stated that the RGB images upon acquisition undergo color space transformation into HSI plane and upon segmentation using K-means clustering the green pixels are masked from the appropriate cluster, and the masked green pixels are removed. Then, the useful segments are obtained. Then, the texture features are extracted using CCM. The classifier used is BPN-FF. But additional color or shape features or both might have improved the efficiency of classification.

Arivazhagan et al. [8] proposed a new method in which the RGB images are converted into HSI plane, and only the hue component is used for further analysis. Then, the green pixels are masked, and the masked green pixels are removed. The useful components are obtained upon segmentation, and only the texture feature is extracted using co-occurrence matrix. Then, the neural network employing SVM classifier is used to detect and classify with the efficiency of 94.74%. Here, only a single feature extraction is employed and the classifier is not that efficient in classifying the disease but effectively detects whether the leaf is diseased or not.

Kaur and Himanshu [9] provided the study of various classifiers, namely, K-Nearest Neighbors (KNN) classifier, Support Vector Machines (SVM) classifier, Backpropagation Neural Network-Feedforward (BPN-FF) classifier, Probabilistic Neural Network (PNN) classifier, and General Regression Neural Network (GRNN) classifier.

Sanjeev et al. [10] provided a new way in which the k-means segmentation is done, followed by feature extraction using GLCM and the classification is done via BPN. Here, only hue component is used and instead of k-means segmentation, other algorithms could have employed to extract lesion more accurately.

Ravichandran and Koteeshwari [11] built android app for the crop prediction which suggests the suitable crop to farmer based on soil parameters, which were entered in application by farmer after soil testing. Here, automatic soil testing is not performed and there is no check on the current crop status.

3 Proposed System

We are proposing a system which is a combination of hardware and software. For obtaining the physical attributes, we use sensors like pH and soil moisture sensors, which are interfaced with controller. Here, we are using raspberry pi controller. Figure 1 shows the overview of proposed system.

Digital camera is used to capture the images from farm and transfer them to computer where image processing is performed in order to detect any diseases on the plant leaves. If starting of disease is detected, then appropriate fertilizer is sprinkled on crop using fertigation pump. As we are having pH and moisture-related data of soil, suggestion can be given to farmer which crop is suitable for that soil. We have gathered the dataset of over 200 high-resolution images of leaves. First, preprocessing of images is performed to remove any noise in image to perform enhancement of image; then, we perform segmentation using the conversion of image from RGB to YC_BC_R plane and feature extraction of where we have considered features such as mean, standard deviation, kurtosis, and skewness for texture features and discrete waveform transformation for the color feature extraction. Classifier utilized is minimum distance classifier which is utilized to arrange obscure picture information to classes which limit the separation between the picture information and the class in multi-include space. The separation is characterized as a record of comparability; so,

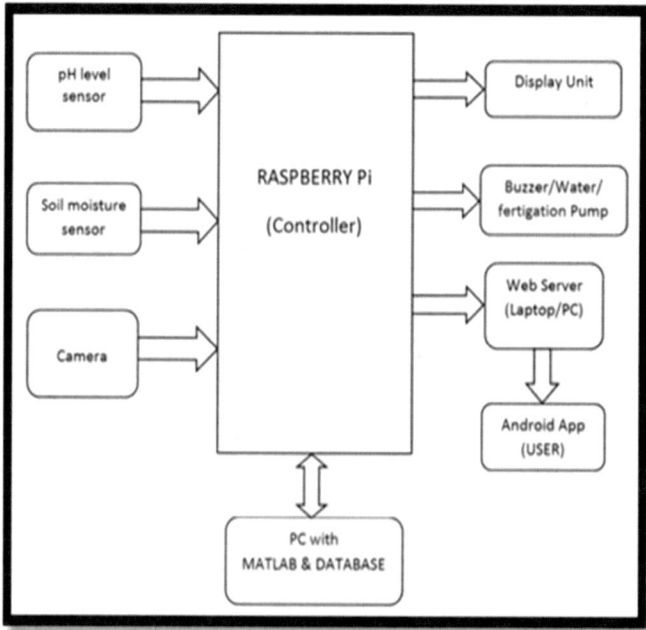

Fig. 1 Block diagram of the system

the base separation is indistinguishable to the greatest closeness with the reference picture, in light of which choice is taken if leaf is having any illness or in solid state.

4   Result Analysis

The result of the proposed system is given below.

After training the dataset, the preprocessing of the input image is done by using the top hat and bottom hat approach. Figure 2 shows the top hat and bottom hat enhancement results. The enhancement is used for enhancing the features of the raw image; after the enhancement, the features of the input image get enhanced and the level of noise gets reduced [12].

After the enhancement, the image segmentation is performed which is used to segment the image into various contours; after the segmentation, the features are extracted. The output of YC_bC_r model is given in Fig. 3.

$$Y = (77/256)R + (150/256)G + (29/256)B \tag{1}$$

$$C_b = -(44/256)R - (87/256)G + (131/256)B + 128 \tag{2}$$

$$C_r = (131/256)R - (110/256)G - (21/256)B + 128 \tag{3}$$

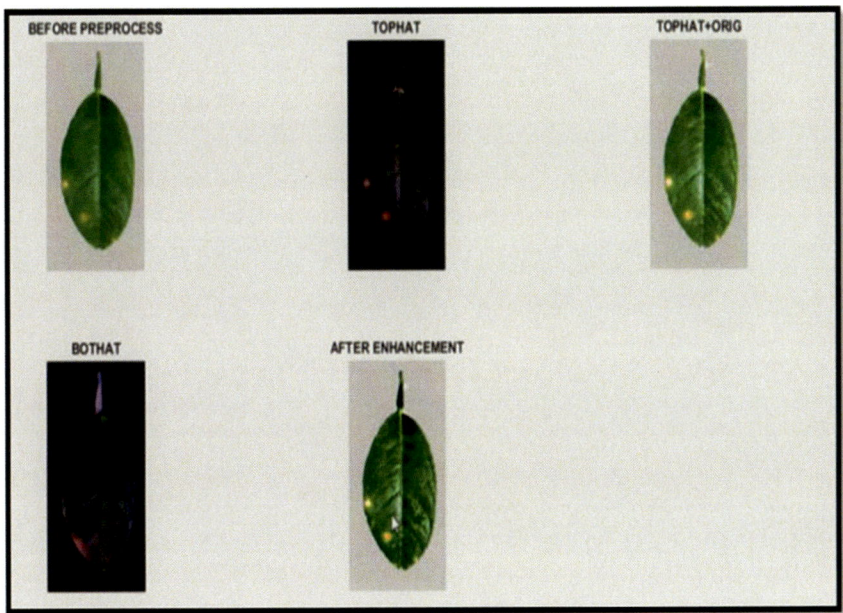

Fig. 2 Output of final enhancement

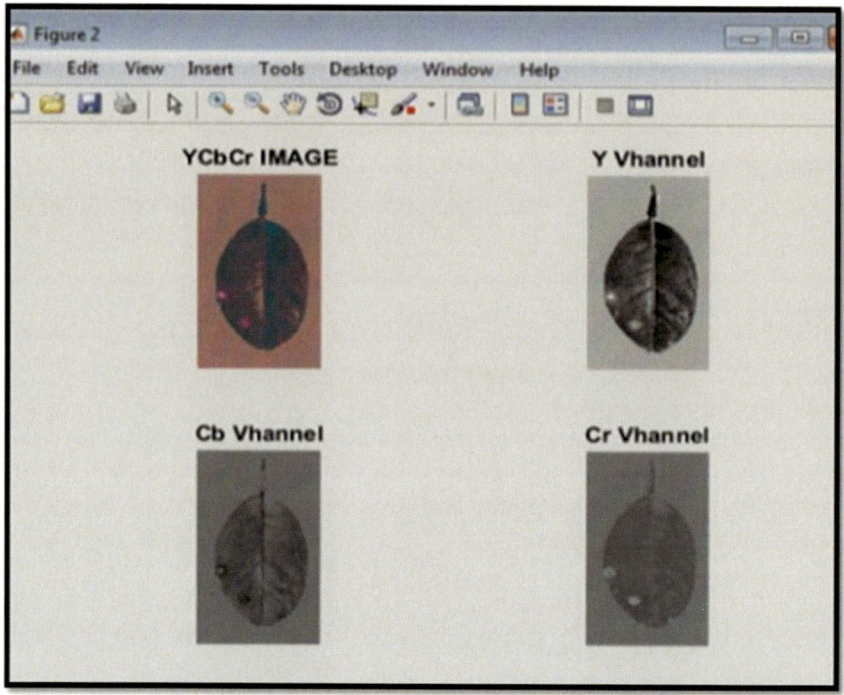

Fig. 3 Output of YC_bC_r model

We have calculated the features of image matrix for example. Mean for a random variable vector A is made up of N scalar observations; the mean is defined as [13].

$$\mu = \frac{1}{N} \sum_{i=1}^{N} A_i \tag{4}$$

The standard deviation is an apportion of how spread out quantities are.

$$y_j = \sigma_j = \sqrt{\frac{\sum_{i=1}^{M} |u_{ij} - \mu_j|^2}{M - 1}} \quad 1 \leq j \leq N \tag{5}$$

Skewness is a measure of the asymmetry of the information around the example mean. In the event that skewness is negative, the information are spread out more to one side of the mean, i.e., to left than to other side. On the other hand, if skewness is certainly positive, the information are spread out to other side. The skewness of the ordinary conveyance (or any flawlessly symmetric dispersion) is zero. The skewness of a distribution is defined below:

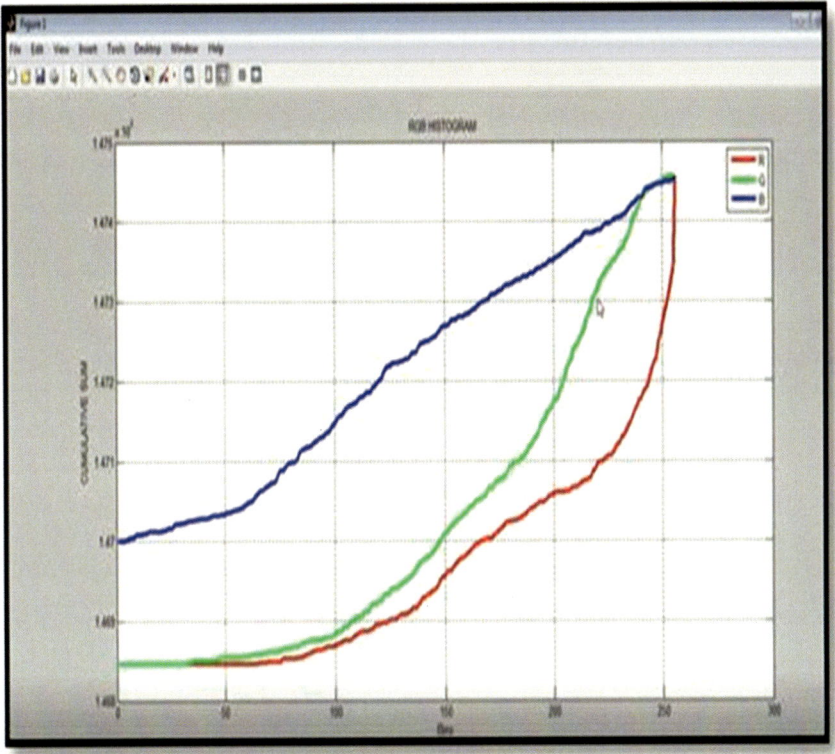

Fig. 4 RGB histogram

$$s = \frac{E(x - \mu)^3}{\sigma^3} \qquad (6)$$

Kurtosis is a measure of how an outlier-prone distribution is. The kurtosis of the normal distribution is 3. Distributions that are more outlier-prone than the normal distribution have kurtosis greater than 3; distributions that are less outlier-prone have kurtosis less than 3. The kurtosis of a distribution is defined as (Fig. 4).

$$k = \frac{E(x - \mu)^4}{\sigma^4} \qquad (7)$$

After the segmentation, the feature extraction is done; here, we are calculating the color feature and color moments using the RGB to HSV conversion; the output of the color feature extraction is shown in Fig. 5. The C_r channel image is multiplied with complete back image for masking and we can see only diseased part of leaf; however, this image is without color; therefore, we again multiply it with enhanced image pixel by pixel and we got below color segmented image (Fig. 6).

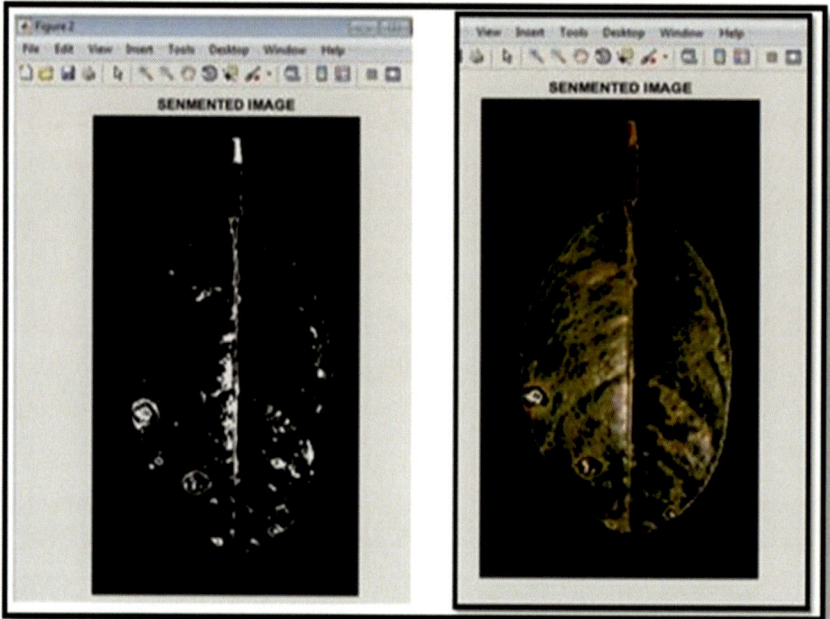

Fig. 5 Color segmented image

After color feature extraction, we move on to extract the texture feature. For this purpose, we make use of the discrete wavelet transform. The discrete wavelet changes state to wavelet changes that the wavelets are disconnectedly evaluated. A change which restrains a capacity both in space and scaling has some important properties contrasted with the Fourier change. The change is fixated on a wavelet lattice, which can be figured more rapidly than the practically equivalent to Fourier framework. Most outstandingly, the DWT [14] is utilized for flag coding, where the advantages of the change are abused to connote a discrete flag in an additional repetitive shape, regularly as a preconditioning for information pressure. The discrete wavelet change has a tremendous amount of uses in science, software engineering, arithmetic, and designing. The mean and standard deviation of the dispersion of the wavelet change coefficients are utilized to build the element vector.

5   Conclusion

To keep up efficiency development in a manageable way, there is a need to move from input concentrated to innovation serious and expertise escalated horticulture. This diminishes the product misfortune by appropriately instructing the ranchers with the harvests points of interest and their prerequisites. The MATLAB ANN Toolbox

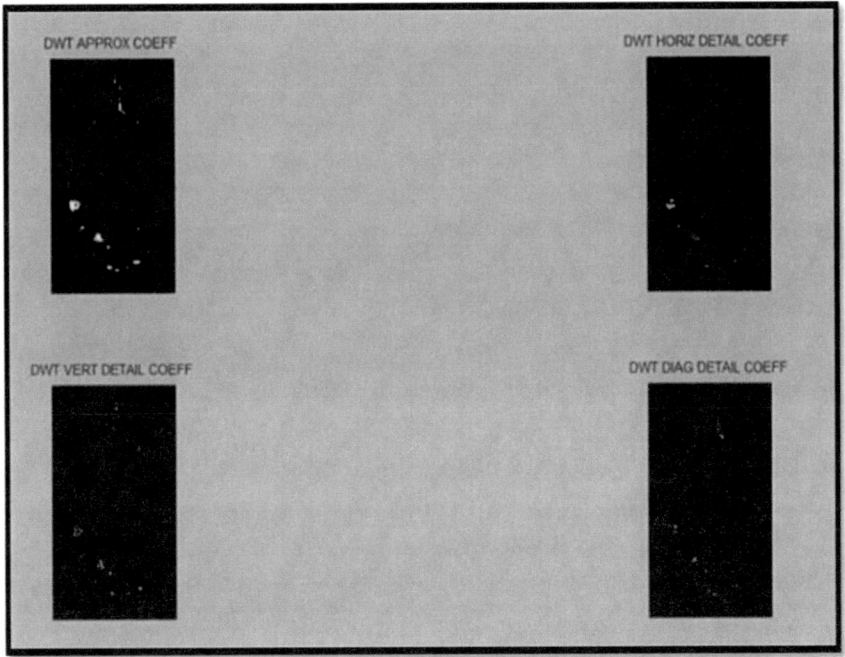

Fig. 6 Output of feature extraction

is the product space in which the expectation framework is constructed and after that imported to Android stage. Most imperative favorable position of utilizing this technique is that the plant maladies can be recognized at the beginning time or the underlying stage.

References

1. Vibhute A, Bodhe SK (2012) Applications of image processing in agriculture: a survey‖. Int J Comput Appl 52(2):34–40
2. Aware AA, Joshi K (2016) Crop and weed detection based on texture and size features and automatic spraying of herbicides. Int J Adv Res Comput Sci Softw Eng 6(1)
3. Aware AA, Joshi K (2015) Wavelet based crop detection and automatic spraying of herbicides. Int J Innovations Adv Comput Sci 4(2)
4. Singh V, Misra AK (2017) Detection of plant leaf diseases using image segmentation and soft computing techniques. Inf Proc Agric 4:41–49
5. Raj Kumar S, Sowrirajan S (2016) Automatic leaf disease detection and classification using hybrid features and supervised classifier. Int J Adv Res Electr Electron Instrum Eng 5(6) https://doi.org/10.15662/ijareeie.2016.0506008
6. Al Bashish D, Braik M, Sulieman B (2010) A framework for detection and classification of plant leaf and stem diseases. In: IEEE international conference on signal and image processing

7. Niket A, Jadhav B, Smeeta N (2014) Detection and classification of plant diseases by image processing. Int J Innovative Sci Eng Technol 1(2)
8. Arivazhagan S, NewlinShebiah R, Ananthi S, Vishnu Varthini S (2013) Detection of unhealthy region of plant leaves and classification of plant leaf diseases using texture features. CIGR J 15
9. Kaur G, Himanshu M (2012) Classification of biological species based on leaf architecture. Int J Comput Sci Inf Technol Secur (IJCSITS) 2(2)
10. Sanjeev S, Vijay S, Nargund VB (2013) Pallavi–Diagnosis and classification of grape leaf diseases using neural network. In: IEEE-2013 4th ICCCNT, July 4–6, 2013
11. Ravichandran G, Koteeshwari RS (2016) Agricultural crop predictor and Advisor using ANN for Smartphones. In: IEEE international conference on emerging trends in engineering, technology and science (ICETETS)
12. Malik S, Lone T (2014) Comparative study of digital image enhancement approaches. In: International conference on computer communication and informatics. http://dx.doi.org/10.1109/iccci.2014.6921749
13. Antoni M, Barlaud M, Daubechies I (1992) Image coding using Wavelet transform. IEEE Trans Image Process 1(2)
14. Kiani1 S, Azimifar Z, Kamgar S (2010) Wavelet based crop detection and classification. In: ICEE 2010, May 11–13, IEEE 2010

Cross Domain Recommendation System Using Ontology and Sequential Pattern Mining

S. Udayambihai and V. Uma

Abstract Recommendation system is very helpful to filter the information according to the user interest and provide user personalized suggestion. Recommendation system is emerging now-a-days in many social networks like Facebook, Twitter, e-commerce etc. Cross domain recommendation system is one of the method to develop the recommendation where we can gather the knowledge from different domains and recommend most similar items related to the user search term. In this work, we try to extend cross domain recommendation by finding semantic similarity of items in Ontology, applying Collaborative Filtering and recommending user preferred items using PrefixSpan algorithm. The similarity between items can be achieved through modified Wpath method. Finally, we can recommend the most preferred items and evaluate using performance measures like F-score.

1 Introduction

Recommendation system is a subclass of information filtering system which is used to check the user 'preference' or 'rating' given to an item. In the e-commerce world, it is used in many fields like movies, music, news, books and research articles. Recommendation system can be implemented using two methods namely Collaborative Filtering (CF) or Content Based Filtering (CBF). Collaborative Filtering (CF) is used to recommend similar items preferred by other users having similar taste. Content Based Filtering (CBF) is alien to CF and is used to match the content according to the user characteristics [1].

Recommendation system is used to retrieve the user preferred information on the internet. By using this, we can increase the average order value and can easily reduce the traffic in services and improve the delivery of relevant content to the user.

S. Udayambihai (✉) · V. Uma (✉)
Department of Computer Science, Pondicherry University, Pondicherry, India
e-mail: itsudaya09@gmail.com

V. Uma
e-mail: umabskr@gmail.com

© Springer Nature Switzerland AG 2019   1885
D. Pandian et al. (eds.), *Proceedings of the International Conference on ISMAC in Computational Vision and Bio-Engineering 2018 (ISMAC-CVB)*, Lecture Notes in Computational Vision and Biomechanics 30,
https://doi.org/10.1007/978-3-030-00665-5_173

Cross domain recommendation systems have the capability to access information belonging to one or more domains [1]. Recommendations can be improved by exploiting the knowledge from source domains and enhancing the recommendations in a target domains. By applying this we can accomplish better accuracy and overcome data sparsity problem [3].

Ontology is used for domain knowledge representation. Ontology represents concepts and relationships between facts in different domains. Ontology is used to describe the individuals, classes, attributes and relations. Ontology can also be represented in the form of Knowledge Graphs (KG) [8].

Semantic similarity is a measure that indicates the similarity between two different entities, sentences, words or documents. Semantic similarity can be applied for topology of ontologies [8]. There are two major approaches for calculating the similarity between words: Pairwise and Group-wise. In pairwise approaches, similarity between two set of words can be calculated by combining the similarity of their words. In group-wise approaches, there are three representation methods viz. set, graph or vector. By using Collaborative Filtering (CF) in recommendation systems, helps to reduce the overloaded information between domains. Collaborative Filtering (CF) when combined with semantic similarity, more accuracy can be achieved with the better recommendations [9].

In this work, ontology is used for representation of knowledge about 2 domains (Music and Books). Semantic similarity is used to measure the similarity between concepts belonging to multiple domains and retrieve the most similar items [11] for recommendation. The proposed cross domain recommendation system, will recommend the items based on semantic similarity measurement [1].

Thus, this proposed work can recommend similar items in E-commerce by involving the user-item-ratings. Section 2 discusses the literature survey. Section 3 explains the system architecture and modules involved. Section 4 briefly explains the experimental setup and Sect. 5 concludes the paper.

2 Literature Survey

Various research works have been carried out related to cross domain recommendation system. Table 1 gives a comparative analysis of some of the related works.

In [1], items are recommended by using User-item rating matrix for Movie-Book domain and by using WordNet based ontology for representing the knowledge. Items are recommended by clustering to form a latent space between different domains. Cosine similarity measure is applied to retrieve the most similar items followed by Collaborative Filtering. But the accuracy is very low due to data sparsity problem. In [3] items are recommended by constructing user-tag-item matrix for Movie—Book domain. Similarity is calculated based on ratio of frequency of occurrence of words and total number of words occurred by applying the k-means clustering and Semantic fuse algorithm the most similar items are recommended. In [8], Semantic Similarity

Table 1 Comparative study of various recommendation approaches

Title and year of publication	Technique used	Evaluation	Knowledge represenation for semantic similarity	Cross domain
[1] Semantic clustering based cross domain recommenda-tion—2014	Clustering to form a common latent space and collaborative filtering	Cosine similarity is used to measure the closeness between source and target domains	WordNet ontology	Movie-book domain for User-item-rating
[2] Personalizing health and food advices by semantic Enrichment of multilingual cross-domain questions—2015	String matching algorithm	Semantic techniques such as OWLIM and sesame RDF	User profile ontology	Health and food domain
[3] Cross domain recommendation using semantic similarity and tensor decomposition —2016	Tensor decomposition, k-means clustering and semantic fuse algorithm	Semantic similarity based on frequency of occurrence of words divided by total number of words occurred	Semantic matrix	User-tag-item for Movie—book domain
[4] Exploiting trust and usage context for cross-domain recommenda-tion—2016	Item based collaborative filtering using association matrix	Computing the cosine similarity of item vectors as the similarity for each item pair	Rating matrix user-item-matrix into item-item matrix association	DVDs, Beauty, Books, Travel and CiaoCafe
[5] Multi-domain collaborative recommendation with feature selection—2016	Feature selection and collaborative filtering	Cosine distance is used to measure the similarity between users using rating distribution and normalization of the ratings	Construction of user-item rating matrix	Two multi-domain product review datasets are used, i.e., Ciao and Epinions
[6] Enhancing cross domain recommendation with domain dependent tags—2016	Cross domain collaborative filtering in matrix factorization	Similarity is calculated using Pearson coefficient	Apply spectral clustering to align domain dependent tags as user-tag and item- tag	User-tags-rating for movie and book

(continued)

Table 1 (continued)

Title and year of publication	Technique used	Evaluation	Knowledge represenation for semantic similarity	Cross domain
[7] Clustering based transfer learning in cross domain recommender system—2016	k-means clustering algorithm and hierarchical agglomerative algorithm	Similarity is measured using cosine of the angle between two vectors	User-item-rating matrix	Tourist and location domain
[8] Computing semantic similarity of concepts in knowledge graphs	Using aspect based sentiment analysis category classification	Similarity is calculated using Wpath semantic similarity. Wpath is measured using shortest path length and contribution of the Least common subsumers of information content	Using knowledge graph to represent ontology	Restaurant domain
[9] A user based cross domain collaborative filtering algorithm based on a linear decomposition model—2017	Collaborative filtering for user-item rating matrix	Semantic similarity can be measured using Pearson correlation with items and ratings on item	Compute the similarity using rich ratings	Book and music domain
[10] Merged ontology and SVM—based information extraction and recommendation system for social robots—2017	Support vector machine (SVM) system	Positive and negative polarities	Merged ontology is the combination of Medical ontology, City ontology and Hotel ontology	Diabetic drugs, hotel and tourism domain
In proposed system, cross domain recommendation system using ontology and sequential pattern mining	Collaborative filtering and prefix span algorithm	Semantic similarity can be measured using modified Wpath method	Two different domain knowledge in ontology	Database for movie from (Movie Lens) and book from (Goodreads dataset)

between different items in the restaurant domain is represented by Knowledge Graph (i.e.) Ontology. Semantic Similarity is calculated using Wpath method,

$$\text{Sem}_{\text{wpath}}\left(c_i, c_j\right) = \frac{1}{1 + \text{length}\left(c_i, c_j\right) * k^{IC(c_{\text{lcs}})}}$$ (1)

By using this method we can get more accuracy comparing with the other existing methods. This method is also evaluated using datasets of Aspect Based Sentiment Analysis and also shown the best performance in terms of accuracy and F-score. In [9], items are recommended by constructing user-item rating matrix. Pearson correlation is applied for calculating semantic similarity for Movie-Book domains. Collaborative Filtering method is also applied for retrieving the most similar items. This method can overcome the cold start problem. This method can also be improved by extending the similarity range from local to total similarities.

In this proposed system, item recommendation is done in e-commerce by using Ontology and Sequential Pattern Mining. Cross domain recommendation is provided for two different domains viz. Movie and Book. Ontology is used for knowledge representation. Modified Wpath method is used to calculate the semantic similarity and to improve the results Collaborative Filtering is used in item-rating matrix. Prefix Span algorithm is used to retrieve the Sequential patterns of most preferred items by the regular users. Hence this method is expected to increase the accuracy compared to the existing methods and reduce the data sparsity problem.

3 System Architecture and Modules Involved

3.1 Architecture

The architecture of the proposed cross domain recommendation system is shown in Fig. 1.

3.2 Modules Involved

Module 1: Calculate Semantic Similarity
Semantic similarity is used to measure the similarity between the input like movie name (i.e.) auxiliary domain and the other items in the target domain like books with the help of generated ontology. It also helps in retrieving the most similar items in the target domain.

Here, semantic similarity measure is calculated by using modified Wpath method wherein, length between the two items and height of the Least Common Subsumer (LCS) are considered. It helps to identify the genre and by using this we can easily

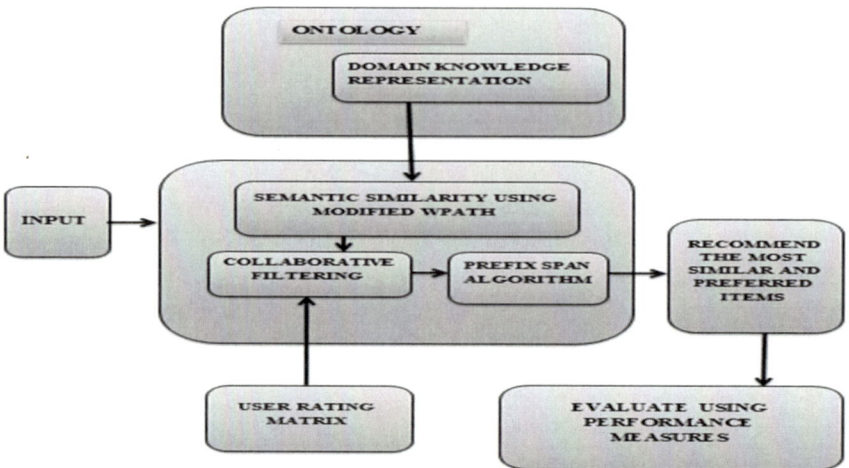

Fig. 1 Proposed system architecture

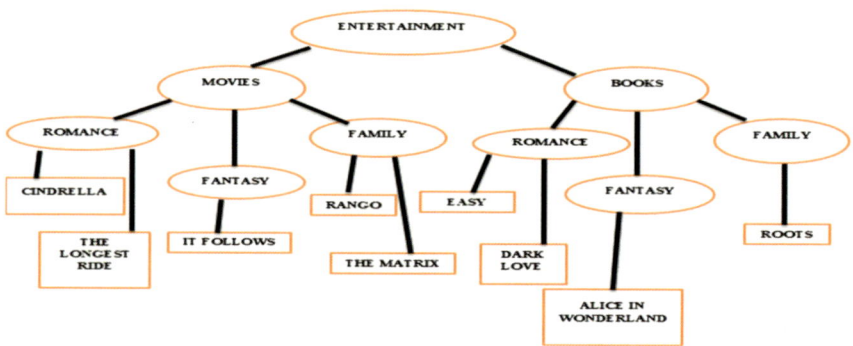

Fig. 2 Fragment of movie and book domain dataset represented as ontology

generate the user rating matrix. Figure 2 represents the movie and book domain in the form of ontology. If Cinderella is given as search term from auxiliary domain then the semantic similarity will be calculated between Cinderella and Easy, Dark love from target domain as both books and movies are from the same genre Romance.

Semantic similarity can be calculated using Eq. (1) by modifying the Wpath method [8].

$$\text{modified Wpath}\,(c, c_i) = \frac{1}{1 + \text{length}(c, c_i) * \text{height(lcs)}} \tag{1}$$

where c is term given by the user in auxiliary domain movies and c_i is the item of the same genre in the target domain books and length (c, c_i) is the shortest path between *c* and c_i and height (lcs) is the maximum height from the least common subsume (lcs).

Module 2: Collaborative Filtering

Collaborative Filtering is used to recommend the items. In item based Collaborative Filtering, similarities are calculated from item-rating matrix. First, write the user-item rating data in the form of matrix. Then create the item-item similarity matrix. The item-item similarity can be calculated using cosine similarity measure.

In this paper, first we have to apply the semantic similarity measure in the Ontology and we have to calculate most similar books that is related to given movie and recommend the item-rating matrix according to the most similar rates.

By using this item-rating matrix apply item based collaborative filtering to find the item-item similarity matrix. To find item-item similarity we can apply cosine similarity measure. The cosine similarity between two item vectors (A, B) can be calculated using Eq. (2).

$$\cos(A, B) = \frac{\sum_{i=1}^{n} A_i B_i}{\sqrt{\sum_{i=1}^{n} A_i^2} \sqrt{\sum_{i=1}^{n} B_i^2}} \tag{2}$$

where A and B are the ratings given by 'n' number of users for 'm' items. By using this item-item rating matrix we can easily retrieve the most similar items.

Module 3: Apply Prefix Span Algorithm

After applying Collaborative Filtering we go for a new pattern growth method for mining the sequential patterns called Prefix Span algorithm. By applying this algorithm, we can recommend the most similar and also most preferred items of the user.

4 Experimental Setup

Dataset

We have used the freely available dataset for Movie domain from Movie lens dataset which contains 100,000 ratings of 9000 movies given by 700 users. The details <User_id; Movie_id; Movie name; rating; genre> are obtained. We have also used the Book domain dataset from Goodreads dataset which contains 411,156 ratings applied to 7000 books by 600 users. From this available dataset we have taken <User_id; Book_id; Book name; rating; genre>.

In Fig. 3, we are comparing No. of Items in Movies, No. of Items in Books, No. of users for Movies and No. of users for Books using the freely available dataset.

In Fig. 4, we are illustrating the user ratings between Movie and Book domain.

The system is to be implemented and evaluated using F-Score.

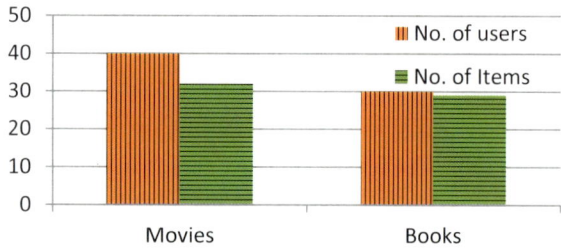

Fig. 3 Illustration between no. of items and users in movie and book domains

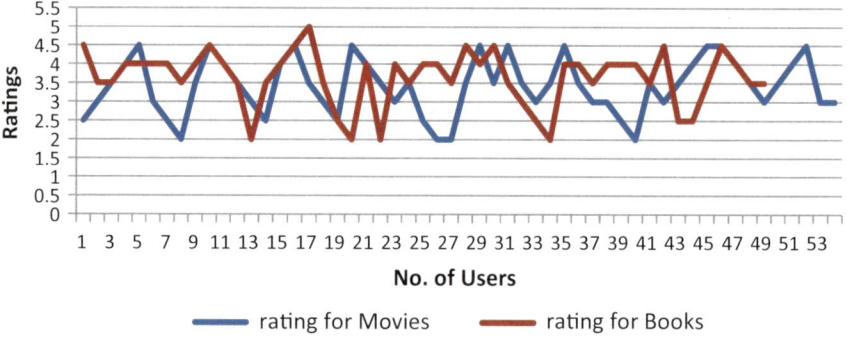

Fig. 4 Illustration of user rating for movies and books

5 Conclusion

By this method we can recommend the most similar items from different domains by combining Ontology with Collaborative Filtering and with the help of user-item-ratings. Modified Wpath method helps in finding the semantic similarity. Prefix Span algorithm is used to retrieve the most preferred items. Hence, Cross Domain Recommendation System using Ontology and Sequential Pattern Mining is expected to achieve more accuracy. Implementing the proposed system and evaluating using the performance measures is left as future work.

References

1. Kumar A, Kumar N, Hussain M, Chaudhury S, Agarwal S (2015) Semantic clustering-based cross-domain recommendation. In: IEEE SSCI 2014–2014 IEEE symposium series on computational intelligence—CIDM 2014: 2014 IEEE symposium on computational intelligence and data mining, proceedings, pp 137–141. http://doi.org/10.1109/CIDM.2014.7008659
2. Al-Nazer A, Helmy T (2015) Personalizing health and food advices by semantic enrichment of multilingual cross-domain questions. In: 2015 IEEE 8th GCC conference and exhibition, GCCCE 2015, pp 1–4. http://doi.org/10.1109/IEEEGCC.2015.7060095

3. Kumar V, Shrivastva KMP, Singh S (2016) Cross domain recommendation using semantic similarity and tensor decomposition. Procedia Comput Sci 85(Cms):317–324. http://doi.org/10.1016/j.procs.2016.05.239

4. Xu Z, Zhang F, Wang W, Liu H, Kong X (2016) Exploiting trust and usage context for cross-domain recommendation. IEEE Access 4:2398–2407. https://doi.org/10.1109/ACCESS.2016.2566658

5. Liu L, Cui J, Song W, Wang H (2017) Multi-domain collaborative recommendation with feature selection. China Commun 14(8):137–148. https://doi.org/10.1109/CC.2017.8014374

6. Hao P, Zhang G, Lu J (2016) Enhancing cross domain recommendation with domain dependent tags. In: 2016 IEEE international conference on fuzzy systems, Fuzz-IEEE 2016, pp 1266–1273. http://doi.org/10.1109/FUZZ-IEEE.2016.7737834

7. Thendral SE, Valliyammai C (2017) Clustering based transfer learning in cross domain recommender system. In: 2016 8th international conference on advanced computing, ICoAC 2016, pp 51–54. http://doi.org/10.1109/ICoAC.2017.7951744

8. Zhu G, Iglesias CA (2017) Computing semantic similarity of concepts in knowledge graphs. IEEE Trans Knowl Data Eng 29(1):72–85. https://doi.org/10.1109/TKDE.2016.2610428

9. Yu XU, Jiang F, Du J, Gong D (2017) A user-based cross domain collaborative filtering algorithm based on a linear decomposition model. IEEE Access 5. http://ieeexplore.ieee.org/abstract/document/8113474/

10. Ali F, Kwak D, Khan P, Ei-Sappagh SHA, Islam SMR, Park D, Kwak KS (2017) Merged ontology and SVM-based information extraction and recommendation system for social robots. IEEE Access 5:12364–12379. https://doi.org/10.1109/ACCESS.2017.2718038

11. Zhang Q, Haglin D (2016) Semantic similarity between Ontologies at different scales. IEEE/CAA J Automatica Sin 3(2):132–140. http://ieeexplore.ieee.org/abstract/document/7451100/

Identifying the Risk Factors for Diabetic Retinopathy Using Decision Tree

Preecy Poulose and S. Saritha

Abstract The role of data mining in healthcare industry is to improve the health systems and to use these data analytics to identify the inefficiencies and to improve care. Accuracy is important when it comes to patient care, and handling this huge amount of data improves the quality of the healthcare system. Diabetic retinopathy is a disease mainly occurring in patients with high sugar level in their blood. The situation occurs when sugar levels in blood are high, thus causing damage to blood vessels. The blood vessels swell and leak and gradually affect the eye vision. This is not detected in early stage of diabetics, even though it affects the eyesight from the beginning. Decision tree classification helps to detect the problem in the initial stage that helps to find the risk factors that cause this disease. Better treatment is provided to those infected patients. A detailed analysis of the result from the decision tree classifier is also presented in this work, and decisive factors for diabetic retinopathy are concluded herewith.

1 Introduction

Diabetic Retinopathy (DR) is a well-known leading cause of preventable blindness, affecting more than 520 million people across the world. DR can be detected at the initial stages by expert doctors through routine analysis of the eye fundus [1]. DR causes serious issues to the retina, and the blood vessels lead to swell and leak. The diabetic patients with prolonged years have more chance to get affected by DR. The patients with high blood pressure and cholesterol, abnormal protein rate in urine, and less oxygen level in retina are more prone to the disease. The rest of the symptoms are scotoma [2] and blurred vision. These symptoms are not identified at the earlier stage. So there is more chance for the patient to get completely blind when not properly treated at the right time. No symptoms are shown at the beginning stage. Later due

P. Poulose (✉) · S. Saritha
Rajagiri School of Engineering and Technology, Kakkanad, Kochi 682039, Kerala, India
e-mail: preecyp@gmail.com

© Springer Nature Switzerland AG 2019
D. Pandian et al. (eds.), *Proceedings of the International Conference on ISMAC in Computational Vision and Bio-Engineering 2018 (ISMAC-CVB)*, Lecture Notes in Computational Vision and Biomechanics 30,
https://doi.org/10.1007/978-3-030-00665-5_174

to prolonged years, the patient feels pain on eyes, swelling on the retina, and blurred vision.

The retina is a well-known source of naturally occurring molecule that enables the early identification of several human diseases, disorders such as high rate in blood pressure, heart diseases, and DR [3]. DR is caused by the damage that has been occurred in the retina due to diabetes. Before 126.6 million people have been diagnosed with DR, and now showing a growth of up to 190 million people by 2030 as per [http://en.wikipedia.org/wiki/Diabetic_Retinopathy], accessed on March 30, 2018.

The algorithms used are random forest and decision tree classifier. This algorithm is used to find out whether the patient will be affected by DR. A simpler way to find the risk factors is using the random forest algorithm. DT is used to identify the decisive factors that affect the DR [4].

The paper is organized into the following sections: Sect. 1 gives the entire introduction of work that is being proposed. Section 2 briefs about literature review of current work. Section 3 describes the proposed system design and architectures. The results and discussions are presented in Sect. 4 and concluded in Sect. 5.

2 Literature Survey

Previously published works were used as a guideline to carry out the survey. Several classifiers have been used for the prediction of risk factors affecting DR. They are classified on the basis of accuracy. The important concepts are discussed below.

In the research work, the non-dilated retina images are taken as an input for diagnosing the diabetic retinopathy [5]. Attribute selection plays a major role in decision tree classifier for classifying the input image data. The input data are then preprocessed and based on the result the images are classified into infected and non-infected images. If the image is infected, it is then given for further processing and classified them into severe infected and moderate infected image. The feature extraction or attribute selection is used for both the train data and the test data. By calculating the accuracy, the performance of the GMM classifier is obtained and identified [5].

DR occurs mostly in both middle-aged and aged people. The diagnosing of the disease at the early stage and providing proper treatment and awareness at the right time can slow down the progression of disease and prevents from complete blindness or vision loss [6]. An automated detection method can be used by the non-experts to detect which patient needs to be referred for an ophthalmologist. It reduces the burden on ophthalmologist, in order to improve the efficiency of diabetic screening program and reduction of cost. PNN classifier [6] is successfully used to detect the exudates in retinal images.

3   Design of Proposed System

The system architecture mainly consists of four modules including data, data pro-cessing, decision tree classifier, and lastly classifier evaluation. Here in the project, the input data are taken as the raw data regarding the medical report and symptoms of diabetic patient. The raw data are then preprocessed and then given to the decision tree classifier, and finally the results are evaluated. The two-third data are given as the training data, and one-third data are given as the test data. Decision tree classifier is implemented on both the data, and finally the result is being compared and evaluated (Fig. 1).

3.1   Data Module

The input data are taken from the medical records provided in the UCI repository. These data include the features and attributes that cause the DR. The data are provided as a text format.

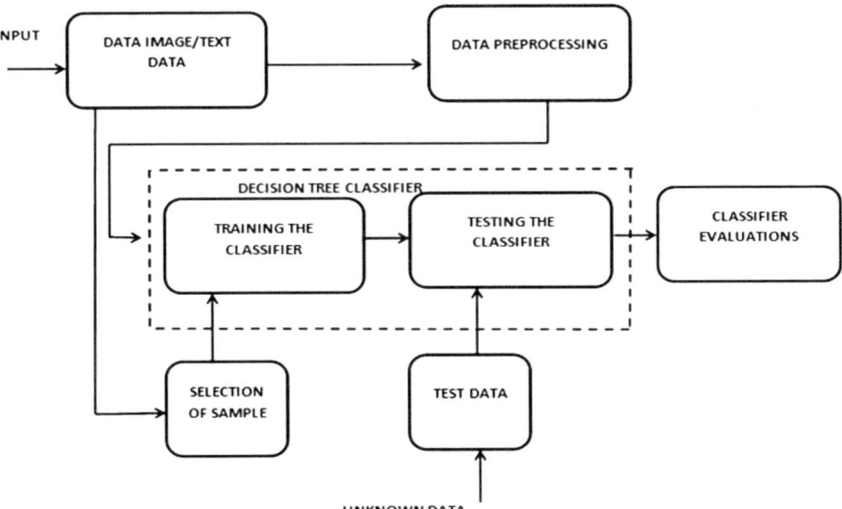

Fig. 1 Overall system architecture of the proposed design

3.2 Data Preprocessing

The raw data are preprocessed to finally give to the classifier. The missing data are filled with the real data, and incorrect data are corrected with real attributes. Some tuples are ignored which have incorrect values.

3.3 Classifier Module

The two-third of data is given for training and one-third is given for testing. Decision tree classifier is used with class labels yes/no.

Train data

The two-third of data from the dataset is taken as an input or train data. It is then given to the decision tree classifier. The algorithm regarding the decision tree will work on these data, and finally it will train the dataset. While training, the last column of attributes is not taken into concern, i.e., the status of the patient, whether the patient will be affected by the diabetic retinopathy or not.

Test data

The one-third set of dataset is given as the test data. And it will then compare the actual result and the predicted result. With this, the final accuracy is being calculated.

Decision tree classifier

The decision tree classifier uses an ID3 algorithm to find out the decisive factors that mainly affect the diabetic retinopathy. The factors are determined by calculating the entropy of each attribute from the dataset. The attribute with the highest entropy is then selected as the decisive factor, and then further classification is provided on this attribute to find the succeeding nodes and finally, the algorithm terminates when the nodes get equal to the class labels.

3.4 Classifier Evaluation Module

The classifier is evaluated based on three categories and they are accuracy, precision, and recall. Two-third of data is given for training and one-third is given for testing. If results are not satisfactory, then training samples are improved.

Random Forest classifier

The advantage of using the algorithm is its problem-solving capability, mainly in regression problems.

Decision tree Classifier

One big advantage of using decision tree classifier is its ability to solve problem, decisions, and outcomes of decision.

4 Results

The dataset includes the severe symptoms that occur in a diabetic patient. These data are in a text format, and the attributes included in the dataset are patient id, age, gender, prolonged year, glucose level before fasting, glucose level after fasting, HbA1c, blood pressure, protein in urine, triglycerides, scotoma, pain on eyes, macular oedema, and oxygen level on retina. These data are then undergone a preprocessing stage that added up the missing values and ignored the incorrect values or tuples. The implementation is done in Python 2.7.9.

4.1 Sample Decision Tree Implementation

See Fig. 2.

4.2 Analysis of Decision Tree

On analysis of the decision tree, the decisive factors that help in identifying diabetic retinopathy are (a) scotoma, (b) pain on eyes, (c) triglycerides, and (d) prolonged

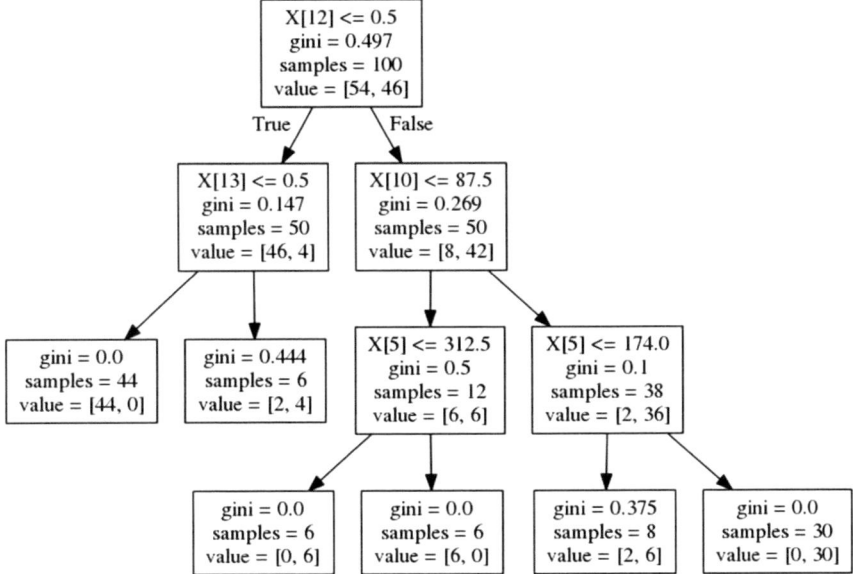

Fig. 2 Decision tree

Fig. 3 Operational mode of decision tree

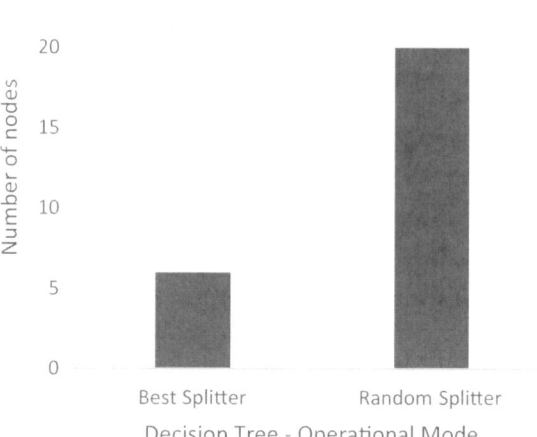

Fig. 4 Accuracy plot for different number of attributes

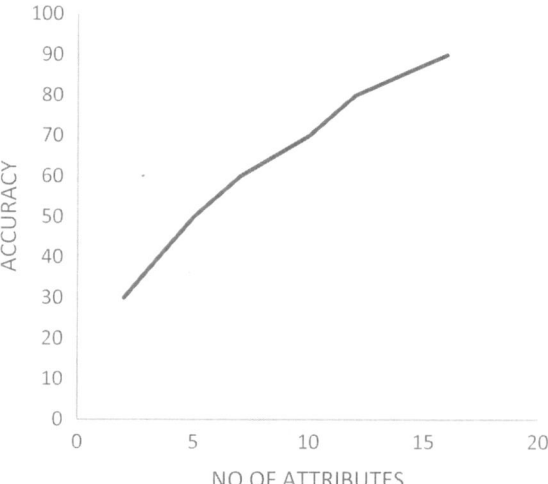

year of diabetes in the order of importance. The factors have been concurred with the aid of a domain expert.

The graphical representation of operational mode of decision tree is given in Fig. 3. As per the splitter being used in the algorithm, the nodes in the decision tree can vary. Best splitter is used to find the best decisive factors that affect DR. While using this best splitter, the number of nodes is reduced in the decision tree.

The accuracy of decision tree obtained is dependent on number of attributes in the input dataset. As the number of attributes increases, the accuracy associated also increases. A graphical representation of the same is presented in Fig. 4 (Table 1).

Number of attributes	Accuracy (%)
6	56.33
8	77.32
10	88.21
12	92
14	93.21
16	97.57

Table 1 Relation between attributes and accuracy

5 Conclusion

This work presented a data mining method to find the occurrence of diabetic retinopathy in diabetic-affected patients in the early stage itself. The factors causing diabetic retinopathy is identified in this work with the help of decision tree classifier. The decisive factors identified from this work were verified with a subject expert and was appreciated as commendable. A detailed analysis of the decision tree obtained is also done in the experiments. As a future scope, the input data of the design can be changed to retina image and appropriate image mining techniques can be applied to obtain better results.

References

1. Zhang Y, Tsai C-W, Hassan MM, Alamri A (2017) Health-CPS: healthcare cyber-physical system assisted by cloud and big data. IEEE Syst J 11(1)
2. Hamsagayathri P, Sampath P (2017) Priority based decision tree classifier for breast cancer. In: 2017 international conference on advanced computing and communication systems (ICACCS-2017), 06–07 Jan 2017, Cancer Detection
3. Senthilnayaki B, Venkatalakshmi K, Kannan A (2013) An intelligent intrusion detection system using genetic based feature selection and modified J48 decision tree classifier. In: 2013 fifth international conference on advanced computing (ICoAC)
4. Sarwinda D, Bustamam A (2017) A complete modelling of local binary pattern for detection of diabetic retinopathy. In: 2017 1st international conference on informatics and computational sciences (ICICoS)
5. Dhanasekaran R, Mahendran G, Murugeswari S, Fargana SM (2016) Investigation of diabetic retinopathy using GMM classifier. In: 2016 international conference on advanced communication control and computing technologies (ICACCCT)
6. Mahendran G, Dhanasekaran R (2014) Identification of exudates for diabetic retinopathy based on morphological process and PNN classifier. In: International conference on communication and signal processing, 3–5 Apr 2014

Heavy Vehicle Detection Using Fine-Tuned Deep Learning

Manisha Chate and Vinaya Gohokar

Abstract Heavy vehicles develop technical snag and traffic jam on streets. Accidents between heavy vehicle and road users, for example, pedestrians often result in severe injuries of the weaker street users. The highway safety and traffic jams can be secured with detection of heavy and overloaded vehicles on the highway to facilitate light motor vehicles like cars, scooters. A model for heavy vehicle detection using fine-tuned based on deep learning is proposed to deal with entangled transportation scene. This model comprises two parts, vehicle detection model and vehicle fine-grained detection. This step provides data for the next classification model. Experiments show that vehicle's make and model can be recognized from transportation images effectively by using our method. Experimental results demonstrate that the proposed detection system performs accurately with other simple and complex scenarios in detecting heavy vehicles in comparison with past vehicle detection systems.

1 Introduction

Traffic congestion is another one of the curses of human driving, and in particular of clumsy driving. Heavy vehicles are still using regular roads and will be subject to congestion and space constraints of the roads. Heavy vehicles develop snag while climbing, especially at bridge. If this happens, vehicular movement gets blocked. Roads are pricey and in many cities there is no room to build new roads. Heavy vehicles on roads just end up with waiting in traffic jams. Much congestion, though, is just caused by heavy vehicles in rush hours. Heavy vehicle detection system provides solutions to avoid accidents and traffic jams. Here, we address the problem of heavy vehicle detection and recognition.

M. Chate (✉)
Amity University, Noida, India
e-mail: mmchahande@amity.edu

V. Gohokar
MIT, Pune, India
e-mail: vvgohokar@mitpune.edu.in

© Springer Nature Switzerland AG 2019   1903
D. Pandian et al. (eds.), *Proceedings of the International Conference on ISMAC in Computational Vision and Bio-Engineering 2018 (ISMAC-CVB)*, Lecture Notes in Computational Vision and Biomechanics 30,
https://doi.org/10.1007/978-3-030-00665-5_175

If fine-grained deep learning approach of heavy vehicle detection is applied in transportation and public security, we can acquire more meta information like vehicle make, model, logo, production year, max speed, acceleration, and so on [1]. By acquiring this information dynamically, we can build a large intelligent transportation system that can monitor the whole city's road. Further, we can analyze the heavy vehicles on the road at different times to find the discipline of people's going out, and then we can schedule transportation rules accordingly; these will make cities more smart and intelligent.

The rest of this paper is organized as follows: Sect. 2 outlines some relevant research on vehicle detection and CNN applications; Sect. 3 provides a meticulous description of the proposed system. Implementation details and experimental observations will be provided in Sect. 4, followed by conclusion in Sect. 5.

2 Related Work

Diverse methods for vehicle detection and classification have been invented recently. Sivaraman and Trivedi [2] use active learning to learn from front part and back part vehicle images, and achieve 88.5 and 90.2% precision on front and rear part vehicle detection. Chen et al. [3] use a measurement-based feature (MBF) and intensity pyramid-based HOG (IPHOG) combined feature set for vehicle classification. Kafai and Bhanu propose a classification approach for rear vehicles [4]. They define a set of features that include tail light and plate position information, and then pass it to a hybrid Bayesian grid for classification.

Various vehicle recognition methods for complex environments have been proposed and made some achievements. Generally, the research on vehicle recognition can be classified into two categories: "handcrafted feature and classifier" based methods and deep-learning-based methods. For example, Chen et al. [5] used the symmetric SURFs feature to detect the matching feature pair in the vehicle and classified the vehicles by the sparse classifier, which obtained better classification accuracy. Li et al. [6] applied HOG numbering your section headings.

Histogram of oriented gradients is used to describe the apparent properties of the vehicle and constructed a top-down and-or model combined with SVM to achieve the vehicle detection and recognition. The accuracy of recognition can be improved obviously with the introduction of and-or model. Lin et al. [7] proposed a 3D modeling method based on the deformable parts model (DPM), which is a classic part-based algorithm. This method can generate the vehicle classification model for fine-grained recognition and achieve good results. The latter category of vehicle recognition research is designed to procure classification model from big data using deep learning. The existing research shows that vehicle recognition is capable of improved performance in a very significant manner more effectively by discovering high-level depth features via big data [8]. Compared with the traditional idea of combining handcrafted features and classifier, recognition accuracy has improved significantly. Heikki et al. [9] adopted the end-to-end structure of five-layer convolutional neural network (CNN) network to achieve the vehicle recognition. Wang et al. [10] abandoned the traditional end-to-end deep structure and applied the feature expression of mid-level layers, combined the SVM classifier with the triplet mining method based on the order and shape condition, which achieved the state-of-the-art recognition results which were obtained on the car-196 dataset.

The deep-learning-based methods for vehicle recognition can ward off the process of handcrafted features design and automatically learn the deep features from big data. The high-level features can provide the important data of the category while suppressing the irrelevant background information. The deep-learning-based methods also have the excellent properties of generalization and robustness. Although these methods overcome the impact of the complex environments on vehicle recognition to some extent and increase the recognition accuracy and efficiency, they do not consider the relationship among the data, but only employ the advantage of CNN structure to obtain the classification model brutally. Inspired by previous research aimed at displaying and activating CNNs to carry out special tasks, we focus on exploring the transfer approach that is commonly used in deeper learning applications. The benefit of the transfer of learning is that the number of images needed for training and training time is highly minimized.

3   Transfer Learning

Transfer learning is considered to be the transfer of knowledge from a task learned to a new task in machine learning. In the context of neural networks, it transfers the features learned from a pretrained network to a new problem. The formation of a convoluted neuronal network from the beginning in any case is not usually effective when there is not enough amount of training data [2, 4]. The common practice in deep learning for these cases is to use a network that is trained in a large dataset for a new problem. While the initial layers of the pretrained network can be fixed, these last layers need to be refined to learn the specific features of the new dataset. Learning transfer often involves faster training times than the formation of a new convolutional neuronal network since it is not mandatory to estimate all the parameters of the new network. Transferable learning is useful when you want to form a CNN in your data set, but for various reasons, the dataset may not be suitable for forming a complete neural network (that is, too small). While data growth is a viable option in many cases, transfer learning has also proved effective. Transfer learning refers to the process of assuming a precompressed CNN, which replaces the completely the last convolutional layer and the formation of such layers in the dataset. By freezing convolutional layers weights, deep CNN can extract image features like edges, while fully connected layers can proceed with this information and use it to categorize data relevant to the problem. In practice, when there is a moderate amount of data, transfer learning is done at the last level and the classification layer is completely connected with a 1/100 learning rate of the value used to obtain the model convergence during training in ImageNet.

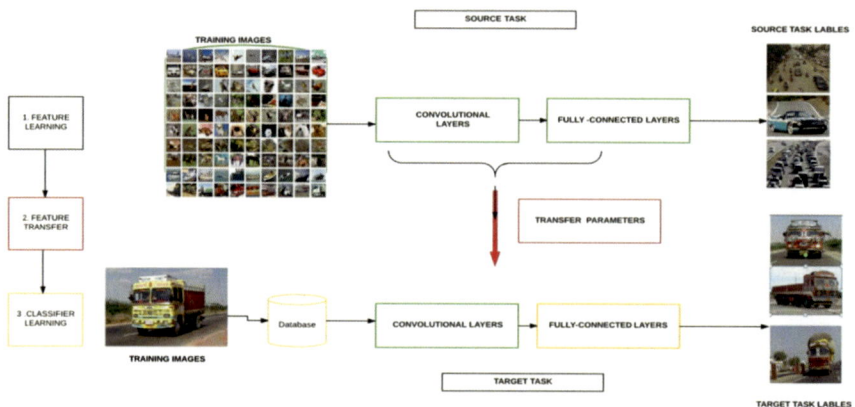

Fig. 1 Network architecture for proposed heavy vehicle detection system

4 Proposed Algorithm

We propose a heavy vehicle detection system based on fine-tuned deep learning. In the initial phase of our system, an input image is applied to pretrained CNN. The image process is from the R-CNN detector. The R-CNN detector only analyzes areas that may contain an object. This greatly minimizes the computational costs incurred while executing a CNN. As soon as transfer learning is used, in the transfer of learning, a network trained in a large collection of images is used as a starting point for solving a new detection activity.

Proposed algorithm can be explained in the following steps:

Step 1: Data collection.

Step 2: Train CNN using CIFAR-10 Data, with initial learning rate of 0.001.

Step 3: Approve CIFAR-10 network training. After the network is trained, it should be validated to ensure that training was successful. First, a swift visualization of the first convolutional layer's filter weights can help determine any immediate issues with training.

Step 4: Input Training Data.

Now that the network is working satisfactorily for the CIFAR-10 classification, the transfer learning methodology can be used to fine-tune the network for heavy vehicle detection.

Step 5: **Load** the ground truth data.

Training data accommodated in a table that contains the image file name and the ROI tags. Each ROI tag is a boundary box around the objects of concern within an image. To train the heavy vehicle detector, only the heavy vehicle image ROI tags are required (Fig. 1).

Step 6: Train R-CNN heavy vehicle detector.

Finally, train the R-CNN object detector using the R-CNN object detector. The training function automatically modifies the original CIFAR-10 network, classified in a network that can classify images in two classes: heavy vehicle and a generic background class.

Step 7: Visualize the feature map superimposed on the test image.

5 Experiments

First, a CNN was previously trained utilizing the CIFAR-10 data set, which consists of 50,000 training images. So this pretrained CNN is tuned for our heavy vehicle with only 41 on road heavy vehicle images. For testing, we have considered heavy vehicle and non-heavy vehicle dataset.

5.1 Detection Results

We are apt to use image instead of video, because it is easy to measure the accuracy. Our system is able to detect the frontal view of each vehicle image accurately. We performed the detection algorithm on all images in the system, and got 100% accuracy with respect to detection accuracy. Thus, our detection algorithm proved to be effective and fast in the desired system. Once the frontal view of a vehicle was obtained, we fed it into trained deep model as input; the output label is used to recognize the vehicle model. In order to keep the environment constant for comparison, we trained the deep architectures for 1, 10, 20, 50, and 100 epochs. During the training, the base learning rate was set to 0.01.

Figure 2 shows the results of number of epochs with accuracy. Table 2 shows the average accuracy of the CNN cascade method and our approach. Our method shows the best result. Figure 3 shows sample images of heavy vehicle database. Figure 4 shows detected images (Table 1).

Table 2 shows the final result of proposed method for heavy vehicle detection. These images were taken from real street CCTV.

Table 1 The result of AUC of heavy vehicle

Method	AUC (%)
Proposed method	100
Multi-scale CNNs [11]	98.31
Cascaded CNN [12]	96.50

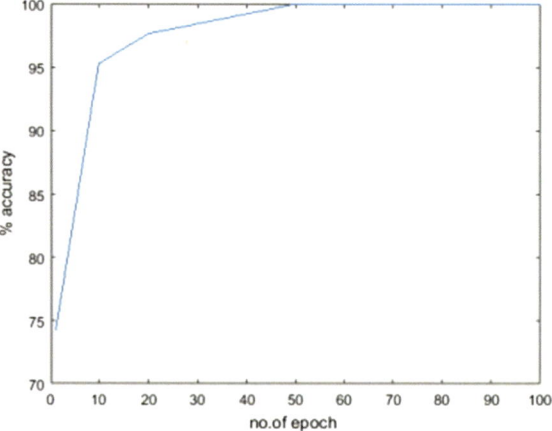

Fig. 2 Loss versus no. of epochs

Fig. 3 Sample images of heavy vehicle database

Fig. 4 Detected images

Table 2 Fine-tuned results on heavy vehicle dataset

Epoch	Time elapsed loss	Mini-batch loss	Mini-batch accuracy (%)	Base learning rate
1	110.65	0.8095	74.22	0.0010
10	1117.10	1587	95.28	0.0010
20	2522.52	0.0479	97.66	0.0010
50	6489.36	0.0074	100.00	0.0010
100	15,698.94	0.0002	100.00	0.0010

6 Conclusion

In this paper, a transfer learning based heavy vehicle detection system proposed. In addition, a reformative CNN structure is presented to enhance recognition accuracy and hugely scale down the computational cost incurred when running a CNN. By introducing transfer learning for proposed system, accuracy rate is significantly improved. Considerable tests have been performed, yielding favorable results.

References

1. Ge Q, Wen C, Duan S (2014) Fire localization based on range-range-range model for limited interior space. IEEE Trans Instrum Meas 63(9):2223–2237
2. Sivaraman S, Trivedi MM (2012) Real-time vehicle detection using parts at intersections. In: 15th International IEEE conference on intelligent transportation systems (ITSC). IEEE, pp 1519–1524
3. Chen Z, Ellis T, Velastin S (2012) Vehicle detection, tracking and classification in urban traffic. In: 15th International IEEE conference on intelligent transportation systems (ITSC). IEEE, pp 951–956
4. Kafai M, Bhanu B (2012) Dynamic bayesian networks for vehicle classification in video. IEEE Trans Ind Inf 8(1):100–109
5. Chen LC, Hsieh JW, Yan Y, Chen DY (2015) Vehicle make and model recognition using sparse representation and symmetrical SURFs. Pattern Recogn 48(6):1979–1998
6. Li B, Wu T, Zhu SC (2014) Integrating context and occlusion for car detection by hierarchical and-or model. In: Computer vision ECCV 2014. Springer International Publishing, pp 652–667
7. Lin YL, Morariu VI, Hsu W, Davis LS (2014) Jointly optimizing 3d model fitting and fine-grained classification. In: Computer vision–ECCV 2014. Springer International Publishing, pp 466–480
8. Xie S, Yang T, Wang X, Lin Y (2015) Hyper-class augmented and regularized deep learning for fine-grained image classification. In: Proceedings of the IEEE conference on computer vision and pattern recognition, pp 2645–2654
9. Huttunen H, Yancheshmeh FS, Chen K (2016) Car type recognition with deep neural networks. arXiv preprint arXiv:1602.07125
10. Wang Y, Choi J, Morariu VI, Davis LS (2016) Mining discriminative triplets of patches for fine-grained classification. arXiv preprint arXiv:1605.01130
11. Sermanet P, LeCun Y (2011) Traffic sign recognition with multi-scale convolutional networks. In: The 2011 International joint conference on neural networks (IJCNN), July 2011, pp 2809–2813
12. Zang D, Zhang J, Zhang D, Bao M, Cheng J, Tang K (2016) Traffic sign detection based on cascaded convolutional neural networks. In: 2016 17th IEEE/ACIS International Conference on software engineering, artificial intelligence, networking and parallel/distributed computing (SNPD), Shanghai, pp 201–206

Using Contourlet Transform Based RBFN Classifier for Face Detection and Recognition

R. Vinothkanna and T. Vijayakumar

Abstract Face is a highly non-rigid object; in such case, face detection and recognition has become an essential part of biometric systems in the majority of the applications. Numerous applications like robots, tablets, surveillance systems, and cell phones revolve around an efficient face detection and recognition technique in the background for access. Human–computer interaction systems like expression recognition, cognitive state/emotional state, etc. are used. Recognizing with the increased need for security and anticipation of spoofing attacks, almost all techniques have been proposed in the past to successfully detect and recognize the face through a single or combination of facial features, which is a challenging task given the complex nature of the background and the number of facial features involved. Here, the proposed work involves a multi-resolution technique, namely, the Contourlet transform along with linear discriminant analysis for feature detection given to an RBFN classifier for effective classification. It could be clearly seen that the proposed technique outperforms the other conventional techniques by its recognition rate of nearly 99.2%. The observed results indicate a good classification rate in comparison with conventional techniques.

1 Introduction

One of the recent booming technologies in the field of security systems is biometrics which involves identification of a person based on physiological or behavioral traits. By unauthorized access to confidential places, data is being reported throughout the globe. Hence, the need for authenticated entry mechanism has become the need of the hour. Biometric systems offer an effective solution in providing access by

R. Vinothkanna
ECE, Lakshmaiah Education Foundation, Guntur, Andhra Pradesh, India
e-mail: rvinothkannaphd@gmail.com

T. Vijayakumar (✉)
ECE, Guru Nanak Institute of Technology, Hyderabad, India
e-mail: vishal_16278@yahoo.co.in

© Springer Nature Switzerland AG 2019   1911
D. Pandian et al. (eds.), *Proceedings of the International Conference on ISMAC in Computational Vision and Bio-Engineering 2018 (ISMAC-CVB)*, Lecture Notes in Computational Vision and Biomechanics 30,
https://doi.org/10.1007/978-3-030-00665-5_176

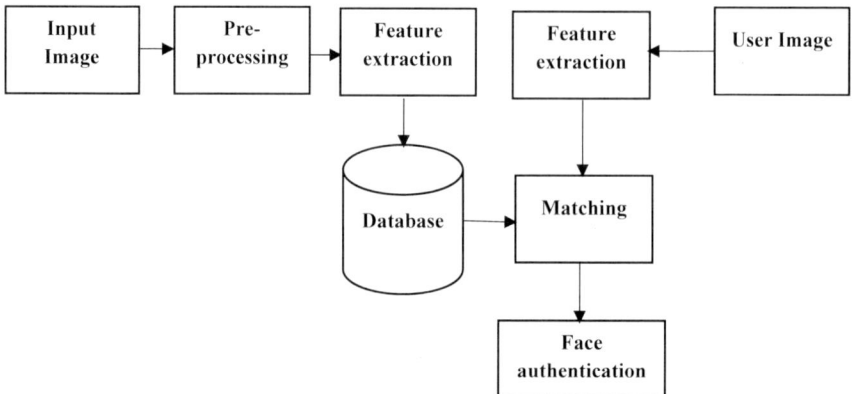

Fig. 1 A simple face authentication system

authenticating the user based on the physical or behavioral characteristics of the particular individual. Physical traits may include facial features [1, 2], fingerprint, palm-print, finger veins, iris, retina, etc., while the behavioral characteristics may include the style of walking referred to as gait of an individual, speech, etc. Since the objective of this paper is to detect an individual based on the facial features, the physical traits are taken into account in this paper. The objective [3] of face detection from face involves several features such as the orientation at the time of capture, the environmental conditions like lighting, etc., the focusing of the image acquisition device, the hair color, skin texture, brow length and arc angle, distance between the brows, the chin protrusion, etc., and the process of face detection involves pre-processing, feature extraction and a classifier as shown in Fig. 1. Any biometric mechanism basically involves two phases namely, the registration and authentication phase. Registration phase refers to the stage where the inputs are fed into the database and trained for the features. Authentication stage refers to the stage where the user is authenticated by comparing the features with the stored features in the database.

The figure illustrates two inputs, namely, the input image and the user image. Input image represents the image used for registration of features into the data base, while the user image denotes the image of user to be authenticated. As shown in Fig. 1, the enrollment phase is where the individuals are registered into the system using the features of their fingerprints. The unique features of each of the individuals are stored in the database. Verification phase takes place before the actual implementation where a known user claims the identity to the access system and verification is done based on the biometric features of a sample taken from the database. In the identification phase, any individual claims identity and access to the system and the biometric system decides on a true or false identity. In the above system, even though the features are unique, their storage in a database always poses a risk of theft leading to spoofing and hacking attacks. Face detection should be performed before recognition system. This is done to extract relevant information for face and facial expression analysis.

2 Related Work

Face detection using Eigen methods is one of the most widely and conventionally used techniques for face detection which comes under the appearance- based approach. They utilize a holistic approach to capture variations in a set of images in combination with principal component analysis (PCA) [4] for feature extraction. The notable demerits are the time consumption for eigen-value computation and also the observed results indicate that they are sensitive to lighting conditions [5]. Euclidean distance has also been used as parameter for feature matching [6] which is observed to detect variations in feature data due to lighting conditions. Recent face detection and classification techniques are mostly based on neural networks [7] due to its feasibility of training a system to capture the complex class of face patterns. Further, they are non-linear and hence the feature extraction step may be more efficient than the principal component analysis. 96.2% accuracy with a classification time less than 0.5 s has been observed [8]. A number of evolutionary variants have been utilized for face recognition like El Bakry's fast neural network [9], Huang's polynomial neural network (PNN), and conventional neural network of Matsugu [10] which has been extensively discussed in [5]. The problem of subject independence and translation, rotation, and scale invariance [11] in facial expression recognition has been addressed in the above variant techniques. With the advent of multi-resolution techniques, there have been more face detection techniques based on wavelet transforms [12]. The wavelets concentrate the energy of image into coefficients and further processing at frequency level causes least visual changes in the human visual perception system. A Gabor wavelet transform [13] addresses the issue of light sensitiveness in previous methods by applying the Gabor wavelets to a local binary pattern (LBP) [14] and experimentations indicated a good acceptance rate in comparison with conventional techniques. Linear discriminant analysis (LDA) techniques have been of late being used for feature extraction and reduction [15, 16]. LDA is a single input analyzer and comes under supervised learning techniques and is a linear technique. It combines the independent feature which leads the largest mean differences between the desired classes.

Hybrid techniques have been developed recently like the works of Sahoolizadeh et al. [17] based on combined Gabor wavelet and ANN feature classifier. The work indicated a drastic reduction in dimensionality, and linear discriminate analysis on down sampled Gabor wavelet faces is found to increase the discriminate ability. A modification was carried out in [18] where Gabor wavelets transform was used in hybrid combination with a feed-forward neural network for finding feature points and extracting feature vectors. The location of feature points contains information about the face in this approach. A correlation- based technique was presented in [19] by estimating areas of candidate of face presence followed by extraction of Gabor wavelets characteristics and neural network classifier. A back-propagation neural network [20] along with a Fourier Gabor filter used color skin to detect face regions and recognition achieved by using BPNN. The conventional support vector machine (SVM) has also been used to determine the match [21].

3 Proposed Work

Face detection is followed by face recognition where the process of face detection involves image preprocessing and feature extraction. A simplified face detection and recognition flow are shown in Fig. 2.

A. PREPROCESSING: Preprocessing is done on the input image to remove noise. In the process of capturing images, distortions including rotation, scaling, shift, and translation may be present in the face images, which make it difficult to locate at the correct position. Pre-processing removes any un-wanted objects (such as, background) from the collected image. It may also segment the face image for feature extraction. Histogram equalization could also be performed to distribute the image intensities throughout the image.

B. FEATURE EXTRACTION AND REDUCTION: Contourlet transform is used for decomposing the input image into sub-bands of differing scales of resolution. A k-level decomposition provides 2^k sub-bands out of which the first half are horizontally oriented and the other half are vertically oriented. The transform is a directional multi-resolution image representation scheme proposed by Do and Vetterli [22] which is effective in representing smooth contours in different directions of an image, thus providing directionality and anisotropy. While the Curvelet transform operates on continuous domain, the Contourlet transform is flexible to be operated in discrete domain. It utilizes a double filter bank; the Laplacian pyramid (LP) is used to detect the point discontinuities of the image and then a directional filter bank (DFB) to link point discontinuities into linear structures. A four-level decomposition of an input image from the ORL data base is shown in Fig. 3.

The Laplacian and directional filter bank are combined to capture the directional information. The texture information is captured in the directional sub-bands and hence the low-frequency sub-bands are ignored. In the proposed work, the LDA is applied to these sub-bands containing the maximum texture features to reduce the dimensionality of the feature vector and capture class-specific features by removing redundant elements. For each directional sub-band, all instances of the same person's face are grouped into one class and the faces of different subjects into different classes for all subjects in the training set, followed by labeling of the instances. Accordingly, within-class and between-class scatter matrices are designed as follows:

$$S_{\text{WC}} = \sum_c \sum_{ci} (\mu_c - \overline{x_i})(\mu_c - \overline{x_i})^{\text{T}} \tag{1}$$

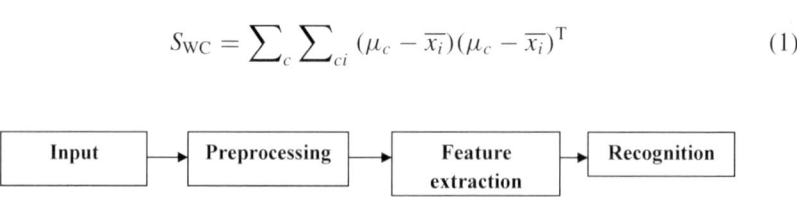

Fig. 2 A simplified face detection and recognition system

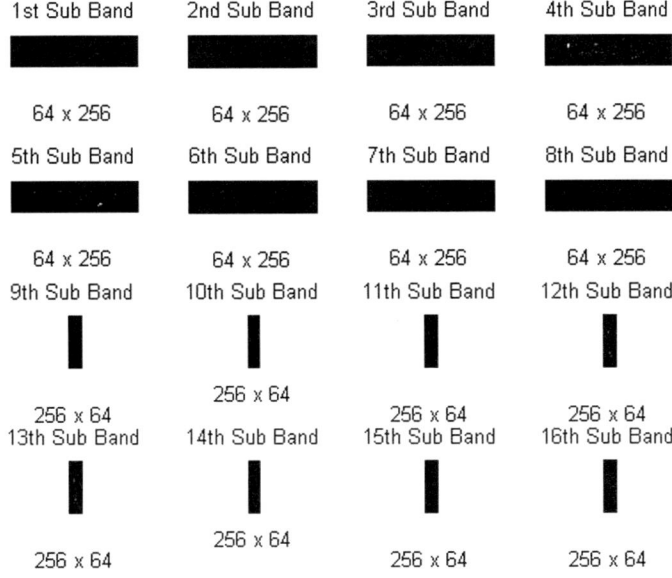

Fig. 3 A four-level Contourlet decomposition of input image from set 1

$$S_{BC} = \sum_c \sum_{ci} (\overline{x_i} - \mu_c)(\overline{x_i} - \mu_c)^T \qquad (2)$$

where S_{WC} and S_{BC} represent the scatter matrices of within class and between class, respectively. C denotes the number of classes, μ_c is the mean of the class, and x_i is the mean of all classes with the objective function defined by

$$J = w^T S_{BC} w / w^T S_{WC} w \qquad (3)$$

C. CLASSIFIER DESIGN: The proposed work utilizes a radial basis feed forward neural network (RBFN) with an input layer, a hidden layer, and an output layer. The input layer of network is set of l units, which accept l-dimensional feature vector. A model of the RBFN is shown in Fig. 4.

In this paper, the output layer has single output unit to determine whether the claimed user face is authentic or not. Connections between the input and hidden layers have unit weights and need not be trained. The output of an RBFN is given as

$$y(o) = g\left(\sum_{k=1}^{m} w_i h_i(o) + w_o\right) \qquad (4)$$

where w_i is the weight and h_i is the activation function taken to be Gaussian in this paper. The number of nodes in the output layer is dependent on the number of classes taken.

Fig. 4 Radial basis feed
forward neural network
model

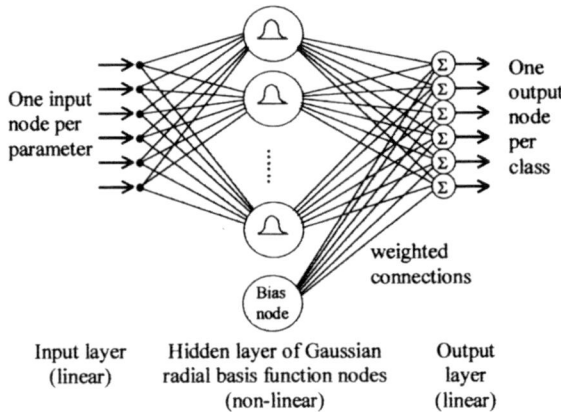

One input
node per
parameter

One
output
node
per
class

weighted
connections

Bias
node

Input layer
(linear)

Hidden layer of Gaussian
radial basis function nodes
(non-linear)

Output
layer
(linear)

4 Results and Discussion

In this paper, the input images have been taken from the ORL database and features
extracted through the Contourlet transform and dimensions reduced using LDA. The
feature set is fed into the RBF neural network and trained to obtain the classification
as authentic or unauthentic. The data-base consists of 10 images of 40 different
individuals taken at different angles, lighting conditions, and facial expressions. The
input images are in .pgm format with sizes of 91×112 pixels taken in frontal and
upright positions. The images taken for experimentation are depicted in Fig. 5.

The feature set obtained from the decomposition and LDA is fed into the neural
network to obtain a high classification rate of nearly 98.8%. The proposed work has
been compared with techniques like eigen-face, PCA, LDA-based face recognition,
and wavelet-based recognition techniques, and results are tabulated. The efficiency
of the proposed system is measured in terms of FAR (False acceptance Ratio] and
false reject rate (FRR). False acceptance rate and false rejection rate are depicted as

$$\text{FAR} = N_F/N; \quad \text{FRR} = N_R/N \qquad (5)$$

where N_F denotes the number of samples that were falsely accepted, while N denotes
the total number of samples which were matched and N_R denotes the number of
samples which were genuinely rejected. 200 images have been used for input out
which 120 have been used for training the RBFNN. During feed forward, each input
node receives an input value and relays it to each hidden neuron, which in turn
computes the activation and passes it on to each output unit, which again computes
the activation to obtain the net output. During training, the net output is compared
with the target value and the appropriate error is calculated. From this, the error
factor is obtained which is used to distribute the error back to the hidden layer. The
weights are updated accordingly. In a similar manner, the error factor is calculated
for units. The output layer contains one neuron. The result obtained from the output

Fig. 5 Input images taken from ORL database

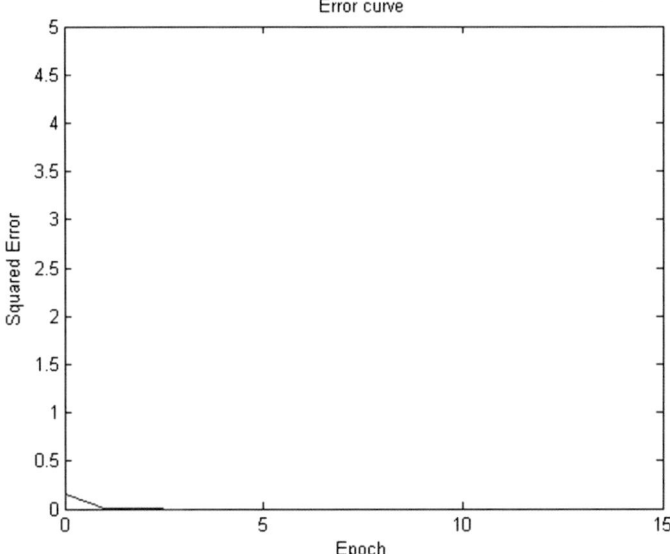

Fig. 6 Convergence of error versus number of epochs

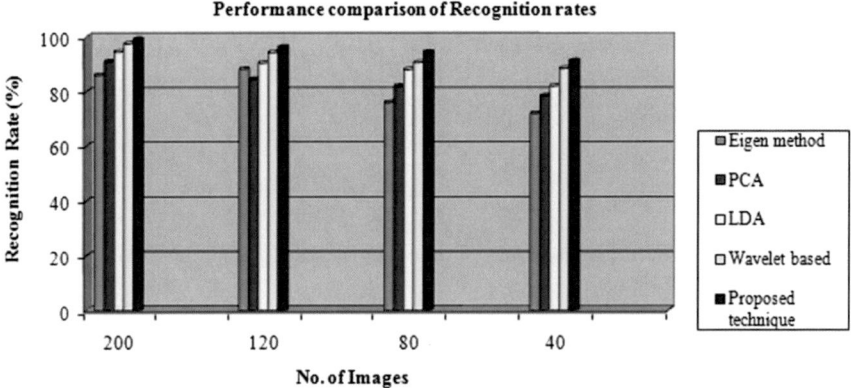

Fig. 7 Performance comparison of recognition rates

layer is given as the input to the RBF. The error incurred during the training phase in terms of epoch is shown in Fig. 6.

The proposed technique has been tested with a set of 200 images and quantified in terms of false acceptance rate and recognition rate. Figure 7 illustrates the comparative analysis. It could be clearly seen that the proposed technique outperforms the other conventional techniques by its recognition rate of nearly 99.2%. Out of the 200 images tested, three were found to be falsely accepted.

Table 1 Comparative analysis of proposed technique

No. of images	Acceptance ratio (%)					Execution time (s)				
	Eigen	PCA	LDA	Wavelet	Proposed technique	Eigen method	PCA	LDA	Wavelet	Proposed technique
200	92.5	93.8	95.14	95.89	97.25	80	78	77	72	50
120	90.1	90.5	91.1	92.4	94.2	64	63	61	57	42
80	86.5	88.4	89.25	90.1	93.1	51	48	47	41	38
40	82.4	86.4	88.4	89.2	90.89	44	40	41	36	32

Another important parameter is the computation time and as already mentioned in the previous sections, computation of eigen-faces takes longer which could be reflected in the observations tabulated in Table 1. Since the proposed technique does not involve computation of eigen faces, a drastic reduction in computation time could be observed. LDA provides the dimensionality reduction on the directional sub-bands, hence reducing the computational steps thus accounting for the reduced computation time. For a sample of 200 images, it could be seen that the classification results could be obtained in a little over 50 s.

5 Conclusion

The proposed frequency- domain- based dominant feature extraction algorithm provides an excellent space–frequency localization, which is clearly reflected in the high within-class compactness and high between-class reparability of the extracted features provided by the linear discriminant analysis algorithm. The use of a multiresolution transform avoids the need to go for an entire image at a single instant to obtain the features. On the other hand, it provides a set of highly informative and directive sub-band to work with. LDA provides a good feature compaction reducing the dimension of the feature vector to over 1×100. The observed recognition rates indicate that the proposed technique is not sensitive to varying facial features due to lighting, environmental factors, and hence plays a key role in discriminating different faces. Moreover, it utilizes a very low-dimensional feature space, which ensures lower computational burden. For the task of classification, a radial-based classifier has been employed and it is found to provide a satisfactory recognition performance, hence reducing the complexity of the recognition network. From our extensive simulations on different standard face databases, it has been found that the proposed method provides high recognition accuracy of images with variations in orientations, lighting. This technique could be extended in future to moving images or images in moving background.

References

1. Bowyer Kevin W, Biswas Soma, Aggarwal Gaurav, Flynn Patrick J (2013) Pose-robust recognition of low-resolution face images. IEEE Trans Pattern Anal Mach Intell 35:3037–3049
2. Amani N, Shahbahrami A, Nahvi M (2013) A new approach for face image enhancement and recognition. Int J Adv Sci Technol 52(1):1–10
3. Al Allaf ONA (2014) Review of face detection systems based artificial neural network algorithms. Int J Multimedia Appl 6(1):1–16
4. Zhang D, Ding D, Li J, Liu Q (2015) PCA based extracting feature using fast fourier transform for facial expression recognition. In: Transactions on engineering technologies, pp 413–424
5. Patel Riddhi, Yagnik Shruti B (2013) A literature survey on face recognition techniques. Int J Comput Trends Technol 5(4):189–194
6. Gao Q, Gao F, Zhang H, Hao X-J, Wang X (2013) Two dimensional maximum local variation based on image Euclidean distance for face recognition. IEEE Trans Image Process 22(10):3807–3817
7. Park Usang, Jain Anil K (2010) Face matching and retrieval using soft biometrics. IEEE Trans Inf Forensics Secur 5:406–415
8. Yang MH, Kriegman DJ, Ahuja N (2002) Detecting faces in images: a survey. IEEE Trans Pattern Anal Mach Intell 24(1):34–58
9. Elbakry A (2002) Further experience with the tubularized-incised urethral plate technique for hypospadias repair. BJU Int 89(3):291–294
10. Matsugu M, Katayama T, Hatanaka K (2004) Method of extracting image from input image using reference image. U.S. Patent 6,766,055, Issued July 20
11. Li Z, Park U, Jain A (2011) A discriminative model for age invariant face recognition. IEEE Trans Inf Forensics Secur 6(3):1028–1037
12. Zheng Y, Zhang C, Zhou Z (2012) A wavelet based recognition for multispectral face. In: Proceedings of SPIE, vol 8401
13. Sharma P, Arya KV, Yadav RN (2013) Efficient face recognition using wavelet based generalized neural network. Mach Learn Intel Image Process 93(6):1557–1565
14. Anuradha K, Tyagi MK (2012) A novel method of face recognition using LBP, LTP and Gabor features. Int J Sci Technol Res 1(5):31–35
15. Moon H-M, Pan SB (2013) The LDA-based face recognition at a distance using multiple distance image. In: Seventh international conference on innovative mobile and internet services in ubiquitous computing, pp 249–255
16. Kaur A, Singh S, Taqdir A (2013) Face recognition using PCA and LDA techniques. Int J Adv Res Computer Commun Eng 4(3):308–310
17. Sahoolizadeh H, Jannesari A, Dousti M (2018) Noise suppression in a commongate UWB LNA with an inductor resonating at the source node. AEU-Int J Elect Comm 96:144–153
18. Kaushal A, Raina JPS (2010) Face detection using neural network & Gabor wavelet transform. Int J Comput Sci Technol (IJCST) 1(1):58–63
19. Abadi Mohammad et al (2011) Face detection with the help of Gabor wavelets characteristics and neural network classifier. Am J Sci Res 36:67–76
20. Bouzalmat A et al (2011) Face detection and recognition using back propagation neural network and Fourier Gabor filters. Signal Image Process Int J (SIPIJ) 2(3):15
21. Zhao Lina, Wanbao Hu, Cui Lihong (2012) Face recognition feature comparison based SVD and FFT. J Signal Inf Process 3:259–262
22. Do MN, Vetterli M (2001) Pyramidal directional filter banks and curvelets. In Image Processing, 2001. In: Proceedings 2001 International Conference on, vol 3, pp. 158–161. IEEE

Breast Cancer Recognition by Support Vector Machine Combined with Daubechies Wavelet Transform and Principal Component Analysis

Fangyuan Liu and Mackenzie Brown

Abstract The method of identifying the abnormal mammary gland tumor images was presented in order to assist the medical staff to find the patients with breast diseases accurately and timely. Db2 wavelet transform and principal component analysis (select the optimal threshold) is used to extract the effective features, support vector machine (set appropriate penalty parameter) is used to classify health and diseased samples, and 10-fold cross-validation is used to verify the classification result. The experimental results show that the method is feasible, the average sensitivity is $83.10 \pm 1.91\%$, the average specificity is $82.60 \pm 4.50\%$, and the average accuracy is $82.85 \pm 2.21\%$.

1 Introduction

The treatment of cancer is one of the most important problems for researchers in the present age. At present, the incidence of breast cancer is getting higher and higher, and the age span is getting wider and wider, which causes unpredictable damage to human health. In the light of breast cancer, experts from various industries combined computer-aided diagnosis with X-ray radiography technology or B-ultrasonic. Researchers can detect breast tumors at early stage, identify malignant and benign tumors accurately, and adopt effective treatment methods in a timely manner. There-

---

The original version of this chapter was revised: The Corresponding Author function has been assigned to Mackenzie Brown and the wrong labeling of images in figure 1 has been corrected. The correction to this chapter is available at https://doi.org/10.1007/978-3-030-00665-5_178

F. Liu
School of Computer Science and Technology, Nanjing Normal University, Nanjing 210023, Jiangsu, China
e-mail: 2191513453@qq.com

M. Brown (✉)
School of Engineering, Edith Cowan University, Joondalup, WA 6027, Australia
e-mail: mackbrown@ieee.org

© Springer Nature Switzerland AG 2019, corrected publication 2023   1921
D. Pandian et al. (eds.), *Proceedings of the International Conference on ISMAC in Computational Vision and Bio-Engineering 2018 (ISMAC-CVB)*, Lecture Notes in Computational Vision and Biomechanics 30,
https://doi.org/10.1007/978-3-030-00665-5_177

fore, it is necessary and urgent to explore a kind of recognition method with good stability and high accuracy.

In response to the detection problem of human breast cancer, researchers have come up with a series of solutions. Xu [1] presented a kernel orthogonal transformation method based on linear orthogonal transform. In the experiment, the vector values of the test samples were compared with the set threshold after the transformation of the nuclear method to determine the malignant and benign breast tumors. By dividing the training–testing sets of samples in different proportions to improve the performance, the final experimental results show that the optimal partition detection rate of 4:1 reaches 100%. Senapati et al. [2] introduced a stable and effective way to identify breast cancer, which combined local linear wavelet neural network (LLWNN) with recursive least squares (RLS). The researchers used RLS in order to select the optimal parameters, including scale, translation, and local linear model. The accuracy of LLWNN-RLS is 97.2%, which is superior to other diagnostic methods. Uzer et al. [3] created two mixed feature selection methods to improve the comprehensiveness of the samples. The workers combined the sequential forward selection (SFS) and the sequential back selection (SBS) with PCA, respectively. Then, the artificial neural network based on SCG algorithm is used to classify the input features, and SBS-PCA+NN achieved 98.57% which is better than SFS-PCA+NN. Azar and El-Said [4] studied six kinds of detection methods based on support vector machine (SVM) for recognizing breast tumors: standard SVM, proximal SVM, smooth SVM, Lagrangian SVM, linear programming SVM, and finite Newton method for Lagrangian SVM. After adjusting the model parameters, the above methods obtained good recognition rate, in which the linear programming SVM performance is the best under each index.

Wavelet transform can extract important texture features of cancer foci. Hence, we used the wavelet transform [5–14] to extract distinguishing features. The principal component analysis was further used to reduce the features. Finally, support vector machine [15–20] was used to help establish the classifier.

We make a detailed exposition of the content after the introduction of the first section in this paper. The second section introduces the dataset and method used in the experiment, the third section analyzes the experimental results, and the fourth section summarizes the practicability and the extension of the proposed method.

2 Dataset and Method

We download public open-access mini-MIAS database [21], which contains 322 mammogram images sizes of 1024×1024. We picked up 200 samples randomly from the mini-MIAS dataset—100 healthy and 100 abnormal breast images, respectively (Fig. 1).

At the first step, we use discrete wavelet transform (DWT) and principal component analysis (PCA) to extract the features of breast images and select the most effective features. DWT is the discretization of the wavelet function in scale and dis-

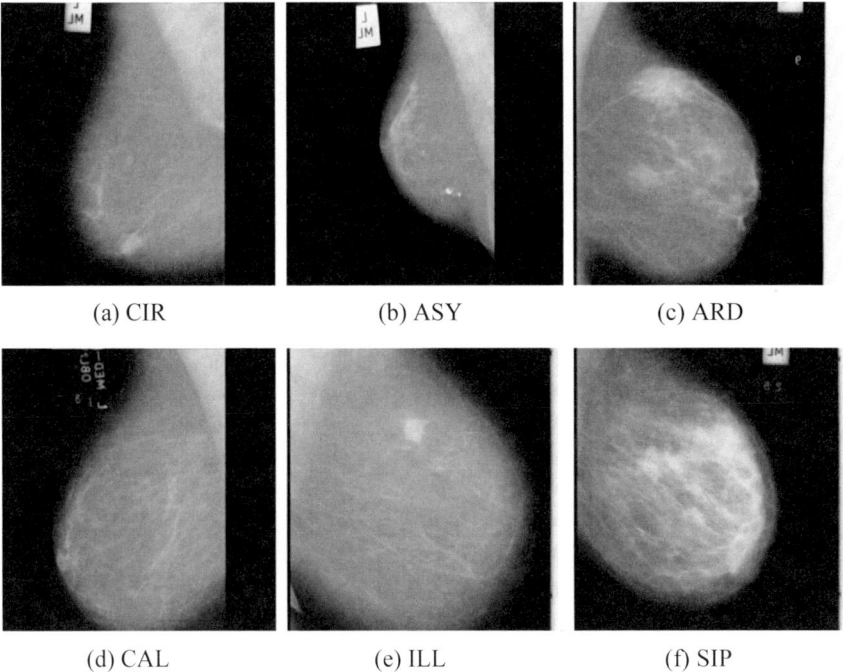

(a) CIR   (b) ASY   (c) ARD

(d) CAL   (e) ILL   (f) SIP

Fig. 1 Six abnormal breasts ((a) CIR = Circumscribed Mass; (b) ASY = Asymmetry; (c) ARD = Architectural distortion; (d) CAL = Calcification; (e) Ill = defined masses; (f) SPI = Spiculated masses)

placement, and the image of the mammary gland is decomposed, which is convenient for computer to process [22–24]. Furthermore, DWT can not only reduce the degree of redundancy of wavelet transform coefficients but also guarantee the integrality of information to some extent [25–29]. The expression of the wavelet basis function is shown in Formula (1), a represents the scale, τ represents the displacement, and t represents time. The Formulas (2) and (3) correspond to the discretization of scale and displacement, respectively. The expression of the discrete wavelet transform for function $f(t)$ is shown in Formula (4).

$$\psi_{a,\tau}(t) = \frac{1}{\sqrt{a}}\psi\left(\frac{t-\tau}{a}\right) \tag{1}$$

$$a = a_0^i, \ a_0 = 2, \ i = 0, 1, 2, \ldots \tag{2}$$

$$\psi_{a,\tau}(t) = \psi(t-\tau), a = 2^0, \tau = k\tau_0, k = 0, 1, 2 \ldots \tag{3}$$

$$WT_f(i, k) = \int_R f(t) \cdot \psi_{i,k}(t)d_t, \psi_{i,k}(t) = 2^{-\frac{i}{2}}\psi\left(2^{-i}t - k\tau_0\right) \tag{4}$$

Fig. 2 Illustration of SVM

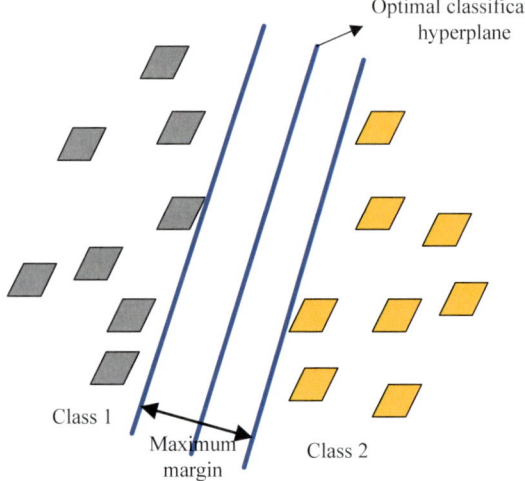

After DWT, the principal component analysis (PCA) was utilized to reduce the feature dimension and to remove the correlation between features [30]. The application of this method is considered because it can decrease computational complexity and improve the accuracy of identifying abnormal breast tumor images. The threshold T is an important parameter in PCA.

In the last step, we presented SVM to classify the breast tumor dataset to determine whether the breast is normal. SVM, as shown in Fig. 2, accurately divides the input eigenvectors according to the determination of the separating hyperplane and the maximum margin classification rule, which is mature in statistical classification and regression analysis. Therefore, SVM is used to identify medical image that can guarantee the reliability of the classification results. We did not use deep learning techniques, such as convolutional neural network [31–34] and autoencoder, because the dataset used in this study is relatively small.

Tenfold cross-validation was used to estimate this "DWT+PCA+SVM" approach. In tenfold cross-validation, nine folds were used as training set, and the rest was used as test set [35]. For a more strict analysis, this tenfold cross-validation repeats 10 times, and the average and standard deviation of the performances will be reported.

3 Experiments and Results

We used db2 wavelet in this study, the threshold of PCA is set to 95%, and the penalty parameter of SVM is set to 10. The classification performance measurement results of this 10×10-fold cross-validation are shown in Table 1. Here, we calculate the sensitivity (Sen), specificity (Spc), and accuracy (Acc) of each run at each fold.

Table 1 Measurement of proposed breast cancer detection at each fold (*Unit %*)

Sen	F1	F2	F3	F4	F5	F6	F7	F8	F9	F10
R1	90	90	80	80	100	90	90	80	70	80
R2	80	70	90	70	90	80	80	70	90	90
R3	70	80	80	100	90	80	90	70	100	80
R4	80	100	80	100	90	70	70	90	90	80
R5	70	90	90	90	90	80	80	70	80	80
R6	90	70	100	90	60	90	80	80	80	90
R7	80	60	60	90	90	80	90	100	100	50
R8	70	90	80	70	90	80	80	90	90	80
R9	80	90	80	90	90	70	100	100	80	80
R10	50	90	90	90	70	80	90	90	100	80

(continued)

Table 1 (continued)

Spc	F1	F2	F3	F4	F5	F6	F7	F8	F9	F10
R1	60	80	90	70	60	70	70	90	70	80
R2	70	90	80	80	90	80	80	100	100	80
R3	80	90	80	70	90	90	100	80	100	80
R4	70	90	90	90	100	90	90	80	70	70
R5	70	90	90	100	60	90	80	80	80	80
R6	60	80	80	70	80	70	90	100	80	70
R7	90	100	100	70	100	70	80	90	90	90
R8	60	70	80	100	80	80	90	70	80	70
R9	90	80	70	70	100	90	90	80	90	90
R10	90	90	90	50	100	80	90	80	100	90

(continued)

Table 1 (continued)

Acc	F1	F2	F3	F4	F5	F6	F7	F8	F9	F10
R1	75	85	85	75	80	80	80	85	70	80
R2	75	80	85	75	90	80	80	85	95	85
R3	75	85	80	85	90	85	95	75	100	80
R4	75	95	85	95	95	80	80	85	80	75
R5	70	90	90	95	75	85	80	75	80	80
R6	75	75	90	80	70	80	85	90	80	80
R7	85	80	80	80	95	75	85	95	95	70
R8	65	80	80	85	85	80	85	80	85	75
R9	85	85	75	80	95	80	95	90	85	85
R10	70	90	90	70	85	80	90	85	100	85

Table 2 Average measure of proposed method (*Unit %*)

Run	Sen	Spc	Acc
1	85	74	79.50
2	81	85	83.00
3	84	86	85.00
4	85	84	84.50
5	82	82	82.00
6	83	78	80.50
7	80	88	84.00
8	82	78	80.00
9	86	85	85.50
10	83	86	84.50
Avr	83.10 ± 1.91	82.60 ± 4.50	82.85 ± 2.21

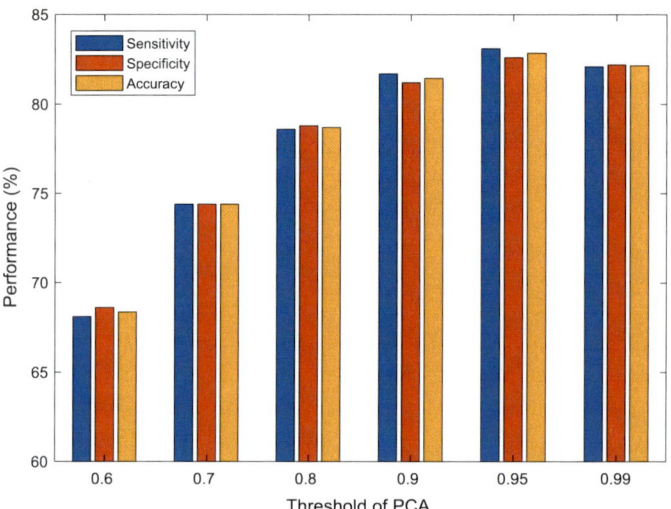

Fig. 3 Choosing optimal threshold of PCA

The average and standard deviation of sensitivity, specificity, and accuracy of our method are listed in Table 2, which indicates that the average sensitivity of our method is $83.10 \pm 1.91\%$, the average specificity is $82.60 \pm 4.50\%$, and the average accuracy is $82.85 \pm 2.21\%$.

In the second experiment, we validated why we set the threshold of PCA to 95%. In Table 3, the effect of different PCA threshold values is shown. We compared the results at PCA threshold T of 60, 70, 80, 90, 95, and 99%, respectively. For a clear view, the comparison results are presented in Fig. 3. Obviously, $T = 95\%$ yields the greatest performance, and hence we chose this value as the optimal one in this study.

Table 3 Effect of different PCA threshold values (*Unit %*)

$T = 0.6$	Sen	Spc	Acc	$T = 0.7$	Sen	Spc	Acc	$T = 0.8$	Sen	Spc	Acc
R1	66	65	65.5	R1	71	75	73.0	R1	74	82	78.0
R2	76	73	74.5	R2	76	70	73.0	R2	84	76	80.0
R3	67	63	65.0	R3	77	74	75.5	R3	79	80	79.5
R4	67	68	67.5	R4	72	80	76.0	R4	82	72	77.0
R5	73	72	72.5	R5	78	78	78.0	R5	75	76	75.5
R6	64	66	65.0	R6	78	73	75.5	R6	83	81	82.0
R7	68	68	68.0	R7	75	75	75.0	R7	81	81	81.0
R8	67	68	67.5	R8	71	78	74.5	R8	75	84	79.5
R9	69	67	68.0	R9	70	70	70.0	R9	74	82	78.0
R10	64	76	70.0	R10	76	71	73.5	R10	79	74	76.5
Avr	68.10 ± 3.78	68.60 ± 3.95	68.35 ± 3.16	Avr	74.40 ± 3.10	74.40 ± 3.50	74.40 ± 2.17	Avr	78.60 ± 3.86	78.80 ± 3.99	78.70 ± 2.06

(continued)

Table 3 (continued)

	T=0.9			T=0.95			T=0.99		
	Sen	Spc	Acc	Sen	Spc	Acc	Sen	Spc	Acc
R1	84	83	83.5	85	74	79.5	84	81	82.5
R2	80	84	82.0	81	85	83.0	79	80	79.5
R3	88	81	84.5	84	86	85.0	81	77	79.0
R4	83	82	82.5	85	84	84.5	78	81	79.5
R5	75	82	78.5	82	82	82.0	84	83	83.5
R6	84	83	83.5	83	78	80.5	85	86	85.5
R7	77	83	80.0	80	88	84.0	80	82	81.0
R8	88	74	81.0	82	78	80.0	82	79	80.5
R9	77	76	76.5	86	85	85.5	88	89	88.5
R10	81	84	82.5	83	86	84.5	80	84	82.0
Avr	81.70±4.52	81.20±3.43	81.45±2.49	83.10±1.91	82.60±4.50	82.85±2.21	82.10±3.11	82.20±3.49	82.15±3.01

4 Conclusion

This study proposes a hybrid breast cancer recognition method based on db2 wavelet transform, principal component analysis, and support vector machine. This proposed method obtained promising results, with sensitivity of $83.10 \pm 1.91\%$, specificity of $82.60 \pm 4.50\%$, and accuracy of $82.85 \pm 2.21\%$.

The future studies contain applying advanced variants of wavelet transform, e.g., the stationary wavelet transform or the wavelet package transform. In addition, we may try to collect private breast image to increase the size of dataset, and then test deep learning methods, e.g., the autoencoders [36–38].

References

1. Xu Y, Zhu Q, Wang J (2012) Breast cancer diagnosis based on a kernel orthogonal transform. Neural Comput Appl 21(8):1865–1870
2. Senapati MR et al (2013) Local linear wavelet neural network for breast cancer recognition. Neural Comput Appl 22(1):125–131
3. Uzer MS, Inan O, Yilmaz N (2013) A hybrid breast cancer detection system via neural network and feature selection based on SBS, SFS and PCA. Neural Comput Appl 23(3):719–728
4. Azar AT, El-Said SA (2014) Performance analysis of support vector machines classifiers in breast cancer mammography recognition. Neural Comput Appl 24(5):1163–1177
5. Zhou XX, Zhang GS (2016) Detection of abnormal MR brains based on wavelet entropy and feature selection. IEEJ Trans Electr Electron Eng 11(3):364–373
6. Yang JQ et al (2016) A novel compressed sensing method for magnetic resonance imaging: exponential wavelet iterative shrinkage-thresholding algorithm with random shift. Int J Biomed Imaging. Article ID. 9416435
7. Sun P (2016) Preliminary research on abnormal brain detection by wavelet-energy and quantum-behaved PSO. Technol Health Care 24(s2):S641–S649
8. Yang M (2016) Dual-tree complex wavelet transform and twin support vector machine for pathological brain detection. Appl Sci 6(6): Article ID. 169
9. Zhou X-X (2016) Comparison of machine learning methods for stationary wavelet entropy-based multiple sclerosis detection: decision tree, k-nearest neighbors, and support vector machine. Simulation 92(9):861–871
10. Lu HM (2016) Facial emotion recognition based on biorthogonal wavelet entropy, fuzzy support vector machine, and stratified cross validation. IEEE Access 4:8375–8385
11. Nayak DR (2017) Detection of unilateral hearing loss by stationary wavelet entropy. CNS & Neurol Disord-Drug Targets 16(2):15–24
12. Wang S-H (2016) Single slice based detection for Alzheimer's disease via wavelet entropy and multilayer perceptron trained by biogeography-based optimization. Multimed Tools Appl. https://doi.org/10.1007/s11042-016-4222-4
13. Li Y, Cattani C (2017) Detection of dendritic spines using wavelet packet entropy and fuzzy support vector machine. CNS & Neurol Disord-Drug Targets 16(2):116–121
14. Li P, Liu G (2017) Pathological brain detection via wavelet packet tsallis entropy and real-coded biogeography-based optimization. Fundam Inform 151(1–4):275–291
15. Chen M (2016) Morphological analysis of dendrites and spines by hybridization of ridge detection with twin support vector machine. PeerJ 4:e2207
16. Chen S, Yang J-F, Phillips P (2015) Magnetic resonance brain image classification based on weighted-type fractional Fourier transform and nonparallel support vector machine. Int J Imaging Syst Technol 25(4):317–327

17. Dong Z (2014) Classification of Alzheimer disease based on structural magnetic resonance imaging by kernel support vector machine decision tree. Prog Electromagn Res 144:171–184

18. Ji G (2013) An MR brain images classifier system via particle swarm optimization and kernel support vector machine. Sci World J (130134)

19. Wu L (2012) Classification of fruits using computer vision and a multiclass support vector machine. Sensors 12(9):12489–12505

20. Wu L (2012) An MR brain images classifier via principal component analysis and kernel support vector machine. Prog Electromagn Res 130:369–388

21. The mini-MIAS database of mammograms (2018) Available from http://peipa.essex.ac.uk/info/mias.html

22. Gorriz JM (2017) Multivariate approach for Alzheimer's disease detection using stationary wavelet entropy and predator-prey particle swarm optimization. J Alzheimer's Dis https://doi.org/10.3233/jad-170069

23. Phillips P (2018) Intelligent facial emotion recognition based on stationary wavelet entropy and Jaya algorithm. Neurocomputing 272:668–676

24. Han L (2018) Identification of Alcoholism based on wavelet Renyi entropy and three-segment encoded Jaya algorithm. Complexity 2018(3198184):13

25. Zhou X-X et al (2016) Combination of stationary wavelet transform and kernel support vector machines for pathological brain detection. Simulation 92(9):827–837

26. Atangana A (2018) Application of stationary wavelet entropy in pathological brain detection. Multimed Tools Appl 77(3):3701–3714

27. Chen Y, Chen X-Q (2016) Sensorineural hearing loss detection via discrete wavelet transform and principal component analysis combined with generalized eigenvalue proximal support vector machine and Tikhonov regularization. Multimed Tools Appl 77(3):3775–3793

28. Chen Y, Lu H (2018) Wavelet energy entropy and linear regression classifier for detecting abnormal breasts. Multimed Tools Appl 77(3):3813–3832

29. Zhan TM, Chen Y (2016) Multiple sclerosis detection based on biorthogonal wavelet transform, RBF kernel principal component analysis, and logistic regression. IEEE Access 4:7567–7576

30. Schimit PHT, Pereira FH (2018) Disease spreading in complex networks: a numerical study with principal component analysis. Expert Syst Appl 97:41–50

31. Zhao G (2017) Polarimetric synthetic aperture radar image segmentation by convolutional neural network using graphical processing units. J Real-Time Image Process https://doi.org/10.1007/s11554-017-0717-0

32. Muhammad K (2017) Image based fruit category classification by 13-layer deep convolutional neural network and data augmentation. Multimedia Tools Appl. https://doi.org/10.1007/s11042-017-5243-3

33. Tang C (2017) Twelve-layer deep convolutional neural network with stochastic pooling for tea category classification on GPU platform. Multimedia Tools Appl. https://doi.org/10.1007/s11042-018-5765-3

34. Lv Y-D (2018) Alcoholism detection by data augmentation and convolutional neural network with stochastic pooling. J Med Syst 42(1):2

35. Gupta A, Kumar D (2018) Beyond the limit of assignment of metabolites using minimal serum samples and H-1 NMR spectroscopy with cross-validation by mass spectrometry. J Pharm Biomed Anal 151:356–364

36. Hou X-X (2017) Seven-layer deep neural network based on sparse autoencoder for voxelwise detection of cerebral microbleed. Multimedia Tools Appl. https://doi.org/10.1007/s11042-017-4554-8

37. Jia W (2017) Five-category classification of pathological brain images based on deep stacked sparse autoencoder. Multimedia Tools Appl. https://doi.org/10.1007/s11042-017-5174-z

38. Jia W (2017) Three-category classification of magnetic resonance hearing loss images based on deep autoencoder. J Med Syst 41(10):165

Correction to: Breast Cancer Recognition by Support Vector Machine Combined with Daubechies Wavelet Transform and Principal Component Analysis

Fangyuan Liu and Mackenzie Brown

Correction to:
Chapter "Breast Cancer Recognition by Support Vector Machine Combined with Daubechies Wavelet Transform and Principal Component Analysis" in: D. Pandian et al. (eds.), *Proceedings of the International Conference on ISMAC in Computational Vision and Bio-Engineering 2018 (ISMAC-CVB),* **Lecture Notes in Computational Vision and Biomechanics 30,** **https://doi.org/10.1007/978-3-030-00665-5_177**

The Corresponding Author function has been assigned to Mackenzie Brown and the wrong labeling of images in figure 1 has been corrected.

The correction chapter and the book has been updated with the changes.

The updated original version of this chapter can be found at
https://doi.org/10.1007/978-3-030-00665-5_177

D. Pandian et al. (eds.), *Proceedings of the International Conference on ISMAC in Computational Vision and Bio-Engineering 2018 (ISMAC-CVB),* Lecture Notes in Computational Vision and Biomechanics 30,
https://doi.org/10.1007/978-3-030-00665-5_178